# ACTIVITY IN RED-DWARF STARS

# ASTROPHYSICS AND SPACE SCIENCE LIBRARY

A SERIES OF BOOKS ON THE RECENT DEVELOPMENTS
OF SPACE SCIENCE AND OF GENERAL GEOPHYSICS AND ASTROPHYSICS
PUBLISHED IN CONNECTION WITH THE JOURNAL
SPACE SCIENCE REVIEWS



VOLUME 102
PROCEEDINGS

# ACTIVITY IN RED-DWARF STARS

PROCEEDINGS OF THE 71st COLLOQUIUM
OF THE INTERNATIONAL ASTRONOMICAL UNION
HELD IN CATANIA, ITALY, AUGUST 10–13, 1982

Edited by

PATRICK B. BYRNE
*Armagh Observatory, Northern Ireland*

and

MARCELLO RODONÒ
*Astrophysical Observatory and University of Catania, Italy*

Springer-Science+Business Media, B.V.

**Library of Congress Cataloging in Publication Data**

CIP

IAU Colloquium (71st : 1982 : Catania, Italy)
Activity in red-dwarf stars.

(Astrophysics and space science library ; 102)
Includes index.
1. Red dwarfs–Congresses. I. Byrne, Patrick, B., 1947–
II. Rodonò, Marcello, 1941– . III. Title. IV. Series.
QB843.R4I27 1982 523.8 83-8698

ISBN 978-94-009-7159-2 ISBN 978-94-009-7157-8 (eBook)
DOI 10.1007/978-94-009-7157-8


Originally published by D. Reidel Publishing Company, Dordrecht, Holland in 1983.
Softcover reprint of the hardcover 1st edition 1983

TABLE OF CONTENTS

Preface xiii

List of Participants ixx

M.RODONO' - Introductory Address 1

R.ROSNER - Magnetic Fields and Activity of the Sun and Stars: An Overview 5

SESSION I: GENERAL CHARACTERISTICS OF ACTIVE DWARFS

B.R.PETTERSEN - Global and Photospheric Physical Parameters of Active Dwarf Stars (Invited Paper) 17

R.F.WING - Mean Colors and Effective Temperatures of K and M Dwarfs 35

J.L.LINSKY - The Quiescent Chromospheres and Transition Regions of Active Dwarf Stars: What Are We Learning from Recent Observations and Models? (Invited Paper) 39

C.JORDAN, T.R.AYRES and A.BROWN - The Structure and Energy Balance in Main Sequence Stars 61

D.R.SODERBLOM - The Distribution of Chromospheric Emission Strengths Among Red Dwarfs 67

S.CATALANO and E.MARILLI - Chromospheric Emission and Rotation of Main Sequence Stars 71

A.C.COLLIER - Late-Type Ca II Emission-Line Stars in the Southern Hemisphere 75

R.DE LA REZA, C.J.BUTLER, C.A.O.TORRES and C.C.BATALHA - The Atmosphere of a Probably Very Young BY Dra-Type Flare Star 77

M.SAXNER - Some Properties of Stellar Transition Regions 79

L.GOLUB - Quiescent Coronae of Active Chromosphere Stars (Invited Paper) 83

H.M.JOHNSON - Einstein and IUE Observations of Nearby Red Dwarfs (Invited Paper) 109

P.C.AGRAWAL, A.R.RAO and B.V.SREEKANTAN - X-Ray Emission from Flare Stars 125

R.A.STERN and M.C.ZOLCINSKI - Stellar X-Ray Activity in the Hyades 131

SESSION II: OBSERVED ACTIVITY IN RED-DWARFS

S.S.VOGT - Spots, Spot-Cycles and Magnetic Fields of Late-Type Dwarfs (Invited Paper) 137

P.B.BYRNE - Optical Photometry of Flares and Flare Statistics (Invited Paper) 157

C.A.O.TORRES, I.C.BUSKO and G.R.QUAST - The ITA Survey on Red Dwarf Variable Stars - Final Report 175

M.RODONO', V.PAZZANI and G.CUTISPOTO - Synopsis of BY Dra and II Peg Light Curves: Variability and Possible Evidence of Differential Rotation 179

G.LA FAUCI and M.RODONO' - Two-Spot Modeling of Synoptic Light Curves of II Peg 185

F.VAN LEEUWEN and P.ALPHENAAR - Variable K-Type Stars in the Pleiades 189

S.L.BALIUNAS - Stellar Activity and Calcium Emission Variability 195

S.L.LIPPINCOTT - EV Lacertae: Is Flare Activity Related to an Unseen Planet-Like Companion? 201

B.N.ANDERSEN - An Unusual Flare on EV Lac 203

S.P.WORDEN - Optical and Ultraviolet Stellar Flare Spectroscopy (Invited Paper) 207

M.S.GIAMPAPA - Results from Optical and UV Stellar Flare Spectroscopy (Invited Paper) 223

P.F.CHUGAINOV, P.P.PETROV and A.G.SCHERBAKOV - Observations of $H_\beta$ and He II λ4686 Lines in the Spectra of Flares of UV Cet-Type Stars 237

B.R.PETTERSEN - The Timescales of Variations in Continuum and Hydrogen Lines During Stellar Flares 239

G.E.BROMAGE, B.E.PATCHETT, K.J.H.PHILLIPS, P.L.DUFTON and A.E.KINGSTON - Far-Ultraviolet and Visible Observations of Flares on dMe Stars 245

C.J.BUTLER, A.D.ANDREWS, J.G.DOYLE, P.B.BYRNE, J.L.LINSKY, T.SIMON, N.MARSTAD, M.RODONO' and V.PAZZANI - IUE Spectra of the BY Dra/Flare Star AU Mic 249

R.E.STENCEL - Flare-Like Activity of Red Giant Stars 251

B.M.HAISCH - X-Ray Observations of Stellar Flares (Invited Paper) 255

D.M.GIBSON - Quiescent and Flaring Radio Emission from dMe Stars (Invited Paper) 273

SESSION III: OBSERVED ACTIVITY IN RELATED OBJECTS

G.M.SIMNETT - Flares on the Sun: Selected Results from SMM (Invited Paper) 289

K.TANAKA - Flares on the Sun: Selected Results from HINOTORI (Invited Paper) 307

R.PALLAVICINI - Solar Flares with SMM and Implications for the Physics of Stellar Flares 321

J.G.DOYLE, J.C.RAYMOND, R.W.NOYES and A.E.KINGSTON - The Interpretation of EUV Spectra of Sunspots 325

L.GOLDBERG - Possible Origins of the 12μ Emission Lines in the Solar Spectrum 327

K.R.LANG - Multiple Wavelength Observations of Flaring Active Regions 331

M.R.KUNDU - Magnetic Development of Flaring Regions at Centimeter Wavelengths 335

J.KUIJPERS, K.F.TAPPING and D.GRAHAM - Very Long Baseline Interferometry of Solar Flares 339

S.CATALANO - RS CVn Stars: Photospheric Phenomena and Rotation (Invited Paper) 343

B.W.BOPP - RS CVn Stars: Chromospheric Phenomena (Invited Paper) 363

S.S.VOGT and G.D.PENROD - Doppler Imaging of Starspots 379

C.BLANCO, S.CATALANO, E.MARILLI and M.RODONO' - Twenty Years of Dedicated Photometry of RS CVn at Catania Observatory 387

G.GIURICIN, F.MARDIROSSIAN and M.MEZZETTI - Periods of Unevolved Late-Type Close Binaries: Evidence of Magnetic Activity 391

L.MILANO, G.RUSSO and S.MANCUSO - Migrating Waves in Solar-Type Short-Period Eclipsing Binaries 393

C.BLANCO, G.BODO, S.CATALANO, A.CELLINO, E.MARILLI, V. PAZZANI, M.RODONO' and F.SCALTRITI - Features of the Wave-Like Distortion in Some RS CVn Binaries 395

P.VIVEKANANDA RAO and M.B.K.SARMA - On the Distortion Wave in the Short Period RS CVn Type Eclipsing Binary WY Cancri 399

Y.W.KANG and F.B.WOOD - Recent Photometry of AR Lacertae 401

K.OLAH - HK Lacertae 403

E.F.MILONE and B.J.HRIVNAK - The Solar-Type Eclipsing Binary System AI Phoenicis 407

IL-S.NHA and J.Y.OH - Light Curves and Ca II Emissions of V711 Tauri During 1981-82 409

M.ZEILIK, R.ELSTON, G.HENSON and P.SMITH - A Flare-Like Event from the Short-Period System XY UMa 411

P.A.CHARLES - RS CVn Systems: The High Energy Picture (Invited Paper) 415

P.A.FELDMAN - Preliminary Results of a Five-Year Survey of Radio Emission from RS CVn and Similar Binaries 429

G.BASRI and R.LAURENT - Activity Correlations in Close Binary Systems 439

A.D.ANDREWS, P.B.BYRNE, C.J.BUTLER, J.L.LINSKY, T.SIMON, N.MARSTAD, M.RODONO', C.BLANCO, S.CATALANO, E.MARILLI and V.PAZZANI - IUE Spectra of RS CVn Stars 443

F.M.WALTER, D.M.GIBSON and G.S.BASRI - Coronal and Chromospheric Structure in AR Lac - II: Physical Characteristics of the Atmosphere 445

A.K.DUPREE - Contact Binary Stars (Invited Paper) 447

G.RUSSO, L.MILANO and S.MANCUSO - The Evolutionary Status of Short-Period RS CVn and Related W UMa Eclipsing Binaries 463

E.BUDDING - Angular Momentum Loss and the Formation of W UMa Systems 465

S.M.RUCINSKI, O.VILHU and J.KALUZNY - Activity of Contact Binaries 469

O.VILHU and S.M.RUCINSKI - Rotation-Activity Connections in Main Sequence Binaries 475

A.P.LINNELL - The VW Cephei System 479

K.E.EGGE and B.R.PETTERSEN - A Bright Spot and a Serendipitous Stellar Flare on the Contact-Binary VW Cep 481

L.XUEFU and L.CHENGZHONG - Numerical Analysis of Orbital Period Variations and a Mechanism for Changes in the Light Curve of VW Cep 485

R.E.GERSHBERG - On Activities of UV Cet-Type Flare Stars and of T Tau-Type Stars (Invited Paper) 487

G.F.GAHM and P.P.PETROV - The Variability of the T Tauri Star DI Cephei 497

W.HERBST - Optical Monitoring of T Tauri Stars 501

F.J.VRBA, A.E.RYDGREN and J.T.SCHMELZ - Periodic Light Variability in Four Late-Type Pre-Main-Sequence Stars 503

T.SIMON, P.R.SCHWARTZ, H.M.DYCK and B.ZUCKERMAN - New Results on the Binary Companion of T Tauri 505

A.BROWN and C.JORDAN - Chromospheric Properties of T Tauri Stars Determined from EUV Spectra 509

V.P.GRININ, P.P.PETROV and N.I.SHAKHOVSKAYA - Quasi-Periodic Variations of Balmer Line Profiles in Spectra of the T Tau-Type Star RW Aurigae 513

M.ŠOLC and J.SVATOŠ - Eruptive Phenomena as a Source of the Observed Variations of Intrinsic Polarization in Long Period and T Tau Variables 515

V.PIRRONELLO, G.STRAZZULLA and G.FOTI - The Interaction of Fast Particles with Frozen Gases in T Tau Nebulae: The Physical Background 519

G.STRAZZULLA, V.PIRRONELLO and G.FOTI - On the Origin of $H_2$ in T Tau Stars 523

SESSION IV: THEORETICAL ASPECTS

D.J.MULLAN - Models of Spots and Flares (Invited Paper) 527

E.R.PRIEST - Magnetic Instabilities in Stellar Atmospheres (Invited Paper) 545

D.S.SPICER - Flaring Processes in Stellar Atmospheres: The Importance of Electromagnetic Coupling (Invited Paper) 559

K.KODAIRA - Empirical Models of Stellar Flares: Constraints on Flare Theories (Invited Paper) 561

G.BELVEDERE - Dynamo Theory in the Sun and Stars (Invited Paper) 579

A.NORDLUND - Photospheric Sources of Magnetic Field Aligned Currents 601

H.U.BOHN - Turbulent Sound Generation in Red Dwarf Stars 605

M.K.DAS and J.N.TANDON - A New Triggering Mechanism for Red Dwarf Flares 609

V.P.GRININ - Theory and Observations of Negative Preflares in UV Cet Stars 613

M.M.KATSOVA and M.A.LIVSHITS - Flares on Red Dwarf Stars as a Result of the Dynamical Response of the Chromosphere to the Heating 617

G.BODO, A.FERRARI, S.MASSAGLIA and R.ROSNER - MHD Thermal Instabilities in Inhomogeneous Atmospheres 621

Y.UCHIDA - Magnetic Accretion Model for the Activities of Very Young Stars 625

Y.UCHIDA and T.SAKURAI - Interacting Magnetospheres in RS CVn Binaries: Coronal Heating and Flares 629

L.PATERNO' and F.ZUCCARELLO - Basic Parameters Determining X-Ray Emission Level in Stars of Spectral Type Later than F5 633

SESSION V: FUTURE RESEARCH DIRECTION AND CONCLUSION

N.O.WEISS - Future Research Directions: Theoretical Approach and Perspective (Invited Paper) 639

G.S.VAIANA - Observational Approach and Perspective (Invited Paper) 651

L.GOLDBERG - Summary of the Colloquium 653

Subject Index 663

# PREFACE

IAU Colloquium No. 71 had its immediate origins in a small gathering of people interested in the optical and UV study of flare stars which took place during the 1979 Montreal General Assembly. We recognized that a fundamental change was taking place in the study of these objects. Space-borne instruments (especially IUE and Einstein) and a new generation of ground-based equipment were having a profound effect on the range of investigations it was possible to make. To extract maximum benefit from these new possibilities it would be necessary as never before to have good communication with colleagues in other disciplines, for instance, with atomic and solar physicists. Similarly, studies of phenomena associated with the outer atmospheres of the late-type stars could now hope to give significant insights into certain aspects of solar activity. So, in view of the wide range of backgrounds of those participating, the meeting had an unusually high proportion of invited reviews while most of the contributed papers were presented as posters. It is gratifying that in the short time since the meeting a good deal of correspondence has been received from participants remarking on the success of this format.

Once the decision had been taken in principle to hold the meeting, a very considerable amount of work fell on the two organizing committees, viz. the Scientific and Local Organizing Committees. The Scientific Organizing Committee was chaired by D.J. Mullan and consisted of A.D. Andrews, B.W. Bopp, A.K. Dupree, R.E. Gershberg, G. Godoli, J.L. Linsky, M. Rodonò, O.B. Slee and G.S. Vaiana. The Local Organizing Committee comprised M. Rodonò (chairman), C. Blanco, S. Catalano, E. Marilli, L. Paternò and S. Serio.

We would like to gratefully acknowledge the cooperation of the IAU and in particular the General Secretary, P.A. Wayman, and Assistant General Secretary, R.M. West, at all stages of preparing for the Colloquium. The Colloquium was sponsored by IAU Commission No. 27 and co-sponsored by Commissions No. 10, 29, 40, 42, 44.

The following chaired the sessions: A.K. Dupree, C. Jordan, J.L. Linsky, D.J. Mullan, D.M. Popper, M. Rodonò, R. Rosner, Y. Uchida, G.S. Vaiana and N.O. Weiss.

There are a number of people we would like to thank whose names do not appear in any of the Committees. Sterling work was done by Fabia



*P. B. Byrne and M. Rodonò (eds.), Activity in Red-Dwarf Stars, xiii–xv.*

Foggi in the months before and both during and since the meeting. A. Calì, S. Del Popolo, G. Gentile, E. Martinetti, M.G. Minaldi, M. Miraglia, M.L. Rapisarda, S. Sardone, M.C. Scuderi, C. Sorge, and V. Stancanelli also assisted tirelessly in various capacities during the meeting. Their faces, if not their individual names, will be known to all who attended. Last, but far from least, our thanks go to Mrs. Eleonora Rodonò, who acted as interpreter, guide, social secretary and even personal confident on a number of occasions!

Vital financial support was provided by the National Research Council of Italy, the Astrophysical Observatory and the University of Catania, the Astronomical Observatory of the University of Palermo, the Ministero della Pubblica Istruzione, the Regione Siciliana, the Comune di Catania and the Amministrazione Provinciale di Catania. Social entertaiments were provided by Azienda Autonoma delle Terme di Acireale, Comune di Acicastello, Camera di Commercio di Catania, Ente Provinciale per il Turismo di Catania, and Azienda di Soggiorno e Cura di Acireale.

This preface would not be complete without commenting on a very special aspect of IAU Colloquium No. 71. Scientific intercourse was made doubly easy by the very successful social programme which accompanied the scientific meeting. Perhaps in this context we should thank one final chairman, viz. Dave Gibson, who presided during the closing dinner.

It has been our task as Editors to bring this large and diverse collection of material to a wider audience. Within each session the invited papers are followed by the related contributed ones. The order is not strictly that in which they were presented. Rather we have grouped the papers according to subject matter.

Discussions which took place after the orally presented papers were recorded on magnetic tape. We have transcribed these and they follow the text of the appropriate paper. There are inevitable gaps due to loss of recording for various reasons. Generally the discussion was transcribed word-for-word but to improve readability repetitions, excessive use of colloquialisms, etc were removed. We are confident that no unacceptable charge of emphasis resulted.

An index follows the text at the end of the book. In preparing it we were conscious of two dangers. The first was that of producing too sparse an index which would make the exercise a trivial one. The second lies in the direction of "safety first", including all references one could possibly imagine. We have tried to take a middle road, bearing in mind the likely use of the book. For instance, in the case of star names, we have not referenced every one appearing in the book but rather those which are discussed at some length.

Finally we would like to thank all of those authors who brought their manuscripts in camera-ready form at the beginning of the meeting as requested, making our editorial duties much easier. Our thanks go to D.Reidel Publ.Co. in the person of Mrs. Pols v.d.Heijden for their excellent cooperation.

30th October 1982 P.B. Byrne and M. Rodonò

1. Agrawal, 2. Linnell, 3. Bopp, 4. La Fauci, 5. Wing Jr., 6. Wing, 7. Bodo, 8. Rodonò, 9. Kameswara Rao, 10. Fischerstrom. 11. Weiss, 12. Scaltriti, 13. Mrs. Scaltriti, 14. Das, 15. Doyle, 16. Gokhale, 17. Goldberg, 18. Pallavicini, 19. Simon, 20. Pirronello, 21. Olah, 22. Bruca, 23. Gimenez, 24. Herrero, 25. Sarma, 26. Lang, 27. Magazzù, 28. Andersen, 29. Pidatella, 30. Butler, 31. Elliott, 32. Tanaka, 33. Paternò, 34. Belvedere, 35. Oskanian, 36. Bohn, 37. Haisch, 38. Giampapa, 39. Walter, 40. Brown, 41. Feldman, 42. Linsky, 43. Charles, 44. Venugopal, 45. Torres, 46. Van Leeuwen, 47. Reglero Velasco, 48. Golub, 49. Cutispoto, 50. Worden, 51. Stencel, 52. Basri, 53. Mrs. Catalano, 54. Catalano, 55. Stern, 56. Sakurai, 57. Kuypers, 58. Kodaira, 59. Angelico, 60. Calvet, 61. Ducati, 62. Baliunas, 63. Priest, 64. Hidayat, 65. Serio, 66. Dupree, 67. non participant, 68. Rucinski, 69. Byrne, 70. Budding, 71. Popper, 72. Peres, 73. Bookmeyer, 74. Mrs. Vaiana, 75. Mrs. Serio, 76. Uchida, 77. Alphenaar, 78. Melikyan, 79. Chugainov, 80. Blanco, 81. Rosner, 82. Mullan, 83. Andrews, 84. Maggio, 85. Vaiana, 86. Sciortino, 87. Ni, 88. Mattig, 89. Gibson, 90. Jordan, 91. De La Reza, 92. Pazzani, 93. Zuccarello, 94. Rapisarda (LOC), 95. Godoli, 96. Del Popolo (LOC), 97. Mrs Rodonò, 98. Micela, 99. Simnett, 100. Tsai, 101. Pettersen, 102. Collier, 103. Collura, 104. Strazzulla, 105. Minaldi (LOC), 106. Scuderi (LOC), 107. Calì (LOC), 108. Foggi (LOC), 109. Nha.

## LIST OF PARTICIPANTS

Agrawal P.C., Tata Institute of Fundamental Research, Homi Bhabha Road, Bombay 400 005, India

Alphenaar P., Sterrewacht Leiden, Huygens Laboratorium, Wassenaarseweg 78, Leiden 2300RA, The Netherlands

Andersen B.N., Institute of Theoretical Astrophysics, University of Oslo, P.B.1029, Blindern, Oslo 3, Norway

Andrews A.D., Armagh Observatory, College Hill, Armagh BT61 9DG, Northern Ireland, UK

Angelico G., Osservatorio Astrofisico, Città Universitaria, 95125 Catania, Italy

Anile M.A., Seminario Matematico, Città Universitaria, 95125 Catania, Italy

Baliunas S.L., Center for Astrophysics, 60 Garden Street, Cambridge, MA 02138, USA

Basri G., Department of Astronomy, University of California, Berkeley, CA 94720, USA

Belvedere G., Istituto di Astronomia, Città Universitaria, 95125 Catania, Italy

Blanco C., Osservatorio Astrofisico, Città Universitaria, 95125 Catania, Italy

Bodo G., Osservatorio Astronomico, 10025 Pino Torinese, Italy

Bohn H.U., Instituteof Astronomy and Astrophysics, Am Hubland, D-8700 Wurzburg, Germany FR

Bookmyer B.B., University of Maryland, European Division, Im Bosseldorn 30, 6900 Heidelberg, Germany FR

Bopp B.W., Department of Physics and Astronomy, University of Toledo, Toledo, OH 43606, USA

Bromage G.E., Building R25, Rutherford-Appleton Laboratory, Chilton, Didcot, Oxon OX11 0QX, UK

Brown A., Department of Physics, Queen Mary College, Mile End Road, London E1 4NS, UK

Bruca L., S.I.S.S.A., International Center for Theoretical Physics, Miramare, 34100 Trieste, Italy

Budding E., Department of Astronomy, University of Manchester, Manchester M13 9PL, UK

Butler C.J., Armagh Observatory, College Hill, Armagh BT61 9DG, Northern Ireland, UK

Byrne P.B., Armagh Observatory, College Hill, Armagh BT61 9DG, Northern Ireland, UK

Calvet N., C.I.D.A., Apartado 264, Merida 5101-A, Venezuela

Castellani V., Istituto Astrofisica Spaziale, Casella Postale 67, 00044 Frascati, Italy

Catalano F.A., Osservatorio Astrofisico, Città Universitaria, 95125 Catania, Italy

Catalano S., Istituto di Astronomia, Città Universitaria, 95125 Catania, Italy

Cellino A., Osservatorio Astronomico, 10025 Pino Torinese, Italy

Charles P.A., Department of Astrophysics, South Parks Road, Oxford OX1 3RQ, UK

Chugainov P.F., Crimean Astrophysical Observatory, P.O. Nauchny, 334413 Crimea, USSR

Collier A.C., Physics Department, University of Canterbury, Christchurch, New Zealand

Collura A., Osservatorio Astronomico, Palazzo dei Normanni, 90134 Palermo, Italy

Cutispoto G., Osservatorio Astrofisico, Città Universitaria, 95125 Catania, Italy

Das M.K., Department of Physics and Astrophysics, S.V. College, Delhi University, Dhaula Kuan, New Delhi 110 021, India

De La Reza R., CNPQ Observatorio Nacional, Rua General Bruce 586, 20921 Rio De Janeiro, Brazil

Doyle J.G., Armagh Observatory, College Hill, Armagh BT61 9DG, Northern Ireland, UK

Ducati J.R., Centre de Donnees Stellaire, Observatoire de Strasbourg, 11 rue de l'Universite, 6700 Strasburg, France

Dupree A.K., Center for Astrophysics, 60 Garden Street, Cambridge, MA 02138, USA

Elliott I., Dunsink Observatory, Castleknock, Co.Dublin, Ireland

Evans D.S., Department of Astronomy, University of Texas, Austin, TX 78712, USA

Feldman P.A., Herzberg Institute of Astrophysics, National Research Council of Canada, Ottawa K1A 0R6, Canada

Fischerstrom C., Stockholms Observatorium, S-13300 Saltsjobaden, Sweden

Gahm G., Stockholms Observatorium, S-13300 Saltsjobaden, Sweden

Gershberg R.E., Crimean Astrophysical Observatory, P.O. Nauchny, 334413 Crimea, USSR

Giampapa M.S., Center for Astrophysics, 60 Garden Street, Cambridge, MA 02138, USA

Gibson D.M., Physics and Astronomy Department, New Mexico Tech., Socorro, NM 87801, USA

Gimenez A., Departamento de Astrofisica, Facultad de Fisicas, Universidad Complutense, Madrid 3, Spain

Giuricin G., Osservatorio Astronomico, Via G.B.Tiepolo 11, 34131 Trieste, Italy

Godoli G., Cattedra di Fisica Solare, Università degli Studi, Largo E.Fermi 5, 50125 Firenze, Italy

Gokhale M.H., Indian Institute of Astrophysics, Bangalore 560 034, India

Goldberg L., Kitt Peak National Observatory, P.O.Box 26732, Tucson, AZ 85726, USA

Golub L., Center for Astrophysics, 60 Garden Street, Cambridge, MA 02138, USA

Haisch B.M., Div.52-54, Bldg 201, Lockeed Palo Alto Research Laboratory, 3251 Hanover St., Palo Alto, CA 94304, USA

Hartmann L., Center for Astrophysics, 60 Garden Street, Cambridge, MA 02138, USA

Herrero A., Instituto de Astrofisica de Canarias, Universidad De La Laguna, Tenerife, Spain

Hidayat B., Bosscha Observatory, Lembang, Java, Indonesia

Johnson H.M., Dept. 52-12, Bldg. 255, Lockheed Missiles and Space Co., 3251 Hanover Street, Palo Alto, CA 94304, USA

Jordan C., Department of Theoretical Physics, University of Oxford, 1 Keble Road, Oxford OX1 3NP, UK

Kameswara Rao N., Indian Institut of Astrophysics, Bangalore 560034, India

Kodaira K., Department of Astronomy, Faculty of Science, University of Tokyo, Bunkyo-ku, Tokyo 113, Japan

Kuijpers J., Astronomical Institute, Zonnenburg 2, 3512 Utrecht, The Netherlands

La Fauci G., Osservatorio Astrofisico, Città Universitaria, 95125 Catania, Italy

Lang K.R., Department of Physics, Robinson Hall, Tufts University, Medford, MA 02155, USA

Linnell A.P., Department of Physics and Astronomy, Michigan State University, E.Lansing, MI 48824, USA

Linsky J.L., Joint Institute for Laboratory Astrophysics,University of Colorado, Boulder, CO 80309, USA

Magazzù A., S.I.S.S.A., Inter.Center Theor.Physics, Miramare, 34100 Trieste, Italy

Maggio A., Osservatorio Astronomico, Palazzo dei Normanni, 90134 Palermo, Italy

Marilli E., Osservatorio Astrofisico, Città Universitaria, 95125 Catania, Italy

Mattig W., Kiepenheuer Institut, Schoneckstr. 6, 7800 Freiburg, Germany FR

Melikyan N.D., Byurakan Astrophysical Observatory, Erevan, 378433 Armenia, USSR

Micela G., Osservatorio Astronomico, Palazzo dei Normanni, 90134 Palermo, Italy

Milone E.F., Physics Department, University of Calgary, Calgary, Alberta T2N IN4, Canada

Monsignori Fossi B., Osservatorio Astrofisico di Arcetri, Largo E.Fermi 5, 50125 Firenze, Italy

Mullan D.J., Bartol Research Foundation, University of Delaware, Newark, DE 19711, USA

Nha Il-S., Department of Astronomy and Meteorology, Yonsei University, Seoul (Private Bag), Korea

Ni W.T., Department of Physics, National Tsing Hua University Hsinchu, Taiwan, Republic of China

Nordlund A., University Observatory, Oster Voldgate 3, 1350 Copenhagen K, Denmark

Olah K., Konkoly Observatory, P.O.Box 67, 1525 Budapest, Hungary

Oskanian V.A., Byurakan Astrophysical Observatory, Erevan, 378433 Armenia, USSR

Pallavicini R., Osservatorio Astrofisico di Arcetri, Largo E.Fermi 5, 50125 Firenze, Italy

Paternò L., Osservatorio Astrofisico, Città Universitaria, 95125 Catania, Italy

Pazzani V., Osservatorio Astrofisico, Città Universitaria, 95125 Catania, Italy

Peres G., Osservatorio Astronomico, Palazzo dei Normanni, 90134 Palermo, Italy

Pettersen B.R., Institute of Mathematical and Physical Sciences, University of Tromsø, O.O.Box 953, N-9001 Tromsø, Norway

Pidatella R.M., Osservatorio Astrofisico, Città Universitaria, 95125 Catania, Italy

Pirronello V., Osservatorio Astrofisico, Città Universitaria, 95125 Catania, Italy

Popper D.M., Department of Astronomy, University of California, Los Angeles, CA 90024, USA

Priest E.R., St.Andrews University, St.Andrews, Fife KY16 9SS, Scotland, UK

Reglero Velasco V., Universidad de Valencia, Facultad de Matematica, Departamento de Mecanica y Astronomia, Valencia, Spain

Rodonò M., Osservatorio Astrofisico, Città Universitaria, 95125 Catania, Italy

Rosner R., Center for Astrophysics, 60 Garden Street, Cambridge, MA 02138, USA

Rucinski S., Max Planck Institut for Astrophysics, K.Schwarzschild Str.1, 8042 Garching bei Munchen, Germany FR

Russo G., Osservatorio Astronomico di Capodimonte, via Moiarello 16, 80131 Napoli, Italy

Sakurai T., Department of Astronomy, University of Tokyo, Bunkyo-ku, Tokyo 113, Japan

Sarma M.B.K., Department of Astronomy, Osmania University, Hyderabad 500 007, Andhra Pradesh, India

Saxner M., Astronomiska Observatoriet, Box 515, 751 20 Uppsala, Sweden

Scaltriti F., Osservatorio Astronomico, 10025 Pino Torinese, Italy

Sciortino S., Osservatorio Astronomico, Palazzo dei Normanni, 90134 Palermo, Italy

Serio S., Osservatorio Astronomico, Palazzo dei Normanni, 90134 Palermo, Italy

Simnett G.M., Department of Space Research, P.O.Box 363, University of Birmingham, Birmingham B15 2TT, UK

Simon T., Institute of Astronomy, University of Hawaii, 2680 Woodlawn Dr., Honolulu, HI 96822, USA

Soderblom D.R., Center for Astrophysics, 60 Garden Street, Cambridge, MA 02138, USA

Solc M., Department of Astronomy and Astrophysics, Charles University, Svedska 8, 150 00 Praha 5, Czechoslovakia

Spadaro D., Osservatorio Astrofisico, Città Universitaria, 95125 Catania, Italy

Spicer D.S., Institute of Astronomy, ETH-Zentrum, 8092 Zurich, Switzerland

Stencel R.E., Joint Institute Laboratory Astrophysics, University of Colorado, Boulder, CO 80309, USA

Stern R.A., M.S. 169-327, Jet Propulsion Laboratory, 4800 Dak Grove Dr., Pasadena, CA 91109, USA

Strazzulla G., Osservatorio Astrofisico, Città Universitaria, 95125 Catania, Italy

Tanaka K., Tokyo Astronomical Observatory, Mitaka, Tokyo, Japan

Ternullo M., Osservatorio Astrofisico, Città Universitaria, 95125 Catania, Italy

Tornambè A., Istituto di Astrofisica Spaziale, Casella Postale 67, 00044 Frascati, Italy

Torres C.A.P.C.O., CNPQ/ON, Observatorio Astrofisico Brasileiro, Caixa Postal 21, 37500 Itajuba MG, Brazil

Tsai C.H., Taipei Observatory, Yuan Shan, Taipei 104, Taiwan, Republic of China

Uchida Y., Tokyo Astronomical Observatory, University of Tokyo, Mitaka 181, Tokyo, Japan

Vaiana G.S., Osservatorio Astronomico, Palazzo dei Normanni, 90134 Palermo, Italy

Van Leeuwen F., Royal Greenwich Observatory, Herstmonceux Castle, Hailsham, East Sussex BN27 1RP, UK

Vazquez M., Instituto de Astrofisica de Canarias, Universidad De La Laguna, Tenerife, Spain

Vennerstrom S., Astronomisck Observatorium, Oster Voldgade 3, 1350 Copenhagen, Denmark

Venugopal V.R., Radio Astronomy Centre, Tata Institute of Fundamental Research, P.O.Box 8, Ootacamund 643 001, India

Vogt S.S., Lick Observatory, University of California, Santa Cruz, CA 95064, USA

Vrba F.J., U.S.Naval Observatory, Flagstaff Station, P.O.Box 1149, Flagstaff, AZ 86002, USA

Walter F.M., Joint Institut Laboratory Astrophysics, Boulder, CO 80309, USA

Weiss N.O., Department of Applied Mathematics and Theoretical Physics, Silver Street, Cambridge CB3 9EW, UK

Wing R.F., Astronomy Department, The Ohio State University, 174 West 18th Avenue, Columbus, OH 43210, USA

Wood F.B., Department of Astronomy, University of Florida, Gainesville, Florida 32611, USA

Worden S.P., Astronomy Department, UCLA, Los Angeles,
CA 90024, USA

Zuccarello F., Osservatorio Astrofisico, Città Universitaria,
95125 Catania, Italy

# INTRODUCTORY ADDRESS

M. Rodonò
Osservatorio Astrofisico and Università degli Studi, Catania, Italy

Mr. Chairman, Official Representatives of the National, Regional and Local Government and Institutions, Ladies and Gentlemen,

The count-down to start our meeting is now over, and it is my great privilege and honour to give the Opening Address of the 71st Colloquium of the International Astronomical Union on "Activity in Red-Dwarf Stars". First of all let me thank all of you for accepting our invitation to attend the Opening Ceremony. Such an impressive gathering of so many distinguished astrophysicists from all over the world is in itself a reward for all of us in the Scientific and Local Organizing Committees for the demanding work which we have been doing. We hope to meet all your expectations for a pleasant and scientifically fruitful meeting.

Many people have made essential contributions at the various stages of the preparation of our Colloquium in its various scientific, organizational and financial aspects. All these combined efforts have made it possible, first to conceive, then to shape, reshape, implement and eventually see the beginning of the first meeting on Stellar Activity approved by the International Astronomical Union and the first I.A.U. Colloquium in Catania.

Other related meetings have preceded the present one in the recent past, such as those held at the Harvard-Smithsonian Center for Astrophysics, in Bonas and some of you are just emerging from the related Symposium in Zurich on Solar and Stellar Magnetic Fields. Actually, an ever increasing amount of work on stellar variability in the past two decades has promoted a widespread interest in stellar activity, both from an observational and theoretical point of view. Clearly, stellar variability is no longer a source of fear such as led our ancestors to label one of the most famous large-amplitude variable stars with the Arabic equivalent of "devil", i.e. Algol. On the contrary, stellar variability



*P. B. Byrne and M. Rodonò (eds.), Activity in Red-Dwarf Stars, 1–3.*

phenomena are closely scrutinized because the pathological situations that trigger variability are an invaluable source of information in the study of stellar physics.

To synthetize all the pieces of information we have been collecting in the recent past, we are here to discuss, to agree - and maybe disagree - on observations and especially on interpretation and theories. Each of us is the depository of single pieces of a highly difficult puzzle, or mosaic, that needs to be put into place. Therefore, this Conference Hall is something like a puzzle box. In the coming days the puzzle box will be opened and, with the various pieces in our hands, we will try to speculate on which scenario is to be built and how to combine all the available pieces. However, we have a reference map to look at, the Sun. Its activity is known in, perhaps, too many details. Nevertheless, problems that were considered exclusively solar have now become central topics for stellar physics. Stellar spots, flares, plages, transition regions and coronae are very recent issues in astrophysics. The macroscopic phenomena we observe on stars are seen in much greater detail on the Sun, so that the microphysics of stellar activity is also available to us. Eventually, solar and stellar astrophysicists have found what a distinguished colleague, who is here with us, has properly defined a fruitful "two-way street", by interpreting the current trend of thought.

Solar-type phenomena in stars other than the Sun were postulated since the beginning of this century. But actual progress has required long-term dedicated observational and theoretical programs. Moreover, enormous stimulus has been given by recent observations with instruments on board artificial satellites that have allowed us to study the outermost layers of stellar atmospheres in previously unexplored spectral bands.

My own view is, and I believe that most of you will agree, that the major achievement of the recent past is the concept that stellar atmospheres are not passive intermediaries between the stellar interior and the outside world: something more or less like opaque walls shaping the stars. Stellar atmospheres are "active" structures exhibiting a whole panoply of spectacular solar-type phenomena. If the "two-way-street" connecting solar and stellar physics is fully exploited, it will be possible to outline a much better picture of both the Sun and the stars. "If it ain't broke, don't fix it" goes and old American saying. But the classical picture of stellar atmospheric equilibrium and energy balance is somehow cracked and needs to be fixed, especially after the fundamental discovery of the important role of stellar magnetic fields, as a result of SKYLAB and EINSTEIN satellite observations. We will hear in the coming days additional exciting results from SMM and HINOTORI as well as from IUE, an apparently inexhaustible source of fundamental discoveries.

What are the motivations for a meeting in Catania on a specialized topic such as stellar activity on red-dwarfs? Several reasons underlie this choice. First of all, both solar and stellar research work have been a long-lasting tradition at Catania Observatory and at the Institute of Astronomy of Catania University: records of solar activity from the last century are on our files; dedicated photometry of several active stars was started in 1963, when I had what turned out to be the good fortune of beginning my thesis work on the now famous RS Canum Venaticorum system, which is one of the best example of stars showing huge photospheric, chromospheric and coronal activity phenomena.

Similar dedicated observations of another interesting "spotted" star, BY Draconis,were independently started in the Crimea soon afterwards. In subsequent years stellar activity observations were carried out at an ever increasing number of places. Apart from a few dispersed stellar flare observations, the sixties and seventies also witnessed a booming of systematic and fundamental research programs at Armagh, Byurakan, Catania, Crimea and McDonald Observatories. This large data base has already made important contributions to our knowledge of stellar flares and spot phenomena.

Lastly, thanks to a collaborative effort involving the Joint Institute for Laboratory Astrophysics (JILA), the Harvard Smithsonian Center for Astrophysics and the Armagh and Catania Observatories, dedicated research programs have recently been extended into the UV and X-ray bands to study in a more systematic way stellar activity in the outermost atmospheric levels. Also, recent theoretical work on stellar dynamos and on activity has been carried out successfully at Catania. I think that the above research activity justifies the choice of Catania for the 71st Colloquium of the I.A.U.. However, I should stress that the enthusiasm and total dedication of many colleagues, and friends, in the study of stellar activity in red-dwarf and related objects made the Colloquium possible.

I hope that at the end of this Colloquium, some of the small pieces of our puzzle will fall into place, so that an easier job will be left to the participants at the next meeting on "Stellar Activity", wherever it may be.

# MAGNETIC FIELDS AND ACTIVITY OF THE SUN AND STARS: AN OVERVIEW

Robert Rosner
Harvard-Smithsonian Center for Astrophysics
Cambridge, MA 02138 USA

## ABSTRACT

I review recent work on the observation and theory of solar and stellar magnetic field activity and its relation to stellar activity, with particular emphasis on those aspects relevant to the problem of activity of red dwarf stars.

## 1. INTRODUCTION

The low-mass stars on the main sequence have long fascinated students of both stellar and galactic structure: it is in this mass range that main sequence stars may become fully convective and attain lifetimes on the main sequence comparable to the age of our galaxy; and it is these stars that provide the dominant component of the luminous mass of our galaxy and very likely constitute the dominant discrete source of the galactic component of the diffuse soft x-ray background. These various characteristics are without doubt of considerable interest in and of themselves; but it is remarkable that a number of these unique features are tied to a rather general problem of astrophysical interest: the connection between magnetic field generation in turbulently-convecting fluids and the presence of "activity".

The physical connection between "activity" on late-type (low-mass) main sequence stars and magnetic field dynamics in such stars is not well-understood. The intimate phenomenological relation between magnetic fields and the flaring activity of *UV Ceti*-type stars, and the similarily close relation between stellar magnetic fields and the optical light modulation associated with *BY Draconis* stars, have of course been long established. However, some of the most basic questions which arise cannot be easily answered: how are the magnetic fields generated in the interiors of these stars, and brought to the surface; what is the detailed relation between magnetic field emergence and stellar surface activity; how is stellar magnetic activity related to the parameters that presumably define the physical state of a star: its mass, composition, age, and rotation rate. These difficulties may be traced to at least three distinct historical problems. First, until recently, direct or indirect measurements of magnetic fields on late-type stars other than the Sun were simply not available. Second, the absence of (sufficiently-sensitive) observations which directly showed evidence for solar-like surface actvity (i.e., observations of *quiescent* emission from a solar-like outer atmosphere) allowed for the possibility that the observed phenomena were in fact rather non-solar in character; hence, the extent to which the solar analogy could be applied was not at all self-evident. Finally, both chromospheric and coronal physics, as well as MHD and dynamo theory, had not advanced sufficiently to be able to predict with any assurance the likely behavior of magnetic fields on low-mass stars, and the consequences for the structure of their outer atmospheres. This situation has now in part dramatically changed. The advent of



*P. B. Byrne and M. Rodonò (eds.), Activity in Red-Dwarf Stars, 5–14.*

the *International Ultraviolet Explorer (IUE)* and the *Einstein Observatory*, and the application of advanced detector and spectroscopic technology at ground-based facilities in both the radio and optical domains, have now allowed study of the quiescent component of the outer atmospheres of low-mass stars and, in the course of these studies, strongly reenforced the notion that what one is observing is indeed analogous to solar surface activity (Noyes 1981; Rosner 1982). In parallel with these advances, solar astronomers have substantially enlarged our knowledge of the fine structure of the solar photosphere (cf. Tarbell 1983), as well as of the solar interior (cf. Deubner 1981); these studies provide the essential fundamental observational constraints on models of magnetic activity, as well as define the interpretive framework for stellar observations of analogous phenomena.

It is remarkable that theory now also finds itself in a period of ferment. In addition to very much increased levels of sophistication of numerical simulations (exemplified by the calculations of Gilman 1982 and Frisch 1983 and collaborators), new analytical techniques have come to the fore, most prominently the systematic application of bifurcation theory to the study of hydrodynamic and magnetohydrodynamic instabilities and dynamo problems (see N. O. Weiss in these Proceedings). The vigorous level of the discussions held at the recent IAU symposium on the subject of solar and stellar magnetic fields in Zurich (Stenflo 1983) are testimony to the excitement currently felt by both observers and theoreticians in this field; the organizers of this Colloquium have asked me to convey some of the flavor of these discussions, and to place them in the context of activity in red dwarf stars. My aim is not to provide a complete overview of magnetic activity in stars (see instead Schussler 1983 and Belvedere in this volume), but rather to call particular attention to some of the major recent results (and question which have subsequently arisen) which in my view signal a new way of looking at the problem of stellar activity.

## 2. ON SOME OBSERVATIONAL QUESTIONS

In much of astrophysics, successful theoretical prediction of new observational effects is a rarity; generally, the theorist is obliged to look toward observations for guidance in defining the relevant physical effects and in picking out the important (= relevant) parameter regimes to study. The student of stellar activity is in this sense hardly at an advantage since the governing physics involves the extreme nonlinearities associated with magnetohydrodynamics and plasma physics; it therefore behooves us to begin with the observations. The problem at hand is to understand the root cause of stellar surface activity; the level of difficulty of this problem is exemplified by the fact that it is by no means obvious whether the observations relevant to the construction of theories can indeed be carried through (even in principle). What then are the new observational facts that have occasioned all the excitement?

Central to the new observational perspective are the realizations, first, that stellar surface activity is common to *all* dwarf stars of roughly solar mass and less; and, second, that the wide range of observed stellar activity levels at any given, fixed spectral type (as manifested in, for example, stellar x-ray emission; Vaiana *et al.* 1981) is not simply related to the stellar properties (composition, mass, and luminosity) which largely define a star's position in the H-R diagram. Indeed, one of the major results of the *Einstein* surveys is that the total x-ray luminosity of late-type stars appears to be only very weakly related to the effective stellar surface temperature (and hence to the level of surface fluid turbulence). What the relevant set of stellar parameters determining stellar activity levels might involve was suggested early on by *HEAO 1* observations of *RS CVn* stars, which indicated that rotation might be a significant determinant of coronal luminosity for these binary systems (Ayres & Linsky 1980; Walter & Boywer 1981). Such a connection between rotation and *chromospheric* activity level was of course long known from observations of chromospheric Ca II emission from nearby solar-type dwarf stars (Wilson 1966; Skumanich 1972); but the recent work of Vaughan and collaborators (Vaughan 1983; Noyes 1983) have placed the question of the detailed correlation of chromospheric activity of late-type dwarfs with spectral type and rotation rate on a far more exacting footing. Furthermore, studies based on the *Einstein*/CfA stellar survey data have established a *coronal* luminosity-rotation correlation for isolated or effectively single late-type dwarf stars, which shows that the level of coronal emission, once a well-developed surface convection zone exists, depends little on the effective temperature of the underlying star; it seems that the x-ray luminosity data is well-described by a power law dependence on the rotation rate (with an exponent in the range 1 - 2), largely independent of spectral type (Pallavicini *et al.* 1981, 1982; Walter 1981). These studies have gained particular force since the analyses of stellar "activity" parameters have moved away from an anectodal approach; thus, substantial effort has recently been made to systematically construct stellar x-ray luminosity functions (Topka *et al.* 1982; Rosner *et al.* 1981; Rosner 1983) and to fully simulate the characteristics (including possible selection effects) of the Ca II data samples (Noyes 1983; Hartmann *et al.* 1983).

Roughly contemporaneously, the extensive modeling of solar data from *Skylab*, *OSO-8*, and the *Solar Maximum Mission (SMM)* have given us a very good idea of how stellar surface activity correlates with magnetic fields, as several papers presented in these Proceedings make clear: chromospheric and coronal activity seems to be very much tied to the emergence of magnetic flux to the stellar surface, as well as to its subsequent evolution on the surface (see, for example, Golub *et al.* 1981). Indeed, the ACRIM observations of solar bolometric luminosity fluctuations (Willson *et al.* 1981) strongly suggest that magnetic fields in the outer convection zones of stars modulate the total stellar luminosity (see also Hartmann & Rosner 1979; Spiegel & Weiss 1980). Furthermore, advances in ground-based observing techniques have led to increased understanding of the small-scale structure of solar surface magnetic fields (Tarbell 1983); and allow study of the solar interior (in particular, of the temperature stratification and rotational state of the convection zone) with the advent of "solar seismology" (cf. Deubner 1981 and references therein). In the latter case, resolution of the spectrum of the "5-minute oscillations" by ACRIM raises the possibility that similar studies applied to stars may be possible, leading to the remarkable prospect that the internal structure of convection zones on stars other than the Sun may be accessible to observational study (Hudson 1982). Thus, both solar and stellar data now give us firm observational grounds for believing that the interaction of magnetic fields with turbulent fluids is central to the problem of the chromospheric and coronal phenomenon on late-type dwarf stars; and, conversely, that one might hope to use observations of stellar surface activity to probe magnetic field dynamics in the outer convection zones of stars.

Now, there is no single, decisive, piece of observational evidence that the solar analogy is apt for stars other than the Sun; instead, at least four independent lines of reasoning converge to strongly support this analogy: first, it has now been possible to show that the classic indicators of activity on the solar surface (e.g., Ca II emission, UV transition region line emission, and x-ray emission) are observed from stars, and do correlate (cf. Pallavicini *et al.* 1982; Zwaan 1983 and references therein). Second, simultaneous ground-based and space observations of the rotational modulation of activity-related emission have shown that stellar chromospheric and coronal emission is spatially correlated with photospheric regions thought to be dominated by strong magnetic fields (analogous to the general association of sun spots with solar active regions; Kahler *et al.* 1981; Baliunas *et al.* 1983; Marcy 1983). Third, detailed modeling of stellar chromospheric, transition region, and coronal emission based on simple extensions of solar "loop" models seem to give a fairly good account of the observations (cf. articles by Dupree, Giampapa, Golub, and Linsky in these Proceedings, and by Linsky 1983 and Vaiana 1983). Fourth, the range of variability in stellar activity-related emission – ranging from short, flare-like transients to rotational modulation and long-term, cycle-like variations – comport with expectations based on solar observations (cf. Noyes 1983; Vaiana 1983). It therefore seems safe to conclude that, at least to first order, "activity" phenomena on late spectral-type dwarf stars may be thought of as a variant of familiar solar phenomena [although the new stellar magnetic field measurement techniques originally applied by Robinson, Worden, and Harvey 1980 yield the remarkable result that active solar-type stars can have both large magnetic field strengths (> 1 kG) *and* large field surface area covering factors (up to ≈ 75% of the visible disk; Marcy 1983) – these are hardly solar conditions].

Recent studies have uncovered two further facets of the data which bear on the question of stellar activity. The first effect of note is the possibility of a "gap" in total Ca II emission strength (the so-called "Vaughan-Preston gap") for any fixed (main sequence) spectral type; that is, there appears to be a range of Ca II emission levels at fixed spectral type (in the B-V range ≈ 0.45 - 1.0) in which there is a relative absence of field stars (Vaughan & Preston 1980). Since such a gap is apparently not seen in either the Li abundance or rotation data (Soderblom 1983), one might ask whether the effect is real (in which case the several contending theoretical accounts already published become relevant; see Durney, Robinson, & Mihalas 1981 and Knobloch, Rosner, & Weiss 1981), or is due to some as yet unrecognized selection effect(s). From a theoretician's perspective, one hopes that the "gap" is vindicated: a dramatic change in chromospheric activity at some fixed rotation rate would add a powerful constraint to theories of stellar magnetic activity; more observational work will be needed to establish this result.

A further interesting new result bears on the question of where dynamo action takes place in a stellar convection zone. It has been argued recently that the production of toroidal magnetic fields in the Sun must take place at large depth, basically because the emergence of magnetic fields due to magnetic buoyancy can be significantly altered only if the flux resides in a stably-stratified region (e.g., the convective overshoot region; Rosner 1980; Spiegel & Weiss 1980; Schmitt & Rosner 1982). A direct test of this idea would be to examine stars which do not have a radiatively-stratified interior: that is, the fully-convective M dwarfs. Such a study has been carried out recently by M. Giampapa (1983), with the remarkable result that main sequence M stars of very late spectral type seem to show (with considerable statistical uncertainty) an absence of Hα emission (an effect which seems to find some corroboration in the available x-ray data for very late dwarf M stars from the *Einstein* data; see L. Golub in this volume). This is a result well-worth of further pursuit.

## 3. ON SOME THEORETICAL QUESTIONS

What is the underlying physical basis for the observed correlations between stellar activity, rotation, and magnetic fields? It is probably fair to say that a complete answer currently exists only at the level of a "cartoon explanation", based on the ideas outlined by Parker (1979): a regenerative magnetic dynamo in a rotating, convecting star produces magnetic fields which inevitably rise to the stellar surface, where these magnetic fields are continually "jostled" by the turbulent convective surface motions; this "jostling" of the emerged fields presumably leads to plasma heating, and hence to a chromosphere and corona.

Can one do better than this qualitative description? Upon closer examination, the theoretical underpinnings of our current understanding of stellar activity dissolve into a myriad of subproblems, each of which is in its own right not well-understood at present: formal kinematic and dynamical dynamo theory, turbulent magnetic field diffusion, magnetic flux tube formation and dynamics, and so forth. A first-principles theory (which starts with the equations of stellar structure and Maxwell's equations, and attempts to predict, for example, coronal emission levels as a function of, say, stellar composition, mass, age, and rotation rate) thus seems well out of reach; a more realistic assessment of the immediate future of theory (at least as regards the dynamo problem) is given by N. O. Weiss in these Proceedings. In the following, I would instead like to briefly explore the current status of some of the major theoretical elements which enter into the discussion of magnetic field-dominated stellar activity.

The theoretical problems which arise in discussing the "rotation-activity" and "magnetic field-activity" connections can be conveniently grouped into two distinct categories, in each of which substantial progress has recently been made:

***(i) Dynamo theory and the "rotation-activity" connection.*** The problem of magnetic field generation in the solar interior has been attacked from both the point of view of full simulations and "model" (non-linear) calculations. The general status of dynamo theory is reviewed in these Proceedings by G. Belvedere (see also Schussler 1983); and the latter (non-linear "model") calculations are covered in some detail in N. O. Weiss' discussion in this volume. I will therefore not dwell further on non-linear "model" calculations, except to point out that such calculations are the best antidote to the impulse (felt by some) to extrapolate the behavior of classic linear (kinematic) dynamo theory to the non-linear domain: as can be seen from, for example, the recent calculations of Cattaneo *et al.* (1983), the solutions to the non-linear MHD equations bear little, if any, resemblence to the solutions of the linear problem.

Because the recent full magnetohydrodynamic (numerical) simulations of convection and dynamo action in spherical shells carried out by Gilman (1982, 1983) would appear to be most relevant to the specific problem of predicting surface magnetic field activity levels as a function of stellar parameters, I would like to focus on these for the moment. Among Gilman's several general conclusions, the most relevant to the present discussion is that the "$\alpha$-effect" (due to non-vanishing mean helicity of motions on fairly large scales, which results in the generation of meridional magnetic fields from toroidal magnetic fields) appears to be far more vigorous than previously suspected (to the point that, in the limit of using standard values for the eddy transport coefficients, one obtains Coriolis force-dominated solutions -- an "$\alpha^2$-dynamo" -- rather than the standard "$\alpha\omega$-dynamo"). As pointed out by Gilman, these results are subject to several major qualifications: first, because of limited spatial resolution (imposed by computational limitations), the formation and dynamics of "flux tubes" cannot be followed; the limited spatial grid resolution is also responsible for the inclusion of eddy transport coefficients (because the scales on which true diffusive behavior occurs cannot be modeled simultaneously with the large spatial scale dynamics; in contrast, see Frisch, Pouquet, & Meneguzzi 1983). Second, the calculations are not consistent in their treatment of compressibility; in particular, magnetic buoyancy is not accounted for. Finally, although turbulent magnetic diffusion is allowed for, the effect of helicity due to motions on small spatial scales (which enters in standard mean-field dynamo theories) is not. Are these limitations fatal to any attempt to use such simulations in understanding stellar activity? The inclusion of compressibility (as for example in the Boussinesq limit proposed by Spiegel & Weiss 1981), and the consistent application of eddy transport coefficients would appear to be straightforward extensions of Gilman's calculations. More problematical is the proposition put forward by U. Frisch that *any* simulation which invokes eddy transport coefficients cannot be rightly viewed as a full simulation; and that simulations which do not appeal to eddy diffusivities cannot be made sufficiently complex, given the forseeable state-of-the-art in large-scale computing, to realistically simulate the solar convection zone and its full dynamo properties. It is not obvious whether this argument will be vindicated; but I suspect that the results of model non-linear calculations (as exemplified by the calculations shown here by N. O. Weiss) suggest its correctness: even relatively simple non-linear systems of equations appear to have an amazingly rich repertoire of behavior, so that it would not be surprising that the full solar dynamo problem (which does not invoke eddy diffusivities) is similarly (if not far more) complex.

Now, more generally, consider the evolutionary stellar spindown problem on the main sequence: in simplest terms, what is the angular momentum loss rate as a function of stellar parameters? Note that the seemingly much simpler question of solar angular momentum loss (and its correlation with solar activity) cannot be easily answered (because present solar wind measurements are largely restricted to the ecliptic). Unfortunately, the position of the Alfven radius $R_A$ and the mass flux at $R_A$ are not observables for late-type dwarf stars (Hartmann 1983); nor does theory readily provide these as a function of stellar parameters (Roxburgh 1983). In fact, even the classic spin-down time scale argument is now in doubt: using the specific angular momentum dependence on mass [log J ~ (2/3)log M] obeyed by early-type stars (Kraft 1967) as an initial condition for solar-type stars, and assuming solid-body rotation, one can constrain the time scale for solar spindown; however, recent observations of global solar oscillations have cast considerable doubt on the solid-body rotation hypothesis (cf. Claverie *et al.* 1981), so that the spin-down time scale is not well-defined. It hence seems that at present, there is no proper theory which can connect stellar spin-down to the level of surface activity, and subsequently to the state of interior rotation (which presumably largely determines the workings of the dynamo processes that lead to spindown itself). That will be a tall order for the future.

*(ii) Flux tube dynamics and plasma heating.* The presence of inhomogeneous magnetic field structures (= flux tubes) at the solar surface seems to be an essential aspect of the inhomogeneity of the solar outer atmosphere. Two basic questions arise: how are these field structures formed (e.g., are they surface phenomena, or do they reflect a basic result of the interaction between turbulent fluids and magnetic fields); and how do they participate in the energetics of the hot outer atmosphere overlying the photosphere. Within the past few years, much work has been done on the question of the formation of thin flux tubes at the solar surface; and until very recently, these studies could be distinguished into two general categories: those calculations in which the flux tube is viewed as the endproduct of an MHD instability at the solar surface (viz., Spruit 1983), and those in which magnetic field concentration is regarded as a consequence of organized flows (i.e., granular and supergranular flows) in the solar convection zone (Galloway, Proctor, & Weiss 1977; Proctor 1983). The past year has, however, seen the suggestion of yet a third possibility: that (in the context of dynamo models in which toroidal flux generation largely takes place in the overshoot region of the convection zone -- the "shell dynamo"; Rosner 1980; Spiegel & Weiss 1980) double-diffusive instabilities lead to flux tube formation at the base of the convection zone (Acheson 1978; Schmitt & Rosner 1982; Hughes 1983; Rosner 1983). Which of these processes really occurs is not at all clear; however, very recent numerical simulations of surface convection and its non-linear interaction with ambient magnetic fields by Nordlund (1983), and their uncanny resemblence to high spatial resolution observations of solar surface magnetic fields and flows shown by Tarbell (1983), seem to suggest that flux concentration in the downflow regions of convection flows is inevitable, even if one were to start with an initially uniform field (see also Galloway *et al.* 1977).

Given the strong spatial intermittency of solar surface magnetic fields, and the spatial correlation between these magnetic fields and enhanced chromospheric and coronal activity (which extends to correlations between the photospheric field and the coronal gas pressure; Golub *et al.* 1980), it is not an unreasonable supposition that the above-mentioned magnetic flux tubes also play a crucial role in the transfer of mechanical energy from the stellar surface (photosphere) to the overlying tenuous plasma. For the theoretician, study of the possible wave modes on flux tubes (which may be involved in this energy transfer process) has proved to be fertile grounds for detailed calculations (cf. Roberts 1983; Spruit 1983). An interesting new idea is that the absorption of wave energy can largely occur in discrete frequency intervals, e.g., that standing modes are set up on flux surfaces, and that these standing waves are damped (largely by viscous forces); if, in addition, it can be shown that the resulting resonance on each flux surface has high Q (i.e., is only weakly damped, so that wave amplitudes are large, and the absorption frequency interval for that flux surface narrow), then the heating rate may be independent of the details of the dissipation process, and the bulk heating properties of coronal structures may be calculated without detailed knowledge of the local heating process (Ionson 1982). Some discussion of this "lumped circuit" approach to coronal flux tube heating in the context of solar flares by D. Spicer can be found in these Proceedings.

## 4. CONCLUSIONS

The above overview represents a very personal outlook on the problem of stellar activity and its relation to magnetic field dynamics in stellar convection zones. Central to the picture I've attempted to sketch is the assertion that one can meaningfully extrapolate our present-day knowledge of solar physics to the stellar domain (indeed, the hope is that the association between hot plasma in the outer layers of the Sun and solar surface magnetic fields goes beyond the relatively narrow confines of solar physics, and may represent a kind of generic behavior of astrophysical objects whose surfaces are turbulent; Vaiana 1981; Rosner, Golub, & Vaiana 1982); the validity of this extrapolation seems to be on fairly secure grounds at least as far as late spectral-type dwarf stars are concerned. Thus, the overall scheme is not only to take advantage of solar observations in order to provide an interpretive framework for discussing stellar observations, but also to use manifestations of activity on stars other than the Sun as an additional observational constraint for exploring the complex interaction between magnetic fields and turbulent fluids which we observe on the Sun. In fact, from the solar perspective, one might hope that the kinds of observations and modeling which will be discussed at this Colloquium will provide additional constraints on both theories of chromospheric and coronal heating and magnetic flux generation (e.g., dynamo theory) which cannot be obtained independently from solar work.

One must however temper these optimistic points-of-view with the caution that it remains unclear to what extent the rapidly-burgeoning new observational and theoretical work has begun to make more solid contact between theory and observations than has been heretofore the case. Over a quarter of a century have passed since the classic dynamo paper of E. N. Parker (1955), which in the immediately-following years had raised the (so far unrealized) hope that the solar activity cycle could be understood from first principles. We think we now know better; dynamo theory and the physics of flux tube formation are now known to be far from well-understood; and the recent new Ca II, UV, and x-ray observations have shown that the behavior of "activity" on stars is substantially more complex than hitherto suspected. It thus appears that a major task for the immediate future will be the problem of understanding how current theory and observations relate.

## ACKNOWLEDGEMENTS

I would like to thank Drs. L. Golub, G. S. Vaiana, and N. O. Weiss for many useful discussions and cogent comments. This work was supported by NASA grants NAGW-79 and NAG8-445 to the Harvard College Observatory.

## BIBLIOGRAPHY

Acheson, D. J.: 1978, *Phil. Trans. Roy. Soc. Lond.* **A**, **289**, 459.
Ayres, T. R., and Linsky, J. L.: 1980, *Ap. J.*, **235**, 76.
Baliunas, S. L., Duncan, D., Noyes, R. W., Vaughan, A. H., and Cronin, P.: 1983, in Stenflo (1983).
Cattaneo, F., Jones, C. A., and Weiss, N. O.: 1983, in Stenflo (1983).
Claverie, A., Isaak, G. R., McLeod, C. P., and Van der Raay, H. B.: 1981, *Nature*, **293**, 443.
Deubner, F.-L.: 1981, in The Sun as a Star, ed. S. Jordan (NASA SP-450), p. 65.
Durney, B. R., Mihalas, M., and Robinson, R. D.: 1981, *P.A.S.P.*, **93**, 537.
Frisch, U., Pouquet, A., and Meneguzzi, M.: 1983, in Stenflo (1983).
Galloway, D. J., Proctor, M. R. E., and Weiss, N. O.: 1977, *Nature*, **266**, 686.
Giampapa, M. S.: 1983, in Stenflo (1983).
Gilman, P. A.: 1982, in Cool Stars, Stellar Systems, and the Sun, ed. M. S. Giampapa & L. Golub.
Gilman, P. A.: 1983, in Stenflo (1983).
Golub, L.: 1983, in Stenflo (1983).
Golub, L., Maxson, C. W., Rosner, R., Serio, S., and Vaiana, G. S.: 1980, *Ap. J.* **238**, 343.
Golub, L., Rosner, R., Vaiana, G. S., and Weiss, N. O.: 1981, *Ap. J.*, **243**, 309.
Hartmann, L.: 1983, in Stenflo (1983).
Hartmann, L., and Rosner, R.: 1979, *Ap. J.* **230**, 802.

Hartmann, L., *et al.*: 1983, in preparation.
Hudson, H.: 1982, private communication.
Hughes, D. W.: 1983, in Stenflo (1983).
Ionson, J. A.: 1982, *Ap. J.*, **254**, 318.
Kahler, S., *et al.*: 1981, *Ap. J.*, **252**, 239.
Knobloch, E., Rosner, R. and Weiss, N. O.: 1981, *M.N.R.A.S. (Comm.)*, **197**, 45P.
Kraft, R. P.: 1967, *Ap. J.*, **150**, 551.
Linsky, J. L.: 1983, in Stenflo (1983).
Marcy, G. W.: 1983, in Stenflo (1983).
Nordlund, A.: 1983, in Stenflo (1983).
Noyes, R. W.: 1981, in **Solar Phenomena in Stars and Stellar Systems**, ed. R. M. Bonnet & A. K. Dupree (Dordrecht: Reidel).
Noyes, R. W.: 1983, in Stenflo (1983).
Pallavicini, R., *et al.: 1981*, *Ap. J.*, **248**, 279.
Pallavicini, R., *et al.*: 1982, **Cool Stars, Stellar Systems, and the Sun**, ed. M. S. Giampapa & L. Golub.
Parker, E. N.: 1955, *Ap. J.*, **122**, 293.
Parker, E. N.: 1979, **Cosmical Magnetic Fields** (Oxford: Clarendon Press).
Proctor, M. R. E.: 1983, in Stenflo (1983).
Roberts, B.: 1983, in Stenflo (1983).
Robinson, R. D., Worden, S. P., and Harvey, J.: 1980, *Ap. J.*, **239**, 961.
Rosner, R.: 1980, in **First Cambridge Cool Stars Symposium**, ed. A. K. Dupree.
Rosner, R.: 1983, in Stenflo (1983).
Rosner, R., *et al.*: 1981, *Ap. J. Letters*, **249**, L5.
Rosner, R., Golub, L., and Vaiana, G. S.: 1982, *Ap. J.* (in press).
Roxburgh, I.: 1983, in Stenflo (1983).
Schmitt, J. H. M. M., and Rosner, R.: 1982, *Ap. J.*, in press.
Schussler, M.: 1983, in Stenflo (1983).
Skumanich, A.: 1972, *Ap. J.*, **171**, 565.
Soderblom, D.: 1983, in Stenflo (1983).
Spruit, H. C.: 1983, in Stenflo (1983).
Spiegel, E. A., and Weiss, N. O.: 1980, *Nature*, **287**, 616.
Spiegel, E. A., and Weiss, N. O.: 1981, Columbia Univ. Preprint No. A10.
Stenflo, J.: 1983, **Solar and Stellar Magnetic Fields: Origins and Coronal Effects**, editor (Dordrecht: Reidel), in press.
Tarbell, T. D.: 1983, in Stenflo (1983).
Topka, K., *et al.*: 1982, *Ap. J.*, in press.
Vaiana, G. S.: 1981, *Inst. Space Astronaut. Sci. (Tokyo)*, **597**, 1.
Vaiana, G. S.: 1983, in Stenflo (1983).
Vaiana, G. S., *et al.*: 1981, *Ap. J.*, **245**, 163.
Vaughan, A. H.: 1983, in Stenflo (1983).
Vaughan, A. H., and Preston, G. W.: 1980, *PASP*, **92**, 235.
Walter, F. W.: 1981, *Ap. J.*, **245**, 677.
Walter, F. W., and Bowyer, S.: 1981, *Ap. J.*, **245**, 671.
Willson, R. C., Gulkis, S., Janssen, M., Hudson, H. S., and Chapman, G. A.: 1981, *Science*, **211**, 700.
Wilson, O. C.: 1966, *Ap. J.*, **144**, 695.
Zwaan, C.: 1983, in Stenflo (1983).

DISCUSSION

Mullan: Can I ask whether there is a connection between flares and coronal heating or do you thing that those topics should be discussed separately?

Rosner: I don't think that they are intimately related. Clearly there is a relation. There is a class of solar flare which occurs in loop geometries in which the physics may be related. But to say something in detail is awfully difficult at this stage.

Kodaira: You used an illustration which showed the relation between Ca II H & K flux and rotation period. Rotation is determined by two methods, the one spectroscopic and the other photometric. Does your ordinate make an allowance for sin i in the data from the former method?

Rosner: Firstly, the illustration in question plotted X-ray luminosity against equatorial rotational velocity not Ca flux. Secondly, in the case of the spectroscopically determined data, it is v sin i which is plotted. So these points will be shifted by a factor sin i.

Kodaira: Stellar astrophysicists are accustomed to looking at plot against v sin i. Looking at this diagram it appears that the scatter is about that expected from the sin i effect. In this case I strongly suspect that the X-ray emission may be confined to the equator or at least to lower latitudes.

Rosner: Yes, that is perfectly plausible.

Gibson: (Part of question lost on tape). I don't think it is quite right to say that. Statistically stars show a rotational velocity which is about half of their equatorial velocity. This is about 0.3 in the log which is smaller than the effect in the diagram.

Rosner: The thing which we do not know is the latitudinal differential rotation rate. It is possible that this is quite large.

Kuijpers: You mentioned the Alvèn radius but did not follow it up. If one takes the angular momentum loss at the Alvèn radius does this give a proper result?

Rosner: I did not have time to address this. The answer is yes, it is reasonable. However one must be very careful when one wants to estimate angular momentum loss using mass loss and scaling the magnetic field strength. There is a coupling back since increasing the magnetic field decreases the mass loss.

Gibson: I would like to return to the question by Mullan about the connection between flaring and coronal heating. There are a number of telling points. Number one: when one measures the energy of the Sun's magnetic fields this is about the same as the energy needed to heat the corona. Secondly, in the case of the most active stars if one measures such parameters as densities or luminosity per unit volume then these are similar to those for small solar flares, making it look as though the stars were covered in by small solar flares. So can this question be made more strongly?

Rosner: Perhaps I did not make myself clear in answering Mullan's question. It depends on in what detail you make the comparison. If the comparison is gross, i.e. is the heating in a flare related to loops in the same way as the quiescent coronal heating is to loops then the answer is yes. If however one asks whether the precise mechanism which leads to flares is the same as that which leads to quiescent coronal heating then the answer is no.

# SESSION I

## GENERAL CHARACTERISTICS OF ACTIVE DWARFS

# GLOBAL AND PHOTOSPHERIC PHYSICAL PARAMETERS OF ACTIVE DWARF STARS

B. R. Pettersen
Institute of Mathematical and Physical Sciences
University of Tromsø, N-9001 Tromsø, Norway.

Abstract
Physical parameters (temperature, luminosity, radius, mass and chemical abundance) of the photospheres of red dwarf flare stars and spotted stars are determined for quiescent conditions. The interrelations between these quantities are compared to the results of theoretical investigations for low mass stars. The evolutionary state of flare stars is discussed. Observational results from spectroscopic and photometric methods to determine the rotation of active dwarfs are reviewed. The possibilities of global oscillations in dwarf stars are considered and preliminary results of a photometric search for oscillation in red dwarf luminosities are presented.

## 1. INTRODUCTION

Flares and spots have been observed in a variety of stars. Extremes are young T Tauri stars and old subdwarfs; luminous giants and faint main sequence dwarfs; very rapid rotators (100 km/s) of the FK Com type and slow rotators (1 km/s) on the main sequence; single stars and multiples of different kinds. In this paper I will concentrate on the lower main sequence, where stars of the UV Cet and BY Dra types are located.

## 2. PHYSICAL PARAMETERS OF FLARE STARS

Broad band photometry and a variety of spectroscopic information on flare stars and spotted stars have been obtained to study the photospheres of these stars in their quiescence.

### 2.1. Effective temperature

Photometric data can be transformed into energy distributions of stars over the observed wavelength interval, and put on a flux scale calibrated in absolute units. By comparing this distribution with that of a theoretical model taking into account the effect of blanketing, one obtains an approximation to the effective temperature of the star. The



*P. B. Byrne and M. Rodonò (eds.), Activity in Red-Dwarf Stars, 17–33.*

theoretical distribution in this comparison may be the flux distribution of a model atmosphere (Mould 1976 a). Unfortunately, no set of models exists for the entire range of temperatures covered by the red dwarf stars. In default of this, black body curves have been used (Veeder 1974, Pettersen 1980). A complicating element with Planck curves is that blanketing effects cannot be dealt with directly. Veeder (1974) used the method of Greenstein et al. (1970) which compensates for blanketing in a subjective manner, while Pettersen (1980) handled the problem by fitting the Planck function to data longwards of 1 μm and simultaneously required that the integral under both distributions be equal. Figure 1 shows the empirical colour-temperature relations obtained by different authors. Also included is the one used by Johnson (1965) which was established from giants with measured diameters. The differences in certain areas of the diagram approach about 250 K. To preserve consistency in the following discussion of flare stars I shall use the results of Planck function comparisons with Pettersen's (1980) method.

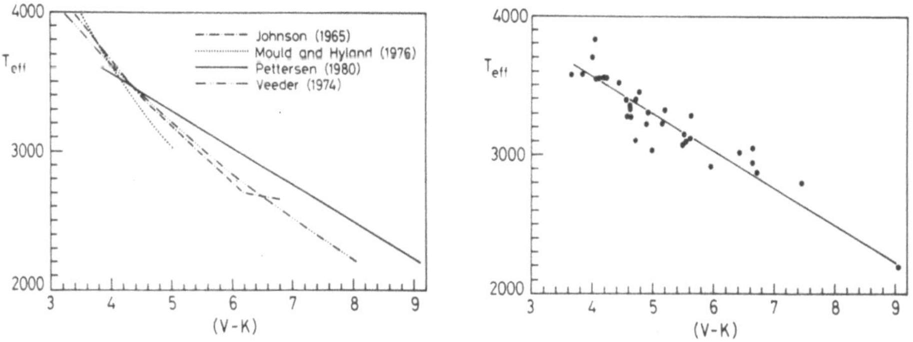


Fig. 1. Left: Empirical colour-temperature relations from different authors. Right: The linear colour-temperature relation of Pettersen (1980) based on analysis of 34 stars.

We find the empirical relations between effective temperature and various colour indices to be linear, as demonstrated in Fig. 1. Results obtained from least square analyses are given in Table 1.

Table 1: Effective temperature and colour indices.

| Colour index | Number of data | Correlation coefficient | Linear fit | Typical scatter |
|---|---|---|---|---|
| B-V | 35 | -0.88 | $T_{eff}$=-1510(B-V)+5738 | ±156 K |
| V-R | 29 | -0.90 | $T_{eff}$=- 645(V-R)+4469 | ±112 K |
| R-I | 29 | -0.95 | $T_{eff}$=- 648(R-I)+4311 | ± 79 K |
| V-K | 35 | -0.93 | $T_{eff}$=- 264(V-K)+4624 | ±120 K |

Several molecules manifest themselves in the photospheric spectra of red dwarfs, with feature strengths in accordance with the temperature of the star. Spectroscopic and photometric observations have revealed temperature sensitivity in TiO, VO, CaH (Jones et al. 1981, Mould 1976 b, Wing 1978), CO, $H_2O$ (Persson et al. 1977), FeH (Cohen 1978), and CaOH (Boeshaar 1976). The response of some of these molecules to a change in temperature is very nearly linear (e.g. $H_2O$, CaH, FeH, and TiO). CO and CaOH show strong non-linear effects (Fig. 3). The temperature dependences are valid for all population types among dM stars and flare stars, so the effects of metallicity on the temperature relations must be small.

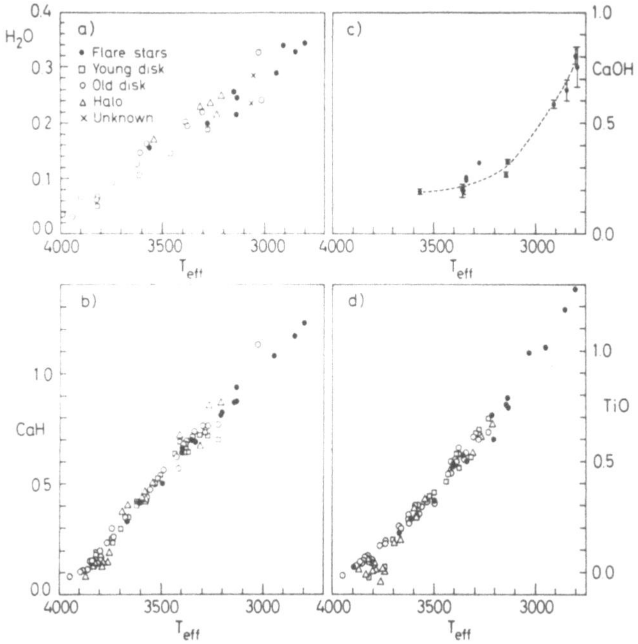


Fig. 2. Empirical relationships between effective temperature and the strengths of molecular features in red dwarf stars. The photometric data used as ordinates are from a)Persson et al. (1977), b) and d) Jones et al. (1981) and Mould (1976), c) Pettersen (unpublished).

## 2.2. Bolometric luminosity and bolometric correction

More than 90% of the flux emitted by red dwarfs is contained within the spectral region from 0.36 μm to 3.6 μm. By integrating the flux distribution over all wavelengths assuming the fitted Planck function to continue on either side of the observed optical and infrared regions, we obtain the bolometric flux emitted by the star. By normalizing to the sun this can be put on the bolometric magnitude scale.

The bolometric correction, defined as $BC=M_{bol}-M_V$, can now be obtained.

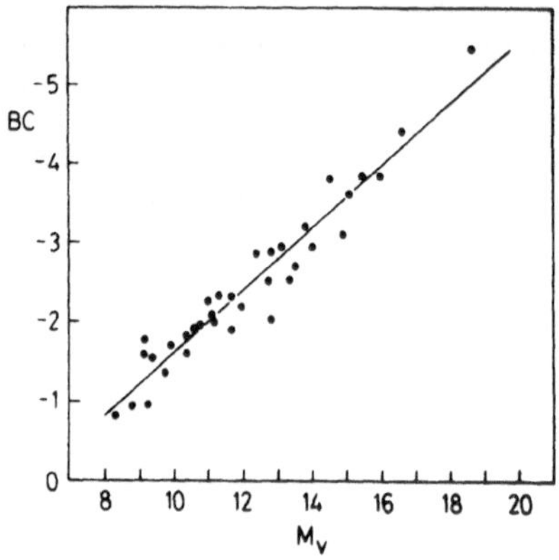


Fig. 3. Bolometric correction versus absolute visual magnitude.

Figure 3 demonstrates that BC is a linear function of $M_V$. A least square analysis yields

$$BC = -0.397\ M_V + 2.386$$

valid for $8<M_V<19$. The standard deviation for a typical value in BC, given a value of $M_V$, is $\sigma=\pm0.27$.

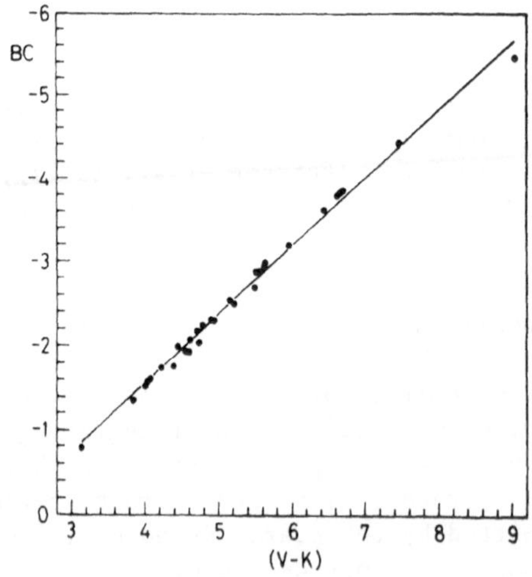


Fig. 4. Bolometric correction versus the (V-K) colour index.

The relations between BC and colour indices also show no deviation from linearity (Figure 4). Table 2 gives the results of the least square analyses. A comparison with the (BC, colour)-relations of Johnson (1965) reveals good correspondance for BC$\geq$-2. For BC<-2 Johnson's (1965) values of BC for a given value of a colour index are consistently smaller than our values.

Table 2: Bolometric correction and colour indices.

| Colour index | Number of data | Correlation coefficient | Linear fit | Typical scatter |
|---|---|---|---|---|
| B-V | 33 | -0.933 | BC=-4.816(B-V)+5.430 | ±0.36 |
| V-R | 26 | -0.988 | BC=-2.267(V-R)+1.689 | ±0.13 |
| R-I | 26 | -0.988 | BC=-2.124(R-I)+0.874 | ±0.13 |
| V-K | 30 | -0.997 | BC=-0.816(V-K)+1.709 | ±0.07 |

## 2.3. Radius

As the effective temperature $T_{eff}$ and the bolometric luminosity L is now determined, the radius R of a spherical star can be found from

$$L = 4\pi R^2 \sigma T_{eff}^4$$

where $\sigma$ is the Stefan-Boltzmann constant. In Fig. 5 we show the empirical radius-luminosity relation for flare stars. Also drawn are theoretical relationships by Ezer and Cameron (1967), Copeland, Jensen and Jørgensen (1970), Hoxie (1970), and Grossman, Hays and Graboske (1974). The relation of the last work fits the observations.

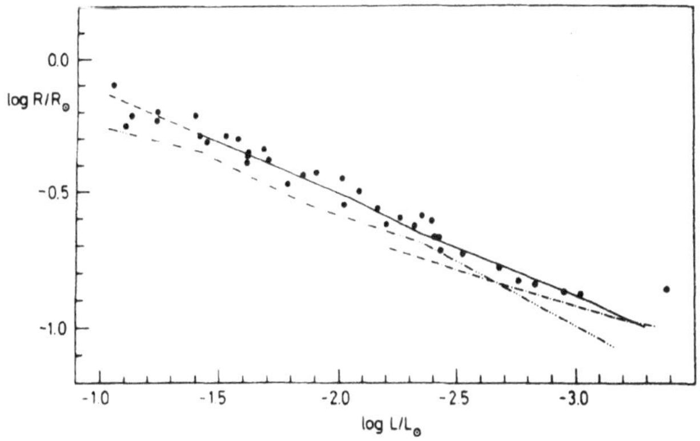


Fig. 5. The radius-luminosity relations of Ezer and Cameron (1967)-•••-, Hoxie (1970)-•-•-, and Grossman, Hays and Graboske (1974)——, compared to empirical results for flare stars. The relation of Copeland, Jensen and Jørgensen (1970) is almost indistinguishable from that of Ezer and Cameron.

## 2.4. Masses and the mass-luminosity relation

For flare stars and a few non-flaring red dwarfs we have compiled astrometrically determined masses. All masses are accurate to 0.05 $M_\odot$ or better, according to the literature sources. The bolometric luminosities of the individual components were determined from multicolour photometry or by using the relationship between bolometric and visual magnitude. Twenty-one stars are plotted and identified in the mass-luminosity diagram in Fig. 6. The scatter of the observed data points is so large that the theoretical mass-luminosity relations of the workers listed in the previous paragraph will all fit the observations. We have drawn the theoretical mass-luminosity relations from Grossman, Hays and Graboske (1974) for models at ages $10^8$ and $10^9$ years, for the mass interval from 0.085 $M_\odot$ to 0.5 $M_\odot$.

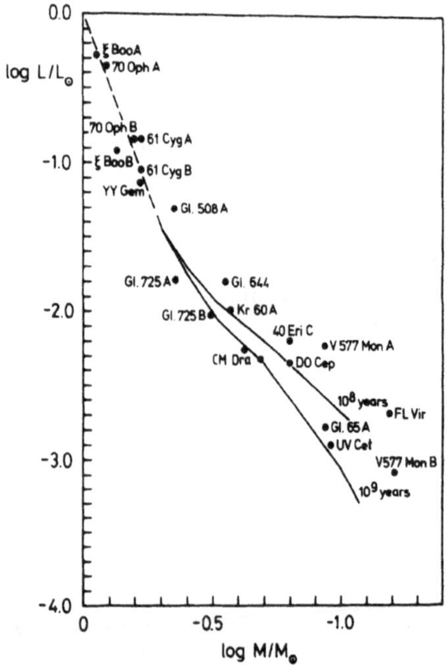


Fig. 6. The mass-luminosity relations of Grossman, Hays and Graboske (1974) for ages $10^8$ and $10^9$ years compared to data for red dwarfs.

## 2.5. Abundances: spectroscopic results

A curve of growth analysis was done by Hartmann and Anderson (1977) from high dispersion spectra in the red at a resolution of about 0.1 Å. They observed the bright flare stars BY Dra and EQ Vir, and for comparison the non-flaring slow rotators 61 Cyg A and B. Using the solar curve of growth for iron and various model atmospheres, their analysis yielded abundances for seven metals. There are no abundance differences between the dMe flare stars and the dM stars in this study, and the abundances found are solar.

Abundance analysis was also done by Mould (1978 b) on five early M dwarfs from Fourier transform spectra between 15000 Å and 25000 Å, with moderate resolution. Two of the stars are absorption line stars (Gliese 15 A and 229), and a third one is 61 Cyg B. A study of the atomic lines leads Mould (1978 b) to conclude that 61 Cyg B and Gliese 229 have solar composition, whereas Gliese 15 A and 411, which he classifies as subdwarfs, are underabundant in metals with respect to the sun.

Rotational lines of the CO and OH molecules are clearly visible in the spectra of M dwarfs. For temperatures below 3500 K one can also identify rotational lines of $H_2O$. Mould (1978 b) measured several lines of CO and OH, and also calculated synthetic spectra for these molecules. He predicted a rather clear-cut abundance sensitivity for OH, and a low one for CO. Yet the measured values of OH in the stars he observed are very similar, and he suggests that oxygen is not underabundant to the extent of the other metals, even in the old disk subdwarfs Gliese 15 A and 411.

The dispersion in the Hertzsprung-Russell diagram for early M type disk population stars is basically a metallicity dispersion. Mould (1978 b) found that metal-rich stars are overluminous with respect to the metal-poor ones.

The light element Li is of particular interest, as its presence in the spectrum of a nearby star is usually taken as an indication of extreme youth. M dwarf flare stars generally do not show the 6707 Å Li-line, with the only exception of the rapid rotator V1005 Ori=Gliese 182 (Bopp 1974, de la Reza et al. 1981). An abundance near the interstellar value of $n(Li)/n(H)=10^{-9}$ was determined by de la Reza et al. (1981). For several other dwarfs an upper limit was set two orders of magnitude lower.

Li has been reported in several rotating dwarfs with spectral types earlier than M0 (see Table 3), both among spotted stars and plage stars. The masses of these stars are between 0.8 $M_\odot$ and 1.2 $M_\odot$. For such stars convective overshoot may produce the required dependence on mass and age, by extending the mixing region deep enough to allow Li burning (Straus et al. 1976).

## 2.6. Chromospheric effects on photometric metallicity determinations

The metal-to-hydrogen ratio can be measured by intermediate band photometry. The Strømgren uvbyβ system uses a line blanketing index dominated by iron lines as a measure of metallicity. The interpretation of such a parameter may not be unique for active stars, however, as the presence of stellar chromospheric activity may affect the photometric index through filling-in of the cores of strong non-magnetic lines as a result of heating in the lower chromosphere. Giampapa et al. (1979) therefore compared measurements of the quiet solar photosphere and an active solar region in the uvbyβ-system. The effect of the active region was to decrease the apparent metal abundance relative to the quiet sun by about 35%. In red dwarfs an apparent low metallicity measurement may thus be the result of active regions present on the star.

## 2.7. Molecular features and metallicity

Superposed on the relationships between temperature and strengths of molecular features is the sensitivity towards other quantities, at least for some of the molecules observed in red dwarfs. MgH is a luminosity discriminator for K stars as CaH is for M stars. Mould (1976) found TiO to be an indicator of metal abundance for the hotter ($T_{eff}$>3600 K) M dwarfs. Halo population dwarfs and old disk subdwarfs can be distinguished from old disk main sequence stars by their weaker TiO for a given temperature (Mould 1978 a).

No photometric investigations of molecular features has yet been published where flare stars are distinguished as a group relative to non-flaring M dwarfs. The observations of Jones et al.(1981) and Mould (1976) permit such a comparison. We have transformed the data to a common system (i.e. that of Mould) for 84 dwarfs, 14 of which are flare stars. In the (CaH, TiO)-diagram (Figure 7) a certain scatter is observed in the slightly non-linear correlation. This dispersion is what one would expect from the main sequence dwarf models of Mould (1976), when the metal abundance is varied between the solar value and one tenth of this. Unfortunately,

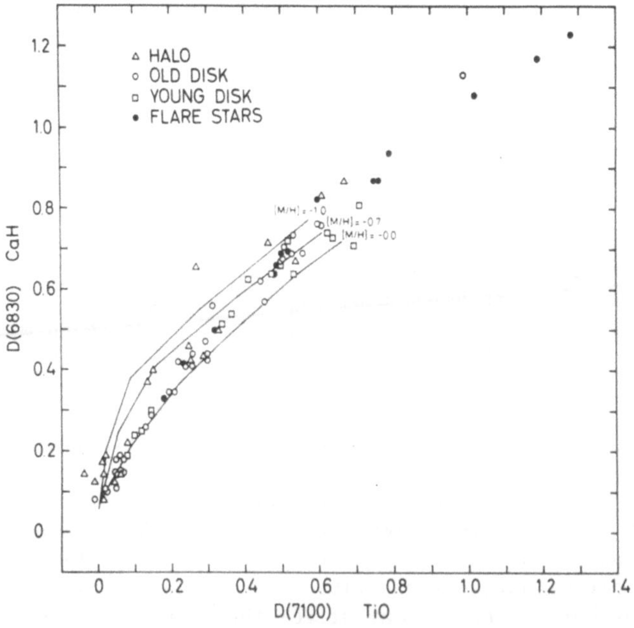


Fig. 7. Photometrically determined quantities for molecular band heads reveal an abundance sensitivity. The lines are theoretical relations from Mould (1976).

no theoretical comparison can be made for $T_{eff}$<3000 K. The halo population stars tend to lie above the disk stars, and the flare stars have no particular characteristic in this diagram. From an abundance viewpoint, they appear to be normal dwarfs.

## 3. STRUCTURE AND EVOLUTION OF RED DWARFS

The Hertzsprung-Russell diagram in Fig. 8 contains those red dwarf flare stars and spotted stars that are single or are single components in binary systems. Also drawn are the theoretical hydrogen main sequences by Ezer and Cameron (1967), Copeland, Jensen and Jørgensen (1970), Hoxie (1970). None of these series of low mass star models fits the observed sequence of stars. The theoretical main sequence that does fit the observations is that of Grossman, Hays and Graboske (1974) with chemical composition X=0.68 and Z=0.03, and with the convective mixing length parameter $\ell/H_p$=1. We have also included the evolutionary tracks for masses between 0.085 $M_\odot$ and 0.5 $M_\odot$.

The stars approach the main sequence during their contraction phase almost vertically in the HR-diagram, as fully convective bodies. Models heavier than 0.3 $M_\odot$ develop a radiative core before they reach the main sequence, at ages between $10^7$ and $10^8$ years. Masses smaller than 0.3 $M_\odot$ remain fully convective also on the hydrogen main sequence. The scatter of the sequence of observed stars is so large that both the $10^8$ and $10^9$ years isochrones are contained within the sequence.

Very recently the question was brought up whether the very low mass stars are actually fully convective. Using Kruger 60 as an observational case for which mass, temperature and luminosity is available, Cox, Shaviv and Hodson (1981) calculated models for the latest molecular opacities available using $\ell/H_p$ as the only free parameter. Their method was to integrate from the surface inwards, and the results are values of $\ell/H_p$ in the surface layer in the range 0.07-0.17, rather than the more conventional values 1-2. Such small values lead to higher temperatures and for some solutions a radiative core develops. The appearance of the radiative core is a very sharp threshold, and depends strongly on the value of $\ell/H_p$. A recalculation of two of the models from Grossman et al. indicates that these models may need revision, using new molecular opacities.

Objects with masses smaller than 0.085 $M_\odot$ never ignite the hydrogen in their cores. Evolutionary tracks for such low masses have been calculated as far as the deuterium burning main sequence. Such objects contract almost vertically in the HR-diagram. Stevenson (1978) developed formulae for the contraction phase and the subsequent degenerate cooling phase. The contraction phase lasts $10^6$-$10^8$ years for black dwarfs with masses 0.01 $M_\odot$ and 0.08 $M_\odot$. The lifetime for black dwarfs in the deuterium burning phase is a few times $10^7$ years (Grossman et al. 1974). At the end of their contraction phase the core becomes degenerate. The black dwarf then cools and gets fainter. The thermal energy is much less than the total internal energy in this phase, and the radius remains almost

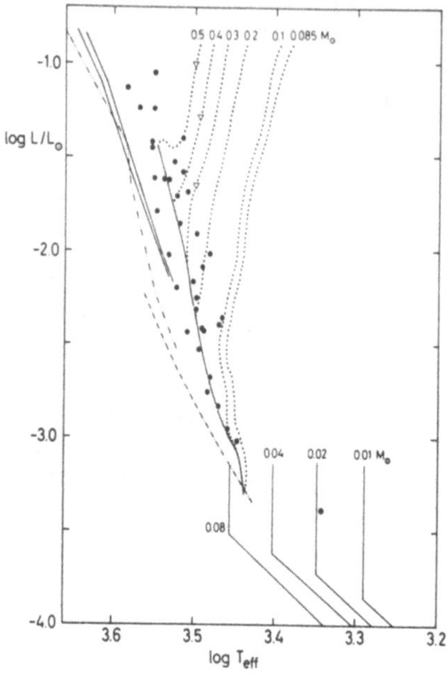


Fig. 8. The Hertzsprung-Russell diagram for nearby flare stars. Theoretical main sequences from Ezer and Cameron (1967)-•••-•••-, Hoxie (1970)-•-•-, and Copeland, Jensen and Jørgensen(1970) — are too warm for a given luminosity. The theoretical main sequence of Grossman, Hays and Graboske (1974) for $0.085 \leq M/M_\odot \leq 0.5$ fits the observations well. The ∇ marks on the pre-main sequence evolutionary tracks indicate where a radiative core develops. Schematic tracks for very low mass black dwarfs are also indicated.

constant. The energy is now conserved in making the electron gas degenerate In Fig. 8 we have indicated schematically the "evolutionary tracks" for $0.01 \leq M/M_\odot \leq 0.08$. The position of Gliese 752 B=VB 10 near the track for a mass of 0.02 $M_\odot$ is remarkable among the flare stars. The star is still contracting towards the degenerate state. Its age is approaching $10^8$ years, and it is well below the deuterium burning phase, where it stayed for about $4 \cdot 10^7$ years.

A particularly interesting point about low mass stars, say $M/M_\odot < 0.3$, is whether they all are frequent flarers. It has been the impression of observers for some time that all stars fainter than a certain absolute magnitude are emission line stars and therefore are likely to flare. The most frequent flarers, like UV Cet and G51-15, are among these low luminosity stars.

Basically, the question of whether all low mass stars are flare stars, has to do with the perseverance of the observers. There are, however, several examples of negative searches for flares in low mass stars.

Gliese 905 showed no flares in the U-filter in 3.2 hours of monitoring, and 29.4 hours of B-filter monitoring by Andersen (private communication) and Shakhovskaya (1974) was also negative. Kunkel (1973) detected no flares on Gliese 752 B in 5.5 hours, and the only report of activity on this star is still the spectroscopic observation by Herbig (1956). Non-emission line stars are also being discovered among the faintest red dwarfs. Liebert et al. (1979) found several stars with no Hα emission in a sample of large proper motion stars with $M_V>15$.

## 4. ROTATION

Observational studies of the rotation of active dwarfs and flare stars have been made by both spectroscopic and photometric methods. The rotational broadening of spectral lines is the basis for the spectroscopic analyses, and the uneven flux distribution across the stellar disk producing a rotational modulation is the interpretative basis for photometric studies.

### 4.1. Rotational broadening of absorption lines

The first attempt to analyze absorption line profiles in flare stars for the purpose of determining their rotation was made by Anderson, Schiffer and Bopp (1977). They observed the bright flare stars EQ Vir (single) and BY Dra (binary), but the spectra were noisy and the expected rotational broadening was only slightly larger than the instrumental width, so their results are quite uncertain. For EQ Vir they determined (v sin i) near 10 km/s, with an uncertainty of at least 2.5 km/s. For BY Dra they put an upper limit to (v sin i) of 10 km/s.

The same method was applied by Vogt and Fekel (1979). They compared an absorption line profile of BY Dra to that of 61 Cyg A. The analysis assumed equal macroturbulence in the two stars and the interpretation is that any excess broadening in BY Dra is due to rotation. Using a rotational velocity of 2 km/s for the comparison star, they found (v sin i)=8.5±2 km/s.

### 4.2. Rotation from a cross-correlation technique

A method has been developed by Benz and Mayor (1981) which utilize a cross-correlation analysis between the spectrum of the star and a reference mask in the spectrograph. This observing technique has long been known as a very efficient way of obtaining accurate radial velocities (Griffin 1967). The position of the correlation dip yields the radial velocity of the star, and the width of the dip reflects the width of the absorption lines in the spectrum. A correlation exists between the width and (v sin i). Lucke and Mayor (1980) applied this method to BY Dra. They obtained for the two components of this binary (v sin i)=8.1±0.3 km/s (primary) and (v sin i)=7.4±1.1 km/s (secondary).

### 4.3. Rotational flux modulation by photospheric spots

A photospheric spot group on a rotating star will modulate the flux received through a broad band filter during photometry in conformity with its rotation period. Analysis of photometric data yields the equatorial rotational period , independent of the inclination of the rotation axis of the star. Twenty-two stars have had their rotational period measured this way, and with radii estimated as described earlier, these quantities can be transformed into equatorial rotational velocities. Spotted stars in Table 3 cover the spectral types from G8V to dM4.5e. More than one half are members of binary systems, and most of them are known to flare.

### 4.4. Rotational modulation by chromospheric plage areas

Narrow band photometry (bandwidth=1 Å) of the Ca II H and K lines relative to nearby continuum windows, has revealed rotational modulation of the Ca flux in several FGK main sequence stars. The situation is analogous to spotted stars with plage areas causing the modulation. Nineteen stars have had their rotational period determined (Vaughan et al. 1981), as given in Table 3. None of the plage stars are known to show classical flare activity.

### 4.5. Rotational velocities of active dwarfs

Among the subset of spotted stars and that of plage stars, none have been measured with both techniques. A comparison between the results of the two methods is therefore not possible at this time. As judged from only two stars, the results of spectroscopic methods and spot photometry are in reasonable agreement.

Examination of Table 3 reveals that for a given spectral type

- members of binary systems tend to be faster rotators than apparently single stars
- spotted stars tend to be faster rotators than plage stars

## 5. GLOBAL OSCILLATIONS IN RED DWARF STARS

The central density of low mass stars is low, $\rho_c/\bar{\rho}<6$ for $M/M_\odot<0.3$. The amplitude of radial adiabatic oscillations increases little from the center of the star to the surface. The destabilizing effects of nuclear reactions can therefore result in vibrational instability.

Because of differences in the constitutive physics of the various star models, results of pulsation calculations are difficult to evaluate. Low mass stars show strong non-ideal gas interactions and these were only recently taken into account (Opoien and Grossman 1974). Large uncertainties are also introduced by the treatment of convection. In stability analyses a linear theory of the time dependence of convection has been used with mixing length type models.

### 5.1 Results for radial oscillations

The most modern calculations of radial pulsations in red dwarfs are those of Gabriel and Grossman (1977), who studied equilibrium models on the hydrogen main sequence with masses between 0.085 $M_\odot$ and 0.5 $M_\odot$. The fundamental mode was investigated by solving the adiabatic wave equation for purely radial motion. They find the models to be pulsationally

Table 3.

ROTATION OF ACTIVE DWARFS

| Star Name or HD | Gliese No. | Spec. type | Dupl. | Type of Activity | Rotational Period | Rotational Velocity | v sin i from spectroscopy | Ref. | Remarks |
|---|---|---|---|---|---|---|---|---|---|
| | 161.1 | dF7 | | Plage | 2.5 d | 24.9 km/s | 22 km/s | 1 | |
| | 654.1 | dF8 | | Plage | 7.6 | 7.3 | 6 | 1 | |
| | 836.7 | dG0 | | Plage | 4.7 | 10.7 | 11 | 1 | |
| 15 Sge | 779 | G1 | | Plage | 14.0 | 3.7 | 3-5 | 1 | Li 6707 Å |
| 9 Cet | 17.3 | G2V | | Plage | 7.9 | 6.3 | 7.0 & 0.7 | 1 | Li 6707 Å |
| HD 26913 | | G3 | | Plage | 7.2 | 6.9 | 6 | 1 | |
| κ Cet | 137 | G5V | | Plage | 8.5 | 5.6 | 15 | 1 | Li 6707 Å |
| | 641 | G8V | | Plage | 11.0 | 4.6 | | 1 | |
| ξ Boo A | 566 A | G8V | | Spot | 10.137 | 4.2 | 3-5 | 2 | Li 6707 Å |
| 12 Oph | 631 | K0V | | Plage | 21.0 | 2.0 | | 1 | |
| VY Ari | 113.1 | dK0 | SB? | Spot | 7.9 or 44 | 6 or 1 | 10 | 2 | Li, flares |
| | 117 | K0V | SB? | Plage | 6.9 | 5.6 | | 1 | Li 6707 Å |
| HD 175742 | | K0e | SB 1 | Spot | 2.898 | 13.5 | | 3 | |
| 36 Oph A | 663 A | K1V | | Plage | 21 | 1.9 | | 1 | |
| 36 Oph B | 663 B | K1V | | Plage | 23 | 1.7 | | 1 | |
| ε Eri | 144 | K2V | | Plage | 11.8 | 3.3 | 15 | 1 | |
| HD 143313 | | K2e | SB 2 | Spot | 9.60 | 4.1 | | 3 | |
| HD 218738 | | K2+K2 | SB 2 | Spot | 3.03 | 12.9 | | 4 | |
| | 688 | dK3 | | Plage | 34 | 1.2 | | 1 | |
| | 233 | dK3+dK5 | SB 1 | Spot | 7.36 | 5.3 | | 5 | |
| | 775 | dK4 | | Plage | 29 | 1.3 | | 1 | |
| | 664 | K5V | SB? | Plage | 17 | 2.0 | | 1 | |
| 61 Cyg A | 820 A | K5V | | Plage | 37 | 0.9 | 2 | 1 | |
| HD 319139 | | dK5e | | Spot | 1.70 | 19.9 | | 13 | Flares |
| TW PsA | 879 | dK5 | SB? | Spot | 10.3 | 3.6 | | 13 | |
| EQ Vir | 517 | dK5e | | Spot | 3.96 | 7.9 | 10 | 6 | Flares |
| 61 Cyg B | 820 B | K7V | | Plage | 48 | 0.7 | | 1 | |
| CC Eri | 103 | K7Ve | SB 2 | Spot | 1.561 | 19.8 | | 6 | Flares |
| BY Dra | 719 | K3V+MVe | SB 2 | Spot | 3.836 | 8.9 | 8.1 & 7.4 | 6 | Flares |
| Vys 124 | | dM0e | | Spot | 8.05 | 4.2 | | 8 | |
| V1005 Ori | 182 | dM0.5e | | Spot | 1.858 | 15.4 | | 7 | Li, flares |
| FF And | 29.1 | dM1e | SB 2 | Spot | 2.17 | 14.2 | | 6 | Flares |
| YY Gem | 278 C | dM1e | SB 2 | Spot | 0.814 | 38.5 | | 6 | Flares |
| | 685 | dM1.5 | | Plage | (9) | (3.2) | | 1 | |
| BF CVn | 490 A | dM1.5e | | Spot | 3.17 | 8.6 | | 9 | Flares |
| FK Aqr | 867 A | dM2e | SB 2 | Spot | 4.083 | 6.6 | | 4 | Flares |
| AU Mic | 803 | dM2.5e | | Spot | 4.865 | 6.2 | | 6 | Flares |
| GT Peg | 875.1 | dM3.5e | EB | Spot | 1.641 | 15.6 | | 10 | Flares |
| CM Dra | 630.1 A | dM4e | SB 2 | Spot | 1.27 | 13.0 | | 11 | Flares |
| EV Lac | 873 | dM4.5e | | Spot | 4.378 | 4.3 | | 12 | Flares |
| YZ CMi | 285 | dM4.5e | | Spot | 2.78 | 5.5 | | 6 | Flares |

**References to the Table:**
1. Vaughan, A. H., et al., 1981, Ap. J. 250, 276.
2. Chugainov, P. F., 1976, Izv. Krymsk. Astrofiz. Obs. 54, 89.
3. Bopp, B. W., Noah, P., Klimke, A., Africano, J., 1981, preprint.
4. Bopp, B. W., Espenak, F., 1977, Astron. J. 82, 916.
5. Bopp, B. W., Hall, D. S., Henry, G. W., Noah, P., Klimke, A., 1981, PASP 93, 504.
6. Bopp, B. W., Fekel, F., 1977, Astron. J. 82, 490.
7. Bopp, B. W., Torres, C. A. O., Busko, I. C., Quast, G. R., 1978, IBVS 1443.
8. Busko, I. C., Quast, G. R., Torres, C. A. O., 1980, IBVS 1898.
9. Pettersen, B. R., 1980, PASP 92, 188.
10. Chugainov, P. F., 1974, Izv. Krymsk. Astrofiz. Obs. 52, 3.
11. Lacy, C. H., 1977, Ap. J. 218, 444.
12. Pettersen, B. R., 1980, Astron. J. 85, 871.
13. Busko, I. C., Torres, C. A. O., 1978, Astron. Astrophys. 64, 153.

unstable for $M/M_\odot \geq 0.1$. The 0.085 $M_\odot$ model is stable, but opacities and the equation of state are not well known for this small mass. Also, the stability analysis is inconclusive for stars with radiative cores, i.e. for $M/M_\odot \geq 0.3$. For some values of the relative perturbation of the pressure scale height stability will occur, while for others there is instability. The period of the fundamental mode is between 14 and 44 minutes for the mass range $0.1 \leq M/M_\odot \leq 0.5$, and the e-folding times are between $3 \cdot 10^7$ years and $2 \cdot 10^8$ years. This is shorter than the corresponding lifetimes of the stars. It could take more than 10 times the e-folding time for the pulsation to manifest itself and the models are therefore marginal candidates for the growth of observable oscillations. The models are also unstable on the deuterium main sequence, and the oscillation period decreases as the star contracts towards the hydrogen main sequence.

## 5.2. Theoretical results for non-radial oscillations

Stability analysis has shown that slightly evolved models for 1 $M_\odot$ star are unstable against some low-order gravity modes. Since stars on the lower main sequence, except the fully convective ones, have essentially the same structure as the solar models, one may expect that they are also unstable, at least during part of the main sequence phase. Noels et al. (1976) made stability calculations for models of 0.5 $M_\odot$ and 0.6 $M_\odot$. Both are found to be unstable for a certain fraction of the main sequence phase. The instability is driven by nuclear reactions, as in the sun. The periods of the adiabatic oscillations are between 36 and 47 minutes for the 0.5 $M_\odot$ star, and between 58 and 67 minutes for the 0.6 $M_\odot$ star.

Ando (1976) made an extensive investigation of non-radial p-mode oscillation with high order spherical harmonics ($\ell \geq 10$). Of interest to us are his coolest main sequence models of 0.8 $M_\odot$ and 1.0 $M_\odot$. He neglected the influence of the core as p-modes with $\ell \geq 10$ are trapped in the envelope of the star. The excitation of p-modes is mainly due to the $\kappa$-mechanism of the hydrogen ionization zone. Unstable modes are found for $10 \leq \ell \leq 10^3$, but the uncertainties are considerable because the coupling between convection and pulsation was ignored in the stability analysis. Observationally, high $\ell$-modes do not give rise to light variability or radial velocity variations, but they show up as a non-thermal velocity field in the stellar atmosphere.

## 5.3. Observations

We have analyzed time series observations of flare stars by a Fourier technique to search for luminosity oscillations. To obtain optimal signal-to-noise ratios and in order to avoid large effects due to flares, we concentrate on photometric data in the R-filter, taken with a time resolution of 9 seconds. The low mass flare binaries UV Cet and FL Vir were observed with the McDonald Observatory 2.1 m telescope. The longest series lasted in excess of 4 hours and the rms deviation in the measurements are generally less than 0.005 mag. We have detected no convincing peaks in the power spectra produced from the time series observations, not even at the frequencies expected from the theoretical results. If red dwarf stars oscillate, their amplitudes are below our present detection limit

## 6. SUMMARY

Based on photometric information the low mass flare stars and spotted stars are shown to be main sequence stars with solar composition. In this respect they are not different from non-flaring disk population M dwarfs. They rotate faster than non-flaring dwarfs, however. Single flare stars have equatorial rotation velocities of 4-20 km/s, those in binaries 7-39 km/s. No red dwarfs have been detected to oscillate yet, although there are theoretical expectations for this.

## REFERENCES

Anderson, C.M., Schiffer, F.H., Bopp, B.W., 1977, Ap.J. 216, 42.
Ando, H., 1976, Publ.Astron.Soc.Japan 28, 517.
Benz, W., Mayor, M., 1981, Astron.Astrophys. 93, 235.
Boeshaar, P.C., 1976, Ph.D. thesis, Ohio State University.
Bopp, B.W., 1974, Publ.Astron.Soc.Pacific 86, 281.
Cohen, J.G., 1978, Ap.J. 221, 788.
Copeland, H., Jensen, J.O., Jørgensen, H.E., 1970, Astron.Astrophys. 5, 12.
Cox, A.N., Shaviv, G., Hodson, S.W.,1981, Ap.J. 245, L37.
de la Reza, R., Torres, C.A.O., Busko, I.C., 1981, MNRAS 194, 829.
Ezer, D., Cameron, A.G.W., 1967, Can.J.Phys. 45, 3429 and 3461.
Gabriel, M., Grossman, A.S., 1977, Astron.Astrophys. 54, 283.
Giampapa, M.S., Worden, S.P., Gilliam, L.B., 1979, Ap.J. 229, 1143.
Greenstein, J.L., Neugebauer, G., Becklin, E.E., 1970, Ap.J. 161, 519.
Griffin, R.F., 1967, Ap.J. 148, 465.
Grossman, A.S., Hays, D., Graboske Jr., H.C., 1974, Astron.Astrophys. 30, 95.
Hartmann, L., Anderson, C.M., 1977, Ap.J. 215, 188.
Herbig, G.H., 1956, Publ.Astron.Soc.Pacific 68, 531.
Hoxie, D. T., 1970, Ap.J. 161, 1083.
Johnson, H.L., 1965, Ap.J. 141, 170.
Jones, D.H.P., Sinclair, J.E., Alexander, J.B., 1981, MNRAS 194, 403.
Kunkel, W., 1973, Ap.J.suppl. 25, 1.
Liebert, J., Dahn, C.C., Gresham, M., Strittmatter, P.A., 1979, Ap.J.233, 226.
Lucke, P.B., Mayor, M., 1980, Astron.Astrophys. 92, 182.
Mould, J.R., 1976 a, Astron.Astrophys. 48, 443.
Mould, J.R., 1976 b, Ap.J. 207, 535.
Mould, J.R., 1978 a, Ap.J. 220, 935.
Mould, J.R., 1978 b, Ap.J. 226, 923.
Mould, J.R., Hyland, A.R., 1976, Ap.J. 208, 399.
Noels, A., Boury, A., Gabriel, M., Scuflaire, R., 1976, Astron.Astrophys. 49, 103.
Opoien, J.W., Grossman, A.S., 1974, Astron.Astrophys. 37, 335.
Persson, S.E., Aaronson, M., Frogel, J.A., 1977, AJ 82, 729.
Pettersen, B.R., 1980, Astron.Astrophys. 82, 53.
Shakhovskaya, N.I., 1974, Inf.Bull.Var.Stars No. 897.
Stevenson, D.J., 1978, Proc.Astron.Soc.Australia 3, 227.
Straus, J.M., Blake, J.B., Schramm, D.N., 1976, Ap.J. 204, 481.
Vaughan, A.H., et al., 1981, Ap.J. 250, 278.
Vogt, S.S., Fekel, F., 1979, Ap.J. 234, 958.
Veeder, G.J., 1974, AJ 79, 1056.
Wing, R.F., 1978, in "The HR-Diagram", eds. A.G. Davis-Philip and D.S. Hayes, p. 451.

DISCUSSION

Haisch: In terms of dynamo theory it seems critically important whether these stars have an interface between the radiative core and the convective zone. Could you comment on the models which suggest that above a certain mass stars should have this interface while below it they are fully convective? Furthermore, what does this mass mean in terms of spectral type?

Pettersen: (Part of reply lost) There is a star observed (van Biesebrock 10) which has a mass of 0.16 solar masses. According to the models of Grossman et al (1974) it is fully convective. New models with revised molecular opacities indicate that there are solutions in which even a star as small as that will develop a radiative core. I have talked within the framework of Grossman et al 1974 models and it may be that better ones are available. Actually, stars with masses about 0.2 - 0.3 $M_{\odot}$ may not be fully convective. In terms of spectral type this occurs at about M4 to M5.

Vaiana: When one observes spotted stars for instance, apart from the long time-scale variations are fluctuations on a shorter time-scale observed?

Pettersen: The technique generally used in observing spotted stars is to measure the star once per night. This makes it difficult to detect variation on a time-scale shorter than one day. I do not know of any published report of such observations.

van Leeuwen: I have been observing stars like this in the Pleiades with periods as short as 5 to 10 hours. The light curves are stable over a period of one year, with amplitudes of about two-tenths of a magnitude.

Baliunas: I would just like to say that in my talk tomorrow I will show some work we have been doing spectroscopically and spectrophotometrically on short time-scale variations in some spotted - and dM-type stars.

Weiss: I was a bit surprised that you discussed G-mode oscillations and rejected P-mode oscillations. In the Sun the G-mode oscillations have only been identified with a period of 2 hrs 40 mins and there is no adequate theoretical interpretation of those. Whereas the P-mode oscillations have been observed in abundance with a 5 mins' period, have been studied in great detail and are regarded as being driven by the convective zone. Therefore I would expect these as being most likely in the late-type stars.

Pettersen: My problem was that the only available observations are photometric and so variations in luminosity are those which are sought.

Linsky: I was interested by your comment that the plage stars do not show flares. The plage stars are believed to be those which have few plages and so exhibit rotational modulation in brightness. Others do not show rotational modulation because they have many plages uniformly distributed on the surface. For instance 61 CygA and B, although they are very old stars, show nice rotational modulation and they are obviously not flare stars.

Soderblom: Earlier you showed a relationship between radius and luminosity. If that relationship is extrapolated to one solar radius it does not correspond with one solar luminosity. Does the relationship turn over before this point or might there be some other problem?

Pettersen: The relationship shown is based on the modelling of Grossman, Hays and Graboske and their model calculations were only done for stars of mass up to 0.5 $M_{\odot}$. These cannot necessarily be extrapolated to reach 1 $M_{\odot}$ .

# MEAN COLORS AND EFFECTIVE TEMPERATURES OF K AND M DWARFS

Robert F. Wing
Astronomy Department
Ohio State University
Columbus, Ohio 43210 U.S.A.

Many applications of multicolor wideband photometry depend upon the existence of a table of mean colors, i.e. a listing of the average values of the color indices B-V, V-R, etc. for each spectral type. Such a table relates the observed colors of unreddened stars to their temperature classes and enables one to draw a reference line representing the main sequence on any color-color diagram.

The table of mean colors provided by Johnson (1966), which is based on Johnson's (1965) own observations of some 40 K and M dwarfs, has been (and still is) extensively used. There is, however, a problem with Johnson's mean colors which arises from the spectral classifications which he took from the literature. These are of widely varying precision and are expressed on several scales having large systematic differences. Hence the meaning of Johnson's mean colors is somewhat ambiguous.

Some time ago the writer drew attention to the need for improved spectral classifications for M dwarfs and suggested that narrow-band photometry of a TiO band be used to establish a consistent set of classification standards (Wing 1973). Subsequently more than 100 M dwarfs — including nearly all those for which Johnson (1965) obtained wideband photometry — have been observed by C. A. Dean and the writer on an eight-color system of narrow-band photometry in the near infrared. The spectral classifications derived from this photometry, based on the 7100 Å band of TiO, have an internal precision of one-tenth of a subclass and are expressed on the same scale as the MK types for M giants.

A spectroscopic study of M dwarfs has been carried out by Boeshaar (1976), who also expressed her types on the giant scale. Several of Boeshaar's classifications are given in the Atlas of Keenan and McNeil (1976), where they are adopted as standards for the MK system. Wing and Yorka (1979) have shown that the spectral types for M dwarfs obtained from eight-color photometry are in excellent agreement with Boeshaar's types and are thus on the MK scale.



*P. B. Byrne and M. Rodonò (eds.), Activity in Red-Dwarf Stars, 35–38.*

Table 1. MEAN COLORS AS A FUNCTION OF MK SPECTRAL TYPE

| Sp. Type | U-V | B-V | V-R | V-I | V-J | V-K | V-L |
|---|---|---|---|---|---|---|---|
| K3 V | 1.94 | 1.02 | 0.92 | 1.50 | 1.95 | 2.68 | 2.84 |
| K4 V | 2.42 | 1.22 | 1.06 | 1.80 | 2.22 | 2.99 | 3.16 |
| K5 V | 2.62 | 1.38 | 1.18 | 2.05 | 2.49 | 3.30 | 3.49 |
| M0 V | 2.68 | 1.46 | 1.28 | 2.28 | 2.76 | 3.61 | 3.82 |
| M1 V | 2.70 | 1.50 | 1.40 | 2.50 | 3.04 | 3.92 | 4.16 |
| M2 V | 2.70 | 1.50 | 1.52 | 2.72 | 3.35 | 4.24 | 4.50 |
| M3 V | 2.74 | 1.56 | 1.66 | 3.04 | 3.74 | 4.58 | 4.92 |
| M4 V | 2.92 | 1.66 | 1.84 | 3.50 | 4.28 | 5.15 | 5.64 |
| M5 V | 3.25 | 1.82 | 2.20 | 4.10 | 5.04 | 5.90 | ---- |
| M6 V | 3.60 | 2.00 | 2.74 | 5.10 | 5.90 | 6.70 | ---- |

Table 1 is a new compilation of the mean colors of K3 - M6 dwarfs, obtained largely from the photometry of Johnson (1965) by sorting the stars he observed according to their new spectral classifications from the eight-color photometry. For each color index, a plot was made of color vs. spectral type and a smooth curve was drawn through the data. With the new spectral types, these curves are very well determined.

There are substantial systematic differences between the new mean colors and those of Johnson (1966), especially toward the later types. For example, the mean colors given by Johnson for an M7 dwarf are nearly the same as those given here for an M5 V star on the MK scale. Thus it is important to use the new tabulation of mean colors whenever classifications on the MK system are employed.

Another use of wideband multicolor photometry is to estimate stellar temperatures, and the compilation of mean colors allows the derivation of a relation between spectral type and temperature. Johnson (1966) listed "effective temperatures" for each spectral type by comparing the calibrated flux curves corresponding to each set of mean colors to a family of blackbody curves. New temperatures based on the mean colors of Table 1 are now being derived with allowance for TiO blanketing at several of the filters, which has been evaluated quantitatively as a function of MK spectral type by Smak and Wing (1979).


Boeshaar, P. C. 1976, Ph.D. thesis, Ohio State University.

Johnson, H. L. 1965, Astrophys. J. 141, 170.

Johnson, H. L. 1966, Ann. Rev. Astron. Astrophys. 4, 193.

Keenan, P. C., and McNeil, R. C. 1976, *An Atlas of Spectra of the Cooler Stars: Types G, K, M, S, and C*, The Ohio State University Press.

Smak, J., and Wing, R. F. 1979, Acta Astron. 29, 187.

Wing, R. F. 1973, in *Spectral Classification and Multicolour Photometry*, ed. Ch. Fehrenbach and B. E. Westerlund (Reidel), p. 209.

Wing, R. F., and Yorka, S. B. 1979, in *Spectral Classification of the Future*, ed. M. F. McCarthy, A. G. D. Philip, and G. V. Coyne (Vatican Observatory), p. 519.

DISCUSSION

Mullan: Could you comment on one of the conclusions which arose from the earlier classification by Joy and Abt namely that all stars considered by them of spectral type M 5.5 or later were emission-line stars? With your revised classification is that conclusion still valid?

Wing: The conclusion remains the same except that maybe instead of calling it 5.5 I would call it 5.0. There is a half spectral class difference in that spectral region between myself and Joy. Generally however the correlation between our two classifications are good. So any general conclusions like that should still pertain.

Linsky: Would anyone else like to comment on that particular point because so far as I know Mark Giampapa has observed some stars of later spectral type still which show no Hα emission.

Mullan: But that is the point of the question. What was the basis of those later spectral types?

Wing: I don't know which stars they were so I am not sure whether I have observed them or not.

Worden: Van Biesbrock 10 is one of the stars which Mark (Giampapa) observed.

Wing: I cannot give a good spectral type for that. I observed it once about 20° from a Full Moon and so cannot give a good spectral type. But I believe it is about M6.

Linnell: In your abstract you had something to say about the temperature scale. Could you expand on that a little?

Wing: What I did was to follow the procedure of Johnson in his 1966 review paper. He compared each set of colours for each spectral-type to an absolute calibration of the photometry and then compared these to black body curves. The result is a temperature index which he called effective temperature. I am not sure that this is the best way of defining effective temperatures but at least it gives a temperature index. I thought that I could do somewhat better by making a quantitative allowance for TiO from my measurements. This work is not finished however and so I do not have figures to hand.

Linnell: Effective temperatures or absolute fluxes which are equivalent can only be determined in a fundamental way through determining of angular diameters or their equivalent. At this end of the scale there are only two good points i.e. γγGem and CM Dra. Between these and the middle A-types, where one can use intensity interferometers, there are only two other good points viz. the Sun and Procyon, each of which has special problems. So the temperature scale is not well established at the M stars.

# THE QUIESCENT CHROMOSPHERES AND TRANSITION REGIONS OF ACTIVE DWARF STARS: WHAT ARE WE LEARNING FROM RECENT OBSERVATIONS AND MODELS?

Jeffrey L. Linsky[1]
Joint Institute for Laboratory Astrophysics National Bureau of Standards and University of Colorado, Boulder, Colorado 80309 USA

ABSTRACT

I will review the rapid progress in our understanding of active dwarf stars, which has been stimulated by recent IUE, Einstein, and ground-based observations, by asking a series of questions. The most fundamental question is the extent to which magnetic fields control nonflare phenomena in these stars. There are a number of aspects to this question:

(1) What is the evidence for large scale magnetic structures similar to solar plages in these stars and how does a plage system differ from a quiescent spectrum?

(2) Can the enhanced heating in these stars be explained by solar-like magnetic flux tubes?

(3) What roles do systematic flows play in active dwarf atmospheres?

(4) What is the relation between heating rates in different layers of these stars?

(5) By what mechanisms are active dwarf chromospheres and transition regions heated?

(6) What are semiempirical models telling us about active dwarf stars?

Recent observations are permitting us to begin to answer these questions.

[1]Staff Member, Quantum Physics Division, National Bureau of Standards.



*P. B. Byrne and M. Rodonò (eds.), Activity in Red-Dwarf Stars, 39–60.*

## I. DO MAGNETIC FIELDS CONTROL THE NONFLARE PHENOMENA IN ACTIVE DWARF STARS?

I suspect that there is nearly complete agreement among astronomers studying late-type dwarf stars that the active phenomena observed in these stars are somehow a direct consequence of strong, variable magnetic fields. I have chosen to review the topic of quiescent, that is nonflaring, chromospheres and transition regions (TRs) of active dwarf stars by pointing out the range of phenomena in these stars that are probably magnetic in character and by suggesting how the magnetic fields likely control these phenomena.

This topic is difficult because there exist only a few measurements of magnetic fields in active dwarf stars and none yet for the M dwarfs. Even so, the existing magnetic field measurements (reviewed by Vogt 1982; Marcy 1982) are averages over the stellar surface, whereas the true fields are likely inhomogeneous and filamentary. Furthermore, theoretical calculations of heating processes and the energy balance in magnetic flux tubes, taking into account magnetic forces and realistic flux tube geometries, are extremely difficult. In view of these difficulties, I cannot be rigorous but must instead by speculative, and will follow solar analogies in trying to explain the interesting results coming from IUE and Einstein. There is a danger, however, in this approach as solar analogy may be a poor guide for explaining phenomena on stars with parameters very different from the Sun.

It is important to ask what the properties of stellar chromospheres and TRs would be for stars without magnetic fields. According to the Vogt-Russell theorem (cf. Lang 1980), the mass, age, and initial chemical composition of a star determine its effective temperature, radius, and gravity, and hence its location in the H-R diagram. Thus, for nonmagnetic stars there are unique values for the luminosity and convective zone parameters at each position in the H-R diagram. Furthermore, the atmospheric structure should be relatively uniform across the surface of these stars, except for small variations due to gravity darkening at the poles, meridional circulation, and the spatial variations due to the cellular nature of convection. The nonradiative heating by acoustic waves or other nonmagnetic waves generated by convective motions should still produce chromospheres, TRs, and coronae in late-type stars, but the heating rates should be relatively small. We therefore expect only quiescent outer atmospheric layers that are weak, steady emitters in the ultraviolet and X-rays, are spatially homogeneous, and show few differences between stars of similar effective temperature and gravity.

By contrast with this rather boring picture of idealized, nonmagnetic stellar atmospheres, we know that magnetic flux tubes in the solar atmosphere are the basic structure responsible for atmospheric inhomogeneity, are regions of enhanced nonradiative heating, are the locations of time-variable phenomena on time scales of seconds to months, are regions of important systematic flows, and are characterized by very different energy balance and atmospheric properties than regions of weak fields. The main theme of this review then is to describe and compare the phenomena that are similar in active and quiescent dwarf stars, and to summarize the various roles that magnetic fields likely play in modifying the chromospheres and TRs of the active stars. Some recent reviews relevant to this topic include Dupree (1981;1982), and Linsky (1980;1981a,b,c,d;1982).

## II. WHAT IS THE BASIC STRUCTURE OF ACTIVE DWARF ATMOSPHERES?

### a) Spatial Inhomogeneity

The best evidence we have for the existence of large scale bright structures in the chromospheres and TRs of stars comes from measurements of periodic variations in the intensity of specific spectral features at their presumed rotational periods. Vaughan et al. (1981), for example, monitored the chromospheric Ca II H and K lines to derive rotational periods of 19 stars, mainly dwarfs of spectral type F6-M0. These data argue conclusively for the inhomogeneous distribution of brightness across the chromospheres of these stars, and solar analogy argues that the structures primarily responsible for this patchiness are bright plage regions that are non-uniformly distributed in longitude and that are long-lived compared to a rotational period. If a star were completely covered with plages or the plages were uniformly distributed in longitude, then there would be no modulated intensity signal to indicate a rotational period. Hallam and Wolff (1981) followed a similar technique using IUE to determine rotational periods of three dwarf stars — 111 Tau (F8 V), ε Eri (K2 V), and 61 Cyg A (K5 V). They observed rotational modulation of intensity in lines of Lyα, Si II 1812 Å, and Mg II 2796 Å. Several such monitoring programs are under way, in some cases with coordinated magnetic field observations to confirm the hypothesis that the plage regions have strong magnetic fields.

Close binary systems with cool components, which are tidally forced to be rapid rotators, typically show bright, ultraviolet emission lines and photometric variability indicative of rotational modulation of dark star spots (e.g. Kunkel 1975; Hall 1981; Vogt 1983). One therefore expects that the ultraviolet emission line flux should vary with rotational phase such that emission line maximum (maximum coverage by plages) corresponds to photometric minimum (maximum coverage by dark star spots). Baliunas and Dupree (1982) tested this correlation for the long-period RS CVn system, λ And (G8 III-IV + ?) and confirmed that the emission lines are strong at photometric minimum and weak at photometric maximum. Marstad et al. (1982) and Byrne et al. (1982) monitored three subgiant RS CVn systems (HR 1099, II Peg, and AR Lac) and the two dwarf binaries BY Dra and AU Mic. They found clear evidence for periodic variability in HR 1099 and II Peg. For HR 1099, the data cover three rotational periods and are consistent, indicating that the plages are long-lived. The II Peg data reveal important clues concerning stellar plages:

(1) The strength of all the II Peg emission lines increases rapidly at orbital phase 0.45, is roughly constant for one-half the period, and then decreases just as rapidly at phase 0.95 (see Fig. 1). This is strong evidence for the rotational modulation of a relatively compact plage group centered at phase 0.70. The compactness of this plage group and the absence of large intensity changes for half a period are important new information.

(2) The increase in emission line strength corresponds to the decrease in photometric brightness (see Fig. 1) measured simultaneously by the FES on IUE, indicating that the plage group overlies dark star spots and thus the plages are regions of strong magnetic fields.

(3) The spectrum of II Peg at plage maximum differs considerably from that at plage minimum (see Fig. 2) in the sense that the high temperature TR lines (C IV 1550 Å, Si IV 1400 Å, and C II 1335 Å) are enhanced by much larger factors than the chromospheric lines (Si II 1812 Å, C I 1657 Å, and Mg II 2800 Å). This phenomenon is seen in solar plages and throughout the IUE data, as we shall see; therefore, it is an important consequence of magnetic fields in stellar atmospheres that requires an explanation.

b) Time Variability

According to our corollary to the Vogt-Russell theorem, nonmagnetic stars emit nearly constant flux, and flux variability, therefore, should be a direct consequence of a quantity not included in the classical theory, namely a magnetic field. Variability on many time scales has already been detected in the chromospheric and TR emission lines in late-type stars. For example, Wilson (1978) discovered periodic variations in the Ca II flux from G-K dwarf stars with periods ~10 years,

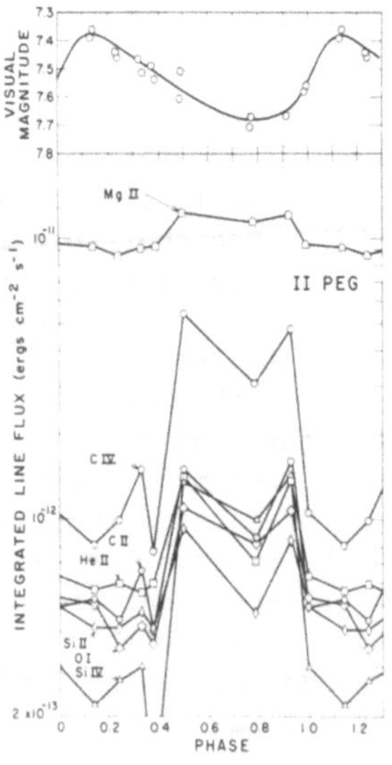


Fig. 1. Observed flux (from Marstad et al. 1982) in the Mg II λ2800, C IV λ1550, He II λ1640, C II λ1335, Si II λ1812, O I λ1304, and Si IV λ1400 features for II Peg as a function of phase. Also given at the top are FES visual magnitudes obtained simultaneously with the IUE spectra.

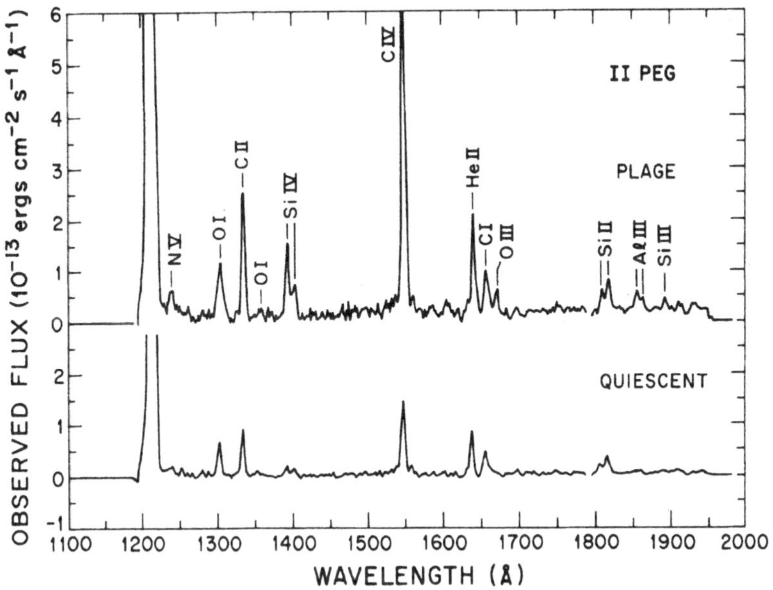


Fig. 2. Composite IUE spectra from Marstad et al. (1982) of the RS CVn-type system II Peg when the plage is on the disk (sum of three spectra) and when the plage is not present (sum of five spectra).

which he ascribed to magnetic cycles. White and Livingston (1981) found similar changes in the solar Ca II line flux between minimum and maximum in the solar cycle, and they argued that these changes are due to differences in the fractional area coverage by plages and not to any changes in the chromospheric network. We believe that the stellar data can be interpreted in a similar way.

Flares, a major topic of this Colloquium, have been detected in IUE spectra of the RS CVn systems UX Ari (Simon, Linsky and Schiffer 1980a) and HR 1099 (Marstad et al. 1982), as well as the dMe flare stars GL 867A (Butler et al. 1981) and Proxima Centauri (Haisch et al. 1982). Such flares probably also are magnetic in character, with the dMe star flares similar to solar flares and the RS CVn flares perhaps involving reconnection between flux tubes of the two stars. In addition to the expected enhancement of the emission lines, Butler et al. (1981) found continuous ultraviolet emission during a flare on GL 867A. One property seen in both the dMe and RS CVn flares is the relative enhancement of the TR lines compared to the chromospheric emission lines. This is illustrated by comparing the IUE spectra of HR 1099 during a flare and at quiescent times immediately before and after the flare (see Fig. 3). This important property will be discussed in §VI below. Also, Byrne et al. (1982) detected emission line variability in BY Dra and AU Mic, on time scales of minutes to hours, and Baliunas et al. (1981) detected similar variability in ε Eri.

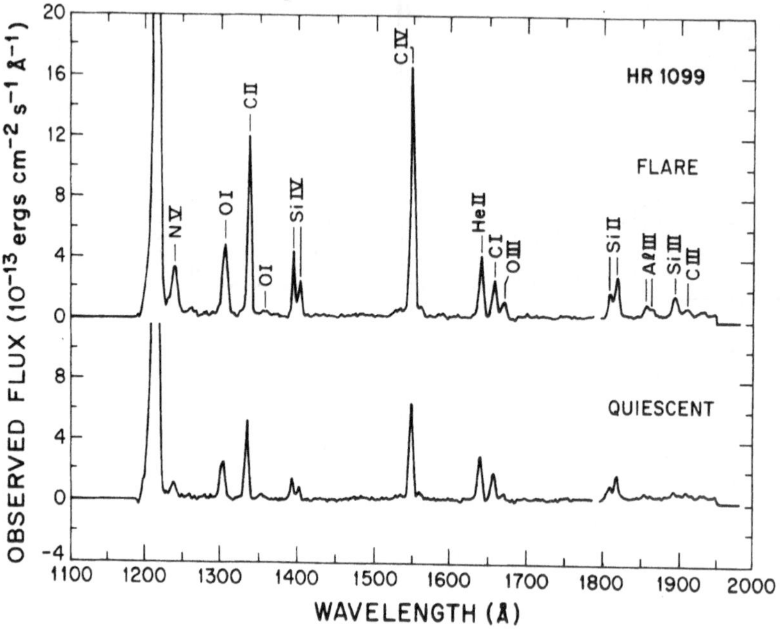


Fig. 3. IUE spectra of the RS CVn-type system HR 1099 obtained during a flare and when the system was quiescent.

## III. CAN THE ENHANCED HEATING IN ACTIVE DWARF STARS BE EXPLAINED BY SOLAR-LIKE FLUX TUBES?

A vital role played by the magnetic field is to enhance the local nonradiative heating rate in the chromosphere and higher layers above that which occurs in regions of weak or no fields. This conclusion is based on the tight spatial correlation of bright chromospheric emission with photospheric magnetic field strength (e.g. Skumanich, Smythe and Frazier 1975). Stein (1981) has summarized the arguments why slow mode MHD waves are the most likely heating mechanism for those regions in stellar chromospheres that have strong magnetic fields.

It is important to go beyond this general conclusion to a quantitative evaluation of the nonradiative heating rates at different atmospheric layers in different types of stars. One way of comparing the nonradiative heating rates is by plotting emission line surface fluxes, the flux per unit area of the star, as a function of mean temperature of formation of each line in different stars. Linsky et al. (1982) compared line surface fluxes for nine active chromosphere dwarf stars and six quiet chromosphere dwarf stars of spectral types G2-M5.6e using IUE observations. The surface fluxes of these stars divided by the mean quiet Sun values are shown in Figure 4. Several conclusions can be drawn from these data:

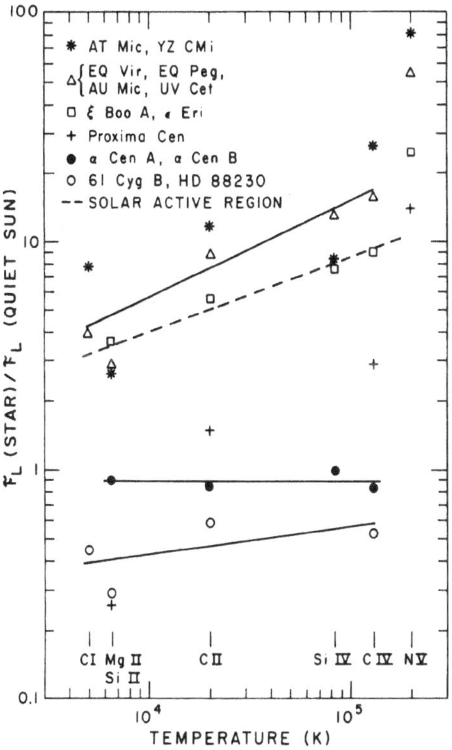


Fig. 4. Ratios of emission-line surface fluxes from Linsky et al. (1982) for six groups of stars and a bright solar active region relative to the quiet Sun. The stars are grouped according to their C IV ratios into three groups of dMe stars (AT Mic and YZ CMi; EQ Vir, EQ Peg, AU Mic, and UV Cet; Proxima Centauri), the dM stars (61 Cyg B and HD 88230), the active G-K dwarfs (ξ Boo A and ε Eri), and the quiet G-K dwarfs (α Cen A and α Cen B). Straight lines are drawn to indicate rough trends in the data for several groups of stars. The data for ξ Boo A and ε Eri are similar to those for a solar active region.

(1) The surface fluxes of the M dwarf stars with the weakest emission lines are about a factor of three below the mean quiet Sun. These stars likely have very weak magnetic fields with few plage regions, and since 61 Cyg B (K7 V) shows rotationally modulated Ca II emission (Vaughan et al. 1981), the plage regions on these stars are intrinsically fainter (implying lower heating rates) than the mean quiet Sun.

(2) The two stars most like the Sun [α Cen A (G2 V) and α Cen B (K1 V)] have surface flux ratios close to unity independent of temperature, suggesting magnetic fields similar to those on the Sun.

(3) The active chromosphere stars, including the dMe stars, have surface flux ratios that increase from 3-8 in the chromosphere to 10-100 at $2 \times 10^5$ K in the TR. Similar trends in the surface flux ratios are detected in other active stars including: (a) G giants and supergiants (e.g. Hartmann, Dupree and Raymond 1982; Stencel et al. 1982a,b) for which the ratios increase from 1 in the chromosphere to about 10 in the TR, and (b) RS CVn systems (e.g. Simon and Linsky 1980) for which the ratios increase from 10-20 in the chromosphere to as large as 600 in the TR. T Tauri stars (e.g. Imhoff and Giampapa 1982) also show very large ratios.

To what extent can we explain these large surface flux ratios by assuming that the active stars are mostly covered by solar-like plage regions? Included in Figure 4 is a dashed line indicating the surface flux ratios for a bright solar plage (Vernazza and Reeves 1978). Since the active M dwarf stars, RS CVn binaries, and T Tauri stars lie above this line, even complete coverage by solar-like plages cannot explain the large surface flux ratios. Furthermore, there is evidence described in the previous section (see Fig. 1) that these active stars also have a patchy distribution of emitting regions.

There is another important point, however, that must be considered. Solar plages are not homogeneous regions, but instead consist of many individual magnetic flux tubes that are not resolved in the existing data as their widths are much smaller than 1 arcsec (Frazier and Stenflo 1973; Stenflo 1973). By comparison, the solar plage data in Figure 4 were obtained by the EUV spectrometer-spectroheliometer on *Skylab*, which had a spatial resolution of 5 × 5 arcsec$^2$ (Vernazza and Reeves 1978). The High Resolution Telescope Spectrograph (HRTS) experiment (cf. Brueckner, Bartoe and Van Hoosier 1977; Basri et al. 1979) obtained ultraviolet solar spectra with a resolution of 1 × 1 arcsec$^2$. These data show a range of intensities between the brightest portions of plages and the darkest quiet regions that is a factor of 100 in the C IV 1548 Å line (Brueckner, private communication). Schindler et al. (1982) used HRTS data to show that the He II 1640 Å line in the brightest plage regions is a factor of 50 brighter than the mean quiet Sun value. Even in these data, the magnetic flux tubes are not resolved, and therefore we cannot yet say whether the observed flux ratios for chromospheric and TR lines in the active dwarf stars, RS CVn systems, and T Tauri stars can be explained by solar-like flux tubes covering portions of their surfaces. It is interesting, however, that the TR line widths observed by Ayres et al. (1982a) are no different in three active dwarf stars ($\chi^1$ Ori, $\xi$ Boo A, $\varepsilon$ Eri) compared with two quiet dwarf stars ($\alpha$ Cen A, $\alpha$ Cen B), implying that there may be no great difference between magnetic flux tubes in the two types of stars except for the fractional surface coverage.

## IV. WHAT ROLES DO SYSTEMATIC FLOWS PLAY IN ACTIVE DWARF STAR ATMOSPHERES?

Perhaps the most exciting and unexpected discovery by *IUE* concerning cool stars is the very recent evidence concerning flows of the $10^5$ K plasma. Stencel et al. (1982a,b) measured line centroid velocities for 18 lines in the SWP high dispersion spectrum of the supergiant $\beta$ Dra (G2 Ib) by fitting least-squares Gaussians to the observed profiles. They estimated the velocity at the base of the chromosphere using eight subordinate or intersystem lines of C I, O I, S I, and Cℓ I. A measurement of the mean velocity of ten high excitation lines of He II, C III, C IV, N V, O III, Si III, and Si IV gave a relative motion of the high and low excitation lines of 20±4 km s$^{-1}$. The TR plasma is thus flowing down into the star, and we are observing a stellar "antiwind."

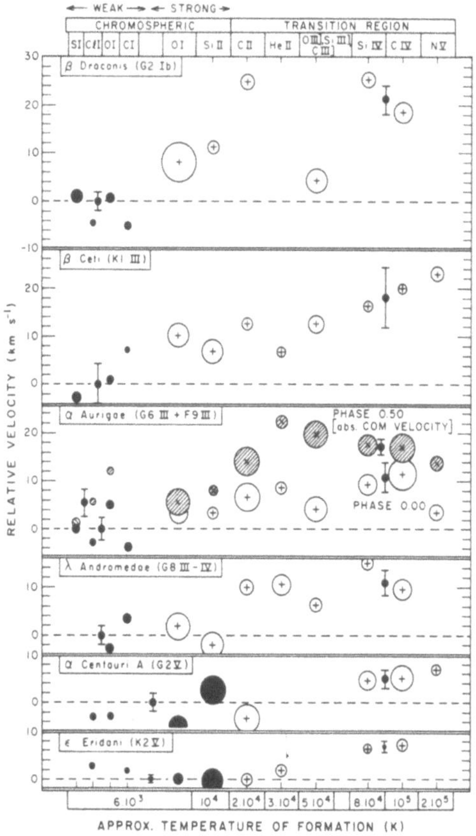


Fig. 5. A comparison of the line-of-sight velocities of high and low excitation lines obtained by Ayres et al. (1982b). The size of the bubbles indicates the total relative flux of the lines of a given ion. The filled circles are narrow chromospheric lines used in obtaining the flux-weighted mean zero velocity, and the standard error of the mean is indicated by the error bars to the left. The open circles indicate the mean velocities of higher excitation lines, and the error bars to the right, indicate the flux-weighted mean velocity of the four C IV and Si IV lines and the error of this mean (including the error of the zero velocity determination). The partially filled bubbles for Capella indicate velocities obtained from a small aperture observation at phase 0.50. These velocities were placed on an absolute center of mass velocity scale by comparison with a platinum lamp exposure obtained after the Capella exposure. Note that the velocity difference between high and low excitation lines is the same for both Capella data sets, but the small aperture data indicate that the chromospheric lines also appear to exhibit a small red shift.

This was an unexpected discovery, because we are accustomed to studying outflows (winds) in luminous late-type stars with mass loss rates that generally increase towards the upper right-hand corner of the H-R diagram. To confirm this result, Ayres et al. (1982b) then re-examined high dispersion IUE observations of active chromosphere stars, and found that β Ceti (K1 III), α Aur Ab (F9 III), λ And (G8 III-IV+?), and ε Eri (K2 V) also show net redshifts of TR emission lines relative to chromospheric lines (see Fig. 5). Even the quiescent dwarf star α Cen A (G2 V) shows a redshift at the 2.5σ level. They also reobserved α Aur Ab during conjunction (when both components of the α Aur A system have the same radial velocity) through the IUE small aperture to obtain an absolute velocity scale, with the result (see Fig. 5) that the chromospheric lines have an absolute redshift of about 5 km $s^{-1}$ and the TR lines have an absolute redshift of 17 km $s^{-1}$.

At first sight, the idea of downflows ("antiwinds") in stars of a wide range of luminosities seems preposterous, but solar downflows of 10-20 km $s^{-1}$ are typically seen in such lines as C IV and Si IV in the OSO-8 (Roussel-Dupré and Shine 1982), Skylab (Doschek, Feldman and Bohlin 1976; Feldman, Cohen and Doschek 1982), and HRTS (Brueckner 1981; Dere 1982) data. These downflows are best seen in observations of the chromospheric network and plages, where magnetic flux tubes are located, but the downflows are detected even in integrated light because the downflowing regions are bright in ultraviolet emission lines and thus make large contributions to the integrated light line profiles. An interesting result seen in the high spatial resolution HRTS data is that the downflow velocities increase with temperature from $10^4$ to $10^5$ K (Dere 1982), but Doschek and Feldman (1977) measured small downflow velocities (≈5 km $s^{-1}$) even for the chromospheric Mg II lines in the solar supergranulation network. Ayres et al. (1982b) detected the same increase in absolute downflow velocities between the chromosphere and TR of α Cen Ab.

Several explanations have been proposed. The flows may be produced by coronal material that is cooling, condensing, and falling back down to the chromospheric footpoints of magnetic loops after an interruption of the internal heating source (Rosner, Tucker and Vaiana 1978). The downflows may be part of a circulation pattern within large flux tubes for which the upleg portion of the circulation is too cool (spicules, for example) to be visible in C IV (Pneuman and Kopp 1977). It is even possible that material is flowing upward at C IV temperatures more rapidly than it is flowing downward, such that the decrease in density required by mass conservative flow will greatly reduce the emission measure of the upward moving gas (since $F_L \sim n_e^2$), resulting in a net redshift (Doschek, Feldman and Bohlin 1976). In any case, the appearance of redshifts is clear evidence for the existence of strong, closed magnetic field structures in stellar atmospheres.

There are important consequences of redshifts, which we henceforth interpret as true downflows of material in magnetic flux tubes:

(1) The existence of downflowing gas implies an enthalpy flux of heat from the corona that must be included in any study of the energy balance of a stellar TR. Pneuman and Kopp (1977), for example, demonstrated that the enthalpy flux in a typical solar downflow exceeds the thermal conductive flux.

(2) Since many stars should have both downflows and upflows and since the downflowing regions probably emit brighter emission lines, one should be especially careful in estimating stellar mass loss rates from integrated disk Doppler shifts. The Sun does have a wind!

## V. WHAT IS THE RELATION BETWEEN HEATING RATES IN DIFFERENT LAYERS OF ACTIVE DWARF STARS?

Using SWP low dispersion observations of 28 cool stars, Ayres, Marstad and Linsky (1981) showed that the emission line fluxes of chromospheric and TR lines are not linearly correlated (see Fig. 6). Instead, as one goes to stars with brighter chromospheric emission (i.e. greater $f_{Mg\ II}/\ell_{bol}$), the TR lines brighten even faster such that $(f_{C\ IV}/\ell_{bol}) \sim (f_{Mg\ II}/\ell_{bol})^{1.5}$ and the coronal X-ray flux brightens faster yet. Walter, Basri and Laurent (1982) and Oranje, Zwaan and Middlekoop (1982) found similar results. This phenomenon was previously noted in the comparison of the II Peg plage to quiescent spectra (Fig. 2), and the flare to quiescent spectra for HR 1099 (Fig. 3) and the dMe star Proxima Centauri (Haisch et al. 1982). It is also seen by comparing solar plage to quiescent spectra and thus must be a general property of stellar atmospheres.

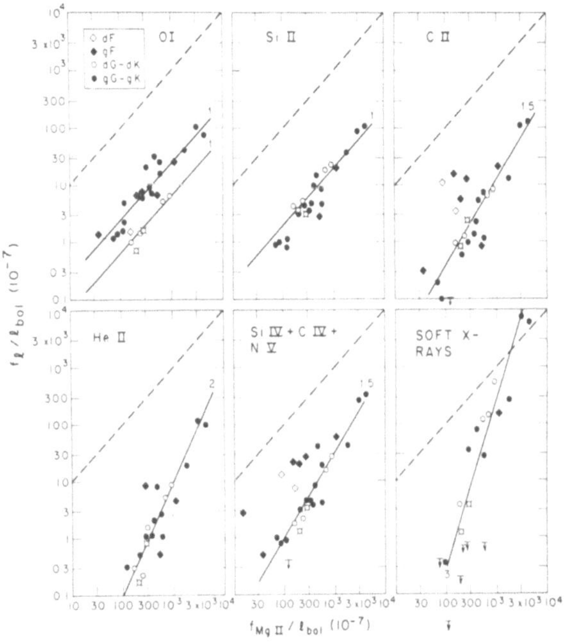


Fig. 6. Correlation plots of chromospheric, TR, and coronal fluxes compared to the Mg II line relative flux (Ayres, Marstad and Linsky 1981).

Hammer, Linsky and Endler (1982) proposed an explanation for this phenomenon. They pointed out that the radiative loss rate in TR emission lines for realistic magnetic flux tube models (e.g. Rosner, Tucker and Vaiana 1978) depends on pressure to a higher power than the corresponding radiative loss rate in chromospheric emission lines (e.g. models in Vernazza et al. 1981). Thus, with increasing mechanical energy flux, the location of the base of the TR (the intersection point of the curves in Fig. 7) moves to larger pressures and the TR lines brighten by a larger factor than the chromospheric lines.

For their sample of ten active dwarf stars (including solar plages), Linsky et al. (1982) showed that the fraction of the stellar flux emitted by the chromosphere increases a factor of 5 as $T_{eff}$ decreases from 5770 to 3200 K, but the fraction emitted by the TR increases a factor of 100. Thus the relative heating rates in different atmospheric layers may depend on $T_{eff}$ as well as the magnetic field.

## VI. BY WHAT MECHANISMS ARE ACTIVE DWARF STAR CHROMOSPHERES AND TRANSITION REGIONS HEATED?

Linsky and Ayres (1978) showed, on the basis of Mg II fluxes, that the chromospheric radiative loss rate per unit surface area of a star shows no dependence on stellar gravity. This implies that the heating rate is also independent of gravity, contrary to computations for the dissipation of shocks produced by nonmagnetic acoustic waves, which imply a $g^{-1}$ dependence and a $T_{eff}$ dependence different than the observed. This result was modified slightly by Stencel et al. (1980),

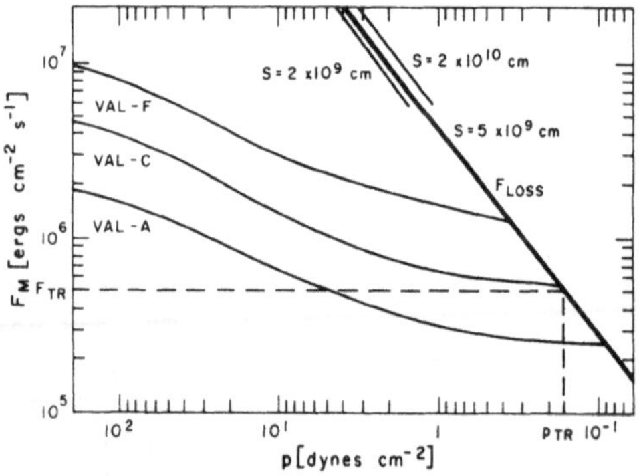


Fig. 7. Total mechanical energy flux $F_M$ as a function of the pressure p for the chromospheric models A, C, and F of Vernazza, Avrett and Loeser (1981). The transition region lies at the intersection point with a curve (drawn heavy) that gives the total energy losses $F_{Loss}$ of the transition region and corona as a function of the base pressure and the semilength S of the coronal loops (cf. Rosner, Tucker and Vaiana 1978). From Hammer, Linsky and Endler (1982).

who showed that IUE observations of cool supergiants are consistent with a small increase ($\sim g^{-1/4}$) in the heating rate as the gravity decreases. These data and the observed large range in heating rates for stars of similar $T_{eff}$ and g, provide important clues on the chromospheric heating mechanism.

Stein (1981) and Ulmschneider and Stein (1982) derived approximate scaling laws for how the emitted flux from a chromosphere ($F_{chromo}$) depends on $T_{eff}$, g, and B for four wave heating modes (acoustic, Alfvén, acoustic slow, and magnetic fast). Only the acoustic slow mode for weak or equipartition magnetic fields predicts a relation

$$\frac{F_{chromo}}{\sigma T_{eff}^4} \approx g^{-0.192}\, T_{eff}^{2.13} \quad ,$$

consistent with the above data. In addition, the chromospheric heating rate will depend on the fractional surface coverage by the magnetic flux tubes, which can explain the range of radiative loss rates in stars of similar $T_{eff}$ and g. Thus stellar observations have played a major role in guiding theoretical calculations by pointing out the important role played by magnetic fields.

In addition to acoustic slow mode heating, the previously discussed enthalpy flux carried by downflows may be an important heating source in the TRs of flux tubes (e.g. Wallenhorst 1980,1981). Also, Cram (1982) has pointed out that the absorption of coronal X-rays in the quiescent chromospheres of dMe stars may be an important chromospheric heating source. In Table 1 we compare surface fluxes in soft X-rays, C IV λ1550 (the largest TR emitter), the Mg II resonance lines (the largest emitters in solar-like chromospheres), and the Balmer

Table 1. Comparison of Surface Fluxes for Active and Quiescent Dwarf Stars

| Star | Spectral Type | Surface Fluxes (ergs $cm^{-2}$ $s^{-1}$) | | | |
|---|---|---|---|---|---|
| | | $F_x$ | $F_{C\ IV}$ | $F_{Mg\ II}$ | $F_{Balmer\ Lines}$ |
| α Cen A | G2 V | 6.6(3) | 5.9(3) | 1.1(6) | -- |
| ξ Boo A | G8 V | 3.1(6) | 8.8(4) | 4.5(6) | -- |
| α Cen B | K1 V | 2.6(4) | 4.6(3) | 9.5(5) | -- |
| ε Eri | K2 V | 4.9(5) | 3.3(4) | 2.4(6) | -- |
| EQ Vir | K5e V | 7.8(6) | 1.4(5) | 2.0(6) | -- |
| AU Mic | M1.6e V | 1.6(7) | 1.3(5) | 8.9(5) | 4.5(6) |
| EQ Peg | M3.7e V | 3.8(6) | 7.5(4) | -- | -- |
| Proxima Cen | M5.5e V | 1.2(6) | 1.9(4) | 1.0(5) | -- |
| UV Cet | M5.6e V | 1.1(6) | 5.9(4) | 1.4(5) | 2.4(6) |

Note: In columns 3-7, the figure in parentheses is the power of ten.

lines (the largest emitters in dMe star chromospheres) from data presented by Linsky et al. (1982), Vaiana et al. (1981), and Helfand and Caillault (1982). Assuming that as much X-ray flux is absorbed in the chromosphere as leaves the star, these data suggest that Cram (1982) may be correct. Since solar coronal X-ray emission is almost entirely confined to magnetic loops, even the nonmagnetic regions of dMe star chromospheres may be heated indirectly by magnetically controlled plasma.

## VII. WHAT ARE SEMIEMPIRICAL MODELS OF ACTIVE DWARF STARS TELLING US?

Until now, we have summarized the different qualitative roles played by magnetic fields. In addition, several investigators have built detailed models of solar and stellar atmospheres for regions in which fields are important. Chapman (1981), Ulmschneider and Stein (1982), and Linsky (1980) have reviewed this work. No models are wholly satisfactory as yet due to gross simplifications in the assumed geometry, treatment of the radiative transfer equation, and the strong dependence of the magnetic flux tube parameters on the assumed filling factor, but I will mention those trends in the models that are likely to be valid:

### a) One-Component Atmospheric Models

The simplest approach is to construct a one-component model atmosphere to match line profiles and fluxes for the brightest regions of a solar plage or the integrated flux of a star with very bright emission lines. Since the data are spatial averages including magnetic flux tubes (the active component) and nonmagnetic plasma (the quiet component) plasma, the derived differences between the active model and the reference quiet Sun or star model are lower limits to the true differences between the properties of magnetic flux tubes and the nonmagnetic regions. Thus we need to extrapolate these trends to the case of complete filling of the aperture by flux tubes.

One-component models of solar plages include the work of Shine and Linsky (1974) using the Ca II lines, Morrison and Linsky (1978) and Kelch and Linsky (1978) using the Mg II lines, Basri et al. (1979) using L$\alpha$, and Vernazza et al. (1981) using a number of continuum and emission line features. These models have a number of elements in common that are illustrated in Figure 8:

(1) The minimum temperature ($T_{min}$) reached between the photosphere and chromosphere is enhanced by several hundred degrees (270 K in the Vernazza et al. Model F and about 200 K in the Morrison-Linsky models) compared to the quiet Sun. There is also considerable enhancement of temperature, and thus nonradiative heating, in the upper layers of the photosphere.

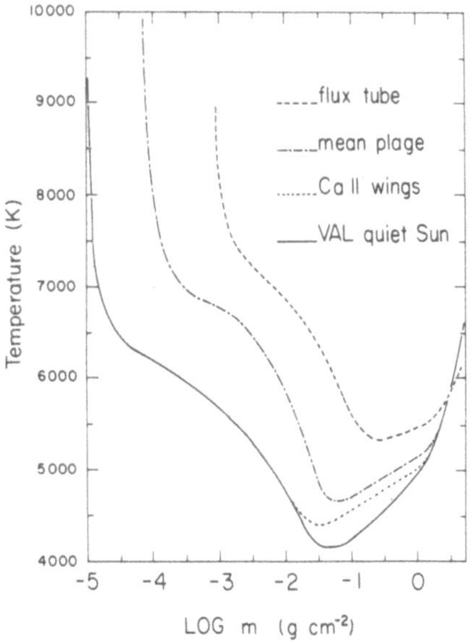


Fig. 8. Plage, flux tube, and quiet Sun models. The solid line is the VAL quiet Sun model. The short dashed lines (Ca II wings) represent a modification of the VAL designed to reproduce the Ca II H and K damping wings. The dash-dot curve is a plage model based on H I L$\alpha$, Ca II K, and Mg II k data obtained by the French (LPSP) experiment on OSO-8. The long dashed (higher) curve represents a flux tube model with a chromospheric portion matching the OSO-8 plage profiles with a 20% filling factor. The photospheric portion ($m > 0.3$ g cm$^{-2}$) is similar to the class of flux tube models advocated by Chapman (1977). From Chapman (1981), courtesy of Colorado Associated University Press.

(2) The location of $T_{min}$ is displaced inward to larger values of column mass density (m); for example, $T_{min}$ is deeper by 40 km and $m(T_{min})$ increases from 0.05 to 0.08 g cm$^{-2}$ between the quiet Sun (Model C) and Model F in the Vernazza et al. grid.

(3) Similarly, the location of the steep rise in temperature beginning at 8000 K and extending through the TR is also displaced inwards to larger values of mass column density. For example, the grid point corresponding to 10,000 K is located 40 km deeper and $m(T = 10^4$ K) increases from $6 \times 10^{-6}$ to $1.2 \times 10^{-5}$ g cm$^{-2}$ between the mean quiet Sun and Model F in the Vernazza et al. grid. In the Shine and Linsky (1974) grid the range in $m(T = 10^4$ K) is nearly a factor of 40. Since the temperature rises steeply in all TR models, the TR pressure ($P_{TR} \approx m(T = 10^4$ K)g) is nearly constant for each model. Thus a sequence of models with increasing $P_{TR}$ typically produces a sequence of emission lines with increasing flux.

(4) Different authors have assumed different functional forms of the temperature structure, T(m), between the $T_{min}$ and the top of the chromosphere ($T \approx 10^4$ K), but a general result is that T(m) and the local electron density at each column mass density are larger for models with greater m($T = 10^4$ K) values. Thus, typical chromospheric lines like Ca II H and K, Mg II h and k, Lα, and Si II λλ1808, 1816 brighten as m($T = 10^4$ K) increases.

(5) Avrett (1981) computed radiative loss rates for the five Vernazza et al. models, the Basri et al. (1979) plage model, a flare model from Machado et al. (1980), and two stars (α Boo and λ And). For the Vernazza et al. models and the plage model, Ca II and Mg II are the dominant emitters, although hydrogen becomes important at the base of the TR. Thus, the observed Ca II and Mg II surface fluxes in the Sun and solar-like stars should be accurate diagnostics of the chromospheric heating rate as had been assumed previously, but, for the dMe stars, the Balmer lines are more important chromospheric emitters (Linsky et al. 1982).

To what extent do models of active chromosphere stars show properties similar to the models of solar plages? Linsky (1980) and Ulmschneider and Stein (1982) have reviewed this question. There are three groups of models that can be used to answer it: models of F-K dwarf stars (Kelch, Linsky and Worden 1979; Simon, Kelch and Linsky 1980b), models of dM and dMe stars (Giampapa, Worden and Linsky 1982), and models of RS CVn-type active subgiants (Simon and Linsky 1980; Baliunas et al. 1979). Evidence for enhanced heating in the upper photosphere and temperature minimum region is clearly shown in the dMe models and perhaps also in the active F-K dwarfs. In addition, the dMe stars and the active G-K dwarfs (see also Ayres et al. 1982a) show broader base widths of the Ca II and Mg II resonance lines, indicating that the column mass density at the temperature minimum is systematically larger in stars with enhanced heating. Finally, all of these active star models have m($T = 10^4$ K) similar to solar plages, so that TR pressures are enhanced over quiet stars with similar $T_{eff}$ and gravity. Thus, the differences between solar plages and quiet models are repeated in the active and quiet late-type stars.

Fosbury (1974) and Cram and Mullan (1979) showed that the Balmer lines are useful diagnostics of chromospheric structure in M dwarfs; in particular, as the amount of chromospheric material in an M dwarf model increases, the Hα line first becomes a deeper absorption feature and then goes into emission. Models of dM and dMe stars computed by Kelch et al. (1979) and Giampapa et al. (1982) to match the Ca II lines show this behavior. In the solar context, Basri et al. (1979) found that the peculiar property of the Hα line in being relatively bright at line center and dark at ±0.5 Å in plages compared to the quiet Sun can be simply explained as a consequence of their different atmospheric structure.

b) Two-Component Atmospheric Models

Chapman (1981) conclusively argued that no one-component model can accurately represent the properties of unresolved flux tubes, but instead the true models for flux tubes must be more extreme versions of the one-component plage models. Three approaches have been followed in estimating flux tube properties:

(1) To estimate the factor by which flux tubes fill the aperture and then to solve for the flux tube parameters assuming that the remaining portion of the aperture is represented by a quiet atmosphere model. Chapman (1981) derived such a flux tube model from OSO-8 plage spectra of the Lyα, Ca II, and Mg II lines assuming a 20% filling factor (see Fig. 8). His solar flux tube model has $T_{min}$ enhanced by 1200 K, $m(T_{min})$ displaced inwards from 0.04 to 0.3 g $cm^{-2}$, chromospheric temperatures enhanced by nearly 2500 K, and $m(T = 10^4$ K) displaced inwards from $1 \times 10^{-5}$ to $1 \times 10^{-3}$ g $cm^{-2}$.

(2) To assume a diverging flux tube in horizontal pressure equilibrium (gas and magnetic forces) with the surrounding nonmagnetic atmosphere. Chapman (1977,1979) matched spatially averaged spectra by assuming a quiet model and solving for the parameters of the flux tubes assuming a value for the base magnetic flux.

(3) To assume a geometry with isolated flux tubes embedded in a nonmagnetic medium and then to solve the transfer equation in two dimensions including the horizontal flow of radiation. Two examples are the work of Stenholm and Stenflo (1977) and Owocki and Auer (1980). This approach increases the computational complexity, but it is probably a necessary complication in the chromosphere where most lines are effectively thin and thus photoexcitation from adjacent structures may be important.

## VIII. SOME SUGGESTIONS FOR FUTURE WORK

I began this talk by mentioning some of the inherent difficulties in this topic, and I suspect that you now appreciate what I meant. While we have made real progress recently in identifying the roles played by magnetic fields in stellar chromospheres and TRs, we now recognize how little we know. Let me conclude by stating how we will probably make real progress in this area:

(1) We must resolve individual flux tubes on the Sun in order to derive their physical properties in a definitive way. I anticipate that this will be the most important accomplishment of the Solar Optical Telescope (SOT) when it flies at the end of this decade.

(2) I suspect that future high spectral resolution stellar observations with IUE and Space Telescope will reveal new phenomena and trends with stellar parameters that will point towards the important physical processes which future theoretical studies must include.

(3) We must elucidate the basic physical processes responsible for downflows in flux tubes.

(4) Even though heating processes will probably remain poorly understood for some time, it is nevertheless important to model magnetic flux tubes properly, taking into account the energy balance, dynamics, and radiative transfer for parameterized heating rates. It is important to study the stability of such models and to look for unique signatures of the location and mechanism of the heating.

This work was supported in part by NASA grants NAG5-82, NGL-06-003-057, and NAG5-199 through the University of Colorado. I would like to thank my colleagues T. R. Ayres, P. L. Bornmann, S. Drake, R. Hammer, N. C. Marstad, M. Schindler, T. Simon, R. E. Stencel, and F. M. Walter for stimulating discussions and for permission to describe unpublished work at this time.

REFERENCES

Avrett, E.H.: 1981, in "Solar Phenomena in Stars and Stellar Systems," eds. R. M. Bonnet and A. K. Dupree (Dordrecht: Reidel), p. 173.
Ayres, T.R., Linsky, J.L., Simon, T., Jordan, C., and Brown, A.: 1982a, in preparation.
Ayres, T.R., Marstad, N.C., and Linsky, J.L.: 1981, Ap. J. 247, p. 545.
Ayres, T.R., Stencel, R.E., Linsky, J.L., Simon, T., Jordan, C., Brown, A., and Engvold, O.: 1982b, in preparation.
Baliunas, S.L., Avrett, E.H., Hartmann, L., and Dupree, A.K.: 1979, Ap. J. (Letters) 233, p. L129.
Baliunas, S.L., and Dupree, A.K.: 1982, Ap. J. 252, p. 668.
Baliunas, S.L., Hartmann, L., Vaughan, A.H., Liller, W., and Dupree, A.K.: 1981, Ap. J. 246, p. 473.
Basri, G.S., Linsky, J.L., Bartoe, J.-D.F., Brueckner, G.E., and Van Hoosier, M.E.: 1979, Ap. J. 230, p. 924.
Brueckner, G.E.: 1981, in "Solar Active Regions," ed. F. Q. Orrall (Boulder: Colorado Assoc. Univ. Press), p. 113.
Brueckner, G.E., Bartoe, J.-D.F., and Van Hoosier, M.E.: 1977, in "Proceedings of the OSO-8 Workshop," eds. E. Hansen and S. Schaffner (Boulder, LASP), p. 380.
Butler, C.J., Byrne, P.B., Andrews, A.D., and Doyle, J.G.: 1981, Monthly Notices Roy. Astron. Soc. 197, p. 815.
Byrne, P.B., Butler, C.J., Andrews, A.D., Linsky, J.L., Simon, T., Marstad, N., Rodono, M., Blanco, C., Catalano, S., and Marilli, E.: 1982, in "Proceedings of the Third European IUE Conference," in press.
Chapman, G.A.: 1977, Ap. J. Suppl. Ser. 33, p. 35.
Chapman, G.A.: 1979, Ap. J. 232, p. 923.
Chapman, G.A.: 1981, in "Solar Active Regions," ed. F. Q. Orrall (Boulder: Colorado Assoc. Univ. Press), p. 43.
Cram, L.E.: 1982, Ap. J. 253, p. 768.
Cram, L.E. and Mullan, D.J.: 1979, Ap. J. 234, p. 579.

Dere, K.: 1982, Solar Phys. 77, p. 77.
Doschek, G.A. and Feldman, U.: 1977, Ap. J. Suppl. 35, p. 471.
Doschek, G.A., Feldman, U., and Bohlin, J.D.: 1976, Ap. J. (Letters) 205, p. L177.
Dupree, A.K.: 1981, in "Solar Phenomena in Stars and Stellar Systems," eds. R. M. Bonnet and A. K. Dupree (Dordrecht: Reidel), p. 407.
Dupree, A.K.: 1982, in "Advances in Ultraviolet Astronomy: Four Years of IUE Research" (in press).
Feldman, U., Cohen, L., and Doschek, G.A.: 1982, Ap. J. 255, p. 325.
Fosbury, R.A.E.: 1974, Monthly Notices Roy. Astron. Soc. 169, p. 147.
Frazier, E.N. and Stenflo, J.O.: 1973, Solar Phys. 27, p. 330.
Giampapa, M.S., Worden, S.P., and Linsky, J.L.: 1982, Ap. J. (in press).
Haisch, B.M., Linsky, J.L., Bornmann, P.L., Stencel, R.E., Golub, L., and Antiochos, S.K.: 1982, Ap. J. (submitted).
Hall, D.S.: 1981, in "Solar Phenomena in Stars and Stellar Systems," eds. R. M. Bonnet and A. K. Dupree (Dordrecht: Reidel), p. 431.
Hallam, K.L. and Wolff, C.L.: 1981, Ap. J. (Letters) 248, p. L73.
Hammer, R., Linsky, J.L., and Endler, F.: 1982, in "Advances in Ultraviolet Astronomy: Four Years of IUE Research" (in press).
Hartmann, L., Dupree, A.K., and Raymond, J.C.: 1982, Ap. J. 252, p. 214.
Helfand, D.J. and Caillault, J.P.: 1982, Ap. J. 253, p. 760.
Imhoff, C.L. and Giampapa, M.S.: 1982, in "Advances in Ultraviolet Astronomy: Four Years of IUE Research" (in press).
Kelch, W.L. and Linsky, J.L.: 1978, Solar Phys. 58, p. 37.
Kelch, W.L., Linsky, J.L., and Worden, S.P.: 1979, Ap. J. 229, p. 700.
Kunkel, W.W.: 1975, in "Variable Stars and Stellar Evolution," eds. V. E. Sherwood and L. Plaut (Dordrecht: Reidel), p. 15.
Lang, K.R.: 1980, "Astrophysical Formulae," 2nd ed. (Berlin: Springer-Verlag), p. 510.
Linsky, J.L.: 1980, Ann. Rev. Astr. Ap. 18, p. 439.
Linsky, J.L.: 1981a, in "Solar Phenomena in Stars and Stellar Systems," eds. R. M. Bonnet and A. K. Dupree (Dordrecht: Reidel), p. 99.
Linsky, J.L.: 1981b, in "Effects of Mass Loss on Stellar Evolution," eds. C. Chiosi and R. Stalio (Dordrecht: Reidel), p. 187.
Linsky, J.L.: 1981c, in "Physical Processes in Red Giants," eds. I. Iben, Jr. and A. Renzini (Dordrecht: Reidel), p. 247.
Linsky, J.L.: 1981d, in "X-ray Astronomy in the 1980's," NASA Technical Memorandum 83848, p. 13.
Linsky, J.L.: 1982, in "Advances in Ultraviolet Astronomy: Four Years of IUE Research" (in press).
Linsky, J.L. and Ayres, T.R.: 1978, Ap. J. 220, p. 619.
Linsky, J.L., Bornmann, P.L., Carpenter, K.G., Wing, R.F., Giampapa, M.S., and Worden, S.P.: 1982, Ap. J. (in press).
Machado, M.E., Avrett, E.H., Vernazza, J.E., and Noyes, R.W.: 1980, Ap. J. 242, p. 336.
Marcy, G.W.: 1983, in "Solar and Stellar Magnetic Fields: Origins and Coronal Effects (Proceedings of IAU Symp. 102, J. Stenflo(ed.)). .
Marstad, N. et al.: 1982, in "Advances in Ultraviolet Astronomy: Four Years of IUE Research" (in press).
Morrison, N.D. and Linsky, J.L.: 1978, Ap. J. 222, p. 723.
Oranje, B.J., Zwaan, C., and Middelkoop, F.: 1982, Astr. Ap. (in press).

Owocki, S.P. and Auer, L.H.: 1980, Ap. J. 241, p. 448.
Pneuman, G.W. and Kopp, R.A.: 1977, Astr. Ap. 55, p. 305.
Rosner, R., Tucker, W.H., and Vaiana, G.S.: 1978, Ap. J. 220, p. 643.
Roussel-Dupre, D. and Shine, R.A.: 1982, Solar Phys. 77, p. 329.
Schindler, M., Kjeldseth-Moe, O., Bartoe, J.-D.F., Brueckner, G.E., and Van Hoosier, M.E.: 1982, Ap. J. (submitted).
Shine, R.A. and Linsky, J.L.: 1974, Solar Phys. 39, p. 49.
Simon, T., Kelch, W.L., and Linsky, J.L.: 1980b, Ap. J. 237, p. 72.
Simon, T. and Linsky, J.L.: 1980, Ap. J. 241, p. 759.
Simon, T., Linsky, J.L., and Schiffer F.H. III: 1980a, Ap. J. 239, p. 911.
Skumanich, A., Smythe, C., and Frazier, E.N.: 1975, Ap. J. 200, p. 747.
Stencel, R.E., Linsky, J.L., and Ayres, T.R.: 1982a, in "Advances in Ultraviolet Astronomy: Four Years of IUE Research" (in press).
Stencel, R.E., Linsky, J.L., Ayres, T.R., Jordan, C., and Brown, A.: 1982b, Ap. J. (submitted).
Stencel, R.E., Mullan, D.J., Linsky, J.L., Basri, G.S., and Worden, S.P.: 1980, Ap. J. Suppl. 44, p. 383.
Stenflo, J.O.: 1973, Solar Phys. 32, p. 41.
Stenholm, L.G. and Stenflo, J.O.: 1977, Astr. Ap. 58, p. 273.
Stein, R.F.: 1981, Ap. J. 246, p. 966.
Ulmschneider, P. and Stein, R.F.: 1982, Astr. Ap. 106, p. 9.
Vaiana, G.S. et al.: 1981, Ap. J. 245, p. 163.
Vaughan, A.H., Baliunas, S.L., Middelkoop, F., Hartmann, L.W., Mihalas, D., Noyes, R.W., and Preston, G.W.: 1981, Ap. J. 250, p. 276.
Vernazza, J.E., Avrett, E.H., and Loeser, R.: 1981, Ap. J. Suppl. 45, p. 635.
Vernazza, J.E. and Reeves, E.M.: 1978, Ap. J. Suppl. Series 37, p. 485.
Vogt, S.S.: 1983, this volume.
Wallenhorst, S.G.: 1980, Ap. J. 241, p. 229.
Wallenhorst, S.G.: 1981, Ap. J. 249, p. 176.
Walter, F.M., Basri, G.S., and Laurent, R.: 1982, in "Advances in Ultraviolet Astronomy: Four Years of IUE Research" (in press).
White, O.R. and Livingston, W.C.: 1981, Ap. J. 249, p. 798.
Wilson, O.C.: 1978, Ap. J. 226, p. 379.

DISCUSSION

Jordan: I'm surprised that you didn't mention that the HeII widths are different in these stars. Other lines have normal widths but HeII is clearly anomalous.

Linsky: Which stars are you specifically referring to?

Jordan: $\alpha$ Cen A, $\alpha$ Cen B, $\xi$ BooA and $\varepsilon$ Eri.

Linsky: Remind me, which way do they go.

Jordan: $\alpha$ Cen A or one of them is distinctly narrower. The formation of HeII is rather more complicated than the other lines. It may be formed partly through recombination from the corona and partly through transient excitation. My explanation would be that one mechanism or the other may be dominant as one goes from one star to another with different temperature gradients. I don't find it very surprising.

Linsky: A plot of HeII (λ1640) flux normalized to $L_{bol}$ against X-ray flux normalized in the same way shows that all of the active dwarfs lie approximately in a straight line. This suggests, but does not prove, that the mechanisms of formation are the same in these stars.

de la Reza: Is there a correlation between the Balmer emission-line strength and photometrically determined rotation period? If there is a correlation then we may expect that the absorption features will not be visible when rotation is high.

Linsky: The Balmer-line observations I referred to were made with a resolution of 120mA. At this the lines were not fully resolved but were obviously broadened. I would interpret this as an optical effect. Is your argument that the lines are rotationally broadened?

de la Reza: If the rotation is high perhaps the predicted flat tops to the emission-line profiles will disappear. I am suggesting that this might be a way to resolve some of the difficulties that exist with photometrically determined rotation periods.

Worden: Perhaps I can answer that. In our data on rapidly rotating stars, like YY Gem and other close binaries, the central reversals are smeared out. Simulations which we have done with L. Cram, M. Giampapa and others also confirm this. Its a very key point that one must take the effect of rotation out.

______ (Some recording lost)

Giampapa: Just to expand on Pete (Worden's) comment, a high pressure transition region will also smear out the central reversal. So as he suggests one would have to make systematic observations of the Hα line to determine any rotational effects on the line profile.

Bromage: Was I correct in seeing a value of 29 ± 3 km/s for the FWHM of high-resolution lines on AU Mic?

Linsky: That's correct.

Bromage: The instrumental profile is equivalent to 20 km $s^{-1}$ and is not well known. How can you then get an error of ± 3 km $s^{-1}$

Linsky: The numbers I have shown refer to the widths after taking out the instrumental profile. We assumed an instrumental correction corresponding to 12000 and took that out by taking the sum of the squares.

Bromage: But the profile is not Gaussian and indeed is not very well known.

Linsky: That's correct and there is considerable uncertainty but I think that there is no doubt that the width of the line is less than it is in stars like α Cen A and ξ Boo A.

Dupree: You suggested that there might be short timescale variations in the UV flux from the dMe stars. I am a little puzzled because some of the variations seem to be of the order of 10-15% while the reproducibility of IUE is generally thought to be 20-25%. Can you comment on the size of the error bars?

Linsky: Those of us have worked with IUE for a long time recognize that there are problems with photometry of the integrated line profiles in low dispersion. These photometric errors get much larger as you go to the weaker lines. However in AT Mic for instance, the C IV line which is one of the strongest shows variations of 70% which appears to be real. On the other hand where we have variations of about 10-20% we would be concerned about its reality.

# THE STRUCTURE AND ENERGY BALANCE IN MAIN SEQUENCE STARS

+ C. Jordan, * T.R. Ayres and + A. Brown
+ Department of Theoretical Physics, Oxford University, UK
* LASP, University of Colorado, Boulder, USA
+ Department of Physics, Queen Mary College, London, UK

ABSTRACT

High-resolution spectra obtained with the IUE satellite have been used to study the structure and energy balance in the main sequence stars $\xi$ Boo A, $\alpha$ Cen A, $\alpha$ Cen B and $\varepsilon$ Eri. The EUV observations are combined with X-ray fluxes to predict the coronal temperatures, the electron pressures and energy lost or transferred by radiation and thermal conduction

## 1. INTRODUCTION

In our IUE programmes we have been studying the brighter main-sequence stars for which high-resolution spectra may be obtained. The aim is to compare the structure and energy input requirements of their chromospheres and coronae and thus elucidate the physical processes operating.

Methods established for interpreting solar emision line fluxes in terms of the atmospheric structure (eg. Jordan and Wilson, 1971), should be applicable to stars of similar gravity. Line fluxes alone do not yield unique models, the electron density must also be measured. Because suitable lines are blended or are in a region of strong continuum high resolution observations become necessary. The line fluxes and subsequent models can be used to calculate the radiative losses and net conductive flux through the atmosphere and hence the energy input required to balance these losses can be deduced. Also, line widths allow the non-thermal energy density to be examined and compared with wave-carried energy fluxes. The methods for carrying out the above analysis have been set out by Jordan and Brown (1981) and have already been applied to a $\alpha$ CMi (F5IV-V). (Brown and Jordan, 1981). At present the results are limited by the lack of reliable temperature measurements for the stellar coronae, in most cases only broad-band fluxes or approximate temperature estimates are available.

In the present work we use new high resolution observations of $\xi$ Boo A and $\varepsilon$ Eri obtained with exposure times of 952 min and 447 min respectively. We use observations of $\alpha$ Cen A and $\alpha$ Cen B that have been published by



*P. B. Byrne and M. Rodonò (eds.), Activity in Red-Dwarf Stars, 61–65.*

TABLE 1.

Stellar Parameters & Results for Pressures, Temperatures & Energy Fluxes

| Star | $R/R_\odot$ | log g | log a | log $P_o$ | log $T_c$ | log $F_{R1}$ | log $F_{R2}$ | log $F_c(T_o)$ |
|---|---|---|---|---|---|---|---|---|
| α Cen A (G 2V) | 1.23 | 4.2 | 17.6 | 14.5 (14.5) | 6.1 | 6.1 | 4.2 | 5.2 |
| α Cen B (K 0V) | 0.87 | 4.5 | 17.6 | 15.0 (15.1) | 6.4 | 6.2 | 4.9 | 6.2 |
| ξ Boo A (G 8V) | 0.98 | 4.4 | 18.6 | 16.0 (15.9) | 6.8 | 7.0 | 6.5 | 7.2 |
| ε Eri (K 2V) | 0.82 | 4.5 | 18.3 | 15.5 (15.5) | 6.5 | 6.6 | 5.8 | 6.5 |

Ayres et al. (1982). X-ray fluxes and temperature estimates are from observations by Vaiana et al. (1980), Walter et al. (1980), Golub et al. (1982) and Johnson (1981). The stellar parameters adopted are given in Table 1.

## 2. RESULTS

The line fluxes have been used to construct the emission measure distributions between 6300K and $2\times10^5$K. These all have essentially the same shape between $\sim2\times10^4$K and $2\times10^5$K. The position of the density sensitive CIII (1909A) and SiIII (1982A) intersystem lines relative to each mean distribution around $5.6\times10^4$K has been used to determine or limit the electron pressure. In ξ Boo A the ratio of the SiIII lines at 1206A and 1892A can also be used to estimate the pressure. (Details of atomic data will be given in the fuller paper to be submitted to MNRAS). Solar abundances were adopted since individual values for the relevant elements are not known.

The similar shape of the emission measure distributions shows that the energy input has the same dependance on $T_e$ in these four stars. If energy input balances radiation losses, in the absence of conduction, the best single power function would be

$$dF_m/dT_e = A\,T_e^{\,x} \qquad \text{where } -1 < x < 0. \tag{1}$$

However the large conductive fluxes found would require a dynamic not a static model; this point will be discussed in the fuller paper.

The X-ray fluxes and temperature estimates for ξ Boo A ($10^7$K) and the α Cen A + α Cen B system ($2.1 \pm 0.4\times10^6$K from Golub et al. 1982) may be used to determine the mean coronal pressure, $P_c$. These pressures ($1.2\times10^{16}$ and $8.9\times10^{14}$ cm$^{-3}$K for ξ Boo A and α Cen B respectively) may be adequately reproduced by adopting an emission measure scaling between $2\times10^5$K and $T_c$, such that $E_m(T_e) = a\,T_e^{3/2}$, as found also in the Sun. (Jordan 1980). The scaling law between $P_o$ (at $2\times10^5$K), a, $g_*$ and $T_c$ is then

$$T_c^{\,5/2} - T_o^{\,5/2} = 1.2\times10^8\, P_o^{\,2}/a\, g_* \tag{2}$$

(Jordan and Brown, 1981). The values of $P_o$ for all four stars may then be determined. (See Table 1). They are in reasonable agreement with the estimates from line ratios. The ratio of the CI (1944A) line, which is pumped by the optically thick CI (1656A) multiplet (Jordan, 1967) to CII (1334A + 1335A) is also density sensitive. The relative density values show remarkably good agreement with the relative values of $P_o$ and are shown in brackets below $P_o$ in Table 1.

The total radiation losses between 6300K and $2x10^5$K ($F_{R1}$) may be found directly from the observed emission measures and power loss calculations (eg. Raymond et al. 1976). Above $2x10^5$K the radiative loss ($F_{R2}$) depends mainly on $E_m(T_c)$. The net conductive flux, $Fc(T_o)$ back at $2x10^5$K is found from the local emission measure and $P_o$. (See Jordan and Brown, 1981 for details).

To match the temperature variation of the non-thermal widths of the optically thin lines, if these do represent a wave-carried energy flux in a region where $P_e$ and B are constant, then the power of the heating function (equation 1) must be $-0.10<x<0$. For acoustic waves $A = 10^{-11}a$. Interpreted as an acoustic wave flux the energy carried by the observed non-thermal motions is comparable with the radiation losses; only very small (~10 gauss) magnetic field strengths would be required to provide sufficient Alfvén wave flux. However the non-thermal motions may be related to the large conductive flux or energy from the magnetic field. The radiative losses and conductive fluxes are given in Table 1. Neither the temperatures, pressures or ratio of radiative and conductive fluxes fit the predictions of the scaling laws proposed by Rosner et al. (1978) or Hearn (1977), which imply a fixed ratio for these fluxes. The total energy losses do scale as $P_o$ in the way predicted by these scaling laws because the total is insensitive to the split between radiation and conduction (Jordan, 1980).

REFERENCES

Ayres, T.R., Linsky, J.L., Moos, H.W., Stencel, R.E., Basri, G.S., Landsman, W, & Henry, R.C., 1982, Ap. J. In Press.
Brown, A. & Jordan, C: 1981, M.N.R.A.S. 196, pp. 757-779.
Golub, L., Harnden, F.R., Pallavacini, R., Rosner, R. & Vaiana, G.S: 1982, Ap. J. 253, pp. 242-247.
Hearn, A.G., 1977, Sol. Phys, 51. pp. 159-168.
Johnson, H.M., 1981, Ap. J. 243, pp. 234-243.
Jordan, C., 1967, Sol. Phys. 2, pp. 441-450.
Jordan, C., 1980, Astron. Astrophys. 86, pp. 353-363.
Jordan, C. & Brown, A., 1981, in 'Solar Phenomena in Stars and Stellar Systems' (Eds. R.M. Bonnet and A.K. Dupree) D. Reidel. pp. 199-225
Jordan, C. & Wilson, R., 1971, in 'Physics of the Solar Corona' (Ed. C.J. Macris) D. Reidel, pp. 211-236.
Raymond, J.C., Cox, D.P. & Smith, B.W., 1976, Ap. J. 204, pp. 290-292.
Rosner, R., Tucker, W.H., Vaiana, G.S., 1978, Ap. J. 220, pp. 643-665.
Vaiana, G.S. et al. 1981, Ap. J. 245, pp. 163-182.
Walter, F.M., Linsky, J.L., Bowyer, G.S. & Garmire, G., 1980, Ap. J. (Letts) 236, pp. L137-141

DISCUSSION

Rosner: I have a point to make about the calculation of emission measures. Several groups (Los Alamos & Shube at Stanford) have made calculations as follows. They take a prescribed temperature gradient and, initially, a Maxwellian distribution. They then do a Fokker-Planck calculation i.e. a particle calculation in which one calculates the electron distribution function as a function of height. The interesting result of this calculation is that when one looks at the electron distribution at low temperatures (i.e. as defined by the ions' thermal distribution e.g. at 20,000°K) one finds a superabundance of electrons in the high-energy Maxwellian tail. This superabundance is by many orders of magnitude and is density and pressure sensitive. So higher excitation species such as C IV, O VI, etc. are locally produced not by thermal electrons but by what are essentially coronal electrons. So if one compares emission-measure curves from these calculations with those derived from the observations one finds that the calculated curves are virtually flat, independent of temperature instead of concave up. This suggests that in a low-density atmosphere, such as the quiet Sun's, one does not have a temperature diagnostic from these lines. Would you like to comment?

Jordan: Yes. It was in fact I who first pointed out that He II was sensitive to this effect and so I was delighted to see that a proper calculation was done. He II is particularly sensitive to this effect. I have looked for it deliberately and systematically in solar work for the other lines but the exponential factor exp (-kT) in the excitation rate is so temperature insensitive for the Li-like resonance lines such as N V, O VI that I would maintain it is not effecting those lines. It is, however, clearly affecting He II and I am glad you pointed this out. There is a huge spread in the He II line λ1640 and this is because it is sensitive to the local temperature gradient. This is true whether or not one has coronal X-rays controlling the ionization balance or the electron tail from the corona. One cannot distinguish between the two from the spread of the He II line. So He II is very sensitive to the temperature gradient. The temperature gradient goes up as the pressure goes up and so (referring to Table 1) He II increases from α Cen B, ε Eri to ξ Boo A because in ξ Boo A the temperature gradient is greater than in ε Eri, for instance. So this effect is controlling the He II emission and perhaps the energy balance lower down but I do not believe it affects lines with small values of (W/kT) in the line factor.

Rosner: Yes, but this also means that the evaluation of the conductive flux cannot be done in the simple-minded way by evaluating $kT^{5/2}$ times the temperature gradient.

Jordan: No, this will give you simply an indication for each star that you have to go back and do a proper evaluation of the turbulent carrying of energy. So I agree with you entirely but to my knowledge nobody is able to do a turbulent, convective, conductive calculation at the moment. This kind of work can tell us those stars where we are going to have to do that.

THE DISTRIBUTION OF CHROMOSPHERIC EMISSION STRENGTHS AMONG RED DWARFS

David R. Soderblom
Harvard-Smithsonian Center for Astrophysics
Cambridge, Massachusetts USA

The survey of Vaughan and Preston (1980, hereafter VP) of Ca II emission among solar neighborhood stars has shown the distribution of chromospheric emission (CE) of these stars. For stars bluer than (B-V)= 1.50, it is possible to transform VP's equivalent width $S$ into a relative flux by use of Middelkoop's (1982) formulae. This enables construction of a chromospheric color-magnitude diagram (illustrated in Soderblom 1982), which shows the same general features as VP's log $S$ vs. (B-V) plot, except that the CE (as a fraction of the stellar luminosity) declines with mass due to the decline of the ZAMS rotational velocity with mass (Soderblom 1983).

The presence of a gap in this diagram is problematical because a number of effects can contribute to produce systematic trends (Soderblom 1983). What is of interest here is the distribution of CE's for K and M dwarfs, i.e., for (B-V)>1.0. The few K-M dwarfs that are extraordinarily strong Ca II emitters are BY Draconis variables or flare stars. The spread in CE for the rest is fairly small: 0.4 dex encompasses all the K-M dwarfs except for a few very weak emitters. As expected, halo population objects have weaker CE on the average than the disk stars do.

(B-V) is a poor temperature indicator for K-M stars. (V-R)'s would be superior but are unavailable for most of VP's stars. To examine the distribution of CE for K-M dwarfs, I have used (R-I) from Gliese (1969). Middelkoop's formulae are not appropriate for such cool stars, so Figure 1 shows log $S$ vs. (R-I). The F and G stars in this diagram are compressed below (R-I)=0.40 ((B-V)$\lesssim$1.0). As before, there are BY Draconis and flare stars in the upper right corner. For (R-I)$\gtrsim$0.70 ($\simeq$dM2, (B-V)$\simeq$1.4), the distribution appears to turn over, so that the very weakest stars exhibit weak CE despite the apparent increase in CE that should result from weaker continua. This turnover may be caused in part by VP's continuum bands, which show unexpected behavior for (B-V)>1.0 (see Fig. 3 of VP). However, this would affect all stars redder than (R-I)=0.40.

The most probable cause of this turnover is the nature of the sample for very cool stars. As Upgren and Armandroff (1981) have shown, our



*P. B. Byrne and M. Rodonò (eds.), Activity in Red-Dwarf Stars, 67–70.*

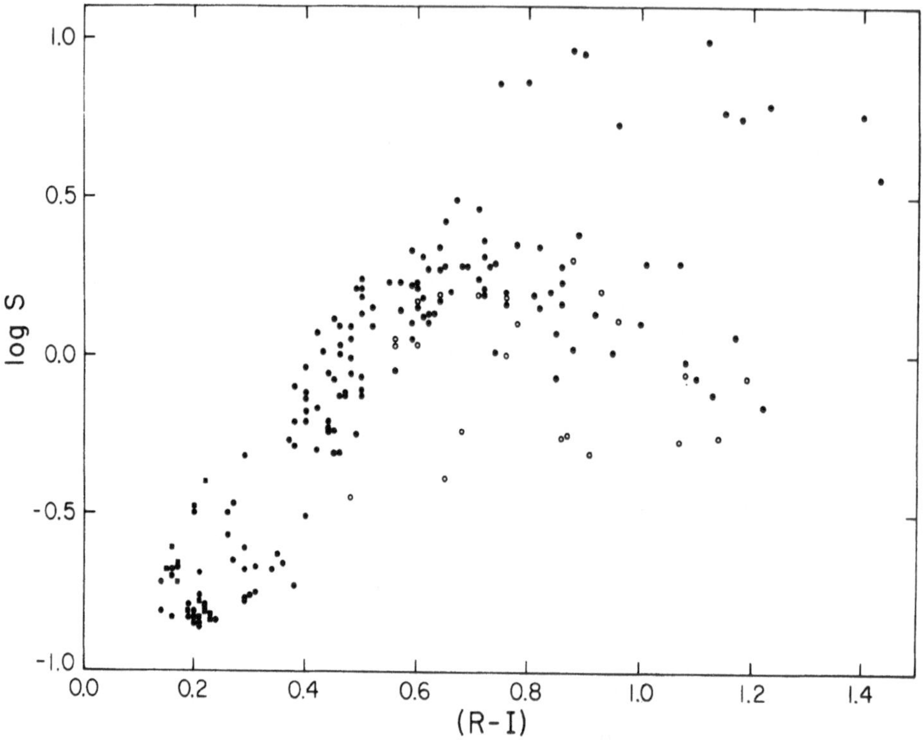


Figure 1. The logarithm of $S$, Vaughan and Preston's (1980) equivalent width of the Ca II H and K emission reversals, against (R-I) (in the Kron system) from Gliese (1969). The squares below (R-I)=0.30 are Johnson colors transformed to the Kron system. Open circles denote old disk and halo objects ($|W+10|>30$ km $s^{-1}$ or $(U^2+V^2)^{1/2}>65$ km $s^{-1}$).

knowledge of the stellar composition of the solar neighborhood is complete only to about (B-V)=1.40 ($M_V$=+9), i.e., (R-I)=0.70. Stars cooler than this tend to be selected on the basis of unusual spectra in objective prism surveys (the BY Dra and flare stars) or because of large proper motions. This latter group tends to be much older than the disk stars. Our knowledge of young to intermediate age stars is poor for low masses.

This conclusion is reinforced by considering the space motions of the stars in Figure 1. If the very active stars are excluded (those with log $S$>0.50), the 47 stars with (R-I)≥0.70 have $W_{rms}$=25 km $s^{-1}$, while $W_{rms}$=16 km $s^{-1}$ for the 43 stars with 0.50≤(R-I)<0.70. This latter velocity is typical of the disk population, but 25 km $s^{-1}$ is appropriate to very old stars (Wielen 1974). Thus a more complete knowledge of K-M dwarfs in the solar neighborhood should turn up stars of moderate CE at all colors.

As Noyes (1983) has shown, the correlation between rotation period and CE is very tight for late-type dwarfs. BY Draconis stars have rotation periods of 2 to 10 days. The longest rotation period found by Baliunas *et al.* (1983) is 48 days for 61 Cygni B (K7V, (R-I)=0.60, log $S$=+0.04). The presence of stars as much as 0.4 dex below this suggests that their rotation periods may be as long as 120 days. However, Noyes' relation works for the mean chromospheric emission. These very cool stars are faint, and so knowledge of their CE variability is lacking. Among the most active stars, the average CE observed tends to be near the high end of the overall range of CE for the star, suggesting that one is unlikely to observe a star in a quiet state. Because the CE of halo stars is very weak, detection of their rotation periods will be difficult. An additional complication is that the low metal abundance of halo stars may systematically change the level of CE appropriate to a given rotation period.

## REFERENCES

Baliunas, S.L., Vaughan, A.H., Middelkoop, F., Hartmann, L.W., Mihalis, D., Noyes, R.W., Preston, G.W., Frazer, J., and Lanning, H.: 1983, Astrophys. J., in press.
Gliese, W.: 1969, Veroeffentl. Astron. Rechen-Inst. Heidelberg, No. 22.
Middelkoop, F.: 1982, Astron. Astrophys., 107, 31.
Noyes, R.W.: 1983, in J. Stenflo (ed.), *Solar and Stellar Magnetic Fields*, (IAU Symp. 102), Reidel, Dordrecht.
Soderblom, D.R.: 1983, in J. Stenflo (ed.), *Solar and Stellar Magnetic Fields*, (IAU Symp. 102), Reidel, Dordrecht.
Upgren, A.R., and Armandroff, T.E.: 1981, Astron. J., 86, 1898.
Vaughan, A.H., and Preston, G.W.: 1980, Pub. Astron. Soc. Pacific, 92, 385.
Wielen, R.: 1974, in G. Contopoulos (ed.), *Highlights of Astronomy*, Reidel, Dordrecht, v. 4, pt. 1, p. 395.

## DISCUSSION

Giampapa: It is important to consider the nature of the chromospheres in these stars. It may well be that the selection effects you mentioned play a role but when you get to the very cool stars,as Linsky has pointed out,Ca II is not the dominant emitter. In fact, the energy in the Balmer lines becomes more important than Ca II. What one would like to do would be to replot your diagram taking into account all of the important emitters. As a second more minor point I should point out that dMe stars are found among both younger and older (kinematically) populations.

Soderblom: Yes, certainly Ca II is only one element in what is going on in these stars' chromospheres. This has the benefit of being a large survey, however. As regards the distribution of dMe stars what you say is also true. What I do not know is the frequency of flaring among stars of different ages. For instance, if you look at kinematically old stars what fraction are you likely to find in flare.

Vaiana: Is this sample volume complete so as to speak?

Soderblom: The stars are taken from the Gliese and Wooley et al catalogues. These catalogues are definitely not complete. Upgren and Armandroff recently concluded that the sample was representative although not complete to (B-V)=1.4. What they did was to examine those stars which lay in the Northern hemisphere and delete known binaries or those with radial velocity variations.

Anon: Can you say something about the gap?

Soderblom: The reason I do not think the gap is real is the following. Firstly I added about 40 stars to the original Vaughan-Preston sample and at least 6 lie in the so-called gap. The second reason is as follows. The upper branch of the area around the gap is about where one finds stars of the age of the Hyades. Now the difference in chromospheric emission between Hyades stars and the Sun is about 0.5 dex in good agreement with what you would expect on the basis of a $t^{-\frac{1}{2}}$ dependence and the relative ages of the Hyades and the Sun. Consider now the difference between the Pleiades and the Hyades. The Pleiades are an order of magnitude older than the Hyades. So on the basis of the above argument one would expect a further 0.5 dex in the chromospheric emission. This is not observed. Instead they are above the Hyades by about 0.1 dex. What is being seen is some kind of saturation phenomenon. From rotational velocity measurements it appears that it too begins to saturate and so the proportionality between the two is maintained. This effect will appear to make a concentration of stars just above where the gap is supposed to be. So there is no physical significance in the gap.

# CHROMOSPHERIC EMISSION AND ROTATION OF MAIN SEQUENCE STARS

S. Catalano[1] and E. Marilli[2]
[1]Istituto di Astronomia, Università di Catania
[2]Osservatorio Astrofisico, Catania, Italia

Here we present a quatitative approach to the problem of the chromospheric emission and rotation in main sequence stars based on a consistent analysis of recent published data of stars from F8 to K5. This analysis has been performed using the following physical parameters:

a) total power emission in the CaII K line, $L_K$;
b) stellar rotation period, $P_{rot}$, from chromospheric emission variability;
c) stellar ages from lithium abundance.

The obtained results are summarized as follows.

- The K line luminosity, $L_K$, follows an exponential law with the rotation period (see the Figure), not dependent on the spectral type,

$$\log L_K = 29.02 - P_{rot}/27.02 \quad (1)$$

- The K line luminosity, $L_K$ of one solar mass stars, follows an exponential decay with the square root of the age

$$\log L_K = -1.485\times10^{-5}\, t^{\frac{1}{2}} + 29.28 \quad (2)$$

The combiantion of relations (1) and (2) leads to

$$V_{rot} = 1.27\times10^{5}\, t^{-\frac{1}{2}} \quad (3a)$$

for stars of about one solar mass and age larger than $2.6\times10^{8}$ years. This result compares fairly well with the Soderblom's observed relation (Soderblom 1981)

$$V \sin i = 1.26\times10^{5}\, t^{-\frac{1}{2}} \quad (3b)$$

- The K line luminosity for stars of the same age follows a power law



*P. B. Byrne and M. Rodonò (eds.), Activity in Red-Dwarf Stars, 71–73.*

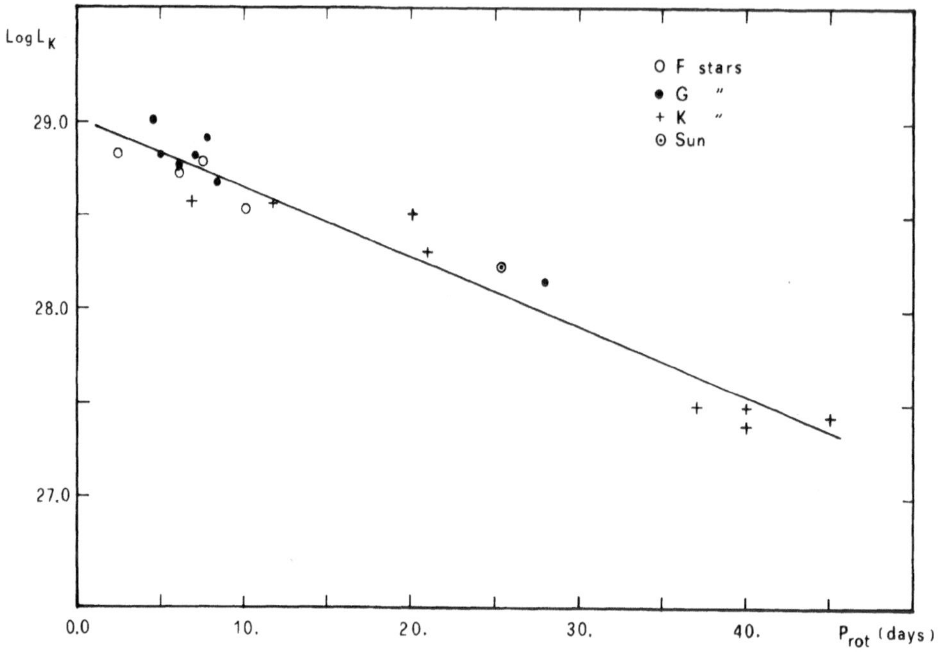


Figure 1. Log $L_K$ versus $P_{rot}$ plot for F-K main sequence stars. The solid line represents a linear fitting for all the stars in our sample, including the Sun.

with the stellar mass

$$\log L_K = 4.97 \log (M/M_\odot) + 28.87 \text{ (Hyades)} \tag{4a}$$

and

$$\log L_K = 5.17 \log (M/M_\odot) + 29.48 \text{ (Pleiades)} \tag{4b}$$

The combiantion of the observed relations (2) and (4) leads to a simple relation for the chromospheric emission of main sequence stars

$$L_K (M/M_\odot, t) = L_K (1,0)\, (M/M_\odot)^\alpha\, e^{-\beta t^{\frac{1}{2}}} \tag{5}$$

where $\alpha$ and $\beta$ are real positive coefficients. Further observations are needed to establish if

$$\alpha = \alpha(t) \tag{6a}$$

and

$$\beta = \beta(M/M_\odot) \tag{6b}$$

If $\alpha$ and $\beta$ are constant, then, relations (1) and (5) lead to

$$P_{rot} = -\alpha' \log(M/M_\odot) + \beta' t^{\frac{1}{2}} + \text{const} \tag{7}$$

So the rotation period of main sequence stars would be determined only by their masses and ages. On the other hand this result would have strong implications for rotation at

$t = 0$ (initial angular momentum)

and

$t = t_o$ (angular momentum at the Zero Age Main Sequence)

REFERENCE

Soderblom, D.R.: 1981, Astrophys. J.

# LATE-TYPE Ca II EMISSION-LINE STARS IN THE SOUTHERN HEMISPHERE

A.C. Collier
Physics Department, University of Canterbury,
Christchurch, New Zealand

## ABSTRACT

The results of a photometric and spectroscopic study of twenty-three late-type southern stars with strong Ca II H and K emission are presented.

The presence of photometric variability, with similar properties to the photometric wave phenomenon seen in the RS Canum Venaticorum binaries, is noted in thirteen of the fifteen stars for which extensive photometry was obtained. Of twenty stars for which high resolution radial velocity measurements were obtained, ten are found to be single-lined and seven are found to be double-lined spectroscopic binaries. The periods range from 0.66 up to 53.9 days.

Two stars with no radial velocity variations but with high rotational broadening ($v_e \sin i \geqslant 40$ km $s^{-1}$) of their spectra are identified as probable new members of the FK Comae group of rapidly-rotating, chromospherically-active late-type giants.

High resolution ($\Delta\lambda < 0.3$ Å) Hα spectra were obtained of twenty-one of the candidate stars. Nine stars show Hα as an emission feature above continuum level. Eight show Hα either absent or unusually shallow, due to filling of the absorption profile with emission. High resolution ($\Delta\lambda < 0.3$ Å) Ca II H and K spectra of ten of the program stars yield line surface fluxes in the range

$$1.6 \times 10^5 \leqslant F(K_1) + F(H_1) \leqslant 1.9 \times 10^7 \text{ erg cm}^{-2}\text{s}^{-1}$$

in the H and K emission cores.

The high proportion of single-lined RS CVn binaries in the sample is discussed in terms of selection effects. The space density of RS CVn systems with mass ratios less than $q \simeq 0.8$ is found to be at least equal to that of the $q > 0.8$ systems. This implies a total space density of at least $10^{-5}$pc$^{-3}$ for the RS CVn binaries.



*P. B. Byrne and M. Rodonò (eds.), Activity in Red-Dwarf Stars, 75–76.*

Differences between the RS CVn binaries and the related short-period, long-period and semi-detached G and K binaries with Ca II H and K emission are discussed in terms of evolutionary processes in close binary systems. A space density calculation based on the evolutionary requirements for membership of the RS CVn group is shown to reproduce closely the mass and luminosity distributions among the RS CVn binaries. The theoretical space density of RS CVn systems in the Galactic plane is found to be $1.9 \times 10^{-5} pc^{-3}$ by this method, in reasonable agreement with the observed value.

# THE ATMOSPHERE OF A PROBABLY VERY YOUNG BY Dra-TYPE FLARE STAR

R. de la Reza[1], C.J. Butler[2], C.A.O. Torres[1], C.C. Batalha[1]
[1]National Observatory-CNPq Rio de Janeiro , Brazil
[2]Armagh Observatory - N. Ireland, U.K.

## ABSTRACT

The flare star Gliese 182 (dM0.5e) seems to be only known single BY Dra type that presents Li in its atmosphere. This characteristic and others, principally activity, and high rotation, indicate that probably this is a very young object. We analyse in this work the possibility to interpret with a single typical model for a dMe atmosphere, some observed lines of Gliese 182 and to predict others. The lines belong to the following atoms; Li I, H I, He I, He II, C I, C II. A relatively good agreement exists only for neutral lines but not for ionized lines. The upper chromosphere and transition region must be studied in more detail. Some comparison are made with the UV observations of the double star BY Dra. Empirical relatively high X-ray fluxes are predicted for both stars. The Li abundance of Gliese 182 is confirmed to be similar to that of the interstellar medium.

This work will be submitted to Monthly Notices of the Royal Academy of Sciences.



*P. B. Byrne and M. Rodonò (eds.), Activity in Red-Dwarf Stars, 77.*

SOME PROPERTIES OF STELLAR TRANSITION REGIONS

*Mikael Saxner*
Astronomiska Observatoriet
Uppsala

## 1. Introduction

Among the strongest lines from stellar transition regions are the resonance lines of N V (1240Å), Si IV (1394,1403Å), and C IV (1549Å), which are all observable by the IUE.
In an earlier paper (Saxner, 1981) it was noticed that there is a correlation between the fluxes in these three lines. In Fig. 1 the ratio of the N V to the C IV flux is plotted against the ratio of the C IV to the Si IV flux. Verbally the correlation may be stated as follows: the Si IV flux and the N V flux are covarying; if one is strong the other also tends to be strong, and vice versa. In all cases the C IV line is the strongest.
The paucity of points above and below the main diagonal in the diagram could be explained if there were objects with only one of the Si IV and N V fluxes too weak to be detected. However, the points along the lower

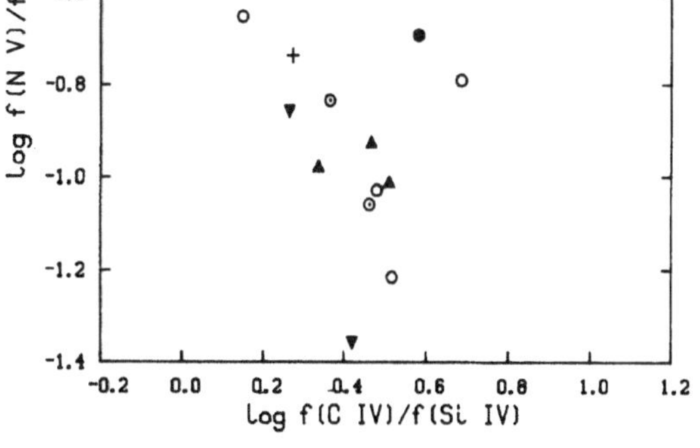

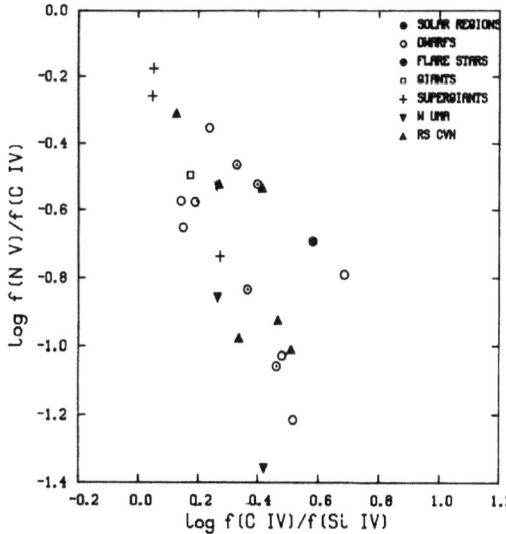


Fig 1. *Observed relation between the transition region emission line fluxes*



*P. B. Byrne and M. Rodonò (eds.), Activity in Red-Dwarf Stars, 79–81.*

part of the diagonal are observed despite having both Si IV and N V weak. Thus the correlation is probably not due to a selection effect.
The correlation also seems to be the same for widely differing types of objects, and is likely to reflect some common property of all stellar transition regions. In this paper we suggest that the correlation is the result of different stellar transition regions having a similar shape, while other properties, such as pressure and geometrical extent, may vary. This together with optical depth effects can explain the observed distribution.

## 2. The model

To investigate the effects of the shape of the transition region we have assumed that the temperature gradient varies as

$$\frac{dT}{dz} = \alpha \cdot T^{\beta} \tag{1}$$

between two temperatures $T_0$ and $T_1$ echosen such that all of the radiation in the three lines is formed within this temperature interval.
The transition region is assumed to be static and to have constant pressure $p_0$. It does not, however, have to be homogeneous.
The ionization equilibria needed in the calculations are taken from Jordan (1969). In accordance with these data we have chosen $\log T_0 = 4.4$ and $\log T_1 = 5.8$ .
Throughout this model transition region the equation of radiative transfer is solved with the source function

$$S_{\lambda} = \frac{n_e C_{12}}{B_{12}} \tag{2}$$

where $n_e C_{12}$ is the electron collision excitation rate to the upper level, and $B_{12}$ is the Einstein coefficient for absorption. This particular form for the source function is valid when self-absorption is negligible compared to collisional excitation, and when spontaneous decay dominates over collisional de-excitation.
The equation of transfer is solved separately for each component of the multiplet and the total intensity in the multiplet is obtained by integration over the line profiles.

## 3. Results

For a given value of $\beta$ different values of the pressure $p_0$ define a curve in Fig. 1. In our computations the pressure was varied to cover the whole range of optical depths in the lines.
The parameter $\alpha$ in Eq. (1) is a function of $\beta$, the stellar gravity, and the fractional amount of mass, x, between $T_0$ and $T_1$ compared to all mass above $T_0$. In the computations we have used $g = 10^4$ and $x = 0.1$ . Other choices for these parameters merely move the point along the given $\beta$-curve.
In Fig. 2a the theoretical curves for different values of $\beta$ are shown together with the observed points. The fit is not very good, especially in the left-hand part of the diagram where the observed points seem to

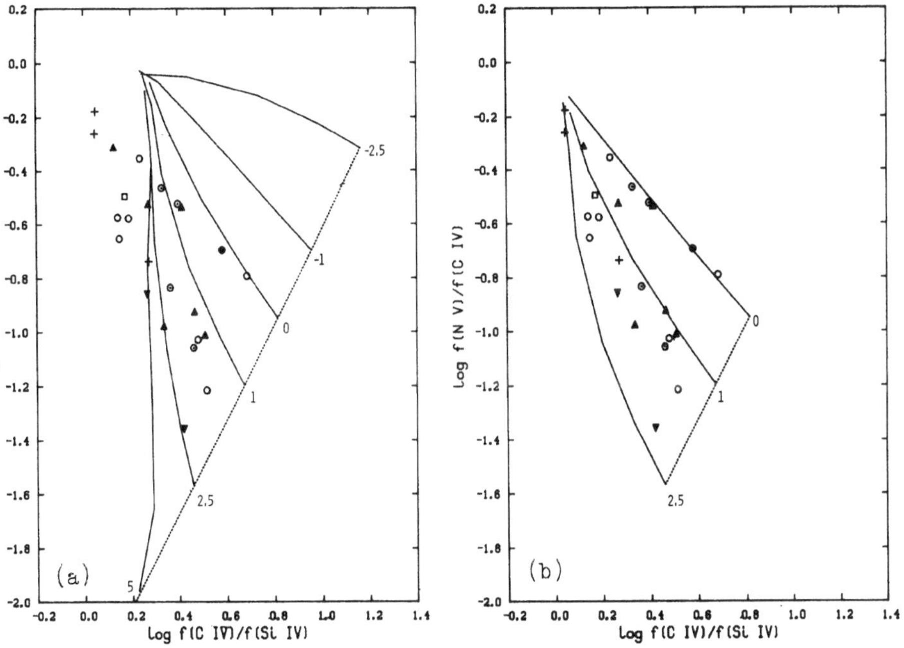


Fig 2. *The solid curves show the theoretical relations without microturbulence (a), and with a microturbulence of 30 km $s^{-1}$ (b). The curves are marked with their value of β. The dotted lines mark the optically thin limit. Also shown are the observations.*

have stronger Si IV emission than the model predicts. In Fig. 2b we show the result if we introduce a microturbulent velocity of 30 km $s^{-1}$. Here the fit is much better, both as regards position and slope. A further increase of ξ to 100 km $s^{-1}$ gives almost identical result. This is not surprising, since at 30 km $s^{-1}$ the microturbulence already dominates over the thermal velocities.

## 4. Conclusions

There are two main conclusions to be drawn from this investigation. Firstly, if we take the observational scatter into account, we find from Fig. 2b that β lies between about 0.5 and 1.5 . As defined here, β is some kind of "integrated" mean over a large temperature interval, and the present result cannot rule out the possibility that β is close to -2.5, the value expected for constant conductive flux, in the higher transition region.

Secondly a comparison of Fig. 2a and b shows that there must exist small-scale turbulence in stellar transition regions with velocities exceeding about 20 km $s^{-1}$ in order to explain the observed distribution.

## 5. References


Jordan, C., 1969, MNRAS, 142, 501

Saxner, M., 1981, Astron. Astrophys., 104, 240

QUIESCENT CORONAE OF ACTIVE CHROMOSPHERE STARS

Leon Golub
Harvard-Smithsonian Center for Astrophysics

ABSTRACT

The *EINSTEIN* Observatory has for the first time provided high sensitivity X-ray measurements of quiescent coronal emission from a large sample of dwarf stars. We now have observed a sufficient number of the nearby M-dwarfs to determine an X-ray luminosity function and we have explored the activity and variability of these stars to the extent of observing, for the first time, X-ray flares with simultaneous ground-based optical and IUE ultraviolet coverage.

The M dwarfs are found to have a much higher degree of variability in X-rays than does the Sun; however, in most cases a quiescent level is definable. We will discuss the quiescent emission from these stars and the changes in quiescent level on time scales from hours to $\sim 1$ year. We have determined coronal temperatures for many of these stars; they are generally hotter than the Solar corona and some of the more active dM stars have $T_{cor} \sim 10^7$ K.

Arguments are presented in support of the hypothesis that M-dwarf coronae are magnetically dominated, as is the Solar corona. We then examine the usefulness of loop model atmosphere calculations in elucidating the coronal heating mechanism and the ways in which observations may be used to test competing theories. The X-ray measurements can be used to predict magnetic field strengths on these stars, with testable implications.

I. OVERVIEW: SOLAR-TYPE STARS.

In attempting to discuss coronal emission from M-dwarf stars, we are exploring a totally new area, for which there has until now been very little data available and which we can therefore approach with an open mind. The observation of X-ray emission from these stars is an especially recent achievement and it is only with the very high sensitivity available from the *EINSTEIN* Observatory that we are beginning to obtain a reasonably complete initial survey of the coronal properties of these stars.



*P. B. Byrne and M. Rodonò (eds.), Activity in Red-Dwarf Stars, 83–108.*

In this review I will summarize the presently available data on coronal emission from dwarf M-stars. In so doing, it will be necessary to broaden the discussion somewhat in order to examine late-type dwarfs in general. The discussion will include not only the Solar-type active chromosphere stars which are similar in many respects to the M-dwarfs, but also the larger class of cool stars which are expected to have outer convective zones and the consequent magnetically dominated outer atmospheres related to the presence of Solar-like dynamo activity.

Since we are finding that the properties of red dwarfs are in many respects only an extension of the behavior observed in Solar-type stars, it will be useful to begin this review by examining some of the Solar data. The reasons for doing so are: first, that the Sun is the only star for which we have been able to see any details of the coronal structure and also, because we now have reason to believe that the coronae of dM stars differ from the Sun mainly in an exaggeration of properties which are present in the Solar context, rather than in the appearance of qualitatively new phenomena.

a. The Solar X-ray Corona

Figure 1 shows a typical image of the Solar corona as seen in high-resolution X-ray observations. The wealth of detail to be found in such data has been discussed at length elsewhere (e.g., Vaiana and Rosner 1978 and references therein), so that we will mention here only the basic point that the X-ray emission derives predominantly from closed loop structures. These are seen in Figure 1 in a range of sizes and brightness levels, corresponding in general to the evolutionary history of the surface magnetic fields which control the coronal structure. The most intense emission occurs in active region cores, which contain the strongest magnetic fields. Further away from these areas and also at locations where older, more evolved active regions are found, the X-ray corona is larger, more diffuse, cooler and weaker.

On the basis of such observations, we may argue for an active participation of the magnetic field in coronal formation and heating (see e.g., Vaiana and Rosner 1978). Thus, the emerged surface magnetic fields not only control the coronal topology, but are now viewed as providing the means for direct mechanical heating of the coronal plasma.

This role of B has been tested in a series of quantitative studies involving direct predictions among observable quantities. The first of these relations was a thermodynamic one, which viewed the isolated closed loop structure as a relatively isolated mini-corona. By taking such a loop as a closed system in hydrostatic equilibrium, Rosner, Tucker and Vaiana (1978) were able to derive the now well-known scaling law

$$T_{max} = 1.4 \times 10^{3} (pL)^{1/3} \quad (1.1)$$

which provides a quantitative test of the hypothesis that the structuring of the atmosphere into closed loops is a fundamental consideration in coronal formation.

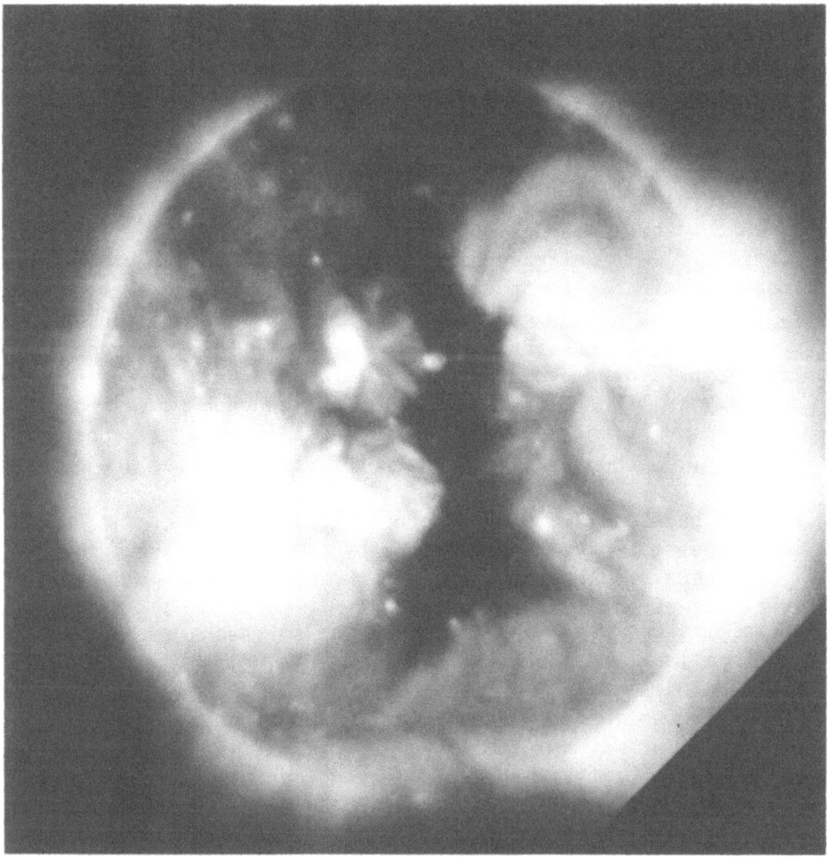

Fig. 1. The Solar x-ray corona, as seen from Skylab on 1 June 1973 (Photo courtesy G. Vaiana).

A test of the direct link to B in the coronal heating process was provided by Golub et al. (1980). Using a simple, general model in which magnetic stresses induced by turbulence in the HCA are transmitted into the corona and dissipated in situ, and employing the RTV relation (1.1), a scaling law involving B is obtained:

$$p = 63 B_z^{3/2} L^{-1/4} V_\phi^{3/2} , \qquad (1.2)$$

where $B_z$ is the average longitudinal magnetic field at the loop footpoint and $V_\phi$ is the effective twisting velocity of the longitudinal magnetic field and is related to the level of surface turbulence in the $B \gtrsim 1$ region of the Solar atmosphere.

Quantitative studies of this kind are continuing and provide the most direct means of testing the role of the magnetic field in the coronal heating process. Using the Solar observations as a starting point, we will examine in § III below the nature of M-dwarf magnetic fields as deduced from the X-ray

b. Extent of Solar-type Coronae along the M-S

Before turning to the M-star data, it is appropriate to review briefly what we have learned about coronal emission from Solar-type stars in general. As already discussed, I will take the view that all magnetically dominated coronae of the Solar type should be viewed together, as long as we can consider that their coronae are due to the emergence and subsequent activity of magnetic fields, presumably due to a Solar-type dynamo operating in the star's outer convective zone. This category is likely to include all main-sequence stars from late A up to and including the M-dwarfs; it may also include all rotating convective stars, if the observations of evolved cool stars are a guide.

What we have learned about the average level of coronal emission from late-type stars may be summarized by two basic observational facts: 1) all dwarf stars of spectral type dF through dM are X-ray emitters at some level between about $3 \times 10^{26}$ and $10^{31}$ erg $s^{-1}$; and 2) the main factor determining the level of X-ray emission for stars later than about F7 is the stellar rotation rate.

These two basic points are summarized in Figure 2. This figure is from a study under way as a follow-up to the Pallavicini et al. (1981) paper, in which we first reported the quantitative connection between $L_x$ and v sin i. Figure 2 shows the rotation dependence of coronal emission for stars of spectral type F7 to M5 and luminosity classes III, IV and V. The empty symbols indicate spectroscopic determinations of v sin i, and the filled symbols indicate close binary systems and spotted stars for which $v_{rot}$ represents the equatorial rotation rate. The straight line indicates a best-fit to the total data sample and is the relation we found earlier, namely that $L_x$ is proportional to the square of the stellar rotation rate. Note especially that all luminosity classes fall on the same line, which is the basis for the statement above that rotating, convective systems in general may behave in the same way as the Solar-type stars.

The influence of rotation is even more striking if we consider that early-type stars do not show any rotation dependence in the coronal X-ray emission. Thus, for stars all the way from early O through Altair at A7, the X-ray emission is found to be proportional to the stars' bolometric luminosity (Pallavicini et al. 1981 and references therein), the proportionality constant being $10^{-7}$. The transition between early- and late-type behavior along the main sequence occurs at spectral type F. It is thus tempting to speculate that the observed sharp rise in X-ray emission at $\sim$ F0 is due to the onset of convection and the consequent beginning of Solar-type dynamo activity.

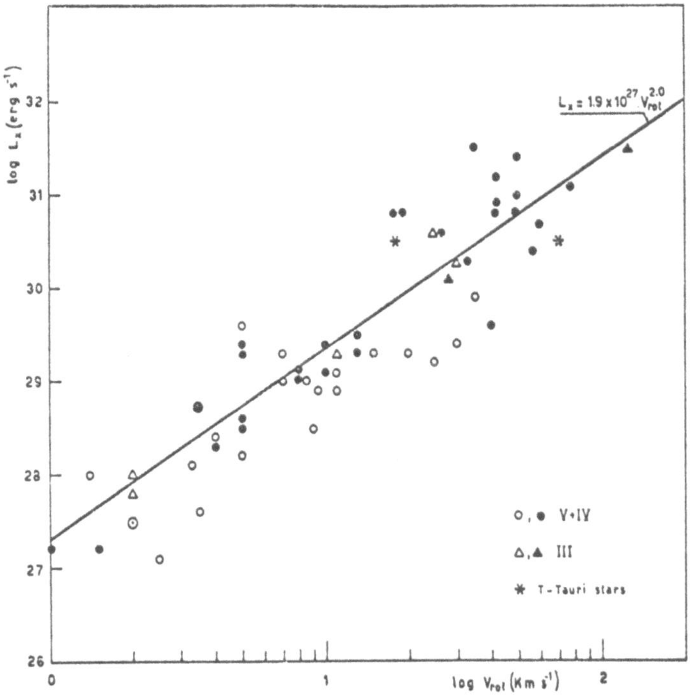


Fig. 2. $L_x$ vs. v sin i for stars detected by the EINSTEIN Observatory.

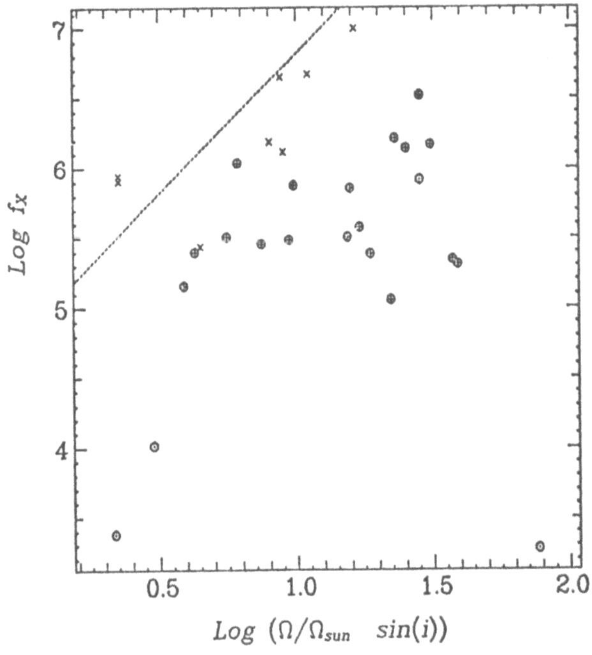


Fig. 3. X-ray surface flux vs. rotation rate for late-A through early G dwarf stars. Circles are A-stars, Circles with plus signs are F0-F5 and Crosses are F6-F9 (courtesy J. Schmitt, CFA).

This idea is being explicitly pursued in a combined theoretical and observational project at the CfA. Figure 3 (courtesy J. Schmitt, CfA) shows preliminary results of a detailed survey of late A and early F star X-ray luminosities vs. rotation rates. The circles at the bottom of the figure show the A-stars, which is mentioned above, have $L_x$ x $10^{-7}$ $L_{bol}$; their X-ray emission is therefore relatively weak and independent of rotation rate.

At the other extreme, the dashed line near the top of the figure is the $L_x$ vs. v sin i relation found for late-type stars (G-, K-, and M-stars. The x's represent F-stars later than F6, and they are close to the dashed line, thus obeying the late-type star behavior. The early and middle F stars fall in between, with a smaller slope. By combining these (and more) data with theoretical calculations, we hope to clarify the behavior of dynamos and possibly to understand in a quantitative manner the relationship between stellar rotation and magnetic field production.

## II. THE M-STAR DATA

The combination of high sensitivity, high spatial resolution and pointing capability offered by the EINSTEIN Observatory have provided a totally new perspective on coronal emission from M-dwarfs. We now have detected steady X-ray emission from over three dozen M stars; in contrast, the pre-Einstein reports of steady emission included only the triple system 40 Eri (Cash et al. 1979) and the likely identification of BY Dra and AD Leo (Ayres et al. 1979), all from HEAO-1. Previous reports also included detection of X-ray flares or transients from such well-known flare stars as YZ CMi, UV Cet and Proxima Cen (Heise et al. 1975; Haisch et al. 1977; Kahn et al. 1979). Even in these cases, however, simultaneous coverage at other wavelengths was generally sparse, making quantitative analysis and comparison with Solar flares difficult (cf. Kahler 1977).

A search of the current literature for the available data on M-dwarf emission yields the list shown in Table 1; sources of these data are indicated. This list includes all of the published data and preprints of which we were aware as of the date of this meeting, as well as a number of previously unpublished observations from our own CfA survey.

We are at this moment in the midst of a very active period of data reduction, and there are sure to be several additions to this list during the next year. However, the number of sources which are already available and the quality of the observations allow us to discuss the general characteristics of X-ray emission from M-dwarfs. Our sample is already representative of this class and we can expect that the preliminary conclusions which we are now able to reach will remain fairly accurate.

An indication of the completeness of the present data set is shown in Figure 4, which is an H-R diagram made up of the stars listed in

Table 1. Summary of M-Dwarf Data

| Star Name | Sp | log $L_x$ | Ref. No. |
|---|---|---|---|
| +43 44 AB | dM1 + dM6 | 27.1T | 1 |
| UV Cet | dM5.5e+dM5.5e | 27.3-27.6T | 1 |
| 40 Eri C | dM4.5e | 27.8 | 1 |
| Ross 47 | M6VI | 27.1 | 1 |
| Ross 986 | dM5e | 27.5 | 2 |
| YY Gem | dM1e-dM1e | 29.6 | 1 |
| YZ CMi | dM5e | 28.6 | 8 |
| AD Leo | dM4.5e | 29.0 | 2 |
| CN Leo | dM8e | 26.6-27.1 | 1,2 |
| DM+44 2051 | dM2e+dM8e | 27.5-28.5T | 1,3 |
| DM + 36 2322 | dM0e+dM4e | 28.9T | 1 |
| Prox Cen | dM5e | 26.6-27.4 | 1,4 |
| Wolf 630AB | dM3.5e | 29.3 | 1 |
| DM + 68 946 | dM3.5 | 26.9 | 2 |
| Barnard's | M5 VI | 26.1 | 1 |
| GL 752 AB | dM3.5e+dM5e | 27.1T | 1 |
| HD202560 | dM0e | 27.2 | 2 |
| HD204961 | dM1 | 26.8 | 2 |
| Krüger 60 | dM3+dM4.5e | 27.4T | 1 |
| L 789-6 | dM7e | 26.9 | 2 |
| EQ Peg AB | dM4e+dM5.5e | 28.8T | 1 |
| Ross 614A | dM7e | 26.9 | 2 |
| CR Dra | dM1.5e | 29.1 | 5 |
| BY Dra | dM0e+dM2e | 29.5T | 5 |
| AU Mic | dM2.5e | 29.9 | 6 |
| Gl 867 AB | dM2e+dM4e | 29.0T | 3 |
| Gl 852 AB | dM4.5e+dM5e | 29.5T | 3 |
| AT Mic | dM4.5e+dM4.5e | 29.3T | 3 |
| Gl 229 | dM2.5 | 28.8 | 3 |
| CC Eri | dK7e | 29.3 | 3 |
| DM+01 2684 | dM0e | 28.2 | 7 |
| Ross 476 | dM6 | 28.0 | 7 |
| DM+40 2208 | dK8e | 28.4 | 7 |
| AC+45 217-363 | dM2+M5 | 27.9 | 7 |
| Wolf 630C | dM5e | 26.4 | 9 |

1 Vaiana et al. 1981
2 Unpublished CfA 1982
3 Haisch & Tsikoudi 1982, preprint (HEAO-1)
4 Haisch et al. 1980
5 Vaiana et al. 1982
6 Caillault 1982
7 Topka 1981
8 Kahler et al. 1982.
9 Swank & Johnson 1982

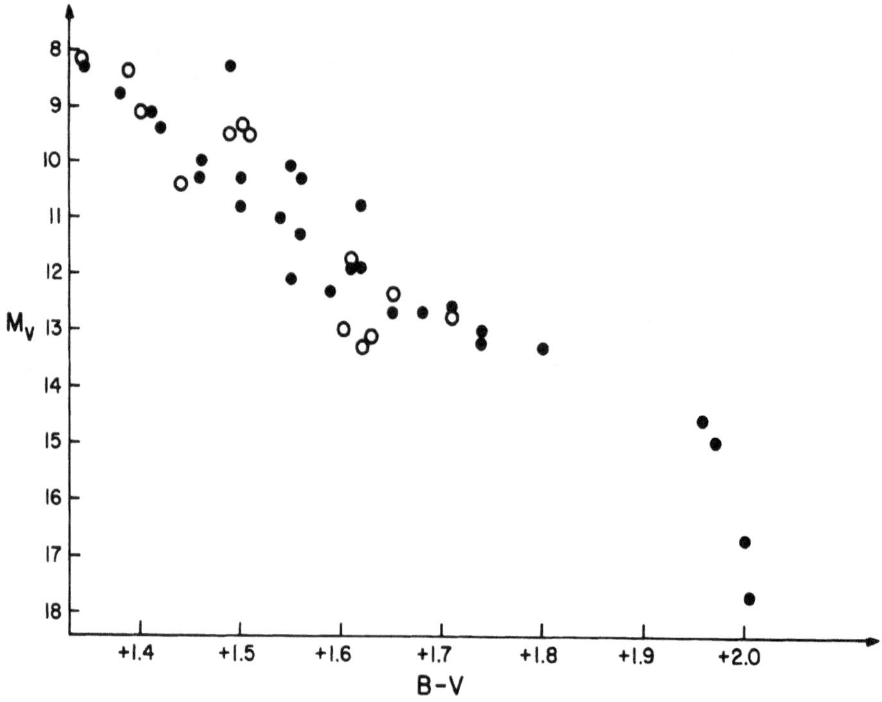


Fig. 4. H-R diagram consisting of M-dwarfs which have been studied in x-rays; filled circles are those seen in the CFA Stellar Surveys and open circles are reported observations from other sources, including HEAO-1 and EINSTEIN.

Table 1. (In some cases $M_V$ or B-V data were not available, and these stars are not listed.) It is clear from the figure that we have good coverage and, aside from selection effects which would not be evident on such a diagram, we are in a position to examine the general properties of coronal emission from M-dwarfs as a class.

a. Quiescent Emission Levels

In order to determine the level of X-ray emission from M-dwarfs we must face the same question which arises in the case of Solar X-ray emission, namely is there any steady emission or is it all due to flares and transients? For the case of the Sun, the consensus view is that there is indeed steady, nonflare coronal emission (Orrall 1981). Likewise, for Solar-type stars such as α Cen, the X-ray emission is often stable to the few percent level for several hours (Golub et al. 1982), thus arguing for a steady and non-transient emission mechanism.

For the case of more active Solar-type stars such as $\Pi^1$ UMa, or cooler dwarfs (late K through M), the level of variability in X-ray

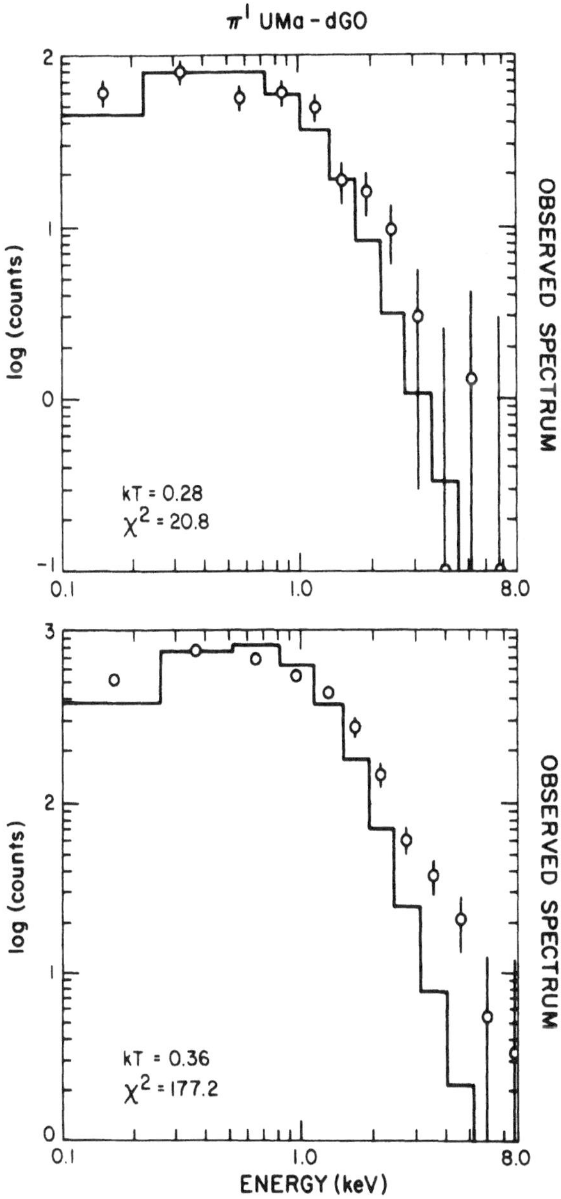


Fig. 5. EINSTEIN imaging proportional counter (IPC) data for the active G-dwarf $\pi^1$ UMa, showing spectral data obtained during a quiescent (top) and an active (bottom) orbit. In addition to a change in the x-ray emission level, there is also observed a change in the best-fit temperature kT and evidence of a second high temperature component during active times.
(Data courtesy G. Vaiana).

emission increases, as illustrated in Figure 5. These data will be discussed in more detail below; for now we show them in order to display the type of variability typically observed in stars with high average surface X-ray flux levels. In practice, we have chosen to exclude only obvious flare events when quoting X-ray luminosities. However, we caution that short-term variability is nearly always present in X-ray emission from active chromosphere stars and it will need to be examined in detail when we discuss measurements of the coronal temperatures.

An example of the way in which the quiescent level is determined in the presence of flare events is shown in Figure 6a, b. The figure shows two observations of Prox Cen from Haisch et al. (1980, 1982) and both containing sizable flares. The dashed lines show the quiescent levels determined; note in Fig. 6b that the 1979 quiescent level was substantially higher than that seen in 1980.

A less clearcut type of variability is illustrated in Figure 7, which shows portions of IPC data for two BY Dra-type stars, CR Dra and BY Dra itself. It is clear that variability is present and, as we discuss below, the spectral fitting procedure reflects this variability, but with short pointings there is no obvious way to find a quiescent level. In some cases we have separated the data into "active" and "inactive" periods (usually determined by satellite orbits); however, most pointings were not long enough to permit such a division and we used the total pointing in determining the coronal parameters.

A graphic summary of the basic data on X-ray luminosity is provided in Figure 8, which shows $L_x$ (0.15-4.0 keV bandpass) vs. B-V for the stars listed in Table 1. Binaries are identified by the asterisk symbols and are plotted with the X-ray flux evenly divided between the two components. Data from HEAO-1 are indicated by open circles and the stars observed more than once are connected by solid lines to indicate the range of $L_x$ found. Upper limits are indicated by arrows.

Two properties of the X-ray emission are immediately apparent from examination of this figure. First, as is the case of G- and K-dwarfs, there is a large spread of 3 orders of magnitude in the observed $L_x$ at any particular value of B-V. This spread is presumably linked to the stellar rotation rate and the sparse data on rotation of M-dwarfs tends to confirm this idea (Pallavicini et al. 1982).

The second major feature in the graph is the total absence of high X-ray luminosity sources beyond a B-V of $\approx$ 1.7. There appears to be a pronounced tailing off of the X-ray luminosity function toward later spectral types; however, a large spread in emission levels is still observed. This observation may be taken as an indication that the activity level is decreasing in low-mass stars, with the implication that magnetic flux production is likewise decreasing.

Recently, Giampapa (1983) reported the results of a high spectral resolution search in the vicinity of Hα for a sample of low mass M-dwarfs.

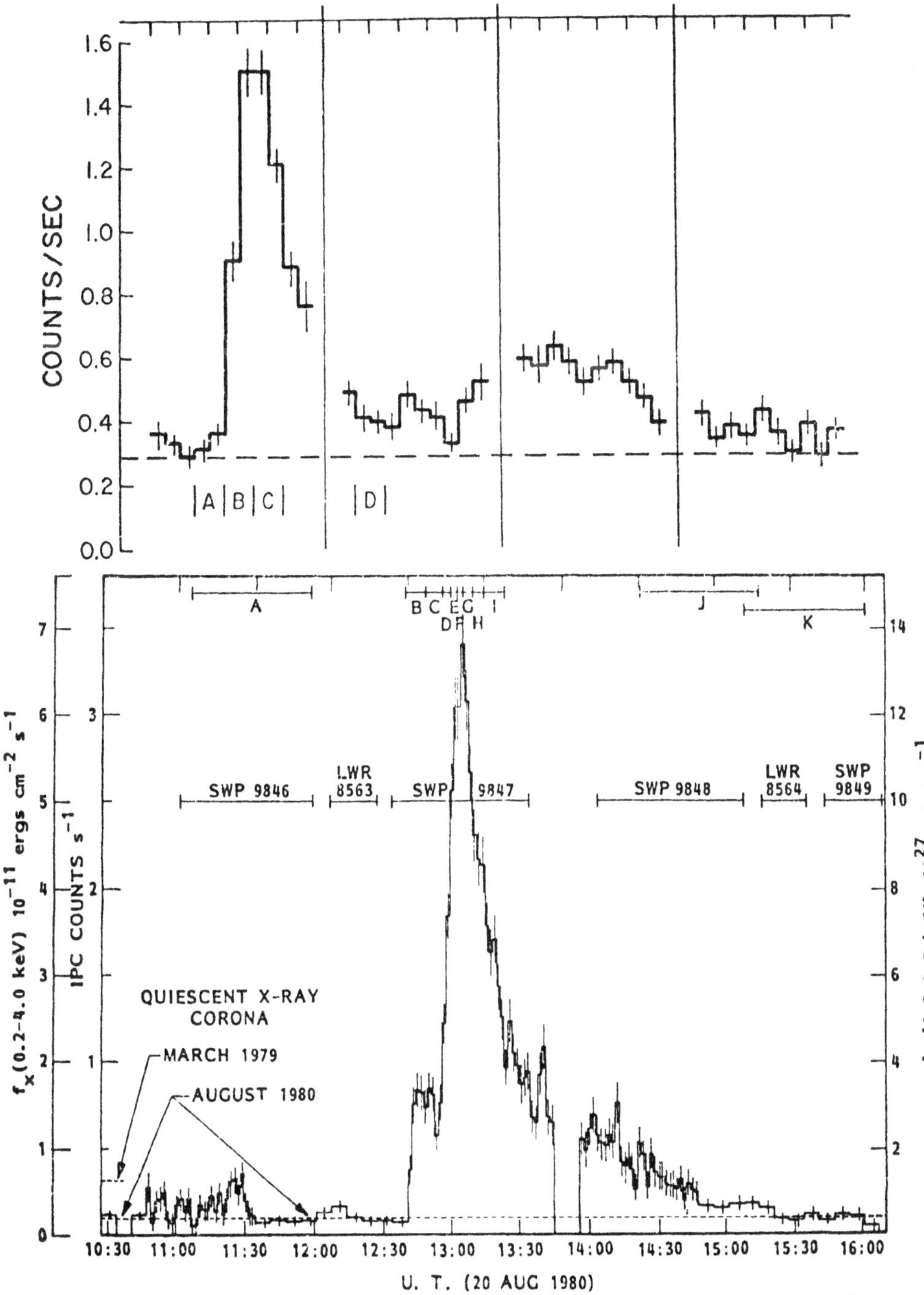


Fig. 6. Determination of a quiescent x-ray level in the presence of a flare event, illustrated by two flares observed on Prox Cen.
a) Dashed line shows quiescent level for the March 1979 event;
b) August 1980 flare, with simultaneous IUE coverage.

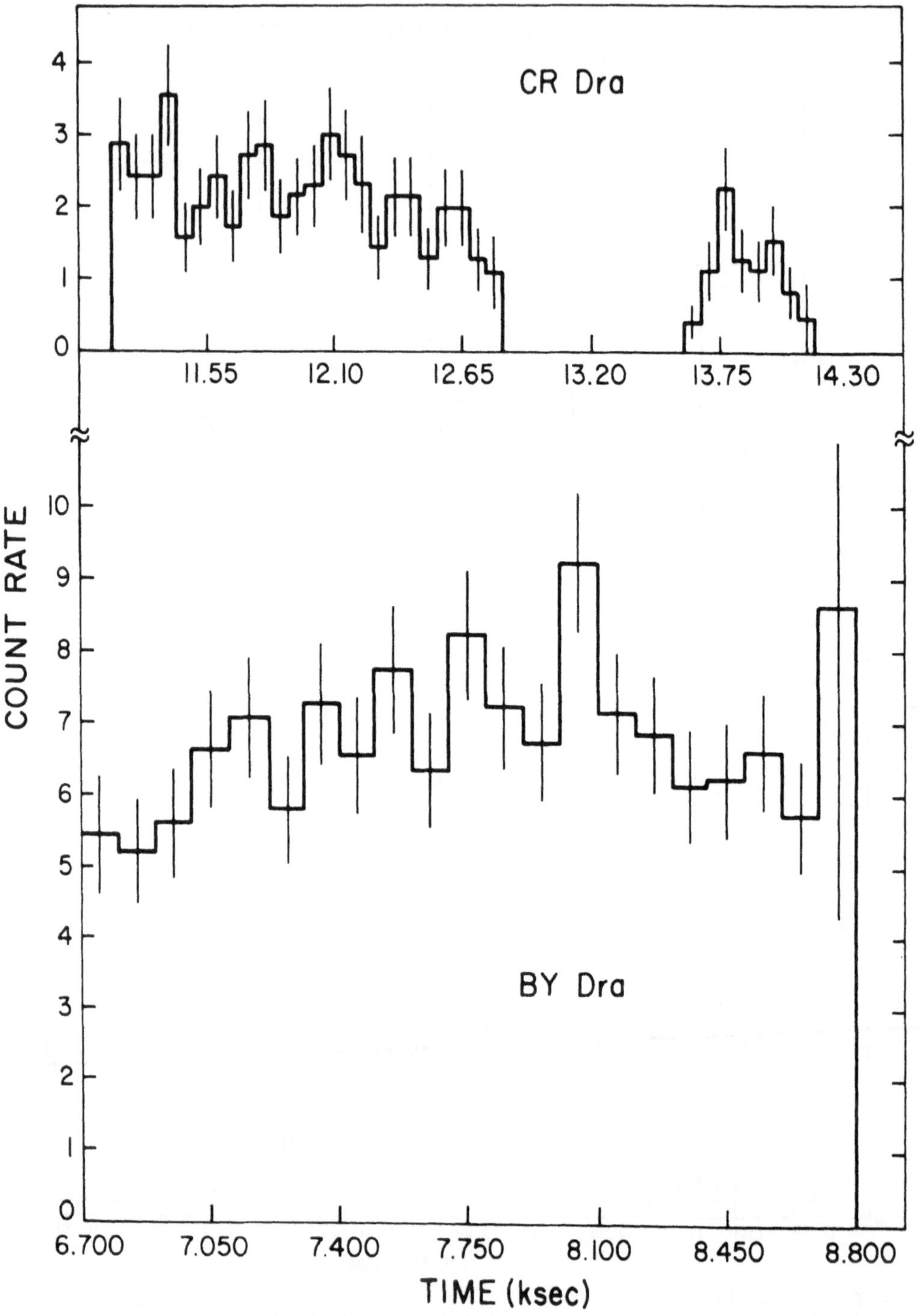


Fig. 7. IPC count rate during portions of two observations, of the BY Dra-type star CR Dra (top) and of BY Dra itself (bottom).

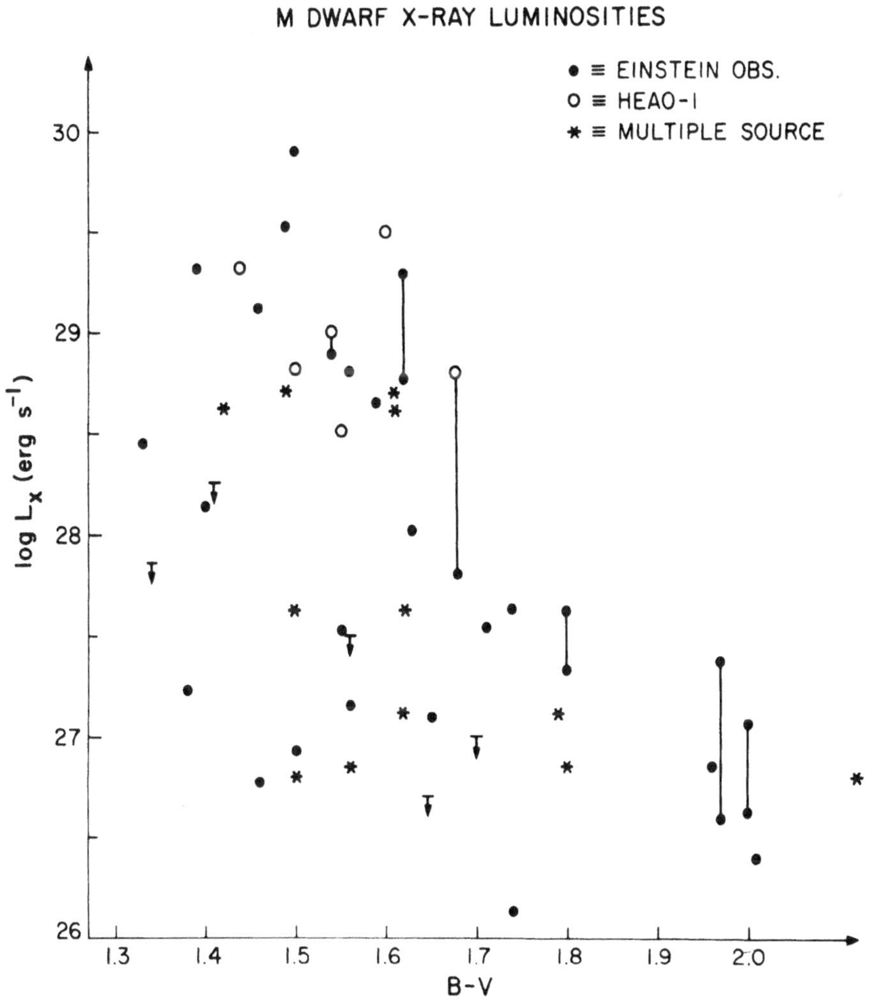


Fig. 8. Summary of all data available to date on x-ray luminosities of M-dwarfs. Closed circles indicate EINSTEIN data, open circles are HEAO-1 data; stars indicate multiple unresolved systems for which the total observed flux has been arbitrarily divided among the known components.

The stars were selected from the Luyten's Proper Motion Survey; the selection criterion used was a determination from the strength of the TiO molecular bands that the stars were later than ∿ dM6 (Liebert 1982). Contrary to prior expectations, these stars showed no Hα emission, implying that the chromospheric indicators of activity can become extremely faint in this sample. If we take into consideration the very high contrast level at which a Solar-strength chromosphere would be detected against a dM6 (or later) photosphere, then we conclude that any coronae or chromospheres on these stars must be extremely weak, even by Solar standards.

Preliminary analysis indicates that the stars observed in this program are OD rather than halo. If this is so, then the likely conclusion is to be drawn from these observations is the same as that from Figure 8, that the level of chromospheric and coronal emission drops for very low-mass dwarfs.

The significance of these results is that we expect that at around dM5 stars will become fully convective (Copeland et al. 1970). Present-day dynamo calculations lead to the expectation that Solar-type magnetic-field generation with a Solar cycle must occur primarily at or near the base of the convective zone (Schüssler 1983, Rosner 1983), although mechanisms also exist to amplify fields throughout the convective zone. In a fully convective star there would appear to be no room for the primary magnetic field amplification to take place, leading at a minimum to a change in the pattern of magnetic flux production. Such Solar-type phenomena as the cycle, polarity reversals and latitude migration of activity could be affected. It will require detailed numerical simulations to determine how the dynamo might change.

However, we show in § III that, although the X-ray luminosity of these stars decreases, the actual amount of magnetic flux produced in the star may be quite high, possibly higher than in G- and K-dwarfs. This apparent contradiction arises from the belief that coronal heating is due to activation of surface magnetic fields by the surface turbulence. In low-mass red dwarfs the surface turbulent velocities decrease, leading to a probable decrease in the amount of heating per unit magnetic flux. Thus, a given level of X-ray emission may imply substantially more magnetic flux on a red dwarf than on a Solar-type star.

Finally, we should note that there may be a selection effect operating in the manner of finding faint red dwarfs, as pointed out by Soderblom (these proceedings). The possibility exists that catalogs utilizing large proper motions of faint red stellar sources will miss the young red dwarfs with low space velocity. Indeed, the stars in Table 1 with the lowest X-ray emission, such as Ross 47 and AC + 79° 3888 are halo stars. However, it is not obvious that the converse statement applies, i.e., that YD stars are the highest X-ray emitters.

A possible resolution of this difficulty may come from the EINSTEIN Observatory "medium survey" (Maccacaro et al. 1982). By searching all medium sensitivity fields for serendipitous sources, this survey finds 3 M-dwarfs in ∿ 50 square degrees of sky, about the same number density as in the Wooley catalogue. If these X-ray sources turn out to be high luminosity, low-mass stars, then the selection effect suggestion will be confirmed. However, if none of them are found to be the "missing" red dwarfs in Figure 8, then it will be more difficult to argue the incompleteness of the optically selected sample. The work needed to answer this question is presently in progress.

A more direct means of testing the possible turnoff of dynamo activity for low-mass dwarfs would be to obtain rotation rates for the stars already observed in X-rays. We know that all dwarfs from F6 through M5 adhere strictly to the rule relating X-ray emission to rotation rate (Pallavicini et al. 1981, 1982; Walter 1982). If the same dependence upon rotation is found for dwarfs later than M5, then we argue that the dynamo has not changed character; if less X-ray emission is found at a given rotation value than expected, then we argue that the efficiency of magnetic flux production has gone down. This type of test should be independent of biases due to incompleteness of the sample.

b. Coronal Temperatures

Using the EINSTEIN Imaging Proportional Counter (IPC), we have thus far determined coronal temperatures for twelve M-dwarfs. In addition, Swank & Johnson have reported a temperature determination for Wolf 630AB using the EINSTEIN Solid State Spectrometer (SSS), and Ayres et al. (1979) reported approximate temperatures for three M-dwarfs using HEAO-1. These data are shown in Figure 9, where we have plotted X-ray luminosity vs. coronal temperature. The observed temperatures fall in a narrow range, from log T = 6.3 to 6.7, with only BY Dra and CR Dra above this range; the latter two stars will be discussed separately below.

For quiescent emission, assumed to derive from a large-scale quiet corona, a simple model based on scaling laws obtained in the Solar context can be used to predict a rather steep $T_c^{5/2}$ dependence between $L_x$ and $T_c$ (Rosner, Golub and Vaiana 1982). The model predictions are indicated on the figure as solid lines. The intersecting, nearly vertical lines are the expected rotation rates, based on the $L_x$ vs. $v_{rot}$ law of Pallavicini et al. (1981). Thus, a low emission level of $10^{27}$ erg $s^{-1}$ is expected to come from M-dwarfs having low coronal temperatures of $\sim 10^{6.0}$K and rotation periods of $\sim$ 30 days. In contrast, emission at $10^{29}$ erg $s^{-1}$, is predicted to come from stars having $T_c \sim 4 \times 10^6$ K and rotation periods $\sim$ 4 days. For a given rotation period lower mass stars will have a lower $L_x$ and a slightly lower coronal temperature.

It appears likely that the temperature determinations for very active stars such as BY Dra and CR Dra are strongly influenced by a hard component due to flarelike activity. We showed in Figure 7 the highly variable nature of the coronal emission from these stars. The spectrum and temperature fit obtained for BY Dra is shown in Figure 10; the result for CR Dra is similar. In contrast to the lower activity M dwarfs, for which the $\chi^2$ of the fit is generally in the range 5-20 with 10 degrees of freedom, we find a very high $\chi^2$ of 188 for this observation. Examination of the figure shows that the problem is due to an inability of a single temperature to fit the total spectrum. We encounter the same problem in, e.g. fitting the YZ CMi or Prox Cen observations if the obvious flares are included in the total fit; removing flaring portions of the data in those cases allowed an

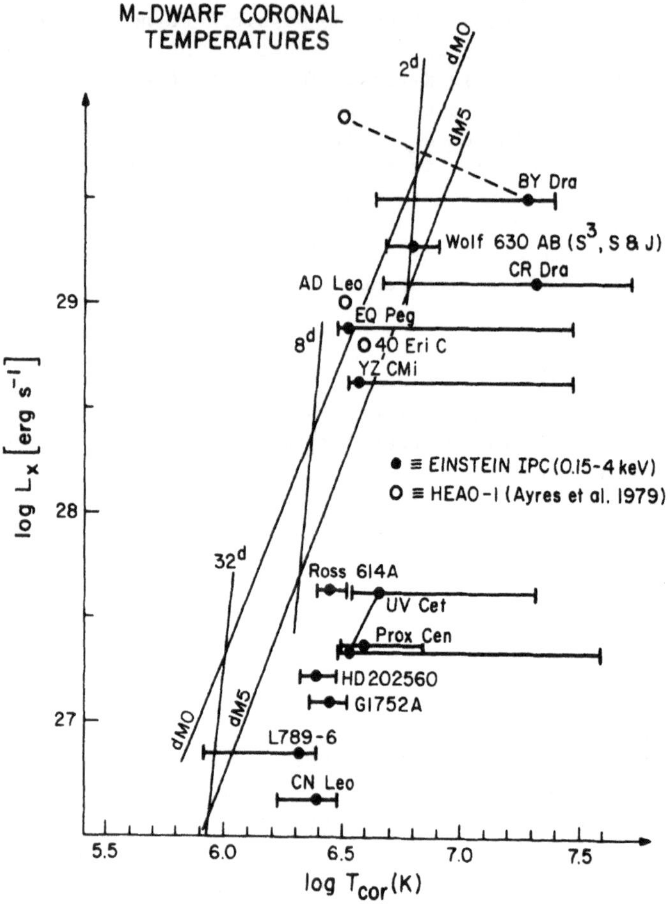


Fig. 9. Level of soft x-ray emission vs. coronal temperature for a sample of M-dwarfs; solid lines are from the theory of Rosner et al. 1982 (see text for discussion).

acceptable (lower temperature) fit to be made. Moreover, SSS observations of Wolf 630 AB by Swank and Johnson (1982) have explicitly found the two temperature components, the lower of which is plotted on Figure 9.

## III. THE ROLE OF MAGNETIC FIELDS

In § I we presented arguments to support our contention that M-dwarf coronae are basically the same as the solar corona, but with some of the controlling physical parameters having exaggerated importance. In this section we will examine these ideas in detail, focusing on the properties of the magnetic fields which control coronal formation and heating.

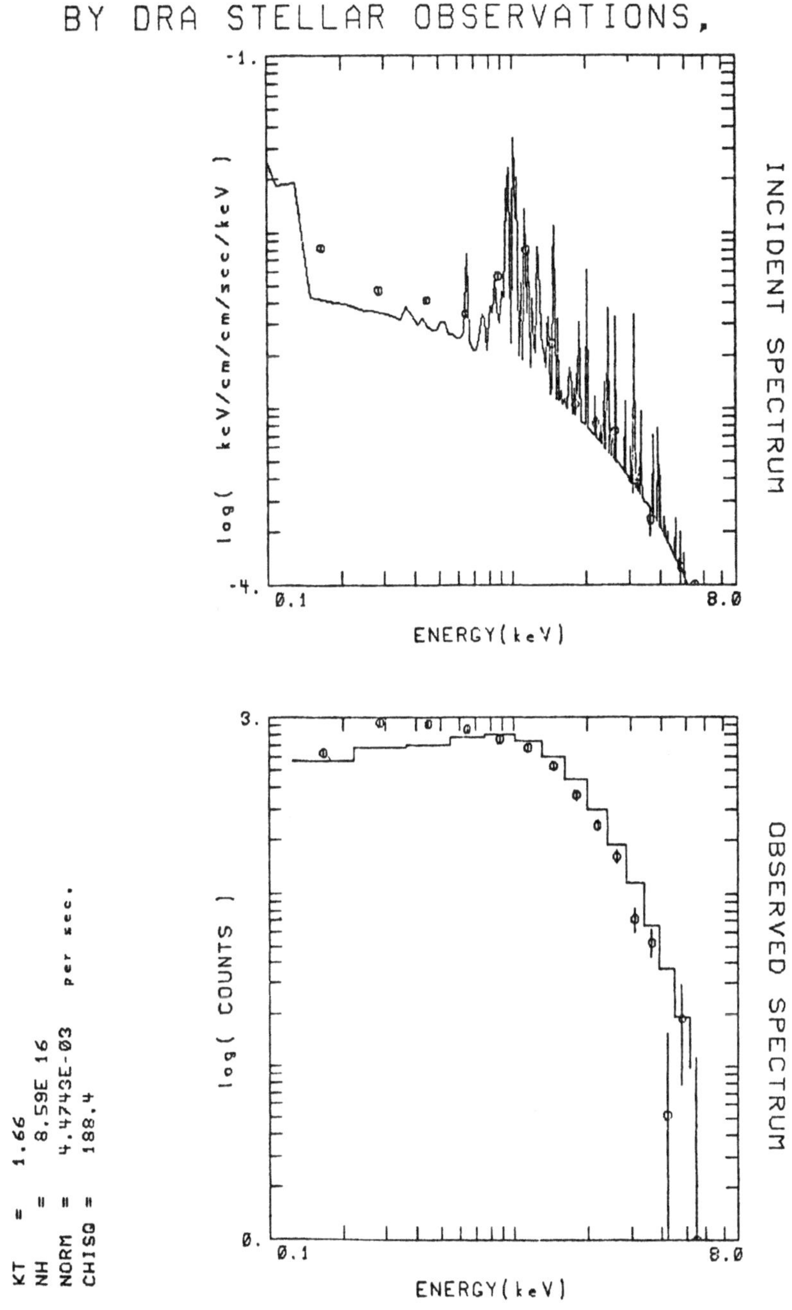


Fig. 10. IPC spectral data for BY Dra, showing best-fit thermal spectrum (bottom) and implied incident spectrum (top). The large value of $\chi^2$ indicates the need for a two-component temperature fit; these data should be compared with $\pi^1$ UMa active data (fig. 5). (Courtesy G. Vaiana and S. Sciortino)

In the present discussion we explicitly adopt the view that previous quantitative studies of Solar coronal properties can be used to help in understanding M-dwarf coronae. Thus, scaling laws relating observable quantities such as coronal temperature and pressure (Rosner, Tucker and Vaiana 1978) and magnetic field strength (Golub et al. 1980) will be utilized; parameters such as surface gravity and surface turbulence velocities appropriate to low mass stars can be included in the calculations. At the same time we will arrive at predictions for quantities such as magnetic field strengths and filling factors, which may soon be testable by direct observation and which will lead to stringent tests of the heating theory which we have derived in a purely Solar context (Rosner et al. 1978; Vaiana and Rosner 1978; Golub et al. 1980, 1982).

a. Loop Model and Scaling Laws

The model which we use to relate the level of X-ray emission to magnetic field strength has been discussed in detail elsewhere (e.g., Golub et al. 1982 and references therein), so that we present only a brief description here. Figure 11 shows the assumed loop topology.

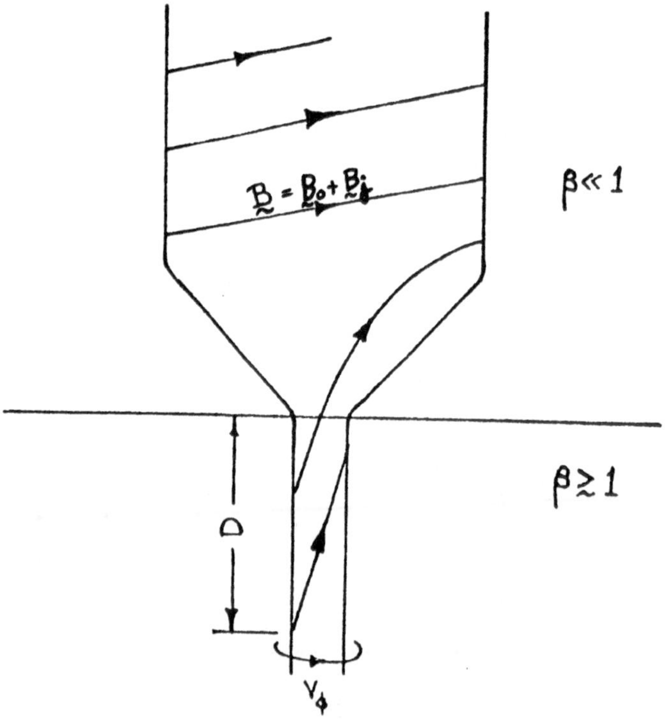


Fig. 11. Schematic representation of the geometry of the magnetic field and labelling of the loop model parameters for the theory described in the text.

The energy available for heating resides in the non-potential component of the magnetic field, which is viewed as being generated from a potential B-field by shear in the $\beta \gtrsim 1$ region of the Solar atmosphere ($\beta \equiv 8\pi p_{gas}/B^2$):

$$B = B_z + B_j, \tag{3.1}$$

$$W_m = B_j^2 \, V/8\pi \tag{3.2}$$

$$B_j = B_z \frac{\partial v_\phi}{\partial z} . \tag{3.3}$$

The coronal portion of such a loop will have a larger cross-section due to the drop in external gas pressure with increasing height above the Solar surface. Because of this expansion, any twist which is generated in the high-$\beta$ region will be transmitted upward and amplified, as discussed by Parker (1974). Viewed from the corona, this transmission of stresses becomes an effective twisting velocity at the base of the loop, which we label $v_\phi$.

The energy available for heating the corona is then

$$W_m = \frac{1}{4} B_j \cdot B_z \, v_\phi \, R^2, \tag{3.4}$$

where R is the cross-sectional radius of the loop in the corona. Combining this result with that of Rosner, Tucker and Vaiana (1978):

$$\frac{W_m}{V} = E_H = 10^5 \, p^{7/6} \, L^{-5/6} \tag{3.5}$$

we obtain the scaling relation

$$p = 2.6 \, B_2^{12/7} \, L_9^{-1/7} \, (\alpha v_4)^{6/7}, \tag{3.6}$$

where the subscripts indicate division by the indicated powers of ten and $\alpha \equiv B_j/B_z$.

We have tested two ways of removing $B_j$ from this relation, since it is not a directly measurable quantity. They amount to taking either $\alpha$ or the coronal $\beta$ ($\beta_j \equiv 8\pi p_{gas}/B_j^2$) as constant. The scaling laws relating <u>measurable</u> quantities which then result are

$$\text{constant } \alpha\text{:} \quad p \propto B_z^{12/7} \, L^{-1/7} \, (\alpha v_\phi)^{6/7} \tag{3.7a}$$

$$\text{constant } \beta_j\text{:} \quad p \propto B_z^{3/2} \, L^{-1/4} \, v_\phi^{3/2} \tag{3.7b}$$

In the Solar context, both of these formulations are found to provide acceptable fits to the available data (Golub *et al.* 1982a).

We may attempt to combine the RTV scaling law (Eq. 1.1) with the magnetic field-related law (Eq. 3.7b) in order to eliminate one of the variables, e.g., coronal pressure p. The result is

$$T_{cor} = 1.2 \times 10^3 B_z^{1/2} L^{1/4} \left(\frac{v_t}{v_\odot}\right)^{1/2} \tag{3.8}$$

where we have normalized the twisting velocity $v_\phi$ to the Solar value in order to take into account the variation of turbulent surface velocities for different stars.

For the case of stars such as the Sun with fairly steady quiescent coronal emission, we may calculate an atmospheric model which allows us to use the X-ray measurements to arrive at a direct estimate of the magnetic field parameters. We will see that direct application of the method to more active stars having large short-term variability leads to problems, which can only partially be overcome at the moment; however, we can still get some idea of the important magnetic field parameters on active chromosphere stars and M-dwarfs in general.

b. Predicted Magnetic Field Values

We have shown elsewhere (Golub et al. 1982) that for a loop atmosphere in which all of the X-ray emission derives from a single type of loop, i.e. in which all of the loops are specified by the same parameters, a measurement of $L_x$ and $T_{cor}$ is nearly sufficient to fully specify the atmosphere. The only additional requirement is a means of specifying either the loop pressure p, or the loop length L, or the coronal filling factor f. Knowledge of any one of these three will allow us to determine the remaining values by utilizing the RTV scaling law relating T, p and L.

If we are modelling quiescent emission from a low-activity star, then our experience with Solar emission shows that it is reasonable to view the stellar emission as coming from large-scale evolved loops. This is because: (i) during the emergence process, loops tend to be very active and variable in their emission properties and (ii) the evolution of surface fields operates in only one direction, i.e. that of diffusion, with a rapid emergence and growth followed by gradual and sustained spreading of the fields; we may expect that quiescent emission involves large, diffused magnetic loops which may (depending on coronal temperature and stellar surface gravity) be larger than the pressure scale height of the coronal plasma. In that case the total X-ray luminosity, which may be represented by

$$L_x = 4\pi R_*^2 H n_e^2 P(T) f, \tag{3.9}$$

can be rewritten with $H = s_p = 5 \times 10^3 T (g/g_\odot)^{-1}$ and $n_e = \frac{p}{2kT}$. We may solve for f, letting $F_x \equiv L_x/4\pi R_*^2$ and also solve Eq. 3.8 for B:

$$f = \frac{3.4 \times 10^{-9}}{P(T)} F_x \left[\frac{T^3}{(g/g_\odot)}\right]^{-1} \tag{3.10a}$$

$$B_{em} = 1.2 \times 10^{-8} \left(\frac{v_t}{v_\odot}\right)^{-1} \left[\frac{T}{(g/g_\odot)}\right]^{1/2} \qquad (3.10b)$$

where the subscript of $B_{em}$ indicates that it is the average field in the regions doing the emitting. The stellar average magnetic field is $\langle B \rangle = f\, B_{em}$, so that the total magnetic flux on the star is

$$\Phi_T = 4\pi R_*^{\,2} \langle B \rangle$$

$$= \frac{4 \times 10^{-17}}{P(T)}\, L_x \left[\frac{T^3}{g/g_\odot}\right]^{-1/2} \left(\frac{v_t}{v_\odot}\right)^{-1}. \qquad (3.11)$$

We have calculated magnetic field values for several stars, based on the X-ray measurements of their coronal luminosity and temperature; the results are shown in Table 2. We have calculated Solar values first as a check of the procedure and the results are quite reasonable. The Solar-type stars α Cen A and B show, not surprisingly, values near those of the Sun; a quiet Sun value for α Cen A and an active Sun value of total magnetic flux for α Cen B.

Table 2. Predicted Magnetic Field Values for Solar-Type Stars

| Star | $F_x$(erg cm$^{-2}$s$^{-1}$ | T(K) | f | $B_{em}$ | $\langle B \rangle$ | $\log\phi_T$ |
|---|---|---|---|---|---|---|
| Quiet Sun | $5 \times 10^4$ | $1.8 \times 10^6$ | 0.25 | 30 | 8 | 23.7 |
| Active Sun | $3 \times 10^5$ | $3.5 \times 10^6$ | 0.20 | 80 | 16 | 24.0 |
| α Cen A | $1.3 \times 10^4$ | $2.1 \times 10^6$ | 0.06 | 90 | 5 | 23.6 |
| α Cen B | $6.1 \times 10^4$ | $2.1 \times 10^6$ | 0.18 | 60 | 11 | 23.8 |
| $\pi^1$ UMa | $1.2 \times 10^6$ | $4.0 \times 10^6$ | 0.22 | 100 | 20 | 24.2 |
| λ And | $1.5 \times 10^6$ | $7 \times 10^6$ | 0.75* | 1110 | 830 | 27.4 |

*Filling factor f redefined to take into account the large ratio of coronal scale height to stellar radius.

The active chromosphere star $\pi^1$ UMa is predicted to have a magnetic field configuration typical of the active Sun. The emission appears to derive from strong field regions which cover ∿ 22% of the surface. The total magnetic flux on the star is only slightly greater than on the Sun during an active period.

We have also calculated a magnetic field value for the active RS CVn type star λ And. The results are markedly different from those obtained for the other stars in the list. λ And seems to be covered over most of its surface (∿ 75%) with very strong field regions ($\gtrsim$ 1000 gauss). Comparison with the results of a direct measurement of the magnetic field on λ And by Giampapa and Worden (reported in these Proceedings) yields surprisingly good agreement, considering the simplicity of our calculation and the difficulties involved in measuring the magnetic field strength. Note that the total amount of magnetic flux which we calculate for λ And is quite high, being more than 1000 times larger than any Solar value yet observed.

If we apply this same procedure to M-dwarfs we will, in general, find that it yields inconsistent results. In particular, the calculated filling factors are consistently larger than unity. This would be acceptable for RS CVn's having lower stellar surface gravity and high coronal temperature, since f > 1 would in those cases be only an artifact of the way we have chosen to define the coronal filling factor. However, this explanation is not viable for M dwarfs which are found to have values of the atmospheric emission scale height generally small compared to the stellar radius. The results of a calculation using Eqs. 3.10 for some M-dwarfs are listed in Table 3 under the columns labelled "$L = s_p$."

Table 3. Predicted M-Dwarf Magnetic Field Values

| Star | $F_x$ (erg cm$^{-2}$s$^{-1}$) | T(x10$^6$ K) | $L = s_p$ | | $B'_{em}$ |
|---|---|---|---|---|---|
| | | | f | $B_{em}$ | |
| BY Dra | 8 x 10$^7$ | 19 | 2.0 | 2300 | 2960 |
| Prox. Cen | 7 x 10$^5$ | 3.5 | 1.8 | 155 | 210 |
| YZ CMi | 1.4 x 10$^7$ | 3.7 | 26 | 195 | 950 |
| L789-6 | 2.5 x 10$^5$ | 2.7 | 3.6 | 175 | 320 |
| CN Leo | 3 x 10$^5$ | 2.5 | 4.1 | 190 | 370 |

We see that, without exception the calculated filling factors are > 1. At the same time we note that the calculated magnetic field values are also quite large by Solar standards; they are all in the 100-200 range. It is clear that our assumption that the emission comes from loops larger than or equal to the pressure scale height is not valid for M-dwarfs. We must look into the case for which the emitting loops are smaller than $s_p$, i.e., more compact, higher pressure, active region type loops.

Unfortunately, there is no obvious bound on L in this case. Certainly the coronal pressure must be smaller than the photospheric gas pressure, but this limit is not very useful. We can obtain a lower limit on the magnetic field strengths by taking the case f = 1. Then we have

$$B'_{em} \geq 7 \times 10^{-13} \left[\frac{F_x}{P(T)}\right]^{1/2} \left(\frac{v_t}{v_\odot}\right)^{-1}. \qquad (3.12)$$

The values of magnetic field strength determined in this manner are listed in the last column of Table 3. They are, of course, larger than the previously determined values and they represent lower limits, which will increase by a factor $f^{1/2}$ if the filling factor is less than unity.

We may draw some tentative conclusions from our first examination of the M-dwarf data and from our comparison with Solar observations. These are:

1. M-dwarfs are essentially all X-ray emitters, as initially reported by Vaiana et al. (1981) and now confirmed by a larger sample. There is some evidence that halo stars are weaker, by at least one order of magnitude;

2. The strong correlation between X-ray emission and stellar rotation rate continues to hold down to late M. For stars later than dM5 there is particular interest in obtaining rotation rates because

3, there is a marked decrease in the observed range of $L_x$ values for dwarfs redder than B-V $\approx$ +1.7, leading to the possibility that we are seeing a decrease in dynamo efficiency toward fully convective stars. A quantitative determination of $L_x$ vs. rotation rate for these stars would provide an unbiased test of this idea;

4. Using quantitative studies relating Solar X-ray emission to magnetic field strength and loop size, we estimate that M-dwarfs in general have active region-strength fields covering most of the stellar surface, or stronger fields over a smaller fraction of the surface; BY Dra stars are similar to the RS CVn's in requiring kilogauss fields and large area coverage;

5. M-dwarfs and active Solar-type stars show substantial variability on short time scales. This is consistent with the results in 4) above, arguing that the X-ray emission is dominated by the continual presence of emerging flux regions, with strong and active magnetic fields covering a large fraction of the stellar surface.

## ACKNOWLEDGMENTS

I would like to thank Drs. G.S. Vaiana and S. Sciortino for use of their data and for helpful discussions, Drs. M. Giampapa and R. Rosner for advice and suggestions, and Dr. M. Rodono' for organizing a wonderful meeting. This work was supported in part by NASA under grant NASW-112, by the Smithsonian Institution Visitors' Program, and by the Secretary's Fluid Research Fund.

## REFERENCES

Ayres, T.R., Linsky, J.L., Garmire, G. and Cordova, F.: 1979, Ap.J.(Letters), 232, L117.
Caillault, J.-P.: 1982, Astron. J., 87, 558.
Cash, W., Charles, P., Bowyer, S., Walter, F., Ayres, T.R. and Linsky, J.L.: 1979, Ap.J.(Letters), 231, L137.
Copeland, H., Jensen, J.O. and Jorgensen, H.E.: 1970, A.&A., 5, 12.
Giampapa, M.S.: 1983, in Proc. IAU Symp. 102, ed. Stenflo, (Reidel: Dordrecht).
Golub, L., Harnden, F.R., Jr., Pallavicini, R.P., Rosner, R. and Vaiana, G.S.: 1982, Ap.J., 253, 242.

Golub, L., Maxson, C.W., Rosner, R., Serio, S. and Vaiana, G.S.: 1980, Ap.J., 238, 343.
Golub, L., Noci, G., Poletto, G. and Vaiana, G.S.: 1982a, Ap.J. 259, 359.
Haisch, B.M., Linsky, J.L., Lampton, M., Paresce, F., Margon, B. and Stern, R.: 1977, Ap.J.(Letters), 213, L119.
Haisch, B.M., Linsky, J.L., Harnden, F.R., Jr., Rosner, R., Seward, F.D. and Vaiana, G.S.: 1980, Ap.J. (Letters), 242, L99.
Haisch, B.M., Linsky, J.L., Bornmann, P.L., Stencel, R.E., Antiochos, S.K., Golub, L. and Vaiana, G.S.: 1982, Ap.J. (in press).
Haisch, B.M. and Tsikoudi, V.: 1982 (preprint).
Heise, J., Brinkman, A.C., Shrijver, J., Mewe, R., Gronenschild, E., den Boggende, A. and Grindlay, J.: 1975, Ap.J.(Letters), 202, L73.
Kahler, S.W.: 1977, Ap.J., 214, 891.
Kahler, S.W. et al.: 1982, Ap.J., 252, 239.
Kahn, S.M., Linsky, J.L., Mason, K.O., Haisch, B.M., Bowyer, C.S., White, N.M. and Pravdo, S.H.: 1979, Ap.J.(Letters), 234, L107.
Maccacaro et al.: 1982 (in preparation).
Pallavicini, R.P., Golub, L., Rosner, R., Vaiana, G.S. Ayres, T.R. and Linsky, J.L.: 1981, Ap.J., 248, 279.
Parker, E.N.: 1974, Ap.J., 191, 245.
Rosner, R.: 1983, in Proc. IAU Symp. 102, ed. Stenflo (Reidel: Dordrecht).
Rosner, R., Golub, L. and Vaiana, G.S.: 1982, Ap.J. (submitted).
Rosner, R., Tucker, W.H. and Vaiana, G.S.: 1978, Ap.J., 220, 643.
Schüssler, M.: 1983, in Proc. IAU Symp. 102, ed. Stenflo (Reidel: Dordrecht).
Swank, J.H. and Johnson, H.M.: 1982 (preprint).
Topka, K.: 1981, Ph.D. Thesis, Harvard University.
Vaiana, G.S. et al.: 1981, Ap.J., 245, 163.
Vaiana, G.S. and Rosner, R.: 1978, Ann. Rev. Astron. & Astrophys., 16, 393.
Vaiana, G.S., Serio, S., Blanco, C., Catalano, S., Marilli, E., Rodono', M. and Golub, L.: 1982, in preparation.
Walter, F.: 1982, Ap.J., 253, 745.

## DISCUSSION

Haisch: (beginning of question lost on tape) ... the energy going into the corona is an additional means of energy dissipation and we have to take this into account apart from the energy going into stellar flares.

Golub: Yes, quite right.

Walter: I have two questions about your plot of $L_X$ vs Temperature. First, the relation is awfully steep. Is that telling us that there is not much of a relation between temperature and luminosity but that coronae exist only at certain temperatures? The second question is what does the relationship look like if you plot, not luminosity, but surface flux? Your diagram plotted both the RS CVn stars and the M dwarfs. These have very different surface areas.

Golub: The first plot that I showed consisted of M dwarf stars only and this was steep. Even the simplest theories are in agreement with this.

Walter: What is the gradient of this relationship?

Golub: I believe it is 5/2.

Jordan: I would like to make a comment. I am alarmed to see that you are building yet again on the scaling law between pressure, temperature and length. That scaling law fails to reproduce the most fundamental property of the solar emission-measure distribution i.e. the minimum at 200 000 K. Moreover it is not generally realized that the boundary conditions used in producing that scaling law fix the ratio of conductive to radiative flux at a constant value of 1.6. This is very similar to the minimum energy-loss method which fixes this ratio at 1. This does not fit the behaviour of the solar atmosphere and there is no reason to think that it will apply to stellar atmospheres. So I think it is time that we stopped using that scaling law and went back to the observations.

Golub: I thought you told me you wouldn't do that! (laughter). Bob (Rosner) might like to comment. For my part I would say that we have results, not just from analytical work but also numerical results from modelling of loop structures, which successfully reproduce the differential emission measure through the transition region and one does get this kind of scaling in the corona. Perhaps Bob (Rosner) would like to say more.

Rosner: I think that this could be a very long discussion.

Jordan: Perhaps we should discuss it privately.

Simnett: The magnetic fields that you infer are very sensitive to the temperature you adopt. A good fraction of the flux observed in the upper energy range of your instrument comes from line emission which is sensitive to abundance. What abundances do you assume?

Golub: We use solar abundances since we are dealing with solar-type stars. We now have the capability in our analyses to vary the abundances but have not done this yet.

Simnett: I thought that we were dealing with cool stars which are not solar-type. Anyway do you know how sensitive you are to these effects?

Rosner: With proportional counter data it is very difficult to distinguish between abundance (next word lost) and temperature. Questions like that are unlikely to be answered by the IPC data but rather by high-resolution spectroscopy.

Simnett: Yes, but abundance is critical in determining coronal temperature and, thence, magnetic field.

# EINSTEIN AND IUE OBSERVATIONS OF NEARBY RED DWARFS

Hugh M. Johnson
Lockheed Palo Alto Research Laboratory

## ABSTRACT

This paper summarizes data for 40 representative nearby stars from one guest observer's coordinated programs with the Einstein observatory or IUE observatory. The coronal X-ray, chromospheric ultraviolet, and auxiliary optical properties of sets of these stars are tabulated or illustrated in several ways. Factors of stellar duplicity are shown to be quite prevalent in presenting the observations. The most luminous X-ray dwarfs below the Sun are strongly prone to binary status. X-ray luminosity, and the ratio of chromospheric flux to X-ray flux, are dependent on photospheric radius. A very long period BY Draconis variable of type dM6e (HH And) is a detected X-ray source, and some presumably quite old (halo) stars are detected.

## 1. INTRODUCTION

As a sequel to an analysis of the nature of the unidentified high-latitude X-ray sources that were cataloged by early 1978 (Johnson 1978) I proposed an Einstein guest-observer program to detect a representative sample from the 100 stars within 6.5 pc, listed by Allen (1973). This sample was extended in 1979, and in 1980 some spectroscopic binaries within 25 pc were added to the total. Activity such as flaring was a neutral factor in the composition, and there is actually some bias against flare stars in the sample because flare stars had been evidently preferred as the prior HEAO-B Consortium selections of late dwarfs (cf. Giacconi et al. 1978).

Johnson (1981) produced Einstein guest observer data for 16 nearby-star IPC targets (several of which are unresolved binaries) and for five HRI images from the same sample. This paper extends earlier Einstein guest observer data of stars nearer than 25 pc by 15 more IPC targets, several of which are again known or suspected binaries that the IPC does not resolve. HRI data for three IPC targets, as reprocessed in 1982, now resolve two of the close binaries and pinpoint one component



*P. B. Byrne and M. Rodonò (eds.), Activity in Red-Dwarf Stars, 109–124.*

of each of them. In order to provide ultraviolet spectral information along with the X-ray data for the IPC targets, IUE observations were made in 1981 of 13 of the targets from the total of 29 candidates. Woolley 319A was also observed with the IUE but not with Einstein. A log of the IUE data sorted by right ascension is included in IUE NASA Newsletter, No. 18 (1982). For the sake of a unified nomenclature the primary identifications of stars here will be according to the catalog of Woolley et al. (1970) where coordinates and many other data may be found. Other observers, notably Weidemann et al. (1980) and Linsky et al. (1982), have independently obtained IUE spectra of four more stars in the present Einstein sample (Woolley 15A, 144, 440, and 820B).

Because red dwarfs are the most numerous species of stars, they are naturally represented most often in the above programs; and because red dwarfs are prototypes for activity such as flaring and the BY Draconis syndrome, active stars are found in these programs. However, other stars are needed and included for comparative purposes. Several known or suspected but unclassified binary companions are also present.

## 2. TABULATIONS OF DATA

Table I requires the following explanation. V and $M_V$ are from Woolley et al. (1970), and $M_{bol} = M_V + BC$, where the bolometric correction BC is the function of MK spectral type according to Johnson (1966). The column of spectral type is the miscellaneous compilation in Woolley et al. (1970). The dM(e) types are updated by some MK types according to Wing and Dean's (1982) 8-color photometric classification in the next column, or (in parentheses) according to Wing and Yorka's (1979) tables of correspondence with MK types. The dK types are updated from Buscombe (1977). One can still find a little disagreement among recent classifications. For example, Boeshaar (1978) and Keenan and McNeil (1976) call Kapteyn's star a subdwarf, and the latter classify TZ Ari M5-V. Wing's system is used in this paper also as needed for dM stars from Johnson (1981). The column of Binary Characteristics refers only to spectroscopic binaries, SB, all of them single-line, followed by the range of velocity (km $s^{-1}$), and the number of measured radial velocities, RV, compiled from the literature. It should be noted that the designation SB based on as few as 2 RV depends on the judgment of spectroscopists who know their equipment and measures. There are outstanding puzzles with some of the stars, e.g. Woolley 905 for which Abt (1973) lists five 1939-44 plates with successive RV = +24.0, -16.4, -70.1, -96.1 and -67.2 km $s^{-1}$ from the Mt. Wilson files. But Abt (1982) has found no publication other than Joy (1947), who gave a mean RV = $-81 \pm 2.5$ km$s^{-1}$ from three (of the same?) plates, and Wilson (1953) who also lists -81 km $s^{-1}$ from four plates! The Table I range is for the five plates, and the designation SB is questioned only because of the foregoing history.

The Binary Characteristics of Table I are compiled to emphasize that the red dwarfs may often be close binaries and this may distort many of the conclusions that might be made of them were they all single

Table I. Some Optical Properties of the Extended Sample

| Woolley | Other Name | Sp. Type | Other Datum | V | $M_V$ | $M_{bol}$ | Binary Characteristics |
|---|---|---|---|---|---|---|---|
| 9066 | TZ Ari | dM5e | M4.5 V | 12.28V | 13.92V | 11.4V | 1 RV |
| 191 | Kapteyn | M0 V | M0.0(V) | 8.85 | 10.89 | 9.8 | non-SB(8), 10 RV |
| 206 | V998 Ori | dM4e | (M3.5 V) | 11.50V | 10.73V | 8.6V | SB(43), 3 RV |
| 293 | EG 56 | DF | | 14.5 | 15.7 | | 0 RV |
| 319A | $+10^{\circ}1857$ | dM1 | (M0.0 V) | 9.68 | 8.71 | 7.6 | prob.SB(16),3 RV |
| 334 | $-8^{\circ}2582$ | dM0 | (K5.0 V) | 9.51 | 9.1 | 8.0 | prob.SB(13),3 RV |
| 402 | Wolf 358 | dM5 | M3.8 V | 11.66 | 12.42 | 9.9 | SB(34), 3 RV |
| 440 | L145-141 | DA | $C_2$ band | 11.48 | 13.05 | | 0 RV |
| 447 | FI Vir | dM5 | M4.1 V | 11.10 | 13.49 | 11.0 | non-SB(10), 7 RV |
| 9400A | SV101251 | dK8 | V=1.5 | 9.77 | 8.3 | 7.3 | prob.SB(11),7 RV |
| 576 | $+6^{\circ}2986$ | K5 | K0 V;M0 V | 9.87 | 8.7 | 8.2 | prob.SB(35),5 RV |
| 628 | $-12^{\circ}4523$ | dM4 | M3.5 V | 10.12 | 12.10 | 9.9 | SB(25), 6 RV |
| 9566 | HD149162 | dK1 | K0 V | 9.3 | 7.4 | 7.2 | SB(37), 5 RV |
| 861 | Wolf1037 | sdK6 | | 14.19 | 13.0 | 12.3 | SB(37), 2 RV |
| 866 | L789-6 | dM6e | M5.5 V | 12.18 | 14.60 | 11.8 | 0 RV |
| 905 | HH And | dM6e | M5.1:V | 12.29V | 14.80V | 12.0V | ?(120), 5 RV |

stars or binaries with defined orbits. Even Woolley 144 ($\epsilon$ Eri) with many RV measures, and a range of only 1.9 km $s^{-1}$ among the observatory means, has been published as a close binary in speckle interferometry (Blazit et al. 1977). An M3 V companion can fit into the interferometric, photometric, and astrometric data; but the RV data may be satisfied only with an orbit of low inclination.

Table II gives $R/R_{\odot}$ and $T_{eff}$ according to Pettersen (1980) or an interpolation of his data plotted as functions of $M_V \geq +9$. Allen's (1973) functions of $M_V < +9$ extend the table up the main sequence. White dwarf R,$T_{eff}$ are from Shipman (1979). Luminosity is estimated as L(R, $T_{eff}$) and again as L($M_{bol}$) for comparison. The galactic orbit eccentricity e and the space velocity $|U^2 + V^2 + W^2|^{1/2}$ from Woolley et al. (1970) provide a basis for estimating the population or age group of each star as young (Y) or old (O) in the absence of other previously published designations, chiefly Veeder's (1974) or Mould and Hyland's (1976). The respective designations for Woolley 447 differ as shown.

Table III compiles the available BY Draconis period or the consistent equatorial rotation information. For Woolley 71 ($\tau$ Cet) only $v_e$ sin i = 2.4 km $s^{-1}$ is known. For Woolley 15B an incomplete photometric cycle was published under the name CQ And (misprinted for GQ And) and questioned as rotationally dependent because all other stars accepted as members of a BY Draconis class by Bopp and Espenak (1977) had periods of less than five days. Woolley 905 is designated HH And and classified as a BY Draconis type variable of 120-day period in Kukarkin et al. (1976), so it raises strong questions about the definition of the

Table II. Properties of the Entire Sample

| Woolley | $R/R_\odot$ | $T_{eff}$ (100K) | L(R,T) (ergs $s^{-1}$) | $L(M_{bol})$ (ergs $s^{-1}$) | Galactic Orbit e | Space Velocity (km $s^{-1}$) | Age Group |
|---|---|---|---|---|---|---|---|
| 15A | 0.41 | 35 | 8.7(31) | 1.1(32) | 0.12 | 51 | OD |
| 15B | 0.19 | 32 | 1.3(31) | 1.9(31) | | | |
| 71 | 0.87 | 51 | 1.8(33) | 1.6(33) | 0.22 | 37 | ? |
| 9066 | 0.19 | 31 | 1.2(31) | 8.3(30) | | 53 | ? |
| 139 | 0.91 | 54 | 2.4(33) | 2.3(33) | 0.36 | 127 | OD |
| 144 | 0.83 | 48 | 1.3(33) | 1.3(33) | 0.09 | 22 | Y |
| 191 | 0.43 | 33 | 7.6(31) | 3.6(31) | | 293 | halo |
| 206 | 0.44 | 33 | 7.9(31) | 1.1(32) | 0.007 | 25 | Y |
| 213 | 0.23 | 32 | 1.9(31) | 2.1(31) | 0.36 | 126 | halo |
| 9193 | 0.012 | 44 | 2.0(29) | | | | ? |
| 293 | | | | | | | ? |
| 319A | 0.65 | 38 | 3.0(32) | 2.8(32) | 0.19 | 54 | O |
| 334 | 0.61 | 37 | 2.4(32) | 1.9(32) | 0.08 | 39 | Y |
| 402 | 0.30 | 32 | 3.3(31) | 3.3(31) | 0.10 | 40 | YD |
| 412A | 0.49 | 34 | 1.1(32) | 1.2(32) | 0.36 | 130 | OD |
| 412B | 0.13 | 29 | 4.1(30) | 1.5(30) | | | |
| 440 | 0.010 | 95 | 3.1(30) | | | | ? |
| 445 | 0.24 | 33 | 2.4(31) | 2.5(31) | 0.27 | 121 | halo |
| 447 | 0.21 | 31 | 1.4(31) | 1.2(31) | 0.09 | 26 | YD,OD |
| 9400A | 0.67 | 41 | 4.4(32) | 3.6(32) | 0.10 | 52 | O |
| 9400B | 0.56 | 36 | 1.8(32) | | | | |
| 576 | 0.65 | 38 | 3.0(32) | 1.6(32) | 0.24 | 93 | O |
| 628 | 0.33 | 32 | 3.9(31) | 3.3(31) | 0.04 | 26 | YD |
| 9566 | 0.73 | 41 | 5.2(32) | 4.0(32) | 0.22 | 75 | O |
| 643 | 0.22 | 32 | 1.8(31) | 1.6(31) | 0.13 | 41 | OD |
| 644A | 0.45 | 34 | 9.3(31) | 9.1(31) | 0.13 | 41 | OD |
| 644B | 0.45 | 34 | 9.3(31) | 1.0(32) | | | |
| 644C | 0.13 | 25 | 2.3(30) | 1.0(30) | | | |
| 702A | 0.89 | 52 | 2.0(33) | 2.1(33) | 0.07 | 29 | Y |
| 702B | 0.73 | 41 | 5.2(32) | 5.3(32) | | | |
| 780 | 1.01 | 58 | 4.0(33) | 4.0(33) | 0.11 | 49 | OD |
| 783A | 0.79 | 45 | 8.8(32) | 1.0(33) | 0.33 | 138 | halo |
| 783B | 0.28 | 31 | 2.5(31) | 2.5(31) | | | |
| 820A | 0.72 | 41 | 5.1(32) | 4.8(32) | 0.28 | 105 | OD |
| 820B | 0.67 | 39 | 3.6(32) | 2.8(32) | | | |
| 860A | 0.35 | 32 | 4.4(31) | 4.0(31) | 0.11 | 32 | OD |
| 860B | 0.24 | 31 | 1.8(31) | 1.2(31) | | | |
| 861 | 0.21 | 32 | 1.6(31) | 3.6(30) | 0.54 | 209 | halo |
| 866 | 0.25 | 30 | 1.7(31) | 5.8(30) | 0.19 | 79 | OD |
| 905 | 0.16 | 30 | 7.2(30) | 4.8(30) | 0.29 | 84 | OD |

type. Kron (1950) reported the 120-day period of Ross 248 (another name

Table III. Period and Equatorial Rotation Velocity

| Woolley | Period (days) | $v_e$ ($km\,s^{-1}$) | References |
|---|---|---|---|
| 15B | > 7 | < 1.4 | Bopp and Espenak (1977) |
| 71 | ≤ 19 | ≥ 2.4 | Soderblom (1981) |
| 144 | 12 | 3.5 | Hallam and Wolff (1981) |
| 702A | 20 | 2.3 | Stimets and Giles (1980) |
| 820A | 35 | 1.0 | Hallam and Wolff (1981) |
| 820B | 45 | 0.7 | Hallam and Wolff (1981) |
| 905 | 120 | 0.07 | Kron (1950) |

for Woolley 905) from the unpublished light curve observed by Gordon and Kron (1982), and noted that the shape of the light curve simulates the secondary variation of YY Gem (Kron 1952). He concluded that the Ross 248 amplitude of 0.06 magnitude originates in the same way, by the rotation of the star, in this case with a long period.

Table IV gives the new X-ray IPC data derived and arranged as for the earlier observations (Johnson 1981).

Table IV. X-Ray Properties of the Extended IPC Sample

| Woolley | $\alpha(1950+\mu_\alpha \Delta T)$ | O–C (s) | $\delta(1950+\mu_\delta \Delta T)$ | O–C (") | $F_x$ ($ergs\,cm^{-2}s^{-1}$) | $L_x$ ($ergs\,s^{-1}$) |
|---|---|---|---|---|---|---|
| 9066 | $01^h57^m30^s.3$ | -0.5 | +12°49′11" | -26 | 1.5(-12) | 4.0(27) |
| 191 | 05 10 00.3 | ... | -45 02 46 | ... | < 1.9(-13) | < 3.5(26) |
| 206 | 05 29 29.6 | +0.2 | +09 47 11 | -10 | 4.8(-12) | 1.2(29) |
| 293 | 07 52 15.7 | ... | -67 38 44 | ... | < 8.9(-14) | < 3.6(26) |
| 334 | 09 04 19.0 | +0.8 | -08 36 22 | -36 | 2.6(-13) | 4.4(27) |
| 402 | 10 48 17.3 | ... | +07 04 39 | ... | < 1.4(-13) | < 8.3(26) |
| 440 | 11 43 10.7 | ... | -64 33 40 | ... | < 1.5(-13) | < 4.2(26) |
| 447 | 11 45 10.3 | -2.4 | +01 05 21 | -02 | 3.2(-13) | 4.2(26) |
| 9400AB | 12 13 25.6 | ... | +05 55 05 | ... | < 1.4(-13) | < 6.5(27) |
| 576 | 15 02 25.7 | +2.8 | +05 50 12 | -03 | 1.7(-13) | 6.3(27) |
| 628 | 16 27 30.8 | ... | -12 32 53 | ... | < 1.5(-13) | < 2.9(26) |
| 9566 | 16 30 21.9 | -0.1 | +03 21 05 | +20 | 1.5(-12) | 1.1(29) |
| 861 | 22 26 17.0 | ... | +05 33 16 | ... | < 2.1(-13) | < 7.8(27) |
| 866 | 22 35 49.9 | -0.6 | -15 34 21 | -43 | 8.0(-13) | 1.0(27) |
| 905 | 23 39 27.3 | -2.0 | +43 54 23 | -23 | 1.5(-13) | 1.8(26) |

Table V gives the reprocessed X-ray HRI data also as arranged for the original HRI observations (Johnson 1981) except that net counts per star are tabulated rather than $F_x$ (thus without intervention of a con-

Table V. Reprocessed HRI X-Ray Data

| Woolley | $\alpha(1950)+\mu_\alpha \Delta T$ | O-C (s) | $\delta(1950)+\mu_\delta \Delta T$ | O-C (") | Net Counts ($10^3$ s)$^{-1}$ | Exp. (s) |
|---|---|---|---|---|---|---|
| 644A 644B | $16^h52^m46^s.87$ | +0.17±0.03 | − 8°15′08″.2 | +0.2±0.5 | 648±26 | 4158 |
| 702A | 18 02 56.12 | +0.12±0.03 | + 2 30 02.1 | −1.9±0.5 | 193±14 | 8912 |
| 702B | 18 02 56.05 | +0.19±0.03 | + 2 30 03.9 | −3.7±0.5 | < 5.3 | |
| 783A | 20 07 56.13 | +0.08±0.17 | −36 14 29.3 | +0.2±2.0 | 5.0±2.2 | 5031 |
| 783B | 20 07 56.58 | −0.37±0.17 | −36 14 33.0 | +3.9±2.0 | < 3.8 | |

version factor). The 1950 SAO catalog coordinates are supplemented with binary orbit information for the comparison O-C of Einstein observed coordinates with the calculated coordinates. The HRI reduction program provides error estimates for the observed coordinates which are quoted after O-C. It is not known how much of the larger values of O-C depends on SAO errors. Except for Woolley 644AB, which is still blended with 0″.2 separation, it is possible to decide that component A of each binary provides the X-ray flux. Errors in the original 1979 HRI data were too large to prefer any binary component over the other. Net counts for Woolley 702B and 783B are estimated as upper limits from the background counts data. These HRI ratios A:B of X-ray flux will be applied to the IPC estimates of $F_x$ and $L_x$ in Johnson (1981) for the following discussion. HRI reduction includes an examination of the time stream of data for variability in these images, but no evidence was found for it. The last column of Table V (Exp.) reports the total time in the reprocessed image.

Tables VI and VII summarize new IUE measures of the flux density of detected line emissions, and also a sample of continuum averaged at λ2660 around λλ2640-2680. The low resolution blends several doublets for which mean laboratory wavelength is listed. However, the Mg II emission doublet is partially resolved in Woolley 412A, 820A, 866, and 860AB; and it appears serrated at the peak in Woolley 576 and asymmetric in Woolley 9066. Mg II in Woolley 9566 is in very slight emission on noisy continuum, and in Woolley 783A the profile shows slight emission within shallow absorption. Geocoronal Lyα is blended with each stellar Lyα, which is seen as a pointlike image within the larger geocoronal image on the IUE Photowrite representations of Woolley 644AB, 783A, 820A, and 860AB. The stellar Lyα image is not seen in Woolley 191. The stars for which only Mg II and continuum λ2660 are listed were not exposed in the IUE short wavelength camera. Mg I 2852 is a very deep absorption feature in Woolley 9566, and there is also a large and broad but unidentified emission between Mg II 2799 and Mg I 2852 that peaks at λ2822. This emission does not appear to be an artifact, but it is definitely out of the ordinary when compared with the earlier spectrum of the Sun or the spectra later than K1 V in this collection.

Table VI. Flux Density of Line Emissions (ergs $cm^{-2}$ $s^{-1}$) and of λ2660 Continuum (ergs $cm^{-2}$ $s^{-1}$ $Å^{-1}$)

| Sp. | Woolley 9066 | Woolley 191 | Woolley 644AB | Woolley 783A | Woolley 820A | Woolley 860AB |
|---|---|---|---|---|---|---|
| H I 1216 | | 8.7(-12) | 1.3-1.4(-11) | 1.6(-11) | Saturated | 1.1(-11) |
| N V 1241 | | | 2.9-3.2(-13) | | 5.6(-14) | |
| O I 1304 | | | 4.2-10 (-14) | | 8.8(-14) | |
| C II 1335 | | | 1.4-2.6(-13) | | 1.0(-13) | |
| Si IV 1394 | | | 5.2-19 (-14) | | | |
| Si IV 1403 | | | 1.0(-13) | | | |
| C IV 1549 | | | 3.4-11 (-13) | | 1.2(-13) | |
| He II 1640 | | | 1.4-3.6(-13) | | 9.9(-14) | |
| C I 1657 | | | 9.1-14 (-14) | | 1.0(-13) | |
| Al II 1671 | | | | | 3.0(-14) | |
| Si II 1808 | | | 3.9-9.0(-14) | 3.3(-14) | 1.3(-13) | |
| Si II 1817 | | | 9.1-12 (-14) | 8.3(-14) | 2.0(-13) | |
| Fe II 2585 -2631 | | | 1.2(-12) | | | |
| Mg II 2799 | 1.5(-13) | 4.2(-14) | 3.1(-12) | 1.8(-12) | 1.8(-11) | 4.0-4.6(-13) |
| Continuum | nil | 1.3(-15) | 8.3(-15) | 9.0(-13) | 1.3(-13) | 3.3(-15) |

Table VII. Flux Density of Mg II 2799 Line Emissions and of λ2660 Continuum as in Table VI

| Woolley | 319AB | 412A | 9400AB | 576 | 628 | 9566 | 866 |
|---|---|---|---|---|---|---|---|
| MgII 2799 | 5.1(-14) | 1.9(-13) | 5.8(-13) | 2.0(-13) | 1.4(-13) | 9(-14): | 2.4(-13) |
| Continuum | 6.4(-16) | 2.1(-15) | 3.7(-15) | 2.0(-15) | 8.3(-16) | 3.8(-14) | 5.6(-16) |

## 3. CORRELATIONS OF DATA

Figure 1 plots log $L_x$ vs. log $R/R_\odot$. Photospheric radius R is one of the fundamental stellar parameters and is directly involved in coronal emission measure for scale. If the coronal radii of stars below the Sun were proportional to R while all other factors of emission remained the same, $L_x \propto R^3$ should appear. The large dispersion of data in $L_x$ shows that other factors perturb but they do not rule out this simple geometry. Relatively little is known about red dwarf coronal temperatures among the other factors. An unexpected discovery from the only Einstein Solid State Spectrometer observations of red dwarfs, AD Leo (Swank et al. 1981) and Wolf 630AB (Swank and Johnson 1982), was the latter's quiescent dominant temperature of about 6.5 x $10^6$ K and an indication of additional emission above $10^7$ K. This is strongly reminiscent of the RS Canum Venaticorum class of coronal properties. Wolf 630AB, which appears in Table II as Woolley 644A and 644B, is a peculiar multiple star of which component B is probably a spectroscopic binary.

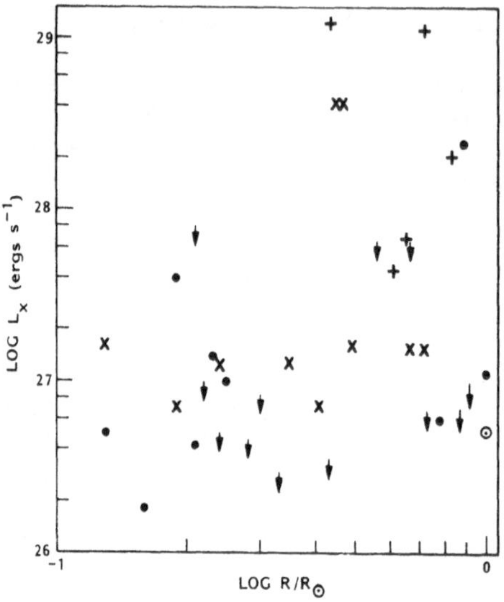


Figure 1: Dependence of $L_x$ on $R/R_\odot$. The symbol "•" plots a star that is a resolved member of a binary or a star that is not known to be a binary; "x" plots a member of a blended binary with half of total $L_x$, so that some of the symbols may be upper limits and some of them may be plotted at half of true $L_x$; "+" plots a primary member of a candidate binary for which the secondary is not plotted because neither its spectrum nor its magnitude is known; and "↓" plots the upper limit of an IPC-undetected star. The average Sun (⊙) is from Pallavicini et al. (1981).

Stellar duplicity may be an important factor in $L_x$. All but three stars in Figure 1 with $L_x > 10^{27}$ ergs $s^{-1}$ are members of visual, interferometric, or suspected spectroscopic binaries. One of these three, Woolley 9066 (TZ Ari), has only one published radial velocity so it is untested. Another, Woolley 213 (Ross 47), is listed by Abt (1970) with four Mt. Wilson velocities (+80.6, +112.3, +104.1, and +98.0 km $s^{-1}$). They do not exclude the star as a potential spectroscopic binary. The third star, Woolley 780 ($\delta$ Pav) is the only one of the three in question that is fairly certainly established as a single star, with $L_x = 1.1 \times 10^{27}$ ergs $s^{-1}$. Although the stars that are binaries may often be close binaries, Woolley 702A (70 Oph A) is the most probable example of a luminous X-ray dwarf that is not a close binary. Batten and van Dessel (1976) found marginal evidence for a short-period variation in the radial velocities of 70 Oph A, but later observations do not support it (Batten 1982).

Figure 2 shows a decline of $L_x$ with $v_e$, the equatorial velocity of rotation. Most of the stars are above or to the left of Pallavicini et al.'s (1981) least-squares fit to partly different data, $L_x = 1.4 \times 10^{27} (v_e \sin i)^2$, and that relation is an upper envelope for $v_e$ dependence.

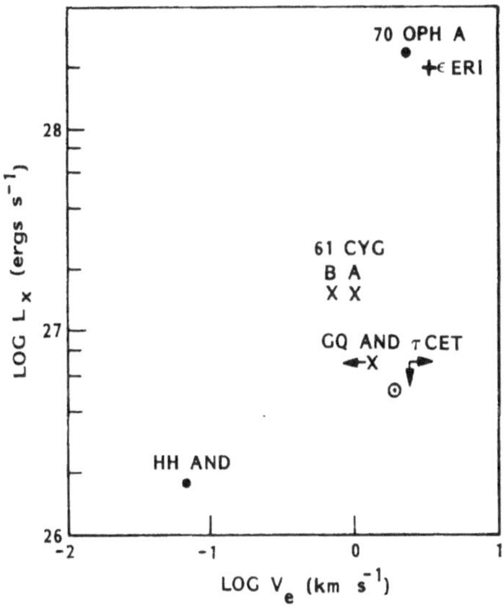


Figure 2: Dependence of $L_x$ on $v_e$, the equatorial velocity of rotation. Plot symbols are the same as in Figure 1. The upper limit on $v_e$ for GQ And results from a lower limit on observed light-curve period. The lower limit on $v_e$ for $\tau$ Cet results from a $v_e$ sin i observation.

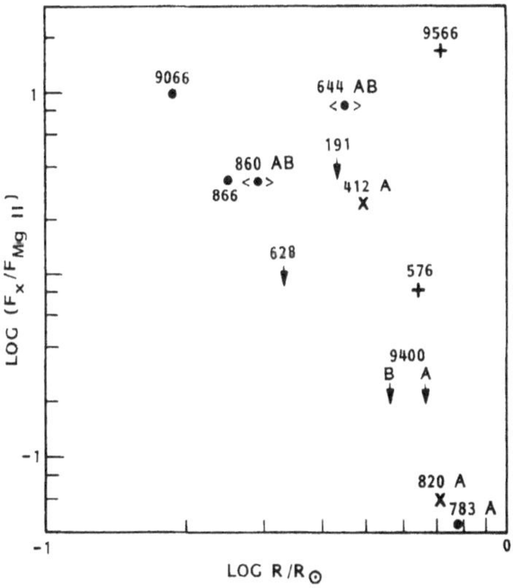


Figure 3: Dependence of the ratio $F_x/F_{Mg\ II}$ (coronal flux in the IPC band to chromospheric flux in the Mg II doublet line) on $R/R_\odot$. Plot symbols are the same as in Figure 1 except that "<>" brackets appear for binaries that are blended in both IUE and <u>Einstein</u> images.

HH And is the most important plot in Figure 2 since it extends the range of $v_e$ over an order of magnitude below previous investigations, and its inferred rotational velocity is only one or two percent as large as that of YZ CMi, a dM4.5e star that is probably single and young. Bopp and Espenak (1977) have maintained that all dMe stars (such as HH And) are subject to BY Draconis variability, but Bopp et al. (1981) clearly would not expect to encounter BY Draconis variability in dMe stars with $v_e < 3$ km s$^{-1}$. The Hα region of the HH And (Ross 248) spectrum is complex (Worden et al. 1981). The star may be completely convective and it obviously deserves much further investigation of photometric variability, rotation period, and the nature of its radial velocity range. The generation of magnetic flux and the related structure of spots in such convective stars has been only briefly discussed theoretically (e.g. Galloway and Weiss 1981; Durney and Robinson 1982). If the 120-day period in the light curve of HH And is truly rotational and not the growth and decay of a polar spot such as Mullan (1974) first supposed for convective dwarfs, or the integrated secular display of many spots, then rapid rotation is not an essential for the BY Draconis syndrome.

Figure 3 shows the relation of coronal flux to a representative measure of chromospheric flux. Coronal flux rises very rapidly with respect to chromospheric flux as radius shrinks in the progression down the main sequence from the Sun, with Woolley 9566 the exception to the rule. If the optically fainter spectroscopic binary component of Woolley 9566 provides dominant $F_x$ and $F_{Mg\ II}$ in the system, the plot might be moved to smaller radius and so to better agreement with the other stars. Unfortunately the measure of the ratio of fluxes for Woolley 9566 is also the least certain of the plots.

## 4. X-RAY FLARING BEHAVIOR

The IPC counts of the detected stars in Table IV were analyzed by standard *Einstein* computer programs for variability. Only one of them,

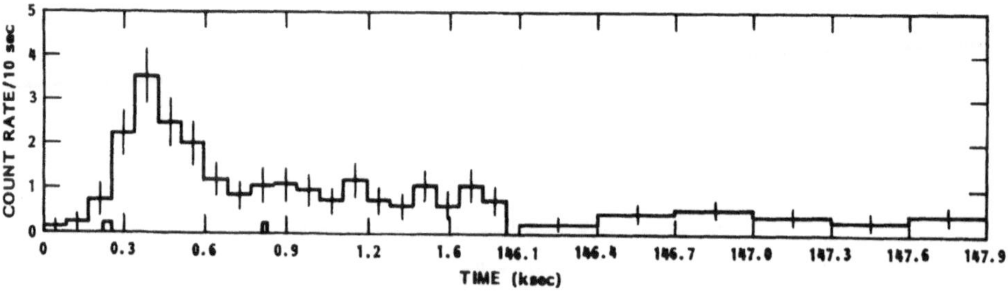


Figure 4: Two major sections of the IPC count rates for TZ Ari. Small cells near zero level mark interruptions in the data stream, for which a computer program has corrected the count rates. Time begins 1980 July 14 at UT 07:38:09.

TZ Ari (i.e. Woolley 9066) exhibited an X-ray flare, while the others remained constant within statistical errors during periods of 1277 s to 4738 s in the integrated image.

Figure 4 shows the X-ray light curve of TZ Ari in two sections 1.7 days apart. The apparent flare duration at half peak level above postflare level is about 4 minutes, and the postflare level continued to decline for 1.7 days. Haisch (1983) reviews the behavior of several stellar X-ray flare events.

Figure 5 shows the results of an Einstein IPC spectral analysis for Wolf 630AB, divided into preflare and flare sections of the light curve (Johnson 1979, 1981). These may be compared with each other and with the Einstein Solid State Spectrometer (SSS) analysis for a quiescent period (Swank and Johnson 1982). Although Raymond thermal spectral fits were also made to the IPC data, their $\chi^2$ values were not so good as the values of 12.7 and 6.7, respectively achieved for preflare and flare sections fitted to exponential spectra (with gaunt factor) over IPC spectral energy bins 4-12 (i.e. 0.5-4.8 keV). These spectra suggest that the temperature approximately doubled from preflare to flare, and that preflare temperature on this model is higher than the low-temperature component (kT = 0.54 keV) of the two-temperature model derived from the SSS observation. The observed total flux density in the 0.5-4 keV spectral range of the SSS observation was $7.9 \times 10^{-12}$ ergs $cm^{-2}$ $s^{-1}$ as compared with the IPC results of $7.3 \times 10^{-12}$ ergs $cm^{-2}$ $s^{-1}$ during preflare state and $1.8 \times 10^{-11}$ ergs $cm^{-2}$ $s^{-1}$ during flare.

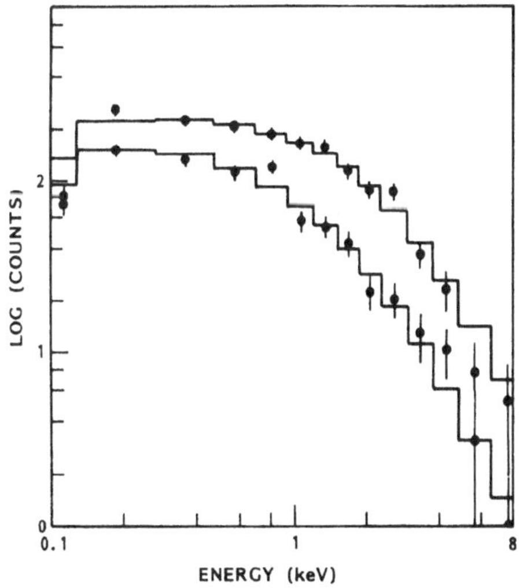


Figure 5: Observed preflare (lower) and flare (upper) spectra of Wolf 630AB fitted to exponential spectra (with gaunt factor) of kT = 1.25 keV and kT = 2.75 keV, respectively. The fits are at minimum $\chi^2$ values. Preflare counts include the first 1.2 ksec of the IPC light curve.

## 5. CONCLUSIONS

Stellar X-ray astronomy first utilized close binary systems with a compact component and its deep potential well to explain a class of bright galactic sources. Then around 1980 binaries were rejected in favor of unexpectedly energetic coronae to explain another larger class of weaker galactic sources. Each new dwarf-star X-ray source since 1979 has been almost exclusively relegated to one or the other class. Yet the abundance of binary stars suggests that caution is in order and that stellar duplicity may be a factor in the coronal X-ray stars. Stellar duplicity as a factor in red dwarf chromospheric activity has long been suspected (cf. Greenstein and Wilson 1969; Abt 1978). It is fairly clear that the factor of duplicity has not been observationally or theoretically investigated very much for the coronal class.

The mechanisms by which duplicity may be important include, again, Roche-lobe overflow or stellar wind to a companion, but not necessarily a very compact companion, interacting coronae of non-synchronous binaries, and tides. The fence that has been erected several times recently in the HR diagram to separate stars having weak or strong stellar winds, e.g. by Ayres et al. (1981), runs out between MK luminosity classes III and IV and leaves the question of red dwarf winds on open range. However, Durney and Leibacher (1973) and Coleman and Worden (1976) have discussed red dwarf winds positively, and Siscoe and Heinemann (1973) have foreseen colliding binary-system winds. There are still almost no quantitative coronal or wind studies dedicated to interacting red dwarf pairs, or a red dwarf with a partially degenerate companion, or a red dwarf with a compact companion so disposed as to be only an inefficient mass gainer and relatively weak X-ray producer. Rappaport et al. (1982) have started in this direction. Systems such as V471 Tau and PG 1413+01 (van Buren et al. 1980) may represent the inefficient kind of red-dwarf, white-dwarf pair with low $L_x$. Such studies might be applicable to many stars in this paper. Tidal effects on RS Canum Venaticorum behavior have been investigated (De Campli and Baliunas 1979; Scharlemann 1982). A corresponding approach to red dwarf binaries might be made. Once the possible binary interactions have been theoretically explored, X-ray astronomy may contribute much more to binary-star statistics as well as to astrophysics.

## 6. ACKNOWLEDGEMENTS

The Einstein guest observer work has been done under NASA contracts NAS8-33332, NAS8-33968, and NAS8-34563, and the IUE guest observer work under NASA contract NAS5-26619. I am indebted to the Center for Astrophysics Einstein group, especially Fred Seward and Dan Harris, for much assistance there. I am also glad to acknowledge the assistance of the IUE observatory staff in the acquisition and reduction of the ultraviolet spectral data at Goddard Space Flight Center. Charles Worley again freely provided binary orbit data. Presentation to this Colloquium was supported by the Lockheed Independent Research Program and by a grant from the U.S. National Committee for the IAU. Finally, I am grateful to the Local Organizing Committee for further support at the meeting.

## 7. REFERENCES

Abt, H.A.: 1970, Astrophys. J. Suppl. 19, pp. 387-505.
Abt, H.A.: 1973, Astrophys. J. Suppl. 26, pp. 365-489.
Abt, H.A.: 1978, in T. Gehrels (ed.), "Protostars and Planets", University of Arizona Press, Tucson, pp. 323-338.
Abt, H.A.: 1982, private communication.
Allen, C.W.: 1973, "Astrophysical Quantities", 3rd ed.
Ayres, T.R., Linsky, J.L., Vaiana, G.S., Golub, L., and Rosner, R.: 1981, Astrophys. J. 250, pp. 293-299.
Batten, A.H.: 1982, private communication.
Batten, A.H., and van Dessel, E.L.: 1976, Pub. Dominion Astrophys. Obs. 14, pp. 345-353.
Blazit, A., Bonneau, D., Koechlin, L., and Labeyrie, A.: 1977, Astrophys. J. Letters 214 pp. L79-L84.
Boeshaar, P.C.: 1978, in A.G.D. Philip and D.S. Hayes (eds.), "IAU Symposium No. 80, The HR Diagram", Reidel, Dordrecht, pp. 21-22 and 405.
Bopp, B.W., and Espenak, F.: 1977, Astron. J. 82, pp. 916-924.
Bopp, B.W., Noah, P.V., Klimke, A., and Africano, J.: 1981, Astrophys. J. 249, pp. 210-217.
Buscombe, W.: 1977, "MK Spectral Classifications, Third General Catalogue", Northwestern University, Evanston.
Coleman, G.D., and Worden, S.P.: 1976, Astrophys. J. 205, pp. 475-481.
De Campli, W.M., and Baliunas, S.L.: 1979, Astrophys. J. 230, pp. 815-821.
Durney, B., and Leibacher, J.: 1973, in S.D. Jordan and E.H. Avrett (eds.), "Stellar Chromospheres", NASA, Washington, pp. 282-285.
Durney, B.R., and Robinson, R.D.: 1982, Astrophys. J. 253, pp. 290-297.
Galloway, D.J., and Weiss, N.O.: 1981, Astrophys. J. 243, pp. 945-953.
Giacconi, R., Forman, W., Schreier, E.J., and Jones, C.: 1978, "The HEAO-B X-Ray Observatory Consortium Observing Program" (CFA/HEA 78-214, Appendix D).
Gordon, K.G., and Kron, G.E.: 1982, private communication.
Greenstein, J.L., and Wilson, O.C.: 1969, in S.S. Kumar (ed.), "Low-luminosity Stars", Gordon and Breach, New York, p. 201.
Haisch, B.M.: 1983, these proceedings.
Hallam, K.L., and Wolff, C.L.: 1981, Astrophys. J. Letters 248, pp. L73-L76.
Johnson, H.L.: 1966, Ann. Rev. Astron. Astrophys. 4, pp. 193-206.
Johnson, H.M.: 1978, in A. Reiz and T. Andersen (eds.), "Astronomical Papers Dedicated to Bengt Strömgren", Copenhagen University Observatory, pp. 339-348.
Johnson, H.M.: 1979, Bull. Am. Astron. Soc. 11, pp. 775-776.
Johnson, H.M.: 1981, Astrophys. J. 243, pp. 234-243.
Joy, A.H.: 1947, Astrophys. J. 105, pp. 96-104.

Keenan, P.C., and McNeil, R.C.: 1976, "An Atlas of Spectra of the Cooler Stars: Types G, K, M, S, and C", The Ohio State University Press, Columbus.
Kron, G.E.: 1950, Astron. J. 55, p. 69.
Kron, G.E.: 1952, Astrophys. J. 115, pp. 301-319.
Kukarkin, B.V., Kholopov, P.N., Fedorovich, V.P., Frolov, M.S., Kukarkina, N.P., Kurochkin, N.E., Medvedeva, G.L., Perova, N.B., and Pskovsky, Yu.P.: 1976, "Third Supplement to the Third Edition of the GCVS", Nauka, Moscow.
Linsky, J.L., Bornmann, P.L., Carpenter, K.G., Wing, R.F., Giampapa, M.S., Worden, S.P., and Hege, E.K.: 1982, Astrophys. J., in press.
Mould, J.R., and Hyland, A.R.: 1976, Astrophys. J., 208, pp. 399-413.
Mullan, D.J.: 1974, Astrophys. J. 192, pp. 149-157.
Pallavicini, R., Golub, L., Rosner, R., Vaiana, G.S., Ayres, T., and Linsky, J.L.: 1981, Astrophys. J. 248, pp. 279-290.
Pettersen, B.R.: 1980, Astron. Astrophys. 82, pp. 53-60.
Rappaport, S., Joss, P.C., and Webbink, R.F.: 1982, Astrophys. J. 254, pp. 616-640.
Scharlemann, E.T.: 1982, Astrophys. J. 253, pp. 298-308.
Shipman, H.L.: 1979, Astrophys. J. 228, pp. 240-256.
Siscoe, G.L., and Heinemann, M.A.: 1973, in C.T. Russell (ed.), "Solar Wind Three", Inst. Geophys. and Planet. Phys., UCLA, Los Angeles, pp. 243-255.
Soderblom, D.R.: 1982, preprint.
Stimets, R.W., and Giles, R.H.: 1980, Astrophys. J. Letters 242, pp. L37-L41.
Swank, J.H., Boldt, E.A., Holt, S.S., Marshall, F.E., and Tsikoudi, V.: 1981, Bull. Am. Phys. Soc. 26, p. 570.
Swank, J.H., and Johnson, H.M.: 1982, Astrophys. J. Letters 259, in press.
van Buren, D., Charles, P.A., and Mason, K.O.: 1980, Astrophys. J. Letters 242, pp. L105-L113.
Veeder, G.J.: 1974, Astron. J. 79, pp. 1056-1072.
Weidemann, V., Koester, D., and Vauclair, G.: 1980, Astron. Astrophys. 83, pp. L13-L15 (erratum 94, p. 206).
Wilson, R.E.: 1953, "General Catalogue of Stellar Radial Velocities", Carnegie Institution of Washington.
Wing, R.F., and Dean, C.A.: 1982, in preparation.
Wing, R.F., and Yorka, S.B.: 1979, in M.F. McCarthy et al. (eds.), "IAU Colloq. 47, Spectral Classification of the Future", Vatican Observatory, pp. 519-534.
Woolley, R., Epps, E.A., Penston, M.J., and Pocock, S.B.: 1970, Royal Obs. Ann., No. 5.
Worden, S.P., Schneeberger, T.J., and Giampapa, M.S.: 1981, Astrophys. J. Suppl. 46, pp. 159-175.

DISCUSSION

Giampapa: I notice from your graph of $L_X$ against MgII flux that these two are comparable in magnitude. Would it be true to say then that if one sums together the Mg flux, the Ca flux, the Balmer and transition region fluxes that these would exceed by perhaps an order of magnitude the coronal X-ray flux?

Johnson: Yes, it seems possible that is true.

Vaiana: The influence of the binary nature on the emission process appears crucial. Would you like to comment on whether this effect might be through higher induced rotation or otherwise e.g. through some kind of physical connection between the two stars? I ask this because on some Einstein images of well-separated binaries both stars are seen in emission.

Johnson: My comment would be that among the RSCVn stars, for instance, duplicity would seem to be a necessary condition. Yet in many of them the stars would not appear sufficiently close for one to directly influence the other. While I realize that enforced synchronism may lead to rapid rotation and thence to activity by mechanism described by others at this meeting, there are active binaries among the non-synchronous rotators. It is possible that the coronae or winds of these stars could interact and produce extra heating. I would say that the field of double star effects has yet to be explored in X-ray astronomy in any depth.

Linsky: There are two important observations which theoreticians need to explain and they are these. Firstly, the energy losses from the chromosphere of the quiet Sun exceed by two orders of magnitude those from the transition region and corona. On the other hand in the dMe stars the coronal X-ray loss, omitting any flows or heat conduction losses, already exceeds radiative losses in their chromospheres and transition regions. So there is a huge difference in the energy balance in the solar atmosphere and the dMe atmosphere in the sense that the Sun's outer atmosphere radiates predominantly in the cromosphere while in terms of energy the corona is irrelevant. Whereas in the dMe stars the corona is dominant and it could even be that the energy radiated by the chromosphere is a result of back-heating from the corona, an idea promoted by Cram. Could you comment on that?

Johnson: I am not a theoretician so I don't aim to comment.

Walter: I would like to make a comment on your comment on rapid rotation vis-a-vis RSCVn stars. RSCVn's are by definition binaries whereas in the dMe stars and stars like $\xi$ Boo we are seeing the same kind of effects in single stars as long as they are rapidly rotating. So it appears that the effect of stellar duplicity is to keep a star rotating fast longer. I have a poster paper on X-ray and IUE observations of AR Lac through eclipse whereby we measure the size of the coronae. They are comparable in size to the solar corona. They are not big enough to interact with one another except perhaps during the large and peculiar flares that one sees on the RSCVn's. These flares have no solar analogues. The RSCVn coronae are small and solar-like and I would presume that you are seeing the same kind of thing in stars like Wolf 630 and the other dMe stars.

Johnson: Wolf 630 is a very complex system and I'm not sure how analogous it is to an RSCVn. Would you say that Capella is rapidly rotating because it is a binary?

Walter: No! Capella is rapidly rotating because it was a rapidly rotating A star which is evolving off the main sequence and there, duplicity doesn't matter. A star like ξ Boo is not rotating rapidly because of duplicity but because it is young. So rapid rotation is important in these objects and not membership of a binary.

Johnson: What you are saying is that rapid rotation has been proved to be important and not proved that duplicity is not important.

Walter: If you pick two stars with the same rotation rate, one in a close binary and one not, they are identical.

# X-RAY EMISSION FROM FLARE STARS

P.C. Agrawal, A.R. Rao and B.V. Sreekantan
Tata Institute of Fundamental Research, Bombay, India

Flare stars are a group of mostly dMe stars, which show intense flaring activity in the optical as well as in the radio and X-ray bands. These stars are characterized by the presence of chromospheric emission lines like $H_\alpha$ and CaII H and K which are present even during the quiescent state. The presence of transition regions and coronae have been inferred from the detection of UV emission lines like NV, CIV, SiIV etc. with IUE and X-ray observations made with the Einstein Observatory. We report here X-ray observations of flare stars made with Einstein to measure their coronal X-ray emission during the quiescent state.

Seven nearby flare stars were observed with the Imaging Proportional Counter on Einstein. Optical properties of the stars and details of the X-ray observations are summarised in Table 1. All are single UV Cet

Table 1.

| Gliese No. of the Star | Other Name | Spectral Type | V | Distance (pc) | Effective Exposer Time (secs) | IPC Counts Rate ($\times 10^{-2}$ Counts $sec^{-1}$) | Log $L_X$ | Log($L_X/L_{Bol}$) |
|---|---|---|---|---|---|---|---|---|
| 229 | BD-21°1377 | dM 2·5e | 8·13 | 5·7 | 10500 | 1·07 ± 0·22 | 26·8 | -5·60 |
| - | PZ Mon | dK 2e | 10·8 | 16 | 5343 | 2·10 ± 0·35 | 28·0 | -3·66 |
| 398 | L1113-55 | dM 4e | 12·61 | 15·2 | 1426 | 7·42 ± 0·86 | 28·5 | -3·17 |
| 493·1 | FN Vir* | dM 5e | 13·34 | 10·1 | 1282 | 2·54 ± 0·69 | 27·6 | -3·55 |
| 729 | V1216 Sgr. | dM4·5e | 10·60 | 2·9 | 3676 | 26·8 ± 0·90 | 27·6 | -3·44 |
| 735 | V1285Aql | dM 3e | 10·07 | 10·9 | 7518 | 26·7 ± 0·60 | 28·7 | -3·57 |
| 791·2 | HU Del | dM 6e | 13·06 | 9·4 | 5265 | 6·32 ± 0·45 | 27·8 | -3·53 |



*P. B. Byrne and M. Rodonò (eds.), Activity in Red-Dwarf Stars, 125–129.*

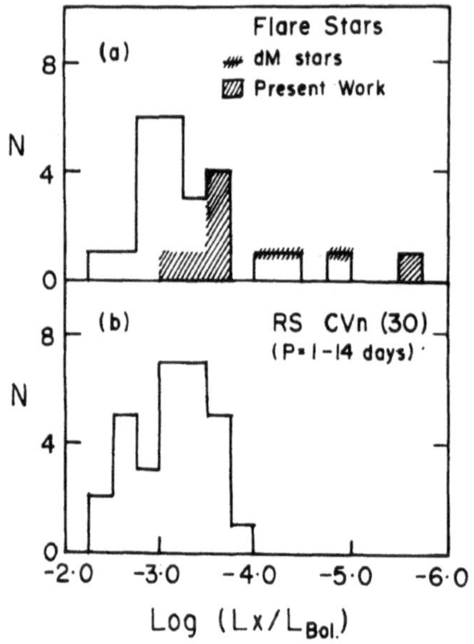


Figure 1. The distribution of $L_X/L_{bol}$ for (a) flaring and non-flaring dM stars and (b) regular period RS CVn stars.

flare stars except FN Vir which is a spectroscopic binary, and all were detected in X-rays. Two stars, Gl729 and Gl735 were detected as bright X-ray emitters with a count rate of 0.27 count $sec^{-1}$. A conversion factor of $1.4x10^{-11}$ erg $cm^{-2}$ $sec^{-1}$ per IPC count $sec^{-1}$ derived from the energy spectra of Gl729, Gl735 and Gl791.2 obtained with the IPC, is used to convert the observed count rate to flux values. Using these we computed X-ray luminosity ($L_X$) and the ratio of X-ray luminosity to bolometric luminosity ($L_X/L_{bol}$) for the observed stars and these are shown in Table 1. With the exception of the star Gl729 from which a weak X-ray flare was detected, there was no compelling evidence for flaring activity in any of the other stars during the X-ray observations. The observed $L_X$ values therefore represent the quiescent state coronal emission of these stars. Except for Gl229, which has a rather low $L_X=6x10^{26}$ ergs $sec^{-1}$, the $L_X$ values lie within a narrow range of $(0.4-5)x10^{28}$ ergs $sec^{-1}$, with a median $L_X=1.8x10^{28}$ ergs $sec^{-1}$. Note that the ratio $L_X/L_{bol}$ agrees within a factor of 3 for these six stars, with a mean value of $3.5x10^{-4}$. The distribution of $L_X/L_{bol}$ for 22 X-ray emitting flare stars, including the 7 from the present work, and 3 non-flaring dM stars reported so far, is shown in fig.1(a). The observed $L_X/L_{bol}$ distribution indicates that coronae of non-flaring dM stars are one to two orders of magnitude fainter compared to those of the flare stars. Although Gl229 is classified as a dMe flare star, its X-ray emission characteristic resembles that of the non-flaring dM stars.

In Fig.1(b) we have also shown the $L_X/L_{bol}$ distribution for the 30 regular period (P=1-14 days) RS CVn binaries for comparison (Walter and Bowyer 1981). The Two $L_X/L_{bol}$ distributions appear to be similar with about the same median value for the two groups of stars. This suggests that coronae of the flare stars are as active as those of the RS CVn stars. This is however not surprising since enhanced coronal ctivity of the RS CVn systems is attributed to the presence of large-scale starspots on their surface and such starspots have also been detected on the photospheres of many flare stars as seen by their BY Dra type variability.

In Fig. 2 we have plotted log $L_X$ vs. log $L_{bol}$ for the flare stars and the regular period RS CVn binaries. The figure strongly suggests linear dependence of $L_X$ on $L_{bol}$. The stright line in the figure corresponds to the relation $L_X=10^{(3.21\pm0.63)}L_{bol}$.

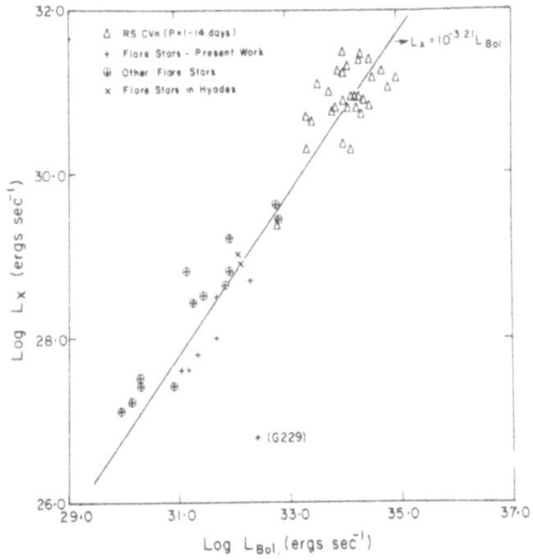


Figure 2. Diagram of $L_X$ vs. $L_{bol}$ for flare stars and the regular period RS CVn stars.

For three of the stars i.e. Gl729, Gl735 and Gl791.2, the number of detected photons was large and therefore we attempted an analysis of the IPC spectral data to estimate temperatures of their coronae. Best fit Raymond-Smith models gave temperature values in the range of (2-5)x $10^6$K which indicate that the spectra of quiescent state coronae of flare stars are somewhat softer than those of the RS CVn binaries (Swank and White 1981).

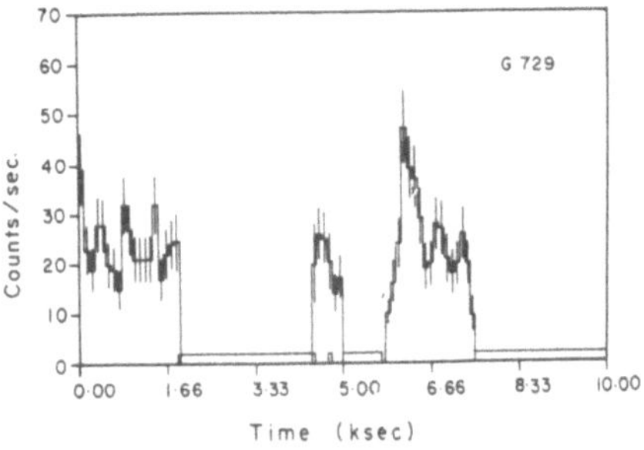


Figure 3. IPC count rate vs.time plot for the flare star Gl729. A moderate X-ray flare is seen at 6k sec after the start of observation.

There is indication of a moderate X-ray flare in Gl729 during the X-ray observations. A plot of counts rate in 100 sec time bins vs.time is shown for Gl729. There is an unmistakable increase in the counting rate starting at 6 ksec after the start of observation which corresponds to 5.063 hour UT on 24th March 1981. The observed increase is at a significance level of 4.9 $\sigma$ above the average count rate before and after the flare. The rise time of the flare is$\sim$50sec, and decay time is $\sim$300 sec. The peak $L_X$ value is $7\times10^{27}$ ergs $sec^{-1}$ and the total senergy release in the X-ray flare is$\sim1\times10^{30}$ ergs. The rise and decay times of the observed X-ray flare are similar to those of the optical flares detected from Gl729 (Cristaldi & Rodonò 1973). Detection of a flare in about one hour of observing time is not inconsistent with the reported optical flare frequency of 0.15-0.35 flare $hour^{-1}$( Feix 1974, Jarrett and Grabner 1974).

A detailed version of this paper will shortly be submitted to Astrophysical Journal.

REFERENCES

Cristaldi, S. and Rodonò, M.: 1973, I.B.V.S. No.835.
Feix, G. : 1974, I.B.V.S. No.943.
Jarrett, A.H. and Grabner, G.: 1974, I.B.V.S. No.968.
Swank, J.H. et al. : 1981, Astrophys. J. 246, 671.
Walter, F.M. and Bowyer, S.: 1981, Astrophys. J. 245, 671.

DISCUSSION

Rosner: Where did the data come from for the integral luminosity function?

Agrawal: From the present work, from a paper by Vaiana et al. from a paper by Johnson et al. and two stars are from HEAO-1 observations.

Rosner: When preparing a luminosity function you have to be careful that you are dealing with a volume-limited sample. Many of the stars in your sample are very far away. When this is so, you bias your sample towards very luminous sources.

Agrawal: All of the flare stars included in the sample are in catalogues of nearby stars and so are all within 25pc. Those included from the surveys of Vaiana et al and Johnson et al are also nearby stars.

Bopp: Caution should be exercised in using the dMe star classification. You referred to Gliese 229 as dMe star while Pettersen specifically referred to this star as having the Balmer lines in absorption. So perhaps we could agree to reserve the dMe designation for those stars with the Balmer series in emission. That may be more than a minor point because the behaviour of Hα may be rather different from CaII. So perhaps Gliese 229 is a plage star rather than a spotted star as defined by Pettersen earlier.

Agrawal: That could be true. I called it dMe because it is classified like that by Kunkel.

Evans: I would like to underline what Bopp has said and ask that the dMe classification be reserved for those stars showing the Balmer lines in emission. Secondly, there are problems with this since there are cases of stars which would at one time have been classified dMe and not at others. An example is BY Dra which has shown considerable variation in the intensity of its Balmer emission.

Agrawal: The point which I was trying to make is this, that the dM stars appear to be two orders of magnitude fainter in X-rays compared to the dMe stars. In the paper by Johnson et al are also some dM stars and these have luminosities of $10^{27}$ ergs $s^{-1}$ or less. So the flare stars which are all dMe stars are two orders of magnitude more luminous in X-rays than the dM stars.

Vaiana: Perhaps I may make a comment here. We should remember what happened to the RSCVn stars. These expanded from a relatively pure definition to include a wider range of stars. Perhaps the same thing is happening to the dMe stars. I personally believe this is no harm as long as one has good reason for it. It is appropriate that this should be discussed at this meeting.

# STELLAR X-RAY ACTIVITY IN THE HYADES

Robert A. Stern[1] and Marie-Christine Zolcinski[2]

[1]Jet Propulsion Laboratory, California Institute of Technology
[2]Physics Department, University of New Hampshire

Observations of the Hyades cluster with the Einstein Observatory and with IUE have uncovered a high level of coronal x-ray emission ($L_X \simeq 10^{29}$ erg $s^{-1}$ for solar-type stars) and similarly high fluxes of chromospheric and transition region line fluxes compared to the Sun (Stern et al. 1981, Zolcinski et al. 1981, 1982). A giant x-ray flare from a spectroscopic binary system in the Hyades has also been reported (Stern, Antiochos and Underwood 1982).

In the original x-ray survey of Stern et al. (1981), roughly a 1-2 order-of magnitude spread was noted in the x-ray luminosities of solar-type stars in the Hyades. This was less than the apparent spread in x-ray luminosities of similar stars (of which there were only about 7) in the Vaiana et al. (1981) stellar x-ray survey. Although some of the difference between the two samples could be due to age effects (see, e.g., the review by Stern, 1982), the Hyades sample stars are all of the same age, requiring some combination of x-ray variability or an intrinsically broad distribution in x-ray luminosities, possibly the result of a similar spread in stellar rotational velocities (see, e.g., Pallavicini et al. 1981). A first step in determining the relative importance of each of these effects on the Hyades x-ray luminosity function is to estimate the range of x-ray variability in the cluster stars.

Here we report preliminary results from monitoring the x-ray emission of about 20 Hyades stars both on short ($\simeq$ 1 day or less), and long (3 months-2 years) time scales. All observations were made with the Imaging Proportional Counter of the Einstein Observatory (HEAO-2), and include a number of relatively short duration (2000 sec) exposures from the original x-ray survey of Stern et al. (1981) as well as a series of longer exposures (10,000 sec) of selected regions in the cluster center. Also, two exposures of 2000 s duration from an uncompleted program to study daily variations in stellar x-ray activity are included.



*P. B. Byrne and M. Rodonò (eds.), Activity in Red-Dwarf Stars, 131–134.*

## Short Term Variability and Flaring.

In the Stern et al. (1981) 2000 s survey exposures, no evidence of x-ray flaring was observed. However, in the 10,000 s (22 - 25000 s duration) exposures of the followup survey, flaring was detected in three objects: the giant x-ray flare ($10^{31}$ erg $s^{-1}$ peak $L_X$) in the G dwarf/K dwarf spectroscopic binary (p = 5.6 d) BD + 16 577 (HD 27130), which had a 1/e decay time of ≃ 40 min (Stern, Antiochos, and Underwood 1982); an increase in flux by about a factor of 2.5-3 in the GO V spectroscopic binary (p = 4 d) BD + 14 690 with a risetime of ≃ 1000 s, and an upper limit to the decay time of ≃ 2000 s (peak $L_X \simeq 10^{30.2}$ erg $s^{-1}$); and a factor of about 2 increase in $L_X$ (peak ≃ $10^{29.6}$ erg $s^{-1}$) for the K dwarf vA 500 over approximately the same time scale as for BD + 14 690.

In addition, there is evidence for a gradual decrease (by about a factor of two) in the observed flux from the dMe flare star vA 288 over the 25000 s duration of one of the follow-up observations.

Given that the total monitoring time on a given Hyades star was typically 7 hours or less (including data gaps), the discovery of three x-ray flares in three different Hyades stars may be an indication that flares at the level of 100 times or more brighter than typical solar flares are common in the stars of the Hyades cluster. We note, however, that two of three stars observed to flare are members of short-period ( 6 d.) binary systems. Also, since the detection threshold for flares is relatively high ( ≃$10^{29}$ erg $s^{-1}$), we have likely observed only the brightest flaring activity in the cluster.

## Long Term Monitoring

For about 20 stars, all F-M dwarfs except for the KO giant θ Tau, we have several observations over a time scale of up to 500 days. Within statistical uncertainties ranging from about 10 - 50%, slightly more than half of the stars show no statistically significant change in x-ray luminosity. θ Tau is among the stars with no evident change over the 1 - 1/2 year period. Except for the flaring activity noted in the previous section, the stars which do vary in $L_X$ do so by no more than factors of 2-3 over the span of the observations. Although the length of our baseline is short compared to the typical 11 year length of the solar cycle, the absence of any very large long term changes in x-ray luminosity is intriguing. Thus coronally active Hyades stars do not show strong cyclic activity on time scales significantly less than a solar cycle. However, much more extensive long term monitoring is required to search for longer duration or weaker cyclic activity in the Hyades.

## Summary

Our preliminary conclusions are that flaring behavior is probably common in the Hyades, but that the influence of binarity on the level of flaring activity needs to be investigated further. The long term observations suggest that intrinsic differences in the level of stellar x-ray luminosity may be required to account for the spread in the x-ray luminosity function for solar type stars observed in the original Stern et al.(1981) survey.

This work was performed under NASA contract to the Jet Propulsion Laboratory, California Institute of Technology. R.A.S. wishes to thank the staff of the Arcetri Astrophysical Observatory, Firenze, Italy for help with the final preparation of this manuscript.

REFERENCES

Pallavicini, R., Golub, L., Rosner, R., Vaiana, G.S., Ayres, T., and Linsky, J.L., Ap. J. 248, 279 (1981).

Stern, R.A., Antiochos, S.K., and Underwood, J.H., Ap. J. (Letters), submitted (1982).

Stern, R.A., Zolcinski, M.C., Antiochos, S.K., and Underwood, J.H., Ap. J. 249, 647 (1981).

Vaiana, G.S., et al., Ap. J. 245, 163 (1981).

Zolcinski, M.C., Antiochos, S.K., Stern, R.A., and Walker, A.B.C., Ap.J. in press (1981).

Zolcinski, M.C., Kay, L., Antiochos, S., Stern, R., and Walker, A.B.C., 1982, to be published in the procedings of the symposium. "Advances in Ultraviolet Astronomy: Four Years of IUE Research", held at NASA Goddard Space Flight Center, March 30 - April 1, 1982.

DISCUSSION

Evans: You mentioned occultations. These show that among the Hyades stars there is an extraordinary proportion which are multiple. In spite of this I am certain that we have not found all of them. So I would advise caution in interpreting the cause of the flaring. Allowance must be made for the possibility of faint companions being the source of activity.

Stern: That is a well taken point. In the case of the large flare I tend to the view that there is a system akin to the RS CVn systems. Equally large flares have been seen in such systems. Bernie Haisch discussed them earlier. This is a special type of binary system. So, unless there is an additional dMe star flaring with an energy in X-rays larger than that typically seen in dMe flare stars' then I would stick to the RS CVn explanation. But I agree that there are probably a lot of undiscovered binaries.

Evans: I would suspect that at least 50% of the bright Hyades stars are binary.

# SESSION II

## OBSERVED ACTIVITY IN RED-DWARFS

# SPOTS, SPOT-CYCLES, AND MAGNETIC FIELDS OF LATE-TYPE DWARFS

Steven S. Vogt
Lick Observatory, Board of Studies in Astronomy
and Astrophysics
University of California Santa Cruz

ABSTRACT

A review is presented of the current state of observational knowledge concerning spots, spot-cycles and surface magnetic fields on active late-type dwarfs. The discussion centers primarily on the physical characteristics of starspots on BY Dra-type stars, including spot sizes, temperatures, structural morphology, migratory motions, and activity cycles. The discussion will also include some references to similar spot phenomena on the RS CVn stars. Observational evidence for surface magnetic fields on these stars, and on chromospherically-active G and K dwarfs, is also reviewed.

## 1. INTRODUCTION

I will attempt to present a review of spots, spot cycles, and magnetic fields in late-type dwarfs, specifically the BY Dra stars. As most of you well know, these are UV Ceti flare stars of dKe and dMe spectral types which, outside of periods of flaring, show periodic variations in light of several tenths of a magnitude attributable to large, cool "starspots" on their surfaces. Starspots are also generally accepted as the explanation for the distortion waves on light curves of the RS CVn stars (Hall 1976). There is no doubt in my mind that the mechanism for the periodic light variations of both the BY Dra's and the RS CVn's is one and the same: starspots. In fact, spots seem to be present on any late-type star with an appreciable convection zone and sufficient angular velocity, regardless of evolutionary state. The BY Dra's are mostly spotted pre-main-sequence dwarfs, whereas the RS CVn's are spotted post-main-sequence subgiants and giants. Recently another group of spotted stars was discovered in the Pleiades - rapidly rotating K stars with periods in the $0\overset{d}{.}4$-$1\overset{d}{.}2$ interval and ranges of 0.04-0.06 in V-magnitude. These will be discussed later in the meeting by others. Their sinusoidal light curves are strikingly similar to BY Dra light curves and they are evidently an even younger form of BY Dra stars still approaching the main sequence.



*P. B. Byrne and M. Rodonò (eds.), Activity in Red-Dwarf Stars, 137–156.*

The observational and theoretical evidence is now overwhelming that spots are a direct consequence of deep convection zones, and rapid angular velocity. While spotted stars occur most frequently in close binaries, this is only because the rotation rate of a convective star in a binary system can be maintained against magnetic braking forces through tidal interaction. It has been shown quite clearly (Bopp et al. 1981; Bopp et al. 1980; Bopp and Fekel 1977; Bopp and Espenak 1977) that relatively rapid rotation ($V \geq 5$ km $s^1$) is the underlying cause for spots on late-type dwarfs, and that membership in a close binary system is only a sufficient, but not necessary condition for spots. Recent V sin i measures by Vogt, Soderblom and Penrod (1982) also support this view by demonstrating that apparently single spotted stars rotate at much higher than normal rates ($V = 5\text{-}15$ km $s^{-1}$).

Many excellent and still current reviews of the spot phenomenon, both theoretical and observational, exist in the literature (Hall 1976; Rodono 1980, 1981; Gershberg 1978; Mullan 1976a,b and refs. cited therein). The field of starspot research has grown enormously since I became involved some ten years ago and it is a hopeless task to do it justice in a mere 40 minute review. Rather, I intend to concentrate primarily on the highlights of our present observational knowledge of the physical character of spots and associated magnetic activity, citing key references where appropriate. I feel compelled to offer the usual apologies to the many hundreds of researchers whose references were not cited explicitly in this work, but upon which this discussion was based.

Almost every paper published these days on starspot phenomena starts with a lengthy defense of the spot hypothesis with all the usual references cited. As recently as two years ago Hall (1980) remarked that "...the case for spots is largely circumstantial or, at best indirect. We should work to find sufficient direct evidence to convince the devil's advocate that such spots exist." Similarly, Rodono (1980) considered the spot model of RS CVn binaries as "a working one probably leading in the right direction ... questions remain to be answered before it can be accepted with confidence." By the end of this review, after I have shown you a movie of spatially resolved images of starspots obtained through a new technique we call Doppler Imaging, I expect to have convinced you that it is time once and for all to put aside our noble skepticism about the reality of starspots. We no longer should feel embarrassed or defensive about the ease with which the spot hypothesis accounts for all the observational results. That is not to imply that these starspots are direct analogs of sunspots. Rather, I will present evidence that starspots on RS CVn and BY Dra stars are actually much more analogous to solar coronal holes and solar complexes than to sunspots as regards sizes, shapes, lifetimes, and migratory motions. If true, they are probably more of a manifestation of global-scale processes occurring deep within the star, than are spots on the Sun.

## 2. PHYSICAL CHARACTERISTICS OF STARSPOTS

Most of our knowledge of the physical characteristics of starspots has been gleaned from modeling the photometric variations. These variations are typically a few hundredths to several tenths of a magnitude in range, and vary in shape, period, phase, and mean light level. They involve small but often detectable color variations, a circumstance which allows us to derive spot temperatures. The light curves often show discontinuities in mean light and phase, easily explainable by subtle changes in the spot area and/or distribution on the star. The light curves of the BY Dra's and the "distortion waves" of the RS CVn's are quite similar in most respects though the evidence for systematic period variations in the latter is much stronger due to the larger body of data available.

### 2.1 Geometric Spot Models

Geometric spot models to date have been of three basic types: single-circular spot, two circular or rectangular spots, and equatorial bands. The single circular spot model approach is only valid in cases where the light curve is symmetrical and where it is known (e.g., II Peg in 1977) from the immaculate light level that the spot passes completely out of view at some phases. In situations where the light curve is asymmetric, two spots (either circular or bounded by parallels of longitude and latitude) separated in longitude are able to provide reasonable fits to the light curves. This approach has been used extensively by Dorren and Guinan (1982a), Guinan et al. (1982), Dorren et al. (1981), Bopp and Noah (1980), and others. Here, there are many more free parameters and solution nonuniqueness problems become much more severe. The third approach (Eaton and Hall, 1979), is to assume (by solar analogy) that spots are distributed nonuniformly in longitude in mid-latitude bands symmetric about the equator. The spot area covered is assumed to follow a simple cosine dependence with longitude.

In all of these approaches, there are definite problems with nonuniqueness as regards spot sizes, temperatures, shapes, and locations. Realistic spot modeling requires, at the minimum, knowledge of the inclination of the system, proper accounting of the behavior of the mean light level and the presence of an unspotted (?) companion, and actual detection of color variations in the light curves (to decouple temperature effects from geometrical ones).

The single circular spot assumption allows a major simplification to be realized in computing light curves to model the observations. In fact the machine computation is fast enough that a dense grid of light curves covering the full range of spot size and latitude can be computed and regions of best fit to the observed light curve can be identified in spot (size, latitude) space. Figure 1 illustrates this approach for the 1977 light curve of II Peg (Vogt 1981a). Here we have plotted the goodness of fit (vertical axis) of synthetic light curves generated at different combinations of spot location (latitude) and size. The best fitting solutions occur farthest above the plane and regions below the third of fourth vertical contour from the plane represent unacceptable fits to the light curve as regards shape, amplitude, or mean light level. Similar planes can also be computed for a range of inclinations and spot temperatures, and the regions of acceptable fit can then be combined onto a single 2-d plot as in Figure 2 to produce a reasonably unique spot model within the bounds of the original single circular spot assumption. Here we have combined regions of acceptable fit from a large number of surfaces like that of Figure 1 spanning the full range of stellar inclination and three different spot temperatures. The pronounced N-S latitude ambiguity in Figure 2 arises because of the unknown inclination.

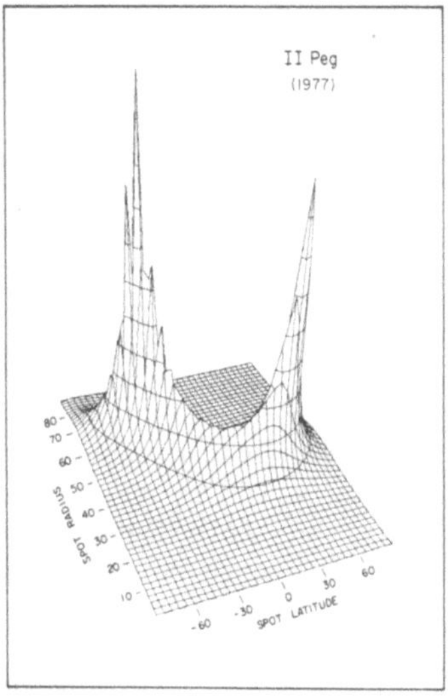


FIGURE 1

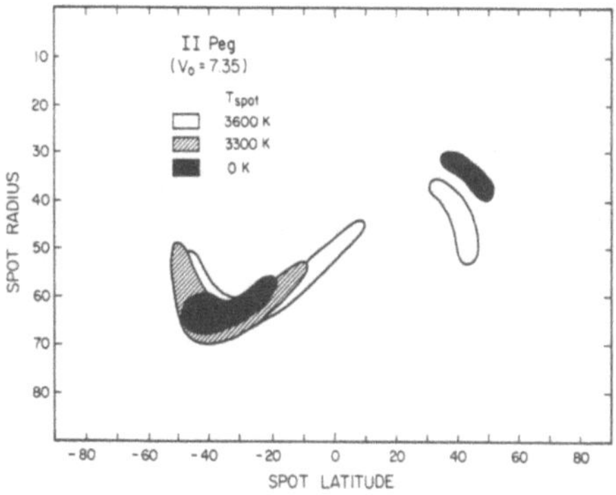


FIGURE 2

In many cases where the light curve is asymmetrical and a two-spot approach is required, the uniqueness problem becomes much more difficult. Often the two spots are assumed to be the same size and to lie at the same latitude, and in several well referenced cases neither the inclination nor the spot temperature was known.

Despite problems of nonuniqueness in the geometrical models, it is clear that starspots, be they circular, triangular, elongated, or whatever shape, are quite large, often covering up to 20% of the stellar surface. This is a simple inescapable fact of the observed amplitudes of light variation and is quite independent of solution uniqueness.

## 2.2 Spot Temperature Modeling

There is a fundamental trade-off between spot temperature and area. Small, cool spots can produce similar light curves to larger, warmer spots. Because of this, any serious model requires knowledge of color variations in the photometry to decouple temperature effects from geometrical ones. Many of the original spot models published simply ignored this effect and their results should not be taken seriously.

Most spot models employ Planckian energy distributions or actual measured stellar energy distributions for star and spot to reproduce color variations. Limb darkening coefficients are taken from published compilations and the coefficient for star and spot are generally assumed to be similar. These analyses generally yield spot temperatures which are 800 to 1200 K cooler than the surrounding photosphere. Unfortunately, the energy distributions of late-type stellar atmospheres are remarkably non-Planckian, and there are problems combining stellar energy distributions accurately on a proper flux scale, requiring integrations of the published distributions through the estimated photometer response functions. To avoid these problems, I developed a technique (Vogt 1981a) for deriving accurate spot temperatures using the Barnes-Evans visual surface brightness relation (Barnes, Evans, and Moffett 1978). This method allows direct determination of the (V-R) color difference and hence temperature difference $\Delta T$ between spot and surrounding photosphere. It consistently yields spot temperatures near 3500 K.

Other methods of determining spot temperatures involve direct detection of spectral features from the spot. Figure 3 shows a low resolution spectrum from 5000-8000Å of the spot on II Peg (Vogt 1981b) which shows a steep rise in flux to the red and pronounced absorption features of the TiO $\gamma$-system. These features led to an estimate of $\sim$M6 for the spot's equivalent spectral type, or T(spot) $\sim$ 3400 K. Ramsey and Nations (1980) also detected enhancement of the $\lambda$8860 bandhead of TiO in HR 1099 as the spot moved into view and derived a spot temperature of $\sim$3750 K or less.

From the set of all published determinations of spot temperature, it might appear (as noted by Bopp and Noah, 1980) that there is no single $\Delta T$ which can be applied to all spotted stars. Published $\Delta T$'s range from a few hundred degrees to about 1200 K, and for the star BY Dra, $\Delta T$ can decrease or even become negative (hot spot) as the cool spot disappears (Vogt 1975, 1981a; Oskanyan et al. 1977). Some of this scatter in $\Delta T$ is undoubtedly a result of solution nonuniqueness however, and some is due to differences in T(star). I have attempted to collect all of the reliable determinations of spot temperatures in Table 1. Only those determinations were used whose spot models were reasonably unique (known inclination, accounting for secondary star and/or mean light level) or where color variations were actually detected.

From this collection of reasonably well-determined spot temperatures, it is clear that most fall very near a value of 3600±200 K. The only significant exception is λ And, though neither the inclination nor the immaculate light level for this star is known and some trade-off between spot temperature and area may be allowable which could further lower T(spot). Also, there is a well-known case, BY Dra, where a large, cool spot dissolved after migrating toward the pole, leaving a plage-like remnant which remained for 5-6 years (Vogt 1981a; Oskanyan _et al._ 1977). A spot temperature of 3900 K was determined for this remnant by Davidson and Neff (1977) and a temperature of 4200-4300 K was derived by Vogt (1975). Clearly these temperatures do not refer to the large cool features we call starspots, but rather to some active remnant. It would be of great interest to follow accurately the change in temperature and area as one of these great spots formed or dissolved. The evidence at hand however points to a fairly consistent, well defined temperature of 3600±200 K for large cool starspots. Table 1 also shows that regardless of the temperature of the star, over the range from 4100 K to 4950 K, the spot temperature remains fairly well fixed, thus giving rise to differing ΔT's noted by Bopp and Noah (1980). This tendency for large cool spots to share a common temperature regardless of star temperature is probably an important clue as to their physical structure.

FIGURE 3

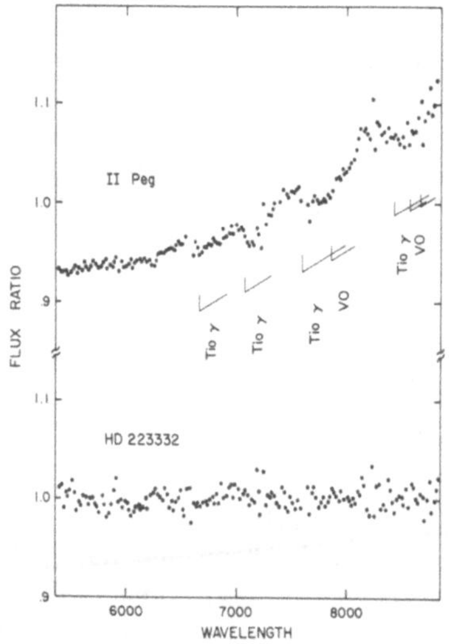


I have also included two columns which list the ratios of T(spot)/T(star) and ΔT/T(star) for comparison with sunspots. In either case, starspots look much more like sunspot _umbrae_ than sunspot penumbrae. If starspots are morphologically similar to sunspots, but just larger, our derived temperatures must refer to the penumbral region, since this contributes 97% of the total flux of a sunspot. That the derived spot temperatures more closely resemble sunspot umbrae is an indication, I believe, that the two are not morphologically similar, but rather that starspots are more like giant umbrae, with little or no penumbrae.

TABLE 1

Temperatures of Stars and their Starspots

| Star | Sp.Type | T(star) | T(spot) | ΔT(star-spot) | T(spot)/T(star) | ΔT/T(star) | Ref. |
|---|---|---|---|---|---|---|---|
| BY Dra | M0 V | 4100 | 3500 ± 450 | 600 | 0.85 | 0.15 | 1 |
| II Peg | K2 IV | 4600 | 3400 ± 100 | 1200 | 0.74 | 0.26 | 1 |
| HR 1099 | K1 IV | 4700 | <3800 | >900 | <0.81 | >0.19 | 2 |
| | | | 3450 | 1250 | 0.73 | 0.27 | 3 |
| | | | <3700 | >1000 | <0.79 | >0.21 | 4 |
| SZ Psc | K1 IV | 4700 | 3500 ± 400 | 1200 | 0.74 | 0.26 | 5 |
| DK Dra | K0 III | 4700 | 3600 ± 150 | 1100 | 0.77 | 0.23 | 6 |
| HD 209813 | K0 III | 4790 | <3840 | >950 | <0.8 | >0.2 | 1 |
| HD 32918 | K2 III | 4950 | 3750 ± 100 | 1200 | 0.76 | 0.24 | 7 |
| λ And | G8 III-IV | 5000: | 4200 | 800 | 0.84 | 0.16 | 8 |
| Sun | G2 V | 6050 | 4240 (umbra) | 1810 | 0.70 | 0.30 | 9 |
| Sun | | | 5680 (penumbra) | 370 | 0.94 | 0.06 | 9 |

1 - Vogt (1981a)
2 - Dorren et al. (1981)
3 - Antonopoulou and Williams (1980)
4 - Ramsey and Nations (1980)
5 - Eaton and Hall (1979)
6 - Guinan et al. (1982)
7 - Collier (1982) - this star is an FK Com-like variable
8 - Bopp and Noah (1980)
9 - Allen (1973)

## 2.3 Spot Shapes

It is often found in the geometrical models that spots are characterized by very large extents in longitude. Spot light curves are only rarely seen to have flat portions, indicative of a spot which passes completely out of view, and the smooth, sinusoidal shape of the light curves in stars of high inclination (like eclipsing systems) argues strongly for great longitude extents, often exceeding 150°. Of course, models invoking spots bounded by parallels of latitude and meridians of longitude (Bopp and Noah, 1980) or bands of spots (Eaton and Hall 1979) seem to favor spots quite elongated in the longitudinal direction, whereas circular spot models by their very nature do not allow for this.

Many of the spotted stars however are at low inclination, and we are often seeing circumpolar spots which can remain in view for much, if not all, of the rotation cycle. Here, it is almost impossible to distinguish elongated spots from circular ones, and the derived spot shape is more a reflection of the model assumptions than anything implied by the light curve. Later in this meeting, I will demonstrate some actual images derived of several spotted stars using a new technique called Doppler Imaging which is quite sensitive to spot shape and shows large spots very near the pole (like polar coronal holes) and spots which are greatly extended in latitude, with almost no extent in longitude. At present then, I can only summarize by saying that there is strong evidence for all three types of shape: spots extended predominantly in longitude, or in latitude, or spots which are roughly circular in shape. If I had to second guess future results of the Doppler Imaging technique, I would say that low latitude spots are preferentially extended in longitude, whereas high latitude spots are preferentially circular or elongated in latitude.

It is also not yet clear whether large spots are single entities, or groups of smaller spots. All existing models provide no information on this question. We expect the Doppler Imaging technique to make some progress in this area.

## 2.4 Spot Locations and Migrations

The spot models show a definite tendency for many spots to form at intermediate latitudes and then to drift either poleward or toward the equator. One of the best examples of this was BY Dra, a spotted star of well known inclination whose photometric behavior from 1965 to 1973 is well-modeled by a large cool spot which formed at an intermediate latitude and drifted poleward before dissolving into a plage-like region. This is illustrated in Figure 4 (a plot similar to Figure 2). Here we see that a large spot, some 40°-50° in radius formed at a latitude of 55° in 1965 and then drifted at constant radius to a latitude nearer 80°, where it then dissolved leaving a bright remnant in 1973. While the spot geometry is admittedly somewhat nonunique and idealized, it is clear that from our viewing angle, we witnessed a poleward motion of the latitude centroid of spot area. This particular episode is in fact strikingly similar to the poleward migration of high latitude solar coronal holes, then resulting in a large polar coronal hole (straddling the pole) which eventually dies away leaving an active region remnant.

FIGURE 4

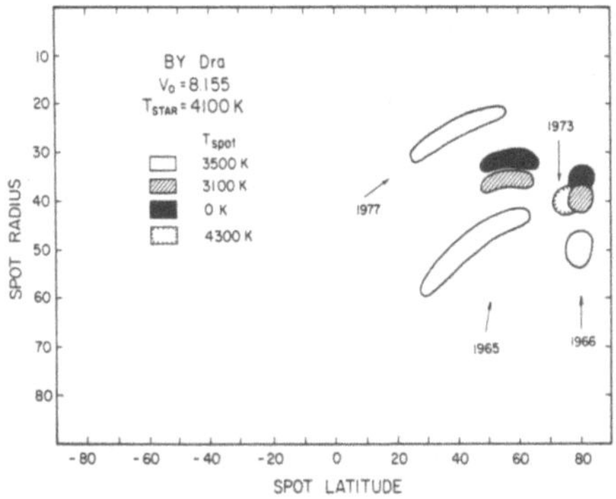


More remarkably, the period change on BY Dra associated with this latitude drift implied a latitudinal shear of the same sign, and very nearly the same value (within 30%) of that observed on the Sun (Vogt 1981a). While the reality of such period changes has been justifiably questioned by Hartmann and Rosner (1979) it is a well-documented effect in the RS CVn's and is becoming increasingly hard to explain away on the BY Dra's in terms of observational uncertainty as the photometric data base improves. The RS CVn stars show longitudinal migrations of spots in both directions, indicative of period changes to values both greater and less than the orbital one (Rodono 1981). The standard interpretation (Hall 1972) is that spotted stars, like the Sun, rotate faster at the equator than at the poles, and thus spots below the latitude of co-rotation rotate faster than the orbit (retrograde migration), and spots above this latitude rotate slower than the orbital motion (direct migration). Occasionally spots drift across the latitude of corotation and spot migration can change from retrograde to direct or vice versa (AR Lac, SS Boo, HR 1099, UX Ari ). While changes in both directions have been well observed on several RS CVn's, only poleward migration has yet been detected on the BY Dra's. Probably migration in both directions is occurring, with all spots above some intermediate latitude migrating poleward, and those below that latitude migrating toward the equator.

Often the migration rate shows discontinuities or sudden accelerations (e.g., Eaton et al. 1980). Such behavior is easily explained by subtle changes in the areal distribution of spot groups, with new spots forming as old ones decay away. Noticeable changes in the areal distribution can occur on time scales as short as five stellar rotations. This makes accurate spot modeling extremely difficult and requires intensive cooperation and coverage by spot photometrists.

## 2.5 Association of Spots with Chromospheric Emission and Flaring

There is now much evidence to suggest that large dark starspots are spatially associated with regions of enhanced chromospheric emission and flaring. Weiler (1978) found a correlation between the strength of Hα and phase of photometric minimum in several RS CVn's. Kodaira and Ichimura (1982) found periodic variations of the emission at Hβ from YY Gem whose maximum corresponded precisely with photometric minimum (dark spot most in view). Dorren and Guinan (1982b) found among a sample of G8-K7V stars an anti-correlation between strength of CaII and Hα emission and mean brightness. Baliunas and Dupree (1982) found for λ And that the phase of maximum spot visibility corresponds with enhancement of both CaII K emission and the ultraviolet transition-region lines. Vogt (1981b) found a strong correlation between Hα strength and spot visibility on II Peg. There are many other reports of such correlations between chromospheric emission and spot visibility.

Generally, the correlation, though certainly present, is much weaker than hoped for, implying that, while the emission is certainly enhanced near the dark spot, it also arises more or less globally. The relatively high mean level of emission on these stars with only subtle modulations also implies that a large fraction of the stellar surface must be covered with plage-like regions. Presumably these plage-like regions, like those on the Sun, have a shallower T(τ) relation than non-active regions and, if area coverage was great enough, could dilute the quiescent photosphere's line spectrum enough to explain the anomalous underabundances often found for the active members of RS CVn systems. This would nicely explain why only the active component ever has "anomalous underabundances", and implies a large areal filling factor for these plages.

The association of spots with flares is much harder to establish, though intriguing correlations have been noticed. Young et al. (1982)

found a strong correlation between the times of flare events and phase of photometric minimum in V471 Tau, an eclipsing binary system consisting of a white dwarf and a K dwarf which shows BY Dra-like spot behavior. Busko and Torres (1978) also found that the flaring rate of AU Mic, a spotted dM0e star, was marginally higher when the dark spot was in view.

## 3. SPOT AND ACTIVITY CYCLES

A fundamental goal of spot monitoring is to search for evidence of periodic cycles or patterns in spot activity. These cycles could be repeated patterns of any observable quantity such as spot area, spot latitude, spot longitude, magnetic field strength, chromospheric emission strength, or others. By analogy with the well-known solar cycle, one expects to find cyclical variations in at least several of these observables which can serve as a probe of the interior structure of the star as well as of the dynamo mechanisms generally thought to be responsible for the activity. Already, many hints of "clocks" running in active late-type dwarfs and subgiants have been detected. The CaII K line monitoring work of Wilson (1978), Vaughan and Preston (1980), Vaughan (1980), and Vaughan et al. (1981) has shown very clear evidence of 7 to 14 year cycles in chromospheric emission from F5-M2 dwarfs. Thus far, only the older stars of the sample show smooth cyclic variations at the K line; these are stars with rotation periods greater than about 20 days. The cycle periods in these stars are apparently uncorrelated with rotation period. There seem to be two distinct levels of activity among the F, G dwarfs: either high or low, with a deficiency of F, G dwarfs exhibiting intermediate K-line emission (the "Vaughan-Preston" gap). This gap has been explained by Durney, Mihalas, and Robinson (1981) as an abrupt transition from a multi-mode to a single-mode dynamo as rotational velocity decreases. Knobloch, Rosner, and Weiss (1981) offer another explanation for this gap in which the convection pattern switches from rolls predominantly parallel to the rotation axis, to normal convection cells as the angular velocity is decreased past a critical rotation period ($\sim$20 days). Whatever the explanation, it seems clear that these cyclic K-line variations are an important tool for understanding the global-scale properties of the star's convection and/or dynamo mechanism.

The RS CVn stars also show evidence of cycles, though the situation is much more complex, and several clocks seem to be present. Haslag (1977) has reported a very striking 30 year cycle in the wave amplitude of RT Lac. Hall (1972) found strong evidence for a 23.5 year cycle of wave amplitude on RS CVn. Bohusz and Udalski (1981) have detected an 8-10 year periodic variation in the phases of photometric minima in II Peg, a transition case between the BY Dra and RS CVn stars. Dorren and Guinan (1982a) found variations in spot area, latitude and longitude for spots on HR 1099, but thus far, none of these patterns have repeated, implying a lower limit of 5 years for any spot cycle period. Rodono (1980) cites evidence of several RS CVn's whose cycles of about 5 years are in evidence. There are numerous other reports of suspected or detected cycles of activity in the RS CVn's. The migration period of a spot in longitude, typically on the order of 5 years or more, might also represent some true cycle of internal activity. The situation is further confused by characteristic several year time scales for the appearance or disappearance of a given spot group which is often misinterpreted as evidence for a cycle.

In short, there is strong evidence for spot activity cycles ranging from as short as 5 years to as long as 30 years on the RS CVn's. At present, the longer cycles seem to me most striking and well defined. Clearly many more years of dedicated photometry will be required to fully identify the various clocks within the star.

The BY Dra stars also show very clear evidence of periodic starspot activity. Lee Hartmann and co-workers have been using the Harvard archival plate collection to study the long-term behavior of the mean light of several BY Dra and RS CVn systems. Phillips and Hartmann (1978) found evidence for a 50-60 year periodic variation in the mean light of BY Dra and CC Eri. Hartmann et al. (1981) found even stronger evidence for a 60 year cycle on Gliese 171.2A, a dK5e spotted star. I present their results in Figure 5 as an example of this type of periodic behavior. Such activity variations may occasionally cease altogether as for II Peg (Hartmann, Londono, and Phillips 1979) where the mean light was essentially constant for about 40 years before beginning to exhibit variations on both short and long time scales. This suggests that the BY Dra and RS CVn stars may undergo periods of relative spot inactivity much like the solar Maunder minimum.

It appears then that clocks of various rates reside in all the groups of stars: the active late-type dwarfs, the BY Dra spotted flare stars and the RS CVn systems and it is interesting to compare these rates and their dependences on spectral type and rotation period with theoretical predictions. For example, Belvedere et al. (1982) developed an $\alpha\omega$-dynamo operating in the convection zone which predicts an increasing dynamo cycle period with advancing spectral type from F5 to M0. When adjusted to fit the 22 year period of the Sun, this model nicely reproduces the 50-60 year periods of the dK5 and dM0 stars Gliese 171.2A and BY Dra. Robinson and Durney (1982) on the other hand predict a dynamo cycle period which decreases with advancing spectral type at a given rotation period for G0 to M5 dwarfs. Their model assumes that the rise time of a magnetic flux tube due to buoyancy is equal to the e-folding amplification time for the dynamo.

For the purposes of comparison with observation, I have assembled in Table 2 and Figure 6 a collection of known or suspected spot cycle candidates from various sources. I make no guarantee of completeness of this compilation, nor of the reality or physical interpretation of the given "cycle" periods, but I have tried to use only the cases where the cycle time (whatever it may be caused by) and rotation periods are accurately known. In addition, there may be some ambiguity by factors of two involving true cycle periods analogous to the 11 year vs. 22 year components of the solar cycle. For the lower group in Table 2, I have simply quoted the time interval between successive minima.

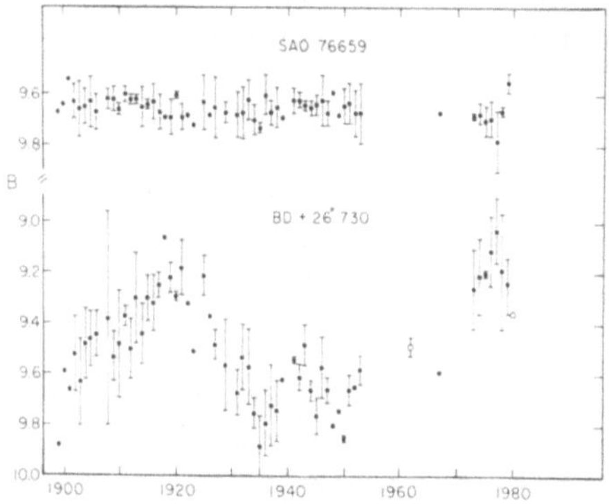


FIGURE 5

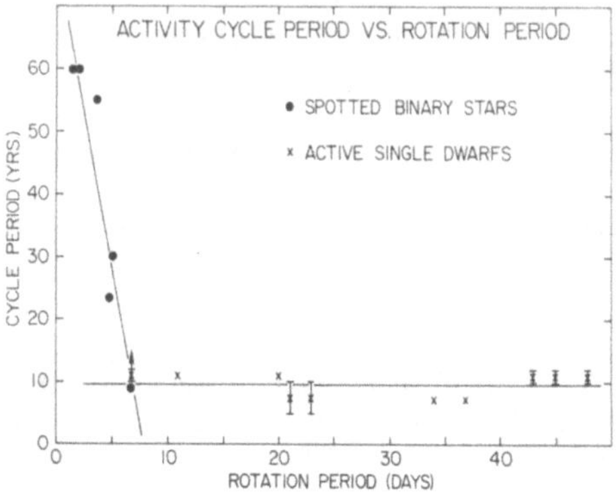


FIGURE 6

TABLE 2

Activity Cycles and Rotation Periods

| Star | Sp. Type | Binary | P(rot)$^d$ | P(cycle)$^{yr}$ | Ref. |
|---|---|---|---|---|---|
| CC Eri | dK7e | * | 1.56 | 60 | Phillips and Hartmann (1978) |
| Gl 171.2A | dK5e | * | 1.9 | 60 | Hartmann et al. (1981) |
| BY Dra | dM0e | * | 3.8 | 55 | Phillips and Hartmann (1978) |
| RS CVn | K0 IV | * | 4.8 | 23.5 | Hall (1972) |
| RT Lac | K1 IV | * | 5.1 | 30 | Haslag (1977) |
| II Peg | K2 IV | * | 6.72 | 8-10 | Bohusz and Udalski (1981) |
| HD 17925 | dK0 | | 6.9 | >10-12 | Vaughan et al. (1981) |
| HD 152391 | dG8 | | 11 | 11 | " |
| HD 4628 | dK4 | | 20 | 10-12 | " |
| HD 155886 | dK1 | | 21 | 5-10 | " |
| HD 155885 | dK1 | | 23 | 5-10 | " |
| HD 160346 | dK3 | | 34 | 7 | " |
| HD 201091 | dK5 | | 37 | 7 | " |
| HD 166620 | dK2 | | 43 | 10-12 | " |
| HD 16160 | dK4 | | 45 | 10-12 | " |
| HD 201092 | dK7 | | 48 | 10-12 | " |

Table 2 and Figure 6 reveal several interesting facts. First, all these stars occur in the G8 to M0 spectral range. Though observational selection effects are certainly present, the mechanism producing such behavior seems to prefer this mix of convection zone depth and rotational velocity. The upper group (spotted binaries) includes both RS CVn's and BY Dra's and shows a very pronounced, essentially linear correlation between cycle time and rotation period in the sense that faster rotation means longer cycle periods. These spotted binaries also show a definite tendency for cycle period to increase with (B–V) though the correlation is not as strong. This latter behavior coincides with the predictions of Belvedere et al. (1982) that cycle period should increase with (B-V). However, these authors assumed that the rotational period was fixed and this is not true among the sample in Table 2, so the comparison should not yet be taken too seriously. Secondly, the single chromospherically active dwarfs of the lower group in Table 2 show no dependence of cycle period with rotation period (as already pointed out by Vaughan et al. 1981), or with (B-V). Instead, these stars all show periods in the 7-12 year range, presumably the analog of the 11 year solar cycle.

From these data, it is clear that some process related to angular velocity turns on abruptly inside the star at a rotation period of about 6-7 days and leads to large spots and cyclic spot activity. Stars rotating slower than this show very little evidence of starspots, and essentially identical chromospheric emission cycle periods very near the solar period of 11 years. Stars rotating faster than this (all binaries in this sample) exhibit cycle periods which increase essentially linearly with angular velocity. Vaughan et al. (1981) point out that, in their sample, 10-12 year activity cycles are found almost exclusively among stars with rotation periods longer than 20 days. However, much of this is probably a selection effect since their monitoring time base is only 13 years now and they are not yet sensitive to the longer activity cycles. As the time base increases, cycles longer than the canonical 11 years length will probably be found on stars with rotation periods below 20 days. Indeed, HD 17925, which has a rotation period of only 6.9 days, shows a steady decline in activity over the 10 year baseline, implying a cycle much longer than 10-12 years.

To summarize, both the single chromospherically active K dwarfs and the spotted RS CVn and BY Dra binaries show clear evidence of dynamo cycles. For rotation periods longer than about 6-7 days the cycle length is essentially constant at $10 \pm 3$ years and is independent of both spectral type (from G8-M0) and rotation period. Stars with rotational periods shorter than this 6-7 day breakpoint show large cool spots and spot cycles whose periods increase linearly with angular velocity. The 6-7 day breakpoint corresponds to an equatorial velocity of 5 km $s^{-1}$ for a K5 dwarf and it is certainly no mere coincidence that this is exactly the critical velocity specified by Bopp and co-workers (Section 1) above which the BY Dra syndrome (spots) turns on. Perhaps, in this latter group, we are witnessing the effects of Coriolis forces which inhibit the buoyant emergence of magnetic flux and thus serve to amplify field strength as well as to increase cycle periods, (Penrod, 1982). The striking difference in Table 2 between starspot cycles of the spotted binaries and the solar-like 10-12 year cycles of the active dwarfs suggests again that starspots are not very close analogs to sunspots, but instead, like solar coronal holes and solar complexes are manifestations of a more global-scale process occurring deeper within the star than the relatively shallow mechanism responsible for spots on the Sun.

## 4. MAGNETIC FIELDS

Much effort over the past 8 years has gone into a search for the magnetic fields suspected on these active dwarfs. The original approach used the time-honored Babcock method to search for longitudinal or line-of-sight fields. The search included work by a number of researchers, including myself and all yielded essentially null results down to the 100-150 gauss level. These noble efforts are only of historical interest now so I will not dwell on them further except to say that, by analogy with the Sun (whose fields are locally bipolar on relatively small spatial scales), it is not surprising that the searches for longitudinal field all failed. Obviously, the small-scale bipolar networks spread uniformly across the disk cancel circular polarization to a very high degree. Recently, Borra and Mayor (private communication) have used the CORAVEL spectrometer at Geneva Observatory and a pockels cell analyzer to push the uncertainties of this technique down to the 3-5 gauss level. To date, after surveying over 30 stars, including RS CVn's, BY Dra's, and active single dwarfs, no evidence of coherent longitudinal fields above the 3 gauss error bars has been detected.

Though circular polarization is largely cancelled by spatially noncoherent fields, the Zeeman splitting of the spectral lines remains and manifests itself as a line broadening which correlates with Landé g factor - the splitting sensitivity of each line. The first convincing detection of this magnetic broadening was reported by Robinson et al. (1980) based on a fourier deconvolution method pioneered by Robinson (1980). Basically the method compares the profiles of two carefully selected lines which are as similar as possible in all respects except for Landé g factor where they are as dissimilar as possible. Generally the magnetically sensitive line is 2.5 times more sensitive to a given field than the insensitive line. The insensitive line is treated, to first order, as an approximation of a totally insensitive line profile to extract the degree of Zeeman broadening from the sensitive line profile. Both the average field strength and the fractional surface area covered by field can be estimated from this technique; the field strength coming from the separation of sigma components in the deconvolved profile, and the area coverage coming from the ratio of sigma to pi component line depths. Using the technique, Robinson et al. (1980) reported detection of a 1900 gauss field covering 10% of 70 Oph A, and a 2500 gauss field covering 20 to 45% of Xi Boo A.

Several groups are now apparently vigorously pursuing magnetic field measures in active dwarfs with similar techniques. One such effort (Marcy 1981), which is now quite well along and with which I am most familiar, is the work of Geoffrey Marcy, a graduate student at Lick Observatory. Marcy's results were presented at the IAU Symposium No.102 in Zurich several weeks ago, and will soon appear in the Astrophysical Journal. I would like to quickly summarize this work for you.

For the past 2 years, Marcy has surveyed a sample of 29 G0V-K5V stars for magnetic activity. This range in spectral type is bounded at the G0V end by decreasing line strength and increasing V sin i, and at the K5V end by line blending and line saturation problems. He uses the 24" coudé auxiliary telescope (CAT) and coudé echelle scanner (Soderblom et al. 1978) to obtain 100:1 signal-to-noise line profiles at 60 milli-angstrom resolution of λ6173 FeI (the magnetically sensitive line: g = 2.5) and λ6240 FeI (the insensitive line: g = 1.0). With the CAT, he can work to V = 5.5 in 3 hours per line or 6 hours per line pair. With the Shane 3-meter telescope, V = 7.5 is about the practical limit.

Figure 7 shows a fairly representative sample of observed line profiles from Marcy's work. The insensitive line in each panel is denoted by the solid line, while the crosses represent the sensitive line.

FIGURE 7

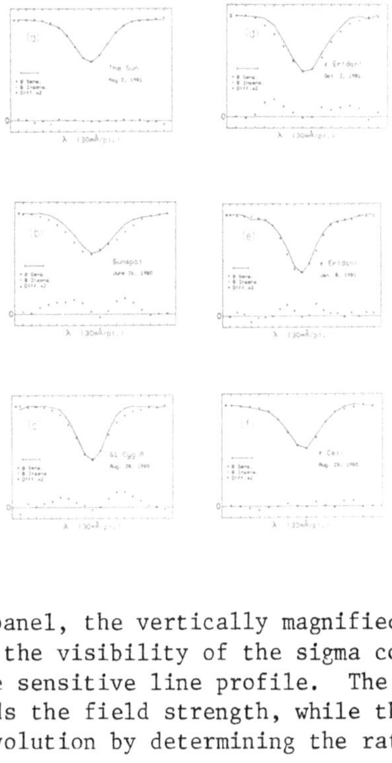

At the bottom of each panel, the vertically magnified difference profile is shown to enhance the visibility of the sigma components in the excess broadening of the sensitive line profile. The separation of these sigma components yields the field strength, while the area coverage is derived in the deconvolution by determining the ratio of sigma to pi component depths.

## 4.1 Results

Of the 29 stars surveyed, 19 show obvious fields; 30% of these in the 800-1000 gauss range with another 35% in the 1000-1500 gauss range. The derived area coverage is somewhat model dependent but area filling factors of 60 to 80% of the visible hemisphere are not uncommon. This is to be compared with filling factors of 1-2% for the Sun. The resultant magnetic surface flux for these stars is then $2\text{-}4 \times 10^{25}$ Mx, a factor of about 20 to 40 greater than the solar flux.

The fields show time variability in several stars in both field strength and area coverage. For example, in Figure 7, the two ε Eri panels show markedly different field strengths as evidenced by the change in separation of the sigma components in the magnified difference. Such variations could be evidence of magnetic flux ropes appearing at the surface and then diffusing away. Monitoring of these field variations should produce some fascinating results in the next few years.

Marcy has attempted to determine the correlation of chromospheric emission at the CaII H, K lines with B and Teff to try to differentiate between acoustic, Alfvén, slow- and fast-mode MHD waves as the origin of this emission. He finds that the corrected H, K emission flux F' is proportional to $\sqrt{BT^4}$, roughly the dependence expected for slow-mode MHD waves, and he is effectively able to rule out Alfvén and fast-mode waves as the dominant chromospheric heating mechanism.

Marcy also finds a good correlation between the X-ray flux [$\log(F_x/F_{BOL})$] and area filling factor. This is in agreement with models which involve large coronal loop structures anchored to magnetic regions at the surface to energize the corona. He also finds that $F_x/F_{BOL}$ is proportional to $B^{-1.5}$ in good agreement with Alfvén waves heating the corona.

A good correlation is also found between the log of the magnetic flux and log of the equatorial rotation velocity: $\Phi \sim V_e^{1/2}\, T_e^{-2.8}$.

Finally, I should mention that in Marcy's sample, fields are detected predominantly for those stars in the upper branch above the Vaughan-Preston gap in the Log S - (B-V) diagram (Vaughan and Preston 1980). Marcy suspects that the area filling factor drops markedly across the gap, making for quite low area coverage of fields along the lower branch of the gap and hence relative absence of detectable fields. The Sun for example has mean surface fields of 1500 gauss, but area filling factors of only 1-2% and would have totally escaped detection in such a survey.

## 4.2 Fields in the Spots and in Flares

Direct detection of fields within starspots has yet to be accomplished. Vogt (1981b) reported the possible detection of a -515 gauss field in the starspot of II Peg, but this result has yet to be confirmed. Geyer and Metz (1977) found for XY UMa an increase in scatter of the Stokes Q, U parameters when the spot passed into view and thus possible evidence of transverse fields within the spots.

Fields have also been indirectly inferred during flaring through circular polarization in radio observations. Various groups have reported circular polarization of up to 20% in the radio at various phases of RS CVn flares, probably attributable to synchrotron emission from the flaring region.

## 5. STARSPOT RESEARCH AT LICK OBSERVATORY

In addition to the magnetic field research of Marcy's, we are developing another technique at Lick for studying spots which should provide some fairly high spatial resolution pictures of spots, and thus greatly improve upon the ambiguous spot solutions obtained from modeling light curves. The method will be presented in greater detail elsewhere in this meeting, but I would like to take these last few minutes to introduce the technique and then to show you a movie of some preliminary Doppler Imaging solutions of UX Ari and HR 1099.

Basically, the technique exploits the direct correspondence between position across the disc of a rotating star and wavelength position across the rotationally broadened spectral line. For V sin i > 40, the rotational broadening dominates and the intrinsic profile can be easily decoupled from the observed profile. Any dark spot on the surface produces an apparent emission bump in any given absorption line profile which propagates across the line as the star rotates. By modeling these bumps simultaneously with the light curve, we can derive quite detailed and unique information on spot sizes, shapes, and locations. Nominally only one dimension of spatial information is present in the line profiles and for inclinations of $90^\circ$ the modeling of the bumps would be terribly nonunique. However, nature was kind enough to give us several stars at intermediate ($35^\circ$-$55^\circ$) inclinations, and on these stars we can actually determine some 2-d information by observing the spot at a range of phases. For example, a spot situated directly on the pole would produce a bump which remained stationary at line center. As the spot is moved to lower latitudes, its velocity excursion about the line center increases accordingly, so accurate latitudes can be obtained. We have even watched circumpolar spots chase each other around the far side of the pole. Photometric modeling of starspot light curves is basically sensitive to longitude extent, whereas our line profile observations are sensitive to latitude extent of the spots. The two methods are thus quite complementary and by combining them into a single modeling formalism, we expect to be able to derive some detailed pictures of starspots. Additionally, we intend to attempt Doppler Imaging in polarized light in an effort to detect fields within the starspots. Our goals are to be able to differentiate between single large spots vs. large groups of smaller spots, to determine the detailed spatial and magnetic structure of a spot (umbra/penumbra) and to follow the development and migration of selected spots through their cycles for determination of differential rotation and spot migration patterns. This work will be published in a more complete form in the Astrophysical Journal.

I wish to thank Don Penrod and Geoffrey Marcy for many interesting and helpful discussions, and Sue Robinson and Gerri McLellan for their expert help in manuscript preparation. This work has been partially supported by grant AST 79-16813 from the National Science Foundation whose support we gratefully acknowledge.

REFERENCES


Allen, C.W.: 1973, Astrophysical Quantities (London: Athlone), p.184.
Antonopoulou, E., and Williams, P.: 1980, IBVS No.1816.
Baliunas, S., and Dupree, A.K.: 1982, Ap. J., 252, 668.
Barnes, T.G., Evans, D.S., and Moffett, T.J.: 1978, M.N.R.A.S., 183, 285.
Belvedere, G., Chiuderi, C., and Paterno, L.: 1982, Astron. Ap.,105, 133.
Bohusz, E., and Udalski, A.: 1981, Acta. Astr., 31, 185.
Bopp, B.W., and Espenak, F.: 1977, A. J., 82, 916.
Bopp, B.W., and Fekel, F.: 1977, A. J., 82, 490.
Bopp, B.W., and Noah, P.V.: 1980, P.A.S.P., 92, 717.
Bopp, B.W., Noah, P.V., and Klimke, A.: 1980, A. J., 85, 1386.
Bopp, B.W., Noah, P.V., Klimke, A., and Africano, J.: 1981, Ap. J., 249, 210.
Busko, I.C., and Torres, C.A.O.: 1978, Astr. Ap., 64, 153.
Collier, A.: 1982, private communication; related results to be published in M.N.R.A.S. (in press).
Davidson, J.K., and Neff, J.S.: 1977, Ap. J., 214, 140.
Dorren, J.D., and Guinan, E.F.: 1982a, Ap. J., 252, 296.
Dorren, J.D., and Guinan, E.F.: 1982b, in Second Cambridge Workshop in Cool Stars, Stellar Systems, and the Sun, in press.
Dorren, J.D., Siah, M.J., Guinan, E.F., and McCook, G.P.: 1981, A. J., 86, 572.
Durney, B.R., Mihalas, D., and Robinson, R.D.: 1981, P.A.S.P., 93, 537.
Eaton, J.A., and Hall, D.S.: 1979, Ap. J., 227, 907
Eaton, J.A., Hall, D.S., and Henry, G.W.: 1980, IBVS No.1862.
Gershberg, R.E.: 1978, Astrophysics, 13, 310.
Geyer, E.H., and Metz, K.: 1977, Ap. Sp. Sci., 52, 351.
Guinan, E.F., McCook, G.P., Fragola, J.L., O'Donnell, W.C., Tomczyk, S., and Weisenberger, A.G.: 1982, A. J., 87, 893.
Hall, D.S.: 1972, P.A.S.P., 84, 323.
Hall, D.S.: 1976, in Multiple Periodic Variable Stars, IAU Coll.29, ed. W.S. Fitch (Dordrecht: Reidel), p.287.
Hall, D.S.: 1980, in Highlights of Astronomy, 5, 841.
Hartmann, L., Bopp, B.W., Dussault, M., Noah, P.V., and Klimke, A.: 1981, Ap. J., 249, 662.
Hartmann, L., Londoño, C., and Phillips, M.J.: 1979, Ap. J., 229, 183.
Hartmann, L., and Rosner, R.: 1979, Ap. J., 230, 802.
Haslag, K.: 1977, M.Sc. Thesis, Vanderbilt University.
Knobloch, E., Rosner, R., and Weiss, N.O.: 1981, M.N.R.A.S., 197, 45p.
Kodaira, K., and Ichimura, K.: 1982, Pub. Astron. Soc. Japan, 34, 21.
Marcy, G.W.: 1981, Ap. J., 245, 624.
Mullan, D.J.: 1976a, Irish Astron. J., 12, 161.
__________.: 1976b, Irish Astron. J., 12, 277.
Oskanyan, V.S.,Evans,D.S.,Lacy,C.,and McMillan,R.S.:1977, Ap.J.,214,430.
Penrod, G.D.: 1982, private communication.
Phillips, M.J., and Hartmann, L.: 1978, Ap. J., 224, 182.
Ramsey, L.W., and Nations, H.L.: 1980, Ap. J., 239, L121.
Robinson, R.D.: 1980, Ap. J., 239, 961.
Robinson, R.D., and Durney, B.R.: 1982, Astron. Ap., 108, 322.
Robinson, R.D., Worden, S.P., and Harvey, J.W.: 1980, Ap. J., 236, L155.

Rodono, M.: 1980, Mem. S.A. It., p.623.
Rodono, M.: 1981, in Photometric and Spectroscopic Binary Systems, ed. E.B. Carling and Z. Kopal (Dordrecht: Reidel), p.285.
Soderblom, D.R., Hartoog, M.R., Herbig, G.H., Mueller, F.S., Robinson, L.B., and Wampler, E.J.: 1978, in Proc. 4th Int. Coll. Ap., ed. M. Hack (Trieste: Trieste Ap. Obs.), p.449.
Vaughan, A.H.: 1980, P.A.S.P., 92, 392.
Vaughan, A.H., Baliunas, S.L., Middlekoop, F., Hartmann, L., Mihalas, D., Noyes, R.W., and Preston, G.W.: 1981, Ap. J., 250, 276.
Vaughan, A.H., and Preston, G.W.: 1980, P.A.S.P., 92, 385.
Vogt, S.S.: 1975, Ap. J., 199, 418.
________.: 1981a, Ap. J., 250, 327.
________.: 1981b, Ap. J., 247, 975.
Vogt, S.S., Soderblom, D.R., and Penrod, G.D.: 1982, in preparation.
Weiler, E.J.: 1978, M.N.R.A.S., 182, 77.
Wilson, O.C.: 1978, Ap. J., 226, 379.
Young, A., Klimke, A., Africano, J., Quigley, R., Radick, R., and Van Buren, D.: 1982, preprint.

DISCUSSION

Anon: Which spectral lines did you use?

Vogt: 6430Å, FeI, we use 6430 and 6439. 6439 is a lot harder, that's Ca. It is difficult because it is almost all CaII in the star and all CaI in the spot. So there is a great change in the line strength when you go from star to spot. This has to be accounted for in the model, so it gets very confusing.

van Leeuwen: There are great problems with these stars. Why are the spots always concentrated to one side of the star? (Next sentence lost). We have been observing K-type stars in the Pleiades and all K-type stars in the Pleiades vary in that way. So why are the spots always concentrated to one side (of the star)?

Vogt: I believe it is a selection effect and that there are many stars which have spots all over them but we can know nothing about them. Their uniform distribution of spots means that they do not vary. They may look a little too cool for their spectral type but other than that I don't think we will ever know.

van Leeuwen: But I have looked at K stars in the Pleiades and they all show varaibility.

Worden: I am worried about the assumption that solar and stellar spots are different in nature. There are two physical facts which must be considered here. The first is that one must examine the depth in the atmosphere at which your lines are being formed. The second relates to the physics of flux tubes. In standard models the flux expands as you go up into the atmosphere. If sufficient flux tubes are put together the brightening or darkening effect according to the models varies. In the Sun for instance if one looks deep enough in even a plage region it looks dark. If a few more flux tubes come together you get a pore. So it is very questionable whether there is a fundamental difference here.

Vogt: I'm not sure which statement you are questioning.

Worden: The statement that there is something different in the physics of what is going on in the sunspot and starspot and that you are seeing differences in umbra, penumbra and plage. I think it all has to do with the flux tube model.

Vogt: All that I am saying is that morphologically they do not seem similar to giant sunspots at all.

Uchida: Takoshi, Sakurai and I have recently proposed a new picture for the starspots. We propose the equivalent of a solar active longitude belt corresponds to the trough of the photometric wave rather than a simple spot. Spots are formed, drift across this belt and disappear.

Vogt: We are saying exactly the same thing.

Uchida: Yes, it is somewhat similar to your view. We have a poster paper on it.

Walter: I just want to say that in the Sun one does see active longitudes. You see a rotational 27 day period in the sunspot number. (Following partly lost) You even may see this in A stars. The other point is that the spots don't appear to go much earlier than A. Some work has been done on post-T Tauri stars and they appear to have spots on them. They are about $2x10^6$ years old, about the same age as the T Tauri stars. So these phenomena appear to be with stars across their entire lifetime.

Vogt: Yes. I don't think it cares about the age of the star but only about the presence of a convection zone and rotation.

Basri: With regard to the definition of a BY Dra star, Chugainov has observed this phenomenon in HD 1835 (?) so apparently a wider range of spectral types exhibit these phenomena than previously thought.

Vogt: That might be.

Basri: There might be A, F and G dwarfs showing these effects.

Vogt: That would be very nice to look for. I did not do a complete literature search to see how early a spectral type has spots. Most of them however seem to be at G8 or later.

# OPTICAL PHOTOMETRY OF FLARES AND FLARE STATISTICS

Patrick B. Byrne
Armagh Observatory, Armagh BT61 9DG, N.Ireland.

## 1. INTRODUCTION

This review discusses optical observations of flares and the results gleaned from these. Excellent reviews of flare activity have been published in the past and one might mention in particular Kunkel (1975), Gershberg (1978) and Gurzadyan (1980). In general these have been more broadly based than the present review incorporating other aspects of flare stars e.g. BY Dra variability, spectroscopic observations, etc. Since these other aspects of flare star behaviour are being covered elsewhere in this volume we will confine ourselves exclusively to optical photometry of the flare phenomenon itself. We will also confine ourselves to the solar-neighbourhood flare stars (i.e. the UV Ceti stars) since other contributors will discuss the T-Tauri, RS CVn, BY Dra and other flaring objects.

## 2. DEFINITION OF A FLARE

The first problem we address is that of defining what we mean by a flare. Fig.1 illustrates what might be called the "classical" flare light curve. It has a fast rise to maximum light followed by a quasi-

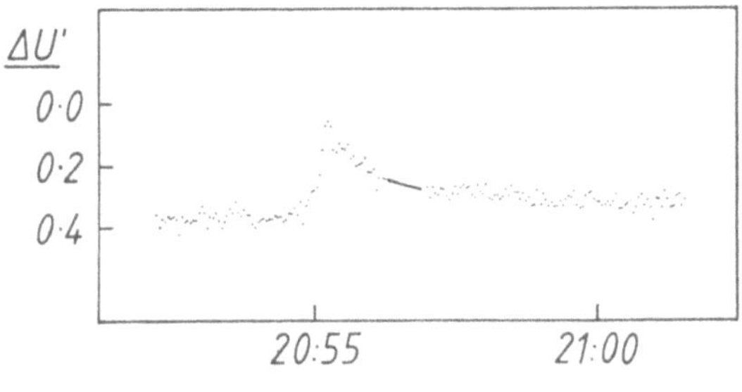


Figure 1. Flare on Gliese 867A observed by Byrne (1979).



*P. B. Byrne and M. Rodonò (eds.), Activity in Red-Dwarf Stars, 157–174.*

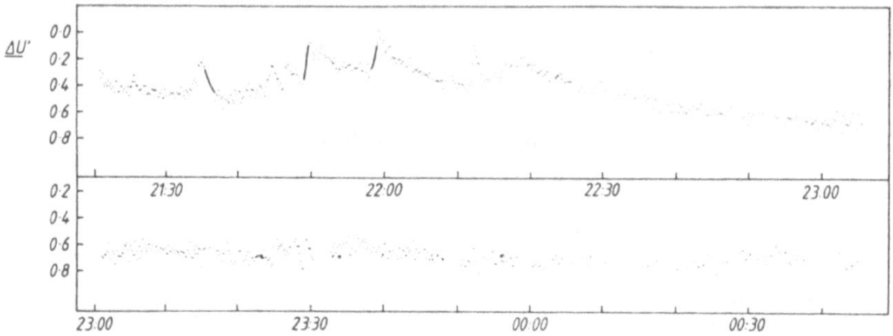


Figure 2. Flare activity on Gl867B observed by Byrne (1979).

exponential decay. The timescale of the rise is generally in the region of 1-1000 seconds, while that of the decay ranges between 1-100 minutes. Such simple flares are not in fact common. The most frequent complication is that of a break or change in slope during decay. When this occurs the initial rate of decay is rapid and then slows considerably. Secondary peaks may occur, some rivalling the primary maximum.

In the most extreme cases a bout of flare activity may be accompanied by a general rise in the "quiescent" stellar background. Superimposed may be many individual flare outbursts. This may continue for several hours before the star settles back to a steady background level (See Fig.2). Fortunately it appears that for statistical purposes counting such outbursts as single "flare events" or as a collection of individual flares does not affect the eventual outcome (see below).

The break in the rate of decay referred to above appears to have a real physical significance. Bopp and Moffett(1973) carried out simultaneous photometry and spectroscopy of flares on UV Ceti and found

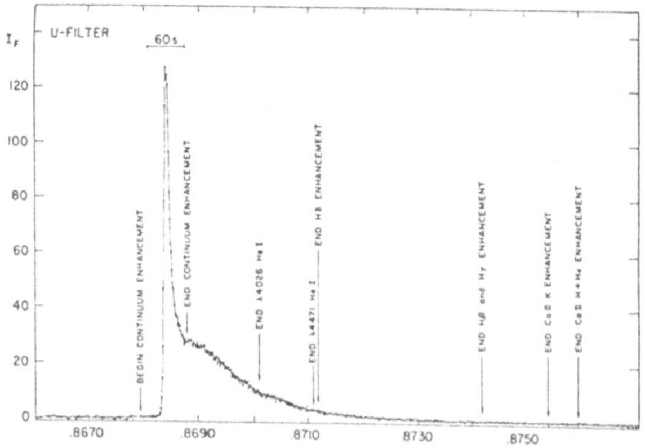


Figure 3. Spectroscopic features of a flare (Bopp & Moffett 1973).

(see Fig.3) that the rapidly varying initial phase is associated with enhancement in both lines and continuum. After the break in slope the continuum enhancement ceased and the excess radiation was provided by emission lines of Ca H + K and the Balmer lines. These results were expanded and reinforced by Moffett and Bopp (1976) and more recently by Mochnacki and Zirin (1980) using a multichannel spectrophotometer.

As the time resolution of flare observations is increased faster changes in the light curves are apparently registered (e.g. see Moffett and Bopp 1976). Kodaira *et al.* (1976) have made simultaneous observations in five wavebands with a time resolution of 0.1 secs. They found however that features on a timescale less than about 5 seconds did not correlate in the different bands suggesting that this may be the physical limit of the time variations of major features of the light curve within at least those particular flares.

## 3. COLOURS OF FLARE LIGHT

It has been obvious for a considerable time that the broad-band colours of flare light are extremely blue. Kunkel (1970) derived mean colours based on UBV photometry of 21 flares of $(\overline{U-B}) = -1.12 \pm 0.15$ and $(\overline{B-V}) = -0.0 \pm 0.22$. Moffett (1974) compiled UBV data for 409 flares on 13 different flare stars and produced (U-B) colours for 153 of them and (B-V) for 77. The lesser number of (B-V) colours is due to the relatively smaller amplitude of flares in the V band. Mean colours from Moffett's data are $(\overline{U-B}) = -0.88 \pm 0.31$ and $(\overline{B-V}) = +0.34 \pm 0.44$, somewhat redder than Kunkel's means but agreeing within the formal errors. The dispersion about these mean values for individual stars and individual flares is considerable. For instance Flare No.1 observed by Flesch and Oliver (1974) had peak colours (U-B) = +0.29 and (B-V) = +0.56. Lacy *et al.* (1976) in analysing Moffett's data derived the following relationships between flare energies in the Johnson bands.

$$E_U = (1.20 \pm 0.08)E_B = (1.79 \pm 0.15)E_V.$$

In terms of flux per unit frequency interval this is quite a flat spectrum increasing somewhat towards the UV.

Several authors have attempted to trace the time-evolution of the colours of flare light. Kodaira *et al.* (1976) investigated two flares of very different amplitudes on EV Lac in five non-standard wavebands in the optical. They found that the spectrum was flat and did not appear to evolve with the progress of the flare. Furthermore although the flares were of very different amplitudes their colours were indistinguishable. Similar conclusions were reached by Cristaldi and Longhitano (1979) from 9 flares on 4 different flare stars observed in Johnson UBV. Walker (1981) observed a large flare also in UBV on Prox. Cen. (B-V) and (U-B) colours remained constant over a considerable duration of the flare and only late in the decay did the colours

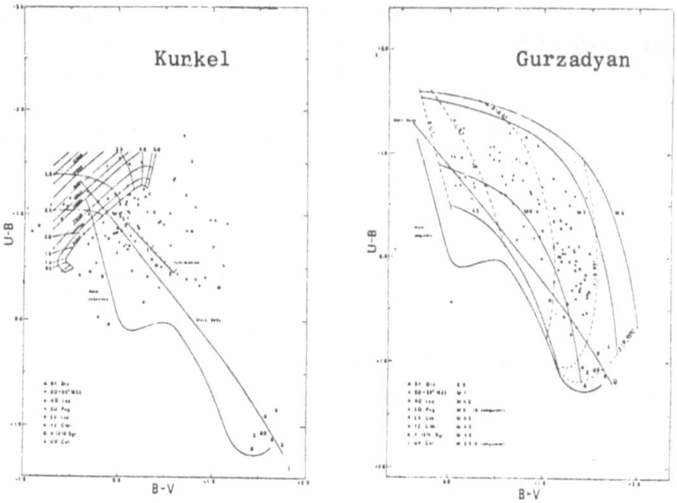


Figure 4. Observed and model flare colours (Cristaldi & Rodonò 1975).

begin to evolve towards the red.

These broadband colours have been used to test the validity of various flare models. Cristaldi and Rodonò (1975) have superimposed grids on the two-colour (U-B)-(B-V) diagram, based on the models by Gershberg (1967), Kunkel (1970) and Gurzadyan (1972). They conclude that the former two are ruled out by their data while Gurzadyan's inverse Compton model can be made to fit the data (see Fig.4). The data of Kodaira et al. likewise eliminate Kunkel's model. Indeed they conclude that Balmer line emission contributes but little to the total flare output. Gurzadyan (1979) has analysed this latter data and shown it to be consistent with his model. These conclusions contrast sharply with that reached by Mochnacki and Zirin (1980) who substantially agree with the model of Kunkel.

If the above discussion illustrates anything it illustrates the wide variety of types of flare observed. This variety is reflected both in the dispersion in the colours of flare light and in the relative importance of the fast and slow components of the light curve. Broadband photometry is probably a very crude tool for making physically meaningful interpretations from the data. This is especially so in the Johnson UBV system where the U-band, in which flares are most easily detected (Kunkel 1973), includes the higher members of the Balmer series, the Balmer continuum, Balmer jump and Ca H & K. Yet photometry has enormous practical advantages over spectroscopy. So attempts to circumvent this problem by using non-standard photometric systems such as those employed by Kodaira et al., Mochnacki and Zirin, and Byrne et al. (1982a) are to be pursued wherever possible.

## 4. TIME DISTRIBUTION OF FLARES

Early work on the time distribution of flares suggested a variety of possible periods (Andrews 1966, Osawa et al. 1968, Chugainov 1969). When effects of aliasing with the observing interval and the daily cycle of observation are taken into account these periods are shown to be spurious. There are nevertheless compelling reasons for searching for such periodicities. This is particularly true of the BY Draconis variables. These stars are believed to possess large starspots (Vogt, this volume). By analogy with the Sun active regions would be associated with such spots and would in turn produce flares. Thus one might expect a higher probability of registering flares when the spot is facing towards the observer.

Oskanyan and Terebizh (1971), Kunkel (1973) and Lacy et al.(1976) all examined large flare data sets and could not distinguish the resulting time distribution from Poisson. This is true whether individual flare peaks or clusters of such peaks (called "flare events" by Moffett (1974)) are considered. All, however, remarked on the possible existence of "precursor flares" i.e. pairs of flares closely spaced in time but whose light curves do not overlap; furthermore the preceding flare is generally the smaller in amplitude by a factor of 3 or more (Fig.5).

Recently Pazzani and Rodonò (1981) have examined this question of "precursor" events. They based their analysis on extensive data on three flare stars observed at Catania, UV Ceti, EQ Peg and YZ CMi. For UV Ceti the times of 424 flare events were analysed. Significant departures from Poissonian behaviour were found for (a) the distribution of time intervals between successive events, (b) the number of events within a fixed time interval and (c) the number of events observed in each complete observing session. The probability that these effects arise by chance lie between 0.1% and 1%. So the physical reality of precursors would appear confirmed.

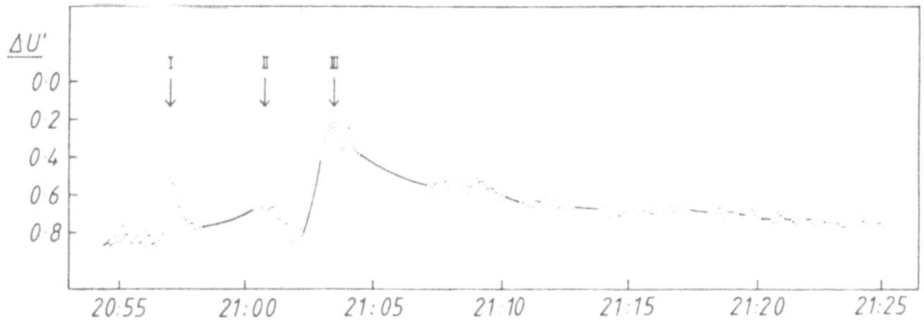


Figure 5. Example of a precursor flare (I) from Byrne (1979).

Lacy et al. (1976) also tested the hypothesis that flare stars behave like relaxation oscillators i.e. that the time between successive flares is related to the energy of the preceding one. They reject the hypothesis on the basis of their data but point out that the result is not necessarily definitive since there may be many active regions on any given star giving rise to flares at any one observing interval. Furthermore, their data was collected over more than one observing season in which time individual active regions would alter or die away completely.

It is of interest to inquire whether on the BY Draconis variables in particular some kind of flare periodicity might reveal itself. Busko and Torres (1978) examined flare activity on a number of known BY Dra stars. They remarked on the unusually large scatter in (U-B) on the star AU Mic which, at the time of their observations, was the largest amplitude BY Dra variable known. A larger scatter in (U-B) than expected on the basis of photon statistics is a common phenomenon of active flare stars (see e.g. Byrne and McFarland 1980) and is generally ascribed to low level flare activity from events not individually distinguishable. The scatter in Busko and Torres (U-B) light curve for AU Mic shows a clear correlation with the mean V light curve in the sense that the (U-B) scatter (= low level flare activity) is greatest at V minimum i.e. greatest spot visibility (Fig.6). Furthermore exceptionally strong flares may occasionally correlate with maximum spot visibility, e.g. the large flare observed by Byrne et al. (1982b) on Gliese 182. So careful examination of the BY Draconis flare stars in particular seems warranted.

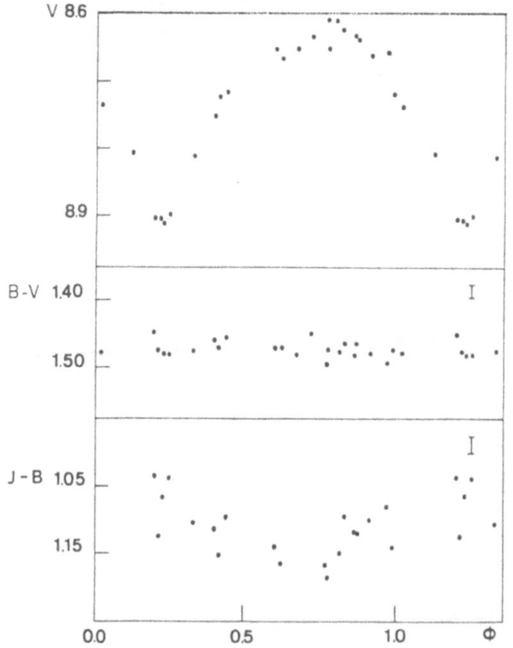


Figure 6. BY Dra-type variability in AU Mic (Busko & Torres 1978).

## 5. THE TIME-AVERAGED ENERGY RELEASED IN FLARES

Two different methods of examining time-averaged activity are currently in use. Kunkel (1973, 1975) postulates a standard form of light curve similar to that in Fig.1. This has the advantage that it is entirely specified by its peak magnitude U(peak) and by the time constant of the quasi-exponential decay, $T_q$. The second method has been developed by Gershberg (1972) and Lacy et al. (1976). They evaluate the integral of the observed lightcurve and refer this to the pre-flare brightness of the star, which is assumed equal to its mean luminosity. Most subsequent work has adopted one of these two forms of analysis and so it is appropriate that we examine them here.

Kunkel (1973) demonstrated that irrespective of aperture there is considerable detection advantage in using the Johnson U-band and this is largely what subsequent observers have used. In analysing his own extensive data Kunkel showed that for a given star the time taken to reach a fraction (1/q) of the flare's peak brightness $T_q$ is independent of U(peak). This introduces a considerable simplification since for any given star a flare's energy is proportional to U(peak) only. In fact the activity of the star can now be specified in terms of a characteristic magnitude $U_0$ such that flares with U(peak) brighter than this will occur at a rate 1 $hr^{-1}$. He then showed that the rate of occurence of flares of U(peak) = U is given by $R(U) = \exp\left[a(U-U_0)\right]$ per hour. The constant a determines the relative frequency of large and small flares and Kunkel showed that it has a value indistinguishable from 1 for all of the stars he observed.

The rate equation can now be integrated between suitable limits to derive the time-averaged energy release in flares. These limits must be finite or the integral will be infinite. So Kunkel referred to his own spectroscopic model of stellar flares (Kunkel 1970) to derive an upper limit. His model indicates mean parameters for stellar flares derived on the assumption of Hydrogen recombination as the dominant source of optical radiation, which in turn imply that U(peak) reflects the area of the flare on the stellar surface. Since a flare cannot be bigger than the visible hemisphere of the star, he adopts the corresponding U(peak) as an upper bound to the rate integral. The lower limit is derived directly from observations of the mean quiescent U magnitude on each night of observation. The scatter in these data are attributed to flares which are individually unobservable and this allows an empirical estimate $U(peak)_{min}$ on the assumption that the constant a above stays close to 1 for the smallest flares.

Kunkel took this analysis one stage further. Using his spectroscopic model he calculated a bolometric correction to the energy of each flare which when corrected for distance could be normalized to the bolometric luminosity of the underlying star. For instance, Kunkel's own observations of UV Ceti yield an estimate of 0.4% of $L_{bol}$ emitted as flare light. In this way Kunkel (1973) was able to suggest that all flare stars emitted less than $1\%L_{bol}$.

Kunkel's treatment has several serious deficiencies. Shakovskaya (1974) showed that the rate of decay in flares is related to their peak magnitude. Therefore in Kunkel's notation a standard value of $T_q$ cannot be used. Furthermore multicolour observations of flares show a wide spread in flare colours which do not agree with Kunkel's model (see section 3) while recent IUE observations of flares (Butler et al. 1981; Bromage, Private Communication) confirm this. So even on a given star flare temperatures may vary widely.

Gershberg (1972) and Lacy et al. (1976) directly integrated the flux under the observed flare light curve and referred to the star's pre-flare brightness to calculate the energy. While there are difficulties with this approach, (Where does the flare begin and end? What if the pre-flare brightness is itself undergoing change (see Byrne 1979; Rodonò et al. 1979)) it does not depend on a chain of inter-relationships as does that of Kunkel. Use is made of cumulative frequency diagrams in determining total flare energy. An example is shown in Fig.7. Gershberg drew attention to the fact that there is a linear portion to these diagrams precisely in the range of flare energies where S/N is best. Above and below this range the slope changes. His proposal was that these changes of slope reflected respectively a saturation effect at the largest energies and the effects of detection threshold at the lower energy end. If the linear relation extended to unobserved flares then the time-averaged flare energy could now be determined as the integral under this line. $E_{max}$ is indicated by the upper turnover and $E_{min}$ as a physical limit of $\sim 10^{26}$ ergs similar to that suggested by Sturrock and Coppi (1966) for solar flares.

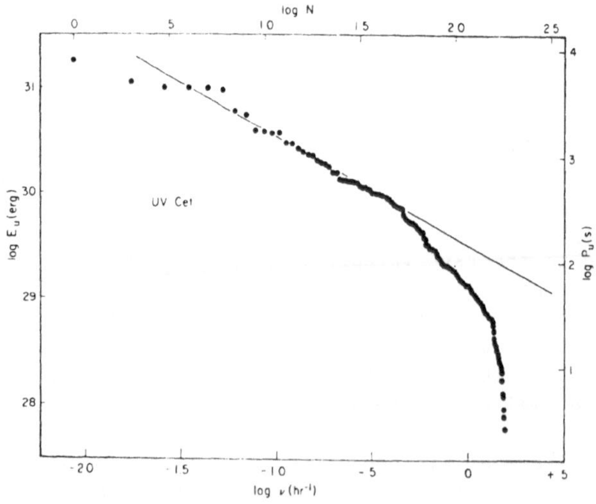


Figure 7. Cumulative frequency diagram from Lacy et al. (1976).

Both Kunkel and Lacy et al. conclude that the absolute energy in flares decreases with the luminosity of the star. Kunkel (1973) suggests that there exists an upper envelope to flare activity defined by

$$M_{U,0,max} = 10.8 + 0.408M_V$$

where $M_{U,0,max}$ is the value of U(peak) expressed as an absolute magnitude. This envelope corresponds to a star expending $\sim 1\% L_{bol}$ in flares.

Lacy et al. find similar relationships for both the mean energy $\overline{E}_U$ of flares and $L_U'$ the time-averaged rate of release of flare energy: both measures are in the U-band. For instance in the latter case Lacy et al.'s relationship takes the form

$$\log L_U' = (0.60 \pm 0.05) \log q_U + (10.3 \pm 1.0)$$

wherein $q_U$ is the quiescent luminosity of the star in the U-band. Several authors have, since then, examined stars not included in Lacy et al.'s original sample and found good agreement with their results (see e.g. Byrne and McFarland 1980). A similar dependence on stellar luminosity was found by Cristaldi and Rodonò (1975).

## 6. OTHER CORRELATIONS WITH LUMINOSITY

Although Kunkel (1973) found that the constant a in his rate equation was close to 1 for all his stars (i.e. that flares of every amplitude contributed equally to the time-averaged flare energy output of the star) Gershberg (1972) and Lacy et al. (1976) found that the relative contribution to total flare energy output of large and small flares varied with spectral type. This relative contribution can be gauged from the slope β of the cumulative frequency diagrams (c.f. Fig.6). Lacy et al. found a loose correlation between β and $q_U$, the quiescent U-band luminosity as follows.

$$\beta = (0.11 \pm 0.02) \log q_U - (3.9 \pm 0.5).$$

The relationship is in the sense that the more luminous stars emit a greater proportion of their time-averaged flare energy in large flares. Kunkel also noted this but attributed it to selection effects. Observations on bright stars, however, with good signal-to-noise have shown it to be real (see e.g. Byrne 1979 and Byrne and McFarland 1980).

Attention has previously been drawn to the effect of mixing data of different detection thresholds together in determining β even for a single star (Byrne and McFarland 1980). As pointed out in the previous section the detection threshold shows itself as a turn-down from the linear portion of the graph. Mixing data will result in several breaks in slope occurring which, if ignored, will lead to an apparent steepening of the cumulative frequency graph. Thus values of β should be determined separately for small sets of data taken under conditions of uniform detection threshold. We illustrate this effect in Fig.8 where different sections of the graph with the same slope (broken lines) are separated by the breaks already referred to. This latter slope is different from the overall slope determined by Lacy et al. (solid line).

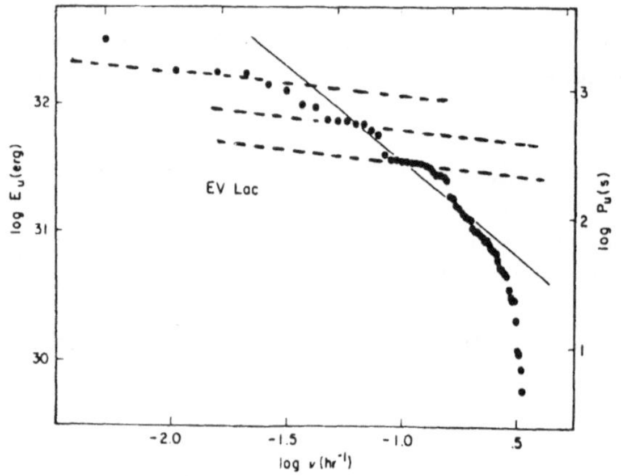


Figure 8. Cumulative frequency diagram from Lacy et al. (1976)(See text).

A relation between the durations of flares and the star's luminosity is well established, in the sense that more luminous stars give rise to flares which on average last longer. Kunkel (1973) quantified this using as parameters the time taken to reach half the peak brightness $T_{0\cdot5}$ and the star's absolute V magnitude $M_V$. He suggests that

$$\overline{\log T_{0\cdot5}} \sim -0.15M_V + (1.61\pm0.17).$$

Results by Busko and Torres (1978) on 4 flare stars are in agreement with this relation.

## 7. VARIATIONS IN ACTIVITY WITH TIME

Since the proposed mechanisms for flare star activity are like those for the Sun it has been natural to seek a periodic behaviour analogous to that of the solar, 11-year spot/activity cycle. The definitions of activity parameters defined in the previous sections may be used for this purpose. The most direct suggestion yet of such a cycle was presented by Kunkel (1975a) for the star Wolf 630 (Gliese 644A). On the assumption that the constant a in Kunkel's (1973) rate equation is 1 data on 120 U-band flares have been used to determine $U_o$, the peak magnitude at which the rate of flaring is 1 $hr^{-1}$. $U_o$ reportedly varies in phase with the star's orbital period of 1.715 years (Fig.9). The effect is only marginally significant however and in view of our discussion in Section 5 should be treated with caution.

Busko and Torres (1978) observed AU Mic in 1974 and compared their cumulative frequency diagram for the star with that from Kunkel's (1973)

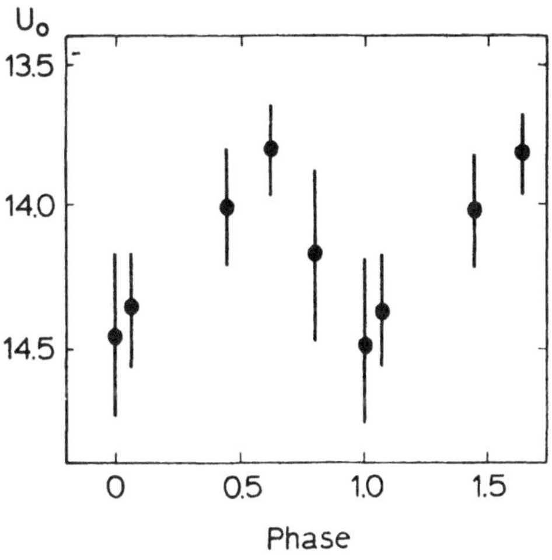


Figure 9. Variations in flare activity for Wolf 630 (Kunkel 1975a).

1970 data. A change in slope between the two is noted but the authors dismiss it as possibly of statistical origin. Byrne and McFarland (1980) compared two season's data on both components of the star Gliese 867 and concluded that no change was apparent in their activity levels.

There are stars within the general population which show flare activity but at a level very much lower than the "standard" UV Ceti stars. Kunkel (1973) drew attention to the low activity of the stars Gliese Nos.54.1, 229 and 752B. For these stars Kunkel measured $M_{U,0}$ = 18.3, 18.4 and an upper limit for Gl 752B of 21.2. Kunkel's own mean relationship for the most active stars would predict $M_{U,0}$ = 15.9, 14.6 and 18.4 respectively. Pettersen and Griffin (1980) have discussed photometric flares on Gliese 15A, 424 and 725A and showed that these too have much depressed flare activity with respect to the most active objects. Byrne (1981) has shown that Gliese 825 should be similarly classified. All of these stars are old disk population stars according to the criteria set up by Veeder (1974) except Gliese 229. It is therefore tempting to conclude that flare activity is a function of stellar age and so in the old disk population activity is reduced. Walker (1981), however, has compared the activity of Prox. Cen (Gliese 551) with the active flare stars and found it too markedly deficient in spite of its well established youth. Byrne et al. (1980) reached a similar conclusion based on chromospheric emission-line strengths measured with IUE.

Membership of a binary system is sometimes cited as a means of enhancing or prolonging the flare active phase in the life of a star. Rodonò (1978) has examined flare activity in 9 binary systems and has been unable to find any correlation with semi-major axis over a range of nine orders of magnitude in separation. Furthermore Gl 182 has been found to be a high-activity star from both IUE spectra and optical flare photometry (Byrne et al., 1982b) in spite of it being old disk population and a single star (Bopp and Fekel 1977). EQ Vir (Gliese 517; Ferraz-Mello 1972) and AU Mic (Gliese 803; Kunkel 1973) are other active single stars. So the link between flare activity and binary nature is not a simple one.

We are forced to conclude therefore that if flare activity varies with time the data presently available do not make it unambiguously clear.

## 8. PRE-FLARE DIPS/INCREASES

For some years pre-flare dips in intensity have been reported in the journals (Rodonò et al. 1979, Cristaldi et al. 1980 and references therein). Examples of two extreme cases are seen in Figs.11 (Giampapa et al. 1982) and 10 (Flesch and Oliver 1974). The former exemplifies a rare large U-band dip. That in Fig.10 is observed only at the red end of the spectrum and indeed corresponds in time to a pre-flare brightening in the U-band. Two dips are also seen in the post-flare decline in Fig.10 and at least one of them corresponds to U-band brightening.

Statistical studies of the occurrence of pre-flare dips and rises have been carried out by Shevchenko (1973), Bruevich et al. (1980) and Cristaldi et al. (1980). Shevchenko and Bruevich et al. have found that dips preferentially precede flares of small amplitude, while rises appear preferred in large amplitude flares. Bruevich et al. also conclude that dips occur more commonly at near IR wavelengths being observed in up to 70% of cases. Cristaldi et al. examined B-band data for 277 flares on 7 stars and observed that pre-flare rises are about 2-3 times more common than dips. They agree that dips occur preferentially on smaller

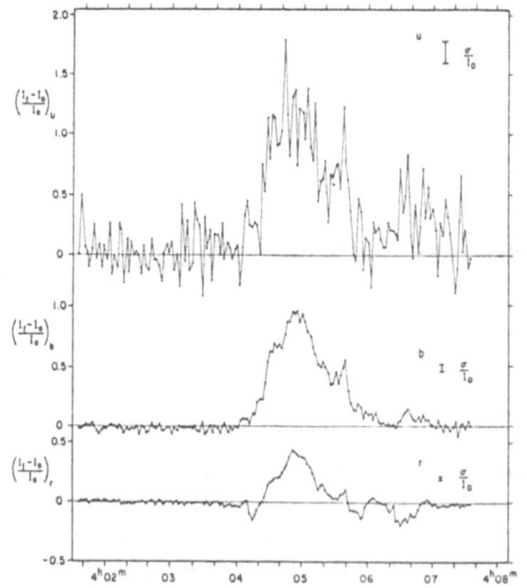


Figure 10. Three-colour data on dips (Flesch & Oliver 1974).

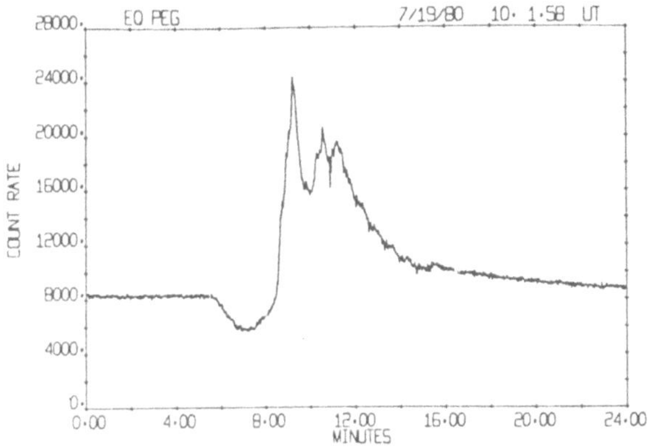


Figure 11. U-band dip from Giampapa et al. (1982).

flares. They also comment that pre-flare rises occur on flares with longer rise-times. This may of course be a result of longer rise times in large amplitude flares. From those flares for which multicolour observations are available it appears that dips are redder than the quiescent star's light (see also Flesch and Oliver), a result in agreement with Bruevich et al.'s findings.

Model predictions of preflare dips have been made by Gurzadyan (1968, 1980), Mullan (1975) and Grinin (1976). We do not discuss the inverse-Compton model of Gurzadyan since it predicts negative flares in the 1-2 μm spectral region which are the inverse of the optical curve and coincident with it in time. Mullan's model proposes that the "red" preflare dips are caused by episodes of Hα absorption. Bruevich et al.'s infra-red filter does not include Hα however and so Mullan's proposal cannot be invoked to explain their data. Grinin has put forward a model based on increased $H^-$ continuum absorption resulting from impulsive heating of the chromosphere by a downflow of material, which when it reaches the photosphere causes the optical flare. This predicts the correct time and spectral sequence for the observed events and deserves further investigation.

The large U-band flare observed by Giampapa et al. (1982) on EV Lac seems to be of a very different kind to those discussed above. It is possible that an episode of increased Balmer absorption similar to that discussed by Mullan could affect the U-band sufficiently to cause this kind of flare.

## 9. POLARIZATION

Since the flare phenomenon is magnetically-related attempts were made quite early on in the history of flare star observing to register polarization both in the quiescent state and during flares. Oskanyan (1964) and Kubicela and Arsenijevic (1966) recorded detections of polarization of the order of several percent. Zappala (1969) later investigated ten stars to a much higher sensitivity without positive effect. Reports of polarization have persisted in the literature however (see e.g. Koch and Pfeiffer 1976). Recently a series of high precision measurements have been described (Pettersen and Hsu 1981; Clayton and Martin 1981) which place upper limits of a few tenths of a percent on any intrinsic polarization present in the flare stars. At present therefore we must regard the presence of polarization in flare stars as unproven at least in the quiescent state.

## 10. CONCLUSION

Although it is true that progress in understanding the physics of flare stars will probably come from the increasingly sophisticated observations outside the optical wavebands, optical photometry is the data which can be gathered with the greatest ease. As a result investigations requiring continuous monitoring over a long timebase will continue to rely heavily on optical photometry. This is probably true also of the need for ground-based data to compliment space-borne instruments (see e.g. papers presented at this meeting by Andrews et al. and Butler et al.). One would wish to encourage, however, a move away from UBV photometry towards somewhat narrower bands with more discriminating power. This is especially important in the U-band region of the spectrum.

## 11. ACKNOWLEDGEMENTS

I would like to express my gratitude to Mr. J.McFarland of Armagh Observatory for his typing of this manuscript under very difficult circumstances.

## REFERENCES

Andrews,A.D. 1966. Publ. Astron. Soc. Pacific,Vol.78,p.324.
Bopp,B.W. & Moffett,T.J. 1973. Astrophys. J.,Vol.185,p.239.
Bopp,B.W. & Fekel,F. 1977. Astron. J.,Vol.82,p.490.
Bruevich,V.V. & 6 co-authors. 1980. Izv. Krymskoi Astrofiz. Obs., Vol.61,p.90.
Busko,I.C. & Torres,C.A.O. 1978. Astron. Astrophys.,Vol.64,p.153.
Butler,C.J., Byrne,P.B., Andrews,A.D. & Doyle,J.G. 1981. Mon. Not. R. astr. Soc.,Vol.197,p.815.
Byrne,P.B. 1979. Mon. Not. R. astr. Soc.,Vol.187,p.153.
Byrne,P.B. 1981. Mon. Not. R. astr. Soc.,Vol.195,p.143.
Byrne,P.B. & McFarland,J. 1980. Mon. Not. R. astr. Soc.,Vol.193,p.525.
Byrne,P.B., Butler,C.J. & Andrews,A.D. 1980. Irish Astron.J.,Vol.14,p.219.
Byrne,P.B., Bakker,R.,Thè,P.S. & Black,E. 1982a. Paper in prep. for Mon. Not. R. astr. Soc.
Byrne,P.B., Doyle,J.G. & Butler,C.J. 1982b. Paper in prep. for Mon. Not. R. astr. Soc.
Chugainov,P.F. 1969. IAU Coll. "Non-Periodic Phenomena in Variable

Stars", Ed.L.Detre. Academic Press, Budapest.
Clayton,G.C. & Martin,P.G. 1981. Astron. J.,Vol.86,p.1518.
Cristaldi,S. & Rodonò,M. 1975. IAU Symp.No.67,"Variable Stars and Stellar Evolution",p.75. Ed.V.E.Sherwood and L.Plaut. D.Reidel, Dordrecht, Holland.
Cristaldi,S. & Longhitano,M. 1979. Astron. Astrophys. Suppl.,Vol.38,p.175.
Cristaldi,S., Gershberg,R.E. & Rodonò,M. 1980. Astron. Astrophys.,Vol.89, p.123.
Ferraz-Mello,S. 1972. IAU Circ.No.2482.
Flesch,T.R. & Oliver,J.P. 1974. Astrophys. J.,Vol.189,p.L127.
Gershberg,R.E. 1967. Astrofizika,Vol.3,p.127.
Gershberg,R.E. 1972. Astrophys. Sp. Sci.,Vol.19,p.75.
Gershberg,R.E. 1978. Astrofizika,Vol.13,p.553.
Giampapa,M.S. & 6 co-authors. 1982. Astrophys. J.,Vol.252,p.L39.
Grinin,V.P. 1976. Izr. Krymskoi Astrofiz. Obs.,Vol.55,p.179.
Gurzadyan,G.A. 1968. Astrofizika,Vol.4,p.154.
Gurzadyan,G.A. 1972. Astron. Astrophys.,Vol.20,p.145.
Gurzadyan,G.A. 1979. Astrofizika,Vol.15,p.335.
Gurzadyan,G.A. 1980. "Flare Stars". Publ. Pergamon Press,Oxford.
Koch,R.H. & Pfeiffer,R.J. 1976. Astrophys. J.,Vol.204,p.L47.
Kodaira,K., Ichimura,K. & Nishimura,S. 1976. Publ. Astron. Soc. Japan, Vol.28,p.665.
Kubicela,A. & Arsenijevic,J. 1966. Publ. Astr. Obs. Beograd,No.12.
Kunkel,W.E. 1970. Astrophys. J.,Vol.161,p.503.
Kunkel,W.E. 1973. Astrophys. J. Suppl.,Vol.25,p.1.
Kunkel,W.E. 1975. IAU Symp.No.67,"Variable Stars and Stellar Evolution", p.15. Eds.V.E.Sherwood and L.Plaut. D.Reidel, Dordrecht, Holland.
Kunkel,W.E. 1975a. IAU Symp.No.67,"Variable Stars and Stellar Evolution", p.67. Eds.V.E.Sherwood and L.Plaut. D.Reidel, Dordrecht, Holland.
Lacy,C.H., Moffett,T.J. & Evans,D.S. 1976. Astrophys. J. Suppl.,Vol.30, p.85.
Mochnacki,S.W. & Zirin,H. 1980. Astrophys. J.,Vol.239,p.L27.
Moffett,T.J. 1974. Astrophys. J. Suppl.,Vol.29,p.1.
Moffett,T.J. & Bopp,B.W. 1976. Astrophys. J. Suppl.,Vol.31,p.61.
Mullan,D.J. 1975. Astron. Astrophys.,Vol.40,p.41.
Osawa,K., Ichimura,K., Noguchi,T. & Watanabe,E. 1968. Tokyo Astron. Bull., 2nd Ser.,No.180.
Oskanyan,V.S. 1964. Publ. Astr. Obs. Beograd,No.10.
Oskanyan,V.S. & Terebizh,V.Y. 1971. Astrofizika,Vol.7,p.83.
Pazzani,V. & Rodonò,M. 1981. Astrophys. Sp. Sci.,Vol.27,p.347.
Pettersen,B.R. & Griffin,R.F. 1980. Observatory,Vol.100,p.198.
Pettersen,B.R. & Hsu,J.C. 1981. Astrophys. J.,Vol.247,p.1013.
Rodonò,M.: 1978, Astron. Astrophys. Vol.66, p.175
Rodonò,M., Pucillo,M., Sedmak,G. & de Biase,G.A. 1979. Astron. Astrophys.,Vol.76,p.242.
Shevchenko,G.G. 1973. Astron. Zirk.,No.792.
Sturrock,P. & Coppi,B. 1966. Astrophys. J.,Vol.143,p.3.
Veeder,G.J. 1974. Astron. J.,Vol.79,p.702.
Walker,A. 1981. Mon. Not. R. astr. Soc.,Vol.195,p.1029.
Zappala,R.R. 1969. Publ. Astron. Soc. Pacific,Vol.81,p.433.

DISCUSSION

Evans: I will be glad to take notice of these criticisms of Lacy, Moffett and Evans. I had better go tell Lacy and Moffett and have a look at what you say. What I really wanted to remark upon, however, was that there are very few observations of correlations between flaring and spot variations. Pettersen and I are currently together some work by a student called Gary Kern in which two stars which showed spot variations were monitored for a considerable length of time for flares. A reasonable number of flares were found and so far as it goes a reasonably good correlation between flare incidence and the appearance of spots was found. There are some complications but that is the general conclusion. As to precursors with regard to which one always feels somewhat biblical, Lacy who was an expert statistician convinced us that there was not anything in it. However I must relate a story about this phenomenon which does not occur and how we actually used it. Bopp, in an earlier manifestation was working with Moffett doing simultaneous spectroscopy and photometry of flares. Moffett working photometrically, when he saw a precursor, would telephone Bopp on the other telescope and tell him to start a spectroscopic exposure. This is in fact how a good chunk of his (Bopp's) doctoral thesis was obtained i.e. using a phenomenon which does not exist. (laughter)

Byrne: So he owes his doctorate to the existence of precursors.

Kodaira: You said that there is no colour evolution of flares. How closely would you place limits on such colour evolution? I, myself, have made five channel simultaneous observations with good time resolution and I noticed that all the energy fluxes remained remarkably constant until very near the end of the flare. It has been said by many broad-band observers and also cited by theoreticians that flare colour evolves in some certain direction in the two-colour diagram in agreement with certain models.

Byrne: One must pay attention to the kind of errors involved in these colour determinations and they can be quite large as you know yourself. For instance, the amplitudes of flares decrease very rapidly as one goes to the red and so the signal-to-noise can also decrease. As a result (B-V) colours can be quite difficult to determine over any length of time. It is obvious that the colour of the flare light must eventually change and evolve towards that of the undisturbed star. The evidence that we have, for istance in Walker's observations, would suggest that within a tenth of magnitude the colours of the flare light over the maximum and for a considerable time after maximum remain constant.

Vaiana: That includes the spike portion?

Byrne: Yes, it does.

Haisch: I have two questions. Based on your knowledge of the frequency of occurrence of flares how often you expect the multiple spikes to be the superposition of flares going off in different parts of the stars. Secondly, do you believe in negative flares? I have seen a few references to these in the literature.

Byrne: On the second question first; if you refer to the written version of my review you fill find a discussion of negative preflare dips. They are almost undoubtedly real. They have been observed very widely.

Vaiana: And frequently?

Byrne: Yes, frequently, particularly in the redder photometric bands. With regard to the occurrence and reality of multiple peaks, this may depend on the star you are looking at. If you are looking at a late-type, very active star which flares frequently then the chances of superposition are very much greater. When however one observes an MO star, which would flare relatively infrequently and with greater energy, then multiple peaks would almost certainly be connected events. I would quote the example of Gliese 182 which I have already mentioned. We have observations of multiple peaks in this star although if flares on average only once every 20 hours.

Oskanian: Recently two papers were published in Astrofizika concerning BY Dra itself. These papers were based on 5 years data gathered at Konkoly Observatory, Budapest and at Byurakan Observatory. We have 20 flares in all. We could not find any correlation between the time of occurrence of flares and the phase of the slow light variations.

Byrne: How large was the amplitude of the BY Dra variability at that time?

Oskanian: About 0.1 or 0.2 magnitudes.

Byrne: I would suggest that it may be best to look at a BY Draconis variable which has the maximum possible amplitude and that this would imply a large concentration of spots and associated flaring active regions on one hemisphere.

Oskanian: At that moment this was the amplitude and it was impossible to ask the star to have a greater amplitude.

Bromage: In deference to those of us who work in wavebands where magnitudes are meaningless and we work in energy, can you say if there is a believable set of relationships which predict the frequency of flares of more than a certain energy in any given band on different stars?

Byrne: Yes, I believe so. The paper of Lacy, Moffet and Evans in particular and some of the follow up works using their methods do in fact allow you to make such predictions in energy terms.

Bromage: Taking into account different observers using different apertures?

Byrne: Yes; taking that into account. We have uniform data sets of our own which were taken under fairly uniform conditions and over a relatively short space of time so that the activity of the star does not vary. These agree with Lacy, Moffett and Evans' results quite well. So we would place a high degree of reliance on some of their relationships.

# THE ITA SURVEY ON RED DWARF VARIABLE STARS FINAL REPORT

C.A.O.TORRES*, I.C.Busko*, G.R.Quast
CNPq-Observatório Nacional
Observatório Astrofísico Brasileiro

From 1971 to 1973, an observational survey on red dwarf stars to search for possible periodic photometric variations was carried out at ITA Observatory - Brazil. Two observing runs at Cerro Tololo in 1974 and 1975 were used as an extension of this survey. We observed a total of 90 stars, as shown in Table I, and almost all stars were measured at least five times during the same season.

TABLE I

| Class | Stars Surveyed | Variables | Suspected |
|---|---|---|---|
| H emission | 20 | 11 | 5 |
| CaII emission | 30 | 2 | 2 |
| Non-emission | 40 | 0 | 0 |

The main conclusions of such survey are;

(a) The hydrogen emission line stars showed a great incidence of variables, with periods in the range of 1.5-8 days. There are some variables that have no detected variation during one season, although they were found to vary on another season. Nevertheless, it is hard to conclude by now that all stars of this class will be periodic variables. As an example, V1054 Oph has shown no variation during all the 3 years spanned by the survey.

(b) The CaII emission line stars showed a much lower incidence rate of variables, and those found have smaller amplitudes and greater periods than the variables of the previous class.

(c) No definite variation was found in non-emission line stars.

* Visiting Astronomers, Cerro Tololo Interamerican Observatory supported by the National Science Foundation.



*P. B. Byrne and M. Rodonò (eds.), Activity in Red-Dwarf Stars, 175–177.*

The suspected and confirmed variable stars are summarized in Table II. Everything said above is consistent in the frame of the evolution of activity in red dwarf stars.

TABLE II

| Gl | Name | Sp | Variability |
|---|---|---|---|
| 103 | CC Eri | K6Ve | 1.561 |
| 182 | V1005 Ori | dM0.5e | 1.86 |
| | DK Leo | dM0e | 7.982 |
| 388 | AD Lec | dM3.5e | YES?2.6? |
| 410 | | dM2e(?) | YES? |
| 494 | DT Vir | dM1.5e | 1.535 |
| | FK Com | G2IIIe | 2.407 |
| 517 | EQ Vir | dk5e | 3.90 |
| | V914 Sco | dM3e+dM4e | 2.69 |
| | V4046 Sgr | dk5e | 1.7035 |
| | FK Ser | K5Ve+K5Ve | 4.54 |
| 729 | V1216 Sgr | dM4.5e | YES? |
| 799 | AT Mic | dM4.5e+dM4.5e | YES? |
| 803 | AU Mic | dM2.5e | 4.8540 |
| | AY Ind | M2Ve | YES? |
| 867A | FK Aqr | dM2e | 4.276? |
| 425 | HD 98712 | K7V | YES 11.56 |
| 566 | ξ Boo | G8V | 10.137? |
| 567 | 131511 | K1V | YES? |
| 879 | TW PsA | K5V | 10.295 |

(d) Au Mic, observed during five years, has shown a variable light - curve from year to year, but all the data may be represented by a single ephemeris of constant period. We do not need to suppose that there is a variable period or a phase shift of the minimum. If we interpret the data as produced by the same activity center, this may be a challenge to the differential rotation hypothesis. It should be noted that the photometric data on the CaII-emission line star TW PsA may be described with a single ephemeris for five years too.

DISCUSSION

van Leeuwen: We have been doing observations of AU Mic in October and November 1981. They show, on combination with published observations by Hoffmann and by Respaju(?), the effect you mention, that is that the minimum of the light curve stays at constant phase but not the maxima. These observations extend over 10 years.

Johnson: It is very interesting if Gliese 644 or Wolf 630 or V1054 Oph is non-variable for 3 years since it is a well-known flare star. I realize that you were looking for BY Dra variability but do I understand correctly that you observed non flares in 3 years?

Torres: Our observations are for the non-flaring variations only. If a flare occurs during our observations that data is removed.

Johnson: So your statement excludes flares.

Torres: Yes.

van Leeuwen: We observed 2 flares during 3 weeks observing AU Mic.

Torres: Yes. We have observed many flares on AU Mic and other stars but this was not the purpose of the search.

Jordan: Is there any periodicity in the flaring as there is in the mean light?

Torres: No.

Mullan: I think that it is worth pointing out that there is an historical distinction between the dMe stars and M dwarfs which do not have the Balmer lines in emission. It may not have any physical significance other than that there may be more chromospheric heating in the emission-line stars. What Cram and I showed was that when one takes a star with an effective temperature of 4000K or 3500K one does not expect to see Balmer lines at all, either in emission or absorption. In order to get formation of Balmer lines even in absorption one needs to have a chromospherere whose energy is a significant fraction of $\sigma T^4$ since the $T^4$ factor falls off so rapidly towards cooler stars. So I think it is misleading to have two separate groups of dM stars. Both those with Balmer lines in emission and those with them in absorption require chromospheric heating which is a significant fraction of $\sigma T^4$.

Linsky: We have observed AU Mic with IUE for 3 days. One thing which is quite obvious from this data is that there is a lot of variability in the UV line strengths. Individual exposures are of duration of the order of 1 hr. Lines such as CIV change by as much as 50% in this time and by as much as a factor of 2 in the course of several hours. There are certainly real. There is no strong evidence of BY Draconis variability. Rather there is a great deal of shorter-term variability. Do you see, superimposed on the optical BY Draconis variability, short-term variability?

Torres: My data would not see this kind of variability since we observed usually only once per night.

van Leeuwen: In observing long-term variability in the Pleiades K stars we see no short-term variability with an upper limit of 2 thousandths of a magnitude.

# SYNOPSIS OF BY Dra AND II PEG LIGHT CURVES: VARIABILITY AND POSSIBLE EVJDENCE OF DIFFERENTIAL ROTATION(+)

M. Rodonò, V. Pazzani and G. Cutispoto
Osservatorio Astrofisico and Università di Catania, Italy

BY Dra (MOVe+MOVe) and II Peg (K2IV-III) are well known non-eclipsing spectroscopic binary systems showing the low-amplitude quasi-periodic photometric variability that is typical of spotted stars.

Since the discovery of their variability (Chugainov 1966, Eggen 1968) additional accurate photometry has been carried out (cf. Rodonò 1982). On account of their highly variable light curves (LC), we have reanalyzed all the available observations and divided the original data into shorter time-interval sets, so that overlapping LCs with different shape could be separated. Additional LCs obtained at Catania Observatory till 1981 were also included.

Notable and systematic changes of both BY Dra and II Peg LCs, from almost pure sinusoidal, to double-peaked to almost flat, on time-scales of the order of one month, are apparent. Representative examples are shown in Fig. 1 and Fig. 2. Wherever possible, the phase of the light minimum ( $\phi_{min}$), and the amplitude and mean level of the LC were obtained by expanding the observed LC with a truncated Fourier series.

For BY Dra, the phase of the light minimum appears to migrate cyclically on the LC in about 5-6 years, i.e. the photometric period ($P_{ph}$) undergoes similar changes with respect to the mean value 3.83624 days, very close to the value given by Chugainov (1966). If the cyclical variation of $\phi_{min}$, or $P_{ph}$, is attributed to latitude migration of spotted areas in a differentially rotating star, a lower limit of $4.4x10^{-10}$ rad $s^{-1}$ $deg^{-1}$ for the latitude shear can be inferred from the gradient of $\phi_{min}$ variation. The latter is accompanied by a sizable, but

(+) Partly based on observations by the IUE collected at the Villafranca Satellite Tracking Station of the ESA.



*P. B. Byrne and M. Rodonò (eds.), Activity in Red-Dwarf Stars, 179–184.*

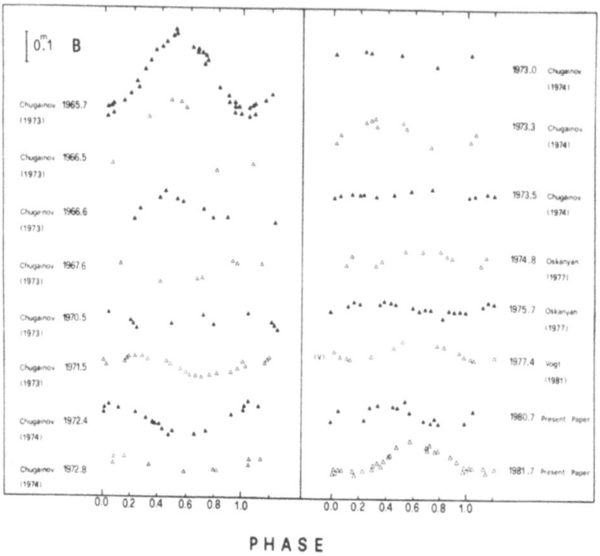


Figure 1. BY Dra light curves. Phases are reckoned from JD 2438983.612 and photometric period $3^d.836$ days.

apparently uncorrelated, variation of the mean luminosity of BY Dra (Fig 3). The long-term variability discovered by Phillips and Hartmann (1978) is confirmed and the most recent observations are consistent with a variability cycle of about 50 years (Fig. 4).

Similar changes are displayed by the LC of II Peg. However, its main features, viz. light maximum and minimum, show quite different migration rates: the maximum migrates towards decreasing orbital phases at a rate several times faster than the minimum, the latter being almost synchronous with the orbital motion (Fig. 5, lower plot). This fact makes the LC of II Peg continuously variable. The magnitudes of II Peg at maximum and at minimum have been decreasing since its variability was discovered (Fig. 5, upper plot). This monotonic light decrease, together with the variation of the LC amplitude, suggest that the activity level of II Peg, as far as spottedness is concerned, has been ever increasing since then. Assuming, as for BY Dra, that the observed migration of the LC features is due to latitude migration of surface inhomogeneities on a differentially rotating star, a lower limit of $1.7 \times 10^{-10}$ rad $s^{-1}$ $deg^{-1}$ for the latitude shear on II Peg is obtained. The values obtained for both BY Dra and II Peg turn out to be somewhat higher than the solar one $0.6 \times 10^{-10}$ rad $s^{-1}$ $deg^{-1}$ (Rodonò 1982).

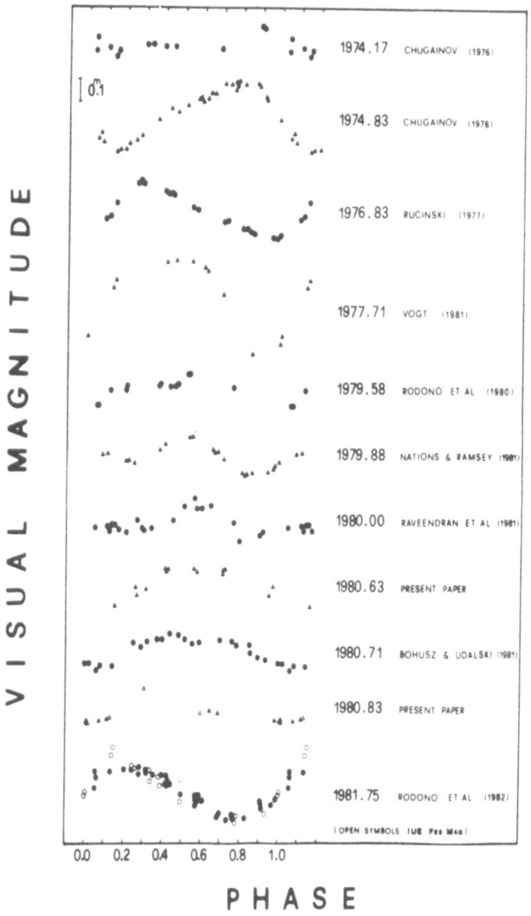


Figure 2. II Peg light curves. Phases are reckoned from JD 2442021.7264 and orbital period 6.724464 days, i.e. from the inferior conjunction given by Raveendran et al. (1981).

In Fig. 6 we also show the II Peg surface fluxes in several UV emission lines obtained from the IUE spectra LWR 8991 and SWP 10328 collected at VILSPA in 1980. At chromospheric and TR levels, II Peg appears to be more active than " *very active*" solar regions, with the higher excitation emission species more enhanced than the low excitation ones. This behaviour is qualitatively similar to the relative enhancement of emission lines in solar active regions with respect to quiescent ones.

An extended version of the present paper will be submitted to Astron. Astrophys.

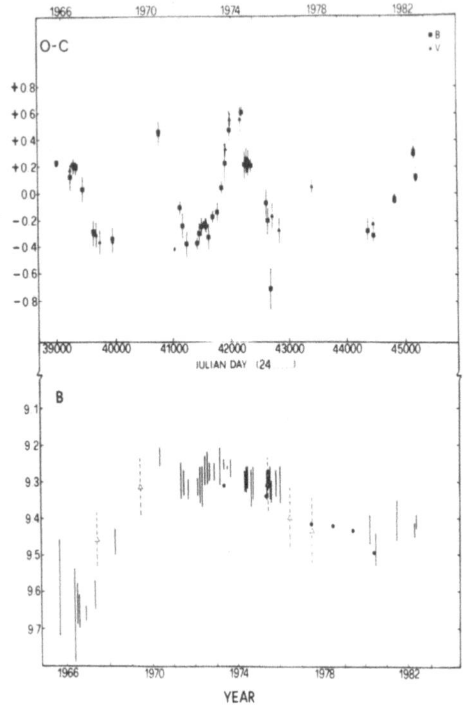


Figure 3. O-C for the phase of the light minimum of BY Dra with respect to the relation $\phi_{min} = -1.682 + 6.338 \times 10^{-5} \times JD$ vs. time (upper inset). B magnitude vs. time (lower inset). Bars: range of variation from pe. observations. Dots: mean seasonal magnitudes (Mavridis et al. 1982). Triangles: Harvard plate magnitudes (Hartmann and Londono 1982).

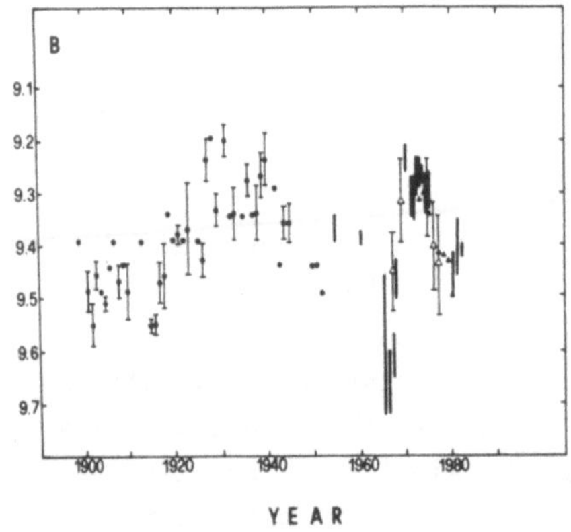


Figure 4. Long term variability of BY Dra from Phillips and Hartmann (1978) and from recent photoelectric photometry. Dots: Harvard plates. Heavy bars: range of seasonal pe. variability. Triangles: seasonal pg. averages.

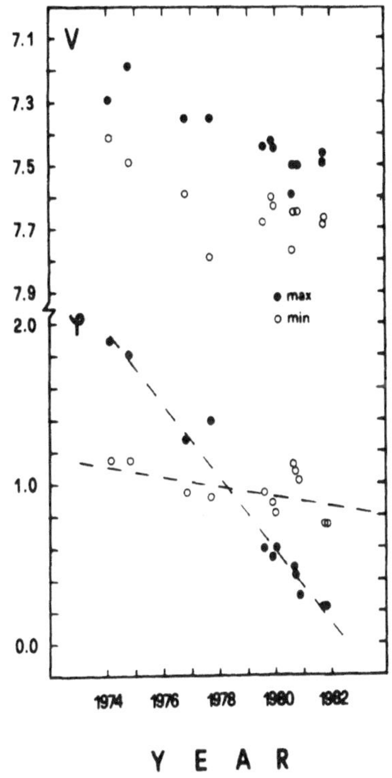


Figure 5. V magnitudes of II Peg at light maximum and minimum (upper plots) and the different migrations of these light curve features with respect to orbital phase (ϕ) (lower plots) versus time. Phases are reckoned from inferior conjunction as in Figure 2.

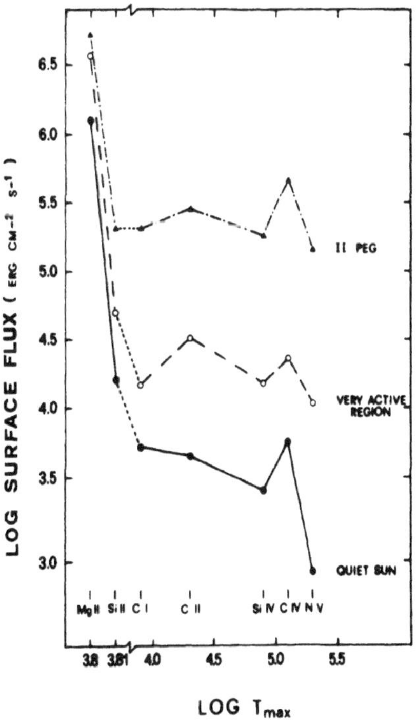


Figure 6. Surface fluxes in UV emission lines from IUE spectra of II Peg compared with those of a "very active" solar region and of the quiet Sun.

REFERENCES

Bohusz, E. and Udalski, A. : 1981, Acta Astron. 31, 185.
Chugainov, P.F.: 1966, IAU Inf. Bull. Var. Stars N° 122.
Chugainov, P.F. : 1973, Izv. Krym. Astrofiz. Obs. 48, 3.
Chugainov, P.F. : 1974, Izv. Krym. Astrofiz. Obs. 52, 3.
Chugainov, P.F. : 1976, Izv. Krym. Astrofiz. Obs. 54, 89.
Eggen, O.J. : 1968, Roy. Obs. Bull. N° 137.
Hartmann, L. and Londono, C. : 1982, private communication.
Mavridis, L.N., Asteridis, G. and Malmoud, F.M.: 1982, preprint.
Nations, H.L. and Ramsey, L.W.: 1981, Astron. J. 86, 433.
Oskanyan, V.S. : 1977, Astrophys. J. 250, 327.
Phillips, M.J. and Hartmann, L. : 1978, Astrophys. J. 224, 1982.
Raveeendran, A.V., Mohin, S. and Mekkaden, M.V. : 1981, M.N.R.A.S. 196, 299.
Rodonò, M., Romeo, G. and Strazzulla, G. : 1980, Proc. Second European IUE Conference, Tübingen, ESA SP-157, p. 55.
Rodonò, M. : 1982, in XXIV Meeting of COSPAR (Ottawa, 1982), Topical Meeting E.3: "Stellar Chromospheres and Coronae", Pergamon Press, in press.
Rodonò, M. et al : 1982, in preparation.
Rucinski, S.M. : 1977, Publ. Astr. Soc. Pacific 89, 280.
Vogt, S.S. : 1981, Astrophys. J. 247, 975.
Vogt, S.S. : 1981, Astrophys. J. 250, 327.

# TWO-SPOT MODELING OF SYNOPTIC LIGHT CURVES OF II PEG

G. La Fauci and M. Rodonò
Osservatorio Astrofisico and Scuola di Specializzazione
in Fisica, Università di Catania, Italy

The light curve (LC) of the single-line spectroscopic binary system II Peg (K2-3, IV-III) generally shows an asymmetrical and highly variable shape (Fig.1, left). Therefore, single circular spot models can be successfully applied only to symmetric LCs, as the 1977 one by Vogt (1981). When asymmetrical or almost flat LCs develop, as appears to be the rule for II Peg, at leasts two-spot modeling is required (Bopp and Noah 1980, Dorren and Guinan 1982).

From classical single-spot models (Torres and Ferraz Mello 1973, Friedman and Gurtler 1975, Bopp and Evans 1973) we have developed a computer code including two separate circular spots of different size which are allowed to assume any relative location on the stellar surface. The model is symmetric with respect to the equator. The spots are assumed to radiate as back bodies with temperature ranging from 100 to 2000 degree lower than the temperature of the unspotted photosphere (4500°K).

Of course the problem of the solution uniqueness still remains a serious one and makes any attempt to study one single LC quite hazardous. However, if several successive LCs are studied with the same method and the obtained results are carefully and critically analyzed, valuable information on the migration and evolution of spotted areas can be drawn. Moreover, if consistent values of the spot temperature, size and location are obtained from the solutions of very different LCs of the same star, the inherent disavantage of not being able to achieve unique solutions might sometime be overcomed.

Our computer code generates a dense grid of synthetic LCs covering a wide range of the location and dimension of two unequal circular spots and of their temperature difference with respect to the unspotted photosphere. Also the inclination of the rotation axis is initially assumed as a free parameter. Successive iterations guided by the goodness of fit between synthetic and observed LCs, allow us to restrict progressively the range of the most probable parameters and



*P. B. Byrne and M. Rodonò (eds.), Activity in Red-Dwarf Stars, 185–188.*

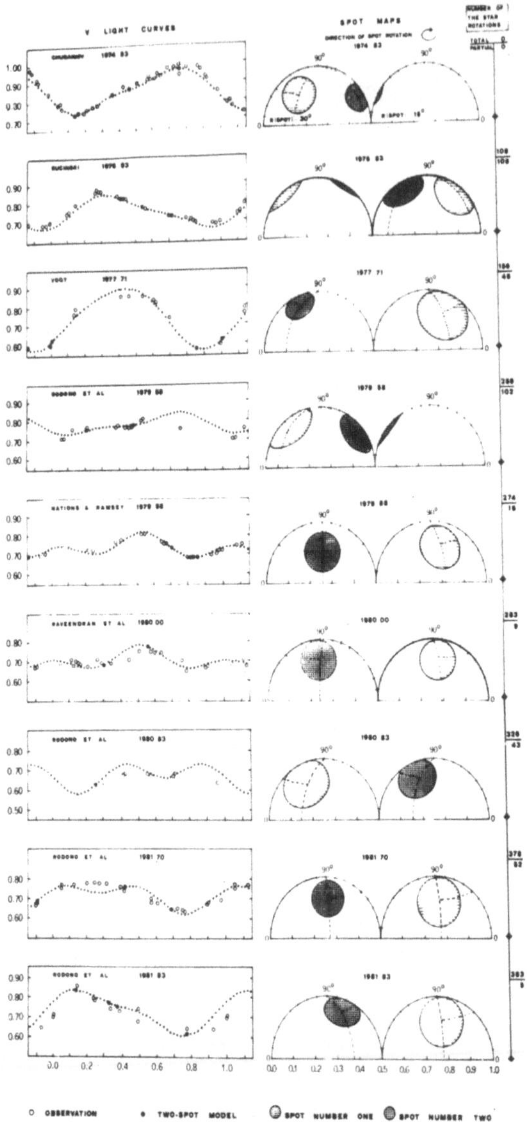

Figure 1 . Observed (open circles) and synthetic (dotted curve) light curves of II Peg from 1976 to 1982 and corresponding spot maps of the northern hemisphere from two-spot model solutions. The two numbers on the left give the number of star rotations from an arbitrary epoch and from the preceeding solution, respectively.

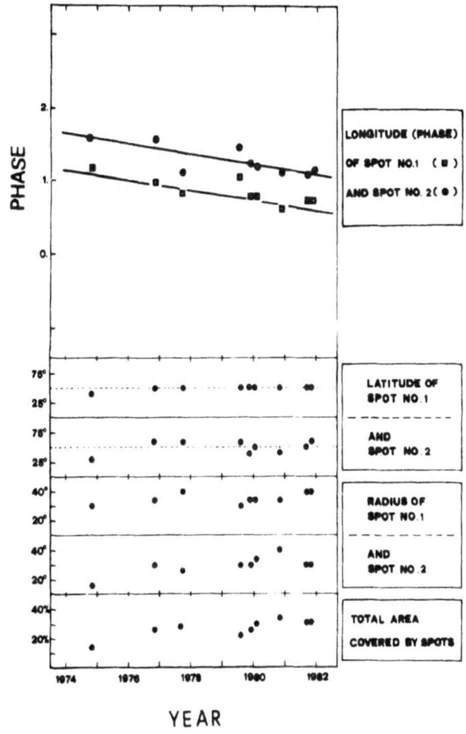

Figure 2. Spot migration, location, dimension and total fractional area from two-spot modeling of light curves in Fig. 1.

eventually to select the best solution. The numerical integration of the resulting flux from the spotted star at a given phase is obtained following the Simpson discrete integration method. A detailed description of our computer code will be published elsewhere. Here we present some results from the analysis of nine LCs of II Peg obtained by various authors from 1947 to 1982 (cf. Rodonò et al 1983).

Since the solutions obtained are symmetric with respect to the equator, only the spot maps of the northern hemisphere are shown in Fig. 1 (right). These maps readily show the large extension of the spotted areas, their relative dimension and location. The best solutions always led to the same values of the inclination of the rotation axis and the spot temperature: 35 degrees and 3300 ± 200°K, respectively. Numerical values of other parameters are plotted in Fig. 2. The upper inset clearly shows that both spots migrate linearly toward decreasing orbital phases, i.e. either the spotted star rotation is not synchronized with the orbital motion, as in several other BY Dra and RS CVn spotted systems, or the spectroscopic period needs to be revised.

Both spots are located near latitude 50° N, the larger spot occasionally extending up the polar region. The dimension of both spots appears to increase in the period covered by the observations, the smaller spot being the most fast-growing one. It appears that the total spotted area has been increasing almost monotonically from 1974 to 1982. This result is consistent with the increase of the median magnitude of II Peg in the same period (Rodonò et al 1983). By using the Harvard archival plate collection, Hartmann et al (1979) have found that, after a fairly long period of inactivity from 1900 to 1940, II Peg is presently exhibiting appreciable light changes. The present systematic decrease of luminosity, i.e. the increase of spottedness, might be part of a regular cycle of variability and therefore be worth of being closely scrutinized both observationally and theoretically.

## REFERENCES

Bopp, B.W., Evans, D.S.: 1973, Monthly Not. Roy. Astron. Soc. 164, 343
Bopp, B.W., Noah, P.V.: 1980, Publ. Astron. Soc. Pacific. 92,717
Dorren, J.D., Guinan, E. F.: 1982, Astrophys. J. 252, 296
Hartman, L., Londoño, C., Phillips, M.J.: 1979, Astrophys. J. 229, 183
Friedmann, C., Gurtler, J.: 1975, Astron. Nachr. 296, 125
Rodonò, M., Pazzani, V., Cutispoto, G.: 1983, This volume.
Torres, C.A.O., Ferraz Mello, S.: 1973, Astron. Astrophys. 27, 331
Vogt, S.S.: 1981, Astrophys. J. 250, 327
Vogt, S.S.: 1983, This volume.

VARIABLE K-TYPE STARS IN THE PLEIADES
(based on observations obtained at the European Southern Observatory, La Silla, Chile, and Lick Observatory, USA)

Floor van Leeuwen, Royal Greenwich Observatory, UK
Peter Alphenaar, Leiden Observatory, Netherlands

Photometric observations in the VBLUW system (Lub, 1979) have been performed during 1980 and 1981 of 19 late G and early K-type members of the Pleiades Cluster, in order to study their variability. All stars showed variations with amplitudes of 0.02 to 0.20 magn. in V. For 12 stars lightcurves were obtained which show periods that range from 0.24 to 1.22 days. The light curves are semi-regular and resemble those of BY Dra stars, although the periods are shorter.

Figure 1 shows three differently shaped lightcurves. The absolute magnitudes in V range from 7.0 to 7.2. The first lightcurve has a 'v' shape, the other two show 'u' and 'n' shapes respectively. The 'v' shape has been observed most often. Comparisons between 1980 and 1981 observations show that the lightcurves can change from one type to the other at constant period. (See Van Leeuwen and Alphenaar, 1982).

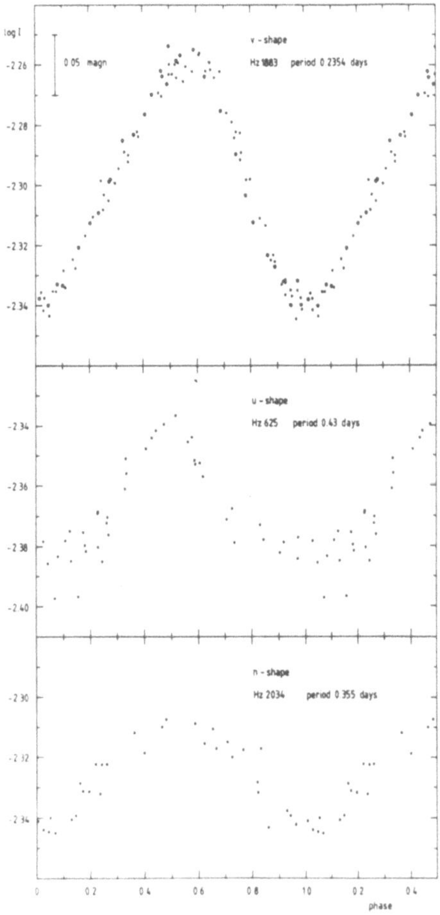


Figure 1.
Lightcurves for three of the Pleiades K stars. These observations are obtained in 1981 except for the dots of Hz 1883, which are 1980 observations.



*P. B. Byrne and M. Rodonò (eds.), Activity in Red-Dwarf Stars, 189–193.*

Figure 2 shows the relation between amplitudes and periods that was found for the variables. The symbols indicate the shapes of their lightcurves. This relation strongly indicates a dynamical background for the variations. There is no indication that a similar relation holds for other BY Dra stars, probably due to their spread in masses and ages. The stars shown in Fig 2 are all of the same age and mass.

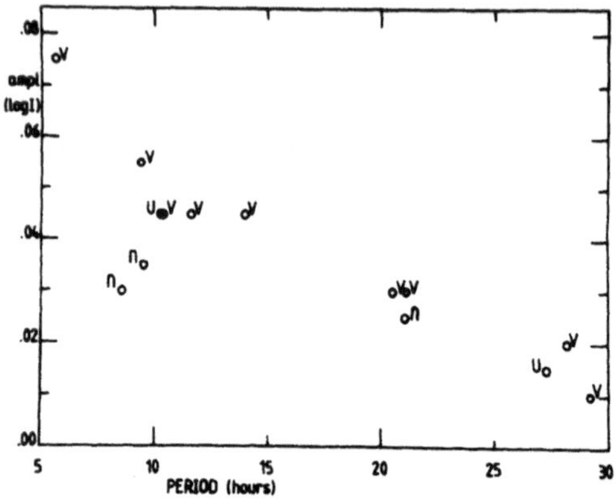

Figure 2. The period-amplitude relation for Pleiades K-stars.

In 1980 and 1981 Dr M F Walker of Lick Observatory carried out spectroscopic observations for two of the variables. He found single stars of spectral type K3Ve. They show very fast rotation: Vsini values of 150 and 75 km $sec^{-1}$ were found for stars with photometric periods of 0.24 and 0.42 days respectively. Such rotational velocities are two orders of magnitude higher than usually observed for K-stars, and bring the rotational periods close to the photometric ones. If they are related, which is generally assumed for BY Dra stars, the variations are due to rotational modulation. The relation of Fig 2 is in that case a relation between amplitude and rotational velocity. Variations in (V-B), the temperature index, have been observed which are proportional to those in V. The temperature variations are probably sufficient to explain most or all of the brightness variations. In the case of rotational modulation this implies a temperature gradient over the stellar surface, of which the amplitude is increased with increasing rotational velocity.

The observed variations decrease in amplitude rapidly for stars brighter than absolute magnitude 6.5 in V. The same happens with their flare activity and the fast rotation (Mayor, priv. comm.). The age of the Pleiades, $10^8$ years (Golay and Mauron, 1982), strongly indicates that these changes in the behaviour of early K stars are due to the last stage in their evolution towards the main sequence.
The amount of angular momentum contained in the Pleiades K-stars is, in case one assumes the photometric and rotational periods to be related, comparable with that contained in the solar system. In relation to their masses, they follow a trend set by O, B and A main sequence stars. This relation is shown in Fig 3, which was presented first by McNally (1965). It probably indicates the initial distribution of angular momentum in star formation. There exists a second relation, set by F, G and K main sequence stars, which shows much lower amounts of angular momentum for these stars. It has been explained as a stability relation, beyond which the stars are not able to keep their initial angular momentum. The Pleiades K stars arrive, according to this explanation, near to the main sequence with most of their initial angular momentum. They are not stable, and loose most of their

angular momentum on a relatively short time-scale. Their photometric variations as well as their flare activity may have much to do with this process.

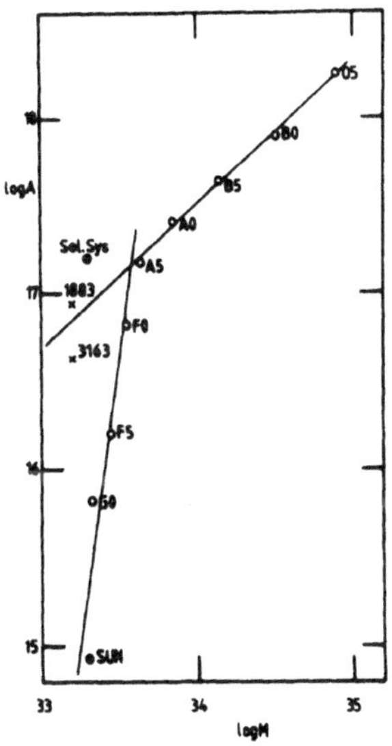


Figure 3.
The relation between the logarithm of the angular momentum per unit mass (A) and the logarithm of the mass (M) for main sequence stars, the solar system, the Sun, and two of the Pleiades K stars (1883 and 3163).

Acknowledgements

We would like to thank Mr J J M Meys, who participated in performing the 1981 observations and reductions.

References

Golay, M. and Mauron, N., 1982: Astron. Astroph. Suppl. Ser. 47, 547.
Lub, J., 1979: ESO Messenger 19, 1.
McNally, D., 1965: The Observatory 85, 166.
Van Leeuwen, F. and Alphenaar, P. 1982: ESO Messenger 28, 15.

DISCUSSION

Rodonò: I notice that you said the light curve was AU Mic-type. I would like to advise you that AU Mic had a v-shaped light curve 5 years ago and since then it has become successively double-peaked and flat and back to sinusoidal. If you say AU Mic-type for a changing light curve I would agree with you. Looking at your light curves (in Fig.1) I would expect that the v-shapes occur when the amplitude is a maximum and the u-type curve occurs at smaller amplitudes.

van Leeuwen: No, in fact the u- and v-shape light curves have similar amplitudes.

Rodonò: What I mean is that I would expect that, sooner or later those stars with v-type curves will change to u- or n-types with smaller amplitudes as we observe on nearby dwarf stars.

van Leeuwen: We have no indication that this is so and, as I have said, the amplitudes of the three kinds of light curves are similar. The changes from n-type to v-type have been observed in both directions.

Rodonò: Yes, but this is over one year, is it not?

van Leeuwen: We have observed 5 light curves over a one year interval and 7 light curves at one epoch only.

Rodonò: This is not enough. You should wait longer.

Rucinski: I would like you to say whether you believe that these are spots or not.

van Leeuwen: No, I do not believe that they are spots. There are many reasons for this belief. One concerns the stability of at least one of the light curves we have observed. It is stable to within 1% over a year or 1500 cycles. There are problems with the relation between amplitude and period. It is also a problem that all of the stars which we have observed to date are variable. I would accept spots if someone can convince me that spots can be always concentrated on one side of the star. Otherwise one would not expect always to see a single large minimum and maximum which stay exactly the same shape.

Rodonò: It was not my intention to prove that these light curves were due to spots. Spots can however reproduce almost any kind of light curve. So it is wrong to quote the kind of lightcurve as evidence against a spot interpretation. I would also like to question the statement that the lightcurve has been stable over a one-year period. We have very beautiful examples of light curves which stayed unchanged for years e.g. BY Dra. Subsequently these began to change their lightcurves. So it is quite possible to reproduce your lightcurves with spots.

van Leeuwen: You have observations as accurate as these?

Rodonò: Yes, I believe so.

Basri: These stars seem very interesting. Are you aware of any X-ray observations or other activity indicators? They would provide an interesting test of the activity-rotation correlations.

van Leeuwen: Since we know of these stars for only 1½ years there are no X-ray observations that I know about.

Basri: It is possible that the Einstein satellite may have observed some of them as part of their survey.

van Leeuwen: It is possible but I don't know of any.

Worden: We had a similar programme to your one at Cloudcraft Observatory New Mexico. We looked at the Pleiades, Hyades and the Malmquist field F, G and K stars. Our results showed, for the Pleiades and Hyades that about 30% were variable. There was a fairly large increase in variability in the K stars. However we did not look for periods, having only observations about a week apart. So assuming that we were seeing the same thing, it is interesting that it continues in the Hyades which are much older.

Hartmann: From Ca II modulation on such stars one can see the same longitudes active for some years at a time. So I think that your expectation that there cannot be active longitudes persisting over one year is not borne out by either experience with RSCVn or BY Dra stars or Ca II variable stars.

Walter: I would like to add to that that even in the case of the Sun one can see active longitudes for very long periods of time, not perhaps for 1500 rotations but perhaps for a year or two at a time.

van Leeuwen: But it is not a case of the stability of a single lightcurve or even of two or three lightcurves but rather of our entire sample. I agree that part of the observations can be explained with spot models. I think you get into problems to explain why all the stars are variable, why and how a relation exists between the amplitudes and periods and why there is so little scatter on that relation. Anything that is interposed between the rotation and its amplitude such as a magnetic field rise to spots would, I believe, give rise to a spread in the relation.

# STELLAR ACTIVITY AND CALCIUM EMISSION VARIABILITY

Sallie L. Baliunas
Harvard-Smithsonian Center for Astrophysics
60 Garden Street
Cambridge, Massachusetts 02138 USA

ABSTRACT. Time series analysis of fluctuations of Ca II H and K chromospheric emission has provided us with much information concerning stellar activity. On all timescales, events which parallel solar behavior can be observed: activity cycles, on timescales of years; rotation of stars and evolution of active areas on timescales of days to weeks; flare-like phenomena on timescales as short as minutes.

We expect that the analogues of solar activity exist on other stars. By studying stellar counterparts to solar activity, we can hope to investigate the physical parameters which are thought to influence chromospheric and coronal activity. The stellar surfaces are usually spatially unresolvable; it is thus difficult to measure directly either small-scale surface inhomogeneities or the associated magnetic fields expected from spatially restricted areas.

On the Sun, however, areas with strong surface magnetic fields show intense chromospheric Ca II H and K emission (Babcock and Babcock 1955; Skumanich et al. 1975). Although indirect, the Ca II H and K features are good indicators of stellar magnetic activity. A major advantage of the Ca II features is their accessibility to ground-based observatories. Long-term synoptic programs are in progress to monitor stellar chromospheric activity, and this paper will highlight ongoing work at Mt. Wilson. Monitoring variations of Ca II H and K chromospheric emission over different timescales can reveal different physical phenomena: (1) Long-term (years) variations corresponding to stellar activity cycles; (2) intermediate term (days-months) variations indicating rotation or evolution of stellar active areas; (3) short-term (minutes-hours) variations resulting from impulsive and flare-like phenomena.

## LONG-TERM BEHAVIOR

In contrast to small luminosity fluctuations in visible light, the sunspot number cycle is prominently marked in disk-integrated Ca II H and K emission which can vary by as much as 40% from solar maximum to minimum (Sheeley 1967; White and Livingston 1978). In a decade-long survey, Wilson (1978) monitored calcium emission from nearly 100 cool dwarf stars at the Mt. Wilson



*P. B. Byrne and M. Rodonò (eds.), Activity in Red-Dwarf Stars, 195–199.*

100" telescope. Wilson (1978) showed that many of the dwarfs in the spectrum type range F-G-K-early M undergo long-term variations. A statistical summary of Wilson's findings is: 40% show smooth fluctuations which appear to be cyclic with periods of 5-10 years or slightly longer. 40% show erratic variations which do not appear to be cyclic over the decade observed. The remainder show little, if any, observable fluctuations.

Since 1978 Wilson's work has been extended at Mt. Wilson on the 60" telescope after the construction of a dedicated frequency-chopping spectrophotometer (Vaughan et al. 1979). The continuation of Wilson's work on activity cycles is important for a number of reasons: First, it is necessary to build up the statistics of the phenomena. In many stars we've seen only one "cycle." Other stars have shown only secular changes which imply periods longer than a decade. Second, it is of interest to define the properties of an average cycle. For different epochs, the solar cycles can be dissimilar--for example, in the extreme case, compare the present-day solar cycle to the Maunder minimum (Eddy 1976). Third, it is possible to explore other stars such as giants, which may allow us to inject the controlled parameter of evolution into the question of activity cycles. Fourth, Vaughan (1980) analyzed the Wilson (1978) survey of dwarfs with activity cycles in the context of the solar neighborhood survey (Vaughan and Preston 1980). There may be a distinction between the morphology of activity cycles for the weak and strong emission-line stars. The old, chromospherically less active stars tend to show a smooth, long-term variation reminiscent of the Sun's activity cycle while the young, active stars show erratic fluctuations, qualitatively different than those observed in the old stars. This dichotomy in cycle behavior has prompted theoretical work in dynamo behavior (Durney et al. 1981; Knobloch et al. 1981).

## INTERMEDIATE TIMESCALE VARIATIONS

Wilson (1978) himself suggested that the "scatter" on a timescale of days to weeks contained in the long-term data trains could be caused by the rotation of long-lived active regions through our line of sight. A concerted effort to monitor intensively these dwarfs nightly bore out this speculation. An important point is that the rotation period is measured directly, rather than the projected rotational velocity. The classical method of determining rotational velocity in late-type stars is notoriously difficult: First, the Doppler line broadening is usually extremely small, and second, the inclination of the stellar rotation axis must be known. Some results of the Vaughan et al. (1981) and Baliunas et al. (1982) quantitative measures of stellar rotation can be summarized

(1) Although previously known from statistical studies of projected rotational velocity in open-cluster dwarfs, (Kraft 1967; Skumanich 1972), we confirm that rotation slows with decreasing Ca II emission strength.

(2) No good correlation exists between rotation period and activity cycle period. A weak threshold effect is apparent, though: stars whose rotation periods are longer than 20 days show morphologically solar-like activity cycles. This finding is consistent with the Vaughan (1980) result that old, relatively inactive stars show long-term fluctuations

similar to those of the Sun. The older stars happen to be the slower rotators, as earlier pointed out from the open cluster studies.

(3) The success rate for determining rotation in the chromospherically more active stars can be over 80%--much higher than for the older stars. This surprising result may be due to two effects:

a) The weak emission-line stars may show changes which are below our experimental sensitivity to such variations, and produce a lower success rate than for the active stars.

b) The lifetime of active regions or active longitudes may be longer in the young stars as compared to the old. As Noyes (1981) has found, the lifetimes of these inhomogeneities may be years.

(4) For a limited range in spectral type, the lengthening of rotation period with decreasing chromospheric emission strength is smooth between strong and weak emission line stars. This evidence is contrary to a discontinuous slaving of rotation as a star ages, an explanation proposed for the lack of intermediate emission-strength stars in the Vaughan and Preston (1980) solar neighborhood survey.

## SHORT-TERM VARIATIONS

Flux variations on timescales of minutes to hours in the solar-type dwarfs are also evident from the long- and intermediate-data sets described above. Both Middelkoop et al. (1981) and Baliunas et al. (1981) described short-term changes for some of the Wilson dwarfs. For HD 22049 (K2V) (Baliunas et al. 1981), 5-10% fluctuations in the emission, which may be akin to flare events, are observed. Not only are brightenings present on a timescale of minutes, but "fadings" of the emission are also seen, especially in Middelkoop et al. (1981, Fig. 3).

The energetics of these flare-like events in chromospheric emission observed in HD 22049 and other G-K stars is similar to those seen in flare stars and large solar flares (Baliunas et al. 1981). It is the poor contrast of the emission variations relative to the nearby, bright photospheric continua in these solar-type dwarfs which masks our impression of flare activity as compared to the dwarf M stars.

## REFERENCES

Babcock, H.W. and Babcock, H.D.: 1955, Ap. J., 121, p. 349.

Baliunas, S.L., Hartmann, L., Vaughan, A.H., Liller, W., and Dupree, A.K.: 1981, Ap. J., 246, p. 473.

Baliunas, S.L., Vaughan, A.H., Hartmann, L., Middelkoop, F., Mihalas, D., Noyes, R.W., Preston, G.W., Frazer, J., and Lanning, H.: 1982, Ap. J., submitted.

Durney, B.R., Mihalas, D., and Robinson, R.D.: 1981, P.A.S.P., 93, p. 537.
Eddy, J.A.: 1976, Science, 192, p. 1189.
Knobloch, E., Rosner, R., and Weiss, N.O.: 1981, M.N.R.A.S., 197, p. 45P.
Kraft, R.P.: 1967, Ap. J., 150, p. 551.
Middelkoop, F., Vaughan, A.H., and Preston, G.W.: 1981, Astron. Astrophys., 96, p. 401.
Noyes, R.W.: 1981, SAO Special Report No. 392, II, p. 42.
Sheeley, N.R.: 1967, Ap. J., 147, p. 1106.
Skumanich, A.: 1972, Ap. J., 171, p. 565.
Skumanich, A., Smythe, C., and Frazier, E.N.: 1975, Ap. J., 200, p. 747.
Vaughan, A.H., Preston, G.W., and Wilson, O.C.: 1979, P.A.S.P., 90, p. 267.
Vaughan, A.H.: 1980, P.A.S.P., 92, p. 392.
Vaughan, A.H., and Preston, G.W.: 1980, P.A.S.P., 92, 385.
Vaughan, A.H., Baliunas, S.L., Middelkoop, F., Hartmann, L., Mihalas, D., Noyes, R.W., and Preston, G.W.: 1981, Ap. J., 250, p. 276.
White, O.R. and Livingston, W.: 1978, Ap. J., 226, p. 679.
Wilson, O.C.: 1978, Ap. J., 226, p. 379.

DISCUSSION

Worden: Regarding your flares, we had some data on ξ Boo A in the U band, which is one of the active stars you have been looking at. We did a Fourier analysis on it and find evidence that there are solar-flare-type events buried in that continuum.

Baliunas: That is very interesting since yours are continuum measurements.

Worden: The light curve looks flat but one can see from the frequency spectrum that it is not just scintillation noise.

Mullan: This matter of the microflares is a very important aspect of your work. Where does one draw the dividing line between stars which are really quiescent and flaring stars? The question has not been addressed here as to how one evaluates what fraction of star's light is coming out in flares This of course is bound up with how much energy is in the microflares. So I think this is a very important point that micro-flares are difficult to see because of the contrast problem.

Belvedere: During your talk you said many times that your results were important for dynamo theory. I would like to know in which sense they are important.

Baliunas: The dichotomy in the activity curves between the old and young stars has spurred a lot of work as to how one can produce long-term cycles which show such a spread in behaviour.

Vaiana: With regard to short-term variability or microflaring we should recall the statement made after Torres paper (by van Leeuwen) that an upper limit of order one thousandth of a magnitude could be placed on variations in the later-type stars. This contrasts with the short-term variability reported here in stars of spectral type G or earlier. In X-rays we find short-terms variability 20-60% in all of the stars for which we have adequate sensitivity. So it may be that for M stars the photosphere does not change more than two thousandth of a magnitude but for G and K stars one has variability such as Worden was reporting.

Baliunas: Well, we have not made continuous measurements.

Evans: You had a diagramme showing the variation in magnitude of a star consequent upon the presence of a spot and its rotation. Part of the curve was level and part showed an oscillation. You explained this in terms of a beat phenomenon. Have you considered the possibility of a spot migrating from a polar region where it would always be in view to a lower latitude where it will be alternately visible or not.

Baliunas: That might also be a reasonable explanation.

van Leeuwen: The variation for the K-stars which we have observed are also very small. But for AU Mic the amplitudes of the flares are not visible at all in Walraven V and B but are very large in L, U and W. We do not see these variations in the visual.

Kodaira: I have been monitoring the Hβ emission of γγGem for 5 years and have found a double wave variation and sometimes a cut off wave. So I would like to ask whether you have found a similar effect in your observations of emission lines or evidence of a migration of the emission-line region?

Baliunas: HD46282 is an interesting example. In our first year of observation we deduced a period of 19 days but later one of the active regions disappeared and the true period turned out to be twice 19 days. So we do see evolution in the light curve.

EV LACERTAE: IS FLARE ACTIVITY RELATED TO AN UNSEEN PLANET-LIKE COMPANION?

Sarah Lee Lippincott
Sproul Observatory, Swarthmore College

ABSTRACT: The presence of an unseen companion to EV Lac with 2 to 4 Jupiter masses is likely. A period of ∿45 years indicates the separation at the time of periastron is ∿2 a.u.

EV Lacertae, BD+43°4305, has been on the astrometric program of the Sproul Observatory since 1938. Yearly coverage during the observing season has yielded 343 nights of observation on plates taken with the 61-cm refractor.

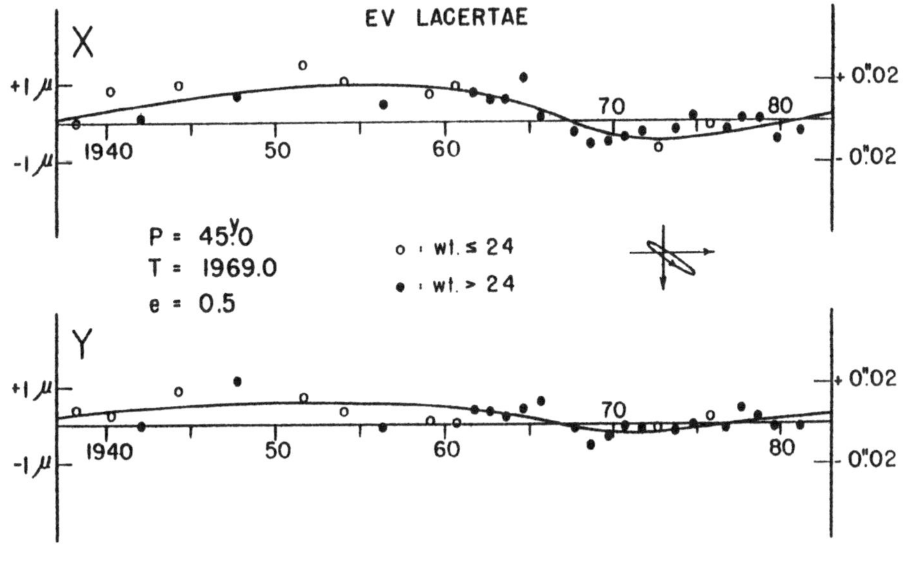


The orbital motion of EV Lac

The normal point positional residuals in X (RA) and Y (decl.) are given with parallax and proper motion removed. The computed displacement curves representing Kepler motion are plotted together in the inset to give the photocentric orbit, in this case the orbit of the primary about the center of mass.



*P. B. Byrne and M. Rodonò (eds.), Activity in Red-Dwarf Stars, 201–202.*

With the elimination of parallactic and proper motion, the yearly mean residuals show trends which can best be explained by Keplerian motion indicating the presence of an unseen companion with a period of ∿45 years and an orbital eccentricity of ∿0.5. Assuming a mass of 0.25 $\mathfrak{M}_\odot$ for the visible component, we find the mass of the companion has a likely range of 2 to 4 times Jupiter's mass, giving the ratio of their masses $\mathfrak{M}_A/\mathfrak{M}_B$ , ∿100 to 1. The range in separation of the two components would be from ∿2 to ∿6 a.u. to be consistent with the values given above. The closest approach, 2 a.u., was ∿1969, and by 1982 the separation is estimated as ∿4 a.u. The interpretation of any long-term changes in the flare activity of EV Lacertae should give some consideration to the likely presence of an unseen planet-like object.

Considering the wide-range magnetic field effects of Jupiter, one might consider comparable effects taking place around the EV Lacertae companion. Also the differential acceleration or tidal effects on EV Lacertae are likely to be 100 times that caused by Jupiter on the Sun. Unfortunately, any change in flare frequency around the time of periastron in ∿1969 from the current frequency is hard to evaluate from inspection of the literature, although some statistical inferences could be made.

There are ∿4000 exposures taken between 1938 and 1981 on the Sproul astrometric program; the current exposure time is 0.5 minutes. From visual inspection there were no flares observed over 0.2 magnitude.

This research is aided by NSF grant AST 81-16514 and past grants.

# AN UNUSUAL FLARE ON EV LAC

Bo Nyborg Andersen
Institute of Theoretical Astrophysics,
University of Oslo, Norway.

## I. INTRODUCTION.

It has long been noted that the characteristics of the intensity variations during a flare on UV Ceti stars depend on the observed spectral region. The intensity increase being larger in the shorter wavelength region. Typical values for the ratioes between the relative intensity increases on EV Lac in the Johnson UBV-bands are $\Delta U/\Delta B \approx 6$ and $\Delta B/\Delta V \approx 3$. (Moffet 1974).

The relative intensity increase decreases further towards the infrared (Bruevich et al. 1979, Pettersen 1982). The differences in sharpness of the intensity peaks has been attributed to different contribution from line and continuum emission. It has been observed that the line emission has a later maximum and slower decline than the continuum emission (Bopp and Moffet 1973, Mochnacki and Zirin 1980, Pettersen 1981). The rare preflare dips are more often observed in the red and near infrared spectral regions than the blue (Flesch and Oliver 1974, Bruevich et al. 1979). However, a very prominent preflare dip was observed in the blue spectral band (Giampapa et al. 1982). The same authors have speculated if the dip is caused by a general increase in the $H^-$ opacity, or by an off limp "disparition brusque", or by a general increase in the Doppler widths caused by MHD waves assosiated with the oncoming flare.

## II. OBSERVATIONS.

During a period in October 1979 the flare star EV Lac was observed photoelectrically with a two-channel photometer mounted on the 30 cm reflector at the Oslo Solar Observatory. The observations were carried out simultainously using a 115Å FWHM centered at Hα and a Schott BG12 filter. The latter approximates a sum of the standard U and B bands. A detailed description of the apparatus is found in Pettersen (1978).

The most interesting of the four flares observed is shown in Figure 1.



*P. B. Byrne and M. Rodonò (eds.), Activity in Red-Dwarf Stars, 203–205.*

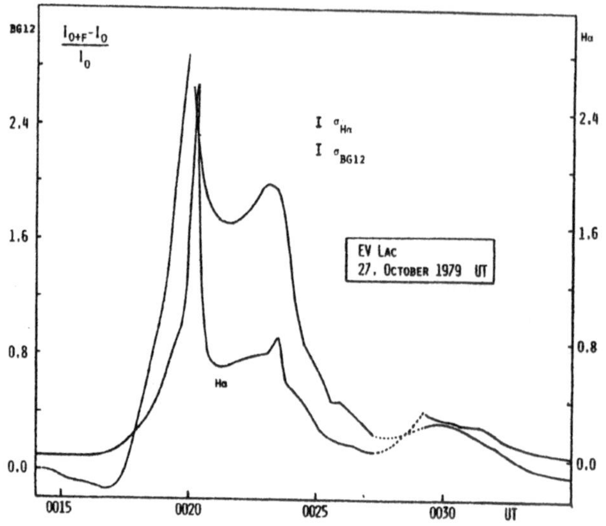


Figure 1. Flare on EV Lac observed in Hα and BG12 filters.

We see a clear preflare dip of 12% in the blue channel while no preflare activity is visible in Hα. The accurate peak intensity in BG12 was not observed as the recorder went off scale. The value in Table 1 is estimated from the intensity gradients.

Table 1. Flare Parameters 27/10-1979.

| Filter | $t_{max}$ | $(I_{o+f}-I_o)/I_o$ | P(min) |
|---|---|---|---|
| Hα | $00^h20^m20^s$ | 2.6 | 5.2 |
| BG12 | $00^h20^m12^s$ | 3.0 | 13.1 |

We note that both the primary and secondary maxima occur ≈10 s later in Hα than in BG12. We also note that the maximum increase in intensity is comparable in the two channels and that the same spiked maximum is shown in both channels.

## III. DISCUSSION.

The most anomalous feature with this flare is the very large and sharply peaked amplitude observed in the Hα filter. Similar observations

(Pettersen 1982, Schneeberger et al. 1982) indicate that our observation in Hα compared to the BG12 filter is an order of magnitude larger than previously observed. Even if the delayed maximum could indicate a significant contribution from the Hα line the sharp peak demonstrates that the emission is continuum dominated. This is in accordance with the wide band Hα measurements by Pettersen (1982). In terms of absolute energy per wavelength unit the emission in theHα filter is ≈ 9 times that in BG12. This is larger than the results indicated by Bruevich et al. (1979).

We could speculate in that the excess radiation in the red has to be seen in connection with the preflare dip in blue. All three mechanisms proposed by Giampapa et al. (1982) for the preflare dip could cause an abnormal increase in intensity at longer wavelengths. As no quantitative theoretical results is available it is not possible to determine if any of the proposed mechanisms provide enough energy to explain our observed phenomena.

REFERENCES:

Bopp B.W., and Moffet T.J., 1973, Astrophys.J. 185, 239.

Bruevich V.V., Burnashev V.I., Grinin V.P., Kilyachkov N.N., Kotyshev V.V., Shakovskaya N.I., and Shevchenko V.S., 1979, Izv. Krymsk. Astr. Obs. 61, 90.

Flesch T.R., and Oliver J.P., 1974, Astrophys.J. 189, L127.

Giampapa M.S., Africano J.L., Klimke A., Parks J., Quingley R.J., Robinson R.D., and Worden S.P., 1982, Astrophys.J. 252, L39

Mochnacki S.W., and Zirin H., 1980, Astrophys.J. 239, L27.

Moffet T.J., 1974, Astrophys.J. Supp.Ser. 29, 1.

Pettersen B.R., 1978, Thesis, University of Oslo.

Pettersen B.R., 1982, Reports from Lund Observatory 18, 114.

Schneeberger T.J., Linsky J.L., McClintock W., and Worden S.P., 1979, Astrophys.J. 231, 148.

# OPTICAL AND ULTRAVIOLET STELLAR FLARE SPECTROSCOPY

Simon P. Worden
United States Air Force, Headquarters, Space Division

## ABSTRACT

As for solar flares, one of the most physically revealing types of data for M-dwarf flares are high-resolution, time-resolved spectra. Due to the intrinsically faint nature of the M-dwarf stars, spectroscopic data has tended to be of low spectral (~ 5 Å) and temporal (~ 5 min) resolution. However, with the development of image intensified spectrographs and fast, efficient digital detectors, the last several years have seen the successful acquisition of both high time and spectral resolution M-dwarf flare spectra. Recent programs have also been successfully conducted using the International Ultraviolet Explorer (IUE) satellite to obtain UV and EUV spectra of M-dwarf flares. These data reveal that dwarf M star flares are remarkably similar to solar flares in all aspects of their spectroscopic phenomenology.

## 1. INTRODUCTION: SOLAR AND STELLAR FLARES

The dwarf M emission line (dMe) flare stars, or UV Ceti-type stars, are a set of late-type dwarf stars which undergo intense, short-lived continuum brightenings, or flares. The basic features of stellar flares have been reviewed by Gershberg (1975), Mullan (1977), and Gurzadyan (1980), as well as in other papers presented at this meeting. As reviewed by Zirin and Ferland (1980), dMe flares are comparable in most respects to solar flares. As with solar flares, one of the most physically revealing types of data are high spectral resolution, time-resolved visible and ultraviolet spectra. In this review, I summarize the spectroscopic data base for dMe flares.

Broad-band stellar flare light curves closely resemble the spatially averaged solar extreme ultraviolet (EUV) 300-1500Å light curve shown in Svestka's(1976) monograph. The transient nature of stellar flares is well documented with a general division into "spike" and "slow" events on the basis of flare rise time in the Johnson U band



*P. B. Byrne and M. Rodonò (eds.), Activity in Red-Dwarf Stars, 207–221.*

(Moffett, 1974). This classification corresponds to a general division of solar flares into those with a "flash" phase and the rarer "spotless" flares (see Zirin and Ferland, 1980; Dodson and Hedeman, 1970). An apparent dichotomy between solar and stellar flares is the time scale for the event. The stellar continuum flare is typically an order of magnitude (minutes) shorter than solar Hα flare durations (hours). However, if the solar continuum flare, either the extremely rare "white light" flare, or the UV continuum flare time scales is used instead, then the timescales for the flash phase of solar and stellar flares are comparable. Moreover, stellar flare line emission persists for hours (Kunkel, 1970) in marked similarity to solar flares. The principle qualitative difference between solar and stellar flares is the presence for the Sun of a substantial photospheric continuum in the visible and near UV.

Energy liberated in solar flares appears to reside in atmospheric magnetic fields, although the physics of the energy release and its partial conversion to electromagnetic radiation is poorly understood (Svestka 1976, Chapter VI). Indirect evidence for substantial magnetic fields on dMe stars comes from both theoretical grounds (Mullan, 1974), and observational, including the presence of photometric variations which may be interpretted as magnetic starspots (Torres and Ferraz-Mello, 1973; Hartman and Rosner, 1979), and the strong polarization, possibly detected in radio observations of stellar flares (Spangler et al., 1974). Direct detection of magnetic fields on dMe stars is very difficult due to the tangled field topologies on their surfaces (Vogt, 1980). Although new techniques relying on high signal-to-noise, and high spectral resolution data (Robinson, 1980: Robinson et al., 1980) may reveal total magnetic flux values for dMe stars, it is unlikely that changes in this flux will be tied to specific flare events. Definitive evidence for changes in solar magnetic field configuration have yet to be detected during solar flares. There is no overriding reason to suppose that stellar flares are fundamentally different from solar flares, although it should be emphasized that both solar and stellar flares are an extremely diverse set of phenomena. Stellar flares can involve much more energy than their solar counterparts, and they are certainly more frequent and represent a much larger fraction of total stellar energy output.

The primary release of solar flare energy apparently occurs in the transition region or low corona. Since this is a nigh-temperature region at greater than $10^6$K (Svestka, 1976), direct radiative flare effects appear in X-ray emission. This spectral region can be studied only with high-spectral resolution X-ray space telescopes, which have not yet yielded sufficient spectral resolution to study the flare fine structure so as to elucidate the primary energy release. However, flare effects in the solar chromosphere and photosphere may be probed through EUV, UV(1500-4000Å) and visible spectra. Observational solar flare studies have centered on the Hα line which is both readily observable and which spans a wide range of levels in the solar atmosphere. In the first portion of this review, I will summarize the principal observational features of solar visible, UV, and EUV spectra.

High spectral resolution observations of line profiles in flare stars could provide data with which to test flare models for both solar and stellar flares. Most data on stellar flares are obtained by broadband visible photometry. However, there is some data on UV and EUV radiation (Haisch et al., 1981, 1982; Butler et al., 1981), X-rays (Kahn et al., 1979; Haisch et al., 1981; Kahler et al., 1982), and radio radiation (Spangler et al., 1974) from flares. However, the intrinsic faintness of the dMe stars has made it extremely difficult to obtain spectroscopic observations with sufficient temporal or spectral resolution to apply fully the diagnostic methods of solar flare research. The development of new digital high speed detectors promises spectral data both in the visible and UV which will be adequate for detailed analyses. These data, although important in isolation, become even more valuable when coordinated with other types of visible, X-ray, UV, and radio observations. In the second part of this review, I summarize the available spectral observations, which in and of themselves suggest approaches for new coordinated programs.

## 2. SOLAR FLARE SPECTROSCOPY

An outstanding review of solar flare spectroscopy in the visible wavelength regions has been presented by Svestka (1972), and the brief summary in this paper will closely follow that review. In general, solar flare visible radiation is emitted in spectral lines, with only extremely rare instances of continuum emission. Depending on the size of the flare, the Balmer lines of hydrogen go into emission. The larger the flare, the higher up the Balmer series the emission extends, up to $H_{16}$ in very large flares. Also in large flares, lines of neutral and singly-ionized metals are excited, suggesting that the flare is progressively heating deeper layers of the atmosphere. Neutral helium lines, which are not normally present in the solar spectrum, go into absorption in small flares, and into emission for larger flares. In the largest flares, He II lines such as $\lambda$ 4686 also appear in emission.

The Balmer spectrum of hydrogen has been the most extensively studied in solar flares. In a following paper, M. Giampapa will discuss how electron density, flare optical thickness, electron temperature, non-LTE conditions, and hydrogen density can be deduced from Balmer line data. Although the central hydrogen emission core has a width comparable to the solar photospheric absorption lines (1 to 2 Å), broad emission wings can extend over 10 Å. This fact, taken with the non-broadening of metallic emission lines, indicates that Stark broadening is the mechanism responsible for the hydrogen line widths (Svestka, 1972). Among the best available solar flare spectra are those published for Hα during the great flares of August 1972 (Zirin and Tanaka, 1973). Figure 1 reproduces a portion of the 7 Aug 1972 flare showing the key features from the maximum Hα emission period. Most of the flare emission is confined to the narrow (~ 2 Å) central peak.

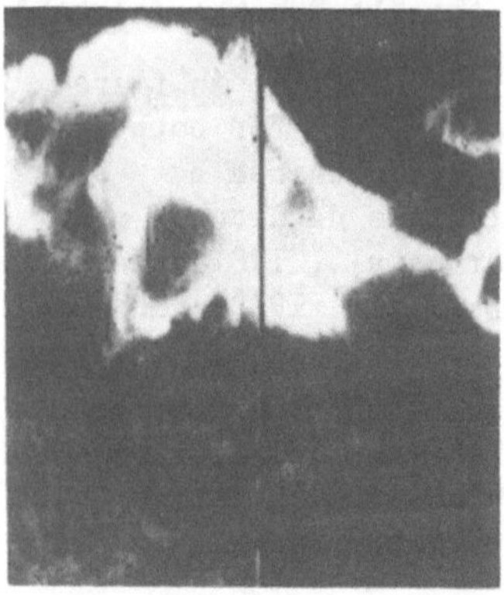

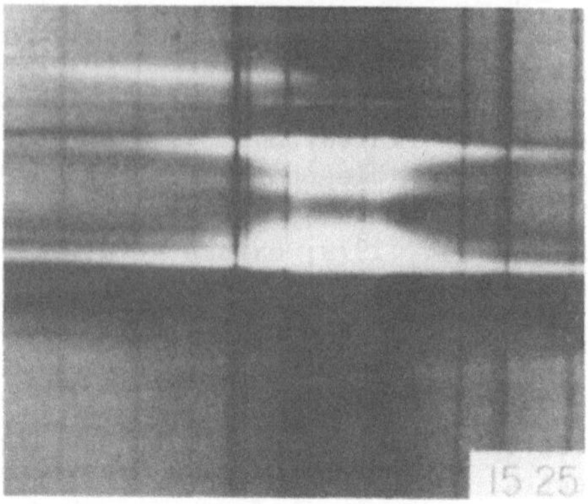


Figure 1. Hα filtergram and spectrum taken by Zirin and Tanaka (1973) for the 7 Aug 1972 flare maximum. The spectrum covers ~ 10 Å; red is to the left.

However, the bright flare kernels show emission extending over at least 10 Å. The strongly redshifted emission feature is also typical of solar flares. Taken as an integrated flare spectrum, the emission asymmetry does not change the center of the emission, but makes one of the wings stronger. At the outset of a solar flare a short-lived blue wing enhancement often occurs. The blue asymmetry disappears within a minute or two and a strong red asymmetry appears, which reaches its maximum roughly concurrent with Hα flare maximum, and then decreases within about five minutes. These phenomena have been interpreted as motions within the flaring region (see Banin, 1965; Severny, 1968); the blue asymmetry is due to a surge during the flash phase, and the red asymmetry is due to condensing material in later flare phases.

The prominent lines of He I, λλ5876, 6678, and 10830 appear in absorption in small flares and in emission during larger ones. The He I lines probe slightly hotter regions ($T_e \sim 2 \cdot 10^4$ K) than the Balmer lines (Te ~ $10^4$K). The He II lines, most conspicuously λ4686, but also such lines as λ5411, probe the region ($T_e \sim 10^4 - 10^5$ °K) between the low-temperature chromospheric flare and the high-temperature coronal X-ray flare ($T_e \sim 10^6$ ° K).

The Ca II H and K lines (λλ3968, 3933) are strengthened during solar flares. These lines are potentially extremely important diagnostics for conditions throughout the solar and stellar chromosphere (see Linsky, 1980). Excellent chromospheric models based on these lines have been developed for all classes of late-type stars, including the dMe stars (Giampapa et al., 1982). The most notable solar flare effect on the H and K line profiles is the disappearance of the central reversal in the emission core.

Metallic lines, both neutral and singly ionized, are probes of the low solar photosphere where $T_e < 10^4$ °K. A thorough study of flare metallic emission lines has been made by Letfus (1964). Metallic lines are typically very narrow, and only occasionally show broadening attributable to material motions (Jeffries and Orrall, 1961). For some events, strong metallic lines in solar flare spectra show a self-absorption (Svestka, 1963). There have been several studies on the sequence with which various lines appear during solar flares, but the results are somewhat contradictory as to whether metallic emission lines appear before or after Balmer emission (Svestka, 1972).

Solar flare EUV emission is a probe of the transition region between the chromosphere and corona. Good data in this spectral region derive primarily from the OSO 4, OSO 6, Skylab, and Solar Maximum Mission Satellites. EUV bursts, both line and continuum, appear simultaneously with the hard X-ray burst, impulsive microwave burst, white light flare, and flare flash phase, but are on the average a few minutes earlier than the Hα and soft X-ray maximum. The EUV burst, lasting a few minutes at most, is much shorter than the Hα or soft X-ray flare, but longer than the hard X-ray burst. Emission lines, representing temperatures of $10^4$ °K to $10^6$ °K, occur in spectral lines ranging from neutral hydrogen and oxygen to highly ionized species such as Fe XVI. Lines arising in the chromosphere ($T_e \sim 10^4$ °K) and transition region ($T_e \sim 10^5$ °K) dominate during the flare impulsive phase, while coronal emission lines ($T_e \sim 10^6$ °K) dominate during later flare phases.

## 3. STELLAR FLARES

Stellar flare spectroscopy is a difficult observational task. Until recently, most stellar flare observations were accidental. Because of the intrinsic faintness of the dMe stars, flare spectroscopic data are usually of low spectral and temporal resolution. However, high-efficiency digital detector spectrometers coupled to large telescopes are making available a growing body of good stellar flare spectra. These data, often taken as part of coordinated flare patrols along with X-ray, UV, EUV, optical photometric, and radio data provide powerful dignostics upon which to base stellar flare modeling.

The main body of dMe stellar flare spectroscopy is summarized in Table I. The first spectrum of a stellar flare on a dMe star was reported by Joy and Humason (1949), who noted greatly enhanced Balmer lines in one 144-minute exposure of UV Cet. They also noted the λλ4471, 4026 lines of He I, as well as the λ4686 line of He II. There were a number of other "accidental" dMe flare spectra taken during the 1950-1965 period. Gurzadyan (1980) notes some of these data; in the interest of brevity, they are not all included in Table I nor in this discussion. Herbig (1956) noted both enhanced Hβ and Ca II emission in BD+4°4048B, and Joy (1958) reported that the Balmer lines in UV Cet doubled in width. However, the utility of flare spectra is greatly enhanced by simultaneous photometry.

Table I. Spectra of Stellar Flares

| Wavelength Range | Wavelength Resol'n | Time Resol'n | Star | Reference and Remarks |
|---|---|---|---|---|
| Blue | | 144min | UV Cet | Joy & Humason (1949). No photometry. One plate had enhanced & broadened H lines. He I and He II emission. |
| 3500-5000Å | 430Å/mm | 4.5 hr | BD+4°4048B | Herbig (1956). Very bright Hβ. Ca II lines brighten less than H. No photometry. |
| Blue | 80Å/mm | 100min | UV Cet | Joy (1958). Visual photometry. H lines double in width. |
| Hα,Hβ, 4226Å ±1000Å | 6-7Å | 1-2min | AD Leo<br>UV Cet | Gershberg & Chugainov (1966,1967). Simultaneous photometry. H lines broaden, Hβ goes from 5.2Å to 10Å. |
| 3500-5000Å | 15Å | 5min | Wolf 359 | Greenstein and Arp (1969). No photometry. Greatly broadened H lines. |
| 3500-7000Å | 6Å | 30min | AD Leo | Gershberg & Shakhovskaya (1971). Red asymmetry in H line wings. Simultaneous photometry. |
| 3500-4900Å | 15Å | 7min | EV Lac<br>YZ CMi<br>AD Leo | Kunkel (1970). Simultaneous UBV photometry. Ca II lines rise slower than H lines. |
| 3800-6000Å | 8Å | 3min | UV Cet | Bopp & Moffett (1973). High-speed photometry. Relative timings of H, He, and Ca II lines. He flashes. |
| all H lines | 3Å | 20min | AD Leo | Kulapova & Shakovskaya (1973). With photometry. Hα width increases to 15Å. |
| 3700-5700Å | 8Å | 3min | UV Cet<br>YZ CMi<br>EV Lac<br>YY Gem | Moffett & Bopp (1976). With high speed photometry. Hβ lags U band by 1-17 min. Complex line response. |
| Hβ±30Å | 10Å | 107sec | UV Cet | Moffett *et al.* (1977). High speed phototometry. Hβ increases much more than continuum. |

Table I. Spectra of Stellar Flares (Continued)

| Wavelength Range | Wavelength Resol'n | Time Resol'n | Star | Reference and Remarks |
|---|---|---|---|---|
| 5300-7300Å | 0.24Å | 2min | AD Leo | Schneeberger *et al* (1979). Has photometry. No width change in Hα. |
| 5090-5350Å | 0.5Å | 5min | YZ CMi | Mochnacki & Schommer(1979). No photometry. Numerous Mg I, Fe I, Fe II emission lines. |
| 3200-7000Å | 20-160Å | 10-30sec | YZ CMi AD Leo | Mochnacki & Zirin (1980). Spectrophotometry. Inverse Balmer decrement during flash phase. |
| 1200-3200Å | 6Å | 5-10min | Prox Cen | Haisch *et al*. (1981). Simultaneous X-ray, *radio*. No photometry. 1200-2000Å increase less than a factor of 2. Possible Mg II emission during X-ray flare. |
| 1200-2000Å | 6Å | 100min | Gl 867A | Butler *et al*. (1981). No other obs. High temperature lines enhanced more than low temperature ones. |
| 4200-5900Å | 4Å | 33sec | YZ CMi | Kahler *et al*. (1982). Simultaneous X-ray, *radio*, optical data. |
| 1200-3200Å | 6Å | 60min | Prox Cen | Haisch *et al*. (1982). Simultaneous X-ray. Numerous transition region lines well observed during flare. |
| 5000-7000Å | 0.24Å | 3min | YZ CMi | Scheeberger *et al*. (1982). Simultaneous photometry. Balmer lines not broadened in 6 flares. Hβ response lags behind Hα. |
| 3500-7000Å | 0.25A-3.8Å | 3min | UV Cet | Giampapa *et al*. (1983). Very large flare (5 mag in U). Broad red-shifted wings with narrow core in H & He lines. |

Gershberg and Chugainov (1966,1967) were the first to study the response of chromospheric emission lines to a flare with coordinated spectroscopic and photometric observations. Since the FWHM of Hα in flare stars is roughly 1-2Å (Worden and Peterson, 1976), low spectral resolution observations cannot reveal detailed information on the response of emission lines to the flare. Moreover, typical flare light curves show that flare phenomena are extremely rapid. Thus, spectral data with high time resolution are essential. Recently, data with spectral resolutions near 0.2Å have become available (Schneeberger et al., 1979, 1982; Mochnacki and Schommer, 1980; Giampapa et al., 1983). The development of efficient digital detectors has also helped in getting higher time resolution data. Using a multichannel scanner, Mochnacki and Zirin (1980) have achieved time resolutions of 10 seconds for a large flare on YZ CMi. In the following summary, I review the principal observational aspects of the spectroscopic data listed in Table I.

### 3.1. Balmer Lines: Balmer Decrement

The ratio of Balmer series line intensities (Balmer decrement) during the flare can be an important indication of physical parameters such as electron density (Zirin and Ferland, 1980). Kunkel (1970) reported an increment, rather than a decrement, where Hβ, Hγ, and Hδ may actually exceed Hα in intensity by up to a factor of two during flare maximum. Higher Balmer lines do show a decrement during flare maximum, with the decrement becoming steeper for the higher series lines. The lower Balmer series reassert the decrement as the flare progresses. Zirin and Ferland (1980) point out that, although poorly studied, there is some evidence of the same behavior in the low Balmer lines during solar flares.

As discussed by Svestka (1972), the highest member of the Balmer series which appears in emission during the flare can provide information on the optical depth within the flaring region. Although there is a paucity of high S/N spectra in the near UV, data such as that reproduced in Figure 2 (Giampapa et al, 1982) show that Balmer emission is present until at least $H_{13}$.

### 3.2. Balmer Lines: Balmer Continuum

In solar flares Balmer continuum observations are applicable for temperature determinations (Svestka, 1976). Solar flare Balmer continuum emission is often absent during flares, but this may be an observational selection effect due to the high background. Kodaira et al.(1976) saw no evidence for Balmer continuum emission in dMe flares. However, Mochnacki and Zirin (1980) did report Balmer continuum emission for a large flare on UV Ceti. Zirin (1982, private communication) has confirmed that Balmer continuum emission is an occasional feature of stellar flares.

### 3.3. Balmer Lines: Widths

Due to low spectral resolution, much of the data in Table I are useless for studying line profiles. Equivalent width results, usually quoted from spectra with resolutions worse than a few Ångstroms, provide little information of value. Since there are no good models of dMe atmospheres, nor good scanner observations of dMe continuum flux, it is very difficult to place equivalent width data on any sort of flux scale. Moreover, the continuum is changing rapidly due to the flare continuum, and probably slowly due to rotation of active regions in and out of the field of view (BY Dra syndrome). Equivalent widths are only meaningful if they can be used to compare emission flux in one line to another. Clearly, due to the problems outlined above, this is not generally possible.

Although there have been numerous reports of Balmer line broadening during flares, the spectroscopic data upon which this is based has marginal resolution to study the broadening. High-resolution Balmer data from Schneeberger et al. (1979, 1982) with resolutions of 0.2Å tend not to show broadening at all. It is apparent from the data in Figure 2 that the Balmer lines (and He I lines) are broadened and asymmetric to the red, while Ca II is unbroadened. A high-resolution Hα profile during a large UV Cet flare from Giampapa et al. (1983) shows that the dMe flare broadening is very similar to the solar flare results in Figure 1. In the Figure 1 spectrum, most of the flare emission is confined to a core no broader than the photospheric absorption width. However, the flare kernels do have an extremely broad profile, with some redshifted, broadened emission as well. The broadening of high series ($n > 10$) Balmer lines discussed by Svestka (1972) has not been reported for dMe flares. Nonetheless, one must conclude that dMe flares exibit Balmer line behavior virtually identical to that of a solar flare.

### 3.4. Helium Lines

Joy and Humason (1949) noted He I λλ4471, 4026 and He II λ4686 in an UV Cet flare. Gershberg and Chugainov (1966, 1967) noted other He I lines in dMe flares, such as λλ4921, 5876. In remarkable similarity to solar flares (see Svestka, 1972), He I lines only appear in the larger dMe flares, and He II lines are present only in the strongest flares (see Gurzadyan, 1980).

### 3.5. Metal Lines

Mochnacki and Schommer (1979) identified numerous Fe I, Fe II and Mg I lines, including the strong Mg I B lines, in a large YZ CMi flare. The fact that these lines are apparent only at high spectral resolution suggests that most absorption lines may have emission phases during large flares. At the limit of their resolution (0.5 Å), Mochnacki and Schommer saw neither line broadening nor displacement from the photospheric absorption rest wavelength. Moreover, none of the lines showed any asymmetries. Mochnacki and Schommer also pointed out that high- resolution line profiles may reveal magnetic field information through Zeeman pattern analysis similar to that proposed by Robinson (1980) for absorption lines. Using a simple form of that analysis, Mochnacki and Schommer set an upper limit of 7.5 kilogauss for magnetic fields within the flaring region.

The Ca II H and K lines (λλ3968, 3933) appear strongly in emission in non-flare dMe spectra and are essential diagnostics for determining low chromosphere models (Giampapa et al., 1982). Enhanced Ca II emission during flares was studied by Gershberg and Shakhovskaya (1971) and Bopp and Moffett (1973). Gurzadyan (1980) notes that Ca II enhancement is generally less than Balmer line enhancement, a fact which is readily apparent in Figure 2. Bopp and Moffett also note a red asymmetry in the Ca II wings during some flares.

3.6. UV and EUV Line Emission

Lines in the UV and EUV region probe the transition region between the chromosphere and corona. Since only the closest dMe stars emit enough flux in the UV or EUV to be detected with the International Ultraviolet Explorer (IUE) Satellite, the best space UV and EUV spectrograph yet available, there are only very limited flare spectra available for this important region. Haisch et al. (1981) were the first to obtain IUE spectra during a time which coordinated ground- and space-based observations indicated a flare was occurring on Proxima Centauri. The IUE data were sufficient to set only an upper limit of a factor of two in the excitation of EUV emission lines, although there was marginal evidence for an enhancement of the Mg II H and K lines (λλ2803, 2793). The enhancement of the Mg II lines during flares should be very similar to the Ca II H and K lines since they share similar excitation, and the Mg II lines are formed at only slightly higher levels in the chromosphere (Gurzadyan, 1980). Two other UV and EUV flare detections have been made with the IUE. Butler et al. (1981) detected a flare on Gliese 867A, but had no coordinated observations to determine other flare parameters. A strong UV continuum was present during the flare along with enhanced lines of N V, O I, C I, CII, CIV, Si II, Si IV, and He II. The lower ionization lines (with the exception of N V) were enhanced by smaller amounts than the higher ionized lines. He II λ1640 was enhanced the least. Recently, Haisch et al. (1982; see also the paper by Haisch in this volume) observed a large flare on Proxima Centauri simultaneously with the IUE and the Einstein X-Ray satellite. One spectrum (1200 - 2000 Å) covering the flare X-ray maximum showed considerable enhancement of many lines (He II, C I-IV, N V, Al III, and Si II-IV), with lesser enhancement the lower the excitation temperature of the line. During the decay phase of the flare only weak C I, C IV, and N V lines were present. By comparison to a well-studied solar flare, the ratio of soft X-ray flux to transition region flux is about 10 times larger for the dMe star, indicating that the radiative energy loss from the transition region compared to the corona is more important than for the un.

### 3.7. Asymmetries and Mass Motions

Line asymmetries, probably indicative of mass motions, have been seen for a number of stellar flares (see Gershberg and Shakovskaya, 1971; also note the red asymmetries in the Balmer and He I lines in Figure 2). Bopp and Moffett (1973) measured shifts which suggest downflow velocities of 1100 km/sec for the Balmer lines, and 600 km/sec for the Ca II K line. As mentioned previously, the redshifted emission may well arise from a small portion of the flare, in similarity to the solar data in Figure 1.

In solar flares, most of the emission is not shifted from the line rest position. Similarly, most stellar flare spectra do not show shifts in the central part of the emission lines. One exception is the spectrogram by Greenstein and Arp (1969) which suggests displacements of the emission lines by tens of km/sec. Careful examination of this spectrum also suggests that the plate shifted during the exposure, based on apparently doubled comparison lines. No reliable evidence exists for overall shifts of emission lines during stellar flares.

### 3.8. Timing and Flare Decay

The relative sequence of line, continuum, radio, and X-ray emission during the flare is extremely important in determining at what levels in the stellar atmosphere the flare occurs, and which physical mechanisms are operating. Although, like the solar data (Svestka, 1972), the stellar results are somewhat contradictory, certain features do stand out.

The most basic fact of line and continuum radiation during the flare is that, although the line emission rises with the continuum flash phase of the flare, the line emission decays considerably slower (Kunkel, 1970). Observations of a large flare on YZ Cmi (Schneeberger et al., 1982) show Hα and Hβ emission not fully decayed to a quiescent level 2.5 hours after the continuum flare. In that work and others (Mochnacki and Schommer, 1980; Mochnacki and Zirin, 1981), the Hα emisson appears to rise simultaneously with the continuum flash flare. However, Gershberg and Chugainov (1966, 1967) show the Hα response delayed by 3 to 5 minutes, and the Hβ response delayed by 5 to 10 minutes. Schneeberger et al. (1982) also found that, while Hα seemed to respond instantly to the flare flash, Hβ emission could be delayed by up to 20 minutes. Bopp and Moffett (1973) reported a general trend of the higher series Balmer lines to decay more rapidly than the lower ones. This effect is consistent with the observation that the Balmer "increment" discussed in Section 3.1. reverts to a decrement after the continuum flash phase. The red asymmetries in the Balmer lines seem to occur only near flare maximum and decay rapidly, more or less in step with the continuum light (Bopp and Moffett, 1973). Mochnacki and Zirin (1980) show that the Balmer continuum enhancement occurs simultaneously with and decays more slowly than the overall continuum flare, but still faster than the Balmer line emission.

Gurzadyan (1980) suggests that flare emission lines follow a general rule in that the higher the ionization potential of the ground state, the later the emission appears, and the shorter it lasts. Helium lines appear later than the H lines and decay more rapidly (for example, see Bopp and Moffett, 1973; Gershberg and Chugainov, 1966, 1967). But Mochnacki and Zirin (1980) indicate that the He I λ5876 line developed in time similarly to the Balmer lines in a YZ CMi flare. He I and He II lines are also observed to appear later and decay faster during solar flares (Svestka, 1972). However, this effect is explained by the fact that the initial response of He I λλ5876, 6678, and 10830 to a flare is to go first into absorption, later go into emission, and then return to absortion in later flare stages before completely disappearing. The time response of other lines to flares has been, as for the Sun, little studied. The available data suggest that Ca II H and K lines appear later, and decay later than the Balmer lines, as well as showing a lower amplitude response to the flare (Gershberg and Shakovskaya, 1971). Mochnacki and Schommer suggest that the same behavior exists for the Mg I, Fe I, and Fe II emission lines.

Clearly, much more coordinated, multi-spectral, time-resolved coverage is needed before a coherent picture emerges of the time development of dMe flares. One area which needs attention is to define what differences, if any, are present for emission line development between slow and fast flares.

## 4. SUMMARY

Through a growing body of high spectral resolution and high time resolution stellar flare spectroscopy, one fact becomes apparent: the better the quality of the data, the more qualitatively similar are solar and stellar flare phenomena. The small differences seem to be associated with the much greater relative importance of the outer atmospheric (and flare) regions to the energy balance of the dMe star. Future opportunities and requirements for flare observations, both for the Sun and for the dMe stars, center on coordinated multi-spectral observations. It is hoped that a major product of this colloquium will be programs for continued and expanded observations of this type.

## REFERENCES

Banin, V.G. 1965, Izv. Kryms. Astrofiz. Observ., 32, 118.
Bopp B.W., and Moffett, T.J. 1973, Ap. J., 185, 239-252.
Butler, C.J., Bryne, P.B., Andrews, A.D., and Doyle, J.G. 1981, M.N.R.A.S., 197, 815-827.
Dodson, H.W., and Hedeman, E.R.:1970, Solar Phys., 13, 401-419.
Gershberg, R.E.:1975, in "Variable Stars and Stellar Evolution," ed. V.E. Sherwood and L. Plaut (Dordrecht: Reidel).
Gershberg, R.E., and Chugainov, P.F. 1966, Astron. Zh., 43, 1168.
Gershberg, R.E., and Chugainov, P.F. 1967, Astron. Zh., 44, 260.

Gershberg, R.I., and Shakhovskaya, N.I. 1971, Astron. Zh., 48, 934.
Giampapa, M.S., Worden, S.P., and Linsky, J.L. 1982, Ap. J., 258, 740-760.
Giampapa, M.S., et al. 1983, in preparation.
Greenstein, J.L., and Arp, H. 1969, Astrophys. Letters, 3, 149-152.
Gurzadyan, G.A. 1980, "Flare Stars" (Oxford: Pergamon).
Haisch, B.M., et al. 1981, Ap. J., 245, 1009-1017.
Haisch, B.M., et al. 1982, Ap. J., submitted.
Hartmann, L., and Rosner, R. 1979 Ap. J., 230, 802-814.
Herbig, G.H. 1956, P.A.S.P., 68, 531-533.
Jeffries, J.T., and Orall, F.Q. 1961, Ap. J., 133, 946-962.
Joy, A.H. 1958, P.A.S.P., 70, 505-506.
Joy, A.H., and Humason, M.L. 1949, P.A.S.P., 61, 133-134.
Kahler, S., et al. 1982, Ap. J., 252, 239-249.
Kahn, S.M., et al. 1979, Ap. J. Letters, 234, L107-L111.
Kodaira, K., Ichimura, K., and Nishimura, S. 1976, Publ. Astron. Soc. Japan, 28, 665.
Kulapova, A.N., and Shakovskaya, N.I.:1973, Isv. Krymsk. Astrofiz. Obs., 48, 31-36.
Kunkel, W.E. 1970, Ap. J., 161, 503-518.
Letfus, V. 1964, Bull. Astron. Inst. Czech., 15, 211.
Linsky, J.L. 1980, Ann. Rev. Astron. Astrophys., 18, 439-488.
Mochnacki, S.W., and Zirin, H. 1980, Astrophys. J. Letters, 239, L27-L31.
Moffett, T.J. 1974, Ap. J. Suppl., 29, 1-42.
Moffett, T.J., and Bopp, B.W. 1976, Ap. J. Suppl., 25, 1-36.
Moffett, T.J., Evans, D.S., and Ferland, G. 1977, M.N.R.A.S., 178, 149-157.
Mullan, D.J. 1974, Ap. J., 192, 149-157.
Mullan, D.J. 1977, Solar Phys., 54, 183-206.
Robinson, R.D. 1980, Ap. J., 239, 961-967.
Robinson, R.D., Worden, S.P., and Harvey, J.W. 1980, Ap. J. Letters, 236, L155-L158.
Schneeberger, T.J., et al. 1982, Ap. J., submitted.
Schneeberger, T.J., Linsky, J.L., McClintock, W., and Worden, S.P. 1979, Ap. J., 231, 148-151.
Severny, A.B. 1968, Nobel Symp., 9, 71.
Spangler, S.R., Rankin, J.M., and Shawhan, S.D. 1974, Ap. J. Letters, 190, L129-L131.
Svestka, Z. 1963, Bull. Astron. Inst. Czech., 14, 234.
Svestka, Z. 1972, Ann. Rev. Astron. Astrophys., 10, 1-24.
Svestka, Z. 1976, "Solar Flares" (Dordrecht: Reidel).
Torres, C.A.O., and Ferraz-Mello, S. 1973, Astron. Astrophys., 27, 231.
Vogt, S.S. 1980, Ap. J., 240, 567-584.
Worden, S.P., and Peterson, B.M. 1976, Ap. J. Letters, 206, L145-L147.
Zirin, H. 1982, private communication.
Zirin, H., and Ferland, G.J. 1980, Big Bear Solar Obs., preprint 0192.
Zirin, H., and Tanaka, K. 1973, Solar Physics, 32, 173-207.

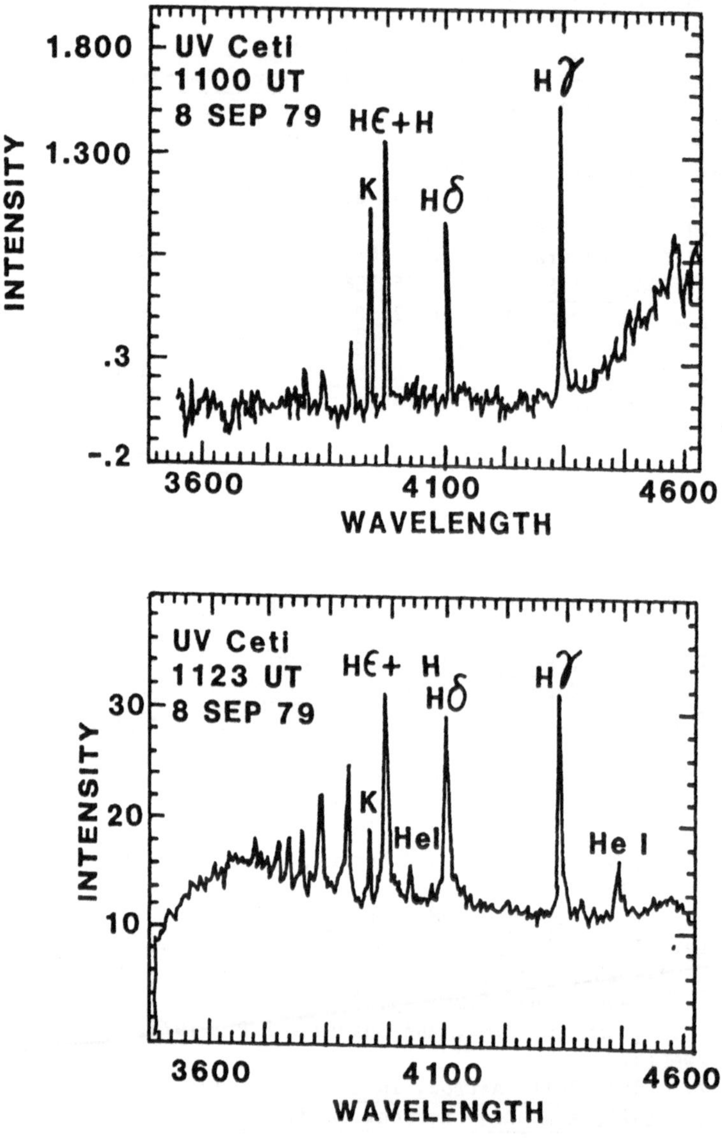


Figure 2. Blue spectra for the dMe star UV Cet from Giampapa et al (1983). The top spectrum shows UV Cet in a non-flare state. The botton spectrum was taken during flare maximum of a 5 magnitde (U-band) flare. Spectral resolution is 3.8Å, and time resolution 3 minutes. Note the broadened and redshifted hydrogen and helium lines.

DISCUSSION

Goldberg: Among your suggestions for future programmes was one for observations in the 500-50A region. How many stars do you expect to penetrate the opacity below 912 A?

Worden: The data which I have seen suggests that we should be able to make observations out to 10-20 pc. This would include most of the flare stars.

Stencel: The science working groups on the Far Ultraviolet Explorer has addressed this question in some detail. In fact some of their results will be published in the proceedings of the recent ESA IUE meeting. The answer is that hundreds of flare stars will be observable below 700 A. I also have a question. In the Sun the broadening of Hα is occasionally very severe. Could you comment on the difficulty of mapping wavelength into velocity to determine physical motions especially in lines like Hα which are Stark broadened.

Worden: Stark broadening is the predominant broadening mechanism. It is difficult to disentangle Stark broadening from that due to mass motion. So one should be careful in interpreting wavelength shifts. There are no observations that are reliable, in my opinion, in stellar flares that show that the central peaks of the hydrogen lines shift nor are there in the Sun. We see extended wings. An important point to remember when you are looking at stellar flares is that you are looking at different areas in different parts of the lines. In the core you are seeing the whole flare apart from those regions which may be undergoing mass motions. In the wings one may be seeing a surge. So it may be very risky to interpret a shift simply as material motion. We have the same problem with a solar surge.

Vaiana: Going back to the question of how far one might see, it would be important to be able to reach at least some of the T-Tauri stars, say in the Pleiades or Hyades. This would be especially so if they could be reached in their quiescent states as well as of course in their flaring state. Then we would have a much better description of these very active young stars.

RESULTS FROM OPTICAL AND UV STELLAR FLARE SPECTROSCOPY

Mark S. Giampapa
Harvard-Smithsonian Center for Astrophysics

ABSTRACT

Optical and ultraviolet spectroscopy can enable the assessment of the physical conditions characterizing a stellar flare atmosphere and thereby potentially elucidate the possible radiative and hydrodynamic transport mechanisms operative during stellar flares. In this review, I present illustrative examples of the spectroscopic diagnostic techniques that can be applied to the analysis of stellar flare spectroscopic data and the resulting inferences concerning stellar flare properties for M dwarf flare events.

INTRODUCTION

High-to-moderate spectral resolution, high temporal resolution spectroscopy of M dwarf stellar flares can yield unique and valuable data that can be utilized to identify the relative roles of lines and continua, and hence the kind of radiative transport mechanisms that operate during the course of flare events. Furthermore, it is possible through spectral line analysis to estimate physical conditions, such as temperature and density, in the flare emitting volume at various depths in the M dwarf flare atmosphere. The atmospheric levels that can be probed depend, of course, on the wavelength coverage of the observations. In this regard, conventional broadband, multichannel photometry can provide extensive wavelength coverage combined with very high (< 1 sec) temporal resolutions. A disadvantage is that the sampling of the flare spectrum is coarse and the relative contributions of lines and continua to the observed flare light is unknown. Thus we cannot rely exclusively on photometry if we seek to thoroughly understand stellar flare activity. For example, the Johnson U and B photometric bandpasses completely encompass the wavelength range of the spectrophotometric flare observation of UV Ceti shown in Figure 1. Thus many of the principal emission lines in the "enhanced chromospheric regions" of the flare are included in these bandpasses. As a result, ambiguity can be introduced into the determination of the operative continuous emission mechanism from photometric colors alone. This



*P. B. Byrne and M. Rodonò (eds.), Activity in Red-Dwarf Stars, 223–235.*

caveat has been repeatedly stressed by flare star observers. As an example, the temporal evolution of the ratio of total line emission flux to total observed flux for the 8 September 1979 UV Ceti flare event (Fig. 1) proceeded from a quiescent value of approximately 0.06 to a value of 0.18 during flare onset, followed by a decline to 0.03 at flare maximum, and then an increase to approximately 0.3 during the decay stage of the flare. This example clearly indicates that stellar flare spectroscopy is imperative for the understanding of flare colors and flare light curves.

OPTICAL FLARE SPECTROSCOPY

Spectroscopic observations of M dwarf flares, such as that of Figure 1, readily reveal the shape of the flare continuum and thus allow us to explore the possible continuous emission mechanisms that may account for the observed flare light. Several investigators have suggested that the optical flare continuum is composed of bound-free and free-free emission from a two-component thermal model characterized by electron temperatures of $T \sim 10^5$ K and $T \sim 2x10^4$ K. I display in Figure 2 the spectral energy distributions of some of the suggested radiative processes that may give rise to flare continua. The wavelength range corresponds to that of Figure 1 and the vertical scale is arbitrary. The curves represent optically thin emission along a ray and are merely used here to illustrate the potential suitability of a given radiative process as a description of the observed wavelength distribution of stellar flare continuous emission. I have assumed LTE for both $H^-$ and Balmer bound-free emission in the construction of Figure 2. For illustrative purposes, I will discuss the various processes represented in Figure 2 within the context of the UV Ceti optical flare event shown in Figure 1.

The shapes of the free-free curves at $T = 10^5$ K and $T = 5x10^4$ K appear to be reasonable fits to the observed flare continuum shown in Figure 1. However, ultraviolet data are required to determine whether these distributions are truly applicable. Each of the free-free curves exhibits a rise toward the blue. In fact, the $T = 10^5$ K distribution would continue to increase monotonically toward 1000 A while the $T = 5x10^4$ K curve would behave similarly, attaining a maximum near 1800 Å, followed by a slow decline in intensity with decreasing wavelength. However, in essence, both curves are quite flat with no more than a factor of 2 variation from maximum to minimum emission between 1000 Å and 5000 Å.

Kunkel (1970) has argued that hydrogen free-bound emission is the principal component of the observed continuous emission in M dwarf flare events. According to Figure 2, the shape of the bound-free curve, like the free-free curves, approximates the shape of the flare continuum displayed in Fig. 1 well in the "red" portion of the flare spectrum but not in the blue. Of course I cannot conclude that free-bound hydrogen emission is not a constituent of the emission; but it may not be the sole component and additional continuum processes originating in higher temperature layers in the flare emitting volume, such as free-free

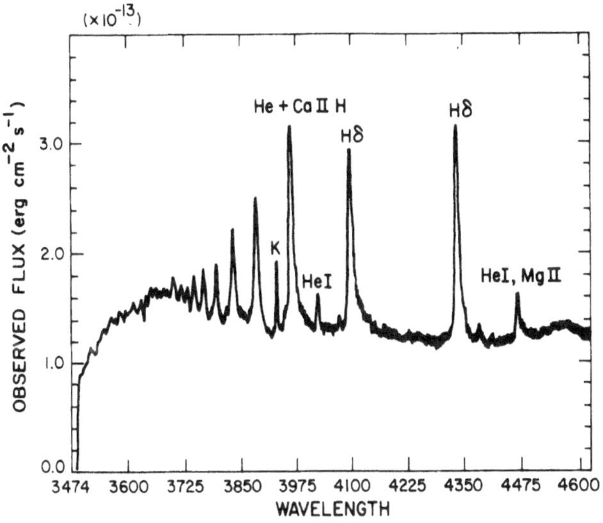


Figure 1: Spectrum during flare maximum for a 5 magnitude (U-band) event observed on UV Ceti (taken from Giampapa et al. 1982).

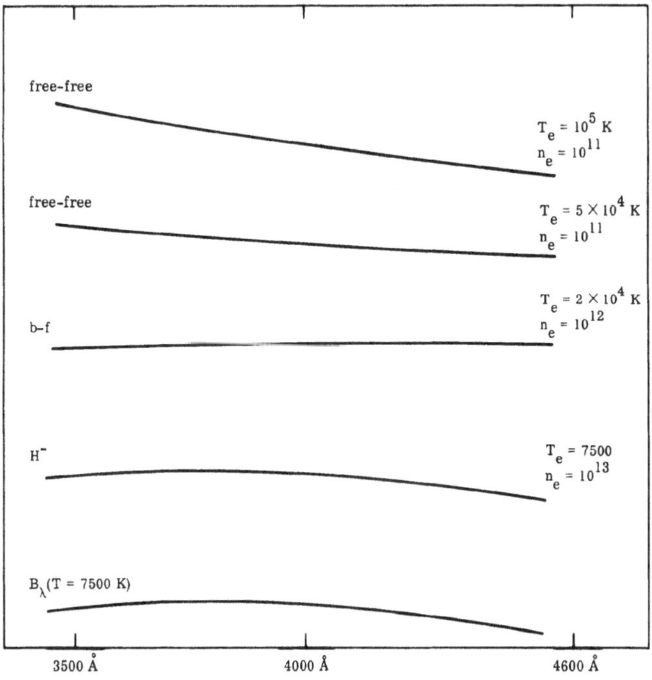


Figure 2: Spectral energy distributions of radiative processes that may give rise to flare continua. The wavelength scale is identical to that of fig. 1. The vertical scale is arbitary.

emissions, are likely present. In this regard, however, I note the caveat that the density of emission lines (especially the higher members of the Balmer series) increases toward the blue and this effect can, in turn, artifically enhance the apparent continuum level in moderate resolution spectra, particularly in the vicinity of the Balmer jump.

An additional mechanism that may participate in the formation of stellar flare continua is $H^-$ emission. Svestka (1966) originally suggested that $H^-$ emission emanating from the temperature minimum region could account for the continuous emission of _solar_ flares as observed in the optical. This suggestion has been recently verified by Neidig (1982 and references therein) in the case of a solar white light flare. Grinin (1976) has applied this suggestion to M dwarf stars to demonstrate that upper photospheric heating by a flare blast wave could enhance the formation of negative hydrogen ions and thereby increase the importance of $H^-$ emission as a component of stellar flare optical continua. There is reasonable agreement between the $H^-$ spectral energy distribution in Figure 2 with the flare continuum shape in Figure 1 in the red region. But, as in the case of hydrogen bound-free emission, there is less agreement in the blue. In fact, $H^-$ emission would decline rapidly toward the UV in those spectral regions currently accessible to the camera on board the _International Ultraviolet Explorer_ (IUE) satellite.

Kunkel (1970) and, more recently, Mochnacki and Zirin (1980) have suggested that stellar optical flare continua can arise from black body emission originating at the footpoints of the flare. This suggestion implies that deep photospheric heating can occur during a stellar flare event. In particular, the point at which $\tau_{5000} = 1$ would be located at higher column mass densities and temperatures in a stellar "flare photosphere" relative to the quiescent stellar photosphere. Utilizing spectrophotometric observations of flare events on the dM4.5e star YZ CMi, Mochnacki and Zirin (1980) concluded that the black body component of the observed continuum emission was the dominant component in the formation of the flare continuum while hydrogen recombination emission was the lesser component. Moreover, these investigators deduced that the black body emission must arise from a projected area that is less than 2% of the area of the visible stellar disk. In a similar analysis, Kahler _et al_. (1982) found that the peak of an optical flare event on YZ CMi was fit well by black body emission characterized by T = 8500 K and extending over an area of 0.5% of the projected stellar disk. This kind of impulsive photospheric heating would result in the ionization of low temperature species, such as molecules and neutral metals, in the upper photosphere-temperature minimum region. Thus absorption line equivalent widths may decline in proportion to the area coverage of the flare footpoints. Unfortunately, a change of less than 2%, as implied by the flare area estimates of Mochnacki and Zirin (1980) and Kahler _et al_. (1982), would be difficult to detect at the high temporal and spectral resolutions required for meaningful stellar flare spectroscopy. Interestingly, Mochnacki and Schommer (1979) obtained a spectrum of

YZ CMi during a flare event in which the photospheric absorption lines were very weak (i.e., filled in by emission or strongly ionized) as compared to the quiescent, non-flare spectrum of YZ CMi in the same wavelength region. In addition, emission lines of Mg I, Fe I and Fe II appeared in the flare spectrum thus indicating (in a schematic fashion) a shift of the chromospheric thermal structure to significantly higher column mass densities at the flare site. The noticeable weakening of the photospheric absorption lines could have resulted from a significant flare area coverage. Alternatively, intrinsically bright emission in photospheric lines arising from small flare areas would have "masked" or "veiled" the photospheric spectrum arising from the undisturbed portions of the stellar surface. Indeed, small flare areas (< 2%) such as those inferred by Mochnacki and Zirin (1980) and Kahler et al. (1982), appear more compatible with the time-scales for M dwarf optical stellar flare events deduced from high-speed optical photometry (see the review by P.B. Byrne in this volume) if the characteristic velocity for a stellar optical flare (i.e., collisional excitation of lines and continuum formation) is of the order of the local sound speed.

## OPTICAL FLARE LINE SPECTRA

An observed stellar flare line spectrum can provide diagnostics of the physical conditions (i.e., $n_e$ and $T_e$) that characterize the line emitting regions. For example, we can utilize the classical Inglis-Teller equation which relates the electron density in the Balmer line emitting layers to the highest resolvable Balmer series member before merging of the lines occurs. According to Kurochka and Maslennikova (1970), the relation is $\log n_e = 22.0 - 7.0 \log m$, where $n_e$ is the electron density and m is the principal hydrogen quantum number for the highest resolvable Balmer series member. Enlarged views of Figure 1 show at least m = 15 for this event and hence $n_e \lesssim 10^{13}$ cm$^{-3}$ in the Balmer line emitting regions. Exploiting these high series members further, we can compare observed line ratios to those computed assuming a Boltzmann (LTE) distribution in the level populations. I display in Table 1 the observed and computed ratios of $f_m/f_{11}$, where $f_{11}$ is the observed flux in the H11 line (the observed fluxes are measured with respect to the nearby flare continuum level). Inspection of Table 1 clearly reveals that the observed flux ratios are in the ratios of the statistical weights. Thus the lines are optically thin and the upper level populations are controlled by collisional processes at the local

Table 1. Observed and Computed Balmer Line Ratios

| m | Observed | Computed |
|---|---|---|
| 11 | 1 | 1 |
| 12 | 0.78 | 0.77 |
| 13 | 0.61 | 0.61 |
| 14 | 0.52 | 0.48 |
| 15 | 0.39 | 0.40 |

electron temperature and density. The physical significance of this result is that it demonstrates that the mechanism for hydrogen line excitation must be of a thermal nature rather than a non-thermal nature.

Following Kurochka and Sitnik (1976), we can use the optically thin H11 line to obtain an estimate of the emission measure. The flux in the H11 line averaged over the stellar surface is $F(H11) = 6 \times 10^4$ ergs $cm^{-2}$ $s^{-1}$. Assuming $T_e = 2 \times 10^4$ K, I find an emission measure of $n_e^2 \Delta V = 2 \times 10^{53}$ $cm^{-3}$ which is a few orders of magnitude greater than the emission measure of the largest solar flares (Moore et al. 1980). The total X-ray luminosity may be crudely estimated from

$$L_x \sim \left[\frac{P(E_1 - E_2)}{n_e^2}\right] (n_e^2 \Delta V),$$

where the quantity in brackets has been tabulated as a function of temperature for optically thin gases of cosmic abundance by Raymond, Cox and Smith (1976). Assuming $T = 3 \times 10^7$ K yields $[P(E_1 - E_2)/n_e^2] = 3 \times 10^{-23}$ ergs $s^{-1}$ $cm^{-3}$ which, in turn, gives $L_x = 6 \times 10^{30}$ erg $s^{-1}$

Mullan (1976) has emphasized the importance of the ratio $L_{opt}/L_x$ as a measure of the fractions of the coronal conductive and radiative losses that participate in the (secondary) heating of the underlying, cooler "optical flare region." For the entire bandpass shown in Figure 1, I find that $L_{opt} = 2 \times 10^{29}$ erg $s^{-1}$ or $L_{opt}/L_x = 0.03$. I regard the ratio as a lower limit since contributions from important lines (Mg II h and k, Fe II, Hα, Hβ, etc.) and continua in the near UV and the optical longward of about 5000 Å are not included in the estimate of $L_{opt}$.

## HELIUM LINES

In addition to the Balmer lines, neutral helium lines are visible in the optical flare spectrum presented in Figure 1. Moffett and Bopp (1976) have noted the appearance of the He I λ4471 triplet line in some M dwarf stellar flare events. The nearby He I λ4026 triplet line was usually not detected except in the strongest U band events reported by Moffett and Bopp (1976). These investigators note that the λ 4026 line is near a strong Mn I absorption feature at λ4030 thus making detection difficult except in the more energetic U band events. According to Robbins (1968), an optically thin recombination spectrum would yield I(λ4026)/I(λ4471) = 0.48. I find for the UV Ceti flare event of Figure 1, an observed ratio of λ4026/λ4471 = 1.0 thus indicating optically thick lines. The He I λ4388 singlet line is also present in this spectrum. The analogous triplet transition is λ4026. The resulting observed triplet-to-singlet ratio is 2.8, which is near the LTE ratio of 3. This kind of triplet-to-singlet ratio is characteristic of active solar prominences with $n_e > 10^{12}$ $cm^{-3}$ and $T_e > 8000$ K (Tandberg-Hanssen 1974; see also Giampapa et al. 1978 for a parallel discussion in the case of the λ6678 and λ5876 singlet and triplet lines).

## HIGH-RESOLUTION BALMER LINE OBSERVATIONS

The Balmer lines are very sensitive to temperatures and pressures in the upper chromospheric ($T > 6000$ K) and transition regions of M dwarf stars while the Ca II H and K lines are strictly lower chromospheric diagnostics (Cram and Mullan 1979; Giampapa 1980; Giampapa, Worden and Linsky 1982). As the region of Hα line formation is more nearly in the vicinity of the presumed site of primary flare energy release, the Balmer lines can be expected to be particularly sensitive to flare activity, especially given their sensitivity to transition region pressures in the M dwarf stars. In fact, the Hα profile in the spectrum of a dMe star exhibits a central reversal (Worden, Schneeberger and Giampapa 1981) which is a non-LTE effect related to the presence of an optical boundary that ultimately leads to Hα source function de-coupling from the local Planck function. However, during a flare the Hα central reversal can vanish (see Schneeberger et al. 1979, their Fig. 3). This observation indicates the presence of a high-pressure transition region at the flare site, or schematically, an inward shift of the chromospheric thermal structure to higher column mass densities. This is essentially a description of the "chromospheric evaporation model," such as those discussed by Machado and Linsky (1975). With regard to the Hα line, chromospheric evaporation models show that hydrogen ionization occurs before the Hα source function can de-couple from the local chromospheric Planck function. As a result, we do not observe a central reversal in the flare Hα profile. In addition, enhanced macroturbulent motions can obscure a central reversal. Furthermore, an increased microturbulent contribution to the Doppler broadening will cause Hα line thermalization at increased temperature and densities in the flare chromosphere.

Finally, the Balmer lines are often considerably more enhanced during a flare than are the Ca II resonance lines. As mentioned previously, the Ca II H and K lines are formed in the lower M dwarf chromosphere in a region that would, during a flare, correspond to the flare footpoints, while Hα is strictly an upper chromospheric-transition region diagnostic. Thus the area of Hα (and Balmer) line emission is greater than that for Ca II emission located at the flare footpoints.

## THE BALMER DECREMENT

Investigators have reported observing a "negative decrement" with $EW(H\beta) < EW(H\gamma) < EW(H\delta)$. Kunkel (1970) has shown that this kind of behavior can be explained if the Balmer lines are driven toward LTE conditions with $n_e \sim 10^{13}$ cm$^{-3}$, $T_e \sim 2 \times 10^4$ K and $\zeta_t \sim 20$ km s$^{-1}$, where $\zeta_t$ is the microturbulent velocity. Recently Drake and Ulrich (1980) computed Balmer line emission from slab geometries for a wide range of conditions. These investigators find $H\beta < H\gamma \lesssim H\delta$ for densities in the range $\log n_e \approx 13$-$14$, $T_e \simeq 2 \times 10^4$ K, and $\tilde{\tau}(L\alpha) \sim 10^5$.

## LINE WIDTHS AND LINE ASYMMETRIES

The review by S.P. Worden (hereafter SPW) noted that broadening of the Hα and Hβ lines is not a general feature of M dwarf flare spectra.

However, in powerful flare events, such as that shown in Figure 1, the base widths of the Balmer lines are clearly greater than that of the pre-flare, quiescent spectrum (see Fig. 2 of the review by SPW). The broadening in Figure 1 is likely due to Stark broadening. In fact, notice that the Ca II K line is narrow compared to the hydrogen lines in Figure 1. The expression for Stark broadening for Balmer lines shortward of Hα is $\gamma = 0.255\ (n_i^2 - n_e^2) n_e^{2/3}$ (Sutton 1978) while that for Ca II is $\gamma = 2.6 \times 10^{-6}\ n_e$ (Raymer 1979). Assuming $n_e \sim 10^{13}$ cm is approximately applicable for both Ca II K and the neighboring H8 line yields $\gamma(\text{CaII K})/\gamma(\text{H8}) = 3.7 \times 10^{3}$. Thus Stark broadening is negligible for Ca II relative to the Balmer lines.

The review by SPW discussed evidence concerning line asymmetries in M dwarf flare spectra. I would add the caveat that blends can mimic asymmetries in moderate or low spectral resolution flare observations. For example, the Ca II λ3968 resonance line can be blended in the wings of the λ3970 Hε line; the He I λ4121 triplet is near Hδ (λ4101); and a Mg II subordinate line at λ4481 resides near the He I λ4471 triplet, as can be seen in Figure 1.

## ULTRAVIOLET FLARE SPECTROSCOPY

Unfortunately a paucity of ultraviolet spectroscopic observations of M dwarf stellar flare events is available at the time of this writing and the published (or in press) observations have been summarized by SPW. A principal feature of the UV observations of a flare on Gliese 867A was the presence of a strong, rather flat UV continuum that is reminiscent of free-free emission at $T \sim 5 \times 10^4$ K or $10^5$ K. However, Butler et al. (1981) attribute the flare continuum to black body emission at $T = 2 \times 10^4$ K. This value is significantly higher than optical black body temperature estimates given by Mochnacki and Zirin (1980) or Kahler et al. (1982).

The flare line emission observed by Butler et al. (1981) exhibits interesting behavior as well. I display in Figure 3 a plot of line surface flux (normalized to the quiet Sun) versus temperature of line formation. I constructed these curves on the basis of data given by Butler et al. (1981) and they should essentially represent the run of emission measure in the flare and quiescent stellar atmospheres. The "flare curve" in Figure 3 is somewhat similar to the "quiet curve." An important difference is a relatively shallow emission measure slope for $T > 10^5$ K. The N V λ1240 line ($T \sim 2 \times 10^5$ K) and the He II λ1640 line were not significantly enhanced as would be expected following an initial X-ray flare outburst. Interestingly, Baliunas and Raymond (1982, private communication) recently observed a UV flare event on EQ Peg that is similar to the Butler et al. (1981) GL 867A observations. In particular, the EQ Peg observations reveal flare continuum emission as well as enhanced line emission but no enhancement of N V or He II was seen. Apparently the flare event was confined to atmospheric levels below the region of C IV line formation, or $T \lesssim 1.1 \times 10^5$ K; that is, these kinds of flares may not be associated with coronal events but may originate in the chromosphere of the M dwarf star.

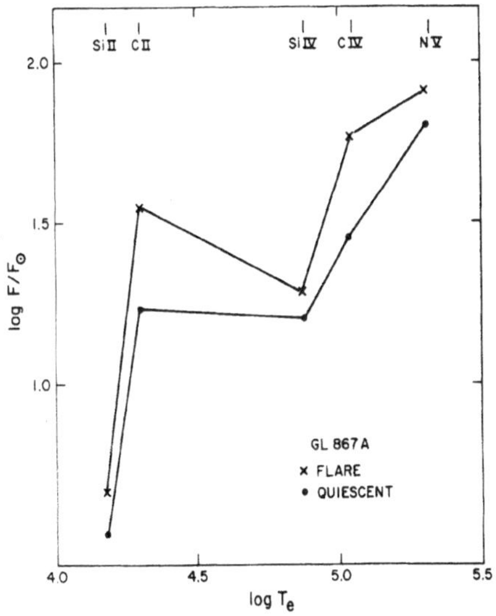


Figure 3: Normalized surface flux versus temperature of formation of the resonance lines of the ions shown during the flare and quiescent state of the dMe star GL867A. Based on data from Butler et al. (1981).

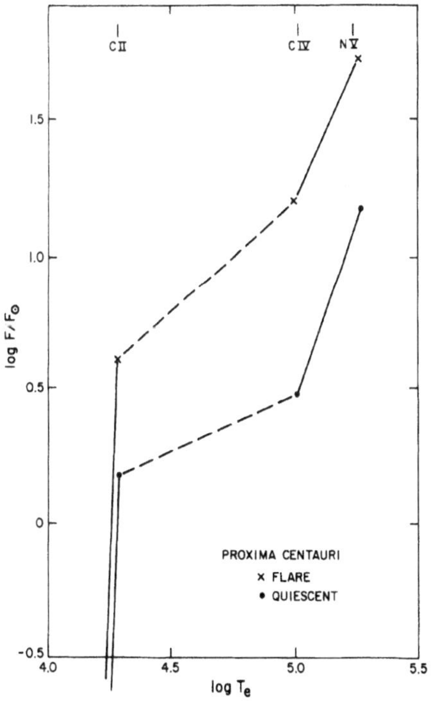


Figure 4: Same as Fig. 3 but for the dMe star Prox Cen. The flare data are taken from Haisch et al. (1982) and the quiescent data are from Linsky et al. (1982).

In juxtaposition to the aforementioned events is the observation of a flare event on Proxima Centauri by Haisch et al. (1982; see the detailed review in this volume). I construct in Figure 4 an "emission measure" curve that illustrates the strong enhancements in the lines (including NV) although the slope between CIV and NV is not quite as steep as in the quiescent spectrum. Haisch et al. (1982) did not observe any UV continuum emission and they concluded that either there was no UV continuous emission during the flare or any continuous emission that occurred was of short duration.

Bromage et al. (1982; see the contribution in this volume) observed a flare event on AT Mic that showed continuum emission increasing toward longer wavelengths in the SWP camera (1100-2000Å) of the IUE. This is reminiscent of bound-free hydrogen emission, $H^-$ emission, or black body emission at $T < 10^4$ K, or, of course, a combination of these radiative processes. A more detailed description of these UV observations appears in this volume (see the contributed paper by G. Bromage and collaborators). In summary, the simultaneous acquisition of additional UV and optical flare data is imperative if we are to progress any further in the understanding of stellar (and solar) flare activity.

As a final comment, it is vital to know the relative importance of the UV and the optical to the total flare energy budget. Unfortunately, simultaneous UV and optical observations are not yet available for M dwarf flares, although solar flares may serve as a guide. The results from the well-observed solar flare of 5 September 1973 reveal that the UV and the optical from 1100Å-8700Å dominated the radiative power output at flare maximum, although soft X-ray observations were unavailable for this analysis (Canfield et al. 1980). Thus the UV and optical may dominate the energy balance in stellar flares as well. However, I note that Wagner et al. (1981) observed that the mass energy flux in a transient associated with the solar flare of 7 April 1980 exceeded the total radiative energy of the flare by at least a factor of 10. Thus we must eventually address the observational problem of measuring the mechanical energy of the material motions in stellar flares in order to discern the relative importance of various flare cooling mechanisms. The answers will ultimately derive from high temporal and spectral resolution spectroscopy.

REFERENCES

Baliunas, S. and Raymond, J. 1982, private communication.

Bromage, G. et al. 1982, in preparation.

Butler, C.J., Byrne, P.B., Andrews, A.D. and Doyle, J.G. 1981, MNRAS 197, 815.

Canfield, R.C. et al. 1980, in Solar Flares, ed. P.A. Sturrock (Boulder: Colorado Associated University Press), p. 451.

Drake, S.A. and Ulrich, R.K. 1980, Ap.J.Suppl., 42, 351.

Giampapa, M.S., 1980, Ph.D. dissertation, University of Arizona, Tucson.

Giampapa, M.S., Linsky, J.L., Schneeberger, T.J. and Worden, S.P. 1978, Ap.J. 226, 144.
Giampapa, M.S., Worden, S.P. and Linsky, J.L. 1982, Ap.J. 258, 740.
Giampapa, M.S. et al. 1982, in preparation.
Grinin, V.P. 1976, Izv. Krim. Astrofiz. Obs., 60, 179.
Haisch, B.M. et al. 1982, Ap.J. submitted.
Kahler, S. et al. 1982, Ap.J., 252, 239.
Kunkel, W.E. 1970, Ap.J., 161, 503-
Kurochka, L.N. and Maslennikova, L.B. 1970, Solar Phys., 11, 33.
Kurochka, L.N. and Sitnik, G.F. 1976, Sov. Astron., 19, 590.
Linsky, J.L. et al. 1982, Ap.J., in press.
Machado, M.E. and Linsky, J.L. 1975, Solar Phys., 42, 395.
Mochnacki, S.W. and Schommer, R.A. 1979, Ap.J. (Letters) 231, L77.
Mochnacki, S.W. and Zirin, H. 1980, Ap.J.(Letters), 239, L27.
Moffett, T.J. and Bopp, B.W. 1976, Ap.J.Suppl. 31, 61.
Moore, R. et al. 1980, in Solar Flares, ed. P. Sturrock (Boulder: Colorado Associated University Press), p. 341.
Mullan, D.J. 1976, Ap.J., 207, 289.
Neidig, D.F. 1982, Solar Phys., 78, 225.
Raymer, M. 1979, Ph.D. thesis, University of Colorado, Boulder.
Raymond, J.C., Cox, D.P. and Smith, B.W. 1976, Ap.J., 204, 290.
Robbins, R.R. 1968, Ap.J., 151, 511.
Schneeberger, T.J., Linsky, J.L., McClintock, W. and Worden, S.P. 1979, Ap.J., 231, 148.
Sutton, K. 1978, J.Q.S.R.T., 20, 333.
Svestka, Z. 1966, Bull. Astron. Inst. Czech., 17, 137.
Tandberg-Hanssen, E. 1974, Solar Prominences (Boston: Reidel), pp. 70-73.
Wagner, W.J. et al. 1981, Ap.J.(Letters), 244, L123.
Worden, S.P., Schneeberger, T.J. and Giampapa, M.S. 1981, Ap.J.Suppl. 46, 159.

## DISCUSSION

Basri: I feel compelled to point out that N V may be a deceptive diagnostic. Results from a large survey of RSCVn stars, where hopefully one does not need to worry about flaring, just about"quiescent" active chromospheres, show a different behaviour in N V from other transition region diagnostic lines. As measured by IUE in low dispersion it behaves more like a chromospheric line than a transition region line. So either there is something strange about the formation of N V or, as Linsky pointed out at the Zurich meeting, it may be heavily contaminated with chromospheric lines. In particular we do not see the same kind of period/ activity relationship for N V as for other transition lines. As I will show tomorrow the slope of the correlation of N V vs other transition region lines is more like that of chromospheric lines vs the other transition region lines. So one would need to be very careful in using IUE data on N V.

Jordan: May I add a comment on that. I had not heard this result before. It is true that there are a lot of C I lines around the wavelength of N V. It is also true that N V has a very long tail to its ionization curve which extends right out to the corona. So that if in the quiescent state the temperature were only in the region of say 500 000 K then N V would be formed over the entire volume of the corona. So that when a flare went off it would only occupy a tiny fraction of that volume and the total emission would not be enhanced as much. I do not know if that would work or not but it is a thought.

Giampapa: That is a really interesting suggestion. I would add that Bromage saw optical events associated with which there were no UV events. So while N V may well be a deceptive diagnostic, it may also be that there are impulsive events which are confined to cooler regions.

Linsky: I have a question and comment. In the Table which you showed from the Canfield reference there is a lot of emission in the region from 1500-2000 Å. So I would like to make a plea that we try very hard to measure the flux in that continuum. I realize that it is very difficult with IUE because of scattered light and its poor sensitivity and also the fact that we are looking at very weak emission. It is important to measure this in order to understand the mechanism whereby the continuum is formed.

Vaiana: In view of the comments on N V presumably a statement about the possibility of having two types of events, one which peaks around C IV and the other in the X-ray region starting from N V, should be put much less strongly. There is another argument against this and that derives from the recent flare seen by Haisch and Todd in which there is the same decrement in N V even though there is very strong X-ray emission.

Giampapa: I did not mean to make a general statement. As we know from the Sun there is an incredible variety of transient phenomena and may be we are not even sure what we should call a flare.

Vaiana: I know of one without X-ray emission but I do not know of one which stops at C IV.

Kodaira: How high was the time resolution at the IUE spectra?

Giampapa: Time resolution is typically 20 or 30 mins.

Kodaira: In that case it is quite probable that the lines come from the decay phase. It is quite important to decide whether you are dealing with the impulsive or the decay phase. So the emission may be dominated by chromospheric lines.

Giampapa: That is an important point. Haisch made this point also. In the optical the continuum emission is of much shorter duration than the line emission. So one may not be able to see the continuum with IUE anyway.

Rosner: I wonder whether anyone has for the solar case tried to compare the X-ray and UV measurements of compact loop flares, which are fairly small structures low down in the atmosphere, and the prominence-associated flares, which are much more disruptive and where there is a lot of motion associated with cool material. These remind me of the kind of events you have been talking about where you see lots of emission in cooler lines.

Giampapa: I am not aware of any such comparison but that is an interesting point.

# OBSERVATIONS OF $H_\beta$ AND HE II $\lambda$ 4686 LINES IN THE SPECTRA OF FLARES OF UV CET-TYPE STARS

P.F. Chugainov, P.P. Petrov, A.G. Shcherbakov
Crimean Astrophysical Observatory, USSR

We have carried out 45.4 hours of continuous spectroscopic and photoelectric B-band observations of AD Leo, DT Vir and YZ CMi. The spectroscopic resolution was about 1 A in wavelength and 5-10 min in time. 6 flares were recorded. Light curves and line profiles are to be presented in the Izv.Crim. Astrophys. Obs., Vol.69. In this paper only the main results are discussed.

## PREFLARE STRENGTHENING OF $H_\beta$ LINE

In 2 flares of AD Leo and 2 flares of YZ CMi, i.e. in 67% of our recorded flares, an increase of the central intensity of $H_\beta$ was observed 10-20 min before the maximum of the star brightness in the B-band. The increase was 10-15% for 3 flares and 50% for one flare in which an increased star brightness was observed 8 min preceding the maximum and a possible negative flare arose 2 min before the maximum. This flare seems to be an unusual one because: 1) in the other 3 flares where the strengthening of $H_\beta$ occured before the maximum the star brightness did not increase at that time; 2) very extensive photoelectric observations of UV Cet-type stars show that such flares with the increase of brightness 10-20 min before the maximum are very rare. Hence one may conclude that the preflare radiation is characterized by an increase of line emission and a strengthening of the continuum emission which only occurs in rare events. At flare maxima it is the continuum emission that shows the greatest enhancement.

## THE $H_\beta$ LINE PROFILES DURING THE FLARES

The spectra of one flare of AD Leo and one flare of YZ CMi definitely indicate the formation of broad wings of $H_\beta$ occuring mainly during flare maximum. These flares surpass the other 4 flares in total optical energy. The line wings are traced ± 15 A for the flare of AD Leo and ± 10 A for the flare of YZ CMi. The intensities of the violet and red



*P. B. Byrne and M. Rodonò (eds.), Activity in Red-Dwarf Stars, 237–238.*

wings are not equal at the same distances from the line center. This effect is especially noticeable on the spectrum obtained near the maximum of the flare of AD Leo where it looks like a 'red asymmetry'. The Stark-effect seems to be the most appropriate explanation of the origin of the wings. Assuming an electron temperature of 10000 K and optical depth at line center of 100-1000 we have found the following estimates of the electron density: log $n_e$=14-15 for the flare of AD Leo and log $n_e$= 14 for the flare of YZ CMi. The asymmetric profiles imply motions with velocities of the order of 100-1000 km/s.

## He II λ 4686 LINE

Special effort was made to search for the presence of this line in the spectra of flare stars in the quiet state. Each star was analysed by 2 or 3 spectra. The averaged spectra showed that the equivalent widths of line emission or absorption do not exceed 0.02 A for AD Leo and DT Vir, and 0.1 A for YZ CMi. But for the case of flares the spectra were not averaged. Thus the upper limit of the equivalent width of the λ 4686 line turned out to be higher than that in the quiet state. It is approximately 0.07 A for flares of AD Leo and 0.2 A for flares of YZ CMi. The λ 4686 line is not revealed in any flare.

The emission in the λ 4686 line can be produced by the cascade recombination of $H_e$ III ions which appear due to the X-ray flux. Assuming the ratios of X-ray, optical and λ 4686 luminosities being L(x)/L(opt)=1, L(4686)/L(x)=0.0012 we have found that the expected equivalent width of the λ 4686 line during flares is not higher than the threshold of our observations. However, equivalent widths of the λ 4686 line of the order of 1 A and greater were observed in 2 flares of UV Cet by Joy and Humason (1949) and Gershberg and Chugainov (1967). These values cannot be explained by the cascade recombination mechanism if the ratio of optical and X-ray luminosities is nearly the same for all flares of UV Cet-type stars.

## REFERENCES

Joy, A.H., Humason, M.L.: 1949, Publ. Astr. Soc. Pacif. 61, pp.133-134.

Gershberg, R.E., Chugainov, P.F.: 1967, Astron. J., USSR 44, pp.260-266.

# THE TIMESCALES OF VARIATIONS IN CONTINUUM AND HYDROGEN LINES DURING STELLAR FLARES

B. R. Pettersen
Institute of Mathematical and Physical Sciences
University of Tromsø, N-9001 Tromsø, Norway.

Abstract
Light curves of major stellar flares have been used to study the behavior of U-B, B-V, and V-R. The majority of the flux transmitted through these filters is continuum radiation, but U and B are affected by emission lines. The variability of Hα and Hβ emission lines were monitored through narrow band filters. The timescales of emission line variability are considerably longer than those for the continuum, and the emission line flare peak occurs a few minutes after the continuum flare maximum. No variability in lines at a timescale of seconds is detected in our data.

## OBSERVATIONS

High speed photometric observations of the stellar flare phenomenon has been made with the 2.1 m Struve telescope at McDonald Observatory. Narrow band filters (Table 1) centered at the Hα and Hβ emission lines were used to monitor the behavior of these lines, and UBVR filters were used to monitor variations of continuum plus lines during flares. One second integrations were taken in each filter. Including the time needed for filter changes, the actual time resolution is 9 seconds in each filter.

Table 1: Characteristics of the narrow band filters

| Designation | $\lambda_{center}$(Å) | FWHM(Å) | Transmission(%) |
|---|---|---|---|
| Hα narrow | 6566 | 9.4 | 69.5 |
| Hα wide | 6572 | 126 | 52.5 |
| Hβ narrow | 4864 | 33 | 30.0 |
| Hβ wide | 4861 | 181 | 67.0 |

## FLARE COLOURS

Figure 1 presents two large U-filter flares on G141-29 (Pettersen 1981) and V577 Mon, together with the time behavior in U-B, B-V, and V-R. It is seen that U-B and B-V remain constant during the flare, except for



*P. B. Byrne and M. Rodonò (eds.), Activity in Red-Dwarf Stars, 239–243.*

increased noise at low flare amplitudes. Average values of U-B=-1.0 and B-V=0.0 are in accordance with Cristaldi and Longhitano (1979) and Moffett (1974). The upper panels of Fig. 1 show that V-R is not constant during the flare. It peaks at U-filter flare maximum and decays according The noise is considerable at small flare amplitudes, but V-R appears constant during the later part of the flares. The U and B filters are contaminated by emission lines during flares, but the V and R filters are little influenced by this. Hα is at the end of the filter windows, and V-R may be a good approximation to a pure continuum index for flares.

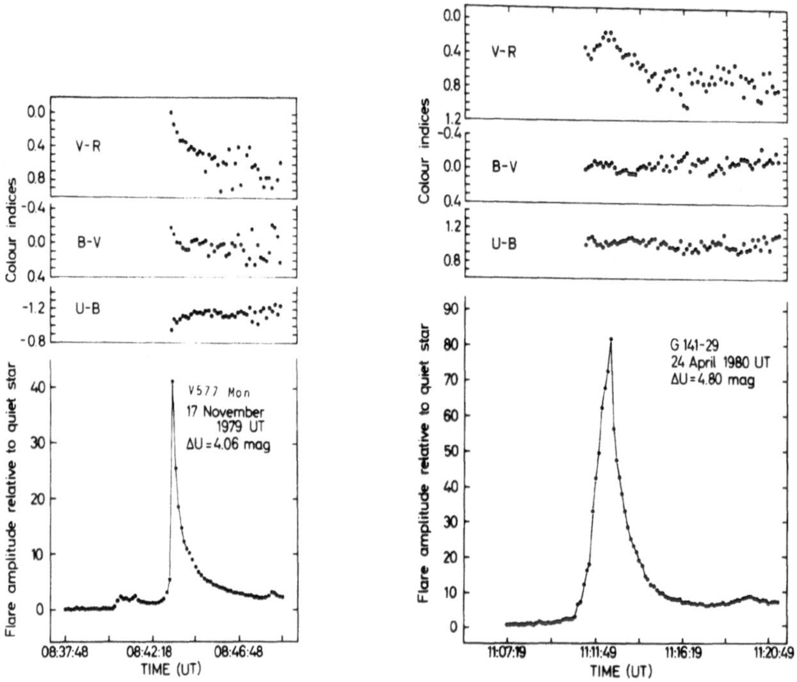


Fig. 1. Flares on V577 Mon and G141-29 with colour behavior of the flare light itself as measured through UBVR filters.

EMISSION LINES

Figure 2 presents variations detected in Hα and Hβ during flares on UV Cet and V577 Mon. The emission line indices are formed by taking the ratio of the fluxes recorded in the narrow and wide filters for each line. This quantity is proportional to the variation of the equivalent width. The timescales involved in emission line variations are longer than those of the flare continuum. This resembles the case for solar flares. Also, there is a time delay between the continuum peak and maximum emission line flux of several minutes.

## DISCUSSION

The presence of Balmer line emission during all phases of a flare indicates a cool flare component ($T_e=10^4$ K) in the chromosphere. The decay timescale for continuum (5-100 seconds) and the values of the flare colours support a hot ($T_e=10^7$ K) coronal flare component (Mullan 1976, 1977).

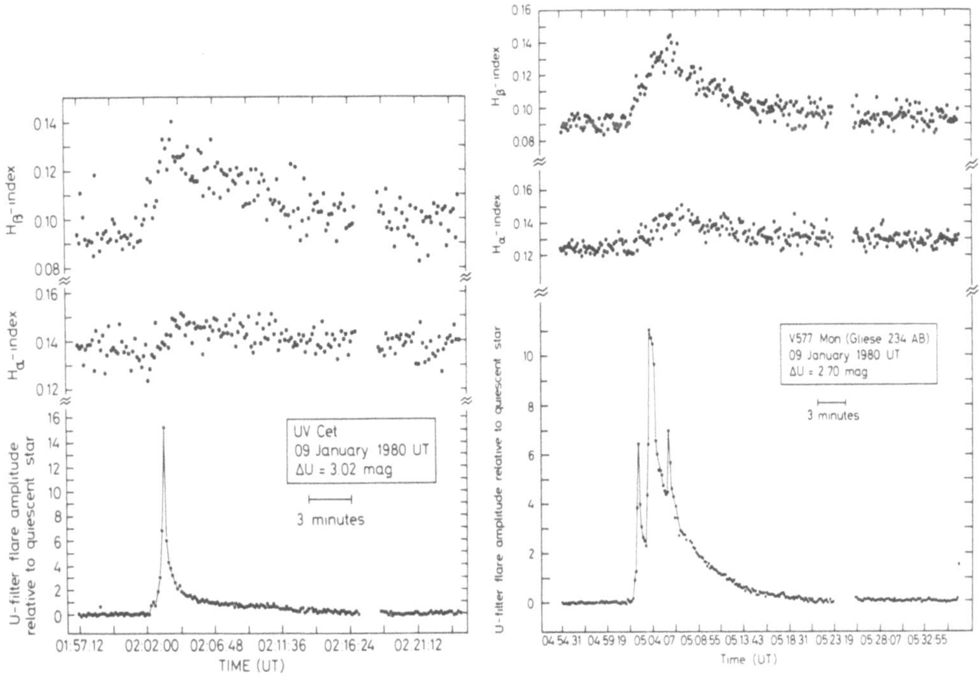


Fig. 2. Flares on UV Cet and V577 Mon with Balmer emission line behavior as measured through narrow band filters.

## REFERENCES

Cristaldi, S., Longhitano, M., 1979, Astron. Astrophys. suppl. 38, 175.
Moffett, T. J., 1974, Ap. J. suppl. 29, 1.
Mullan, D. J., 1976, Ap. J. 210, 702.
Mullan, D. J., 1977, Ap. J. 212, 171.
Pettersen, B. R., 1981, Astron. Astrophys. 97, 199.

DISCUSSION

Linnell: If the flare is the result of increased brightness then I would expect to see a change in (U-B) which is larger than that in (V-R). On your first slide you showed no change in (U-B) at all. Could you explain that please?

Pettersen: I am not sure that I understand your point. This observational conclusion simply says that the flare amplitude in U will be larger than in B, which in turn will be larger than in V and so on. So I don't understand your point as to why (U-B) would not be constant.

Linnell: If the flare is broadband then you expect it to represent a higher surface temperature within the flare region. So you would expect a colour change.

Pettersen: Well, I think that I should say that the (U-B) and (B-V) colours are difficult to interpret because they are influenced but not dominated by emission lines. So what one sees in (U-B) is partly a measure of the continuum and partly of lines. In (V-R) one sees continuum only. There might be some influence from Hα but this is at the edge of the spectral window. My suggestion is that (V-R) is really a continuum index for flares.

Linsky: I think that that is a very important point. Perhaps I make a comment. I would urge people working at wavelengths longer than 3000 A not to use the term ultraviolet. The term ultraviolet nowadays pertaints to the region below 3000 A. It is a question of a nomenclature which can lead to confusion.

Mullan: May I ask you to put up your second to last viewgraph with the spiky flare on it (Fig.2, flare on UV Cet)? It is obvious from this that the Hα and Hβ producing gas is varying much more slowly than the continuum-producing gas. The most direct inference you can draw from this is that the gas producing the continuum has a much smaller inertia than that producing the lines. I believe that this is a strong indication that the continuum arises from the corona i.e. from a region of low density which can respond much more quickly than the Hα and Hβ. This is a very significant result.

Gershberg: It would be desiderable to calibrate your Hα and Hβ indices to equivalent width to compare this result with the previous results by Chugainov and his new spectrographic results. Have you made such a calibration?

Pettersen: I am just in the process of analysing these data. I started with the large flares because I expected to see effects. I have data available which will provide a rough calibration when I compare spectroscopy and photometry. The accuracy of that will be of the order of 1 Å

Gershberg: But is it not a sufficient calibration if you know the filter response?

Pettersen: I know the filter responses very accurately. When I do a calibration from these I get the same results as I do from an empirical calibration. The thing which limits the accuracy of my calibration is the fact that the photometry and spectroscopy were not simultaneous. So average values of the equivalent widths of emission lines were used in the calibration.

# FAR-ULTRAVIOLET AND VISIBLE OBSERVATIONS OF FLARES ON dMe STARS

G E Bromage, B E Patchett, K J H Phillips
Astrophysics Group, Rutherford Appleton Laboratory, Didcot, UK
and
P L Dufton, A E Kingston
The Queen's University, Belfast, UK

ABSTRACT

Four large flare events - one on each of the dMe stars UV Cet, AT Mic, EV Lac and EQ Peg - have been witnessed during a total of 17½ hours of far-UV ($\lambda\lambda$1150-1950) IUE exposures. Some observational characteristics of these four events are compared. Two showed strong enhancements of chromospheric and transition-region line strengths. The other two did not, even though their visible flares were intense ($\Delta U \sim 2$ mag). The brightest UV flare spectrum (EQ Peg) is contrasted with that of the largest solar flare seen from 'Skylab'.

## 1. FREQUENCY OF FAR-UV FLARE DETECTIONS

In aggregate, more than 150 hours of IUE exposure time have been devoted to far-ultraviolet spectroscopy of nearby dMe stars. Many "quiescent state" spectra have been recorded (see for example: Hartmann et al 1979, Haisch and Linsky 1980, Butler et al 1981). Detected line fluxes for the prominent C IV $\lambda$1550 transition-region emission are in the range from 1 to 5 $.10^{-16}$ $Wm^{-2}$ for a sample of 9 flare stars in the solar neighbourhood. Corresponding stellar surface fluxes (20-150 $Wm^{-2}$) are similar to, or somewhat higher than, those of solar active regions.

Occasionally, stellar flares have also been detected by IUE, as transient enhancements of line fluxes. We define an event as one showing enhancements of at least 50% (or else at least 3$\sigma$) for at least 2 lines, when integrated over the typical time-resolved element of ½ hour, or a measured flare continuum flux exceeding $10^{-17}$ $Wm^{-2}$ Å$^{-1}$ over more than 100Å. We are aware of only six such positive detections (Butler et al, 1981; Bromage, 1981; and those reported at this Colloquium by Haisch, by Linsky, by Worden for Baliunas and Raymond; and in the present paper). Thus, IUE observers might now expect on average only one such flare per day. This is rather lower frequency than that predicted from U-band statistics and extrapolations below



*P. B. Byrne and M. Rodonò (eds.), Activity in Red-Dwarf Stars, 245–248.*

3000Å. However, as described below, stellar flares with intense visible flashes but rather insignificant far-UV fluxes may be fairly common.

## 2. FOUR OBSERVED FLARES

The table summarises some contrasting features of the four strong flare events witnessed by the present authors during four IUE shifts, in 1980 September 17 and 19 and 1981 September 2 and 3. These flares are described and discussed in detail elsewhere (Bromage et al 1983).

| | Star | Optical Event | | Far UV Event | | | |
|---|---|---|---|---|---|---|---|
| | $M_V$, dist. (mag, pc) | time of max. (UT) | ΔU max. (mag) | IUE exp. time (min) | C IV λ1550Å enhance factor | C IV λ1550Å flare flux* | other lines enhanced |
| 1. | UV Cet (15.8, 2.7) | 17.9.80 22.22 | 2.4 | 55 | <1.5 | <1 | - |
| 2. | AT Mic (11.1, 7.5) | 19.9.80 16.40:? | ? | 30 | 5±1 | 13±3 | C II, Si IV |
| 3. | EV Lac (11.7, 5.1) | 3.9.81 17.48 | $\gtrsim$2# | 30 | 1.5: | 2: | - |
| 4. | EQ Peg (13.4, 6.5) | 3.9.81 22.55:? | ? | 53 | 9±2 | 18±3 | N V, He II, C II, Si II, C III 1176 (Mg II, Fe II) |
| | | (+22.30:?) | ? | (+24) | | | |

* units of $10^{-16}$ $Wm^{-2}$ averaged over 30 minutes
\# estimate based on: observed ΔB of 0.7 mag (Chugainov, private communication)

The flares on AT Mic and EQ Peg were detected by IUE, but had incomplete ground-based optical coverage due to cloud, and no visible flares were seen. Thus, for these events ΔU values are not known, although statistically it is highly unlikely that ΔU exceeded 3 mags. Also, the times of flare maxima can only be estimated to within about 20 minutes.

A strong continuum is present on the AT Mic flare spectrum, amounting to $4.10^{-16}$ $Wm^{-2}$ $Å^{-1}$ at λ1700-1900Å, smoothed over an assumed effective duration of 100 s. The continuum can be fitted by (inter alia) a black body with T $\sim$ 13000K.

The spectrum of the EQ Peg flare has the highest line fluxes so far reported for dMe stars, even though these are averaged over the whole 53 minute exposure. The estimated peak C IV surface flux in the flare is 3000 $Wm^{-2}$, which is several hundred times that from the Quiet Sun. The far ultraviolet flare spectrum also shows a weak continuum longward of λ1700Å. In addition, just before this spectrum was taken, a mid-ultraviolet (λλ1950-3250Å) spectrum was obtained, and this also shows large line flux enhancements - of Mg II and Fe II multiplet UV1 in this case. From the known position and orientation of the EQ Peg double star in the IUE large aperture at the time of observation, we deduce that the flaring star was component B. It is quite likely that a multiple flare event was witnessed. An effective electron density exceeding $10^{10}$ $cm^{-3}$ is implied by the presence of C III λ1176 and absence of C III λ1909 and Si III λ1892 lines.

Turning to the UV Cet event, simultaneous UBV and optical spectroscopy were obtained with the SAAO Sutherland 0.5 and 1.9 m telescopes. No detectable far UV line enhancements were seen, even though an IUE exposure was optimally timed. Similarly, only small enhancements were recorded during the EV Lac event. This latter was discovered during simultaneous B monitoring by P F Chugainov (private communication).

## 3. COMPARISON WITH A SOLAR FLARE SPECTRUM

The far ultraviolet spectrum of an early stage in the largest solar flare observed from Skylab (Cohen et al 1978, Cohen 1981) has been compared with the IUE stellar flare spectra. Extracts from Cohen's Atlas were digitised, calibrated, and then degraded to the IUE low resolution.

During this Colloquium, the many similarities between dMe and solar flares have been repeatedly emphasised. (See also the report by Kahler et al, 1982). It is therefore worth pointing out here some glaring spectral differences, between this particular solar flare spectrum and that of our EQ Peg event. Relative to the C II and C IV fluxes, we note for example that He II λ1640 and N V λ1240 appear to be an order of magnitude stronger in the stellar case. However, the Skylab solar flare spectrum does bear a somewhat closer resemblance to our AT Mic (and the Butler et al (1981) Gl 867A) weaker stellar flare spectra. A more detailed comparison of these stellar and solar spectra will be reported elsewhere (Bromage et al 1983).

## REFERENCES

Bromage, G E: 1981, *Observatory* 101, p.41.
Bromage, G.E. et al : 1983, in preparation
Butler, C J, Byrne, P B, Andrews, A D, and Doyle, J G: 1981, *MNRAS* 197, p.815.
Cohen, L: 1981, *An Atlas of Solar Spectra, NASA Ref Publ 1069.*

Cohen, L, Feldman, U, and Doschek, G A: 1978, *Astrophys J Suppl* 37, p.393.
Haisch, B M, and Linsky, J L: 1980, *Astrophys. J* 236, p.L33.
Hartmann, L, et al: 1979, *Astrophys J* 233, p.L69.
Kahler, S, et al: 1982, *Astrophys J* 252, p.239.

DISCUSSION

Rodonò: Did you attempt any time-trailing during the observations?

Bromage: No. But normally we made two exposures in the large aperture giving 30 minutes time resolution.

Evans: (Question lost).

Bromage: Because of the position angle with respect to the centre of gravity of the aperture.

Evans: You really can resolve them properly.

Bromage: Yes.

Evans: I noticed on your trailed image that you had in fact two things on top of each other. You have a continuum trace which goes right through emission lines and the emission lines on top.

Bromage: It looks like a continuum but it may just be a lot of emission lines.

Mullan: Your plot of intensity against wavelength was of the intensity per unit wavelength. On a plot using intensity per unit frequency interval it would be almost flat. That's the kind of spectrum you would expect from free-free coronal emission and so that is not an argument in favour of a black body.

Bromage: I am not arguing in favour of a black body.

Linsky: I would like to make two comments. The first is that anyone who uses the continuum in IUE data from flares ought to be very careful about scattered light. It is essential to correct for this. If this is not done then the continuum increases both to short and to long wavelengths. The theoreticians would have fun in interpreting such a spectrum. The second remark is that there is a poster paper (by Butler et al) that shows time-trailed spectra of AU Mic that show evidence of flares. So it is possible to use IUE in its time-trailed mode to give reasonably high time resolution.

# IUE SPECTRA OF THE BY Dra/FLARE STAR AU Mic

C.J.Butler[1], A.D.Andrews[1], J.G.Doyle[1], P.B.Byrne[1], J.L.Linsky[2], T.Simon[2], N.Marstad[2], M.Rodonò[3] and V.Pazzani[3]

[1] Armagh Observatory, Armagh, N.Ireland
[2] JILA/NBS, University of Colorado, Boulder, USA
[3] Osservatorio Astrofisico and Università di Catania, Italy

A coordinated series of ground-based optical and IUE observations of BY Dra variables was undertaken to follow the spectral variation of these stars over one cycle. In the first series 20 LWR and 19 SWP trailed spectra were taken of AU Mic over a three day period 4-6 August 1980 .

In Figure 1 we show the mean integrated fluxes for the strong emission lines in the SWP spectra of AU Mic over the observed phase interval of 0.14 to 0.8 together with an approximate V light curve determined by the FES on IUE. From comparison of the emission line intensities and FES magnitudes in Figure 1 several points emerge.

(a) The light curve at this time had at least three and possibly four minima. If this is to be interpreted as due to rotation of a spotted star, several spotted regions would be required,distributed in stellar longitude.

(b) Repeated flaring of AU Mic makes it difficult to perceive any clearly defined modulation of the emission line intensities due to plage regions in the vicinity of the spots.

(c) Following flare activity during SWP 9695 and SWP 9698 it appears that the intensity of the HeII and SiII lines has remained high for some time after the drop in intensity of the CIV line.

In order to further investigate point (c) we have attempted to refine the time resolution of SWP 9694, 9695 and 9696 using the STAK and TRAK programs on STARLINK devised by Giddings and Settle (1980). With a time spacing of about 25 minutes a flare is evident in each of these spectra with optical flares coincident with the CIV peaks in SWP 9694 and 9696 observed by Barbier and by Touhy respectively (private communications). In Figure 2 it again appears that in all three of these spectra the lower excitation lines peak later than CIV by some tens of minutes. An even finer time resolution for SWP 9695 shown in Figure 3 indicates that in this, the largest of the three UV flares, the emission lines of CI, CII, SiII and HeII all peak around 20 minutes after CIV reaches its maximum intensity. This unexpected result may possibly be due to a high proportion of the CI, CII, SiII and HeII originating from irradiation by soft X-rays.

## REFERENCE

Giddings, J. and Settle, J. (1980) SRC IUE Newsletter No 5,11.



*P. B. Byrne and M. Rodonò (eds.), Activity in Red-Dwarf Stars, 249–250.*

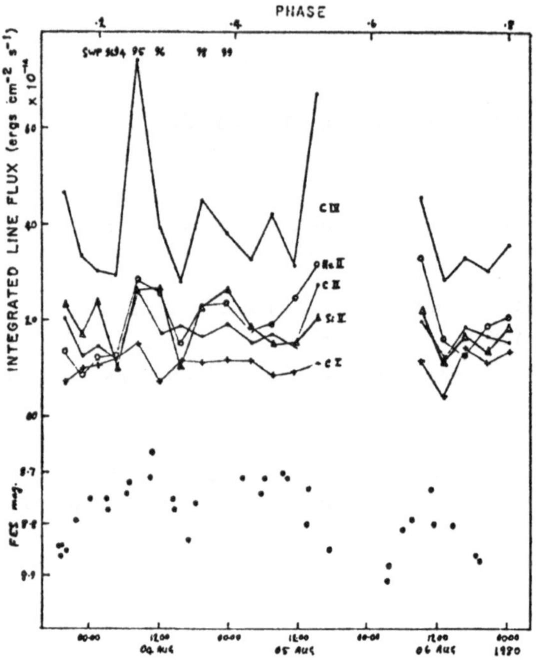


Figure 1. Top, Mean fluxes in strong UV emission lines of AU Mic Bottom, V magnitude of AU Mic determined from Fine Error Sensor

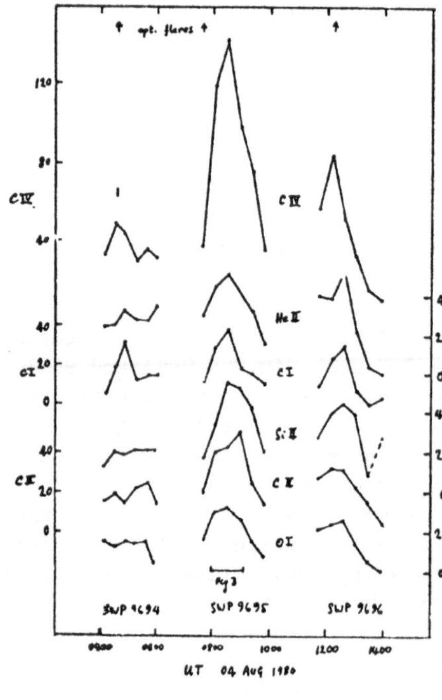


Figure 2. Fluxes of UV emission lines in SWP 9694, 9695 and 9696 with time resolution of approximately 25 mins.

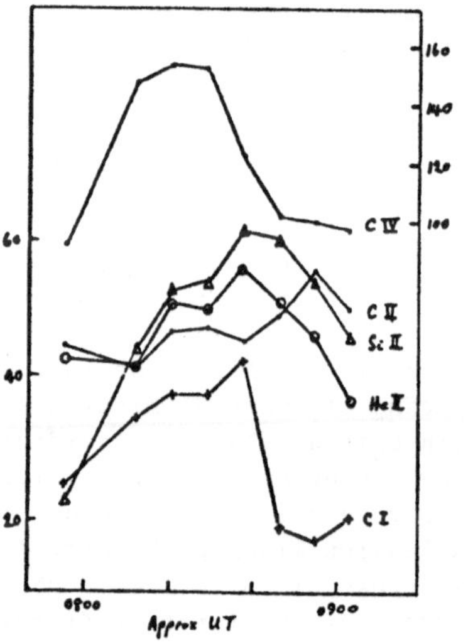


Figure 3. Emission line fluxes with time resolution of 8 mins approx. from SWP 9695.

FLARE-LIKE ACTIVITY OF RED GIANT STARS

Robert E. Stencel
Joint Institute for Laboratory Astrophysics, University of Colorado and National Bureau of Standards, Boulder, CO 80309

ABSTRACT

Evidence for magnetic surface activity among cool stars of low gravity is discussed.

## I. INTRODUCTION

In this colloquium dedicated to surface activity on red dwarf stars, it may be useful to comment on evidence for flare-like activity among higher luminosity cool stars: red giants and supergiants. Although the canonical view is that the rotational angular momentum has been dissipated from the envelopes via massive stellar winds, both their rapid evolutionary timescale and core dynamics might be capable of substantial dynamo regeneration of magnetic fields. Emerging magnetic flux means surface activity for cool stars, regardless of the surface gravity.

There is a profound difference in the structure of the outer atmospheres of red giants compared to red dwarfs. Main sequence stars are known to be solar-like, with geometrically thin chromospheres and thick coronae, plus spots and flares. Red giants and supergiants, in contrast, appear not to have any coronal temperature plasma, but are enveloped by a vast chromosphere which extends out several stellar radii. There is now spectroscopic and direct evidence for these geometrically extended chromospheres surrounding red giant stars (cf. Stencel 1982). The large mass loss/stellar winds of the red giants may be analogous to open magnetic field regions on the Sun.

## II. MAGNETIC ACTIVITY?

Is there evidence for flares on red giants? Among red giants, observations of Ca II H&K (Reimers 1977), Mg II h&k (Mullan and Stencel 1982) and He I 10830 Å (Zirin 1982) lines have revealed profile and flux variations. The timescales for variations are as short as $10^4$ sec, while others occur between observing seasons ($10^7$ sec). Synoptic observations (Baliunas et al. 1981) will be required. In the Mg II study with Mullan, we observed several cool giants and supergiants repeatedly during three years with IUE, and found variation in total emission



*P. B. Byrne and M. Rodonò (eds.), Activity in Red-Dwarf Stars, 251–254.*

strength, or strength/velocity of circumstellar Mg II absorption. The current statistical sample is small and no conclusion about extremes of such events can be drawn at present. Even so, the inferred energy release probably is comparable to that of a modest red dwarf flare, but on a longer timescale.

Boice et al. (1981) reported a radio flare on the red giant Alpha Ceti, which peaked at ten million times the surface flux of a large solar flare. Hayes (reported by Goldberg et al. 1982) has measured polarization changes in the red supergiant Alpha Ori which were correlated with changes in the brightness distribution of the extended chromosphere revealed in narrow band (H-alpha) speckle interferometry. These limited synoptic data suggest that among red giants and supergiants, flare-like events occur on timescales of $10^4$-$10^7$ sec, with associated luminosities of $10^{30}$ to $10^{36}$ ergs/sec. While the luminosity changes are flare-like, the timescales seem much larger than for red dwarf stars.

These events can be interpreted using magnetic topology arguments (Mullan 1982). High gravity stars are capable of maintaining stable, closed magnetic loops which aid production of coronal temperatures. Low gravity stars do not permit emerging magnetic flux loops to find stable, closed configurations, and flux continues to emerge, experiencing reconnection and subsequent forcing out of plasma in a stellar wind. The associated "flares," which may have velocities similar to the circumstellar material (e.g. 50 km/sec) on timescales mentioned, thus have characteristic lengths of appreciable fractions of giant and supergiant stellar radii. Observed line variations, speckle interferometry and polarization argue for atmospheric inhomogeneities in red giant chromospheres which are comparable to the stellar radius. Synoptic photometry and spectroscopy will help to clarify this intriguing situation, and whether or not similar flare physics is involved in giants and dwarfs.

REFERENCES

Boice, D. C., Kuhn, J. R., Robinson, R. D., and Worden, S.P.: 1981, Astrophys. J. 245, p. L71.
Baliunas, S., Hartmann, L., Vaughan, A., Liller, W., and Dupree, A.: 1981, Astrophys. J. 246, p. 473.
Goldberg, L., Hege, E. K., Hubbard, E. N., Strittmatter, P. A., and Cocke, W. J.: 1981, in "Second Cambridge Workshop on Cool Stars, Stellar Systems and the Sun," ed. M. S. Giampapa and L. Golub, S.A.O. Special Report 392, p. 131.
Mullan, D. J.: 1982, Astron. & Astrophys. 108, p. 279.
Mullan, D. J. and Stencel, R. E.: 1982, Astrophys. J. 253, p. 716.
Reimers, D.: 1977, Astron. & Astrophys. 57, p. 395.
Stencel, R. E.: 1981, in "Second Cambridge Workshop on Cool Stars, Stellar Systems and the Sun," ed. M. S. Giampapa and L. Golub, S.A.O. Special Report 392, p. 137.
Zirin, H.: 1982, Astrophys. J. (submitted).

DISCUSSION

Feldman: Whitworth(?) and Hughes once observed a 0.25 Jy flare in R Aql and then spent the next 3 years trying to find another, unsuccessfully. So I would suggest that these specifically radio events are quite rare.

Stencel: That would be consistent with other reports and indeed that is why the information I report is so stetchy.

Popper: I am having a little trouble with terminology. In using the term "red giant" you lump together stars of very different radii and luminosity. For istance, you talked about α Cet, which most astronomers would call a giant, and about α Sco or α Ori,which we would call supergiants. It is my impression that giants are, in terms of atmospheric structure, more like Main Sequence stars then they are like supergiants. When you talked of a flare on α Cet you then talked about the sizes and structure of the chromospheres and supergiants. So giants and supergiants are very different and should be clearly distinguished.

Stencel: Absolutely. Certainly the spectroscopic criteria which distinguished luminosity classes are well founded. In X-rays however there is not yet a way of distinguishing between red giants and supergiants. They both have an almost complete lack of coronal gas and both possess extended chromospheres. So in that sense they are similar. So please excuse me using a common notation. I should have specified giants and subgiants in all cases.

Oskanian: Did I understand properly that the flares you showed us last for 1 day or even more? How then could you decide that these were flares and not some other kind of changes on the stars?

Stencel: In the case of α Ori, for example, the light curve indicates an event of several days duration. In other cases the flickering was on a timescale of hours but its overall shape was not adequately defined in the published data.

Oskanian: I can add to this. On the star μ Cep, a supergiant, Asenievich in Belgrade has observed a real flare lasting several minutes. I have seen the traces and it is a real flare in V. I also had the opportunity to observe once on a giant star, that is, a luminosity class III star, and it flared twice. There are also other giant stars on which I know other people have observed flares of duration a few minutes. I have also observed longer timescale changes on these types of star which may not be flares but other kinds of changes. So we need more proof in order to call these phenomena flares. We know the characteristic of flares well enough.

Stencel: I wish that more of these observations were reported in the widely circulated journals. At the present they simply comprise rumours and stories. The importance of these variations, whatever their true nature, solarlike flares or otherwise, is that they evidently propagate through the entire atmosphere of the stars from a disturbance low in the atmosphere. This latter disturbance could be related to the release of magnetic energy.

Jordan: Can I suggest that you do not call these variations flares because these variations, which may be seen in the optical to radio wavelengths, are not accompanied by any changes in the transition region fluxes, such as C IV, or in the X-ray flux. Schwarzschild and others showed quite some time ago that the scale of the supergranulation on these giants and supergiants is vast compared to that on the Sun. In fact one convective cell could cover almost the entire visible hemisphere of such a star. It is quite plausible that whatever weak magnetic field is there concentrates into the boundary regions of such cells and that these magnetic fields periodically rearrange themselves. There it seems reasonable to me that these events should release small amounts of energy into the low chromospheres. That is an entirely different matter from what we would normally call a flare in which the whole corona and chromosphere is involved. So it is not that I do not believe that what you are seeing is interesting but you should be very careful about calling them flares.

Stencel: Data from Reimers on variations in the transition regions lines of hybrids may contradict your suggestion that there are no variations of that sort.

Jordan: You hadn't mentioned hybrids. You only mentioned the very cool giants and supergiants with extended atmospheres. I recall that the hybrids have steep transition regions.

Collier: Last year in July a programme of monitoring RS CVn candidates at 6 cm wavelength was carried out. Among them was an object HD 101379 which is a K IV giant. Photometric investigations have derived a radius of about 40 solar radii and a rotation period of about 60 days. On one day out of 8 days of monitoring it showed an enhancement of about a factor of 3 in its radio emission at 6 cm up to about 70 mJy. Simultaneous high resolution spectroscopy at Mt.Stromlo showed an enhancement in the red wing of Hα at the same time. This may be symptomatic of the same kind of thing. The time scale was of the order of 8 hours to 1 day.

Stencel: Please publish your results.

# X-RAY OBSERVATIONS OF STELLAR FLARES

Bernhard M. Haisch
Lockheed Palo Alto Research Laboratory
Palo Alto, California, USA

## ABSTRACT

The history of stellar X-ray flare observations prior to EINSTEIN is reviewed. X-ray light curves as measured by the IPC are then presented for all time resolved flare events discovered as of July 1982 in the EINSTEIN data set. These light curves are analyzed in terms of solar-like loop models to derive densities, temperatures, loop lengths, magnetic field strength lower limits, etc. The failure of the model to adequately represent the observations in the case of the YZ CMi flares is discussed. The relationship of X-ray to optical emission and X-ray to UV emission is considered from both an observational and a theoretical viewpoint. It is concluded that the characterization of a flare by a single, time averaged ratio, $L_x/L_{opt}$, is not physically significant.

## I. HISTORY OF X-RAY FLARE OBSERVATIONS BEFORE EINSTEIN

### A. Astronomical Netherlands Satellite (ANS)

The first stellar X-ray flare ever observed was the 19 October 1974 event on YZ CMi (Gl 285; dM4.5e; d=5.99 pc) seen by the low energy (0.2-0.28 keV) and medium energy (1-7 keV) detectors onboard the Dutch ANS satellite as reported by Heise et al. (1975). Figure 1 shows the weighted and summed low (LED) plus medium (MED) energy count rates converted to a luminosity, $L_x$. The background was extremely high and the sensitivity quite low by present standards, as can be seen by the high and fluctuating background in comparison to the quiescent coronal emission ($\sim 3 \times 10^{28}$ ergs/s) measured by EINSTEIN in 1979, discussed below. The short duration of the event ($\sim$ 2 min.) is not too surprising since it must represent only the very peak of a quite energetic flare, $L_x$(LED+MED) $\sim 4 \times 10^{30}$ ergs/s, in comparison to the 1979 EINSTEIN flare, $L_x$(IPC) $\sim 10^{29}$ ergs/s.

During the first 48 s of the flare, the ratio $L_x$(MED)/$L_x$(LED) was about 25, implying a temperature, $T \sim 10^7$-$10^8$ K, although this is only



*P. B. Byrne and M. Rodonò (eds.), Activity in Red-Dwarf Stars, 255–272.*

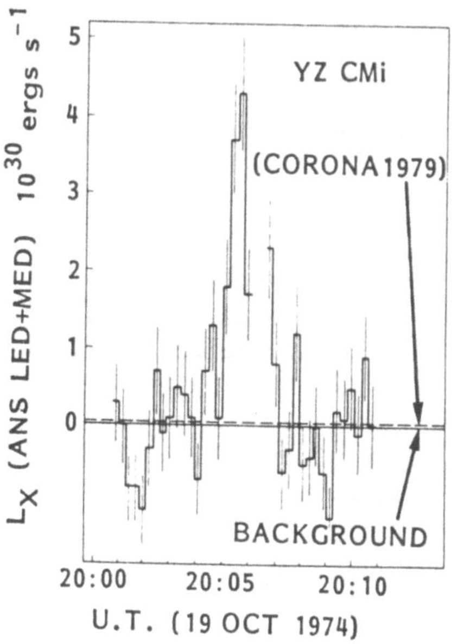


Figure 1. ANS Flare on YZ CMi.

suggestive. Making a crude allowance for radiation outside of the combined passbands, the peak luminosity was $L_x \sim 10^{31}$ ergs/s, and the total energy radiated was $E_x \sim 5 \times 10^{32}$ ergs. No optical data were available for this event.

A second flare observed by the ANS was the 8 January 1975 event on the flare star pair L726-8 + UV Ceti (Gl 65AB; dM5.5e + dM6e; d=2.62 pc; a=2″.06), seen by the LED only. The peak luminosity was $L_x$(0.2-0.28 keV) $\sim 10^{29}$ ergs/s; but if $T \geq 10^7$ K, the total X-ray luminosity would be $L_x \sim 4 \times 10^{30}$ ergs/s (these X-ray "bolometric corrections" are based on emission models of Mewe and Gronenschild (1981), Rosner et al. (1978), and Raymond et al. (1976) among others).

For this event there were simultaneous U and V band observations, from which Haisch et al. (1977) conclude that $L_{opt} \sim$ 5-6 $\times 10^{31}$ ergs/s; we therefore estimate that $L_x/L_{opt} \sim 0.1$.

B. Apollo-Soyuz EUV Telescope

Proxima Centauri (Gl 551; dM5e; d=1.31 pc), a very distant member of the Alpha Cen system (sep. $\sim 2^{\circ}.2$!), was detected during a brief scan by the Apollo-Soyuz EUV telescope in the Parylene filter (44 - 190 Å) passband as reported by Haisch et al. (1977); no optical observations were carried out at that time. In retrospect this paper is more important for its discussion and clarification of the "X-ray to Optical Luminosity Ratio" issue than for the EUV data, which have been superceded by the 1979 and 1980 flares observed by EINSTEIN, discussed below, motivated by this early apparent detection.

C. MIT Satellite SAS-3

YZ CMi was the object of a coordinated X-ray, optical and radio observing program from 30 November - 3 December 1975 using SAS-3 (Karpen et al. 1977). Numerous optical and radio flares were recorded, but no X-ray flares were detected. The brightest optical flare during a time of X-ray monitoring had an estimated luminosity, $L_{opt} \sim 10^{31}$ ergs/s (but note that this is based on an assumed optical emission model by Kunkel as discussed in Haisch et al. 1977). The upper limit $L_x(0.15-0.8\ \mathrm{keV}) < 10^{29}$ ergs/s corresponds at $T \geq 10^7$ K to an upper limit on the total luminosity, $L_x < 2 \times 10^{29}$ ergs/s; we therefore estimate that $L_x/L_{opt} \leq 0.05$.

Prox Cen was the target of a second coordinated X-ray, optical and radio observing program from 16-18 May 1977 (Haisch et al. 1978). The brightest U-band flare yielded $L_{opt} \sim 10^{30}$ ergs/s, again based on Kunkel's emission model. The upper limit on the total X-ray luminosity, assuming $T \geq 10^7$ K, $L_x < 10^{29}$ ergs/s, results in the estimate, $L_x/L_{opt} \leq 0.1$.

D. HEAO-1

The flare star pair AT Mic (Gl 799AB; dM4.5e + dM4.5e; d=8.2 pc) was scanned by HEAO-1 for six days late in 1977. An increase (>9σ) in the LED (0.15-2.5 keV) count rate and an increase (>5σ) in the MED (2-18 keV) count rate occurred on 25 October 1977 (Kahn et al. 1979). A spectral fit resulted in a temperature, $T \sim 4 \times 10^7$ K, a luminosity, $L_x \sim 1.6 \times 10^{31}$ erg/s, and a lower limit for the total flare energy, $E_x > 5 \times 10^{32}$ ergs.

Two suspected flare events on AD Leo (Gl 388; dM3.5e; d=4.85 pc) were inferred by Kahn et al. from a shifting of the centroid of emission away from an unidentified source near that star, and toward AD Leo; this occurred twice on 22 November 1977. Luminosities were estimated to be $L_x \sim 1.5 \times 10^{30}$ ergs/s for both events; however these flare "detections" must be regarded as suggestive only.

## II. THE EINSTEIN OBSERVATIONS OF FLARE STARS

In the two and one half years of EINSTEIN operation (November 1978 - April 1981), 40 of the 70 nearby flare stars in the lists of Pettersen (1976) and Kunkel (1975) were targeted for observation by various X-ray instruments, but primarily by the IPC (in fact 57 of the 67 flare star observations utilized the IPC). Despite this substantial number of observations only four significant X-ray flares and five low level minor events have so far (July 1982) been discovered in the data among the nearby flare stars; in addition one major flare was witnessed in the Pleiades (dK?), one in the Hyades involving a (G0 V? + K0 V?) binary, as well as three minor enhancements on other Hyades stars, only

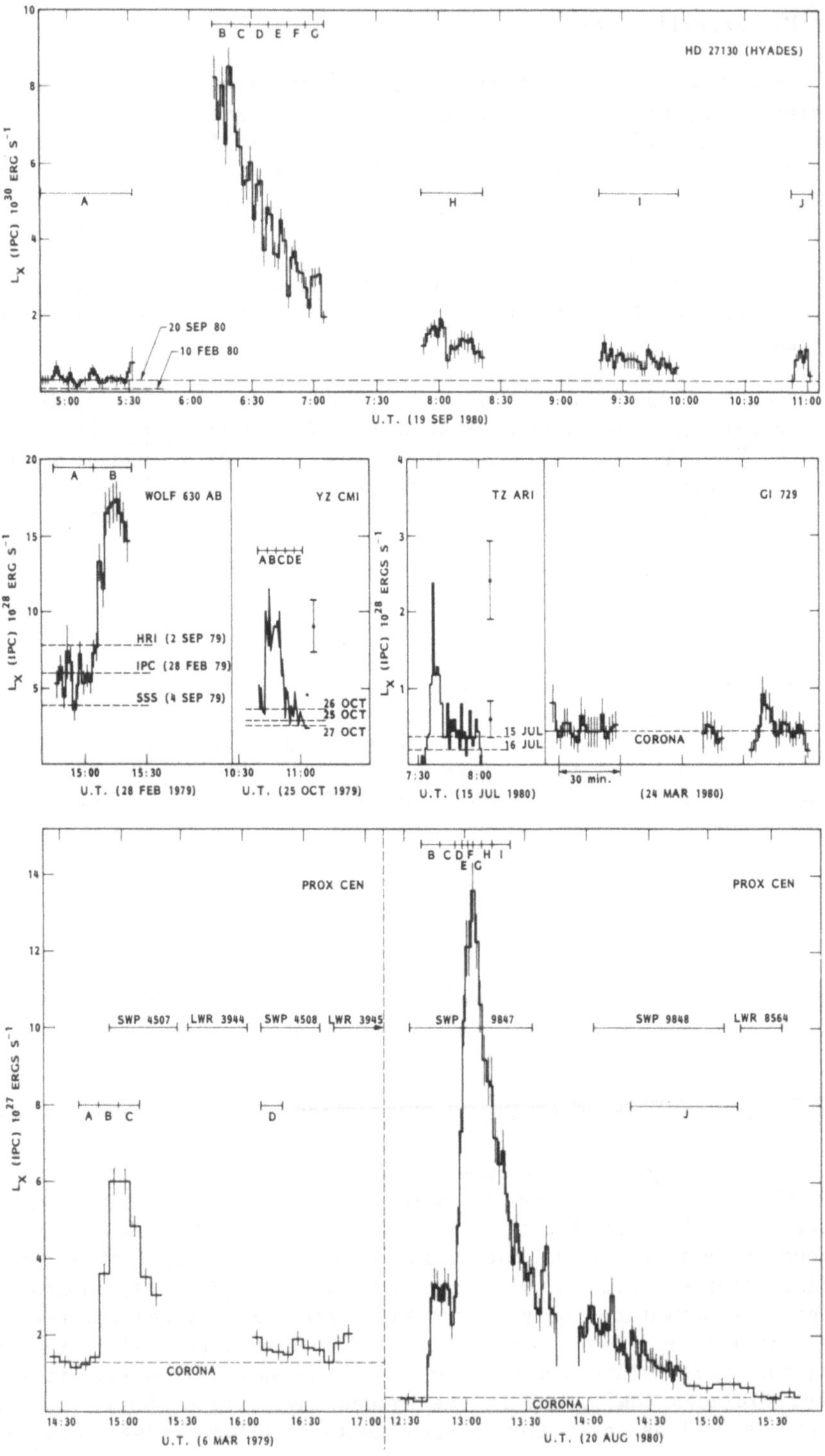


Figure 2. Flare events observed with EINSTEIN

one of which involved a known flare star. These data are summarized in Table 1.

Not all of the EINSTEIN observations targeted on flare stars have yet been thoroughly scrutinized to identify flare events using various timing analysis procedures; nor have serendipitous observations of flare stars lying in the IPC fields of other targets all been identified and analyzed. Furthermore, the ongoing re-processing of IPC data should increase somewhat the effective lengths of many of the observations.

X-ray IPC light curves are shown in Figure 2 for all noteworthy flare events observed by EINSTEIN. Since all of the stars are at known distances, the X-ray luminosities, $L_x$ (IPC), are presented in this figure. Note the change in scale by a factor of 1000 between the Hyades flare and the Prox Cen flares; the time scale, however, is the same for all the light curves. Various quiescent coronal luminosities are also indicated on these plots, and it is clear that there is secular variation in the quiescent coronae, as one would expect from the behaviour of the solar corona.

Table 1. SUMMARY OF X-RAY FLARES OBSERVED BY EINSTEIN

| Star Name | | Sp. Type | Obs. Date | $\frac{Max}{Min}$ | Comments | References |
|---|---|---|---|---|---|---|
| I. Nearby Stars | | | | | | |
| Wolf 630AB[a] | Gl 644AB | dM3e + dM4e | 28 Feb 79 | 3 | Rise, Peak only | Johnson 1981 |
| Prox Cen[b] | Gl 551 | dM5e | 6 Mar 79 | 4 | Rise, Peak, part of Decay | Haisch et al. 1980, 1981 |
| WX UMa[c] | Gl 412B | dM5.5e + dM2 | 24 May 79 | 3 | No time resolution | Johnson 1981 |
| YZ CMi | Gl 285 | dM4.5e | 25 Oct 79 | 3 | Compl. flare; Rel. minor | Kahler et al. 1982 |
| TZ Ari | Gl 83.1 | dM5e | 15 Jul 80 | 4 | Compl. flare; Rel. minor | Johnson, these proc. |
| Prox Cen[b] | Gl 551 | dM5e | 20 Aug 80 | 34 | Compl. flare; Rel. major | Haisch et al. 1983 |
| V1216 Sgr | Gl 729 | dM4.5e | 24 Mar 81 | 2 | Minimal time resolution | Agrawal et al. 1983 |
| EQ Peg | Gl 896AB | dM4e + dM5e | | <2 | Very weak | Golub 1982 |
| HD 24196 | | dK5 + dM3 | | <2 | Very weak | Golub 1982 |
| II. Hyades Stars | | | | | | |
| HD 27130[d] | +16°577 | G0V ? + K0V ? | 19 Sep 80 | >20 | Major Event; Post Peak only | Stern et al. 1981, 1982 |
| HD 27691AB[e] | +14°690 | G0V + ? | | 2-3 | No time resolution | Stern and Zolcinski 1982 |
| vA 500 | VR 17 | dK? | | 2 | No time resolution | Stern and Zolcinski 1982 |
| vA 288[f] | VR 6 | dM2-3? | | 2-3 | Weak; Long Decay only | Stern and Zolcinski 1982 |
| III. Pleiades Star | | | | | | |
| HZ 1136 | | dK? | | 10 | Peak (?); Part of Decay | Caillault et al. 1982 |

[a] Component A is V1054 Oph; both stars may flare; sep. = 0.218" = 1.3 AU.

[b] α Cen C; sep. from α Cen AB = 7850" = $10^4$ AU; probably a bound member of the α Cen system (cf. Walke 1979).

[c] Component A (dM2) is not known to flare; sep. = 28".

[d] Eclipsing Spectroscopic Binary, p = 5.61 d (cf. McClure 1982); analogous to RS CVn Systems.

[e] Spectroscopic Binary, p = 4 d.

[f] Flare Star.

## III. ANALYSIS OF FLARE X-RAY LIGHT CURVES: LOOP MODELS

The behaviour of the thermal soft X-ray plasma during solar flares has been outlined by the Moore et al. (1980) summary of the SKYLAB data. And although SMM studies are presently underway which will revise some of our current concepts, the following flare properties are now widely accepted: (1) soft X-ray emission originates from a flare loop or cluster of loops having maximum temperatures of 10-30 x $10^6$ K, with T peaking early in the event just prior to the peak of the X-ray emission, and remaining above $10^7$ K well into the decay phase; (2) the lengths of the loops, L, range from a few times $10^8$ cm for compact flares, to a few times $10^9$ cm for subflares, to $10^{10}$ cm or more for the largest flares; (3) densities at flare maximum are roughly inversely proportional to loop length, ranging from $10^{11}$-$10^{12}$ $cm^{-3}$ in compact flares to $10^{10}$-$10^{11}$ in large flares; (4) emission measures, EM, range from less than $10^{48}$ $cm^{-3}$ to greater that $10^{50}$ $cm^{-3}$ for the largest flares; (5) peak soft X-ray luminosities lie in the range $10^{26}$ to more than $10^{28}$ ergs/s; and (6), coronal magnetic fields above active regions, where flares generally occur, are of order, $B \sim 100$ G.

### A. Flare Peak and Initial Decay Phase Analysis

As discussed by Moore et al., near flare maximum there are theoretical as well as observational reasons for believing that the observed 1/e decay time, $\tau_d$, the radiative cooling time, $\tau_R$, and the conductive cooling time, $\tau_C$, are all about equal,

$$\tau_d = \tau_R = \tau_C = 2N_e(\tfrac{3}{2}kT)/N_e^2P(T) = 2N_e(\tfrac{3}{2}kT)/(10^{-6}T^{7/2}/L^2). \qquad (1)$$

This equality of the time scales near flare maximum results in two relations among the four variables, $\tau_d$, T, $N_e$ and L (the loop length) as derived by Stern et al. (1982), if one assumes a known temperature dependence for P(T); we take $P(T) \sim 10^{-26.2}\, T^{-1/2}$. If in addition we assume that the loops giving rise to the X-ray emission are of constant cross section with area, $A = (L/10)^2$, then the loop volume is $V = L^3/100$. We thus arrive at the following relations among $\tau_d$, T, $N_e$, L and the emission measure, $N_e^2V = N_e^2L^3/100$,

$$T = (4 \times 10^{-5})(N_e^2V)^{1/4}\tau_d^{-1/4} , \qquad (2a)$$

$$N_e = 10^9\ (N_e^2V)^{1/8}\tau_d^{-9/8} , \qquad (2b)$$

$$L = (5 \times 10^{-6})(N_e^2V)^{1/4}\tau_d^{3/4} . \qquad (2c)$$

We now apply these relations to the seven EINSTEIN flare light curves shown in Figure 2 and the ANS light curve in Figure 1. Temperatures have been determined from spectral fitting at selected intervals (A, B, C, etc.) for some of the data as indicated on the figure and discussed in the original references; the ANS derived temperature was discussed in § I. Since all temperatures are in the range $10^7$-$10^8$ K, about 50-80% of the total soft X-ray flux will fall within the IPC passband (cf. Table 3 in Haisch and Simon 1982). $L_{max}^{tot}$ is a best estimate of the total X-ray luminosity from which the emission measure may be derived. From eqns. (2) we then calculate T, $N_e$ and L as tabulated for each flare in Table 2.

The flares seem to fall roughly into three groups as shown in Table 2, with the Wolf 630 flare falling somewhat between the properties of groups one and two as discussed below.

In the first group we note the following properties: (1) the calculated and observed temperatures are in fair agreement and are solar-like; (2) the inferred loop lengths are comparable to solar subflares, and are a fraction of the stellar radius, however the corresponding densities are as high as those found in solar compact flares; (3) EM and $L_x$ are comparable to the largest solar flares; and (4), the decay times are shorter than on the Sun, $\tau_d$(Sun) ~ 1000-4000 s.

Overall, these flares seem to be quite similar to moderate sized flares on the Sun except that the densities seem to be an order of magnitude or so higher, which one might expect in the high gravity envi-

Table 2. RISE AND DECAY TIME LOOP MODEL ANALYSIS

| | $\tau_d$ (s) | $L_{max}^{tot}$ (ergs/s) | $T_{max}^{obs}$ ($10^6$K) | EM[a] | $T^{calc}$ ($10^6$K) | $N_e$ ($cm^{-3}$) | L (cm) | $R_\star$[b] (cm) | $\tau_r$ (s) | P ($d/cm^2$) | B (G) | $\tau_{Alf}$ (s) |
|---|---|---|---|---|---|---|---|---|---|---|---|---|
| Prox Cen #1 | 1000 | 7(27) | 17 | 3.5(50) | 30 | 8(11) | 4(9) | 1.0(10) | 300 | 6.6(3) | >400 | <400 |
| Prox Cen #2 | 1000 | 2(28) | 27 | 1(51) | 40 | 9(11) | 5(9) | 1.0(10) | 300 | 9.8(3) | >500 | <450 |
| Gl 729 | 300 | 7(27) | -- | 3.5(50) | 40 | 3(12) | 2(9) | 1.5(10) | 100 | 3.5(4) | >900 | <150 |
| TZ Ari | 200 | 3(28) | -- | 1.5(51) | 65 | 6(12) | 2(9) | 1.3(10) | 100 | 1.0(5) | >1600 | <100 |
| Wolf 630 | (600)[c] | 1.5(29) | (30)[d] | 7.5(51) | 75 | 2(12) | 6(9) | 3.1(10) | 250 | 4.2(4) | >1000 | <350 |
| YZ CMi #2 | 120 | 1.2(29) | 20 | 6(51) | 100 | 1(13) | 2(9) | 2.6(10) | 40 | 3.5(5) | >3000 | <100 |
| YZ CMi #1 | 80 | 1(31) | 10-100 | 5(53) | 350 | 3(13) | 4(9) | 2.6(10) | 20 | 3.2(6) | >9000 | <100 |
| HD 27130 | 3400 | >1(31) | 30-100 | >5(53) | >140 | >5(11) | >6(10) | 6-7(10) | --- | >1.9(4) | >700 | <2750 |

[a]Assuming P(T) ∿ 2 x $10^{-23}$ erg $cm^3$ $s^{-1}$.
[b]Data from Pettersen (1980).
[c]Extrapolation from post-peak turnover.
[d]Not a true thermal spectrum fit.

ronment of dM stars, and as a result of the high densities EM and $L_x$ are scaled up from the Sun. The decay phase is thus probably controlled by radiative cooling and would thus be shorter than for the same size flare on the Sun by virtue of the higher density.

The two YZ CMi flares constituting the second group are rather different. The observed and calculated temperatures are not at all in agreement, which probably invalidates any resulting estimate of $N_e$ and L; however the extremely short observed decay times do argue in favor of strong radiative cooling and hence high densities as the analysis suggests; EM and L are higher than on the Sun. The Wolf 630 flare also has a discrepancy in the temperatures, although not as great as in the case of YZ CMi (but note also that the temperature derived by spectral fitting of the IPC data is not based on a true thermal spectrum as discussed by Johnson in these proceedings). The Wolf 630 and YZ CMi #2 flare are quite similar in EM and L, but the Wolf 630 $N_e$ is lower although this depends on the poorly known decay time of the Wolf 630 event. By and large the Wolf 630 flare seems to lie somewhere between the first and second group; overall the YZ CMi flares suggest much denser and perhaps more energetic phenomena than are seen on the Sun based on this analysis. In § V however, we present arguments that indicate that this type of analysis is not particularly appropriate for the case of YZ CMi.

Lastly, the Hyades flare (HD 27130), the most energetic of the events, appears to represent a very large scale event with $L \sim R_*$ and a long cooling time suggesting moderate, solar-like density.

B. Flare Rise and Magnetic Field Strength Analysis

The currents giving rise to coronal magnetic structure are primarily deep in the atmosphere and below, with only a small fraction of the field arising from currents in the corona itself; it is the annihilation of these coronal currents which is thought to energize the flare plasma. Thus the magnetic pressure at the time of the flare will still be greater than the gas pressure,

$$\frac{B^2}{8\pi} > P . \qquad (3)$$

We therefore use the results of the previous analysis to calculate lower limits on B, as presented in Table 2. These values are intrinsically of interest because they suggest considerably stronger magnetic fields than on the Sun, especially for YZ CMi, although of course in the case of those events the previous analysis is questionable in the first place.

In addition, the magnetic field strengths also provide a crude theoretical constraint on the flare rise time, $\tau_r$. The lower limits on B give us a lower limit on the Alfven velocity,

$$v_A = B/(4\pi\rho)^{1/2} \quad . \tag{4}$$

The limit on $v_A$ translates into a constraint on the theoretical flare rise time, $\tau_A$,

$$\tau_A = L/\varepsilon v_A \quad , \tag{5}$$

where $\varepsilon v_A$ is the propagation velocity for the magnetic instability. There are theoretical reasons for believing that $\varepsilon \leq 0.1$ (Antiochos, priv. comm.) and since we have only a lower limit on $v_A$ we take $\varepsilon = 0.1$ to derive $\tau_A$ as tabulated in Table 2. Although there are too many uncertainties to place much credence in agreement between the upper limits on $\tau_A$ and the observed $\tau_r$ in any individual case, the general trend seems to be in the right direction and is encouraging. It is of course unfortunate that $\tau_r$ is unknown for HD 27130 since this event would have apparently provided a critical test; all we know is that $\tau_r < 2500$ s.

C. Flare Energy Analysis: The "Continued Heating" Problem

According to Moore et al., in the large two-ribbon solar flares heating of the thermal plasma continues far into the decay phase, as manifested by the continuous creation of new, higher, $T \sim 10^7$ K, loops, presumably by filling up via chromospheric evaporation; the compact flares show less evidence for continuous heating.

We again address the issue of similarity between the solar and stellar flare phenomenon by asking whether there is any evidence for continued heating in these eight flares. From the previous analysis we have estimates of T, $N_e$ and L at flare maximum; the total plasma energy after the primary heating phase is therefore,

$$E = 2N_e(\tfrac{3}{2}kT)\,L^3/100 \quad . \tag{6}$$

The total amount of energy radiated away, $E_r$, can be estimated from the light curves with an appropriate "bolometric correction". If $E_r < E$, there is no need to invoke additional heating (although of course a detailed model of the time dependence of the energy balance allowing for conduction and mass motions could still suggest heating even if $E_r < E$, but the present data do not warrant this level of modeling); whereas $E_r > E$ would definitely require additional continued heating. In Table 3 we present best estimates of E and $E_r$ for the eight flares.

We find evidence for continued heating in both of the Prox Cen flares and in the Wolf 630 event; however the Wolf 630 thermal plasma parameters are too uncertain to be very convincing. On the other hand the Prox Cen results, especially for the second (20 Aug) flare, do provide credible evidence for continued heating; the determination of

Table 3. ESTIMATED THERMAL ENERGY AND RADIATED ENERGY

| Star | E (ergs) | $E_r$(ergs) |
|---|---|---|
| Prox Cen #1 | 6(30) | 1.2(31) |
| Prox Cen #2 | 2(31) | 3.5(31) |
| Gl 729 | 4(30) | 1.5(30) |
| TZ Ari | 1.3(31) | 5(30) |
| Wolf 630 | 1.3(32) | >1.6(32) |
| YZ CMi #2 | 3(31) | 3(31) |
| YZ CMi #1 | 3(33) | 5(32) |
| HD 27130 | >6(34) | >3(34) |

$E_r$ for the second flare ($E_r \sim 3.5 \times 10^{31}$ ergs) is based on detailed bolometric corrections possible as a result of the numerous temperature determinations during the flare alluded to in Figure 2 and discussed in Haisch et al. (1983); the loop model analysis leading to the estimate, $E \sim 2 \times 10^{31}$ ergs, is fairly well substantiated by the agreement of the calculated and observed maximum temperatures for that flare.

## IV. RELATIONSHIP OF SOFT X-RAY TO OPTICAL EMISSION

The determination of the ratio, $L_x/L_{opt}$ for a given flare as a test of various models has received a great deal of attention in the past several years, especially as a result of the scaling law analysis and the resulting predictions for various flare stars by Mullan (1976). In the above review of observations we have cited several events characterized by $L_x/L_{opt} < 1$; in other papers, ratios $L_x/L_{opt} > 1$ are derived for a given flare (cf. Kahn et al. 1979; Haisch et al. 1981). In fact, as Kahler et al. (1982) clearly and correctly point out, the ratio $L_x/L_{opt}$ varies dramatically throughout the course of a flare event: in the case of their YZ CMi flare, ranging from ~ 0.08 to infinity!

In order to make sense of this it is necessary to again look to the Sun for guidance. The three principal ingredients in a solar flare are: Hα emission, soft X-ray emission and hard X-ray bursts. Unfortunately, there is no universally accepted model regarding the interrelationships of these phenomena, but recent SMM data are providing important new results.

The Hα event itself consists of two distinct phenomena: the rapid, spiky brightenings seen in the Hα kernels and the later appearance of bright, relatively stable Hα loops spanning the magnetic neutral line. As discussed by Zirin et al. (1981) the soft X-ray emission of the thermal phase is closely connected with the formation of the system of Hα loops; whereas the rapid variability seen in Hα kernels (and in the brightenings of transition region lines) is similar to the temporal behaviour of the hard (> 20 keV) X-rays of the impulsive phase (Leibacher, private communication).

As a result of SMM, there is now evidence, from direct chromospheric observations, of chromospheric evaporation driven by two mechanisms: heating by non-thermal, flare accelerated electrons during the impulsive phase; and heating by thermal conduction during the thermal phase (Acton et al. 1982; Antonucci et al. 1982).

We thus arrive at the following general scenario for the phenomenology of a flare. Hard X-ray bursts result from the deceleration of non-thermal electrons as they bombard the dense chromosphere, with the burst size and timing presumably reflecting the magnetic energy release sequence. Chromospheric material is immediately heated and evaporated, and the Hα and transition region line brightenings reflect the rapidly varying heating function. There is a maximum temperature for the heated flare material predicted by dynamic loop models $T_{max} \sim 30 \times 10^6$ K (Pallavicini, private communication); the evaporating material quickly reaches a temperature of this order, but it takes some time for the loops to fill up as conduction and downward irradiation redistribute energy (and there may be additional heating going on as well as discussed in § III), all of which smooth out the temporal behaviour of the soft X-ray emission, which presumably is coming mainly from the tops of the hot loops. Higher and higher loops are filled up; as the lower loops cool, we see loop structure in Hα.

In this context, the spiky nature of the U-band flare reported by Kahler et al. for the 1979 YZ CMi flare is interpreted as a manifestation of the impulsive heating phase, since the U-band brightening can be associated with chromospheric enhancement at temperatures of 7500-9500 K, as shown by the flare spectrophotometry of Mochnaki and Zirin (1980), which we take as evidence of chromospheric heating and evaporation due to fast electron bombardment. The first and peak U-band "burst" coincides exactly with the onset of the soft X-ray, thermal event. In other words, we take the U-band spikes as a proxy indicator of hard X-ray bursts; with this interpretation it is instructive to compare Figure 2 of Kahler et al. (1982) with Figure 1 in Zirin et al. (1981).

We conclude that the characterization of a flare by a single, time averaged ratio, $L_x/L_{opt}$, is not of any particular significance; instead, observations of the detailed time dependence of various diagnostics within the context of a flare model are required. Cram and Woods (1982) have done preliminary studies of the responses of certain such spectral signatures to various candidate energy transport processes in models of stellar flares.

We note that observation of the hard X-ray, impulsive phases of stellar flares has heretofore been impossible since the sensitivity of even such instruments as the EINSTEIN Monitor Proportional Counter (MPC) has been restricted to energies less than 20 keV. However, the forthcoming EXOSAT mission will be sensitive to 50 keV with a reasonably high collecting area using the Medium Energy Experiment, and to 80 keV with the much smaller effective area Gas Scintillation Proportional Counter (GSPC) experiment. Hard X-ray flare measurements should provide an important new diagnostic for stellar flare models.

## V. CRITIQUE OF THE YZ CMI LOOP MODEL RESULTS

In § III a detailed analytical procedure was outlined based on a simple solar model as discussed by Moore et al. (1980) and Stern et al. (1982); the results of this analysis appeared to be self-consistent for some of the flare events, but questionable for others, particularly for the YZ CMi flares. We now assume, as discussed in § IV, that the U-band spikes observed at the onset of the YZ CMi flare are indeed indicative of impulsive heating at the footpoints of loops and are characterized by the chromospheric flare temperatures observed by Mochnaki and Zirin (1980).

In fact, a continuum spectrum was obtained by Kahler et al. (1982) at 10:43 UT coinciding with the first, peak U-band spike which is fairly well fit by a Planck function at $T \sim 8500$ K. The projected area can then be derived from the known surface flux, $\pi B_\lambda$ ($T = 8500$ K), and the observed flux at the earth in the U-band; Kahler et al. derive a projected area, $A\mu \sim 10^{19}$ cm$^2$. If this emission arises at the footpoints of one or more loops which are in the process of being filled up by chromospheric evaporation, and if the cross section of the loop or loop cluster is $(L/20)^2$ for loops of length L (i.e. diameter one-tenth the length), then we find $L \sim 3.5 \times 10^{10}$ cm. This is in sharp contrast to the loop model analysis result of § III, wherein it was estimated that $L \sim 2 \times 10^9$ cm for this event. This dramatic change in L would of course significantly alter the estimates of $N_e$, T, B, etc. The moral here is that as attractive as simple loop models and scaling laws such as those outlined in § III might be as analytical tools, they may lead to quite misleading results when general characteristics of flares on the average are applied to individual cases.

## VI. RELATIONSHIP OF SOFT X-RAY TO UV EMISSION

As indicated on Figure 2, simultaneous IUE and EINSTEIN observations were carried out during both of the Prox Cen flares (cf. Haisch et al. 1980, 1981, 1983 for details). Only quiescent chromospheric and transition region emission was observed during the 1979 event (Haisch and Linsky 1980); however during the 1980 event an IUE short wavelength, low dispersion flare spectrum was obtained coinciding with the peak of the X-ray flare. The total flare energy in the transition region lines of He II, C II, C IV, N V, Al III and Si IV was $1.2 \times 10^{30}$ ergs; the total X-ray energy during the time of the IUE exposure was $2.5 \times 10^{31}$ ergs, or in other words, $E(\text{X-ray})/E(\text{TR}) \sim 20$.

This single energy ratio averaged over the impulsive phase and most of the thermal phase of the flare suffers from the same deficiencies as does the single, average $L_x/L_{opt}$ ratio discussed in § IV; and unfortunately given the sensitivity of the IUE and the low flux levels of even the brightest transition region lines, detailed time-resolved UV flare spectra will probably have to await the coming of the Space Telescope.

## ACKNOWLEDGEMENTS

I wish to thank Drs. H.M. Johnson, S.K. Antiochos, P.C. Agrawal R.A. Stern, L. Golub and J.-P. Caillault for useful discussions and for allowing me to include results of theirs prior to publication. I would also like to acknowledge my gratitude to Drs. J.W. Leibacher, L.W. Acton and M.E.C. Bruner. Special thanks go to Dr. J.B. Reagan for his support and sponsorship making attendance at this Colloquium possible, and to Ms. K. Robbins for her professional assistance in the preparation of this manuscript.

## REFERENCES

Acton, L.W., et al.: 1982, Ap.J., in press.
Agrawal, P.C., Rao, A.R. and Sreekantan, B.V.: 1983, Ap.J., submitted.
Antonucci, E., et al.: 1982, Solar Phys., 78, p. 107.
Caillault, J.-P., Helfand, D.J. and Ku, W.: 1981, BAAS, 13, p. 811.
Cram, L.E. and Woods, D.T.: 1982, Ap. J., 257, p. 269.
Golub, L.: 1982, private communication.
Haisch, B.M., et al.: 1977, Ap.J. (Letters), 213, p. L119.
Haisch, B.M., et al.: 1978, Ap.J. (Letters), 225, p. L35.
Haisch, B.M., et al.: 1980, Ap.J. (Letters), 242, p. L99.
Haisch, B.M., et al.: 1981, Ap.J., 245, p. 1009.
Haisch, B.M., et al.: 1983, Ap.J., submitted.
Haisch, B.M. and Linsky, J.L.: 1980, Ap.J. (Letters), 236, p. L33.
Haisch, B.M. and Simon, T.: 1982, Ap.J., to appear in Dec. 1 issue.
Heise, J., et al.: 1975, Ap.J. (Letters), 202, p. L73.
Johnson, H.M.: 1981, Ap.J., 243, p. 234.
Kahler, S., et al.: 1982, Ap.J., 252, p. 239.
Kahn, S.M., et al.: 1979, Ap.J. (Letters), 234, p. L107.
Karpen, J.T., et al.: 1977, Ap.J., 216, p. 479.
Kunkel, W.E.: 1975, in IAU Symp. No. 67, Variable Stars and Stellar Evolution, ed. V.E. Sherwood and L. Plaut (Dordrecht: Reidel), p. 67.
Mewe, R. and Gronenschild, E.H.B.M.: 1981, Astr. Ap. Supp., 45, p. 11.
McClure, R.D.: 1982, Ap.J., 254, p. 606.
Mochnacki, S.W. and Zirin, H.: 1980, Ap.J. (Letters), 239, p. L27.
Mullan, D.J.: 1976, Ap.J., 207, p. 289.
Moore, R., et al.: 1980, in Solar Flares, ed. P.A. Sturrock (Boulder: Colorado Associated University Press), p. 341.

Pettersen, B.R.: 1976, Catalogue of Flare Star Data, Inst. of Theoretical Astrophysics, Oslo, Rept. No. 46.
Pettersen, B.R.: 1980, Astr. Ap., 82, p. 53.
Raymond, J.C., Cox, D.P. and Smith, B.W.: 1976, Ap.J., 204, p. 290.
Rosner, R., Tucker, W.H. and Vaiana, G.S.: 1978, Ap.J., 220, p. 643.
Stern, R.A., Underwood, J.H. and Antiochos, S.K.: 1981, in Proc. Sec. Cambridge Workshop on Cool Stars, Stellar Systems and the Sun, ed. M.S. Giampapa and L. Golub, SAO Rept. No. 392.
Stern, R.A. and Zolcinski, M.-C.: 1982, private communication.
Stern, R.A., Underwood, J.H. and Antiochos, S.K.: 1982, Ap.J. (Letters), submitted.
Walke, D.G.: 1979, Ap.J. (Letters), 234, p. L205.
Zirin, H., Feldman, U., Doschek, G.A. and Kane, S.: 1981, Ap.J., 246, p. 321.

DISCUSSION

Vaiana: Of course loop modelling should not be applied to flares. One of the basic hypotheses of loop models is hydrostatic equilibrium. It is true that Pallavicini has found that scaling applies to the less-than-impulsive phase of several stellar flares. Ignoring this basic difficulty for a moment, however, and, assuming that there is some way in which scaling can be applied, I wonder whether there is a way round the difficulty in the case of YZ CMi. If I understand this difficulty it is that when you do a total volume analysis you come up with loop lengths which are an order of magnitude greater than those you derive from radiative cooling times for the flare. Given that the quiescent corona of YZ CMi is very different from that of Prox Cen (the temperature is higher giving a higher volume coverage as well as, perhaps, higher pressure) I wonder whether a way round the difficulty, assuming you want to adhere to the loop model, which I do not, would be to have the same kind of intermittency in loops as in the Sun. If one has, for istance, 10 loops emitting at the same time then one has the required area coverage, all of them will be at the high density required by the cooling rate while the total emission will be compatable with your other analysis based on volume factors. That is, instead of having one loop of $10^9$cm in length you will have 10 of them all having essentially the plasma parameters that are advocated by time changes.

Haisch: I think I agree with everything you have said but I am not sure that it makes a great deal of difference as to what kind of corona the flare is embedded in. The flare analysis depends only on equating various time-scales i.e. the conductive cooling time, the radiative cooling time and the observed cooling time. So what kind of loops might be present in the atmosphere before, during or after the flare is not relevant.

Vaiana: But in the one you are analysing you find loop lengths of a few times $10^{10}$cm.

Haisch: But that was based on the observation of a certain amount of flux in the U-band and the fitting of the spectrophotometry to a black body at a certain temperature.

Jordan: I think we are getting into too much detail for general discussion.

Haisch: Basically I think we are in agreement (laughter).

Serio: I would like to comment on your suggestion that there might be continuous heat deposition after the maximum of the flare. We have analysed quite a lot of SMM data with the aid of a numerical lodel for flares in loops and found no evidence for heat deposition after the temperature maximum. Your result may be explained if after the maximum there is continuous input of hot matter evaporated from the chromosphere. So the comparison of the radiated energy with the energy content at any time may be misleading.

Haisch: All I am saying is that if I take the integrated energy emitted during the flare, especially for the two Prox Cent events, then, having carefully applied realistic bolometric corrections, I see that 12 x $10^{30}$ ergs was radiated by the first flare and somewhat more by the second. So these are well determined numbers. But I do not know how much energy was there in the plasma to begin with except in so far as I have estimates from the application of the loop model and so on. If one accepts those loop models then I see that there is twice as much energy being radiated as was stored in the plasma. Furthermore, I know that there are additional losses besides radiation. For istance, half of the radiated energy is radiated downward and there is energy conducted away. So if the energy losses are larger than the energy stored then that would seem to be evidence for continuous heating. On the other hand the estimate of the energy stored is based on a model calculation and it is not clear to what extent one should believe that. There could be a factor of 2 uncertainty or there could be a factor of 10.

Jordan: May I make a comment? By equating the radiated and the conducted energy losses through the cooling time you are, in effect, adopting a condition of minimum energy loss. That method will always underestimate the total energy loss.

Haisch: By how much?

Jordan: I don't know. It depends on what else you may be doing, but it will certainly tend to under rather than overestimate the losses.

Mullan: I have 3 comments. First, the comparison of radiative and conductive cooling in the Sun was done many years ago by Moore and Atlow and they showed that the conductive cooling was much more important right at the flare maximum. Secondly, any assumption of black body emission from a flare seems totally irrelevant. I cannot imagine any circumstances in which the chromospheric emission from a flare would have anything to do with a black body. My third and major point concerns the bolometric corrections. It is well known that the corrections to the optical radiation from a flare is a very difficult problem. As far as I know it is subject to errors at least as large as 200, depending on which model you take for the flare emission. I am therefore not very upset by errors of a factor of 2 in the estimated radiated losses. I also do not believe that values of $L_X / L_{opt}$ evaluated only within certain bands mean anything. You might as well say that I went to the USA last week and did not see you there. So you were not there (laughter). The point is that there is much emission from the chromosphere which has not been seen. (Rest of comment lost).

Haisch: Let me reply to those 3 comments if I can remember them all. The equating of radiative and conductive cooling times at flare maximum I took from Ron Moore's summary of the Skylab data. That was based on both empirical data and theoretical consideration from solar flares. If one pictures a flare as a loop filling up with plasma, this hot plasma arises from conductive flux heating material lower down. Once the radiation begins to exceed conduction as a cooling mechanism then less conduction takes place, leading to a fall-off in the supply of hot plasma and so the light curve turns over. There are theoretical considerations but I believe that there are also Skylab data, which support this balance of conductive and radiative timescales at flare maximum. The second point concerns the use of black body curves. Well this is justified by the observations of Zirin and Mochnacki. They carried out spectrophotometry of flares on YZ CMi and found that the radiation could be fitted to black body distributions at temperatures of 8000-10 000 K. If one has flares characterized by such temperatures I do not see how you can get bolometric corrections $\sim$ 200. If one is looking in the U and B optical bands then one expects to pick up more than 0.5% of the total radiation. Perhaps if one had, a 10 000 000 K radiation temperature this could be so.

Priest: I think that when you are quoting solar results from Skylab you should be very critical of them. In particular, when people calculate conductive cooling times there are many uncertainties. It is especially true that when dealing with flare plasma it is likely to be highly turbulent and so one should use turbulent conductivity in calculations rather than "classical" turbulence. Another problem is that of knowing what the correct length scale is. You see on the Sun one may see some kind of elongated structure in X-rays. That does not mean that the field lines run along its length. They may indeed be directed across it. So great care is needed.

Haisch: I agree with what you are saying. What is needed is more observations and better analysis of solar flares so that we can develop a more sophisticated approach. All I wanted to do here was to show what could be done using our present knowledge and it is obvious that strange results occur as a result.

Jordan: Having restrained myself thus far I would like to make a comment. If you are using the scaling law which you have used you will probably get the electron temperature about right but the electron pressure will always be overestimated. The scaling law is in fact the condition that the pressure be a maximum for the solution as Tony Hearn has pointed out. You can show that very easily on a curve of temperature against pressure. The scaling laws give the locus of critical solutions. This is similar to the work that Hood and Priest have done on thermal stability. The critical point, which is the peak on the curve, is always the maximum pressure point but the critical temperature is never far from the real temperature. So if you want a good estimate of the temperature but a bad estimate of the pressure then the scaling law approach may be a reasonable one.

Haisch: If one always overestimates the density then that would agree with what I find here.

Jordan: You really do not need to estimate the pressure however. If you have the X-ray luminosity and have measured temperature and know the radius of the star, you can find the mean coronal density or the mean pressure. Since these are high pressure objects the pressure will be roughly constant down to the transition region. You can then work from that point and model the pressure. I would encourage that approach.

Haisch: That's right. In fact the system is overdetermined. There was a temperature predicted and a temperature observed and you can compare those two or you could just as well compare the densities.

Linsky: There are two aspects of the data which you did not mention for lack of time. The first is that the temperature peaks before the soft X-ray luminosity by a few minutes. I think that this has been seen in several flares and is typically seen in solar flares. Secondly, there is some evidence for absorption after flare maximum. Would you like to comment on these apsects.

Haisch: I did not plant Jeff (Linsky) in the audience but I am glad he mentioned these points. During this event as well as during the 1979 event (on Prox Cen?) we found that the peak temperature did in fact preceed the X-ray luminosity by 2-3 minutes. When we carried out the standard Einstein processing on the decay part of the 1980 flare we had two free parameters, viz the temperature and the mass column density in the line of sight. During a short time interval in the decay we found a temporary increase in this column density to a significant number i.e. $10^{20} cm^{-2}$. The mass column density derived for the rest of the flare was negligible. So we attribute this to mass ejection during the flare and draw an analogy with the solar Hα two-ribbon flare.

# QUIESCENT AND FLARING RADIO EMISSION FROM dMe STARS

D. M. Gibson
Dept. of Physics
New Mexico Tech
Socorro, NM USA

## 1. INTRODUCTION

In reviewing radio studies of M-dwarf flare stars one is struck by the curious way in which the field developed. Indeed, that is twenty years old may be the greatest surprise, for if one imagines the largest solar flares occurring at distances comparable to the nearest stars the expected flux densities would be < 10 mJy. Yet, despite the fact that detection thresholds in 1963 were about two orders of magnitude higher than the expected value, Lovell et al. (1963) made extensive observations and reported the detection of UV Ceti. This remarkable discovery was followed immediately by detections of V371 Ori (Slee et al., 1963) and EV Lac (Lovell et al., 1964). One might have thought that these unexpected discoveries would have spurred significant interest in this new field but they did not.

Why? Three reasons seem most apparent: 1) The "mainstream" of astronomers in the late 60's and early 70's never fully appreciated these data. Cross-fertilization especially with solar astonomers was virtually non-existant. 2) The data were obtained at very slow rates; e.g., Lovell (1972) reported an average of only one flare of a few Jy amplitude (at 240 MHz) on UV Ceti per 35 hours observation. In an era of radio studies of "glamorous" new objects as quasars and pulsars such an investment of time was difficult to justify. 3) Credibility. As Lovell and his colleagues wrote, "(detections) were difficult because of the sporadic and transient nature of the phenomena, and the danger of interference . (In general, interference monitors were not used). In several thousand hours of observation there are only a few cases where the existence of individual flares of long duration has been unambiguous" (Davis et al., 1978). I cannot over emphasize the problem interference causes. My own experience with total power records showed many "events" (later shown to be interference by simultaneous interferometric observations) which had levels similar to flares reported earlier and their "typical" fast-rise, slow-decay morphology. To add to the feelings of mistrust it seemed that observations made with different instrumentation (e.g., telescopes, frequencies, bandwidths,



*P. B. Byrne and M. Rodonò (eds.), Activity in Red-Dwarf Stars, 273–286.*

integration times and polarizations) yielded qualitatively different results.

The path toward a resurgence in radio studies of flare stars was paved in 1976 and 1977 by the Jodrell Bank group using an interferometer successfully for the first time (cf. Davis et al., 1978). Later, groups at New Mexico Tech (cf. Fisher and Gibson, 1982), Colorado (cf. Gary and Linsky, 1981) and Caltech (cf. Topka and Marsh, 1982) began extensive microwave observations of flare stars using the recently completed VLA. These interferometers provided unambiguous interference and background rejection as well as unparalleled sensitivity. Not only was quality data available but complementary studies at X-ray, UV, and optical wavelengths provided the radio observations a new context within the framework of the new field of stellar activity.

I will discuss the meter-wave (Section 3) and the microwave (Section 4) flaring separately for several reasons. In general, the former were obtained with single-dishes and are subject to confusion with background objects and interference, whereas the latter were obtained with interferometers which reject both. The reported meter-wave flares differ from the microwave flares in that flux densities are typically several orders of magnitude greater than their microwave counterparts. As is the case for the un, there may well be a difference between microwave and meter-wave phenomena. In addition to the flare levels and morphology I will also review the extremely important spectral and polarization observations which are the true keys to developing adequate theories for the emission mechanisms. I will also discuss correlated observations at other wavelengths (primarily optical) in section 5. In section 6, I will review the observations and theory of quiescent stellar radio emission which has recently been detected on UV Ceti (Gary and Linsky, 1981) and several other objects. Like the detection of flaring in 1963, the detection of quiescent emission was quite unexpected. Nevertheless, it has been put to advantage as a sensitive probe of the coronal environment in these objects.

## 2. STELLAR RADIO-ASTROPHYSICS

Although the total luminosity of stellar radio emission is insignificant compared to $L_{bol}$ it is probably the most sensitive probe we have of the coronal environment. Thus, a few preliminary paragraphs are needed before we review the observations to enable the reader to better evaluate the data. Radio flux density $S_\nu$ measurements are made in Jansky's (1Jy = $10^{-23}$ erg $s^{-1}$ $cm^{-2}$ $Hz^{-1}$). Spectral indices are logarithmic ratios between the flux densities at different observing wavelengths

$$\alpha \equiv \frac{\ell n(S_{\nu_1}/S_{\nu_2})}{\ell n(\nu_1/\nu_2)}$$

Thus, negative (-) spectral indices indicate non-thermal phenomena. Polarizations are usually measured either in circular-mode or linear mode. By convention the percent polarization

$$\pi \equiv \frac{S_1 - S_2}{S_1 + S_2}$$

where 1, 2 correspond to R, L or $||$, $\perp$, as appropriate. But the total flux density is defined as $S = (S_1 + S_2)/2$. The early radio astronomers could not conceive of high percents of polarization and, thus, we have apparent situations where the polarized flux density is greater than the total flux density. Finally the brightness temperature $T_B$ is the equivalent of the electron temperature $T_e$ in an optically-thick thermal source. It is related to stellar parameters as

$$T_B = 2.12 \times 10^7 \ S_\lambda \lambda^2 \ d_*^2/R_*^2$$

where $\lambda$ is the observing wavelength (in cm), $d_*$ is the distance to the star (in pc), and $R_*$ is the source radius (in $R_\odot$). Thus a solar-type star could be detected at a distance of 10 pc by today's most sensitive instrument (the VLA operating at 6 cm) provide $T_B > 8 \times 10^6$K.

One can infer a significant amount about the source conditions and emission mechanism(s) by simply examining $S_\nu(t), \alpha, \pi$, and $T_B$ and by remembering that $n_e \lesssim 1.1 \times 10^{13} \lambda^{-2}$ cm$^{-3}$ ($<10^{11}$ cm$^{-3}$ for the type of sources we discuss here). Several rules of thumb apply:

a) If the source varies rapidly ($\tau < 10$ minutes) it is probably non-thermal. Conversely if it remains at about the same level for days-to-months, then it is probably thermal or at least the emission mechanism controlled by the thermal environment.

b) If $\alpha < -2$ or $\alpha > +2$ the emission mechanism may be coherent. In the range $0 < \alpha < +2$ the source is probably optically-thick but it is difficult to distinguish if the source is thermal or non-thermal without other information such as rapid time variability.

c) If polarization is observed the emission mechanism is non-thermal. Very high percentages of circular polarization ($\pi_c > 50\%$) usually require a coherent emission process.

d) If $T_B \gtrsim 3 \times 10^8$ K for reasonable estimates of the true source size -- which can be based on either the timescale for variability (assuming the propogation of some excitation phenomena has a characteristic velocity) or the percent of circular polarization (assuming the magnetic field structure is tangled or radial and you know something about the electron energy distribution) -- then the emission mechanism is probably coherent. Wild variability coupled with very high percentages of circular polarization almost certainly indicate the emission mechanism is coherent. Brightness temperatures $>10^{12}$ are incompatible with incoherent emission mechanisms.

## 3. METER-WAVE FLARES IN dMe STARS

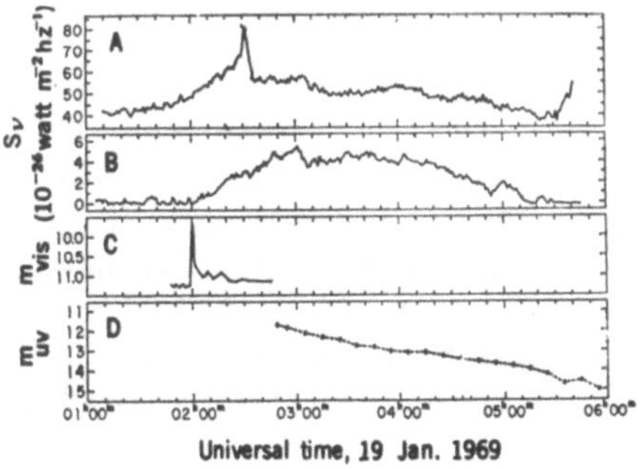


Fig. 1. A large flare on YZ CMi observed simultaneously at 240 MHz (A), 408 MHz (B), the visual (C), and near UV (D) (after Lovell, 1969).

It is very difficult to describe a "typical" meter-wave flare on a dMe star because the observations to date indicate the most believable flares are anything but typical. A classic example is shown in Fig. 1, where total-power flux densities at 240 and 408 MHz obtained at Jodrell Bank by Lovell (1969) are plotted together with simultaneous visual and near-UV photometry by Andrews (1969) and Kunkel (1969), respectively. The difficulty establishing a good baseline is evident in these data. This long three-hour flare had a $T_B^{max} \sim 2 \times 10^{15}$ K and spectral indices ranging from $-5 < \alpha < -1$. The time delay between the optical and radio peaks suggest phenomena similar to those occurring in solar moving bursts but the peak in the 240 MHz emission would be expected to occur after that at 408 MHz. A good interpretation of this event remains to be offered.

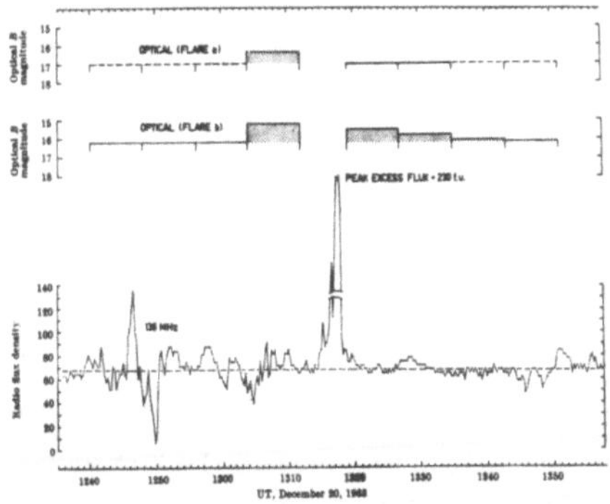

Fig. 2. A "monster" flare presumed to have occurred on one of the nebular variables on the Orion aggregate. $T_B^{max}$ $\sim 10^{20}$ K! (after Slee *et al.*, 1969).

The most spectacular event which has been reported is shown in Fig. 2. Slee *et al.* (1969) detected a gigantic flare from what they think may be a flare star in the Orion aggregate. $T_B^{max} \sim 10^{20}$ K! Two optical flares were seen in different parts of the aggregate at about the time of the radio flare, so it is impossible to identify on which star the flare occurred. The emission would have to be coherent and fully 1% of the total flare energy was estimated to have appeared in the radio band.

The best low-frequency single-dish observations of meter-wave flares on dMe stars are those of Spangler and his colleagues (cf. Spangler *et al.*, 1974b) which were made at Arecibo during 1973-1975. A "typical" observation is shown in Fig. 3 in which a flare on Wolf 424 AB lasting about 30 s is shown together with simultaneous U-filter observations (Spangler and Moffett, 1976). The $T_B^{max}$ ~2 x $10^{14}$ and $\alpha \simeq -4$ again suggest a coherent emission mechanism. The most important of the Arecibo results were obtained for AD Leo (Spangler et al., 1974a) and are shown in Fig. 4. 430 MHz observations using dual circularly-polarized feeds revealed the large polarization of a (small) flare for the first time. $T_B^{max}$ was ~4 x $10^{13}$ K and the degree of right-hand circular polarization varied from 50% to 80% as the flare evolved. The high degree of circular polarization together with linear polarization in the range of 10-20% and the high $T_B$ imply the circular polarization is not a propogation effect and that the emission mechanism is probably not a coherent plasma process.

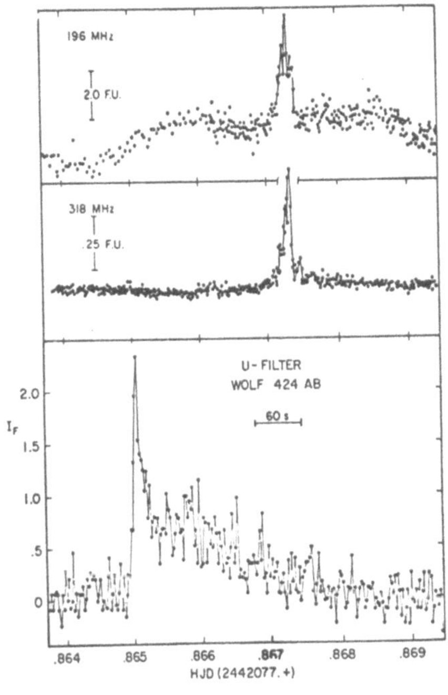


Fig. 3. A typical flare (on Wolf 424) observed at Arecibo, following an optical flare observed at McDonald Observatory (after Spangler and Moffett, 1976).

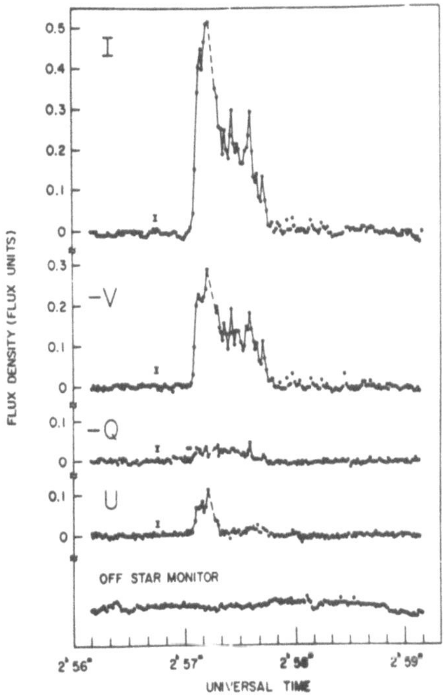


Fig. 4. The first detection of the remarkable polarization seen in flare stars. This flare on AD Leo was as much as 80% right-hand circularly polarized at 430 MHz (after Spangler et al., 1974).

Very real improvement in the low-frequency detection of flares on the dMe stars came at Jodrell Bank with the use of the Jodrell Bank-Defford Radio-link Interferometer and is being continued today with the new MERLIN array. Two "typical" events on the YZ CMi are shown in shown in Fig. 5 (Davis et al., 1978). The event of 1977 Dec. 18 had a

$T_B^{max} = 3 \times 10^{12}$ K and, conceivably, could be incoherent if the source size is $\gtrsim 2R_*$. However, lacking spectral or polarization information there is little physical information in these data except to comment that these are weakest meter-wavelength flares reported to date and still they are $10^4$ times more energetic than a large solar flare (importance 3) in the same band.

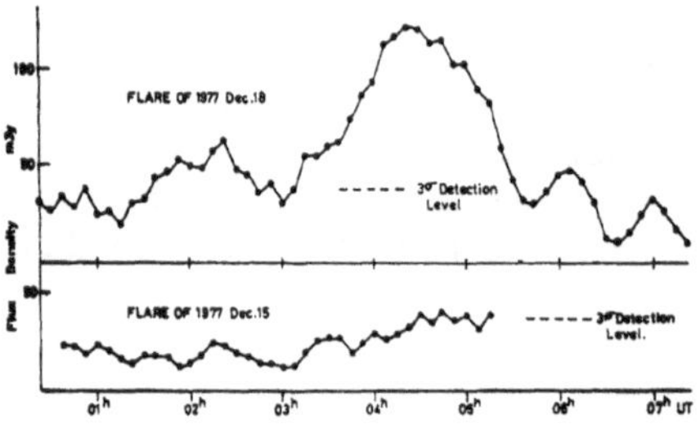


Fig. 5. Two flares on YZ CMi at 408 MHz observed by the Ma IA. Defford interferometer a Jodrell Bank. These are two the weakest low-frequency flares ever observed (after Davies et al., 1978).

## 4. MICROWAVE FLARES IN dMe STARS

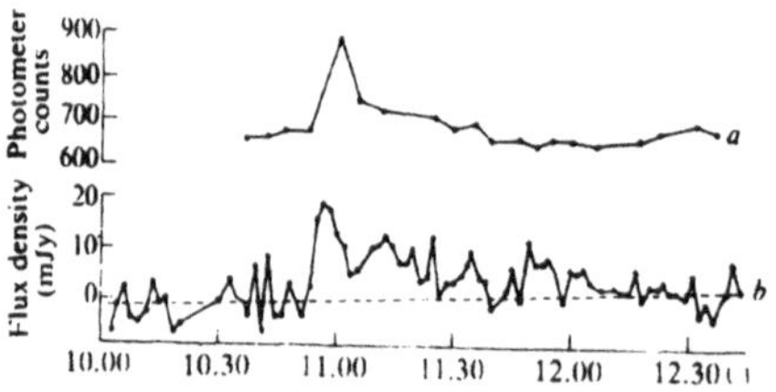


Fig. 6. A weak flare on AT MIC observed at 5000 MHz at Parkes. This flare occurred simultaneously at optical (above) and radio (below) wavelengths (after Slee et al., 1981).

Though studies of the microwave (here defined as λ's < 20 cm) flaring emission from dMe stars have a much shorter history than those at longer wavelengths the body of data is much more useful because of better sensitivity (~10 mJy at Parkes; < 1 mJy at the VLA), less confusion by nearby sources, fewer problems with interference, and, in the case of the VLA, multi-frequency, multi-polarization capabilities. The latter is extremely important because a weak flare, such as that shown in Fig. 6, can have $T_B^{max} < 10^{12}$K and still require a coherent emission mechanism as stated in the "rules-of-thumb" mentioned before. Thus, the studies by Fisher and/or Gibson and Gary/Dulk/Linsky, are the best for the purpose of physical analyses of microwave flares.

To date, only six objects have been detected to flare at microwave frequencies. Variations in measured flux densities have ranged from a few mJy (Fig. 7) to about 60 mJy (Fig. 8), and flare durations have ranged from a few minutes, for relatively simple events (Fig. 8), to several hours for complex events (Fig. 9). None of the flares at 20 cm shown in Figures 7, 8, or 9 were detected at 6 cm, but since $\alpha > -2.5$ one can only conclude the events were non-thermal. The only time an event was detected at both frequencies simultaneously is shown in Fig. 10. One cannot be sure if the peaks observed first at 6 cm and then 80 minutes later at 20 cm are in fact due to the same event since we simply do not have enough data to do a proper statistical analysis.

If they are related, then this event is reminiscent of moving solar bursts but is occurring at a higher frequency than those observed on the Sun. This modified solar analogue would be quite apt since the coronal densities in these systems are higher than in the un.

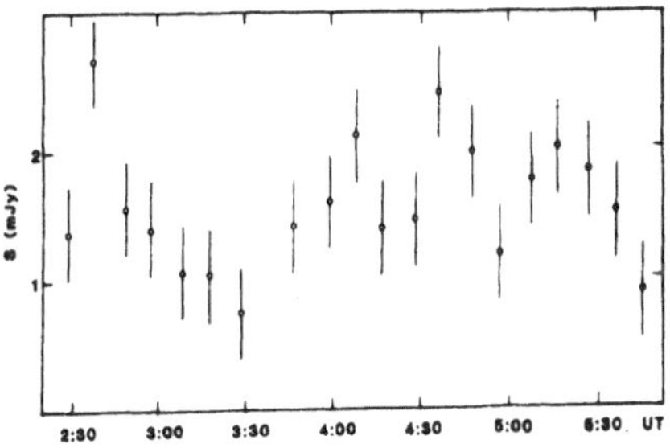


Fig. 7. Very weak flaring on UV Ceti observed at 20 cm with the VLA. The quiescent level of this star is 1.1 mJy (after Fisher and Gibson, 1983).

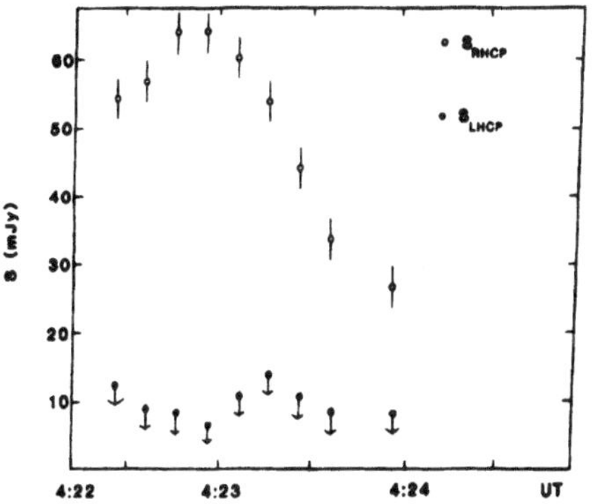


Fig. 8. A short, highly circularly polarized ($\pi_c$>85%) flare on UV Ceti observed at 20 cm (after Fisher and Gibson, 1983).

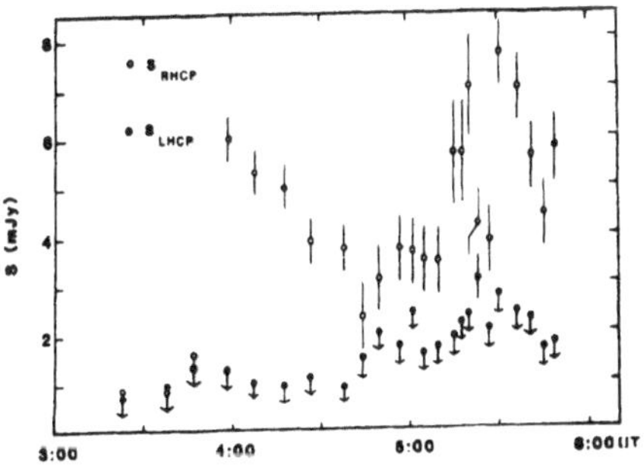


Fig. 9. A complex, highly polarized event on L726-8 A (UV Ceti's companion) observed at 20 cm (after Fisher and Gibson, 1983).

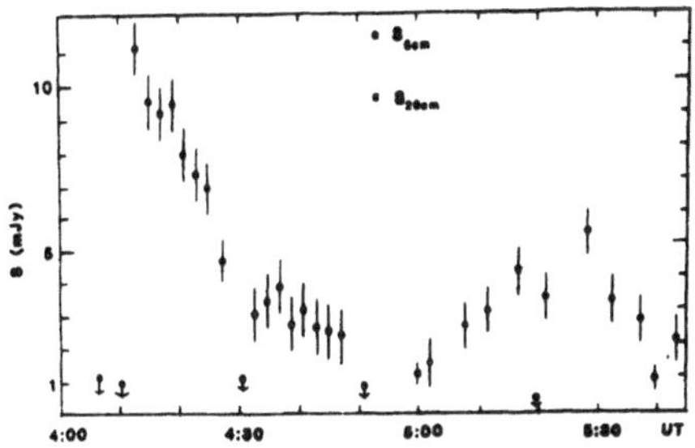


Fig. 10. A 6 cm flare on L726-8 A is followed by a flare at 20 cm. If related, this could be an analogue of a moving Type IV solar burst (after Fisher and Gibson, 1982)

All of the 20 cm bursts observed with the VLA have been highly-circularly polarized, $\pi_c \gtrsim 50\%$, and in some cases $\pi_c \approx 100\%$. The 6 cm flare of L 726-8 A (UV Ceti's companion) shown in Fig. 10 was not significantly polarized, however. As Melrose and Dulk (1982) have pointed out, large circular polarizations ($\pi_c$ > 30%) and high brightness temperatures ($T_B > 10^9$ K) are not mutually compatable within the framework of incoherent emission mechanisms. Thus, even the relatively weak flares seen here may result from coherent processes. Melrose and Dulk (1982) favor a cyclotron maser mechanism which they estimate can achieve $T_B^{max}$ ~$10^{16}$ to $10^{17}$ K and, of course, $\pi_c$'s~100%. The population inversion (or loss cone anisotropy) to drive the maser develops when electrons injected at the top of a flux tube travel down the flux-tube to regions of higher magnetic field and ambient particle density. The low pitch-angle electrons are thermalized before they can be "mirrored" back leaving only electrons which develop a one-sided loss cone.

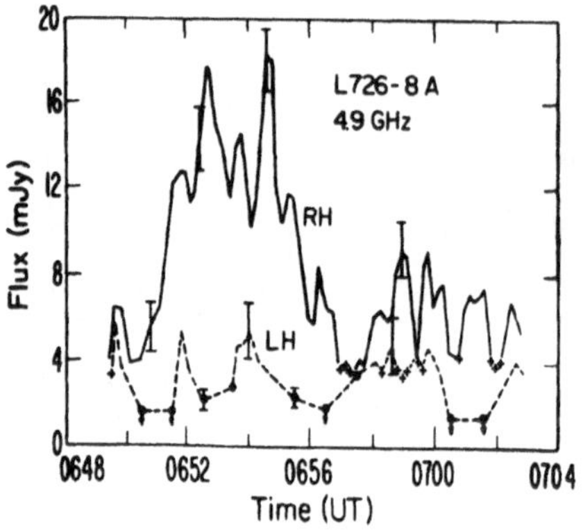


Fig. 11. Possible quasi-periodic oscillations in L726-8 A seen predominantly in r.h.c.p. (after Gary *et al.*, 1983).

The electron-cyclotron maser may have is best application in explaining quasi-periodic 56 sec oscillations seen in L726-8 A (see Fig. 11) by Gary *et al.* (1982). The key datum here is not the oscillation itself, which remains unexplained but could be due to flux-tube oscillations driven by Alfven waves which modulate the emission. Rather, the very rapid variations suggest the excited region is very small and thus $T_B >> 10^{12}$ K. A cyclotron maser theory is very attractive because the same electron population that produces the radio flare could also account for flaring in the hard and soft X-rays, the UV line emission, and optical continua as discussed below.

## 5. CORRELATION OF RADIO FLARING WITH PHENOMENA AT OTHER BANDS

It is most difficult to assess the frequency with which radio flares are accompanied by flares at other bands since the coordinated efforts made to date have varied widely in the bands covered, their sensitivities, and the stars observed. Only the coordinated observations of Proxima Centauri by Haisch *et al.* (1981) provided more-or-less simultaneous coverage at all four major bands -- radio, optical, UV, and X-ray -- and only two others succeeded in providing simultaneous radio, optical, and X-ray observations of YZ CMi (Karpen *et al.*, 1977; Kahler *et al.*, 1982).

The impression that most radio events were accompanied by quasi-simultaneous optical flares was widely held until the mid-1970's primarily because the single-dish observations had no good way of distinguishing radio flares independently. However, since many of these simultaneous events were among the largest ever observed the probability that strong radio flares have optical counterparts remains high. The best observed events of this type (e.g., Fig. 1) show significant time-delays between the optical and radio peaks, suggesting that the radio events could be stellar counterparts of solar type IV-dm bursts. Much more sensitive low-frequency radio/optical studies by Spangler and Moffett (1976) showed the simultaneous occurrence of radio/optical flares about 70% of the time, and occasionally the radio flare preceeded the optical event. A larger survey at still lower-frequencies by Nelson et al. (1979) showed only about 30% of the radio and optical flares were correlated, the radio flares were generally of long duration (>30 min), and all followed the optical flares.

At microwave frequencies the correlation between optical and radio events is virtually non-existent for all stars except Proxima Cen which shows correlated events about 25% of the time (cf. Slee et al., 1982). Otherwise only 2 or 3 correlated events have been reported out of 9 radio and 41 optical events recorded during some 50 hours of coordinated observation in the USA (Fisher and Gibson, 1982) and Australia (Slee et al., 1982).

Two events probably best represent the relation of radio phenomena to those at other bands. At microwave frequencies all coordinated events have shown the optical and radio flares to be co-temporal (cf. Fig. 6) whereas at longer wavelengths the radio emission follows other flare phenomena (see Fig. 12). These two examples support the idea that there are at least two different types of radio flare emission, the former on which the radiating electrons are of the same population as those which "create" phenomena in other bands, and the latter in which the emission probably results from excitation of regions above the flaring loop during the propogation of some outward moving disturbance.

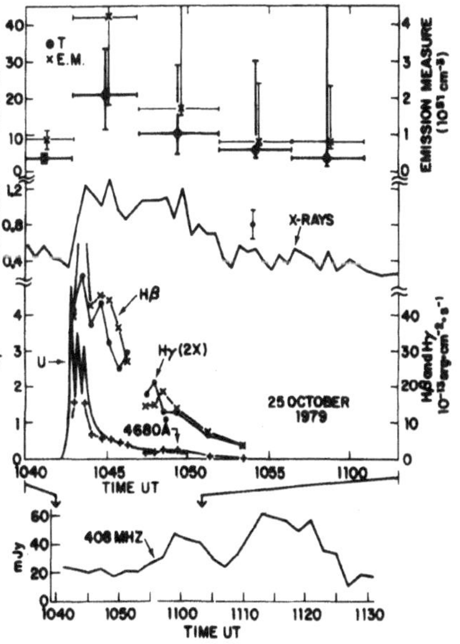


Fig. 12. Multi-frequency observations of a flare on YZ CMi. Note the time delay before the onset of the low-frequency radio emission (after Kahler et al., 1982).

## 6. QUIESCENT EMISSION FROM FLARE STARS

More surprising to this author than the detection of flaring emission from dMe stars was the detection of quiescent emission (cf. Gary and Linsky, 1982; see Fig. 13). The quiescent solar atmosphere only becomes optically thick at cm-wavelength in the outer chromosphere where temperatures are typically ~20000 K. This implies, by analogy, that the expected flux density from a typical dMe star would be <1 μJy, still about 2 orders of magnitude below today's capability. What is required to make them detectable is that the source be both larger and optically-thick at a higher temperature ($T > 10^7$ K). Gary and Linsky (1981) suggest that gyroresonance absorption can make the coronae about these stars optically-thick at frequencies a few harmonics above the gyrofrequency (see Fig. 14). This implies that field strengths in the coronae must be few hundred Gauss. To be consistent with the Einstein X-ray results the temperatures determined at both bands should be the same. Thus, if the temperatures are few x $10^7$ K as suggested by Einstein, the source size inferred by the radio flux densities which are measured imply $R_{radio\ corona} > 3R_*$! This may be pushing the thermal bremsstrahlung/gyroresonance absorption interpretation a bit far. One needs to determine source spectra and polarization characteristics before this model can achieve a full measure of acceptability.

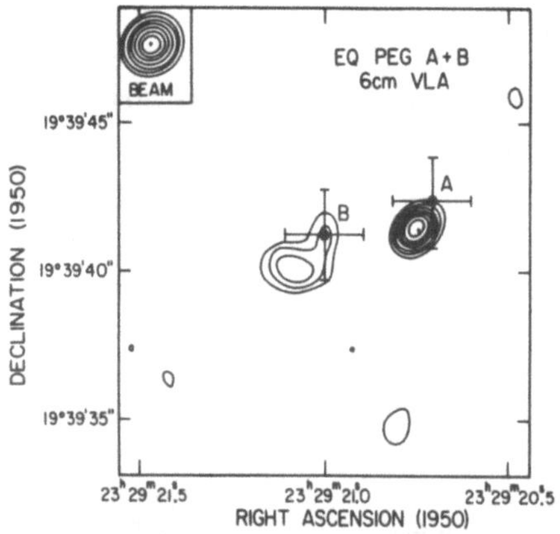


Fig. 13. Quiescent emission from EQ Peg A (0.6 mJy) and B (0.4 mJy) observed by the VLA (after Topka and Marsh, 1982).

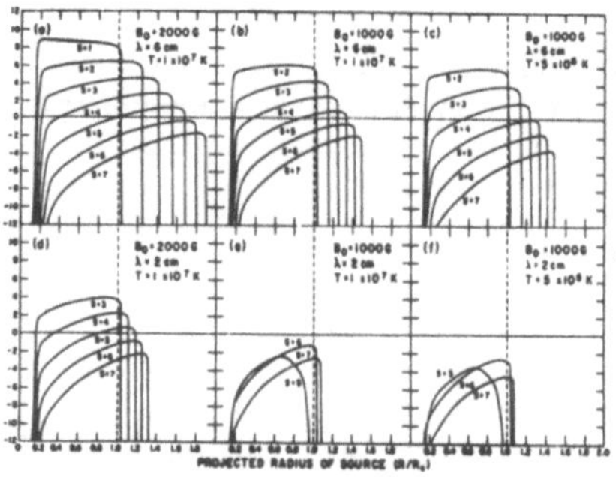

Fig. 14. Optical depths due to gyroresonance absorption as a function of height, magnetic field, wavelength, and temperature. These graphs show it may be possible to make dMe stellar coronae optically thick to a few stellar radii (after Gary and Linsky, 1981).

That the radio corona is thermally controlled, however, is virtually certain. Flux density measurements of UV Ceti show the same levels of emission on timescales of a few hours to 1.3 years. Thus, long-term monitoring of quiescent microwave emission may be a very good way to investigate them for synoptic (solar-cycle type) variations.

ACKNOWLEDGEMENTS

The author would like to thank Paul Fisher, Dale Gary, and George Dulk for helpful comments in preparing this manuscript. I would also like to thank the above and Bruce Slee for permission to use figures in this review which may appear in advance of publication.

REFERENCES

Andrews, A.D. 1969, IBVS No. 325.
Davis, R.J., Lovell, B., Palmer, H.P., and Spencer, R.E. 1978, Nature, 273, 644.
Fisher, P.L. and Gibson, D.M. 1982, in Second Cambridge Workshop on Cool Stars, Stellar Systems, and the Sun (eds. M.S. Giampapa and L. Golub), Vol. II, 109 (SAO:Cambridge).
Gary, D.E., Dulk, G.A., and Linsky, J.L. 1982, submitted to Ap.J.
Gary, D.E. and Linsky, J.L. 1981, Ap. J., 250, 284.
Haisch, B.M., et al. 1981, Ap. J., 245, 1009.
Kahler, S., et al. 1982, Ap. J., 252, 239.
Karpen, J.T. et al. 1977, Ap. J., 216, 479.
Kunkel, W.E. 1969, IBVS No. 325.
Lovell, B. 1969, Nature, 222, 1126.
Lovell, B. 1971, Quart. J.R.A.S., 12, 98.
Lovell, B., Whipple, F.L., and Solomon, L.H. 1963, Nature, 198, 228.
Lovell, B., Whipple, F.L., and Solomon, L.H. 1964, Nature, 202, 377.
Melrose, D.B. and Dulk, G.A. 1982, Ap.J., in press.
Nelson, G.J. Robinson, R.D., Slee, O.B., Fielding, G., Page, A.A., and Walker, W.S.G. 1979, MNRAS, 187, 405.
Slee, O.B., Allen, W.H., Coates, B.W., Page, A.A., and Quinn, P.J. 1982, preprint.
Slee, O.B., Higgins, C.S., Roslund, C., and Lynga, G. 1969, Nature, 224, 1087.
Slee, O.B., Solomon, L.H., and Patson, G.E. 1963, Nature, 199, 991.
Slee, O.B., Touky, I.R. Nelson, G.J., and Renie, C.J. 1981, Nature, 292, 220.
Spangler, S.R. and Moffett, T.J. 1976, Ap. J., 203, 497.
Spangler, S.R., Rankin, J.M., and Shawhan, S.D. 1974a, Ap. J. (Lett.), 194, L43.
Spangler, S.R., Shawhan, S.D., and Rankin, J.M. 1974b, Ap. J. (Lett.), 190, L129.
Topka, K. and Marsh, K.A. 1982, Ap. J., 254, 641.

DISCUSSION

Kuijpers: In the case of UV Ceti did you ever observe a flare going off in between the two stars or are they always concentrated on a single individual star?

Gibson: No. As far as we can tell they are all on individual stars.

Venugopal: If one observed the Pleiades at meter wavelengths, then scaling from flare intensities observed in Orion, one should see much higher intensities than are in fact observed. We observed the Pleiades region in 1973 and 1974 for a total of over 200 hours without recording a single flare of intensity greater than 0.75Jy at $\lambda$ = 92 cm i.e. f = = 327 MHz. During November 1981 we had a further 60-70 hours of observation and again no flares.

Gibson: That is an interesting result and it again points out the probable problems with the early radio data.

Rodonò: It is quite a difficult problem to achieve coincident observation at more than one wavelength. The very high rate of flaring both in optical and radio on the active objects makes it very difficult to attribute an individual optical flare to a particular radio flare. So I would suggest that future observations should not be confined to the very active objects like YZ CMi, for example, but that the many less active objects be observed which will permit a better determination of optical/radio and other wavelength coincidences.

Gibson: I think that the problem with that approach is that we have observed less active stars and we do not see flares at all, at least in the kinds of total observing times available on an instrument like the VLA. Zirin and I have made attempts at simultaneous observation previously and we do not really know how to resolve this problem. It is obviously necessary to have at least optical observations simultaneous with the radio data if we are to put the flares into a proper perspective. However at present only stars which flare as frequently as YZ CMi and UV Cet yield a sufficient number of flares to make it worthwhile undertaking a project like this given that the total amount of observing time is of the order of 100 hrs $yr^{-1}$.

Evans: I would like to address two points. The first is that one of the biggest difficulties in arranging simultaneous observations is that of scheduling time at separate institutions. You mentioned the simultaneous radio and optical work by Spangler and Moffett. In that case there appeared to be a convincing argument in favour of a connection between flares in the two regions of the electromagnetic spectrum whereby the radio flare followed the optical flare by 5-8 minutes. The second point concerns the effects of duplicity and you talked about ringing between one star and another. Well, the duplicity characteristics of flare stars vary a great deal, extending from YY Gem with a period of less than a day to UV Cet whose period has been revised from 250 to 26 years. We did an experiment to see if there was an effect of separation on the rate of flaring. The early observations by Moffett happened to be done near periastron. We then observed the star again about 3 years later and found the same rate of flare energy release. There was some suggestion that the star might not have been flaring in quite the same way. Even at their closest however those stars are pretty far apart and if one were looking for mutual interaction one would be better to search for it in YY Gem or CC Eri.

Gibson: Well I would like to thank you for bringing up the correlation between optical and long-wavelength radio which seems well established. There seems to be correspondence about 50% of the time with a distinct time delay. This would seem to agree with models which invoke the optical flare going off at low level and then a disturbance propagating up into the corona where radio emission can escape from the star and be detected. This may be fundamental different from the kinds of things which we are seeing at higher frequencies.

Kodaira: You have said that the correlation between low radio frequency flares and optical flares is a loose one whereas the high frequency flares come "hand-in-hand" with the optical. Yet your results and those from my survey have failed to produce real coincidences. An exception is the Parkes 64m single-dish observations at 5 GHz.

Gibson: There is only one flare which we have detected simultaneously in the optical and with the VLA and here we only saw the beginning of the flare.

Kodaira: But this was impulsive microwave emission.

Gibson: Most of the rest of the data does come from the Parles data, primarily on Prox Cen and there they seem to get good "hand-in-hand" data.

Kodaira: Yes but this data shows effects which only occur in one integration bin and it only barely exceeds the $3\sigma$ level. Do you credit these data?

Jordan: I think we are getting a little detailed.

Kodaira: Yes but it is important whether they have really detected microwave bursts or whether the effect is too marginal to credit at this stage.

Uchida: My question relates to the propagation of the low-frequency disturbance. Has anyone tried to estimate the velocity of the low-frequency disturbance. I ask this because in the Sun there are two types of propagation radio sources. The first, Type II, is identified with the magnetohydrodynamic shock which travels at typically about 1500 km $s^{-1}$. The second, Type IV, is associated with the expanding electron cloud moving with typically 300 km $s^{-1}$ velocity. The blast-type burst might be particularly interesting because of the possibility of sympathetic flaring in binaries.

Gibson: The assumption was made by Spangler for istance in analysing his delays that he was dealing with the shock rather than the expanding cloud. He used more or less arbitrarily in this analysis parameters which are typically solar because nothing better was available at that time. Perhaps these need to be looked at again and the calculations reworked.

# SESSION III

## OBSERVED ACTIVITY IN RELATED OBJECTS

: THE SUN

: RS CVn BINARIES

: CONTACT BINARIES

: T Tau AND OTHER ACTIVE STARS

FLARES ON THE SUN : SELECTED RESULTS FROM SMM.

G.M.Simnett
Department of Space Research
University of Birmingham
Birmingham B15 2TT

## 1. INTRODUCTION

Observationally the study of solar flares has reached the stage where intensity-time distributions of emission over broad and resolved regions of the electromagnetic spectrum are obtained for spatially resolved parts of the flare. Polarization measurements add an important diagnostic tool in some wavebands but we shall not report on these here. In the optical band good ground based observations have been available for many years, whereas in the UV, soft X-ray and hard X-ray (> 5 keV) bands recent spacecraft have greatly extended the data base. Good high resolution maps are being made in the microwave region with the ground based VLA. We are now at the point where significant progress into understanding the flare problem has been made, and will continue to be made, during the current solar maximum. This coincides with the development of soft X-ray instruments sensitive enough to detect transient and quiescent emission from flare stars, particularly red dwarfs in the solar neighbourhood (e.g. Kahn et al,1979, Haisch et al, 1980) which previously had only been detected in the optical and radio wavebands. In the light of this important milestone it is appropriate to review the current advances being made with the Solar Maximum Mission (SMM) data set on solar flares and to examine how they can further the interpretation of flare star behaviour. We first indicate the physical parameters which can be readily deduced from the observations, and then show how SMM provides sufficiently well co-ordinated measurements for real advances to be made.

## 2. THE PHYSICAL FRAMEWORK OF FLARES

A typical solar flare starts with a slow increase in temperature and emission measure from the subsequent flare site. The energy source is in the strong magnetic field associated with the sunspot groups within which flares almost invariably occur. Generally a bipolar magnetic flux tube is close to the source of energy release, and observations of the post-flare state often indicate an arcade of loops stretching into the low corona. Following the pre-flare increase by some minutes is the



*P. B. Byrne and M. Rodonò (eds.), Activity in Red-Dwarf Stars, 289–305.*

main impulsive phase, when the hard X-rays are emitted. In this phase the bulk of the flare energy resides in the kinetic energy of non-thermal electrons > 10 keV and, perhaps more significantly, ions; the energy is relatively invisible to us if the latter dominates. Electrons which encounter a high density medium such as the chromosphere, immediately radiate via non-thermal bremsstrahlung. The resultant X-ray spectrum generally suggests a power law distribution of electron energies, $dJ(E)/dE \propto E^{-\gamma}$, where E is the energy in keV. The relationship of the index $\gamma$ to the observed X-ray spectrum depends on the thick or thin target nature of the emission, the former appropriate for an electron beam hitting the chromosphere and the latter appropriate for an electron beam trapped in, or escaping from, the corona (Lin & Hudson, 1976). Where electrons encounter suitable magnetic fields in the relative absence of matter, such as at the top of a magnetic arch, they produce microwave radiation. Similarities between hard X-ray and microwave intensity-time profiles suggest that these emissions are from different fractions of the same population of electrons. The ions, if sufficiently energetic ( > few MeV) may produce nuclear gamma rays, but otherwise they remain invisible, contributing only to the thermal energy content of the ambient medium. During the impulsive phase intensity fluctuations are rapid ( < 1s) which makes unambiguous interpretation of the flare difficult.

Both electron and ion populations create a hot thermal plasma as they stop. Moore et al (1980) showed that the most significant fraction of the total flare energy resides in this plasma during the slow decay phase. The plasma is contained in a magnetic flux tube thought to be approximately in pressure equilibrium; therefore the energy is dissipated either through conduction or radiation. Moore et al (1980) derive the characteristic cooling times, $\tau$, of a plasma in the $10^7$K region :

$$\text{conduction dominated} \quad \tau_c \cong 4 \times 10^{-10}\ L\, n_e^{2}\, T^{-5/2} \ \text{s} \qquad (1)$$

$$\text{radiation dominated} \quad \tau_r \cong 10^{7}\ T\, n_e^{-1} \ \text{s} \qquad (2)$$

where T is the temperature (K), $n_e$ the electron density ($cm^{-3}$) and L the length (cm) of the flux tube. In the decay phase temperatures and emission measures may be derived from the observations . The emission measure Y is given by :

$$Y = \int n_e\, n_i\, dV \cong \int n_e^{2}\, dV \qquad cm^{-3} \qquad (3)$$

where $n_i$ is the ion density, assumed equal to $n_e$, and V is the volume ($cm^3$). In some instances the differential (with respect to temperature) emission measure can be derived from line spectra; however analysis of the latter to determine flare parameters is a complex and evolving science, beyond the scope of this review. At higher energies, ≳ 10 keV, the thermal bremsstrahlung continuum dominates, and Crannell et al (1978) derive the following :

$$\frac{dN(E)}{dE} = 1.3 \times 10^3 . Y . e^{-E/aT} . E^{-1.4} . T^{-0.1} \quad (4)$$

where N(E) is the measured photon spectrum and aT is the temperature in keV. Thus the measured photon spectrum can be used to determine the temperature and emission measure of the thermal plasma. Consideration of (1)-(4) shows that the situation, as described, is close to being defined; the uncertainties are the choice between (1) and (2) as to the cooling mechanism, the definition of the volume, and the observational data. In practice other factors such as additional energy input and lack of stability in the principal flare loop are a complication.

One parameter which may link the impulsive and thermal phases is the total energy in the flare. During the impulsive phase integration of the electron spectrum, derived from the hard X-ray spectrum, gives a lower limit to the energy (the balance being in the ions), while in the thermal phase the energy of the plasma, Q, is given by:

$$Q = 3 \, Y \, k \, T \, n_e^{-1} \text{ erg} \quad (5)$$

This immediately sets a constraint on the magnetic field, for if the plasma is to be contained, the magnetic pressure $B^2/8\pi$ must exceed the thermal pressure $n_e kT$. Microwave emission is a function of the electron spectrum and the magnetic field. It can be seen from (1)-(5) that if the observations are of high quality and co-ordinated, then some real limits may be placed on a number of flare parameters. The thermal phase is better understood, and it is this phase alone that can, at the current time, be studied in flare stars.

In so far as they are observed, flares in nearby stars involve energy of a different order of magnitude than even the largest solar flares - $10^{35}$ erg compared with $\approx 10^{32}$ erg. The largest estimate for a stellar flare in the optical band is $6 \times 10^{34}$ erg from YZ CMi on Jan 19, 1969 (Kunkel, 1969). Despite the high energy, they have such a short rise time that they have to be local phenomena, rather than sudden expansions of the star's envelope. For example, UV Cet increased its output by a factor of 420 in 31s, or $2^m.8$/s, during a flare on Sept 22, 1974 (Jarrett and Gibson, 1975). They do not radiate strongly in X-rays, and only recently have soft X-ray flares been detected from some of the nearer red dwarfs, ATMic (dM4.5e), ADLeo (dM3.5e) and Prox. Cen (dM5e) (Kahn et al, 1979; Haisch et al, 1980).

If the understanding of stellar flares is to benefit from solar flare studies, we must discover how to scale from solar conditions to those in the dwarf. Kahler and Shulman (1972) used the ratio of the soft X-ray (1-20 Å) luminosity to optical luminosity as seen in a "typical" solar flare. However, this is quite inappropriate for stellar flares; in UV Cet the soft X-ray luminosity is an order of magnitude too low when compared with the Sun. Mullan (1976) showed that the relative power in the X-ray/optical wavebands is related to the radiated/conducted energy loss from the hot plasma. Under the assumption that conductive losses exceed radiative losses, and taking relative dimensions into account, he argues that X-ray luminosities on red dwarfs should be an order of magnitude, or more, lower than solar equivalents. However, we must remember that the theoretical treatments leading to (1) and (2) are relatively imprecise.

The e-folding decay time of a flare appears to be constant for a given star (Kunkel, 1975). This is not true for solar flares, although some simple compact flares have a rapid, $\lesssim$1 m, decay time for the soft X-rays. In more complex flares, which are generally larger in area, the decay time at a given wavelength is very inhomogenous. Fig.1 shows the 3.5-8.0 keV X-ray intensity-time history of a flare seen by the Hard X-ray Imaging Spectrometer (HXIS) on SMM. The upper left panel shows the complete flare, with the gradual onset before this impulsive phase inset on an expanded intensity scale. The other panels show the time line for different parts of the flare. In all cases the decay of specific parts is much faster than the decay of the whole. The rise time also shows wide differences. This is a caution regarding the use of simple radiation and conduction theory for a complex flare as the energy is redistributed within the flare rather than escaping to an infinite heat sink (the photosphere for conductive losses and space for radiative losses). The uniformity of stellar flares may point to a simple, dominant flare scenario on the red dwarfs, paradoxically involving larger magnetic structures (to explain the high total energy). On intrinsically brighter stars, the magnetic structures are on a smaller scale and are more complex, perhaps leading to very efficient energetic particle acceleration and correspondingly higher hard X-ray luminosity. On the Sun the impulsive hard X-rays are produced primarily by non-thermal electrons not by high temperature thermal plasmas.

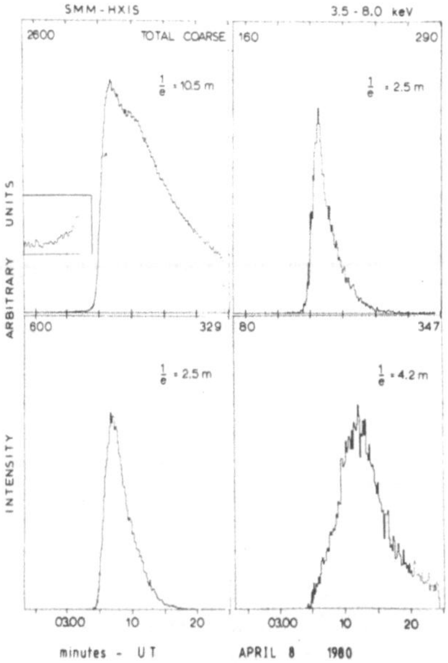


Figure 1. Soft X-ray intensity-time histories from a flare on April 8. Clockwise from upper left: the total flare with the onset magnified inset; 20" NE of the flare core; the flare core; 20" SW of the flare core. The number in the upper left corner of each panel is the intensity at the top of that panel.

## 3. THE RESULTS FROM SMM

NASA launched SMM on Feb 14, 1980 from Cape Canaveral, Florida. It represented the most comprehensive set of instruments ever assembled to study the active Sun. They monitored electromagnetic radiation from the optical to the $\gamma$-ray part of the spectrum, with high time resolution, good spectral resolution and in some regions high spatial resolution (10" or better). The payload is described in Solar Physics 65, 1980; the first results were published in Ap.J.Lett. 244, 1981; and a review of the significant scientific achievements during the period of full operation, up to December 1980 was given by Simnett (1981).

The strength of SMM was its ability to study the impulsive phase simultaneously with the full complement of instruments. It is not surprising, therefore, that the most definitive new results have emerged from these studies. Prior to launch it was suspected that the hard X-rays, >20 keV, did not exhibit faster variations than about 1 s (Hoyng et al, 1976). The concept of an elementary flare burst was introduced to account for this cut-off (de Jager and de Jonge, 1978). However, a study by the Hard X-ray Burst Spectrometer (HXRBS) at time resolutions down to 10 ms has revealed significant structure at this level (Kiplinger et al, 1982). Fig.2 shows 22 s of data from a short, intense event on March 22, 1981; the points in Fig.2a are at 128 ms resolution and features are expanded in Fig.2b,c at 30 ms resolution. Spike 1 has a FWHM of 45 ms while spike 4 shows an abrupt ($4\sigma$) fall in 20 ms. These X-ray variations are the fastest yet seen from the Sun, although ms fluctuations have been reported at microwave frequencies (Slottje, 1978). Kiplinger et al (1982) note that fast events are rare, with only $\sim$10% of suitable events examined showing fast spikes with durations $<1$ s. There is lack of such spikes in major events, although this could be a masking effect caused by lack of spatial resolution or an intense, more gradual component. The more intense fast spikes are seen in the presence of a dominating slowly varying component.

One of the most controversial issues regarding the impulsive phase is the thermal/non thermal origin for the hard X-ray bursts. Short decay times place severe constraints on a thermal model; (2.1) and (2.2) show that fast decays are achieved best if either we have a small dimension L in a conduction dominated plasma or a high density in a radiation dominated plasma. In a non-thermal model fast fluctuations are interpreted as bremsstrahlung from electron beams of short duration, conceptually easy to achieve; however the energy conversion efficiency is so small that large numbers of electrons are needed. Nevertheless, SMM results are supporting the non-thermal hypothesis more than the thermal.

Fig.3 displays the intensity-time history of a fast spike event on May 7[1], imaged by HXIS in the 16-30 keV band with 1.5 s resolution. There are two broad peaks, each with significant fine structure at the 1.5 s level; this event also exhibited 60 ms time structure above 29 keV (Kiplinger et al, 1982). Spatial intensity contours are shown at various times within the bursts. The first bursts in the first peak originate in

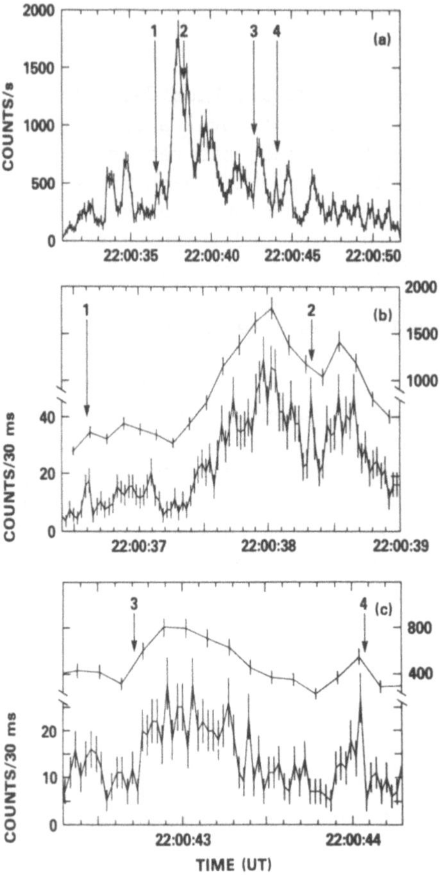


Figure 2. Fast hard X-ray time variations on March 22, 1981. (Kiplinger et al, 1982)

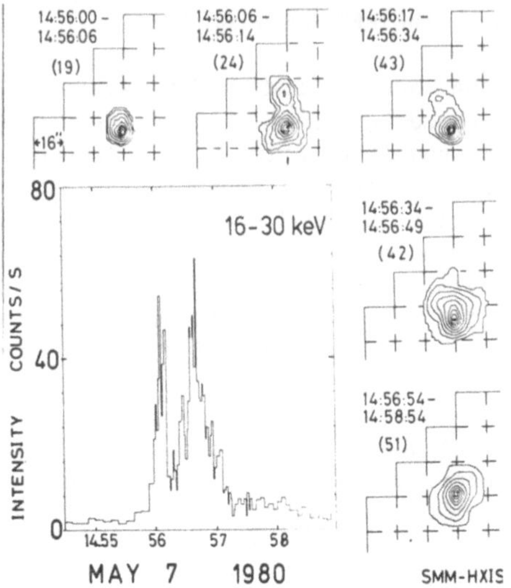

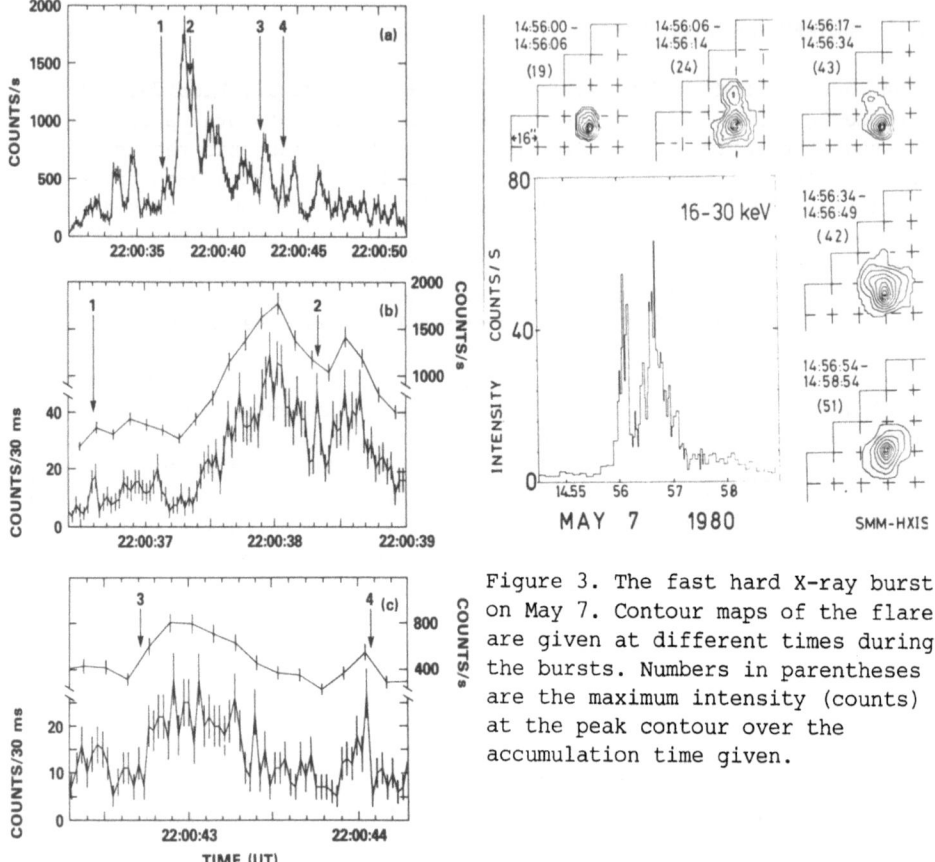

Figure 3. The fast hard X-ray burst on May 7. Contour maps of the flare are given at different times during the bursts. Numbers in parentheses are the maximum intensity (counts) at the peak contour over the accumulation time given.

a compact, <6" FWHM, source, which is co-located with the southern-most of two bright $H_\alpha$ kernels visible just before the flare. The picture from 14.56.06-14.56.14 covering the last part of the first peak shows an additional component to the north. These two bright points are interpreted as footpoints of a magnetic loop. The small bursts between the two main peaks are from the south footpoint. From 14.56.34-14.56.49, covering the second peak, the emission is diffuse. The halo around the centroid could be Compton backscattering off the photosphere from a hard X-ray source rising in altitude within the loop. During the decay the centroid shifts to midway between the footpoints, consistent with the looptop. Data from the X-ray Polychromator (XRP) (Acton et al, 1982) is interpreted as chromospheric evaporation from the south footpoint, driven by the impact of non-thermal electrons rather than a thermal conduction front.

One of the principal arguments advanced by proponents of thermal models for hard X-ray bursts is that for many flare situations, emission from a thermal plasma is more efficient than bremsstrahlung for a non-thermal beam. However, Kiplinger et al (1982) have made a detailed analysis of the May 7 event and have shown there to be no thermal advantage, and have ascribed a non-thermal origin to the fast spikes.

The Gamma Ray Spectrometer (GRS) has observed over 100 flares with emission above its 0.3 MeV threshold with some 30 or so showing evidence for unresolved prompt nuclear de-excitation lines from $^{16}O$ and $^{12}C$ at 6.13 and 4.44 MeV respectively (E.Rieger, 1982, private communication). A subset of the latter also show the 2.223 MeV neutron capture line $^{1}H\ (n,\gamma)\,^{2}H$. This is a delayed line for which the neutrons must first slow down and then be captured by ambient protons. Both processes are density dependent, with the capture time dominant. The delay of the 2.223 MeV line from the prompt de-excitation lines is typically $\gtrsim$100 s, which implies (Wang and Ramaty, 1974) that the mean density at which the neutrons are captured is $10^{17}$ H $cm^{-3}$, or just within the photosphere. Neutrons are produced in the first instance from break-up of the $^{4}He$ by protons $\gtrsim$25 MeV, with contributions from spallation of heavier nuclei.

Events such as May 7 show that electron acceleration up to sub-relativistic energies is impulsive. It transpires from other events that this also applies up to at least a few MeV for electrons and many times this energy for ions. In the June 7 event the time history of the gamma ray intensity at the energies of the prompt de-excitation lines discussed above was equally impulsive as, and in phase with, the hard X-ray intensity (Chupp et al, 1981). Detailed analysis, taking into account the time history of the 2.223 MeV line also, showed that energetic protons ($\gtrsim$50 MeV) and relativistic electrons (1 MeV) were accelerated in a similar time scale, quite contrary to earlier beliefs.

There have been further developments following a limb flare on June 21,01.19 UT, when neutrons up to 600 MeV were detected by GRS (Chupp et al,1982). The timing of the neutrons at 1 A.U.showed that protons up to 1 GeV must have been accelerated coincident with, or within a few seconds of, the impulsive phase. Ion acceleration in this flare must have been efficient, for the 511 keV positron annihilation line was seen. This results from nuclear interactions which produce radio-isotopes such as $^{11}C$, $^{13}N$, $^{15}O$, $^{19}Ne$, etc. The positron line has been observed in very few events. High energy neutrons have recently been detected from the June 3,1982, X8 flare (Forrest, D.J.,private communication) showing that the June 7 event was not unique in this solar cycle.

From an overview of the gamma ray events two things emerge: 1) many events are sub-flares when seen in $H_\alpha$; 2) de-excitation lines are seen more often than the 2.223 MeV neutron capture line, which itself is more common than the 511 keV line. Thus acceleration of electrons up to 0.5 MeV and protons to a few MeV is both common and part of the impulsive phase. The difference between small gamma ray flares and major $H_\alpha$ flares is that in the latter particle acceleration in the impulsive phase is more efficient and may extend to higher energies. The concept of a two stage acceleration process no longer seems necessary to explain the highest particle energies reached in flares.

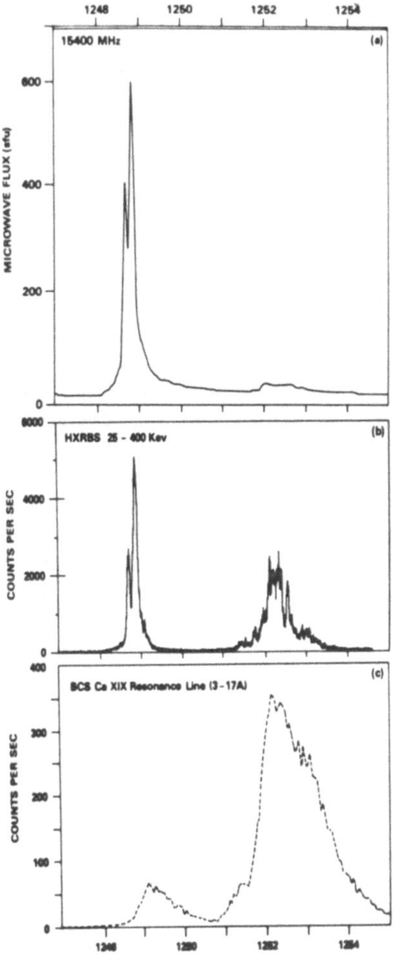


Figure 4. The double flare at 12:48 August 31 (Strong et al, 1982)

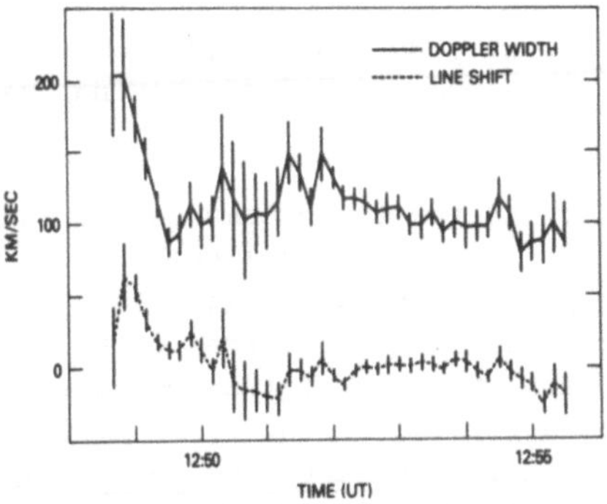


Figure 6. Doppler broadening and blue shift during the double flare on August 31 (Strong et al, 1982)

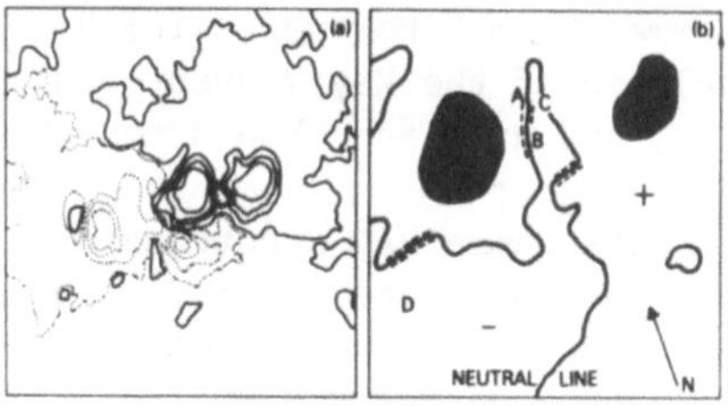


Figure 5 a) the transverse magnetogram on August 31; b) the position of the initial flare brightenings A,B,C. (Strong et al, 1982)

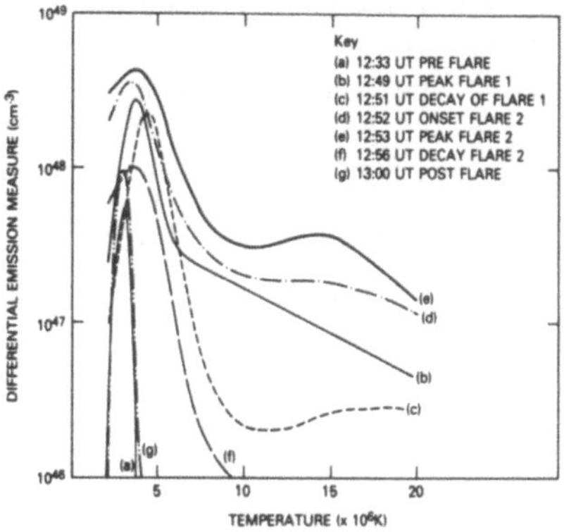


Figure 7. Differential emission measure analysis ($cm^{-3}/10^6 K$) for the double flare on August 31. (Strong et al, 1982)

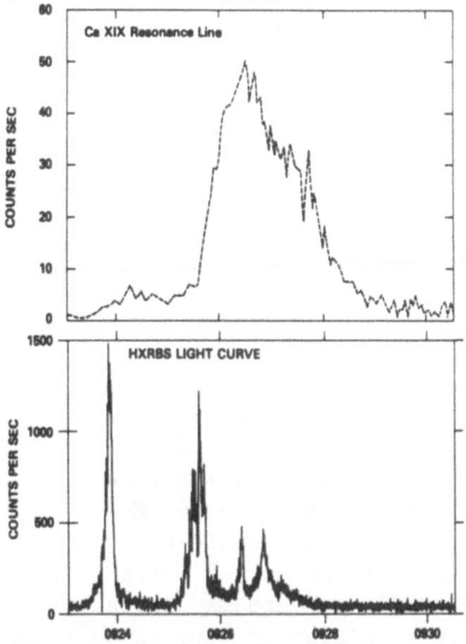


Figure 8. The double flare at 9:24 August 31. (Strong et al, 1982)

We turn now to a compact event, consisting of two closely separated flares, which is illustrative of the co-ordinated analysis possible from SMM and ground based observations. Fig.4 shows the intensity at 15.4 GHz, 25-400 keV and the Ca XIX resonance line from 12.47-12.55 on Aug 31 (Strong et al, 1982). The Ca XIX line is monitored by the Bent Crystal Spectrometer (BCS) of the XRP. The doubly impulsive first flare is very hard with a photon number spectrum obeying $dJ/dE \propto E^{-3}$ up to 150 keV. The microwave profile mirrors that of the hard X-rays (a typical feature) indicating that electrons had access to a region from which microwave radiation was (a) produced and (b) escaped. The soft X-rays (Fig.4c) reached maximum 20-30 s after the hard X-ray peak and exhibited a rapid decay (e-folding time 35 s). Fig.5a shows the longitudinal magnetogram of the flare region where a -ve inclusion is evident between two major +ve sunspots. Fig.5b sketches the region between the +ve spots, showing the locus of the magnetic neutral line; the Ultraviolet Spectrometer and Polarimeter (UVSP) identified the Fe XXI emission at 1354Å at points A, B and C during flare onset. In the impulsive phase of the first event HXIS imaged the 16-30 keV X-rays at two opposite sides of the core of the soft X-ray emission. This is consistent with a flare contained in a magnetic arch, with soft X-ray emitting plasma at the top and hard X-rays produced by electron beams at the footpoints. This model is supported by observations of metric frequency U-bursts.

In the second event a few minutes later the hard X-ray burst is made up of many fast spikes; also the peak in the soft X-ray intensity is co-located with the hard X-ray peak at a point displaced 5" NE of B. The X-ray spectrum from the whole flare has the same slope as for the first flare, but now there is a substantial low energy, thermal component.

The BCS can determine the turbulent velocity of the plasma from the Doppler broadening of the Ca XIX line, and also the bulk motion from the shift in line centre. These parameters are shown in Fig.6. Turbulent velocities reach 200 km/s during the first flare but do not change during the second. The line is quite strongly blue shifted during the first flare but not during the second. Strong et al (1982) concluded that in the second flare, because of the increased density from the earlier event, the non-thermal electrons lose the majority of their energy within the loop, rather than at the footpoints. The density is probably non-uniform throughout the loop so the bulk of the energy dissipation may appear close to a footpoint, from which the plasma will expand to fill the whole loop. If the loop is unresolved, this will manifest itself by a shift in the centroid of the emission, which is in fact seen. It is difficult to distinguish between lateral and vertical motion as the flare was at N12 E31 and the apparent motion is $\sim$NE.

The Flat Crystal Spectrometer (FCS) of the XRP investigated the multithermal nature of the coronal plasma throughout the flares (Strong et al, 1982). A differential emission measure analysis is shown in Fig.7 for seven times during the flares. The plasma state both pre-flare and post-flare appears isothermal while during the flares there is an enhanced high temperature component which has a significantly higher emission measure for the second flare than for the first.

A few hours before the Aug 31 event a weaker but similar pair of compact flares had occurred in the same region. The hard and soft X-ray time histories are shown for comparison in Fig.8. The common features are the first fast impulsive event, with low soft X-ray emission and a second multiple spike hard X-ray burst associated with a more dominant soft X-ray peak. The decay times are similarly fast. The main difference is that the soft X-rays peak significantly after the hard X-ray burst; this is interpreted as a lack of significant expansion of material into the loop following the first event, thus depriving the loop of sufficient stopping power for the second hard X-ray burst.

Events following this pattern may be very common, as may be seen from Fig.9, which shows HXIS data from a flare on October. 20. The initial impulsive hard X-ray peak is followed 1 m later by a longer, weaker peak. At 3.5-5.5 keV the slowly rising weak first event merges with the more intense second event, which like Aug 31 peaks approximately simultaneously with the hard X-rays. The soft X-rays have a fast, 30 s e-folding decay. Inset in Fig.9 is the hard X-ray contour map at the time of the first peak. The two clearly resolved features separated by ≈20" are suggestive of magnetic loop footpoints. During the second peak only the easterly point is bright, both at 3.5-5.5 keV and 16-30 keV. This event is similar to that on May 7 (Fig.3) with visible footpoints during the first burst and emission from one point during the second burst. Projection effects due to the easterly location of the October 20 flare mean the coincidence of the second burst with the easterly footpoint could also arise if the emission was from the looptop.

To summarise, there is a class of flare with two major energy releases in a region containing a compact magnetic loop; the hard X-ray data show considerable structure within each energy release, the first of which often causes chromospheric evaporation. The subsequent evolution depends on how the effects of the first release change the plasma state within the loop. The second release always appears dominant at soft X-ray wavelengths and rarely is resolved as a second footpoint.

Until SMM the existence of electron beams at the onset had not been proved. One of the first major results (Hoyng et al,1981) identified the points of impact of such beams on the chromosphere from a large 2-ribbon flare on May 21. UVSP observed the flare in the Mg II line at 2795Å, and demonstrated coindicence in space and time between the hard X-ray and UV bright points and $H_\alpha$ kernels. After the impulsive phase the centroid of the X-ray emission at both hard and soft wavelengths was at the looptop between the footpoints. Duijveman et al (1982) have analysed this and two other large events for which footpoint emission was seen in the impulsive phase. They show that the near simultaneous brightenings cannot be accounted for by an MHD disturbance triggering one source from the other for any reasonable values for magnetic field and plasma density. The number of electrons in each beam >16 keV in these large flares is in the region $10\text{-}30\times10^{36}$, and the energy in each beam is $2\text{-}9\times10^{29}$erg. This is small compared with the total flare energy estimated at the peak of the thermal phase. Some large flares also show the double structure summarized above for the small compact flares.

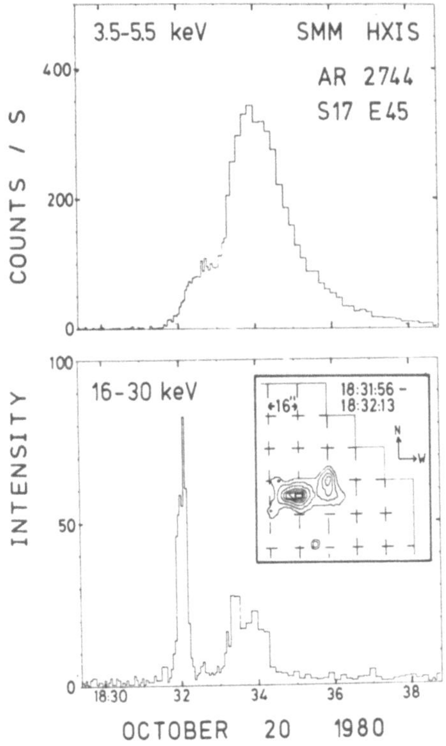

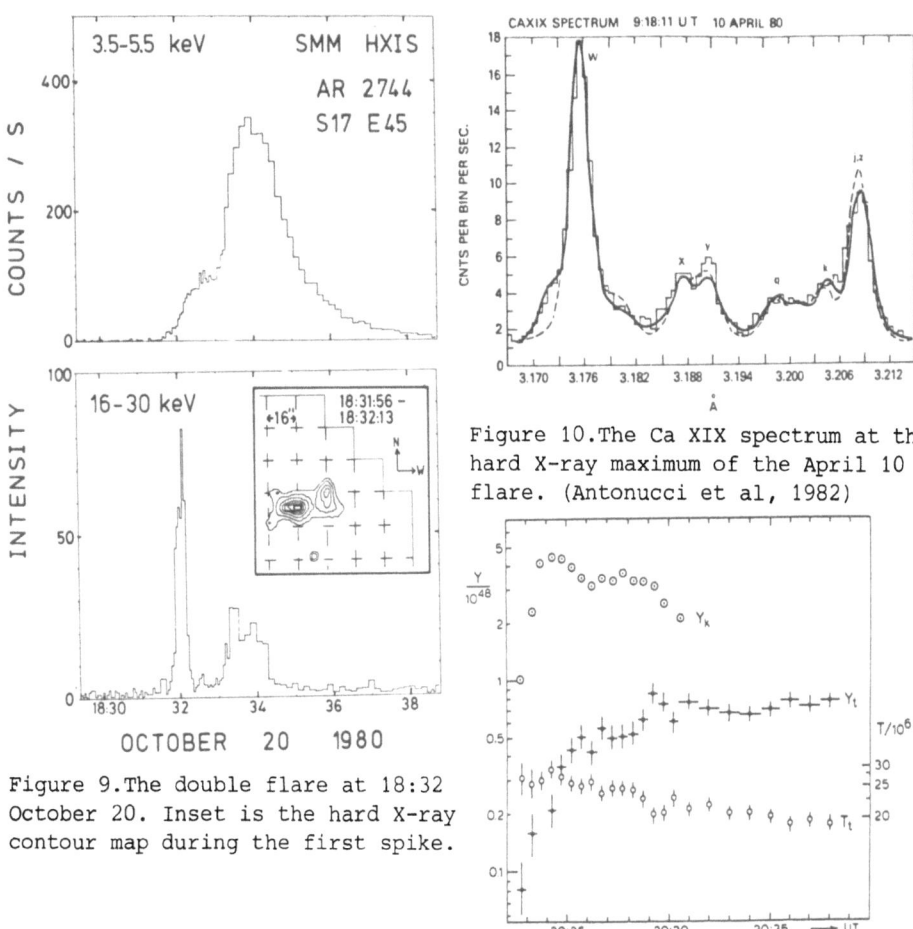

Figure 9.The double flare at 18:32 October 20. Inset is the hard X-ray contour map during the first spike.

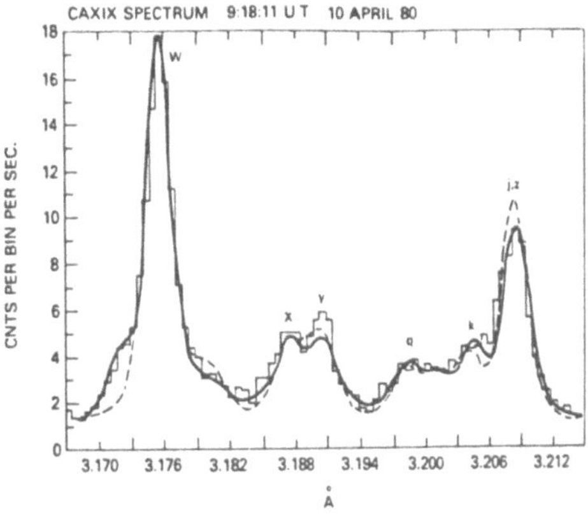


Figure 10.The Ca XIX spectrum at the hard X-ray maximum of the April 10 flare. (Antonucci et al, 1982)

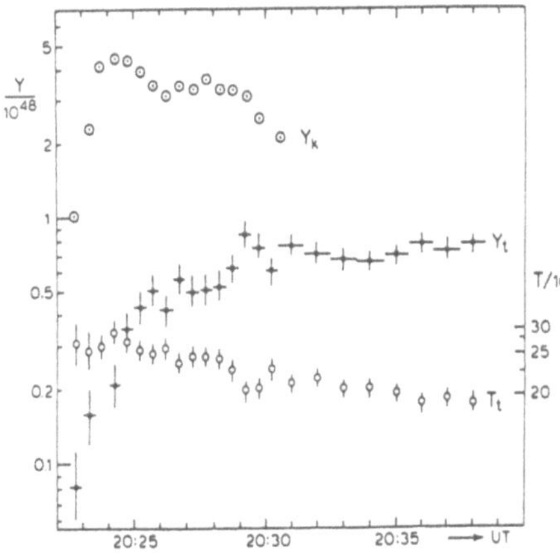


Figure 11. The emission measure of the kernel, $Y_k$ and the tongue, $Y_t$ of the limb flare at 20:25 April 30. Also shown is the electron temperature $T_t$ of the tongue (de Jager et al, 1982)

SMM has studied the dynamics of the flare plasma in unprecedented detail. During the impulsive phase of large flares, the plasma exhibits characteristic turbulence and upward motion, the latter frequently reaching 300-400 km $s^{-1}$ (Antonucci et al,1982). Fig.10 shows the Ca XIX spectrum during the impulsive phase of the April 10 flare. The spectrum shows not only a non-thermal broadening, best seen in the resonance line w but also a well defined blue wing. The theoretical stationary spectrum is shown dotted, and the solid line is obtained by adding a blue shifted component to it. Antonucci et al have computed the heat content and kinetic energy of the upward moving chromospheric plasma and have shown this to be sufficient to account for the energy contained in the flare loop during the main phase. It is well in excess of the energy in the electron beam deduced from an analysis of the hard X-ray burst using a thick target hypothesis (Duijveman et al,1982). The additional energy is probably supplied by ions accelerated with the electrons. The ions are invisible until they reach energies of several MeV/nucleon, when they start exciting nuclei, with resultant gamma ray emission; yet the spectrum below this threshold can easily contain the bulk of the flare energy.

Poland et al (1982) and Woodgate et al (1982) have reported coincidence between the impulsive hard X-rays and brightening in the OV line (1371Å) formed primarily in the transition zone at $2.5 \times 10^5$K. This is seen for both limb and disc flares, and leaves little doubt that the emissions come from loop footpoints. The good spatial resolution of UVSP allowed impulsive excitation from the limb event at 18:23 on June 29 to be tracked up to 30,000 km into the corona, indicating that rapid evaporation follows the impulsive phase (Poland et al,1982).They saw material ejected into the flare loop and a more open coronal structure. The X-ray bursts with the hardest spectra appeared in coincidence with features within the closed loop, where the density is relatively high.

Unlike disc flares, the morphology of limb flares can be examined unambiguously by the imaging instruments. De Jager et al (1982) have shown in the thermal phase of the April 30 flare that while the majority of the X-ray and UV emission was confined to a relatively compact low lying loop, electrons escaped from this loop to populate a much larger structure extending to at least 80000 km into the corona. Fig.11 shows the time history of the emission measure from the main flare and the coronal "tongue", and the electron temperature of the tongue, which initially is higher than the flare kernel. This indicates that only the higher energy electrons escape from the main loop.

Limb flares regularly show large coronal structures which provide a link between widely separated active regions. As the density decreases such structures tend to become invisible, so the observations only give a lower limit to their true altitude. Following the Nov 6 east limb flare (Svestka et al,1982a) an arch was visible to $\approx 1.5 \times 10^5$km; a similar large arch was also imaged in soft X-rays after the May 21 flare (Svestka et al,1982b) and the June 29 flare. In June the link between the active regions, 2522 and 2530 ($\approx 1.5 \times 10^5$km apart) was established while they were still well on the disc. Some flares from these regions produced time coincident (within ≈1m) sympathetic events in the other region, regardless as to which region flared first. At quiet times there was frequent evidence for a soft X-ray bridge between the two regions. Other regions, such as 2370 and 2372 in April were similarly connected, so this must be regarded as a normal state on the Sun.

Flares which occur within such a connection are likely to populate the high arch with trapped particles. Such events are therefore invariably associated with metric radio noise storms from the trapped electrons. One of the most interesting features on Nov 6 was a quasi-periodic variation (≈20 m) in soft X-rays with no response in the chromosphere ($H\alpha$) and little in the O$\underline{V}$ transition region line (Svestka et al,1982a). Thus the energy source was not from fresh small flares; it was presumably from dumping of trapped particles from the corona to a denser part of the atmosphere, similar to a terrestrial aurora. In the solar case the ions are the more likely species to cause the X-ray emission via local heating at the top of the chromosphere.

The corona is playing an ever important role in flare studies. During the early stages of some flares, Benz et al (1983) found that decimetric and soft X-ray emission was observed prior to both the impulsive flash phase and the gradual pre-flash phase as seen in X-rays >26 keV. Significantly, the locations of the Type III bursts in these events are not all the same, indicating that energetic electrons are present prior to the main flare. In some events there is evidence for a reverse drift metric burst prior to the flare, indicating that some electrons are moving downwards. Benz et al conclude that in such events energy carried by the downward moving particles may provide the flare trigger.

There are other interesting aspects to the SMM data. One feature of many active regions, such as AR 2372 in April, is the production of homologous flares, indicating a certain stability in the dominant magnetic structure, despite the instabilities inherent in the flare process. In fact, even flares from different regions may have many similarities as we have seen. The general magnetic field topology may be identical for all these flares and differences between flares may represent trivial variations in dimension and field magnitude that alter the final appearance of the flare out of all proportion.

In sunspot studies Gurman et al (1982) have reported transition region oscillations in the umbra with typical periods in the 2-3m region. Peak intensity is in phase with the maximum blue shift and they interpret the oscillations as upward propagating acoustic waves. The energy flux, however, at $\lesssim 2\times10^{3}$ erg $cm^{-2}s^{-1}$ is $10^{7}$ too small to account for the missing sunspot radiative flux. The global effect of the latter has been demonstrated by the Active Cavity Radiometer (Willson et al,1981). They showed the Solar Constant to have variations as high as -0.2%, due apparently to the presence of large sunspot groups. This energy must ultimately be released, probably in white light faculae; however, lack of significant high frequency fluctuations, <7 days, suggests that the missing energy must be stored in the convection zone for $\gtrsim$7 days.

The remaining instrument on SMM was the Coronagraph/Polarimeter, which did not observe the Sun directly. However, in terms of energy release in flares, it is evident that the corona plays a major role. Detailed analysis by Wagner et al (1981) concluded that the mechanical energy in a coronal transient may exceed the total radiated flare energy by over an order of magnitude. Thus the least visible of solar events turns out to be the most energetic! If this is true in flare stars, then individual events release a prodigious amount of energy. As the transient moves outwards it becomes supersonic, generating an interplanetary shock, and indeed it has been shown that most, if not all, of such shocks are initiated by a coronal transient.

## 4. SUMMARY

The advances made in the study of solar flares could help us to a better understanding of red dwarf flare stars. Before drawing too many parallels, however, we must remember that differential solar rotation and the dynamics of the convention zone are intimately related to sunspot and flare activity; flares are almost perfectly correlated with sunspots and there have been times in history when sunspots disappeared for long period Very little is known about the corresponding phenomena on flare stars, so the flare mechanism might be quite different. Scaling from solar flares to red dwarf flares may be inappropriate at worst, difficult at best. A decisive datum would be the observation of the separate impulsive phase in a stellar flare, but for this we must await more sensitive hard X-ray instrumentation.

ACKNOWLEDGEMENT

The author expresses his deepest thanks to members of the SMM experiment teams for supplying material used in this review.

NOTE[1] Unless otherwise stated all flares are 1980

REFERENCES

Acton, L.W. et al, 1982: Ap.J. (to be published)
Antonucci, E. et al, 1982: Solar Phys. 78, 107
Benz, A.O. et al, 1983, Solar Phys. 83, 267.
Chupp, E.L. et al, 1981: Ap.J.Lett, 244, L171
Chupp, E.L. et al, 1982: Submitted to Ap.J.Lett.
Crannell, C.J. et al, 1978: Ap.J. 223, 620
de Jager, C and de Jonge, 1978: Solar Phys. 58, 127
de Jager, C. et al, 1982: Submitted to Solar Phys.
Duijveman, A. et al, 1982: Solar Phys. 81, 137.
Gurman, J.B. et al, 1982: Ap.J. 253, 939
Haisch, B.M. et al, 1980: Ap.J.Lett. 242, L99
Hoyng, P. et al, 1976: Solar Phys. 48, 197
Hoyng, P. et al, 1981: Ap.J.Lett. 246, L155
Jarrett, A.H. and Gibson, J.B. 1975: IBVS, No.979
Kahler, S. and Shulman, S. 1972: Nature Phys. Sc. 237,101
Kahn, S.M. et al, 1979: Ap.J.Lett, 234, L107
Kiplinger, A.L. et al, 1982: Submitted to Solar Phys.
Kunkel, W.E., 1969: Nature, 222, 1129
Kunkel, W.E. 1975: I.A.U. Symposium No.67, 15
Lin, R.P. and Hudson, H.S. 1976: Solar Phys. 50,153
Moore, R.L. et al, 1980: In "Solar Flares", Colorado Ass.Univ.Press (Ed.P.Sturrock
Mullan, D.J. 1976: Ap.J. 207, 289
Mullan, D.J. 1976: Ap.J. 207, 289
Poland, A.I. et al: 1982: Solar Phys. 78, 201
Simnett, G.M. 1981: Proc. 17 Int.Conf.on Cosmic Rays,Paris, Vol.12,205
Slottje, C. 1978: Nature Phys. Sc. 275, 520
Strong, K.T. et al, 1982: Submitted to Solar Phys.
Svestka, Z. et al, 1982a: Solar Phys. 75, 305
Svestka, Z. et al, 1982b: Solar Phys. 78, 271.
Wagner, W.J. et al, 1981: Ap.J.Lett. 244, L123
Wang, H.T. and Ramaty, R, 1974: Solar Phys. 36, 129
Willson, R.C. et al, 1981: Science, 211, 700
Woodgate, B.E.et al, 1982: Submitted to Ap.J.

DISCUSSION

Stencel: I would like to ask about the flare events. In the second peak always the brightest and if so could this be a reflection with argumentation? Secondly, does the 30 msec timescale reflect something fundamental about spatial structure?

Simnett: The relative brightness of the peaks in double flare events is a function of wavelength. At microwave frequencies and in hard X-rays the first peak is almost always the biggest (Figure 9). Indeed there are some flares where the first peak is only seen in hard X-rays. Vice versa is true for soft X-rays. The model I outlined agrees quite well with this. The energy release from the first flare goes into particles which then gradually heat the medium, evaporate material which they continue to heat. This heated material then expands producing the soft X-rays. Whether the 30 msec interval is fundamental or not is not clear at this stage. You may recall the results of de Jager, de Jong and the Utrecht group based on the TD-1 data which concluded that there was no significant variation on timescales shorter than 1 second. This was due to their poorer time resolution. This result now revises this conclusion. But I should warn that the time resolution here is only 10 msec and variations are seen down to this timescale. So it is not clear whether the 30 msec timescale is fundamental or not. I should add to that the microwave people see variations at the msec level and so I would suspect that, when X-ray time resolution gets down to that, they may see the same thing.

Hartmann: When you say that solar flares are longer in duration than stellar flares, are you comparing events of comparable energy?

Simnett: No. Of course I cannot compare events of comparable energy because most of the stellar flares for which I have data are considerably in excess of a few time $10^{32}$ergs which is the energy of the largest solar flares. The largest flares tend to go for an hour or so and I believe that this is a fairly general rule.

Venugopal: Are you sure about the reduction of the solar constant during a large spot? Is there a quantitative relationship between the two effects?

Simnett: No, there does not appear to be a quantitative relationship between the two. I say this because there were times when there were large spots on the Sun and the reduction in intensity was not very significant. What I am saying is that the two biggest dips in overall solar intensity were coincident with the passage of large spots. I leave you to interpret that how you will.

Venugopal: Can you elucidate the response characteristics of active (word lost) radiometer?

Simnett: Let me discuss this with you later. It is a bit detailed for now.

Linsky: One thing which is very hard to measure in stellar flares is velocity fields or the kinetic energy of gas motions. Could you tell us what SMM tells us about velocities of expansion and the fraction of the total energy going into kinetic motion of gas in the case of solar flares? Do you know what the kinetic energy is and can you compare it with kT?

Simnett: Let me answer that in a slightly different way. One of the instruments from which I did not show any data is the coronograph which has seen frequent coronal transients. Calculations of the energy in these measured at a couple of solar radii is typically a factor of 10 greater than the total radiated energy of the corresponding flare. So to answer your question, the dominant energy transfer would appear to be in the mass motion associated with the coronal transients.

Vaiana: I think it is also important to know what fraction of the flares seen by SMM have coronal transients associated with them. Skylab results suggested that there were in fact two types of solar flares. In one type there was no need to invoke mass motions except in the filling of a loop.

Simnett: Perhaps I can answer this in a qualitative way since a great deal of analysis still remains to be done. It looks as though the compact flares, which are intimately related to large scale magnetic structures linking the flaring region to another separate region, are not associated with coronal transients. On the other hand, the flares, even relatively minor ones, which are associated with much bigger structures often seen linking several active regions, were seen by the coronagraph polarimeter to frequently give rise to coronal transients. So you are correct. To say there are two types of flare only may be over-simplifying the matter but it is certainly true that there are at least two types of flare. In one type coronal transients are generated, and these are very common, while the other type are the compact flares.

Vaiana: (Part of question lost). I do not remember the exact statistics but I thought that from Skylab data mass-loss events were relatively rare although they were very significant in terms of total energy.

Simnett: Well, all of the SMM data has not been processed yet so I think we must wait perhaps another year before we can discuss the statistics with confidence. It is actually quite difficult for the coronagraph to observe transients from flares which are well into the disk.

Gibson: I have two questions. Were there any radio observations simultaneous with the pulsational gamma-rays?

Simnett: Yes.

Gibson: Did they show pulsations also?

Simnett: They did yes.

Gibson: This causes me at least to rethink the possibility that in stars the pulsations might be due to the particle distribution rather than some narrow-band phenomenon. I believe it is important that we begin to look at ways of generating particles in pulses rather than some peculiar arrangement of the fields such that a narrow beam passes through the line of sight.

# FLARES ON THE SUN : SELECTED RESULTS FROM HINOTORI

Katsuo Tanaka

Tokyo Astronomical Observatory, University of Tokyo
Mitaka, Tokyo, Japan

ABSTRACT

The X-ray observations of solar flares by Japanese astronomy satellite Hinotori (ASTRO-A) are reviewed. Detailed results from the X-ray telescope and the soft X-ray spectrometers are given. The hard X-ray images in the 17-40 keV range show a wide variety of source structures. Examples of compact single source, stationary coronal source, and double sources are presented. High resolution spectra in the range 1.7-2.0A indicate strong turbulence and blue-shifted components in the beginning of flares. From the FeXXV and FeXXVI spectra two kinds of thermal plasma are shown to appear. The cooler component of $Te=15\text{-}20\times10^6$K, which increases in the impulsive phase of the hard X-ray burst, is suggested to be produced by the dissipation of the electron beams. The hotter component of $Te=30\text{-}50\times10^6$K and $Ne=10^{11}\text{-}10^{12}\ cm^{-3}$ increases towards the flare maximum. This component is also responsible for emitting 17-40keV continuum, and emitted from a very compact source. The results are discussed in comparison with flare loop models. It is shown that evolutionary changes in the loop density realize various source structures in the hard X-ray under the unitary electron beam hypothesis.

## 1. INTRODUCTION OF HINOTORI SATELLITE

A Japanese astronomy satellite ASTRO-A was launched on February 21, 1981, from Kagoshima Space Station (KSC) by the Institute of Space and Astronautical Science (ISAS). This satellite, nicknamed Hinotori (Japanese for Fire Bird), is intended to observe various aspects of solar flare X-rays with good spatial, spectral and temporal resolutions. The physical instruments aboard Hinotori for flare studies are (1) Solar X-ray Telescope (SXT) with Solar X-ray Aspect Sensor (SXA), (2) Soft X-ray Crystal Spectrometer (SOX), (3) Soft X-ray Flare Monitor (FLM), (4) Hard X-ray Flare Monitor (HXM) and (5) Solar Gamma Ray Detector (SGR). Table summarizes these instruments.



*P. B. Byrne and M. Rodonò (eds.), Activity in Red-Dwarf Stars, 307–320.*

Table Characteristics of Hinotori Instruments for Flare Observations

| | Detector | Energy Range | Resolutions | |
|---|---|---|---|---|
| SXT | 113cm$^2$ NaI Scint. | 17-60keV/5-10kev | 38"(7")/~8sec | (~4sec) |
| | 113cm$^2$ NaI Scint. | 17-60keV/5-10keV | 30"(7")/~8sec | |
| SXA | Fine Solar Aspect Sensor | | 5" | |
| HXM | 57cm$^2$ NaI Scint. | 17-40keV | 7.8ms (HXN-1) | |
| | | 40-340keV | 125ms (HXM-2-7) | |
| SGR | 62cm$^2$ CsI Scint. | 210-6700keV | 128channels/2sec | |
| FLM | 0.5cm$^2$ Xe Gas | 2-25keV | 128channels/4sec | |
| | Scint.Prop. | counts in L/H bands | 125ms | |
| SOX | $SiO_2$/NaI Scint. | 1.72-1.95Å | 2mÅ/~8sec | |
| | $SiO_2$/NaI Scint. | 1.83-1.89Å | 0.15mÅ/~8sec | |

Hinotori is a spinning satellite (about 4rpm) with its spin axis being off-set from the Sun center by 1.2°±0.5°. X-ray image and line spectrum are obtained utilizing this attitude and spin. The data processor aboard Hinotori judges a flare automatically and about 20 minutes data of a flare, preferentially for large flare, are stored in the data recorder. Real time data and recorded data are transmitted at KSC in about 12 minutes for each of five visible orbits per day. In a year and five months operation Hinotori observed 675 flares including 31 X-class flares. The largest flare was a X12 flare that occurred at 16:30UT of June 6, 1982. Early results have appeard in the proceedings of Hinotori Symposium (1982 January at ISAS). Here I review some selected results from this symposium as well as other preliminary results obtained by now.

## 2. RESULTS FROM THE SOLAR X-RAY TELESCOPE

Solar X-ray Telescope (SXT) is a rotational modulation collimator ("Oda" collimator) consisting of two orthogonal bigrid collimators and NaI scintillation counters. Flare position is determined in terms of optical lenses and collimators system (SXA) co-aligned with the X-ray collimators. Informations necessary to reconstruct the X-ray image from modulation patterns are obtained every quarter of the spin period or about 4s. For the image reconstrucion two kinds of methods : arithmetric reconstruction technique (ART) and maximum entropy method (MEM) have been used. Spatial resolution is determined nominally by FWHM of the triangular beam pattern (about 30"). However, due to the sharpness of the pattern edge SXT has a response to a structure less than 7" (Makishima 1982 HS-abbreviation for the proceedings of Hinotori Symposium). For a year after the launch the two collimators were mainly used for the hard X-ray range (17-40keV) with one collimator being switched to soft X-ray range (5-10keV) only for the later phase of flares.

General survey of some dozen flares performed by S. Tsuneta, T. Takakura and K. Ohki has revealed a great variety of the hard X-ray source structures such as compact single source, large loop-like structures and multi-component structures. Evolutionary change from one type to the other and/or mixtures of several kinds of sources in a flare have also been found. Publications of individual cases are in preparation by the SXT team. Here, I introduce some of their preliminary results or results before publications, together with some my own comments (for example, proposed flare types).

An example of compact single source is shown in Fig.1 (Tsuneta p.c. -private communication). The source size is very small, about 10" (7000km) or less and tends to decrease in the post-maximum phase. This kind of source is characteristic of particular flares marked by high X-ray intensity of short duration, structureless time profile and very "soft" hard X-ray spectrum (type A, see Fig.5). Although the flare showing this characteristic from the initial phase is rather rare, similar characteristics of the spectrum are often witnessed after the middle phase of so-called impulsive flares, which are characterized by spike components in the initial phase (type B). Later phases seem to correspond to the compact, single (and sometime shrinking) source. From comparisons with the Hα emissions, the compact source is suggested to be located at the top of the loop (Tanaka, preliminary results).

A large loop-like structure is reported by Takakura et al. (1982 HS) in a long-duration burst of May 13, 1981 which occurred near the limb. Superposition of hard X-ray images on the Hα pictures indicates that X-ray source is located nearly along the tops of the loop arcades which would connect the Hα two ribbon emissions (Hiei et al. 1982 HS). This global structure is rather stationary during a long lasting gradual burst (type C ). Existence of the stationary coronal source has been confirmed in several limb flares, too. Fig.2 shows a case of large limb flare of X5.5 (Tsuneta et al. 1982 HS). Although the source size appears to increase slightly during the initial phase, its center of gravity remains unchanged, at the height of about 15000km above solar photosphere. The soft X-ray image (Tsuneta, preliminary result) is nearly concentric to the hard X-ray image and about double in linear size. It may imply that the hard X-ray source is confined in much smaller portion of the coronal loops than the soft X-ray source.

Impulsive bursts of short duration (type B) show a variety such as point-like, arch-like, double, and extended multi-structures (Takakura, Ohki, p.c.). Some impulsive flares show rapid variations of the source structure. From the imaging of limb flares it is suggested for some flares that the sources are low in height, presumably chromospheric (Ohki, p.c.). An example of double sources which appear cospatial with the two Hα emissions (Tsuneta, Tanaka, Zirin, and others, preliminary result) is shown in Fig.3 (top). The double structure is separated by 47000km, and the weaker component soon disappeared. In later phase, as the intensity variation becomes more gradual the hard X-ray source is shifted to accord with the top or slightly above the top of a bright Hα loop which appeared in this phase and connected the two ribbon emissions. In other limb flare Ohki (p.c.) has reported that the main source stays at about $10^4$km above the photosphere from the beginning, while a weak

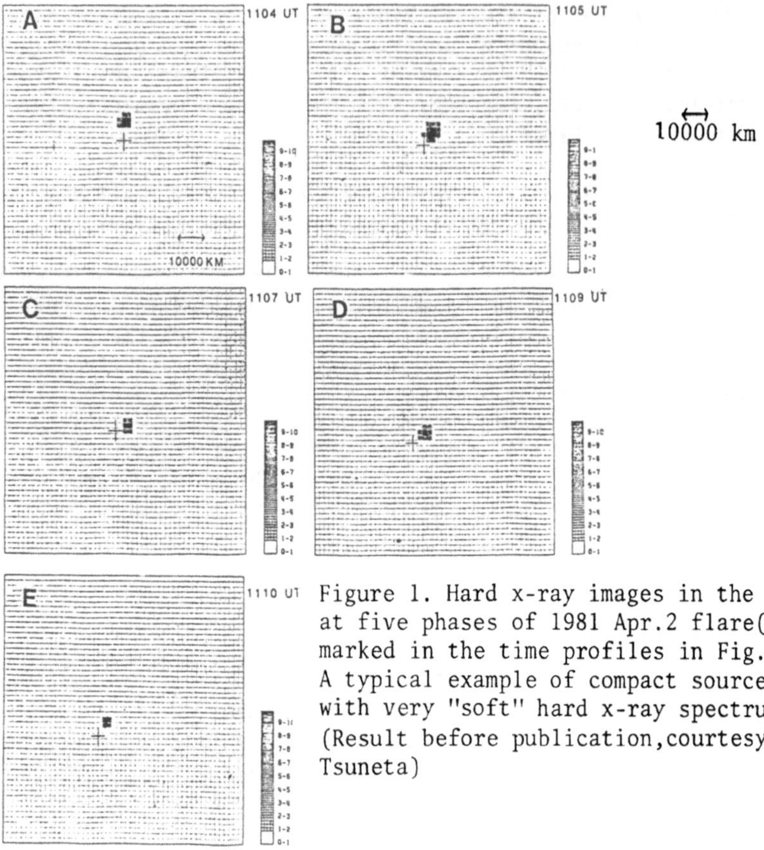


Figure 1. Hard x-ray images in the 17-60 keV at five phases of 1981 Apr.2 flare(times are marked in the time profiles in Fig.5).
A typical example of compact source structure with very "soft" hard x-ray spectrum (type A) (Result before publication,courtesy of Mr.S. Tsuneta)

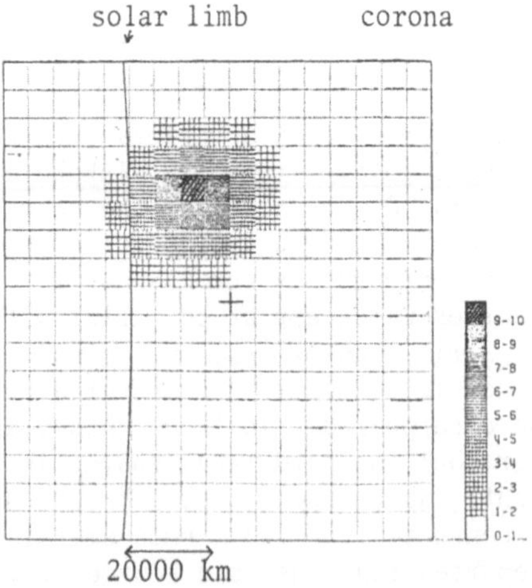


Figure 2. Hard x-ray image in the 17-60 keV of 1981 April 27 flare (limb flare). An example of coronal source. (Tsuneta et al. 1982HS; result before publication)

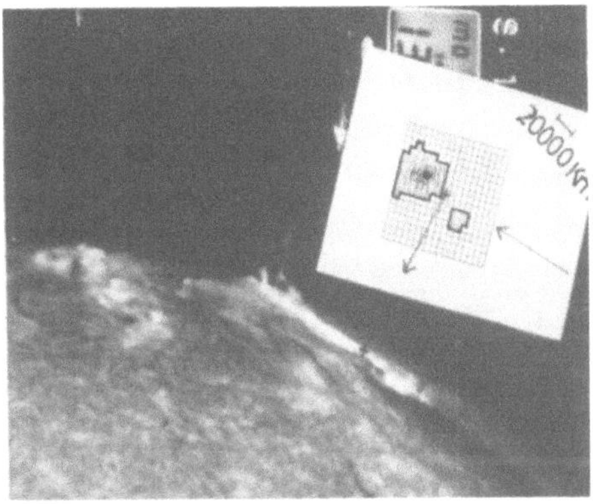


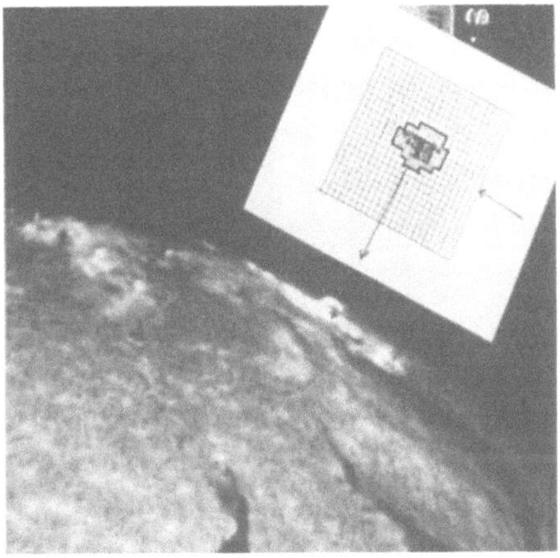

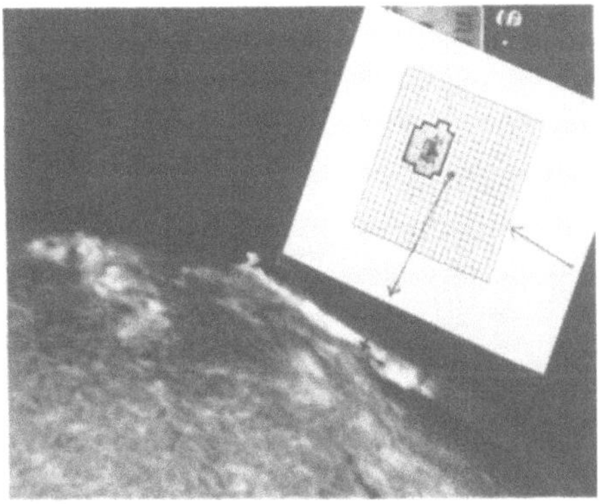

Figure 3. Hard x-ray images in the 17-40 keV and Hα pictures(BBSO) of 1981 July 20 flare.Coordinate centers of hard x-ray images are marked by cross in Ha. Example showing change from double to single sources. (Preliminary result, Tsuneta et al.)

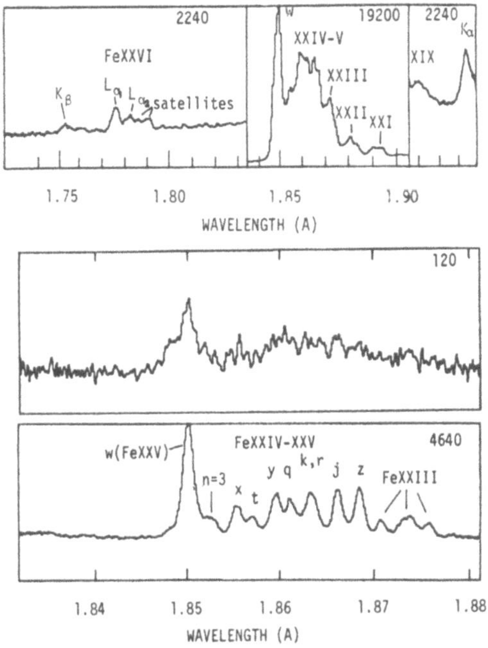


Figure 4. Examples of soft x-ray iron line spectrum in 1.73-1.94A (top) and high dispersion spectra in 1.83-1.88A(lower two). Note the line broadening shown in the middle which is a typical spectrum in the flare start.

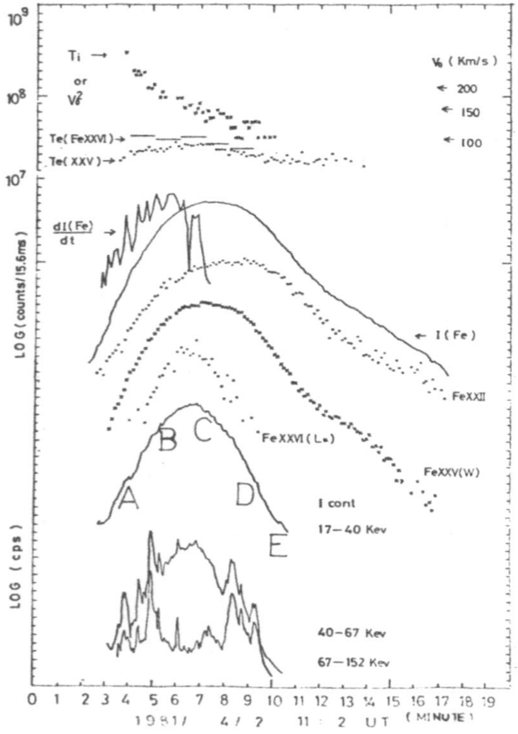


Figure 5. (1981 Apr.2-X2.3)

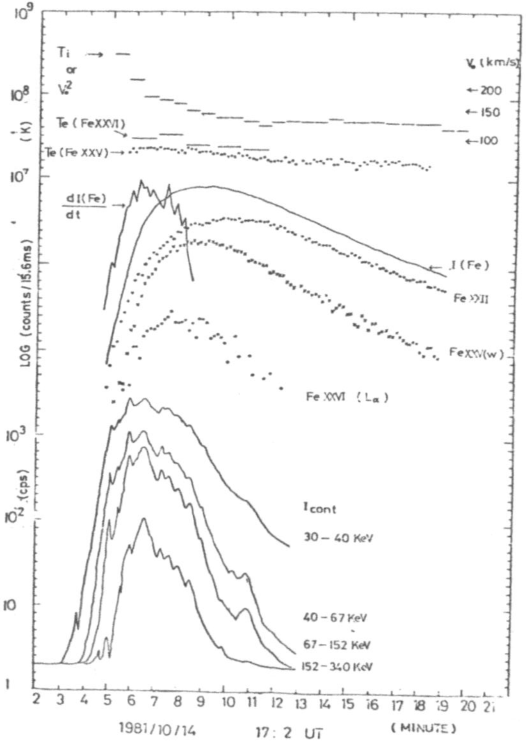


Figure 6. (1981 Oct.14-X3.0)

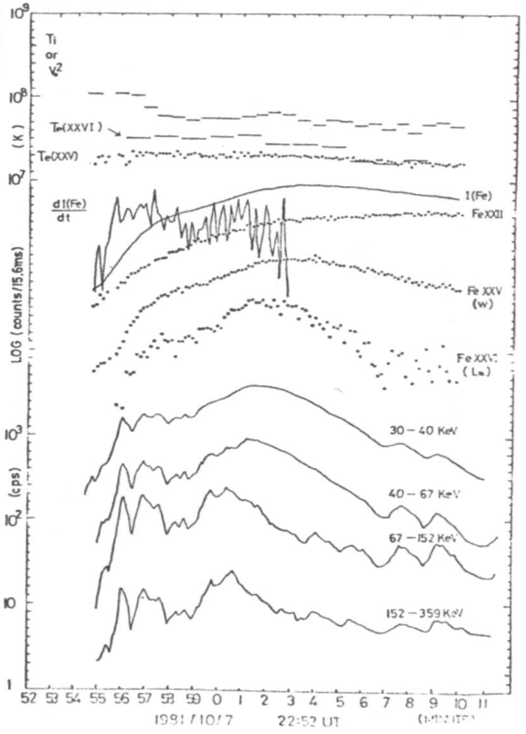


Figure 7. (1981 Oct.7-X3.6)

Time profiles of various plasma parameters derived from line spectra and intensity time profiles for three types of flares( Fig.5 for type A, Fig.6 for type B and Fig.7 for combined type A and B).
From top: ion temperature(turbulence), electron temperature (Fe XXVI), electron temperature(Fe XXV), time derivative of total line emissions, total line emissions:I(Fe), intensities from Fe XXII, Fe XXV(w), Fe XXVI(La), continua of 17-40 keV,40-67 keV, 67-152 keV and 152-350 keV.Ordinate in the same log scale.

sub-source, which might be a footpoint of the loop, appears in the initial phase.

## 3. RESULTS FROM THE SOFT X-RAY CRYSTAL SPECTROMETERS

The soft X-ray rotating crystal spectrometers (SOX) consist of two flat $SiO_2$ crystals and Na(Tl) scintillation counters. The wavelength scan is made by the satellite spin. Two wavelength regions about 1.72-1.95A and 1.83-1.89A are scanned in half a spin period (7-10s) with resolutions of about 2mA and 0.15mA, respectively. About 10 X-class flares have been analyzed (Tanaka et al. 1982a, b, HS). Fig.4 shows an example of the spectra. Various plasma parameters have been derived such as electron temperatures, ionization temperatures, ion temperature or turbulence, and mass motion. Resolved spetrum of FeXXVI (Lα and associated satellites) has provided the electron temperature of the hottest plasma using the line ratio.

Time profiles of obtained plasma parameters are shown, together with intensity time profiles from the soft to hard X-ray ranges in Fig.5-Fig.7. Fig.5 refers to an intense flare of "soft" spectrum which is shown in Fig.1 (type A), Fig.6, a short impulsive flare (type B) and Fig.7, an extended burst which showed spiky components at the early phase and gradual component after the middle phase. We have found remarkable line broadening in the initial few minutes of all flares. The line widths, if interpreted as thermal broadening, give ion temperatures exceeding $10^8$K, which are much higher than the electron temperature. It should rather be attributed to the large turbulence (150-300$kms^{-1}$). An evidence is given that large broadening exists prior to the impulsive hard X-ray burst though it is accompanied by a gradual rise in the hard X-ray. In the very beginning of the line broadening phase small (10%), blue-shifted components with V=150-400 $kms^{-1}$ are often seen in the disk flares. In the limb flares the blue- shifted component cannot be found.

The electron temperature derived from the FeXXV resonance to di-electronic satellite line ratio (B. Dubau et al. 1982) increases rapidly to about $20x10^6$K in the period of large turbulence. Then, it increases gradually (upto $25x10^6$K) or remains constant and decreases slowly in the decay phase. Initial temperature increase occurs in a very short time (10s-1min.) associated with the increase in the emission measure. The electron temperature derived from the FeXXVI line ratio (Dubau et al. 1981) is equal to $30\text{-}40x10^6$K near the maximum, which is much higher than the electron temperature derived from the FeXXV line ratio but in decay phase it becomes closer to the latter. Since results obtained conflict the assumption of uniform temperature, which is adopted in the line ratio method, we searched for solutions which allow for two temperature structure using the same set of lines and assuming the ionization equilibrium. Fig.8 shows an example of temperature and emission measure diagram thus derived. It shows a clear separation of cooler component ($15\text{-}20x10^6$K) from hotter component ($30\text{-}50x10^6$K). Evolution of the

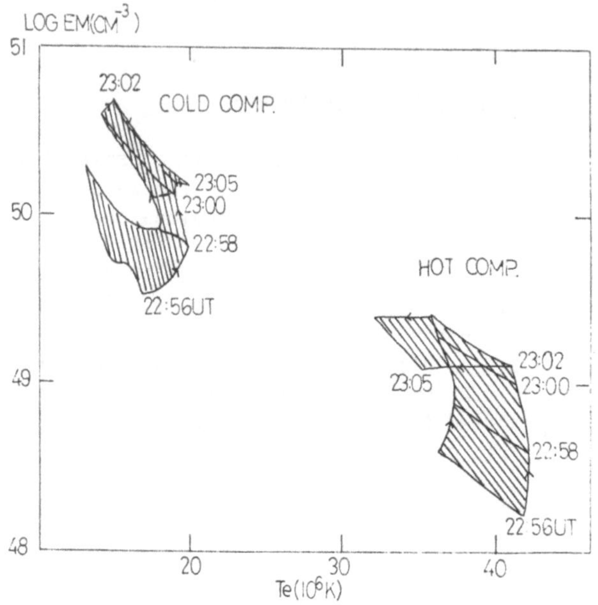


Figure 8. Evolution of temperatures and emission measures of two thermal component (For flare of Fig.7).

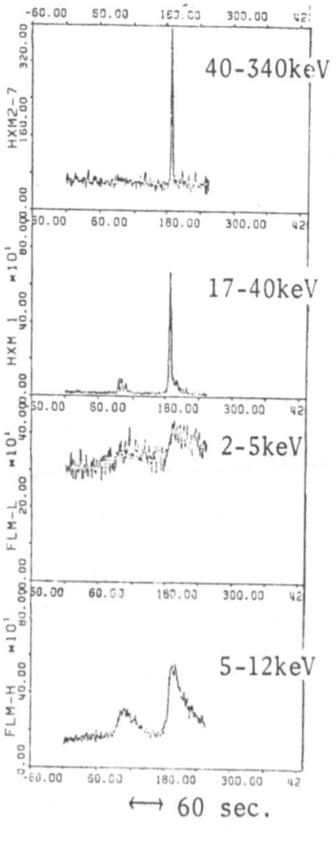


Figure 9. Time profiles of a very short flare.

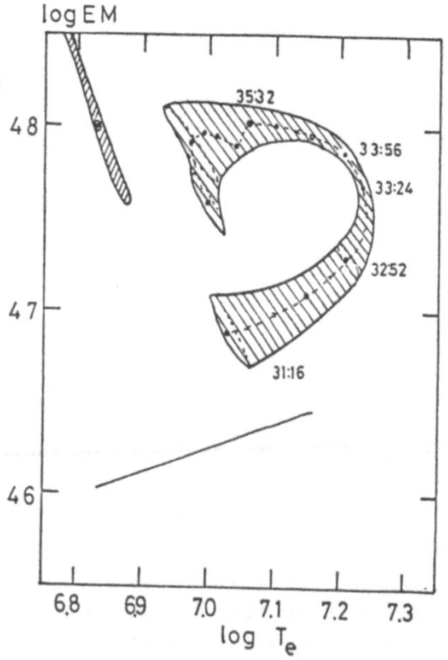


Figure 10. Evolution of temperature and emission measure in a simple ,short flare.(Watanabe et al.1982HS) A straight line shows a case of adiabatic compression.

cooler component is reflected in the intensity profiles of lines from FeXXV or lower ionization, typically the total iron emission : I(Fe), while the intensity profile of FeXXVI L$\alpha$ line represents evolution of the hotter component. Generally the two intensity profiles are different. Comparisons of the line intensity profiles with those of the hard X-ray continuum reveal two facts. Firstly, the increase rate of I(Fe) is propotional to the global intensity variation of the hard X-ray spike components, especially for the harder range above 40keV. It indicates close relationship between the productions of the high energy electrons and the cooler thermal plasma. In particular it has been shown that thick target electron power input as derived from the hard X-ray spectrum explains the increase of the thermal energy content for the 10-20 million degree plasma. The two quantities agree in absolute scale with Ne= $10^{11}$-$10^{12}$cm$^{-3}$, which implies that the electron beams are responsible for the production of the hot plasma by heating the transition region plasma of above densities. Presumably the observed blue-shift which is predominantly vertical to the surface would represent expansion of the heated plasma into the loop. The low height in the hard X-ray source of some impulsive flares may suggest thick target hypothesis, too. These characteristics seem typical for the impulsive phase of flares (type B phase).

The second fact is that the hard X-ray time profiles, in the low energy range of 17-40keV, tend to become smooth near the maximum of flares as well as that their intensity profiles become similar to the FeXXVI L$\alpha$ intensity profile. (see Fig.7) This tendency is most pronounced in type A flare. One can see in Fig.5 the similarity of the two profiles from the first. In short impulsive burst (type B) the tendency is less pronounced. For type A flare the emission measures and temperatures of the hotter thermal component explain the hard X-ray flux in the 17-40keV range. We may safely conclude that, for this kind of flare, the observed compact hard X-ray source in this energy range is identical with the volume of the hotter thermal plasma. Using the hard X-ray source size and the emission measures of the hotter component we obtain Ne=$10^{11}$-$10^{12}$cm$^{-3}$. The high density is consistent with the observed cooling time (150s for $\Delta$Te=$10^7$K) of the hotter component assuming the radiation cooling. It assures also the ionization equilibrium. It should be remarked that H$\alpha$ loop, very bright at the top, appears almost at the maxima of the smooth hard X-ray component and FeXXVI L$\alpha$ (BBSO, flare of Fig.7). It is likely that the highly condensed 30-50 million degree plasma cools down rapidly at the top of the loop. Above results indicate that previously unknown plasma component of Ne= $10^{11}$-$10^{12}$cm$^{-3}$ and Te=30-50x$10^6$K is formed in the middle phase of impulsive flare or from the initial phase of some particular flares. This is considered to be characteristic phenomenon of the type A phase.

## 4. SIMPLE FLARES

A simple flare, which consists of a single spike of very short duration, may be considered as elememtary burst, and useful to estimate elementary flare process. Such burst, as shown in Fig.9, indicates a hard X-ray burst of 5s FWHM duration and associated soft X-ray burst of 40s FWHM duration. Time integration of the hard X-ray intensity is precisely proportional to the rising flux of the soft X-ray burst, and the thick target electron power input above 20keV is shown to be consister with the thermal energy content of the soft X-ray emitting plasma with $Ne=10^{12}cm^{-3}$ (Tanaka et al. 1982 HS). The radiation cooling time of 15 million degree plasma at this density is consistent with the observed decay time of the soft X-ray burst. These facts suggest production of the thermal plasma by the electron beam.

Thermal histories of simple flares have been derived from continuum spectrum in the range of 2-12keV which is obtained from the scintillation proportional counter (FLM). This sensitive detector resolves Fe line and other lines from the continuum (Inoue et al. 1982) which enabled the analysis of pure continuum. The emission measure-temperature diagram of a simple flare as derived from the $\chi^2$ minimum condition is shown in Fig.10 (Watanabe et al. 1982 HS). In small flares like this only cooler thermal component is produced. It is found that the temperature increases till the end of the hard X-ray burst, while the emission measure continues to increase for a minute after that. The temperature, thereafter, decreases with constant emission measure. If radiation cooling is assumed, $Ne=10^{11.5}cm^{-3}$ is obtained. The initial increase rates of the temperature and emission measure happen to satisfy, in this flare, a relation (shown by a line ) which is expected in the adiabatic compression with the conservation of total mass. However this relation is not observed in other simple flares, excluding this process as elementary heating mechanism.

## 5. DISCUSSIONS - FLARE LOOP MODEL

Present knowledge of the hard X-ray images is still fragmentary, and it is still premature to perform self-consistent classification. However, to promote comparisons with models, I tentatively classify the reported hard X-ray images into three categories : stationary coronal source (type C phenomenon), low-chromospheric footprint-source (type B phase), and compact-coronal-source (type A phase). The coronal source suggests dissipative model of the hard X-ray burst (Brown et al. 1979; Smith and Auer 1980) or trapped electron model. Long-enduring electron trap may be realyzed only for low coronal density. Under the assumption of thin target emission, Tsuneta et al. (1982HS) derived $Ne=10^{9}cm^{-3}$ as the lower limit of the ambient density from the observed hard X-ray spectrum and source size for the flare shown in Fig.2, and concluded that the collision time of 20-30keV electron in this density appears to be too short compared with the burst duration. Therefore, continued injections of electrons are needed. If the injection occurs in a form of electron beams, however, the beams will dissipate immediately at this low density due to the ion-acoustic instability of the return current associated with the electron beam. (Brown and Melrose 1977).

In the dissipative model a high temperature (over $10^{8}$K) region will

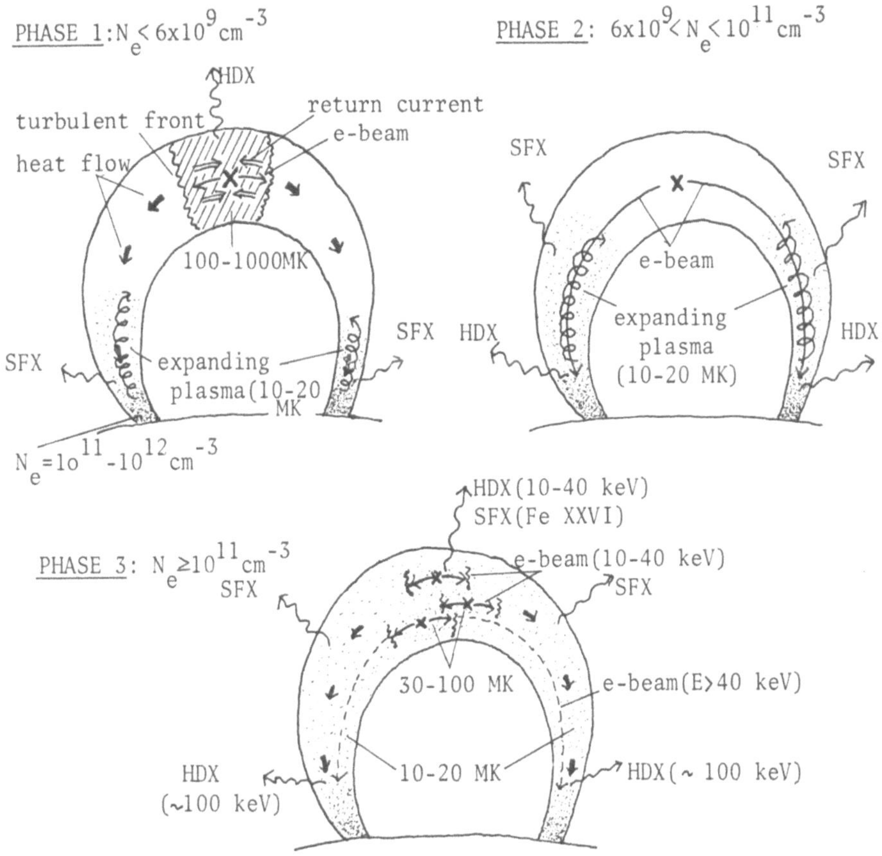


Figure 11. Proposed model of flare loop evolution. As the coronal density increases due to the chromospheric evapolation, hard x-ray source shifts from top of a loop to foot points and again to top. The cross represent energy release site(beam production). HDX:hard x-ray, SFX:soft x-ray MK:million degrees.

be formed at the energy release site, presumably at the loop top. If the dissipation continues, the hot region expands to the whole loop immediately by conduction. The limitation of the conduction due to the anomalous conductivity or due to saturation of conduction does not increase this time scale beyond 10s (Smith and Auer 1980). This situation apparently contradicts the stationary coronal source which is confined in small region. It seems necessary to consider that a heating of short duration (less than 10s) occurs successively in the nearby loops which cannot be resolved by present resolution.

The low (chromospheric) source or double source, on the other hand, suggests thick target emission of hard X-ray at the footpoints of the loop. Electron beams would be responsible for this case. Electron precipitation and subsequent heating of the dense plasma of low-temperatur appears to be consistent with the soft X-ray results ( time profiles of 10-20 million degree plasma and blue-shifts).

While we cannot discriminate among several energy release modes in the dissipative model, we may note that three types of the burst source can be explained in terms of the electron beam hypothesis as results of evolutionary changes in the ambient density. To illustrate this, we consider continued injection of the electron beams at the top of the loop (Fig.11). When the initial density of the loop is low enough ($Ne<6X10^{9}cm^{-3}$, Tanaka et al. 1982HS), the beam dissipates as noted before and a high temperature (over $10^{8}K$) region bounded by the turbulent front will be realyzed for a few seconds (phase 1). Conduction from the hot region will cause ablation of dense transition region, producing 10 million degree plasma. This plasma will expand into the loop and increases slightly the loop density. Then, a condition for the electron beam to stream freely ($Ne>6.10^{9}cm^{-3}$) may be realized, which results in strong beaming down to the footpoints of the loop (phase 2). The thick target hard X-ray emission will occur at the footpoints, together with efficient heating of the transition region or chromophere(type B phase). The heated plasma will expand to fill the loop as the beaming continues, and make the loop density equal to the transition region density or $10^{11}$-$10^{12}cm^{-3}$. In this phase (phase 3) low energy electrons (10-40keV) in the newly produced beam will be thermalized immediately at the source region due to the collision. Actually the 20keV electron decays within 3000km at $Ne=10^{11}cm^{-3}$. Thus, we have compact, dense hot regions which may consist of plasma of $Te<10^{8}K$ (temperature is expected somewhat low because of low efficiency of heating in the high density) at the top of the loop. This phase may correspond to the type A phase. In actual cases this evolution may occur progressevely in unresolved different loops. Then, we may see coronal source and chromospheric source at the same time. Due to varieties in the initial density of loops, duration of individual heating as well as in the number and size of loops there may occur various types of flare in which particular phase(s) of above sequence would be enhanced. The type C phenomenon may be considered as successive realizations of phase 1 or phase 3 in different unresolvable loops. If the initial density is higher than $10^{10}cm^{-3}$, phase 1 will be skipped and footpoint sources are seen first. Also if the loop size is very small, the expansion time is short and phase 3 may be realyzed from the early phase. This case may explain type A flare.

Although above scenario may explain qualitatively plasma heating under the unitary electron beam hypothesis, other mechanisms cannot be excluded. Acceleration of protons almost simultaneous with electrons has been suggested from the $\gamma$-ray observation from Hinotori (Yoshimori et al. 1982HS), which I have not referred to. Plasma heating as the production of low energy protons might be the other possibility (Enome. p.c.). It seems important to explain the observed line broadening to clarify the heating mechanism.

ACKNOWLEDGEMENTS

Hinotori was designed and fabricated at ISAS by the ASTRO-A team under the direction of Prof. Y. Tanaka, the manager of the Hinotori Project. Operation of Hinotori has been carried out by many scientists and thechnicians from various related institutes. I wish to thank all of those who have contributed the Hinotori Project. Particular thanks are to the SXT and HXM team : Prof. T. Takakura, Dr. K. Ohki, Messrs. S. Tsuneta, N. Nitta, Drs. K. Makishima, T. Murakami and Professors M. Oda and Y. Ogawara for permitting to use their results before publications. I am indebted to Mr. S. Tsuneta for providing preliminary results and to Prof. H. Zirin for providing BBSO Hα data. Finally I wish to thank Professors F. Moriyama, K. Nishi and Z. Suemoto for their contributions to the SOX experiment as well as Dr. T. Watanabe and Mr. K. Akita for the computer software developements.

REFERENCES

Bely-Dubau,F., Dubau,J., Faucher,P., and Gabriel,A.H. 1982, M.N.R.A.S 198,239.
Brown,J.C. and Melrose,D.B. 1977, Solar Phys. 52, 117
Brown,J.C., Melrose,D.B. and Spicer,D.S. 1979, Ap.J.228, 592.
Dubau,J. et al. 1981, M.N.R.A.S., 195, 705.
Inoue,H., Koyama,K., Mae,T., Matsuoka,M., Ohashi,T., Tanaka,Y. and Waki,I. 1982, Nuclear Instruments and Methods 196, 69.
Smith,D.F. and Auer,L.H. 1980, Ap.J. 238, 1126.

Tanaka,K., Watanabe,T., Nishi,K., and Akita,K. 1982, Ap.J. (letters), 254, L59.
Tanaka,K., Akita,K., Watanabe,T., Miyazaki,H., Kumagai,K., Miyashita,H., Nishi,K. and Moriyama,F. 1982, Annals of Tokyo Astr. Observatory 2nd Series, Vol.18, 237.

DISCUSSION

Kodaira: Do you have microwave coverage for the smooth hard X-ray compact of your observations? Because microwave data usually has the same kind of smooth profile.

Tanaka: Yes. 35$GH_z$ shows a slow gradual flare with the same structure as the hard X-ray.

Pallavicini: You apparently explain all of your hard X-ray observations using non-thermal electrons. Can you exclude the possibility of a thermal model for some of these flares?

Tanaka: From the hard X-ray imaging we see that these impulsive bursts are very low in the atmosphere, in the chromosphere. So from that point of view a non-thermal model is better. However the spectrum itself is exponential which is very strange for this impulsive flare.

Uchida: So the possibility of a thermal model exists.

Simnett: Can I make a comment about the last question? If one considers the ion acceleration derived from the gamma-ray results it is fairly clear that these cannot be explained on a thermal hypothesis. There are certainly many flares where one needs non-thermal acceleration of particles and I think the best proof of this is the gamma-ray observations. Looking purely at the X-ray observations alone it is possible for "non-believers" to "wriggle out" of the non-thermal hypothesis. It is extremely difficult to do that with the gamma-ray result.

# SOLAR FLARES WITH SMM AND IMPLICATIONS FOR THE PHYSICS OF STELLAR FLARES

R. Pallavicini
Arcetri Astrophysical Observatory, Florence, Italy

ABSTRACT. XUV flare observations from SMM are discussed and a comparison is made with recent X-ray observations of stellar flares.

## I. INTRODUCTION

Observations of flares from the Solar Maximum Mission have provided significant insights into the physics of transient phenomena in the solar atmosphere. The high spatial and spectral resolution obtainable in the solar case are potentially useful for the interpretation of spatially unresolved observations of stellar flares. Two problems relevant for the physics of flares on stars are discussed here: a) the role of non-thermal electrons in the overall flare energetics; b) the process of chromospheric evaporation during the thermal phase of flares.

## II. NON-THERMAL PROCESSES DURING THE IMPULSIVE PHASE.

It is well known that many flares comprise two phases, an impulsive one early in the event, followed by a more gradual phase later on. The impulsive phase is usually interpreted as due to non-thermal electrons accelerated at the flare onset. Observations obtained with the Hard X-Ray Imaging Spectrometer on SMM have given strong support to this notion (Hoyng et al. 1981). The hard X-ray emission observed in the early flare phase appears to be concentrated at the footpoints of magnetic arches connecting regions of opposite magnetic fields and to be spatially coincident with impulsive kernels of optical emission in the chromosphere. Moreover, observations with other instruments on SMM (HXRBS and UVSP) have shown a close temporal correlation between spikes of hard X-ray emission and localized impulsive UV brightenings in the transition region (Cheng et al. 1981).

These observations indicate the presence of beams of accelerated



*P. B. Byrne and M. Rodonò (eds.), Activity in Red-Dwarf Stars, 321–324.*

electrons travelling along magnetic field lines and damping energy through collisional losses in the dense regions at the footpoints of coronal arches. An important question is whether the energy deposited by accelerated electrons is sufficient to explain the total radiative output of flares during the gradual phase. Preliminary energy estimates based on SMM observations and standard thick-target calculations show that this is unlikely, unless the spectrum of accelerated electrons is extended arbitrarely to very low energies. Moreover, the short duration of the non-thermal phase with respect to the rise-time of the thermal phase at all temperatures, including the highest ones observed with SMM, argues against this possibility.

## III. THERMAL PHASE AND FLARE HYDRODYNAMICS

An alternative view is to regard the flare phenomenon as mainly thermal in nature and due to the sudden heating of the coronal portion of a magnetic arch. The acceleration of electrons to suprathermal energies is considered as a second order effect which is not directly related to the main flare phase. In this model the key role is played by the process of heat conduction from the loop top to the footpoints, and by the dynamical response of the chromosphere to the energy input.

A numerical hydrodynamic code which solves the full set of mass, energy and momentum equations for a magnetically confined arch has been developed and used to predict the temporal behaviour of X-ray spectral lines formed over a wide range of coronal temperatures (see Pallavicini et at. 1982 for details). The results have been compared with spectroscopic observations obtained by the X-Ray Polychromator experiment on SMM. A variety of different initial conditions as well as different spatial and temporal dependences of the heating function have been used in the numerical simulations. The model reproduces correctly the observed temporal profile of X-ray spectral lines as well as their relative intensities, with the high temperature lines peaking earlier and decaying faster than low temperature lines. The time behaviour of the highest temperature lines observed with XRP is a good indicator of the time duration of the heating process.

An essential feature of our model is the process of chromospheric evaporation which occurs whenever the dense chromospheric layers receive by conduction more energy than can be radiated away. Chromospheric evaporation causes the coronal arch to be filled with high density material and produces the observed delay of the density peak with respect to the temperature peak. The upflow velocity predicted by model calculations is in agreement with line blueshifts observed by the XRP during the flare rise phase. In the decay phase the flare cools by conduction and radiation, the latter process becoming more and more important as the temperature decreases and the density increases as a consequence of the evaporation process.

## IV. IMPLICATIONS FOR STELLAR FLARES

How peculiar to the Sun are the above processes? Recent observations of stellar flares at XUV wavelengths -in particular X ray observations from the EINSTEIN Observatory- show many similarities between solar and stellar flares. These include the presence of a gradual and an impulsive phase, coronal temperatures of the order of $10^7$ K and inferred coronal densities of the order of $10^{11}$ $cm^{-3}$ , a time delay between the peak temperature and the peak emission measure, a ratio Lx/Lc -and hence the importance of radiative vs conductive cooling- increasing in the decay phase (Haisch et al. 1981, Kahler et al. 1982).

The observed analogies indicate the possibility of applying the detailed knowledge gained in the solar case to stars in general, an extrapolation further suggested by the existence around dM stars of quiescent X-ray emitting coronae similar to the solar corona. If the solar phenomena discussed above are common to stars in general, we may draw the following preliminary conclusions: a) non-thermal electrons, although likely present in stars as in the Sun, may not supply the total energy radiated by flares at all wavelengths; b) the temporal profile of X-ray coronal emission is mainly determined by dynamic processes involving the entire atmospheric structure from the photosphere to the corona.

## REFERENCES

Cheng, C.C. et al. :1981, Astrophys. J. Letters 248, pp. 39-43.
Haisch, B.M. et al :1981, Astrophys. J. 245, pp. 1009-1017.
Hoyng, P. et al. :1981, Astrophys. J. Letters 246, pp. 155-159.
Kahler, S.W. et al. :1982, Astrophys. J. 252, pp. 239-249.
Pallavicini, R., Peres, G., Serio, S., Vaiana, G., Acton, L., Leibacher, J. : 1982, submitted to the Astrophys. J.

## DISCUSSION

Priest: These simulations are very impressive to me. In order to get the best fit to the observations what is the location and duration of the source of the heating?

Pallavicini: We put a heating perturbation at the top of the loop. We also tried putting it at the foot points. The first result is that we did not find much difference numerically between putting the heating at the foot points and at the top of the loop. The only difference is in the early phase of the flare when the loop is empty. So we cannot see very much then. The difference would only be a redshift at the beginning of the flare splutter. We would not expect to see this because the loop is essentially empty. The duration of the event is taken to be about 100 to 200 seconds which was chosen because of computer time. If you need to reproduce exactly a particular flare we need to choose a particular form of heating function. It may for instance have a tail to reproduce the tail observed after the peak of the event.

# THE INTERPRETATION OF EUV SPECTRA OF SUNSPOTS

J. G. Doyle
Armagh Observatory
J. C. Raymond and R. W. Noyes
Harvard-Smithsonian Center for Astrophysics
A. E. Kingston
Queen's University of Belfast.

## 1. INTRODUCTION

We report here on EUV observations of a sunspot observed by the Harvard instrument on Skylab. The observational data used here have been presented in a previous paper by Noyes et al. (1982), in which line identifications and intensities for the wavelength region 350 - 1350 A were given. Several electron density sensitive line ratios suggest a constant density, rather than constant pressure, emitting region, while temperature diagnostic line ratios of several ions yield temperatures below the temperatures expected in ionization equilibrium.

## 2. SUNSPOT PARAMETERS

### a) Electron density

For the temperature range $0.6 - 3 \times 10^5$ K a constant density of $10^{10}$ cm$^{-3}$ fits the data, while for the $10^6$ K plasma the density has decreased by an order of magnitude and a constant pressure approximation may be a valid assumption. The line ratios used in the analysis were from the ions Si III, C III, N IV, O IV, O V and Mg VIII.

### b) Ionization balance

Analyses of transition region emission from the Sun generally assume ionization equilibrium, but departure from equilibrium can affect the interpretation of observed line intensities. We used temperature diagnostic line ratios of N III, O III, S IV, O IV, O V and Ne VII. All but Ne VII, which is formed at a much greater temperature than the other ions, show temperatures lower than the equilibrium temperature and lower than the temperatures observed in the averaged quiet Sun.



*P. B. Byrne and M. Rodonò (eds.), Activity in Red-Dwarf Stars, 325–326.*

c) Model

It has been suggested that the energy radiated at transition zone temperatures is derived from the enthalpy of cooling, downflowing gas from active regions in general (Pneuman and Kopp 1977) and for sunspot plumes in particular (Foukal 1976). The observed departures from ionization equilibrium is thus consistent with those expected for a radiatively cooling gas. In our model we assume a constant density ($10^{10}cm^{-3}$), constant velocity flow (7 km/s) beginning in ionization equilibrium at log T = 5.8. The inferred slope of the emission measure from the model is substantially shallower than the observed, although photoionization can account for some of the discrepancy. The model however can match the overall energetic needs for the transition region, and it accounts for the observed shift away from ionization equilibrium. Other modifications which might account for the remaining discrepancy include random sudden reheating of the cooling gas (Raymond and Foukal 1982) or a geometry in which the line of sight intersects the cooling flow where the gas is at a few x $10^5$K, but much of the lower temperature cooling occurs outside the instrumental field of view.

References

Foukal, P.V.: 1976, Astrophys.J. 210, 575.
Noyes, R.W., Raymond, J.C., Doyle, J.G., and Kingston, A.E.: 1982, Astrophys.J. (submitted).
Pneuman, G.W. and Kopp, R.A.: 1977, Astron.Astrophys. 55, 305.
Raymond, J.C. and Foukal, P.V.: 1982, Astrophys.J. 253, 323.

# POSSIBLE ORIGINS FOR THE 12μ EMISSION LINES IN THE SOLAR SPECTRUM

Leo Goldberg*
Kitt Peak National Observatory, USA**

Braut and Noyes (1982,1983) have reported the detection of about 40 unidentified emission lines near 12μ in the solar spectrum. The strongest lines, at 811.578 $cm^{-1}$ and 818.062 $cm^{-1}$, respectively, appear as broad, shallow absorption lines, less than 3% deep, with central, emission reversals projecting 5-10% above the continuum. The emission lines strengthen at the limb and over spot penumbrae but seem to be absent over spot umbrae. The full width at half-intensity of the emission lines is about 5 km/sec, but the absorption widths are more than 10 times as broad. Over spot penumbrae, the Zeeman splitting of the emission lines is striking. The lines have the appearance of a Zeeman triplet; the central component is nearly absent at the center of the disk but is very strong near the limb where the field is viewed perpendicularly to the line of sight. The splitting over spot penumbrae is about 10 times the width of the central component, and is consistent with that of a spectral line with a Landé g-factor of unity in a magnetic field of 1500 gauss. Braut and Noyes (1982, 1983) point out that the 12μ lines are a potentially powerful tool for magnetic field measurements in stars. Further observational details will be found in their referenced papers.

The great widths of the absorption features in the strongest lines probably offer a strong clue to their identifications. We consider two possible sources for the line broadening, the first being abundance broadening and the second autoionization. At first glance, it may seem surprising that such shallow absorption lines could be abundance broadened, until we realize that the wavelengths are far in the infrared, where lines formed by pure absorption have very small central depths (Mihalas 1978). For example, a line at 800 $cm^{-1}$ formed in the sun according to the Milne-Eddington approximation will have a saturated

*This paper contains additional results and conclusions derived after the paper was presented.

**Kitt Peak is operated by AURA, Inc. under contract with the National Science Foundation.



*P. B. Byrne and M. Rodonò (eds.), Activity in Red-Dwarf Stars, 327–330.*

central depth of a little less than 2% of the continuum intensity. A calculation made purely for illustrative purposes shows that the observed absorption component of the solar line at 811.578 $cm^{-1}$ can be satisfactorily represented in the LTE approximation by the following parameters: $l_0/\kappa = 10000$, $a = 0.02$, $\Delta\nu_D = 2 \times 10^8$ $sec^{-1}$, where $l_0/\kappa$ is the ratio of line to continuous opacity at the line center, a is the ratio of damping to doppler broadening and $\Delta\nu_D$ is the doppler width in frequency units corresponding to a velocity of 2.5 km/sec. Note that the central line absorption coefficient, which is inversely proportional to the doppler width, is at least 20 times larger at 12μ than in the visible. This circumstance favors the appearance of a sharp, central emission, which, in the abundance broadening model, would be formed in the low chromosphere in the presence of a negative temperature gradient. As to the specific element or elements responsible for the 12μ lines, such abundant atoms as Fe I, Mg I and Si I should be looked at in the laboratory. It is possible that the reason the wavelengths cannot be calculated from existing tables of energy levels is because the transitions may involve configurations with outer f-electrons, which require infrared laboratory spectra for their detection.

A second possibility is that the lines may arise from transitions between doubly-excited levels, one or both of which is autoionizing. Since the lines are stronger in spot penumbrae than in the photosphere, one thinks in terms of an abundant, neutral atom in which the first excited state of the ion has a relatively low excitation potential. An obvious candidate is neutral calcium, which is known to possess a large number of doubly-excited levels converging on the first excited level 3d $^2D$ of Ca+. For example, a recent compilation (Sugar and Corliss, 1979) lists about 200 such levels, nearly all of them with J = 1 and belonging principally to the odd configurations 3dnp and 3dnf, with n-values as large as 58. Only odd levels with J = 1 may combine with the ground level of Ca I and thereby be relatively easily observed in absorption in the laboratory. Fourteen levels of the even configurations 3d4d, 3d5d and 3d6d are also listed by Sugar and Corliss as lying above the first ionization limit of Ca I. Inspection of the Sugar-Corliss table shows that a large number of transitions with $\Delta n = 0$ and $\pm 1$ occur throughout the infrared spectrum. However, no exact coincidences in frequency could be found for the three strongest solar lines at 12.2μ. This is not surprising, considering that most of the observed levels are of odd parity with J = 1 and relatively few even levels are known.

If the transitions involve autoionizing levels, it is not obvious how the central emission reversals would be produced. One possibility is that the lines are optically thick and are formed near the photosphere-chromosphere temperature minimum, analogously to the H and K lines of Ca+. Here, a rough calculation demonstrates that there are far too few excited Ca I atoms along the line of sight through the region in which the temperature of the low chromosphere is increasing outward. Part of the problem is that, unlike profiles with Doppler cores, the pure Lorentzian absorption coefficient expected for lines arising from autoionizing levels is relatively small at the line center and decreases very slowly away from the center. If due to Ca I, the emission lines would have to be formed in the photosphere, which would be possible, in principle, by laser action. According to the selection rules for autoionization in LS coupling, radiationless transitions between a term in a doubly-excited configuration and a virtual term in the continuum can only occur when the parity and the L-value of the term are unchanged. This means, for example, that certain terms of a given configuration are autoionizing, while other are not, as follows:

| Configuration | Terms | |
|---|---|---|
| | Autoionizing | Non-Autoionizing |
| 3dnp | $^{31}FP$ | $^{31}D$ |
| 3dnd | $^{31}GDS$ | $^{31}FP$ |
| 3dnf | $^{31}HFP$ | $^{31}GD$ |

Autoionizing levels are closely coupled to the continuum and therefore their populations relative to the ground state of the ion will be close to those given by the Saha-Boltzmann equation. In other words, the factor $b_n$, defined as the ratio of the true level population to that in thermodynamic equilibrium, will be unity or close to it. Levels that do not autoionize and are populated chiefly by radiative capture will be underpopulated as compared with LTE - their $b_n$-factors will be less than unity. Thus, for a transition between a lower non-autoionizing level and an upper autoionizing level, we may have a situation in which $b_L < b_U$, where $b_L$ and $b_U$ are the departure coefficients of the lower and upper levels, respectively. If $b_U > b_L$, the line absorption coefficient may become negative and amplification of the line intensity may occur. The condition is that $(b_U/b_L)\exp(-hc\tilde{\nu}/kT) > 1$ (Goldberg, 1966). Taking $\tilde{\nu} = 800$ cm$^{-1}$ and $T = 4000°$ K, we find that the condition for amplification is $b_U/b_L > 1.3$, or, with $b_U = 1$, $b_L < 0.75$, a value that seems not unreasonable for Ca I levels in the 3d4d or 3d5d configurations, which lie about 0.7 eV below the 3d ionization limit. Model calculations should be performed as a test of the Ca identification and laser mechanism hypothesis.

It should also be possible to test the proposed identification by direct measurements of doubly-excited energy levels in the laboratory. As mentioned earlier, the 3dnd levels are of even parity and do not combine directly with the ground state. In recent years, however, physicists have made a major breakthrough in the study of excited atomic and molecular states through the development of two-photon, laser spectroscopy. For example, the singly-excited 4sns and 4snd levels have been observed to very high quantum numbers by this technique (Esherick, Armstrong, Dreyfus and Wynne, 1976), and it would be of considerable astrophysical importance if the measurements could be extended to the doubly-excited even levels as well.

## REFERENCES


Esherick, P., Armstrong, J. A., Dreyfus, R. W., and Wynne, J. J. 1976, *Physical Review Letters*, **36**, 1296.

Brault, J. and Noyes, R. W. 1982, to be submitted to *Ap. J. Letters*.

Brault, J. and Noyes, R. W. 1983, *paper presented at IAU Symposium No. 102, J. Stenflo (ed.)*.

Sugar, Jack and Corliss, Charles 1979, *Journal Phys. Chem. Ref. Data*, **8**, 865.

Goldberg, L. 1966, *Astrophys. Journal*, **144**, 1225.

Mihalas, D. 1978, *Stellar Atmospheres, 2nd Edition, p313, W. H. Freeman and Company, San Francisco*.


## DISCUSSION

Mattig: In your sunspot spectra the shifted ($\sigma$) lines are much broader than the unshifted ($\rho$) one. Can you explain this?

Goldberg: My guess is that we are looking close to the limb and therefore almost horizontally. So I think you are seeing the radial gradient of the magnetic field in the penumbra, which drops off very rapidly with distance and that is why those components are so broad. The central component is not affected.

Worden: With current detector technology it is highly unlikely that these features will be observed in stars. Do you have an idea for a suitable target stars?

Goldberg: I think that Noyes and Brault are planning to look at some stars. They should stay away from red supergiants which have very broad lines. I calculated that at this wavelength the Zeeman splitting for a field of the size mentioned would be about the same as the Doppler width. So I think one should go to bright giants and Capella would be my first choice.

# MULTIPLE WAVELENGTH OBSERVATIONS OF FLARING ACTIVE REGIONS

Kenneth R. Lang
Department of Physics, Tufts University, Medford, MA 02155

Abstract. The radio emission of quiescent active regions at 6 cm wavelength marks the legs of magnetic dipoles, and the emission at 20 cm wavelength delineates the radio wavelength counterpart of the coronal loops previously detected at X-ray wavelengths. At both wavelengths the temperatures have coronal values of a few million degrees. The polarization of the radio emission specifies the structure and strength of the coronal magnetic field ($H \approx 600$ Gauss at heights $h \approx 4 \times 10^9$ cm above sunspot umbrae). At 6 cm and 20 cm wavelength the solar bursts have angular sizes between 5" and 30", brightness temperatures between $2 \times 10^7$ K and $2 \times 10^8$ K, and degrees of circular polarization between 10% and 90%. The location of the burst energy release is specified with second-of-arc accuracy. At radio wavelengths the bursts occur within the central regions of magnetic loops, while the flaring Hα kernels are located at the loop footpoints. Coronal loops exhibit enhanced radio emission (preburst heating) a few minutes before the release of burst energy. The radio polarization data indicate magnetic changes before and during solar bursts.

Displacements of the radio emission with respect to their optical wavelength counterparts indicate that sunspot-associated sources at 6 cm lie at altitudes of $h \approx 4 \times 10^9$ cm above the photosphere. The brightness temperatures of the 6 cm sources are a few million degrees, and the high degree of circular polarization ($\rho_c \approx 70\%$) is attributed to gyroresonant absorption at the third harmonic of the gyrofrequency ($H \approx 600$ Gauss at 6 cm). The 20 cm emission delineates coronal loops which join regions of opposite magnetic polarity in the underlying photosphere. It is attributed to optically thick bremsstrahlung of a hot plasma trapped within magnetic loops with semilengths $L \approx 5 \times 10^9$ cm, maximum electron temperatures $T_e(\max) \approx 3 \times 10^6$ K and electron densities $N_e \approx 10^9$ cm$^{-3}$ (or optical depth $\tau \approx 2.5$).

For all cases in which we could compare the positions of radio bursts with optical features, the radio emission originates near the central regions of magnetic loops, rather than at the footpoints. In Figure 1



P. B. Byrne and M. Rodonò (eds.), Activity in Red-Dwarf Stars, 331–334.

The degree of circular polarization of the radio emission can increase to 100% about 10 min to 1 hour before the eruption of solar flares, indicating changes in the structure of the coronal magnetic field which may trigger bursts and provide the source of their energy. Changes in both the sense and degree of circular polarization during solar bursts indicate complex magnetic changes and/or propagation effects within the burst plasma.

In Figure 3 a sequence of 10 s snapshot maps at 6 cm wavelength are provided. They were made before, during and after the impulsive burst (second map). The impulsive burst ($\gtrsim$ 8" in size and < 15% circularly polarized) is spatially separated from both the preburst radio emission and the gradual decay component of the burst (10" in size and 30% left circularly polarized). The absence of circular polarization in the impulsive component suggests that this source is located near the apex of the loop where the longitudinal component of the magnetic field is small, whereas the polarization detected in the gradual decay component suggests an origin in a predominantly longitudinal magnetic field of one polarity, most likely in one leg of the loop.

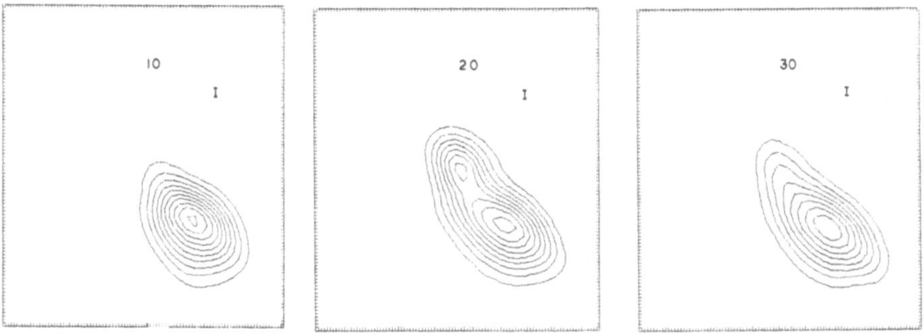


Fig. 3. A series of 10 s snapshot maps before, during and after the impulsive phase (second map) of a 6 cm burst (angular size $\lesssim$ 8").

ACKNOWLEDGEMENTS

Radio interferometric studies of solar active regions at Tufts University are supported under contract F 19628-80-C-0090 with the Air Force Geophysics Laboratory and grant no. INT 8006066 with the National Science Foundation. The National Radio Astronomy Observatory (V.L.A.) is operated by Associated Universities, Inc., under contract with the National Science Foundation. All of this work was done in collaboration with Robert F. Willson.

The complete manuscript of this paper is being published in 1983 as part of the Proceedings of the XXIV COSPAR (Committee on Space Research) Meeting.

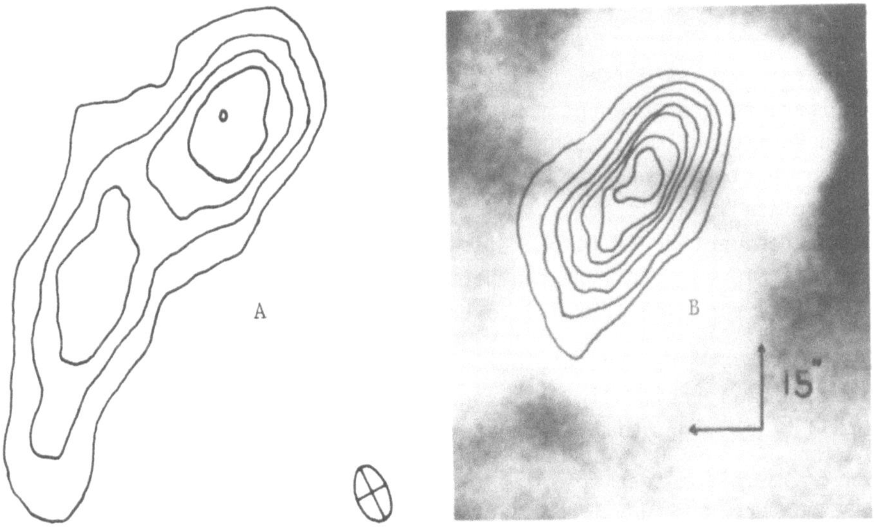


Fig. 1. V.L.A. synthesis map of the preflare (A) and impulsive phase (B) of a burst detected at 6 cm wavelength.

we compare a 10 s snapshot map of the impulsive phase of a 6 cm burst (B) with a map of the preburst radio emission (A: three minutes before the burst) and the flaring Hα kernels (B: at the time of the radio burst). The preburst radio emission was contained within a looplike structure which joins the sites of subsequent Hα emission. The peak brightness temperature of the preburst emission was $T_B \approx 6 \times 10^6$ K as compared with the peak burst brightness temperature of $T_B \approx 4 \times 10^7$ K. The brightness temperature for several ten minute intervals around one hour before the burst was only $T_B \approx 2 \times 10^6$ K, which is typical of quiescent emission of both plage and sunspot-associated sources at 6 cm. The loop plasma was therefore "warmed up" before the release of burst energy (i.e. preburst heating).

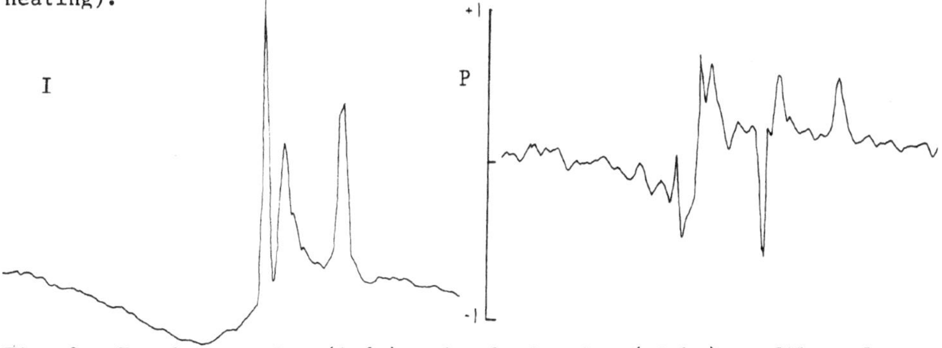


Fig. 2. Total intensity (left) and polarization (right) profiles of a multiple spike burst at 20 cm wavelength. Polarization changes occur before burst emission as well as from burst to burst within the same source of $\approx$ 15" in size.

DISCUSSION

Kuijpers: You said that the bursts which were observed were resolved. I believe that at present you cannot say that. About a year ago we carried out a VLBI experiment using two instruments with a resolution of 0.06 arc sec to settle the problem of whether the energy release in a solar flare takes place in small pockets (Kuijpers, Tapping & Graham, this volume). This experiment needs to be followed up to observe large flares with VLBI techniques before we can settle this.
(Part of discussion lost due to break in recording).

Kuijpers: ... that in the impulsive phases it might be that might be several small regions in which acceleration takes place. Secondly, one should be cautions with averages over 10 seconds.

Lang: It is the case in solar radio astronomy that the bigger the dish used the more one sees. The VLA can see structures much smaller than the ones we resolve i.e. of order tenths of arc sec and they do not detect any signal. So I am surprised that the VLBI observations detects any signal from solar flares. Have you detected signals from solar active regions with resolution of, say, 0.1 arc sec?

Kuijpers: Yes, I think so.

Lang: Well, that is surprising because I know of many experiments which have tried unsuccessfully to detect structure on those angular scales.

Kuijpers: What I got from those results is that they did not have much flaring activity.

Lang: That's possible. But Marshall Cohen and Bernie Burke(?) both tried and did not see anything.

Dupree: I would like to make a comment. I was involved in the Burke attempt with the VLBI. We never had an optimum configuration and a flare. The only flare which occurred as we were changing tapes. There was no occasion when there was a substantial flare which we should have seen.

Lang: I agree that we cannot rule out the possibility that when a strong flare occurs you will see small-scale features.

# MAGNETIC DEVELOPMENT OF FLARING REGIONS AT CENTIMETER WAVELENGTHS

M. R. KUNDU
Astronomy Program, University of Maryland, College Park, MD

It has been known for many years that the flare build-up manifests at centimeter wavelengths (2-6 cm), in the form of increased intensity and increased polarization of the active region. The flare-associated bursts originate in these intense sources, and the probability of occurrence of bursts increases with the increasing intensity of these narrow bright regions. With the availability of arc-second resolution using the VLA it has been possible to study the nature of this build-up from two-dimensional synthesized maps over short periods before the start of a flare. For a hard x-ray associated impulsive 6 cm burst observed on June 25, 1980 (Kundu, Schmahl, and Velusamy 1981), we produced several 15-minute synthesized maps in total intensity (I) and polarization (V) just before the flare onset (Kundu 1981). Figure 1 shows the central 1.′1x1.′6 regions of 15 minute synthesis maps over the period 14:45-15:45 UT. As can be seen from these Figures, the region is very complex, consisting of numerous components many of which are bipolar. These components have brightness temperatures of $6\text{-}9 \times 10^6$ K during the hour before the flare. The burst source was located close to the neutral line of these oppositely polarized regions near B. The burst maximum is identified with a "+" and the burst extent averaged over the period 1551-1600 UT is shown by the dotted contour in the last map.

There was a definite trend for the active region undergoing brightness and polarization changes. The central component B intensified at 1515-1530 UT and increased in polarization slightly. In the last map (1530-1545) several new components appeared with polarizations of 40-80%. However, the most remarkable feature is the change of the sense of polarization of component B; also the component on the northern side of B greatly increased in polarized intensity, with polarization of 80-90%. This might imply the emergence of a flux of reverse polarity at coronal levels. (The photospheric magnetograms show little or no change.) We believe that this reverse polarity may be caused by the expansion of a pre-existing flux tube in which twisting increases its coronal magnetic field; at the same time there must be some heating of the loop. Alternately a previously existing loop of opposite polarity



*P. B. Byrne and M. Rodonò (eds.), Activity in Red-Dwarf Stars, 335–337.*

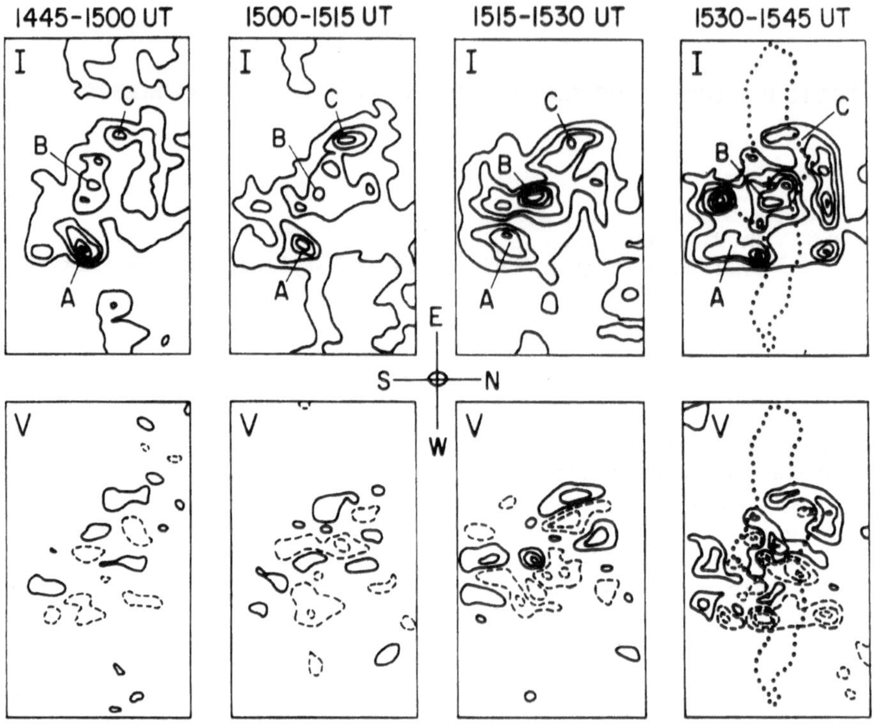


PRE-FLARE 6 CM MAPS 25 JUNE 1980 Fig 1

which was not observable at 6 cm due to its weak magnetic field became observable due to a sudden onset of currents in the loop.

Kundu (1981) and Kundu et al. (1982) discussed a set of 6 cm VLA observations (resolution ~ 2") that pertains to changes in the coronal magnetic field configurations that took place before the onset of an impulsive burst observed on 14 May 1980 (Figure 2). The burst appeared as a gradual component on which was superimposed a strong impulsive phase (duration ~ 2 minutes) in coincidence with a hard X-ray burst. The pre-flare region showed intense emission with peak $T_b \sim 10^7$ K extended along a neutral line situated approximately in the east-west direction. A burst source of intense emission with $T_b \sim 4x10^7$ K, appeared initially. The most remarkable feature of the burst source evolution was that an intense emission extending along the north-south neutral line (line of zero polarization at 6 cm) appeared (Fig. 2), just before the impulsive burst occurred. This north-south neutral line must be indicative of the appearance of a new system of loops, possibly due to reconections. In the 20 seconds preceeding the impulsive peak ($T_b \sim 1.1x10^9$ K) the arcade of loops (burst source) changed and ultimately developed into two strong bipolar regions or a quadrupole structure whose orientations were such that near the loop tops the field lines

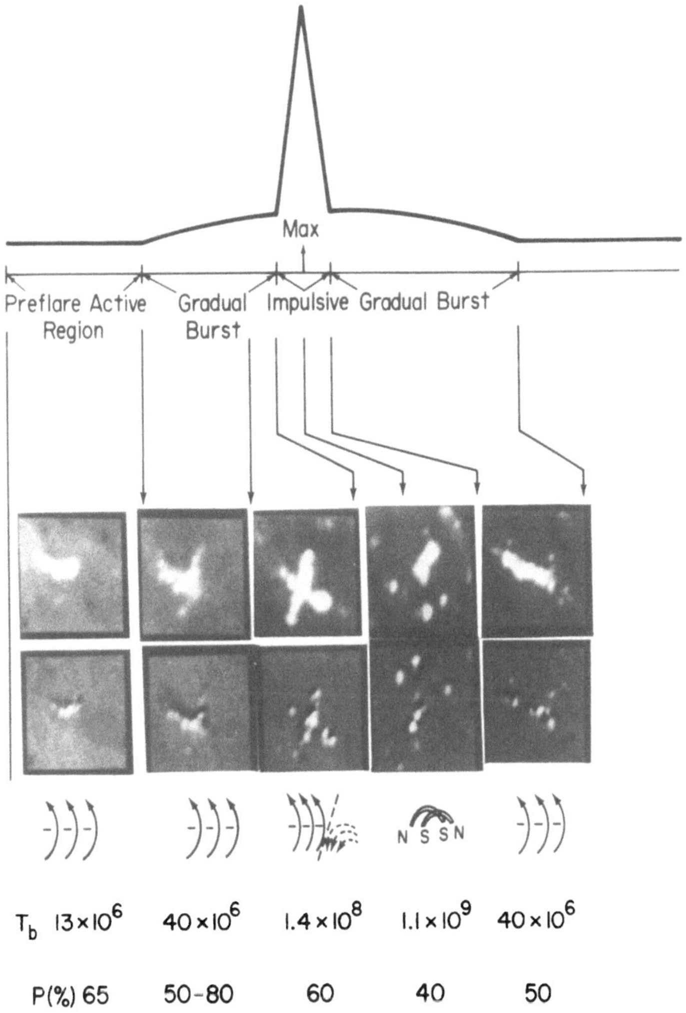


were opposed to each other. This quadrupole field configuration is reminiscent of the flare models in which a current sheet develops at the interface between two closed loops. The impulsive energy release must have occurred near one of the centrally located neutral lines. The bright compact bipolar source is obviously related to the region of energy release by some kind of magnetic reconnection of the field lines originating from the two bipolar regions between which this compact region is located.

References

M. R. Kundu, E. J. Schmahl, and T. Velusamy, Ap. J. 253, 963, (1982).
M. R. Kundu, SMY Workshop, Crimea, p. 24 (1981).
M. R. Kundu, E. J. Schmahl, T. Velusamy, and L. Vlahos, Astron. Astrophys., 108, 188, (1982).

# VERY LONG BASELINE INTERFEROMETRY OF SOLAR FLARES.

J. Kuijpers[1], K.F. Tapping[2], D. Graham[3]
[1] Sterrenkundig Instituut, Zonnenburg 2, 3512 NL UTRECHT, The Netherlands.
[2] Herzberg Institute of Astrophysics, NRC, KIA OR6 Ottawa, Canada.
[3] Max Planck Institut für Radioastronomie Auf dem Huegel 69, D-5300 Bonn, FR Germany.

The aim of the VLBI experiment was a search for the occurrence of sub-arc-second microwave emission centres as tracers of the initial energy release in solar flares. The observations extended over the period April 28 to May 3, 1981 and were performed with the 25 m telescopes at Onsala (Sweden) and Dwingeloo (Netherlands) at a wavelength of 18 cm (1663 MHz). The baseline was 619 km long giving a minimum angular width of the interference fringes of 0.06". This corresponds to a spatial scale of 40 km at the solar disc centre.

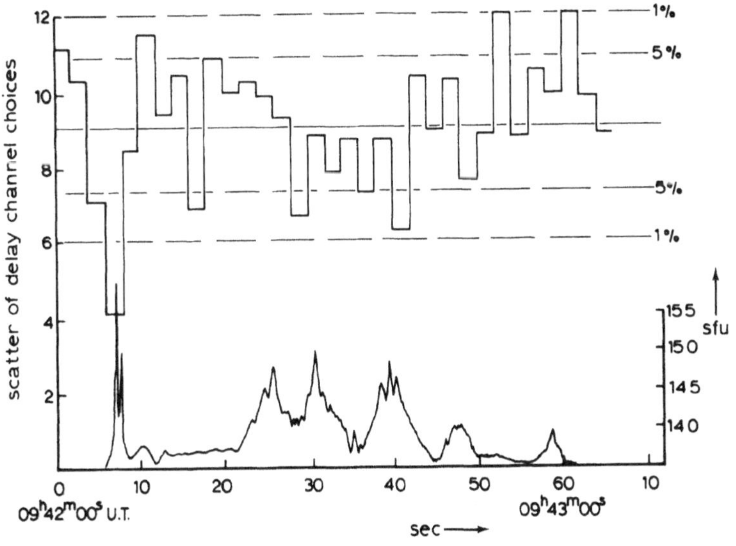


Figure 2. Total flux at 18 cm on 29 April 1982 (below) (1 sfu = $10^{-22}$ $Wm^{-2}Hz^{-1}$) and scatter in channel number (above).



*P. B. Byrne and M. Rodonò (eds.), Activity in Red-Dwarf Stars, 339–341.*

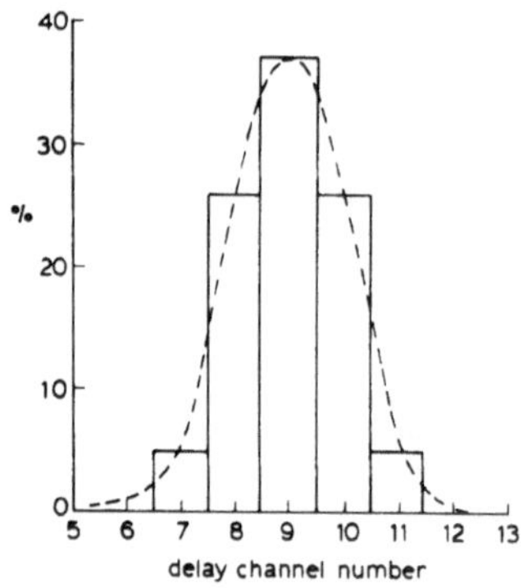


Figure 1. Distribution of standard deviations in channel no. of maximum correlation in absence of bursts.

We used the data recording system of the European VLBI network. The pre-correlation bandwidth was 250 kHz. The data were processed at the MPIfR in Bonn. The shortest available integration time was 0.2 sec and is dictated by the basic correlator cycle-time. The output from the correlation process consisted of a listing for each 0.2 sec time interval of the correlated signal amplitude and fringe phase for each of 31 delay channels (separation 2 μsec ) centered around the delay calculated for the estimated source position and baseline orientation. The annual solar motion and the solar rotation were taken into account by a differential fringe rate correction.

During the observing period three weak outbursts occurred. No large correlations were observed. However a strong indication was found for an unresolved source with a signal to noise ratio of order unity during the impulsive bursts preceding the main phase of one event (Fig 2.). During the impulsive bursts the spread in number of the channel

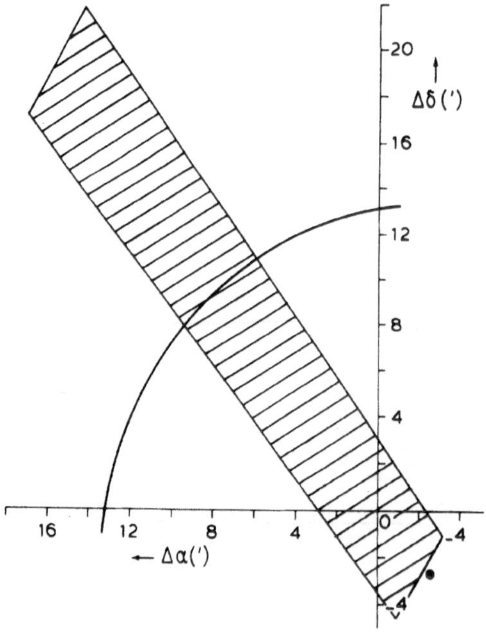


Figure 3. Probable error box for the source position.

which showed the maximum correlation amplitude was reduced significantly below the value for random behaviour (99.5% confidence, see Fig. 2). The spreads were determined for intervals of 2 sec containing ten data-points; their distribution in the absence of bursts is shown by the histogram in Fig. 1, for 190 intervals, together with the expected normal distribution. At the time of the impulsive bursts Fig. 2 shows a clear reduction in the amount of scatter from the expectation value for the standard deviation of 9. The derived brightness temperature is of order $10^{12}$ K (flux of ~ 1 sfu). At lower frequencies a small group of type III bursts occurred; no Hα flare was observed (NOAA, Boulder). Both other outbursts showed only gradual flux variations in time.

From the offsets in delay and fringe frequency from the values for the estimated source position the actual position can be determined. The error box in Fig. 3 overlaps Hale region 17609. The curve represents the solar limb, the dot the solar centre and the estimated position is at the origin.

It is important to repeat this experiment for a larger flare.

The Onsala telescope is operated by the Onsala Space Observatory of the Chalmers University of Technology. The Dwingeloo telescope is operated by the Netherlands Foundation for Radio Astronomy (SRZM) with the financial support of the Organization for the Advancement of Pure Research (Z.W.O.).

# RS CVn STARS: PHOTOSPHERIC PHENOMENA AND ROTATION

S. Catalano
Institute of Astronomy, University of Catalania, Italy

## ABSTRACT

This review presents a summary of observed photospheric phenomena on RS CVn stars: the amplitude, shape, evolution and migration rate of the photometric wave in relation to the rotational and orbital motion.

The main points considered are: 1) the activity level (maximum amplitude, short and long timescale variability) versus rotation period; 2) the activity cycles as inferred from changes in the wave migration rate and direction and from the variation of its amplitude; 3) the detection of differential rotation; 4) the connection between the orbital period variation and activity.

## INTRODUCTION

This is just the twentieth year since we started observing RS CVn at Catania (Blanco et al 1982a) and, it is a great pleasure for me to have the opportunity of discussing the photometric peculiarity and properties of those stars that have since become a special group of binaries (Hall 1972).

Among the phenomena characterizing "Stellar Activity"like that we observe on the Sun are spots, flares, chromospheric emission, variable radio and X-ray emission, coronas, and so forth. Large-scale dark spots appear as the most conspicuous manifestation on RS CVn binaries compared with the Sun itself. The case for spots on the stellar surface is largely indirect. Direct evidence however has recently been provided by spectroscopic investigation of the TiO band (Ramsey and Nations, 1980).

It is generally believed now that photometric variability in the RS CVn binaries and in the BY Dra variables outside eclipse are due to an uneven distribution of large-scale spotted areas. The RS CVn binaries exhibit light variations with amplitudes typically 0.1-0.2 mag. of the combined light of the system, which may become as large as 0.7 mag



*P. B. Byrne and M. Rodonò (eds.), Activity in Red-Dwarf Stars, 343–361.*

when the variation is referred to the spotted star only, i.e. after the contribution of the unspotted companion has been taken into account (Catalano, Frisina and Rodonò 1980). This means spot areas up to 30-40% of the visible hemisphere, i.e. three to four order of magnitude more extended that the largest solar spotted areas.

It is now generally accepted that chromospheric and coronal activity on single and binary stars is essentially determined by the rotation speed (Skumanich 1972, Duncan 1981, Pallavicini et al 1981, Catalano and Marilli 1982). Many people are becoming convinced (Bopp and Espenak 1977, Walter and Bowyer 1981) that rapid rotations is of major importance in determining the high rate of occurrence of active stars in binary systems rather than the binary nature itself. There is however no definite evidence of photometric variability in single stars with similar rotation speed (Blanco et al 1979) and with definitely ascertained chromospheric variability (Wilson 1978, Vaughan et al 1981).

The general properties of chromospheric and coronal activity and theoretical problems on RS CVn binaries are being reviewed at this meeting by Bopp, Charles and Vogt. Therefore I will deal with photometric characteristics related to photospheric activity, spots, spot cycles and rotation.

Reviews on the general properties and problems have been recently published (Hall 1981, Milano 1981, Rodonò 1981 , 1982).

## PHOTOMETRIC ACTIVITY LEVEL AND ROTATION

Many different parameters can be considered as defining the activity level from a photometric point of view, like the maximum wave amplitude, the time scale of amplitude changes, flare events, long-term variability of the median brightness.

If rotation is the main parameter in determining the level of activity we would expect some interdependence among these parameters or at least between some of them and the rotation period. However a number of complications arise from such a kind of analysis. The wave amplitude is not only determined by the total area and temperature of the spots but also from the asymmetric distribution of spots on the stellar surface, their latitude, and the inclination of the rotation axis to the line of sight. The detection of optical flares is strongly dependent on the spectral type of the star (Byrne 1982) and rotation periods may reflect evolutionary status.

Hall (1976) suggested that a distinction can be made between short-period ($P \leq 1$ day), regular (i.e.intermediate), and long-period ($P \geq 15$ days) RS CVn binaries. Even if it is difficult to define a specific domain of $P_{orb}$, to segregate certain sets of physical properties and the evolutionary stage of the components, a broad subdivision can be made. Short-period group members have both components on the main sequence (Milano 1981) with primaries (hotter) of spectral type near G0V and smaller secondaries in the range G5 V to M2 V. Regular and long-period RS CVn's are post-main sequence binaries, in which one star is a subgiant or a giant evolved off the main sequence, but not yet filling its Roche lobe (Popper and Ulrich 1977, Morgan and Eggleton 1979).

A) The Maximum Wave Amplitude

Although the maximum observed wave amplitude is smaller in short-period systems than in the regular and long-period groups, the values, corrected for the dilution effect caused by the light contribution of the companion, turn out to be comparable (on the average 0.3 - 0.4 mag.) in all the groups. This is because in the short-period group the spotted stars are always the faintest in the system, sometimes by a factor of more than three (Table 1), and the correction factor is very large. In regular and long-period systems both components have comparable luminosity or the spotted, larger, giant component is more luminous than the smaller main-sequence companion so that the correction factor is generally small. There are few exceptions to this but RS CVn is one of them. Its maximum observed wave amplitude $\Delta V$=0.22 mag. leads to a large value for the corrected amplitude of 0.67 mag.

TABLE 1. Maximum wave amplitude in short-period RS CVn binaries

| System | Sp.Type | P(orb) | $L_c/L_{SP}$ | $(\Delta V)_{max}$ | $(\Delta V)_{cor}$ |
|---|---|---|---|---|---|
| RT And | F8V+K0V | $0^d.629$ | 4.4 | $0^m.06$ | $0^m.37$ |
| SV Cam | G0V+K4V | 0.593 | 8.3 | 0.06 | 0.75 |
| WY Cnc | G5V+M2V | 0.829 | 0.005^ | 0.08 | 0.08 |
| CG Cyg | G1 +G9V | 0.631 | 2.4 | 0.11 | 0.43 |
| UV Psc | G2V+K0V | 0.861 | 3.0 | 0.08 | 0.36 |
| XY UMa | G2V+G5V | 0.479 | < 0.4 ^ | 0.20 | 0.29 |
| ER Vul | G0V+G5V | 0.698 | 2.9 | 0.04 | 0.17 |

^Spots on the early type component.

Since the spotted stars are of different spectral types and luminosity classes, for system of different period any dependence of the maximum amplitude on the orbital period may not be obvious. In order to attempt to separate the two effects, distributions of the maximum wave amplitude for each spectral type have been considered.

The distribution of the corrected maximum wave-amplitude in K0-K1 IV components does not show any dependence on the orbital period (Figure 1). The largest value in Figure 1 refers to the spotted component of RS CVn. For W UMa binaries Budding and Kadauri (1982) found evidence of a correlation of the wave amplitude and the orbital period. The shorter is the period, the larger is the wave amplitude. W UMa binaries may not be considered true RS CVn binaries, but spot characteristics have been detected (Binnendijk 1970, Eaton, Wu and Rucinski 1980). Hall (1976) discussed them in the context of stellar spots and Russo et al (1982) considered their evolutionary stage in relation to the short-period RS CVn binaries.

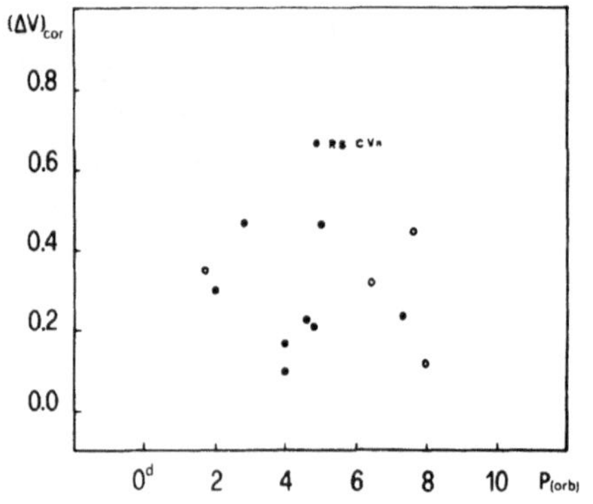


Figure 1. Maximum values of wave amplitude, for RS CVn stars of spectral types K0-K1 IV corrected for the light contribution of the unspotted component, versus the orbital period. Open circles (o) refer to systems for which the luminosity ratio of the two components has been deduced from their spectral types.

B) Time Scale for the Wave Amplitude and Median Luminosity Changes

Waves of a wide variety of shapes have been observed in RS CVn binaries. For any one system shape and amplitude changes can obviously result from changes in the spot sizes and temperatures or redistribution of the spots on the stellar surface (Bartolini et al 1982). Synopses of light-curve evolution based on many years of observation are being published now. Let me summarize some of the most important. RT And (P=0.629 days) (Milano et al 1981), in addition to wave-like distortions migrating towards increasing orbital phases on a timescale of 3.86 years ($\approx$2250 $P_{orb}$) (Blanco et al 1982b), shows changes outside eclipses by as much as 0.1 mag. over a period of 1-2 months (Zeilik et al 1982a), variations in the depths of primary and secondary minima from night-to-night and a changing dip near the end of primary eclipse. UV Psc

(P=0.86 days) showed a nearly constant sinusoidal wave migrating toward decreasing orbital phase between 1976 and 1978 (Vivekananda Rao and Sarma 1981). The wave disappeared in 1979 and 1980 and reapparead in 1981 (Zeilik et al 1981, 1982b). In SV Cam (P=0.593 days) the wave is not clearly visible due to the rounded maxima, which indicates a large proximity effect (Van Woerden 1957), but the large changes outside eclipse agree with a wave migrating toward decreasing orbital phases. Variation in the average light at maxima and the primary minimum occurred in SV Cam between 1969/70 and 1976/77, while the magnitude at the middle of the secondary eclipse remained unchanged. Hilditch et al (1979) found that the cooler K4 V star exhibits a BY Draconis variability, confirmed by the detection of flares (Patkos 1981). XY UMa (P=0.48 days), even if its inclusion in the RS CVn binaries is questionable, shows the most remarkable light curve changes of the short-period systems. Geyer (1980) has collected an extensive series of observations showing changes from symmetric to asymmetric light curves with a periodicity of 3.5-4 years, flickering and brightness spikes dominate the light curve for 1-2 weeks and variations of the average brightness of the system on a time scale of 25 years. For this binary as well as for WY Cnc (Awadalla and Budding 1979) the active spotted star appears to be the hotter component in the system. WY Cnc shows periodic variations in the median brightness in about ten years (Sarma 1976), while CG Cyg is undergoing a steady increase in the median light amounting to about 14% since 1965 (Milone et al 1979).

Collection of data for regular and long period systems can be summarized as follows. RS CVn appears to have the most stable wave, its shape has undergone very little changes during the last 20 years (Blanco et al 1982a). In 1976 (Ludington 1978), as in 1981, the wave showed two minima of different depths. Small cyclic variations on the wave amplitude have been detected with a period of 4.5 years (Rodonò 1981). Variations of the median brightness by 0.1 mag. occur on a time scale of 15-20 years (Reglero 1982). II Peg has the most variable light curve. The shape and amplitude of the wave both change within a few months. Generally the curve is strikingly asymmetric, the rise and decline having markedly different slopes, while the median brightness is now secularly decreasing (Rodonò et al 1982). A similar decline was observed between 1940 and 1950 (Hartmann et al 1979). UX Ari (P=6.438 days) in less than ten years (1972-1980) has shown alternate periods of small and large amplitudes of the wave (Guinan et al 1981). The largest amplitude values observed till now occurred when the system was at the maximum and the minimum of its median brightness. The last event took place between March and November 1980 leading to an amplitude increase from 0.06 to 0.17 mag. The light curve obtained during late 1978 and early 1979, appears to have two maxima and two minima of unequal brightness (Boyd et al 1981). A similar behaviour, but with a reversed temporal sequence was observed in the light curve of V711 Tau (Bartolini et al 1982). The large-amplitude single wave in late 1978-early 1979, developed in late 1980-early 1981 into a light curve showing two maxima and two minima (Figure 2). For V711 Tau the median brightness remained unchanged so that Bartolini et al (1982) reasonably supposed the total spotted area remained also unchanged. The wave amplitude variation results from a redistribution in longitude of the spotted area. The appearance of two distinct minima in the light curve and the decrease of the wave amplitude to a minimum both coincide with the maximum longitude separation of the spot groups. Light curve fitting of V711 Tau by Dorren et al (1981) required a theoretical model of two circular spot well separated in longitude before the two maxima and minima became visible in the light curve.

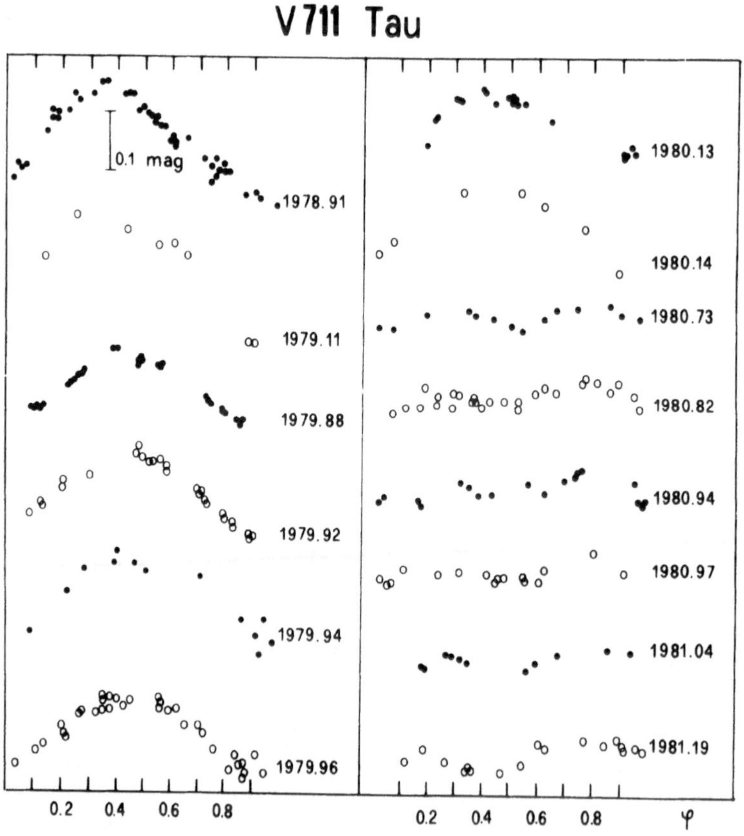


Figure 2. Synthesis of the observations of V711 Tau from 1978 to 1981 (Bartolini et al 1982), showing the evolution of the light curve and the double wave structure.

As in V711 Tau an amplitude increase from 0.07 to 0.13 mag. between 1977 and 1979 was followed in 1980 by the appearance of two maxima and two minima in the wave of AR Lac (Ertan et al 1982). This phenomenon which appears to be transient in many systems and develops on a time scale of one-two years seems to be characteristic of the light curve of some long-period system like λ And (Landis et al 1978), and HD 185151 (Bopp et al 1982).

Quasi-regular wave amplitude changes are observed in SS Boo (Blanco et al 1980) and in RT Lac (Shore and Hall, 1980) with periods of about 10 and 30 years. The matter of periodic variability will be discussed in more detail later on in the context of detection of stellar activity cycles.

No clear evidence has been gathered of optical flare detection in regular and long-period RS CVn systems. Guinan et al (1979) reported flare events on V711 Tau which are not confirmed by observations on the same nights by Bartolini et al (1982). On the other hand flare-like activity has been detected several times in V711 Tau at radio wavelength (Gibson et al 1975, Feldman et al 1978).

Sporadic events have been observed, at the primary eclipse of AR Lac, during totality. The V light at mid-eclipse changed by more than $0^m.05$ between August 6th and August 28th 1979, while smaller changes were observed during the 1978 observations season.

From this short analysis on the best observed systems we can see that regular and long-period RS CVn binaries frequently show nearly cyclical variation of the wave amplitude, remarkable changes in the light curve shape on timescales of months to years, and in a very few cases long-term variation of the median brightness. Short-period systems exhibit, large variability from night to night, frequent optical flare events and well defined, in some cases cyclic variations of the median brightness.

All these latter symptoms are characteristic of the BY Dra stars. This, together with the fact that the components of the short-period RS CVn binaries are both main sequence stars, suggests that they be included in the BY Dra variables rather than in the RS CVn (i.e. if any distinction within the spotted stars must be done). In this context II Peg could be better classified as a BY Dra rather than as a RS CVn member.

It is interesting to recall that the active stars in the short-period systems WY Cnc and XY UMa appear to be the earlier-type components of spectral types G5 V and G2 V respectively, i.e. typical solar-type stars.

## ACTIVITY CYCLES

Spot cycles a decade or so in duration are expected in RS CVn stars both on the basis of the solar analogy and of theoretical investigations. Stellar magnetic cycles have been estimated from a generalization of the dynamo mechanism (Shore and Hall 1980) and overlapping of successive cycles along the main sequence towards later spectral type have been predicted (Belvedere et al 1980a).

Observationally, the detection of cycles it is not an easy task, even if different approaches are possible. The most obvious are to study the variations in amplitude of the wave and or the median brightness of the star as a function of time. A third approach, based on the solar experience, is to study the periodicity in the wave migration direction relative to the orbital phase. If differential rotation and latitude drifts of spots operate on stellar surfaces, giving rise to a butterfly diagram, we would expect direction changes in the wave migration from increasing to decreasing orbital phase as the centre of the spotted area drifts from high latitudes (starting of a new cycle) to the equator (end of the cycle). Correlated amplitude variations would be expected (see the schematic diagramme in Figure 3). Various complications, (Hall 1981, Rodonò 1982) make a proper interpretation of the amplitude changes and of the migration direction variation non-trivial. The presence of two spot groups, which as we saw in the previous paragraph, is very frequent in RS CVn binaries, could cause the wave amplitude to diminish even though the area of those two groups remained the same. In addition the two groups may lead to minima in the light curve migrating in different directions

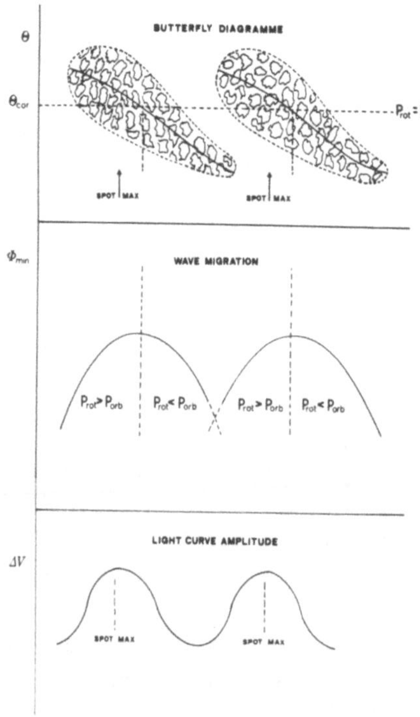


Figure 3. Schematic behaviour of the migration and amplitude variation of the wave in a solar-like cycle of activity.

as in V711 Tau (Bartolini et al 1982). Furthermore, if reference is made to the minimum or to the maximum, the rate and direction of the migration may be completely different, as in II Peg (Rodonò et al 1982) and HK Lac (Percy and Welch 1982).

Bearing in mind these difficulties tentative cycles deduced from the median brightness variations, from amplitude variations of the wave and from wave migration curves, are given in Table 2. The cycles in Table 2 range between 5 and 100 years with values around 40 years. Some evidence of correlation between wave amplitude variation and migration direction seems to be detected in SS Boo and RT Lac (Figure 4).

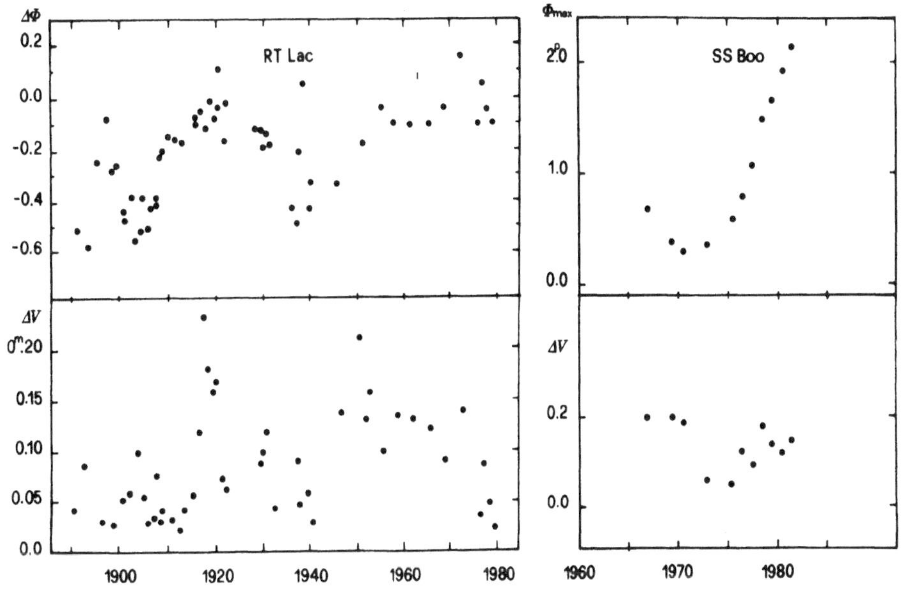


Figure 4. Observed migration and amplitude of the wave for RT Lac (adapted from Hall 1981) and SS Boo (Blanco et al 1982b) showing possible activity cycles.

DETECTION OF DIFFERENTIAL ROTATION

The wave migration with respect to orbital phases in RS CVn binaries indicates that the synchronization between rotation and orbital motion is not perfect. Among Algol binaries, cases of high asynchronism are well documented but, with very few exception, RS CVn binaries show a low degree of asynchronism. Typically their photometric and orbital periods differ by less than 1%. This asynchronism causes the minimum

TABLE 2. Activity Cycles for RS CVn Binaries, and BY Dra Stars

| System | Sp Type | P(orb) | $P_{cy}$ (Years) Bright. | Ampl. | Migrat. | References |
|---|---|---|---|---|---|---|
| UX Ari | G5V+K0IV | 6ᵈ44 | | | 16 | Blanco et al.(1982b) |
| SS Boo | dG5+dG8 | 7.61 | | 12 | 20 | Blanco et al.(1982b) |
| WY Cnc | G5V+M2V | 0.83 | 10 | | | Sarma (1976) |
| RS CVn | F4IV+K0IV | 4.78 | | 4.5? | 100 | Blanco et al.(1982b) |
| RT Lac | G9IV+K1IV | 5.07 | | 30-35 | 35-40 | Hall (1981) |
| AR Lac | G0IV+K0IV | 1.98 | | | 30 | Blanco et al.(1982b) |
| HK Lac | FIV+K0III | 24.43 | | | 22^ | Blanco et al.(1982b) |
| II Peg | K2-3 V-IV | 6.72 | ~ 40 | | | Hartmann et al.(1979) |
| V711 Tau | G5V+G5V | 2.84 | | | 5 | Blanco et al.(1982b) |
| XY UMa | G2V+G5V | 0.48 | 25 | | | Geyer (1980) |
| BY Dra | dM0e+dM0e | 5.97 | 50-60 | | | Phillips & Hartmann (1978) |
| CC Eri | K7Ve | 1.56 | 50-60 | | | Phillips & Hartmann (1978) |
| BD 26°730 | K5V | - | 50 | | | Hartmann (1981) |

From the average photometric period of 24.96 days.

of the light curve outside-eclipse to migrate with respect to orbital phase. However changes in the migration rate (Figure 5) and reversal of the direction of migration (Figure 4) does suggest relative motions of the spotted areas on the stellar surface, which may be ascribed to a differential rotation and a latitude drift as on the Sun. Since migrations both toward increasing and decreasing orbital phases are observed (i.e. rotation period larger and smaller than the orbital period) an equatorial acceleration, as in the Sun, may explain the observation if there exists a latitude away from the equator which is rotating with the orbital period. Therefore the observed cyclic changes of the phase of wave minimum ($\phi_{min}$) reflect the latitude drift of the spots toward the equator (advancing $\phi_{min}$ corresponds to spots at latitude larger than the co-rotating latitude $P_{rot}>P_{orb}$, regressing $\phi_{min}$ corresponds to spots at latitude smaller than the co-rotation latitude $P_{rot}<P_{orb}$).

So both advancing and regressing $\phi_{min}$ have been observed, but in general the migration rate proves to be variable in almost all systems,

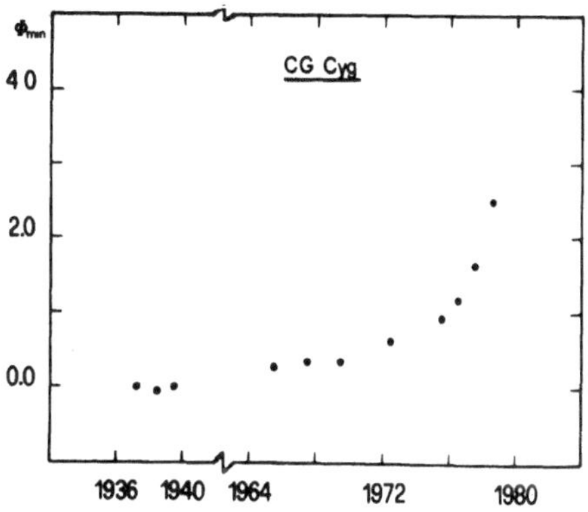


Figure 5. The phase of the minimum of the wave for CG Cygni (data from Blanco et al 1982, and sources referenced therein).

provided enough observational material is available (Blanco et al 1982a). On 30 systems for which reliable information are available, 6 show advancing $\phi_{min}$, 16 regressing $\phi_{min}$ and 8 are variable.

This highly non-uniform distribution of the frequency of advancing (27%) compared to regressing (73%) migration(if the systems of variable migration direction are excluded) may result from different life-times of the migration in the two directions during a cycle. The frequency ratio therefore may give information on the co-rotating latitude relative to the maximum latitude of spot appearence for a new cycle.

Actually a completely repeated cycle has been observed to date only for RT Lac (Hall 1981) so it is not known how smooth the change in $\phi_{min}$ is from one cycle to the next and if the $\phi_{min}$ curve has a degree of asymmetry consistent with the distribution of migration directions.

In this interpretation the difference of the maximum values of the direct-to-reverse migration rate would be a measure of the differential rotation. Typical values of $\Delta\Omega/\Omega$ are $1.4\times10^{-3}$ for short-period and $8.2\times10^{-3}$ for long-period RS CVn binaries. $\Delta P/P$ values plotted versus the orbital period in Figure 6 show a power law dependence on the rotation period with exponent nearly equal to 1, i.e. $\Delta P/P \propto P$ (Blanco et al 1982b). Expressing this in terms of the rotational frequency one obtains $\Delta\Omega \propto (-\Omega^2/\Omega^2)$=const., while theoretical models of differential rotation predict $\Delta\Omega \propto \Omega$ (Belvedere et al 1980b). This is a puzzling result that can be understood in terms of forced synchronization of the stars by tidal action or by the braking effect of interacting magnetic loops (Simon et al 1980). The Sun appears to have a larger differential rotation (values for two different latitudes are given in Figure 6). An alternative possibility is that the interaction of the companion forces the spots to form closer to the co-rotating latitude where the strongest interaction is.

## ORBITAL PERIOD VARIATION AND ACTIVITY

Many eclipsing binaries are known to have orbital period variations

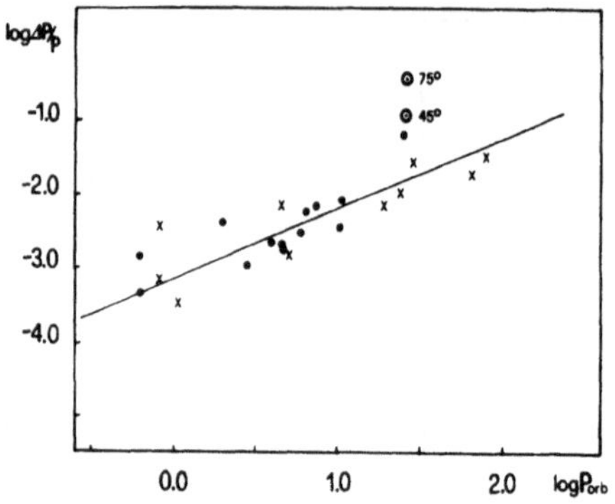


Figure 6. Maximum rate of the wave migration ΔP/P plotted against the orbital period of the system. Crosses (X) refer to systems for which only the difference ΔP between the photometric and the orbital period is known. Values for the rotation of the Sun at two different latitudes with respect to the equator are indicated. The solid line represents the power law relation $\log(\Delta P/P) = \log P + C$.

not due to apsidal motion or to a third companion. This phenomenon which appears to be fairly well understood in semidetached Algol binaries in terms of mass transfer, is difficult to explain in the RS CVn binaries, virtually all of which are clearly detached. The most comprehensive discussions of the period variation of RS CVn systems are those of Hall, Kreiner and Shore (1980) and Hall and Kreiner (1980). They found that of the known eclipsing RS CVn binaries around 40% have variable orbital periods. The observed long-term period changes are period decreases, while short-term alternating periodic changes (Hall, 1972) have been revealed to be spurious, the O-C residuals being due to a light curve asymmetry during the eclipses produced by the photometric distortion wave outside eclipses (Catalano and Rodonò 1974). Hall and Kreiner (1980) explained the long-term period changes with a model in which the active star loses mass in a convectively driven stellar wind which co-rotates with the star out to the Alfvèn radius. Average magnetic fields of about 1 k Gauss and mass losses of about $10^{-9}$ $M_{\odot}$/year are required. Due to the strong longitudinally asymmetric spot activity, they allow a possible anisotropy in the wind. De Campli and Baliunas (1979) found that mass loss rates of $10^{-6}$ $M_{\odot}$/year and high velocity ejection are required for the rocket effect to explain the apparent alternating short-term period changes. Magnetic braking can lower this required rate if the surface magnetic fields are $\geq 1$ k Gauss. This could imply a different spin-orbit coupling which involves the interaction of the magnetic fields of the two components. De Campli and Baliunas have made an important first step computing the synchronization time (about $10^3$ years for RS CVn and AR Lac) on the basis of the effect of magnetic reconnection in non-synchronously rotating binaries suggested by Bahcall et al (1973). Since this synchronizing magnetic torque may be stronger than the de-synchronizing magnetic braking torque, large orbital period changes may be allowed on that time scale.

Scharlemann (1981, 1982) considering tidal coupling in differentially rotating binary stars, found that the coupling torque is sufficient to explain the nearly synchronous rotation, but may not be able to explain the very long observed periods for the migration of the photometric wave. Better agreement is found on the hypothesis of angular momentum loss directly from the orbit.

It would be very interesting to tackle further this complex problem in order to try to explain the period changes of those systems like SS Boo, SV Cam, RS CVn, CG Cyg, RT Lac and AR Lac which exhibit both period increases and decreases. Anyway in this context a much closer analysis would be welcome in order to see if there exists any relation between the period changes and the activity parameters of RS CVn binaries.

Now I would like to mention two cases for which both the wave migration and the period variation seem to show correlated cyclic variation. These are AR Lac (Figure 7 ) and RS CVn (Blanco et al 1982a). In both systems rotation periods smaller than the orbital period (wave migrating toward decreasing phases) correspond to an orbital period decrease and viceversa.

RT Lac, for which data on the wave migration and period variation are available for about 80 years, does not seem to show such a behaviour. Unfortunately there are not enough observations on the wave migration for the other systems exhibiting cyclic period changes. Therefore let me call for new continuous observations on this group of RS CVn systems.

Finally let me emphasize a number of facts arguing against the hypothesis that the migration of the wave with respect to the orbital period may represent a differential rotation of the spotted star. They can be summarized as follows:

1) The relation between the wave migration rate $\Delta P/P$ and the orbital period can result from a forced synchronization.
2) The excess of migration of the waves toward decreasing phases and the corresponding excess of decreasing periods.
3) The correspondence on the cyclic variation of the orbital period and the wave migration cycle.
4) The discordance between the latitude drift of the spotted areas deduced from light curve modelling and from the wave migration curve.

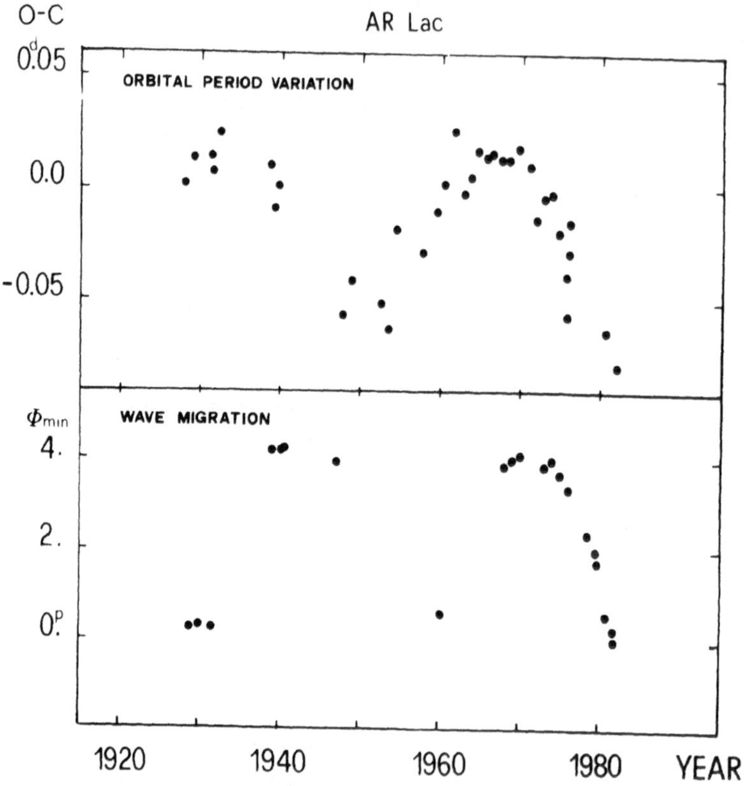


Figure 7. O-C residuals of the orbital period variations (Top) and the observed phases of the wave minimum $\phi_{min}$ (Bottom) for AR Lac (Blanco et al 1982b) . O-C residuals are averaged values from Hall and Kreiner (1980).

5) The difficulty in explaining traveling times of the wave on the light curve as long as 10 years (Uchida and Sakurai 1982).

ACKNOWLEDGEMENTS

I should like to thank the many Colleagues that helped me sending published and unpublished material, the Catania Colleagues C. Blanco, E. Marilli and M. Rodonò for allowing me to include in this review several data of a common work on RS CVn stars before publication and finally P. Brendan Byrne for a critical reading of the manuscript. This work has been partially supported by the CNR-Gruppo Nazionale di Astronomia under contract N. 81.00932.02.

REFERENCES

Awadalla, N.S., Budding, E.: 1979, Astrophys.Space Sci 63, 479

Bahacall, J.N., Rosenblath, M.N., Kulsrud, R.M.: 1973, Nature 243, 27
Bartolini, C., Blanco, C., Catalano, S., Cerruti-Sola, M., Eaton, J.A., Guarnieri, A., Hall, D.S., Henry, G.W., Opkins, J.L., Landis, H.J., Louth, H., Marilli, E., Piccioni, A., Renner, T.R., Rodonò, M., Scaltriti, F.: 1982, Astron. Astrophys., Submitted
Belvedere, G., Paternò, L., Stix, M.: 1980a, Astron. Astrophys. 91, 328
Belvedere, G., Paternò, L., Stix, M.: 1980b, Astron. Astrophys. 88, 240
Binnendijk,K.L.:1970, Vistas in Astronomy 12, 217
Blanco, C., Catalano, S., Marilli, E.: 1979, Astron. Astrophys. Suppl. Series, 36, 297
Blanco, C., Catalano, S., Marilli, E., Rodonò, M.: 1980, Mem. Soc. Astron. Ital. 51, 673
Blanco, C., Catalano, S., Marilli, E., Rodonò, M.: 1982, (in preparation)
Blanco, C., Catalano, S., Marille, E., Rodonò, M.: 1983, This volume.
Bopp, B.W., Espenak, F.: 1977, Astron.J. 82, 916
Bopp, B.W., Fekel, F.C.jr., Hall, D.S., Henry, G.W., Noah, P.V., Africano, J., Wilkerson, M.S., Beavers, W.I.: 1982, Astron J. (in the press)
Boyd, J., Chambliss, C.R., Eaton, J.A., Fried, P., Hall, D.S., Henry, G.W., Landis, H.J., Louth, H.: 1981 (preprint)
Budding, E., Kadouri, T.H.: 1982, (preprint)
Byrne, P.B.: 1983, This volume.
Catalano, S., Frisina, A., Rodonò, M.: 1980, I.A.U. Symposium N° 88, 405
Catalano, S., Marilli, E.: 1983, This volume.
Catalano, S., Rodonò, M.: 1974, Publ. Astron. Soc. Pac. 86, 390
De Campli, W.M., Baliunas, S.L.: 1979, Astrophys.J. 230, 815
Dorren, J.D., Siah, M.J., Guinan, E.F., Mc Cook, G.P.: 1981, Astron. J. 86, 572
Duncan, D.K.: 1981, Astrophys. J. 248, 651
Eaton,J.A., Wu, C.C., Rucinski, S.M.: 1980, Astrophys.J.239, 919
Ertan, A.Y., Tumer, O., Tunca, Z., Ibanoglu, C., Kuruta, C.M., Evren, S.: 1982, (preprint)
Feldman, P.A., Taylor, A.R., Gregory, P.C., Seequist, E.R., Balonek, J.J., Cohen, N.L.: 1978, Astron.J. 83, 1471
Geyer, E.H.: 1980, Symposium 88, 423
Gibson, D.M., Hjellming, R.M., Owen, F.N.: 1975, Astrophys. J. 220, L99
Glatzmaier, G.A., Gilman, P.A.: 1981, in "Solar Phenomena in Stars and Stellar System", eds. Bonnet, R.M., Dupree, A.K., Reidel, Dordrecht, p. 145
Gorza, W.L., Heard, J.F.: 1971, Publ. David Dunlap, Obs. 3, 107
Guinan, E.F., Dorren, J.D., Siah, M.J., Koch, R.H.: 1979, Astrophys. J., 229, 296
Guinan, E.F., Mc Cook, G.P., Fragola, G.L., O'Donnell, W.C., Weisenberger, A.G.: 1981, Publ. Astron. Soc. Pac. 93, 495

Hall, D.S.: 1972, Publ. Astron. Soc. Pac. 84, 323
Hall, D.S.: 1976, I.A.U. Colloquium N° 29, 287
Hall, D.S., Kreiner, J.M.: 1980, Acta Astr. 30, 387
Hall, D.S., Kreiner, J.M., Shore, S.N.: 1980, I.A.U. Symposium N° 88, 383
Hall, D.S.: 1981, in "Solar Phenomena in Stars and Stellar Systems" Bonnet, R.M., and Dupree, A.K. eds. Reidel, Dordrecht, p.431
Hartmann, L.: 1981, in "Solar Phenomena in Stars and Stellar Systems" Bonnet, R.M. and Dupree, A.K. eds. , Reidel, Dordrecht, p.487
Hartmann, L. Londono, C., Phillips, M.J.: 1979, Astrophys. J., 229, 183
Hilditch, R.W., Harland, D.M., Mc Lean, B.J.: 1979, Montly Not. Roy. Astron. Soc. 187, 797
Landis, H.J., Lovell, L.P., Hall, D.S., Henry, G.W., Renner, T.R.: 1978, Astron. J. 83, 176
Ludington, E.W.: 1978, Ph.D.Thesis,University of Florida, Gainesville, Florida
Mílano, L.: 1981, in "Photometric and Spectroscopic Binary Systems", Carling E.B., Kopal, Z. eds., Reidel, Dordrecht, p. 331
Milano, L., Russo, G., Mancuso, S.: 1981, Astron.Astrophys. 103, 57
Milone, E.F., Castle, K.G., Robb, R.M., Swadson, D., Burke, E.W., Hall, D.S., Michlovic, J.E., Zissell, R.E.: 1979, Astron. J. 84, 417
Morgan, J.C., Eggleton, P.P.: 1979, Montly Not. Roy. Astron. Soc. 187, 661
Olah, K.: 1979, Inf. Bull. on Var. Stars N° 1717
Pallavicini, R., Golub, L., Rosner, R., Vaiana, G.S., Ayres, T.R., Linsky, J.L.: 1981, Astrophys.J. 248, 279
Patkos, L.: 1981, Astrophys. Lett. 22, 1
Percy, J.R., Welch, D.L.: 1982, J. Roy Astron. Can. 76, 185
Phillips, M.J., Hartmann, L.: 1978, Astrophys. J. 224, 182
Popper, D.M., Ulrich, R.K.: 1977, Astrophys. J., 212, 131
Ramsey, L.W., Nations, H.L.: 1980, Astrophys. J. 239, L31
Reglero, V.: 1982, Astrophys.Space Sci. (submitted)
Rodonò, M.: 1981, in "Photometric and Spectroscopic Binary System", Carling, E.B., Kopal, Z. eds., Reidel, Dordrecht, p. 285
Rodonò, M.: 1982, in Proc.XXIV COSPAR, Topical Meeting on "Stellar Chromosphere and Coronae", Pergamon Press, London, in press
Rodonò, M., Pazzani, V., Cutispoto, G.: 1983, This volume.
Russo, G., Milano, L., Mancuso, S.: 1983, This volume.
Sarma, M.B.K.: 1976, Bull. Astron. Inst. Czechoslovakia 27, 335
Scharlemann, E.T.: 1981, Astrophys. J., 246, 305
Scharlemann, E.T.: 1982, Astrophys. J. 253, 298
Shore, S.N., Hall, D.S.: 1980, I.A.U. Symposium N° 88, 389
Skumanich, A.: 1972, Astrophys. J. 171, 565
Simon, T., Linsky, J.L., Schriffer, F.H.: 1980, Astrophys.J. 239, 911

Uchida, Y., Sakurai, T.: 1983, This volume.
Van Woerden, H.: 1957, Ann.Sternw.Leiden 21, p.3
Vaughan, A.H., Baliunas, S.L., Middelkopp, F., Hartmann, L.W., Mihalas, D., Noyes, R., Preston, G.W.: 1981, Astrophys.J. 250, 276
Vivekananda Rao, P., Sarma, M.B.K.: 1981, in "Photometric and Spectroscopic Binary Systems", Carling, E.B., Kopal, Z. Eds., Reidel, Dordrecht, p.305
Walter, F.M., Bowyer, C.S.: 1981, Astrophys.J. 245, 677
Wilson, O.C.: 1978, Astrophys.J. 226, 379
Zeilik, M., Elston, R., Henson, G., Smith, P.: 1981, Inf.Bull. on Var. Stars n° 2006
Zeilik, M., Elston, R., Henson, G., Schmolke, P., Smith, P.: 1982a, Inf. Bull. on Var.Stars n° 2090
Zeilik, M., Elston, R., Henson, G., Schmolke, P., Smith, P.: 1982b, Inf. Bull. on Var.Stars n° 2089

## DISCUSSION

Belvedere: I can add a theoretical argument in favour of your different interpretation of the migration of the minimum. If we accept an interpretation based on differential rotation then the rates of differential rotation implied are too small to give rise to an adequate dynamo action. This would not be in keeping with the levels of activity observed.

van Leeuwen: I would like to make a few remarks. These are with respect to the measurements presented by Torres on AU Mic, which showed that the minimum of that star stays at the same place. On your interpretation this would mean that spots would always arise on the same side of the star. Is that right?

Catalano: This would mean that this star is very highly synchronous.

van Leeuwen: But AU Mic is a single star!

Catalano: Can I ask to what you refer the positions of the maxima or the minima? If you measure periods using the maxima and the minima you get different results. One is considered constant and the other changing. This is similar to what we see in the case of II Peg.

van Leeuwen: If one calculates a period from the minima of AU Mic one gets a very good period fit while using the maxima the fit is very poor. So I wonder what keeps the minima at the same place in the spot model?

Catalano: I don't know. The problem for those modelling the RS CVn and

BY Dra stars is why there is this preferred longitude like the Red Spot on Jupiter. On the Sun there are symmetries in the longitude distribution of the spots which are not so strong as in the case of these stars. This is a general problem and perhaps the answer can come from this meeting.

van Leeuwen: There is one other thing I would like to mention. You found no correlation between amplitude and period whereas such a relationship is obviously present in the observations of the Pleiades K-type stars. I believe that the reason that you do not see it is the spread in the ages and the masses of the stars that you took as your sample. If you take stars of the same age and mass then you find a very clear relation between amplitude and period.

Catalano: This may be a statistical problem. We have only about 6 known RS CVn stars. Of those only a few have been observed for a long enough time to see changes in their light curves. Such relationship may also be hidden by other effects. The approach we have adopted is a photometric one. Perhaps spectroscopic observations for instance can give us a better view.

van Leeuwen: Have you considered the possibility of precession of the stars' rotational axis as being the cause of the amplitude changes?

Catalano: No. It would be very difficult to have this effect. You need to have the rotational axis inclined with respect to the normal to the orbital plane. This is a complication which I did not consider.

Vogt: The spots we see appear relatively stable over long periods of time, longer than you would expect if you consider the kind of differential rotation which we see in the Sun. They show the same kind of stability as solar complexes. This would suggest that we are seeing the effect of one internal, radial dependence of rotation rate rather than the latitudinal shear which we observe on the Sun. The other thing I wish to say is that the mass loss rates from these regions derived by Hall is too large to be believable. The analysis of Baliunas and De Campli involving magnetic fields suggest a much lower mass loss rate and still provide the observed period changes. Is not this what you would expect if these spots were large-scale unipolar regions showing the same kind of mass loss that occurs in solar unipolar regions? Does not this provide evidence that these large spots are just filled-in analogues of solar complexes rather than something similar to sunspots?

Catalano: Baliunas considered two case of period change. One is on a short timescale. Their result is that the required mass loss is as great as Hall and Kreiner's model . They conclude that these short period changes are due to a rocket effect and may not be real. The longer period changes they treat very briefly. They just try to calculate the synchronization times for systems like RS CVn or AR Lac. They found a 1000 year timescale for synchronization. If we have the two effects, i.e. the braking effect on the spotted star and the reaction on the other star, then period change would be expected. So on this view we would expect some relation between the change of the period and the rotation of the star that we deduce from the periodic light variation.

Feldman: In superimposing your data on the diagram of Vogt (this volume, Fig.6) I noticed that there were a few points that had activity cycles of order 10 years for rotation periods less than 7 days. Could you comment on those particular systems?

Catalano: It was already pointed out by Vogt in his paper that there appeared to be some kind of relationship between the cycle period and the rotation periods less than about 7 days. You must ask Vogt for details of these. I believe the cycles were observed in variation of the K-line for rotation period larger than 7 days. But as I showed in Table 2 cycles deduced from median brightness, amplitude and wave migration variations in some case are different one another. Some os the systems you are referring to is the same plotted by Vogt, but the cycle is deduced from a different parameter.

Rucinski: I would like to point out that we have an interesting problem here. With regard to the changes in the migration rate, the short period systems seem to be better synchronized. On the other hand we have heard that from a theoretical point of view that these stars cannot generate enough magnetic field to account for the observed activity. This is important, since your period dependencies looked most impressive.

# RS CVN STARS: CHROMOSPHERIC PHENOMENA

Bernard W. Bopp
University of Toledo

ABSTRACT: The observational information regarding chromospheric emission features in surface-active RS CVn stars is reviewed. Three optical features are considered in detail: CaII H and K, Balmer H-alpha and HeI λ10830. While the qualitative behavior of these lines is in accord with solar-analogy/rotation-activity ideas, the quantitative variation and scaling is very poorly understood. In many cases, the spectroscopic observations with sufficient S/N and resolution to decide these questions have simply not yet been made. The FK Com stars, in particular, present us with extreme examples of rotation that may well press our understanding of surface activity to its limits.

## INTRODUCTION

The intense chromospheric activity characteristic of the RS CVn stars is perhaps most clearly indicated by the presence of optical and ultraviolet emission lines. In a complete discussion, features such as CaII H and K; H-alpha; MgII h and k; Ly-alpha; HeI λ10830, λ5876; and UV resonance lines of OI, CI, CII, and SiII might be included. Constraints of time will force me to cover only a few of these in depth. In addition, I will discuss some chromospherically active objects that I believe to be of particular importance to our understanding of stellar activity. My biases will be plain: I will concentrate heavily on the optical chromospheric spectrum, since this is my area of expertise, and I anticipate that the UV will be emphasized by others with greater eloquence at this meeting.

## CALCIUM II H AND K

Historically the CaII H and K lines were the first emission features to be recognized as indicating the presence of a stellar chromosphere. The lines are unquestionably our most valuable and intensively observed optical chromospheric signature. Our association



*P. B. Byrne and M. Rodonò (eds.), Activity in Red-Dwarf Stars, 363–377.*

of CaII emission and RS CVn stars is so intimate that it is almost a definition; certainly it is part of Hall's (1976) by now classic definition of the group: "...the orbital period is in the range one day to two weeks, the hotter component is of spectral type F or G and luminosity class V or IV, and strong H and K emission is seen outside of eclipse." Despite our wealth of new observational data in the last 6 or 7 years, this definition has held up exceedingly well. If we were to substitute the word "rotational" for "orbital" and perhaps increase the two week limit, I'd be very content with this. Bear in mind that I use the term "RS CVn" a bit loosely throughout this review.

A problem that immediately arises is how to discuss the "strength" of the CaII emission. Certainly the RS CVn's have reversals much stronger than those seen in ordinary (usually single) giants or subgiants, but it is important to quantify this statement. Wilson has published extensive lists of eye-estimated emission intensities, on a scale from 0 (no emission present) to 5 (strongest emission); the latest compilation is Wilson (1976). These estimates are from photographic spectra, with all of their limitations. In addition, any intensity estimate (or even a measurement from a tracing) is strongly dependent on resolution: if dispersions significantly lower than that used by Wilson are employed, the CaII emission will appear weaker with respect to the surrounding continuum. Lastly, this intensity scale is not terribly useful for the extremely active RS CVn's, as it "saturates" at the high end: nearly all the RS CVn's of type G-K have emission intensities of 5, but the lines are not equally strong nor are the stars equally active.

More quantitative (photoelectric) measures of H and K flux in giants have been recently published by Wilson (1982). He defines the flux at a wavelength $\lambda$ by:

$$F_\lambda = E_\lambda / E_C$$

where $E_\lambda$ is the energy in the 1 Å wide data channel and $E_C$ is the energy in a nearby continuum bandpass. Wilson includes his earlier photographic intensity measurements in this paper, and we can thus obtain a rough calibration of estimated and measured flux for G-K III-IV stars (Table I). It is perhaps of interest that sigma, the scatter in the flux measurement, increases as the flux increases, but whether this indicates real chromospheric variability or merely difficulties with accurately estimating I(CaII) is not clear.

However, most of the RS CVn's have CaII emission much stronger than I=4. In the case of $\sigma$Gem, for example, the CaII reaches nearly to the surrounding continuum (Figure 1). This star has I=5 on Wilson's scale (as does V711 Tau), but this is hardly descriptive enough for emission so intense. The best and most astrophysically significant description of CaII emission is in terms of absolute surface flux. The most extensive compilation of such data is that of Linsky et al. (1979). The technique requires reasonably high resolution and S/N, but the

Table I
I(CaII) vs. $F_K$
F-G III-IV

| I(CaII) | $F_K$ | σ | N |
|---|---|---|---|
| 1 | 0.143 | 0.018 | 57 |
| 2 | 0.152 | 0.027 | 38 |
| 3 | 0.188 | 0.033 | 25 |
| 4 | 0.236 | 0.045 | 19 |

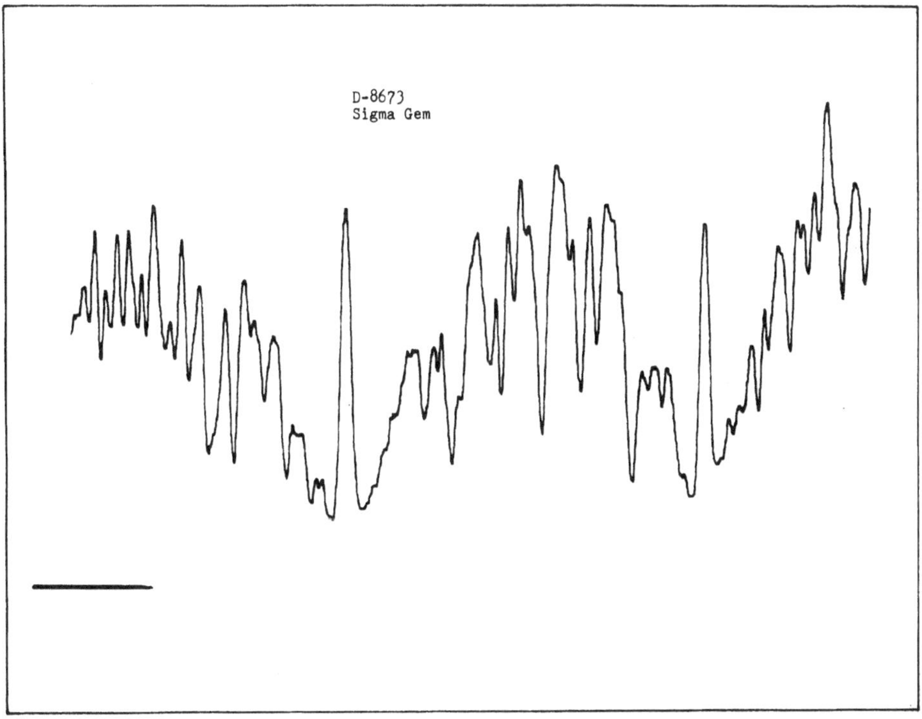


Figure 1: Density tracing of the CaII H and K region in the active-chromosphere giant σ Gem. Original dispersion 13 Å/mm. The clear plate level is indicated by the horizontal line.

information that can be derived is invaluable in the construction of model stellar chromospheres. The surface flux at H and K also yields a measure of emission/chromospheric-activity that is not biased by the color of the star. For example, Linsky et al. note that the surface fluxes at K in αSco (M1Ib,I=4) are actually two orders of magnitude less than in αCom AB (F5V,I=1).

A few years ago we began a project at Toledo to obtain CaII flux measurements of RS CVn stars and look for correlations between chromospheric radiative losses and rotation rate. We obtained data at KPNO, and were able to borrow and analyze a number of spectrograms obtained at Lick and McDonald by Popper and Fekel. Some of the new results are presented in Table II. A comparison of our values for those stars also observed by Linsky et al. shows agreement in the fluxes to about 10%. I caution that due to possible errors in spectrophotometry, estimates in V-R, and the problems due to the continuum contribution from a secondary star, the flux measurements are good to perhaps 50%.

Table II
CaII Surface Fluxes of RS CVn Stars

| Star | V-R* | $F(K_2)$ ergs/cm$^2$/s | $F(H_2)$ ergs/cm$^2$/s | N |
|---|---|---|---|---|
| RS CVn | (0.72) | 1.2(+6) | 9.7(+5) | 1 |
| TY Pyx | (0.55) | 3.2(+6) | 3.1(+6) | 1 |
| LX Per | (0.72) | 1.3(+6) | 8.6(+5) | 1 |
| SZ Psc | (0.70) | 3.3(+6) | 3.6(+6) | 1 |
| WW Dra | (0.72) | 2.2(+6) | 1.6(+6) | 1 |
| AR Lac | (0.77) | 1.4(+6) | 1.2(+6) | 1 |
| RT Lac | (0.77) | 4.2(+6) | 3.9(+6) | 1 |
| UX Ari | (0.72) | 4.3(+6) | 3.4(+6) | 1 |
| σ Gem | 0.90 | 7.0(+5) | 5.7(+5) | 2 |
| λ And | 0.78 | 1.7(+6) | 1.7(+6) | 1 |
| ζ And | 0.84 | 8.7(+5) | 1.1(+6) | 1 |
| V711 Tau | (0.77) | 2.9(+6) | 2.6(+6) | 12 |
| HR 4665 | (0.77) | 2.7(+6) | 2.3(+6) | 4 |
| HR 6469 | (0.47) | 6.4(+6) | 5.2(+6) | 1 |
| HR 7275 | (0.77) | 2.7(+6) | 2.1(+6) | 1 |
| HR 8703 | (0.90) | 7.2(+6) | 6.0(+6) | 1 |
| HD 81410 | (0.80) | 2.9(+6) | 3.0(+6) | 1 |
| HD 82210 | 0.64 | 2.0(+6) | 1.8(+6) | 2 |
| HD 86590 | (0.54) | 1.1(+7) | 7.5(+6) | 1 |
| Sun | 0.53 | 4.6(+5) | 3.8(+5) | |

*Estimates in parenthesis

Two results from these data are worth noting:

1. There was no significant variation in the CaII flux of V711 Tau on 12 spectrograms.

2. No strong correlation was found between chromospheric loss rate at H and K (or H-alpha) and rotational velocity among the RS CVn stars.

A new and extensive spectroscopic study of southern hemisphere RS CVn stars has recently been made by Collier (1982a) who finds a correlation (with much scatter) of CaII surface flux and rotation rate for eight objects.

It is of great importance to determine whether CaII intensity or flux in RS CVn's is modulated by stellar rotation period, implying concentration of emission in active regions. This has been assumed by many to be the case for the RS CVn's; I find the evidence for this behavior compelling, but I am not convinced it is always so. Certainly H and K flux measures have been used effectively to derive rotation periods in late-type MS stars (Vaughan et al. 1981), but Vaughan and colleagues found no clear evidence for periodic modulation of H and K flux in 6 giants.

One of the first investigations of CaII variability in RS CVn's was by Weiler (1978). His oft-quoted result was that there was a correlation between emission intensity and phase in UX Ari, RS CVn, and Z Her. However, Weiler also states that this correlation was only "moderately strong" for RS CVn itself, and marginal for UX Ari and Z Her. Three other binaries showed only random emission intensity variations.

Perhaps the best data, and the best evidence for periodic CaII variations are the results for $\lambda$And given by Baliunas and Dupree (1982). Their observations showed that at light minimum the relative flux at K was larger by 25-100% than at light maximum. In addition, flux changes were accompanied by complex profile changes, with $V/R < 1$ occuring only near light maximum. The behavior of the MgII h and k asymmetries as a function of photometric phase was consistent with that of CaII.

Possibly complicating this analysis are the observations of rapid variability of CaII by Baliunas et al. (1981). Their interesting results showed fluctuations of up to 7% in the H and K emission of $\varepsilon$ Eri, a young active K2V star. K-line variations of as much as 10%, on a time scale of 10 minutes, were seen in $\lambda$ And. The authors conclude that much of the rapid variability is related to flaring activity; the brightenings observed at H and K in these stars is, however, several orders of magnitude larger than the total radiative output of typical solar flares. Apparently large portions of the chromospheric emission come from coherent active regions, of size ~1% of a stellar radius.

## H-ALPHA

The Balmer H-alpha line is perhaps the most accessible chromospheric diagnostic in late-type stars. With the use of red-sensitive detectors (the peak QE of S-20 image tubes, Reticons, and CCD's is very near H-alpha) it is particularly easy to observe, and a reference continuum level may be readily established (unlike the situation near the CaII H and K lines). We associate the presence of H-alpha emission in late type dwarfs, giants, and subgiants with strong chromospheric activity. For example, all the dMe stars (where the "e" denotes the presence of Balmer emission) show stellar flare activity, and the quiescent H-alpha emission strength is well correlated with the frequency of photometric flaring (Gershberg and Shakovskaya 1971).

The situation, however, is not as clear in the case of the RS CVn binaries. I know of only four RS CVn's that show H-alpha as a pure emission feature above continuum at all times: V711 Tau (HR 1099), UX Ari, II Peg, and DM UMa (BD + 61° 1211). Although these are clearly among the most active RS CVn's, there are other systems, with comparable CaII H and K flux and orbital period (e.g., AR Lac) which show the H-alpha feature in absorption. In fact, emission at H-alpha is rare in RS CVn's: the two dozen systems that we surveyed at 1.5 A resolution (Bopp and Talcott 1978) showed only five emitters, and in three of these (HK Lac, SZ Psc, HD 86590) the emission was present only 10-20% of the time.

Certainly the H-alpha feature is enhanced during radio-flaring intervals. During what we now recognize as a rather small (~150 mJy) radio flare on V711 Tau in September 1976, Weiler _et al_. (1978) reported an increase in H-alpha EW of about a factor of three. During a very strong radio flare (peaking near 1000 mJy) in February 1978, both Popper (1978) and Fraquelli (1978) reported H-alpha EW enhancements of 3-5X. The H-alpha emission profile during the 1978 flare of V711 Tau was very broad (~400 km/s total width) and showed a pronounced redward asymmetry (Furenlid and Young 1978, Hearnshaw 1978). Fraquelli (1982), using H-alpha data obtained with concurrent radio monitoring over an interval of several years, demonstrates that there is a good correlation between the H-alpha flux and the log of the radio flux from V711 Tau.

It is significant that during radio outbursts the H-alpha emission behavior tracks with the radio flux, and is not modulated by orbital or rotational period. During the 1976 event, the H-alpha line rose to maximum EW in unison with the radio (and Ly-alpha) flux, then slowly faded on a time scale comparable to that of the radio decay time (several days). The H-alpha emission showed no modulation with the 2.8 day rotational/orbital period, as if it was produced over a very large area of the star, and was not concentrated near the active region that was presumably the site of the outburst.

Regarding quiescent intervals, if solar analogies are invoked, we might expect H-alpha emission in RS CVn stars to vary in antiphase with the V-band measures (or at least to show some modulation with orbital period). The observational evidence for this effect is limited. Fraquelli (1982) finds no correlation between H-alpha flux in V711 Tau and distortion wave minimum, and similar results (with more limited data and lower resolution) are reported by Bopp and Talcott (1978). In that same paper, we also showed a phase-dependent variation of H-alpha emission in UX Ari where V minimum was seen near emission maximum. This result is often referred to as support for "classical" ideas about the distribution of H-alpha emission in active regions. I certainly stand by our data, but I would like very much to re-do the UX Ari observations with higher spectral resolution and S/N--the contribution of the H-alpha absorption profile in the hotter primary of the system needs to be taken into account properly.

Certainly there are (quiescent?) intervals when the H-alpha emission in RS CVn's is "well behaved": in 1979 the EW of the line in V711 Tau did vary in antiphase with the distortion wave (Nations and Ramsey 1980) and phase dependent variability of H-alpha has been seen in II Peg (Bopp and Noah 1980). However, these are more the exception than the rule. In general the emission behavior is stochastic rather than regular, with variability present on time scales of at least a few minutes. The narrow-band H-alpha photometry of V711 Tau by Dorren et al. (1981) presents perhaps the best overview: their observations over several observing seasons showed no phase dependence of the emission in 1977-78 but some evidence of coherent variations in 1979. Overall, however, their H-alpha index was characterized by large scatter, not observational in origin. Flares which increased the index by 0.03-0.04 mag. were seen, and significant variability on a time scale of minutes was frequently present. This last point concerning rapid variability has been very neglected by spectroscopists. About the extent of the spectral data in this regard is my suspicion (Bopp 1979) that V711 Tau showed rapid (~20 minute) changes in H-alpha EW of ±20%, but this is in need of confirmation. I did not have sufficient resolution to look for rapid profile changes; do they exist?

In those RS CVn's that show sporadic H-alpha emission (that is, where emission above continuum is seen perhaps 10-20% of the time) the behavior of the feature is every bit as puzzling. The two systems I single out for discussion are SZ Psc (P = 3.97 days, sp. K1IV+F8V) and HK Lac (P = 24.4 days, sp. K0III+F?). In the case of SZ Psc, the H-alpha line in the active K-star is usually a weak absorption feature, presumably partly filled by chromospheric emission. During a single month in 1978, however, the line profile changed to a broad (~300 k/s) double peaked emission (Bopp 1981). I was unable to explain this in the context of surface activity; in particular, I felt the time scale of the behavior (weeks) ruled out a flaring origin. Instead, I proposed that the emission was circumstellar in origin, produced from a ring or shell of gas around the K-star, ultimately originating in a transient mass-transfer event. Ramsey and Nations (1981), however, observed another H-alpha outburst in this system, similar in duration, and were able to fit their data into the general context of a surface outburst. Still, it was necessary to postulate transient ring or disk structure in the system.

HK Lac, a system with a markedly different orbital configuration, may be capable of producing such structures also: one spectrogram obtained in 1977 (Bopp and Talcott 1980) showed a double peaked H-alpha emission line, identical in appearance to the SZ Psc feature. In the 1980 paper, we attempted, perhaps not entirely convincingly, to ascribe this profile in HK Lac to a sporadic prominence-like ejection of material. But the resemblance of the profiles is so striking that I wonder if shell/ring structures are not a possibility in HK Lac also.

Finally, I should not fail to mention the H-alpha absorption line behavior; this after all is the type of profile seen most commonly in RS CVn's. The low resolution surveys suggested that for those stars for which H-alpha is seen in absorption, it is indeed a normal feature, not filled in by emission, and not showing any EW variability (to a ±20% level) with phase (Bopp and Talcott 1978). However, a survey that we did a few years ago with 0.3 Å resolution showed some interesting results (Smith and Bopp 1982). We found all the RS CVn stars to have abnormal H-alpha profiles when compared with appropriate MK standards. In each case, the H-alpha core intensity was elevated relative to that of the standard (Figure 2). The H-alpha absorption lines of all the

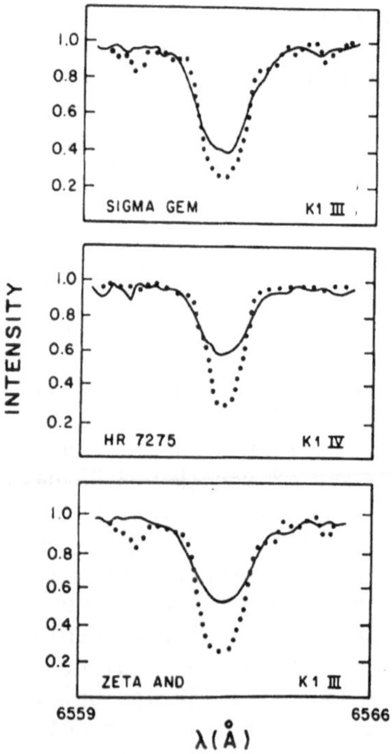


Figure 2: H-alpha profiles of RS CVn-like stars (solid lines) superimposed on those of MK standards (dotted lines).

surface-active stars are thus partially filled by chromospheric emission, though this subtlety was not noticeable at lower (~1 Å) resolution. Once again, we have no information on whether the partial filling of the H-alpha profile is variable with rotational phase, or if profile changes are evident (our high resolution observations were part of a survey program only). However, extensive, high quality H-alpha observations have been obtained by Fekel (1982) in the northern hemisphere and by Collier (1982) in the southern. I look forward to the publication of their results.

## HELIUM I λ10830

The λ10830 line is potentially one of the most important chromospheric (and perhaps even coronal) diagnostics. It is also the least-studied feature that I will discuss. The line is present as an absorption feature in the spectrum of many G and K stars. Of course in an ordinary late-type stellar atmosphere the line should not be present at all, since the metastable lower state of the transition that produces the feature lies 20 eV above the ground state. Clearly the HeI feature is produced in an overlying high temperature region, but the mode of excitation is not clear. There have been a variety of mechanisms proposed to explain the existence of λ10830 in cool stars: Athay (1965) invoked direct excitation by electron collisions in a hot chromosphere; Shine, Gerola and Linsky (1975) proposed diffusion of the helium into a high temperature region where it could be collisionally excited; Zirin (1975, 1976) argued that 10830 is excited by coronal soft X-ray emission, implying that the emission measure of stellar coronae may be determined by measurement of the line.

The only detectors that are presently capable of recording the λ10830 line with adequate resolution are S-1 image tubes and bare (unintensified) Reticons. The published Reticon data have excellent S/N; O'Brien and Lambert (1979) illustrate impressive spectra showing variable HeI emission in Arcturus, where the emission intensity is only a few percent above continuum. Definite variations in the HeI emission were seen on a time scale of about a week, and more rapid variability was suspected. In addition, O'Brien and Lambert observed variable P-Cygni structure in the line, implying variable mass outflow. They mention, but do not discuss, additional detections of HeI emission in γ Aql (K3II) and γ Dra (K5III); strong λ10830 absorption was observed in β Dra (G2II) and β Cet (K1III).

The vast majority of the observations of the λ10830 line have been made by Zirin using S-1 image tubes with photographic plates. The resolution and S/N with the S-1 are considerably lower than comparable Reticon figures, but the image tube is much faster, and is well suited for survey work. The initial photographic results were presented by Vaughan and Zirin (1968) and Zirin (1976). The latest compilation (Zirin 1982) presents data on 455 stars, including many with CaII K-line intensity estimates and known X-ray sources.

With this large a sample, a number of statistical tests are possible. Zirin finds, for example, that binary stars have much higher λ10830 absorption EW's than single stars, but the EW is not correlated with period, except that binaries with periods >200 days show no enhancement. In addition, λ10830 EW is strongly correlated with the X-ray flux of the star relative to its visual luminosity. This suggests that λ10830 measurements may be a convenient way of surveying coronal emission measures: a kind of "poor man's EINSTEIN". Zirin has also compared λ10830 EW and Wilson's (1976) K-line intensity estimates, and the correlations are reasonably good for G and K stars.

Despite these encouraging correlations, the behavior of HeI λ10830 in individual stars, and particularly RS CVn's, is bewildering. The line in the RS CVn's is nearly always seen as an absorption feature according to Zirin (1982), though weak emission of the sort seen by O'Brien and Lambert would not have been detectable photographically. The line is quite variable; changes of factors of 2-3 in EW are not uncommon. Though other stars with extended atmospheres/shells show λ10830 emission, the only RS CVn to ever show strong emission was HR 5110. The observed emission EW was 200 mA on two occasions, two years apart. A third observation a year later showed no HeI feature of any kind! Additionally, the correlation between HeI EW and CaII intensity appears to break down badly for the most extreme CaII emitters: both V711 Tau and HD 115043 have I(CaII)=5, yet the HeI absorption EW's are only 200 mA for V711 Tau (the Sun has an EW of 75-100 mA) and 0 mA for HD 115043.

Finally, it may well be that binarity may influence the behavior of this feature more than we suspect. Zirin (1981) has obtained spectrograms of several eclipses of Algol in the wavelength region of λ10830, where the spectrum is dominated by the X-ray source Algol B. He finds the EW of the HeI absorption feature to be an incredible 6000 mA, the strongest yet measured in a G-K star.

## THE FK COMAE STARS

The FK Comae stars are rapidly rotating G-K giants, all having v sin i ~75-120 km/s. The stars show no evidence of duplicity; I have set limits of <±3 km/s for the velocity variations of HD 199178 over a four year interval, and the recent velocity data on FK Com, obtained by McCarthy and Ramsey (1982), set a limit of <±5 km/s for this object. The observational data are summarized in Table III; a more detailed description of their properties may be found in Bopp(1982).

The late spectral type and rapid rotation certainly suggest that these stars should have very strong chromospheric emission features, and this is indeed the case. Bopp and Stencel (1981) report IUE observations that show FK Com and HD 199178 to exceed the "classical" RS CVn's by up to an order of magnitude in the flux from emission lines such as OI λ1300, SiII λ1530, and CII λ1335. The CaII H and K lines are similarly strong. Though the reversals do not appear particularly strong to the eye, again the appropriate quantity to examine is the absolute surface flux. I obtain, from two KPNO spectrograms taken in March 1979, concurrent with Rucinski's (1981) photometry:

$$F(K_2) = 7.1 \times 10^6 \quad \text{ergs/cm}^2\text{/s}$$
$$F(H_2) = 4.9 \times 10^6$$

These values are higher than any given by Linsky et al. (1979) and are 2-3x greater than our measured surface fluxes for the cool component of V711 Tau. I did not see any significant variability in the flux between my two spectrograms, obtained two nights apart, at photometric phases 0.15 and 0.95 when the star was at about the same visual brightness. Bolton (1977), however, finds the emission lines of H and K to be highly variable in intensity on a number of his spectrograms. The existence of phase-dependent variability of H and K in the FK Com stars is as yet uninvestigated, however.

Table III
Properties of FK Com Stars

| Name | V | Sp. | v sin i (km/s) | Photometric Period(days) | Velocity Var. (km/s) |
|---|---|---|---|---|---|
| FK Com | 8.2 | G2III | 120 | 2.400 | < ± 5 |
| HD 199178 | 7.3 | G5III/IV | ~80 | 3.337 | < ± 3 |
| UZ Lib | 9.2 | K0III | ~80 | 4.75 | < ± 25: |
| [1]HD 32918 | 8.6 | K2III | 50 | 9.55 | < ± 3 |
| [1]HD 36705 | 6.9 | G8III | 70 | 0.51 | < ± 5 |

[1]Data from Collier (1982b).

The CaII H and K emission profiles appear to be dominated by rapid rotation in all the objects. In particular, measures of H and K on my coudé spectrograms of FK Com yield values of 120 km/s for the broadening, identical to that seen in the metallic absorption features and in the MgII emission (Bopp and Stencel 1981).

I must note that the evolutionary status of the FK Com objects is puzzling. We have suggested (Bopp and Rucinski 1981) that the FK Com stars represent the further evolution of contact binary systems, evolved to coalescence (e.g., Webbink 1976). Though alternate binary models for FK Com itself have been proposed (Walter and Basri 1982), I believe the new velocity data effectively rule these out.

A disturbing aspect of the chromospheric behavior of the FK Com stars is the velocity behavior of the CaII emission. It is almost an article of faith among spectroscopists that the velocities one obtains for the chromospheric CaII emission and the photospheric absorption lines do not differ (at least at a 2-3 km/s confidence level). However, the velocities I obtain from CaII in all the FK Com stars show the lines to be shifted by ~15km/s to the red of the absorptions. I have no ready explanation for this, but it is not due to any profile asymmetry in CaII.

But far and away the biggest challenge to spectroscopists is to explain the H-alpha profile of FK Com. The line is a nightmare. Figure 3 shows a KPNO CCD observation (resolution ~0.5A) of the red region that we obtained in April 1982. The emission profile shows a

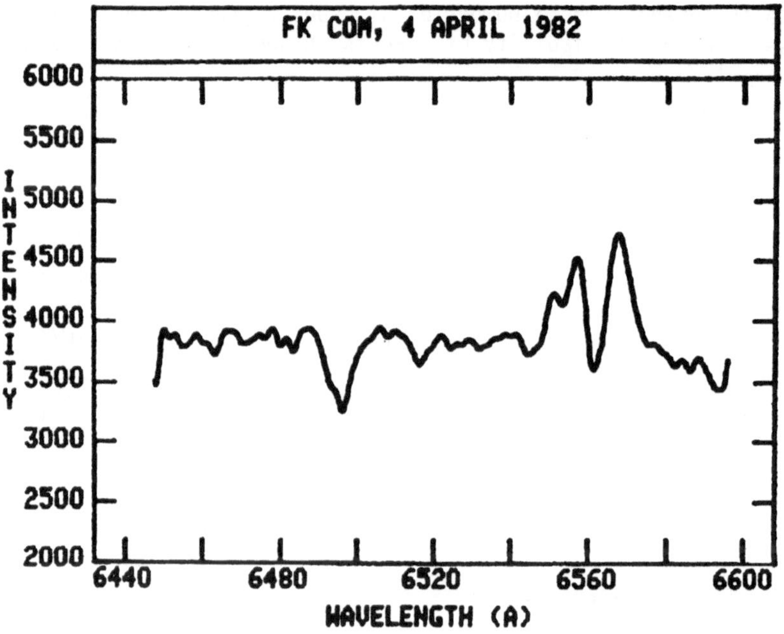


Figure 3: The red region of FK Com, from a KPNO CCD scan. Note the 20Å wide H-alpha emission feature.

full width of ~1000 km/s, a factor of two larger than the (flare enhanced) H-alpha width of V711 Tau. The line is variable in profile and intensity on a night to night time scale. In addition, H-alpha flares, where the emission EW has increased three-fold in the course of about two hours, have been reported by Walter and Basri (1982) and Ramsey and Nations (1981). At least at certain epochs, the H-alpha profile is seen to vary regularly, in the sense that the V/R emission ratio and H-alpha EW are correlated with photometric phase (Ramsey et al. 1981, Walter and Basri 1982).

With its extreme rotation, FK Com provides us with an extreme test of our rotation-activity ideas. It may well be, however, that the H-alpha profile of FK Com is telling us little about the chromosphere of this object and much about an extended atmosphere or mass-loss mechanism. Even if it is so that "one man's chromosphere is another man's extended atmosphere", an understanding of the H-alpha profile of FK Com is a spectroscopic task of the highest priority.

## ACKNOWLEDGEMENTS

I am grateful to the many colleagues that submitted copies of their latest work for inclusion in this review; I am only sorry that every paper could not be discussed. I acknowledge support from NSF (AST 81-15098) and NASA (NAGW-229).

## REFERENCES

Athay, R. G.: 1965, Astrophys. J. 145, 784.
Baliunas, S. L. and Dupree, A. K.: 1982, Astrophs. J. 252, 668.
Baliunas, S. L., Hartmann, L., Vaughan, A. H., and Dupree, A. K.: 1981, Astrophys. J. 246, 473.
Bolton, C. T.: 1977, private communication.
Bopp, B. W.: 1979, Inf. Bulletin Var. Stars, No. 1669.
Bopp, B. W.: 1981, Astron. J. 86, 771.
Bopp, B. W.: 1982, in Second Cambridge Workshop on Cool Stars, Stellar Systems and the Sun, Vol. I, p. 207.
Bopp, B. W., and Noah, P. V.: 1980, Publ. Astron. Soc. Pacific, 92, 333.
Bopp, B. W., and Rucinski, S. M.: 1981, in IAU Symposium 93, Fundamental Problems in the Theory of Stellar Evolution, ed. D. Sugimoto, D. N. Schramm, and D. Q. Lamb (Dordrecht: Reidel), p. 177.
Bopp, B. W., and Stencel, R. E.: 1981, Astrophys. J. (Letters), 247, L131.
Bopp, B. W., and Talcott, J. C.: 1978, Astron. J., 83, 1517.
Bopp, B. W., and Talcott, J. C.: 1980, Astron. J., 85, 55.
Collier, A. C.: 1982a, Ph.D.Thesis, University of Canterbury.
Collier, A. C.: 1982b, Mon. Not. Roy. Astron. Soc., in press.
Dorren, J. D., Siah, M. J., Guinan, E. F., and McCook, G. P.: 1981, Astron. J. 86, 572.
Fekel, F.: 1982, private communication.
Fraquelli, D. A.: 1978, Astron. J., 83, 1535.
Fraquelli, D. A.: 1982, Astrophys. J. (Letters), 254, L41.
Furenlid, I., and Young, A.: 1978, Astron. J., 83, 1527.
Gershberg, R. E., and Shakovskaya, N. I.: 1971, in IAU Colloq. No. 15, Veroff. Der Remeis-Sternwarte Bamberg, BD. IX, NR. 100, p. 126.
Hall, D. S.: 1976, in IAU Colloq. No. 29, Multiple Periodic Variable Stars, edited by W. S. Fitch, (Dordrecht: Reidel), p. 287.
Hearnshaw, J. B.: 1978, Astron. J., 83, 1531.
Linsky, J. Worden, S., McClintock, W., and Robertson, R.: 1979, Astrophys. J. Suppl. 41, 47.

McCarthy, J. K., and Ramsey, L. W.: 1982, in preparation.
Nations, H. L., and Ramsey, L. W.: 1980, Astron. J., 85, 1086.
O'Brien, G., and Lambert, D. L.: 1979, Astrophys. J. (Letters), 229, L33.
Popper, D. M.: 1978, Astron. J., 83, 1522.
Ramsey, L. W., and Nations, H.: 1981, Publ. Astron. Soc. Pacific, 93, 732.
Ramsey, L. W., and Nations, H. L.: 1982, Second Cambridge Workshop on Cool Stars, Stellar Systems and the Sun, Vol. I, p. 225.
Ramsey, L. W., Nations, H. L., and Barden, S.: 1981, Astrophys. J. (Letters), 251, L101.
Rucinski, S.: 1981, Astron. Astrophys. 104, 260.
Shine, R., Gerola, H., and Linsky, J.: 1975, Astrophys. J., 202, L101.
Smith, S. E., and Bopp, B. W.: 1982, Astrophys. Lett., in press.
Vaughan, A. H. and Zirin, H.: 1968, Astrophys. J., 152, 123.
Vaughan, A. H. et al.: 1981, Astrophys. J., 250, 276.
Walter, F., and Basri, G.: 1982, Astrophys. J., in press.
Webbink, R. F.: 1976, Astrophys. J., 209, 829.
Weiler, E. J.: 1978, Mon. Not. Roy. Astron. Soc., 182, 77.
Weiler, E. J., et al.: 1978, Astrophys. J., 229, 919.
Wilson, O. C.: 1976, Astrophys. J., 205, 823.
Wilson, O. C.: 1982, Astrophys. J. 257, 179.
Zirin, H.: 1975, Astrophys. J. (Letters), 199, L63.
_____. 1976, Astrophys. J. 208, 414.
_____. 1981, preprint.
_____. 1982, preprint.

DISCUSSION

Dupree: I would like to ask that caution be exercised in the use of the 10830 Å line as a coronal diagnostic. First of all, the line is very difficult to measure since it is measured using photographic plates and it is confused by blends. Secondly, the case for X-ray excitation of the line was developed only for the Sun which is a dwarf star without an extended atmosphere. It is not at all certain that it would apply to giants or subgiants. Now I have a question. A potentially interesting application of the 10830 Å measurement is in the width of the line. O'Brien's measurements suggest that the terminal velocities of these lines for supergiants and, possibly, supergiant stars agree with the terminal velocities measured from circumstellar absorption lines. My question is whether the spread in the width of the lines in the compilation of Zirin's give any indication of outflows i.e. are they broad, do they show any phase dependence, can we get any information on the wind from looking at the profiles?

Bopp: I certainly agree with your comments that this line is not yet a reliable coronal diagnostic. But I was impressed by Zirin's latest compilation and would encourage additional observations. He did not however attempt to measure velocities or widths. The 10830 feature as recorded in RS CVn and similar systems has a width which appears to be instrumental with the exception of Algol B.

Linsky: After listening to this talk I would urge observers to measure quantities that contain useful information. For instance in the Sun the Hα line is very complicated both observationally and theoretically. In a solar plage for instance the line centre is brighter than in the surroundings but about 0.5 Å off centre it is darker. So if one has a composite Hα spectrum you don't know what you are looking at. In the dMe star we know, both theoretically and observationally, that as you go from the least to the most active stars first Hα goes dark and then it gets bright. So Hα is very confusing and if we spend our valuable observing time watching this thing it may contain no information. Secondly, the Ca II line is a better line to use but may have very little contrast. In particular in the Sun we know that the rotational modulation of the Sun is small whereas in the UV and X-ray lines it is very large. I would like to show a viewgraph to stress this point (Fig.3, Andrews et al., this volume). In the case of II Peg, which has a large photometric variability, if one looks at a chromospheric line such as Mg II it will show only a very small variation of the order of 40-50%. Higher temperature lines however show variation by a factor of 5. So I think it is important to look at lines which show large variation if one wishes to study a plage.

Goldberg: With regard to He I 10830 there is a line which can tell us whether 10830 is being radiatively or collisionally excited i.e. the singlet counterpart at 20582 Å, I believe it is. It may be necessary to have a star with the right radial velocity to get the line away from $CO_2$ absorption. I wonder if anyone has tried to observe that line?

Bopp: No I am sure not. I think there would be problems with appropriate detectors. The 10830 line requires at least 0.5 Å resolution to pick it up.

Goldberg: I think the instrumentation is good enough to do it with Fourier Transform Spectrometers.

Basri: I would just like to say that I presented results at the Zurich meeting (IAU Symp. No. 102) which show no correlation between MgII and activity. I would like to caution however that using the Linsky technique i.e. using (V-R) to get CaII fluxes in RS CVn stars seems very dangerous because of contamination from the other star. I urge people to try to observe the flux directly to try to answer this question directly.

Bopp: I certainly agree. In the text of my talk I caution that the tabulated values are probably only good to 50% or so.

# DOPPLER IMAGING OF STARSPOTS

Steven S. Vogt and G. Donald Penrod
Lick Observatory, Board of Studies in Astronomy and Astrophysics
University of California Santa Cruz

## ABSTRACT

We discuss a newly-developed technique for spatially resolving starspots on some of the more rapidly rotating RS CVn stars. Basically, the method uses high resolution, very high signal-to-noise spectral line profiles and exploits the Doppler velocity correspondence between position across the stellar disk and wavelength position across a rotationally broadened line profile to synthesize an image of the star, showing the location, sizes, and shapes of its starspots. Though still in a developmental stage, the technique is already yielding information about the structure and general appearance of starspots. Examples of Doppler Imaging observations of HR 1099 will be presented, along with a movie showing the behavior of synthetic line profiles generated from a computer spot model.

## 1. INTRODUCTION

Photometric variations attributed to starspots are seen in several classes of fast-rotating late-type stars. In principle, these variations contain a great deal of useful information on the sizes, locations, and temperatures of the spot groups. In recent years, a number of attempts to extract this information have been made, with considerable success (Vogt 1981a; Dorren and Guinan 1982; Dorren et al 1981; Bopp and Noah 1980; Eaton and Hall 1979). While the studies have shown that suitably chosen spot distributions can indeed reproduce the observed light curves, all have been relatively unable to handle the problem of solution uniqueness pointed out by Vogt (1981a). The derived solution may well fit the observations, given the modeling assumptions, but a family of often very different solutions may fit just as well.



*P. B. Byrne and M. Rodonò (eds.), Activity in Red-Dwarf Stars, 379–385.*

In view of these difficulties, we have been pursuing an entirely new line of modeling of the spot distributions which, when combined with simultaneous photometric observations, may largely eliminate the problem of non-uniqueness and reveal important new information on spot shapes and movements. We call this technique Doppler Imaging.

## 2. TECHNIQUE

The Doppler Imaging technique exploits the fact that, in the spectrum of a rapidly rotating star, there exists a one-to-one correspondence between wavelength position across a spectral line profile and spatial position across the stellar disk. Lines of constant radial velocity are chords across the stellar disk parallel to the star's projected rotation axis. Thus, a unique mapping occurs in one dimension between position across the rotating stellar disk and position across the line profile. Any dark or bright region on the stellar surface will produce an associated bump or depression at the corresponding location in the line profile. As the region is carried across the stellar disk by rotation, the associated bump propagates across the line profile. Such bumps were first noticed by Fekel (1980) as line profile asymmetries in HR 1099 whose shape correlated with the phase of the photometric wave.

Figure 1 illustrates the phenomenon. Here we show a star divided into five zones of equal area, each with a characteristic and approximately constant radial velocity. The star is assumed to be rotating sufficiently rapidly that the rotational broadening is several times the width of the intrinsic profile. The left-hand sequence of Figure 1 shows the formation of a rotationally broadened spectral line on an unspotted star , as the sum of five specific intensity profiles each appropriately shifted by the radial velocity of its corresponding zone. The resultant sum produces a greatly broadened line profile whose width merely reflects the star's projected rotation velocity.

The right-hand sequence of Figure 1 illustrates the formation of a line profile on a rapidly rotating, spotted stellar surface. A black, zero flux spot is assumed to cover half of the area of Zone III. The summation of the specific intensity profiles now produces a normal rotationally broadened profile with an apparent "emission" bump in the center. Obviously, this bump does not represent true emission, but rather a lack of line absorption (in real energy units) at the radial velocity in the spectral line corresponding to the spotted Zone III. This effect is simply a geometric one, and bumps would appear almost identically in all of the star's absorption lines. The spot also produces a lowering of the continuum by 10%, but high dispersion spectroscopists rarely determine absolute continuum fluxes. Rather, they usually normalize to the continuum surrounding the line profile, as shown in the bottom frame of the right-hand sequence of Figure 1.

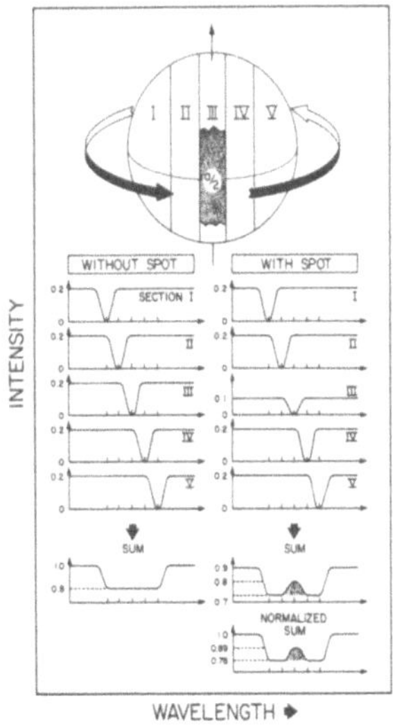


FIGURE 1

In real life, the situation is rather more complex than the idealized case of Figure 1. In general, the continuum flux and line profile from the spot itself must also be included in the computation, along with the center-to-limb dependence of both the line profile and the continuum flux (limb darkening). While these factors complicate the analysis, they can be exploited to yield more detailed spatial and temperature information on the spot groups.

Any single observation provides only one-dimensional spatial information across the disk. A series of observations, with good phase coverage, is necessary to accurately model the true two-dimensional spot distribution. Obviously, a dark spot at the pole will produce an "emission" bump at line center at all observed phases, while spots at progressively lower latitudes will exhibit progressively greater excursions across the line profile, and progressively greater rates of crossing the line profile. In practice, perhaps 6-10 observations uniformly spread across the rotation period of the star are needed to produce a reasonably valid map of the spot distribution.

The effective resolution of the technique is limited more by the rotation velocity of the star and the width of the intrinsic profile than by our instrumental resolution. Our typical instrumental resolution is about 0.15 Å, or 7 km $sec^{-1}$, while the line intrinsic widths are of order 10 km $sec^{-1}$. A projected rotation velocity of 40 km $sec^{-1}$ thus yields 7-8 resolution elements across the stellar disk.

## 3. MODELING

The actual construction of a spot distribution is an iterative process. A distribution is initially assumed, and a series of theoretical profiles are generated for comparison with the observed profiles. The spot distribution, built up from as many as 50 small circular spots, is then modified until a satisfactory fit is achieved.

The theoretical profiles are generated by program PROFILE. Specific intensity profiles are computed at 30 limb angles for both the immaculate stellar photosphere and the spotted areas, and the stellar disk is divided into a rectangular grid of typically 10,000 zones. Each zone is assigned a radial velocity (a combination of rotation and a random element of radial-tangential macroturbulence) and an appropriate specific intensity profile, which depends both on its limb-angle and whether or not it falls within the boundaries of our spot groups. The shifted profiles are then added to produce the integrated profile, which is then convolved with an instrumental profile to produce the final result. This modeling often fits our high-resolution, very high signal-to-noise spectra to better than 0.5% at all resolution elements.

## 4. OBSERVATIONS

Most of our observations are obtained with the double-pass echelle spectrograph (Soderblom et al. 1978) and an image-tube image-dissector scanner (Robinson and Wampler 1972) mounted at the coudé focus of the Lick Observatory Shane 3-m telescope. The spectrograph was typically used with an entrance slit of 0.85 arc-second, yielding a spectral resolution of 100 mÅ, each resolution element being over-sampled at 8 points. Since the detector covers only about 6.5 Å of spectrum in the 9th order, generally only one spectral line can be obtained at each setting.

Additional observations are obtained with the Reticon system (Vogt 1981b) and coudé spectrograph of the 3-m telescope. The detector is an 1872 channel unintensified silicon photodiode array, mounted at the focus of the 40" camera. The entrance slit is normally set to 0.75 arc-second, giving an effective spectral resolution of 200 mÅ, with 3 sample points per resolution element. This set-up provides about 140 Å of spectral coverage.

The spectral lines chosen for most of the observations were Fe I 6430.9 and Ca I 6439.2. These lines are relatively unblended even at the high rotation velocities of our double-lined binary program stars, and allow us to exploit the excellent red response of both detectors.

## 5. RESULTS

As the spot distributions can change on time scales of a few months, it is vital to obtain observations in a relatively short period of time. The autumn-winter cloud cover at Lick Observatory has made this a most difficult prospect during the last two seasons. Our most convincing data set is a set of 8 phases taken of HR 1099 (V711 Tau) during September-October of 1981, shown in the left-hand panel of Figure 2. This star is a well-known bright RS CVn binary with a 2.838-day orbit (Bopp and Fekel 1976). As previously noted, it was on this star that Fekel (1980) observed asymmetries in all of the stellar line profiles, which led to the development of the Doppler Imaging technique. The inclination of the rotation axis was taken to be 33° (Fekel 1982), while the projected rotation velocity was determined to be 38 km $sec^{-1}$. The stellar effective temperature was assumed to be 4700 K, appropriate for its K1 IV spectral type, and the spots were assumed to have a temperature of 3500 K (Vogt 1981a).

Our derived spot distribution (preliminary), shown in the right-hand panel of Figure 2, can be characterized as two spot groups. One, which crosses line center at phase 0.55 (determined from the ephemeris J.D. = 2442766.080 + 2.83774E), is centered at +12° latitude and is roughly circular, covering about 6-7% of the observed hemisphere. The second group consists of a large, almost circumpolar clump and a narrow filament which descends to about +30° latitude. This spot group crosses line center at roughly phase 0.25, about 110° in longitude distant from the first group, and covers about 8% of the hemisphere.

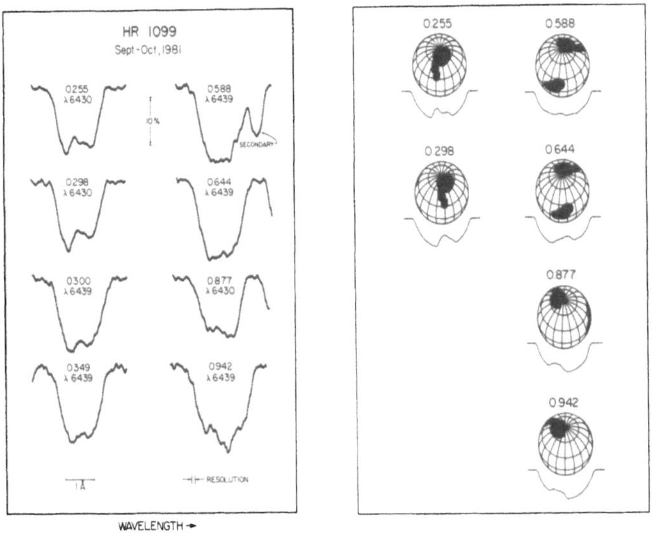


FIGURE 2

A more convincing representation of this spot distribution can be seen in our movie of the evolution of the line profiles, which will be shown by one of us (SSV) elsewhere during the colloquium. The movie is particularly valuable as a demonstration of the relative ease of finding the spot latitudes from their rates of motion across the line profiles.

## 6. DISCUSSION

A comparison of our derived spot distribution with photometric observations will be presented more fully elsewhere. The most remarkable result of this work is the derived shape of the polar spot group. Clearly, any sort of circular spot model is a poor approximation of this group. In fact, the shape of this group is remarkably similar to that of solar coronal holes, which are the most visible and spectacular features associated with the solar complexes of activity (Bumba and Howard 1965). Indeed, the similarities of many physical parameters of "starspots" and those of solar complexes are striking. Both exhibit almost rigid-body rotation, along with a slow poleward drift with time. In each case, the activity is largely confined to "active longitudes", although the starspot (starstripe) on HR 1099 is rather narrower than its solar counterpart. A solar complex dies by drifting to the pole and eventually decaying into bright plage regions. This is remarkably similar to the scenario previously suggested by Vogt (1981a) for the disappearance of a spot on BY Draconis.

The great difference between the solar complexes and starspots is the filling factor of dark spot umbrae. Presumably this simply reflects the vastly greater dynamo activity on the rapidly rotating stars. A solar complex typically has a magnetic flux of $10^{23}$Mx (Galloway and Weiss 1981). If spot fields scale with photospheric pressure, the maximum spot fields on HR 1099 should be of order 1500 G. A typical spot group would then have a magnetic flux of 2-5 x $10^{25}$Mx, hundreds of times the value of its solar analogue.

Hence, we suggest that starspots are analogous more to solar complexes than to either active regions or individual sunspots. Further observations are clearly needed to confirm this suggestion on a broader sample of stars. A more complete discussion of these results (Vogt, Penrod, and Fekel 1982) will be forthcoming in the Astrophysical Journal.

This work was funded by grant AST 79-16813 from the National Science Foundation, whose support we gratefully acknowledge.

REFERENCES

Bopp, B. W., and Fekel, F.: 1976, A. J., 81, 771.
Bopp, B. W., and Noah, P. V.: 1980, P.A.S.P., 92, 717.
Bumba, V., and Howard, R.: 1965, Ap. J., 141, 1502.
Dorren, J. D., and Guinan, E. F.: 1982, Ap. J., 252, 296.
Dorren, J. D., Siah, M. J., Guinan, E. F., and McCook, G. P.: 1981, A. J. 86, 572.
Eaton, J. A., and Hall, D. S.: 1979, Ap. J., 227, 907.
Fekel, F.: 1980, Bull. A. A. S., 12, 500.
_____. 1982, preprint.
Galloway, D. J., and Weiss, N. O.: 1981, Ap. J., 243, 945.
Robinson, L. B., and Wampler, E. J.: 1972, P. A. S. P., 84, 162.
Soderblom, D. R., Hartoog, M. R., Herbig, G. H., Mueller, F. S., Robinson, L. B., and Wampler, E. J.: 1978, in Proc. 4th Int. Coll. Ap., ed. M. Hack (Trieste: Trieste Ap. Obs.), p. 449.
Vogt, S. S.: 1981a, Ap. J., 250, 327.
_____. 1981b, S. P. I. E., 290, 70.

# TWENTY YEARS OF DEDICATED PHOTOMETRY OF RS CVn AT CATANIA OBSERVATORY

C. Blanco, S. Catalano, E. Marilli and M. Rodonò
Osservatorio Astrofisico di Catania and
Istituto di Astronomia, Università di Catania, Italy

We present a complete set of yearly seasonal light curves (LC) of RS CVn obtained from 1963 to 1982 at Catania Observatory, integrated with two LCs obtained by Ludington (1978) in 1975 and 1976. This unique observation set vividly shows (Figure 1) the migrating outside of eclipse "photometric wave" (PW), which was first evidenced by Catalano and Rodonò (1967) and it is now considered the distinctive photometric feature of RS CVn binaries as well as of other spotted stars. By attributing the PW to surface inhomogeneities, or spots, whose visibility is modulated by the spotted star rotation, the present PW migration toward decreasing orbital phases indicates that the angular rotation of the KOIV spotted component is slightly lower than the orbital one. An inspection of Figure 1 readly shows that the almost sinusoidal PW has been fairly stable from 1963 up to 1981, when a double-peaked wave, still present in the 1982 LC, developed. Most probably the "unusual" 1949 LC by Keller and Limber (1951) and the 1976 one by Ludington (1978) were obtained at a similar activity phase.

As apparent from the two upper panels in Figure 2, the migration rate of the PW has been ever increasing: from 0.06 (1963) to 0.29 (1981) orbital periods/year. Also evidence of a past migration reversal has been recently given by the present authors (Blanco et al 1982). In the same paper we have shown that the observed migration behaviour of the PW - advancing, receding and increasing rate - is qualitatively consistent with a solar-type spot-cycle drifting toward the equator in a differentially rotating star.

The highest observed value of the PW migration rate ($0^{P}.29$/year) gives a lower limit of $7.4\times10^{-3}$ for the differential rotation ($\Delta\Omega/\Omega$) between the average latitudes of spots with the lowest and the highest angular velocity. On the other hand, the fairly constant median magnitude of RS CVn outside-of-eclipses does require a remarkably stable



*P. B. Byrne and M. Rodonò (eds.), Activity in Red-Dwarf Stars, 387–390.*

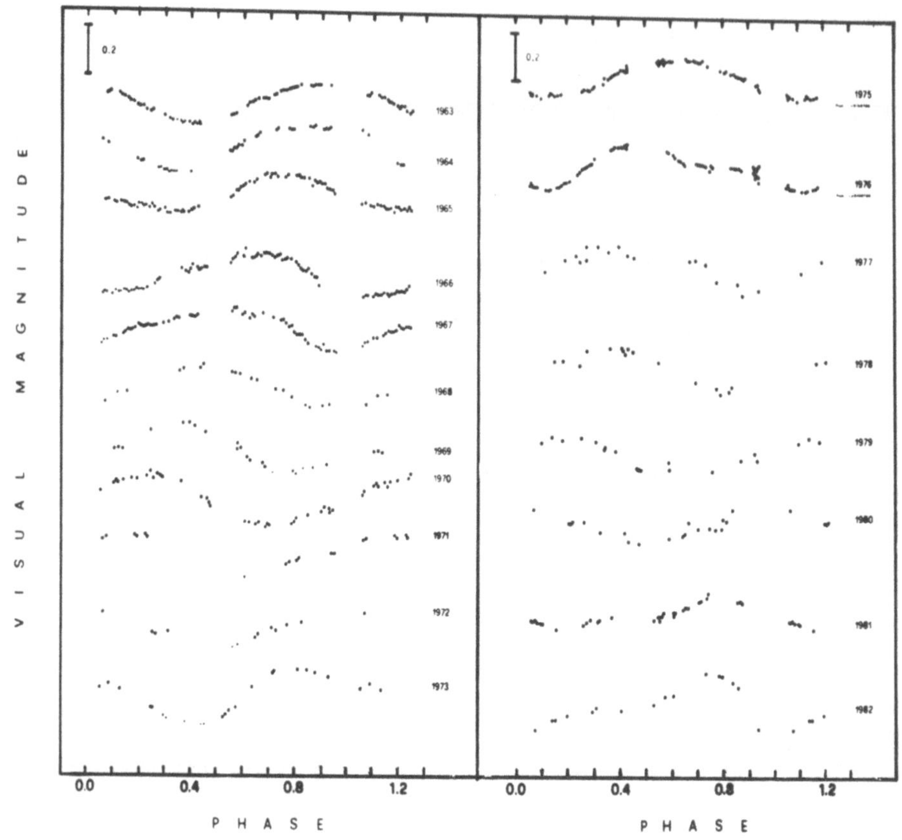


Figure 1. Outside-of-eclipse V light curves of RS CVn obtained at Catania Observatory from 1963 to 1982. The 1975 and 1976 observations are from Ludington (1978).

average spot activity level from 1963 to 1982, though some indication of a 5-year spot cycle has been presented (Rodonò 1981, Reglero 1982).

Finally, the cyclic variation of the orbital period (lower panel in Fig. 2) seems to be connected with the possibly cyclic direct and reverse migration of the PW, i.e. some dynamical interaction between the orbital motion and the differential rotation of the KOIV star producing the migration of the active regions on its surface is likely to occur (see also Rodonò 1981 and Catalano 1982).

We have summarized the main results and questions raised by the systematic observations of RS CVn carried out at Catania Observatory. Twenty years of dedicated photometry have allowed us to establish the short-term behaviour of this interesting binary and to give some indication on its long-term behaviour. As expected, observations ever

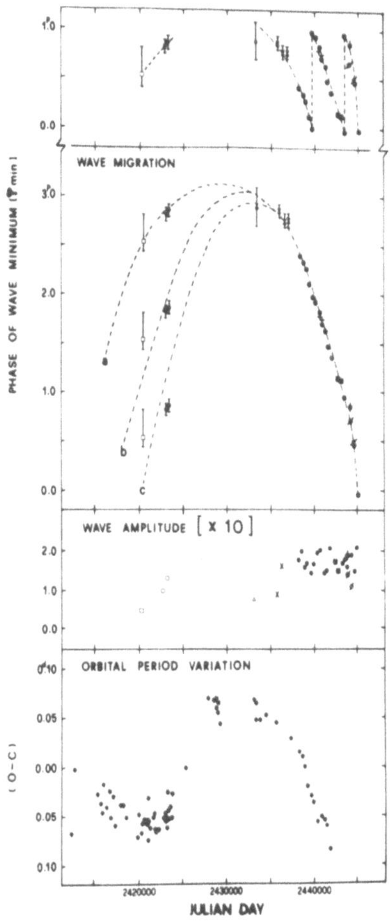


Figure 2. First panel (from top): orbital phase of the wave minimum on the light curves of RS CVn vs. Julian Date. Second panel: successive passages of the wave minimum on the light curve; three possibilities (a,b,c) are indicated. Third panel: outside-of-eclipse wave amplitude vs. JD. Fourth panel: cyclic variation of the orbital period about the mean value 4.797855 from Catalano and Rodonò (1974). The different symbols indicate different sources.

rise questions related to their interpretation and our present ones cannot be an exception. However, we are now beginning to grasp some clues on the RS CVn phenomenon and we hope that several RS CVn systems will be observationally "adopted" by other observers. Actually, only systematic long-term programs will enable us to understand the true nature and physics of RS CVn binaries and of other spotted stars.

REFERENCES

Blanco, C., Catalano, S., Marilli, E., Rodonò, M.: 1982, Astron. Astrophys. 106, 311.
Catalano, S.: 1983, This volume.
Catalano, S. Rodonò, M.: 1967, Mem. Soc. Astron. Ital. 38, 345
Catalano, S., Rodonò, M.: 1974, Publ. Astron. Soc. Pacific 86, 390
Catalano, S., Frisina, A., Rodonò, M.: 1980, IAU Symp. 88, 405

Keller, G., Limber, D.N.: 1951, Astrophys. J. 113, 637
Ludington, E.W.: 1978, Ph.D. Thesis, University of Florida, Gainesville, Florida
Reglero Velasco, V.: 1982, Astrophys. Space Sci., submitted.
Rodonò, M.: 1981, in : Spectroscopic and Photometric Binary Systems", Eds. E.B. Carling and Z. Kopal, Reidel Publ. Co., Dordrecht, p. 285

PERIODS OF UNEVOLVED LATE-TYPE CLOSE BINARIES: EVIDENCE OF MAGNETIC ACTIVITY.

G. Giuricin, F. Mardirossian, and M. Mezzetti
Astronomical Observatory of Trieste, Trieste, Italy.

ABSTRACT

Our study of the period distribution of about 200 FGKM-type unevolved close binaries has revealed a strong deficit of short-period systems.This finding may be connected with the occurrence of a very efficient mechanism of orbital angular momentum loss via magnetic braking by stellar wind, in the earliest evolutionary phases.

INTRODUCTION

Reliable clues to early evolutionary processes in close binaries may come from the distribution of the orbital periods of close binaries, provided that, of the binaries catalogued,one considers only the systems substantially unevolved. Through a survey of the literature we have gathered together the available data on the periods and primary spectral types of 79 FGKM-type eclipsing binaries for which there is sufficient information (i.e., basically, lightcurve analyses must be available) to regard these binaries as unevolved (we have in essence considered nearly main sequence detached systems). Furthemore, we have selected 120 non-eclipsing spectroscopic binaries (with periods < 1000 days), which have a high enough orbital eccentricity e ($e \geq 0.1$) or estimated mass ratio q close enough to unity ($0.7 \leq q \leq 1.3$) to be regarded as probably unevolved.

RESULTS AND DISCUSSION

A period-spectrum plot of our 79 unevolved eclipsing binaries shows a strong deficiency of short-period eclipsing pairs (with periods $\lesssim$ 1



*P. B. Byrne and M. Rodonò (eds.), Activity in Red-Dwarf Stars, 391–392.*

day ). Besides, the lower envelope to the observed periods considerably departs from the lower theoretical limits corresponding to ZAMS contact binaries with members of equal mass. Interestingly enough, this gap, which cannot be attributed to observational selection effects, markedly increases as we go from the G to the M spectral type. A similar behaviour is shown by the period-spectrum diagram of our 120 unevolved spectroscopic binaries, although in this case the real existence of the gap is not well established, since severe observational limitations act against the detection of short-period non-eclipsing spectroscopic pairs.

A reasonable explanation of our findings is that a very efficient mechanism of angular momentum loss via magnetic braking by stellar wind occurs in the earliest evolutionary phases of late-type binaries, the youngest of which are known often to display strong magnetic activity. Thus, regardless of whether the late-type contact pairs form in the pre-main sequence evolutionary phase or in later phases - this latter view seems to be favoured by preliminary theoretical calculations of close binary formation (Popova et al., 1982) - , late-type detached binaries of small enough initial separations are probably rapidly drawn into contact by magnetic braking, which is probably more efficient for systems of small mass and separation. Hence, the resulting systems are observed as contact pairs (W UMa stars) and, therefore, are excluded from our binary sample.

The complete paper will be published in the Astrophysical Journal (Supplements).

ACKNOWLEDGEMENT

The authors thank Mr. A. Janesich for valuable technical assistance.

REFERENCE

Popova, E.I., Tutukov, A.V., Shustov, B.M. and Yungelson, L.R., 1982, in "Binary Stars and Multiple Stars as Traces of Stellar Evolution", IAU Colloquium No. 69 , Z. Kopal and J. Rahe eds. (Dordrecht:Reidel)

# MIGRATING WAVES IN SOLAR-TYPE SHORT-PERIOD ECLIPSING BINARIES

L. Milano, G. Russo, S. Mancuso
Capodimonte Astronomical Observatory - Napoli, Italy

One of the properties of RS CVn-like binaries is the presence of cyclic fluctuations, sometimes called "migrating waves", in the V and other broad-band lightcurves. These fluctuations, perhaps due to spots, vary in amplitudes and periods, in the sense that, the longer the orbital period of the RS CVn system, the higher the amplitude of the "wave", which may be up to 30-35% of the total light variation. Therefore, in short-period (less than one day) RS CVn binaries, these fluctuations are generally difficult to be detected. We propose to use the Wilson (1979) method for lightcurve synthesis to separate the fluctuations from the light variation caused by geometric and photometric effects in the binary system, thus allowing a precise measurement of the amplitude and periodicity of the fluctuations which can be correlated to rotation and magnetic fields. The procedure, already applied to RT And (Milano, Russo, Mancuso, 1981) is here extended to UV Psc, WY Cnc, SV Cam and is being applied to CG Cyg, ER Vul, BH Vir. As an example, we give in Fig. 1 the results of the 1965 V light curve of WY Cnc (upper) and the "migrating wave" isolated by means of this technique (bottom), and in Figs 2 and 3 the results for RT And. The "migrating wave", isolated in this way, can then be used for further numerical computations,

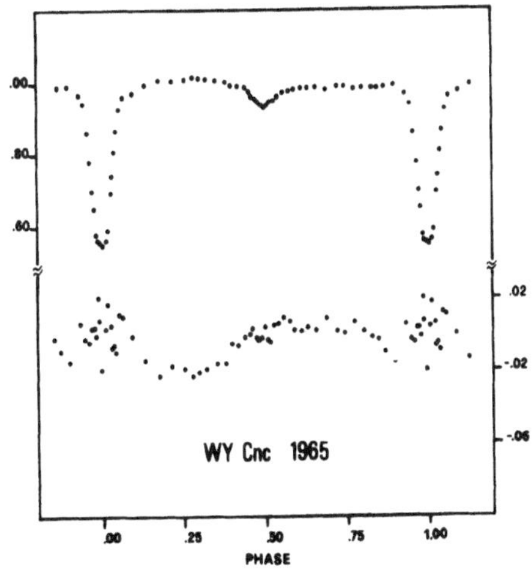


Figure 1. Lightcurve (upper) and "photometric wave" for WY Cnc.



P. B. Byrne and M. Rodonò (eds.), Activity in Red-Dwarf Stars, 393–394.

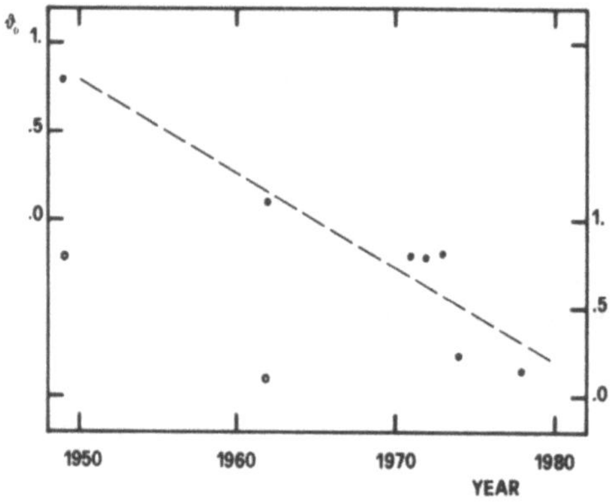


Fig. 2. Phase of "wave" maximum along the time.

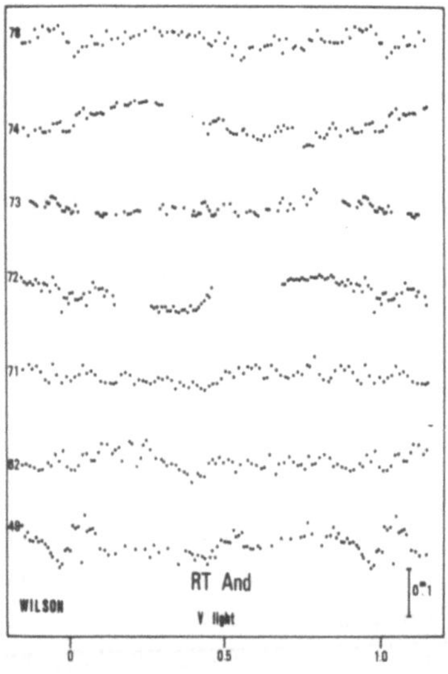


Fig. 3. The "photometric waves" for RT And.

for example for detailed modeling based on the canonical starspot model of Hall (1972), as already applied to the non-eclipsing RS CVn system DM UMa by Kimble, Kohn and Bowyer (1981) to get information on the area covered by starspots and their temperature difference with respect to the photospheres. These starspots should be due to strong coronal and chromospheric activity, enhanced by forced rapid rotations (Linsky, 1980), which can be inferred from modulation in the Ca II H and K lines (Middlekoop et al., 1981). Therefore, as an observational basis of our programme, we are performing a continuous photometric monitoring in the UBV system, and sporadic spectroscopic observations (medium to high resolution) for all the seven short-period systems already mentioned.

The complete discussion of the data we obtained will be published elsewhere.

REFERENCES

Hall, D.S.: 1972, Pubbl. Astron. Soc. Pac. 84, 323.

Kimble, R.A.; Kahn, S.M., and Bowyer, S.: 1981, Ap. J. 251, 585.

Linsky, J.L.: 1980, Ann. Rev. Astr. Ap. 18, 439.

Middlekoop, F., Vaughan, A.H., and Preston, G.W.: 1981, Astron. Astrophys. 96, 401.

Milano, L., Russo, G., and Mancuso, S.: 1981, Astr. Astrophys. 103, 57.

# FEATURES OF THE WAVE-LIKE DISTORTION IN SOME RS CVn BINARIES

C.Blanco [1], G.Bodo [2], S.Catalano [1], A.Cellino [2], E.Marilli [1], V.Pazzani [1], M.Rodonò [1] and F.Scaltriti [2]

[1] Osservatorio Astrofisico di Catania and Istituto di Astronomia dell'Università di Catania

[2] Osservatorio Astronomico di Torino, Italy

One of the outstanding features of RS CVn-type binaries is the changing shape of the so called "wave-like distortion" or "photometric wave" (PW) and its phase migration on the light curve (LC). Here we present a preliminary analysis of the observations of the following objects: CQ Aur, RU Cnc, VV Mon and SZ Psc. For each system we have derived the amplitude and phase of the PW maximum and its migration rate.

Typically, the PW of RS CVn systems can be represented by: $\ell = \ell_0 + \ell_w \cos(\phi - \phi_{max})$. However, this equation applies to an ideal situation, while other effects, such as ellipticity and reflection, modify the LC of binary systems. In general we may adopt the truncated Fourier expansion: $\ell = A_0 + A_1 \cos\phi + B_1 \sin\phi + A_2 \cos(2\phi)$, but higher order terms cannot be excluded. The coefficient $A_2$ allows for the usual ellipticity effect and for a possible asymmetry of the PW, whereas $A_1$ is due, at least in part, to the reflection effect. Therefore, if we want to find out the features of the PW, we need to assume preliminary estimates of ellipticity and reflection effects. The reflection effect has been estimated theoretically from the spectral type of the components. Owing to uncertainties in this determination we have considered several possible values. In all cases the general behaviour is not significantly altered. We have assumed symmetric PWs, i.e. the coefficient $A_2$ is due only to ellipticity. This hypothesis was confirmed "a posteriori", as generally concordant $A_2$ values for all the available LCs of each binary were found. Therefore, we adopted the average values of the ellipticity coefficients found in different years. The results of our analysis for each system follow.

a) CQ Aur is an example of a binary whose LC is modified both by the PW and by reflection and ellipticity effects (Fig. 1). In this case the Fourier expansion gives concordant values of $A_2$ for all years (~ 0.03), whereas a rather small value (~ 0.001) was found for the reflection



P. B. Byrne and M. Rodonò (eds.), Activity in Red-Dwarf Stars, 395–398.

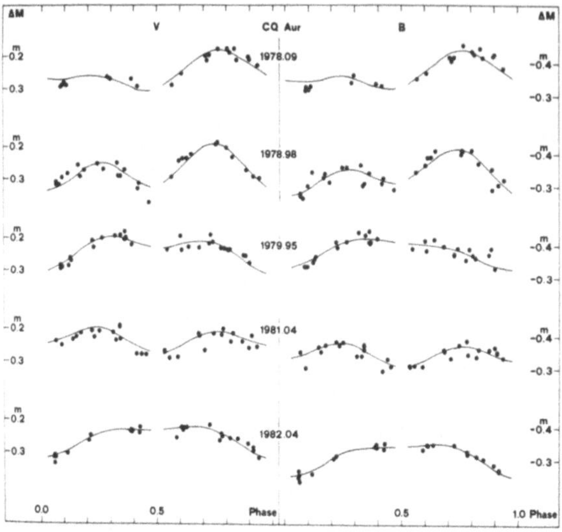


Figure 1. Seasonal V and B light curves of CQ Aur

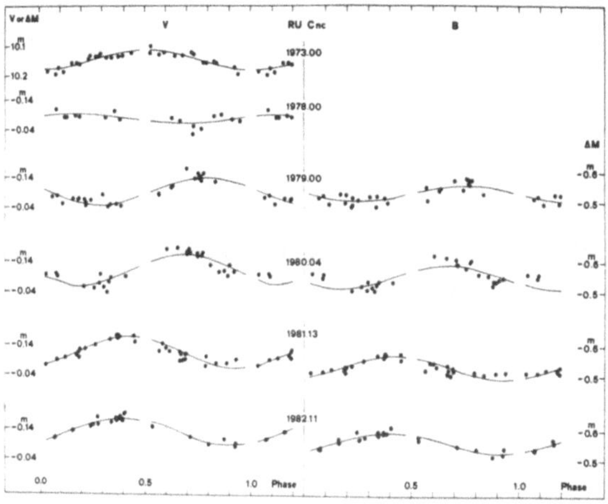


Figure 2. Seasonal V and B light curves of RU Cnc

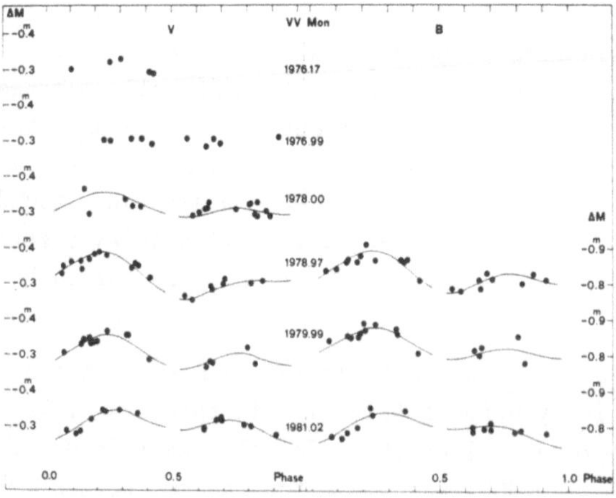


Figure 3. Seasonal V and B light curves of VV Mon

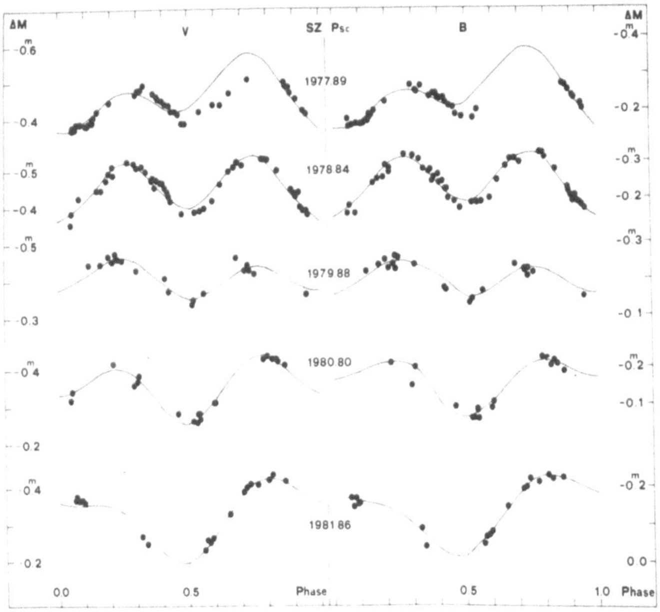


Figure 4. Seasonal V and B light curves of SZ Psc

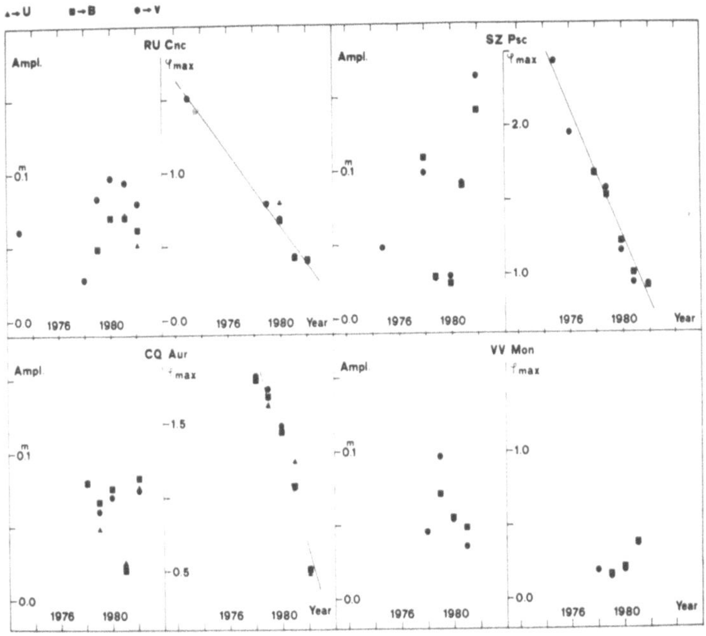


Figure 5. Amplitudes of the photometric wave and migration of the maximum phase ($\phi_{max}$) on the light curve of RS CVn binary systems versus time.

coefficient. By assuming for the latter an higher value (-0.004) we find a variation of $0^p.01$ - $0^p.02$ in $\phi_{max}$ and of only a few thousands of magnitude in the amplitude. In the interval covered by our observations the amplitude of the PW varies in a rather erratic way, while its backwards motion leads to a migration period of 3.0 ± 0.2 years.

b) The RU Cnc out-of-eclipse LC closely resembles that of RS CVn itself. In this case the PW is probably superimposed to a nearly flat maximum (Fig.2). A Fourier expansion including $A_2$ leads to small values of this term with a mean value around zero. Therefore, being negligible the reflection and ellipticity effects, a pure sinusoidal PW was adopted. We found variable amplitude attaining 0.1 mag at most and being smaller in blue than in yellow light. Actually the PW disappeared in 1977-78 leaving a nearly flat LC. Assuming a uniform wave migration towards decreasing orbital phase, we find P(migr) = 7.9 ± 0.4 years (Fig. 5).

c) VV Mon represents a different case with respect to CQ Aur and RU Cnc because neither pure sinusoidal PW nor definitely distorted rounded maximal are apparent on the available LCs (Fig.3). The reflection coefficients we adopted were -0.050 and -0.010, whereas the Fourier analysis of the LCs gives concordant values of $A_2$ for all observing seasons. The average value $A_2$ = -0.025 was adopted. The uncertainty we found by changing the reflection coefficients are ~ 0.03 in $\phi_{max}$ and ~ 0.01 in the PW amplitude. No definite trend in the variation of the PW amplitude is apparent, whereas the phase shift of its maximum seems to indicate a change in the sense of migration. Hence, it is not possible to estimate the migration period.

d) The case of SZ Psc is similar to that of CQ Aur. The adopted reflection coefficients are -0.002 and -0.004, whereas the coefficient $A_2$ turns out to be fairly constant, but in the season 1978-79. Again we adopted an average value for $A_2$ (-0.045). Combining our data with other previous results, we have obtained a backwards motion of the PW with a migration period of 4.6 ± 0.3 years (by assuming a uniform rate) whereas its amplitude varies in a more complicated way. It is interesting to notice that the minimum of the PW amplitude occurred at the time of the Hα outbursts observed by Bopp (1981). The available LCs are shown in Fig. 4.

Further photoelectric observations are planned in order to achieve a more complete picture of the LC variations and of the physical phenomena involved.

REFERENCE

Bopp, B.W.: 1981, Astron.J. 86, 771.

# ON THE DISTORTION WAVE IN THE SHORT PERIOD RS CVn TYPE ECLIPSING BINARY WY CANCRI

P. Vivekananda Rao and M.B.K. Sarma
Centre of Advanced Study in Astronomy,
Osmania University, Hyderabad-500007, India

The eclipsing binary WY Cancri (G5 V, M2 V), having a period of $0^{d}.82937122$, has been classified as a short period RS CVn type binary (Hall, 1976). Chambliss (1965), Oliver (1974), Sarma (1976), Awadulla and Budding (1979) have reported observations of this system. In order to determine a more reliable period of the distortion wave, its source, nature and period of the overall luminosity change from one observing season to the next, we have included this star in the observing schedule of RS CVn binaries at our observatory. Observations were made in standard UBV filters using the 48-inch reflector telescope of the Japal-Rangapur Observatory. The stars BD+28° 1672 and BD+27° 1701 were used as a comparison and check stars respectively. The light curves in yellow and blue colours for 1973, '76, '77 and '78 are shown in figure 1 which shows the following features:

(i) Changes in overall luminosity outside the eclipses from one observing season to the next.
(ii) Presence of the distortion wave outside the eclipses.
(iii) Variation in the depth of the primary minimum which is not very prominent compared to the secondary.

The light outside the eclipse ($\theta_e$=26°) for all the years was represented by a truncated Fourier series of four terms. The Fourier coefficients of the cos$\theta$ terms comprise of two parts: (i) due to the reflection and eclipticity effects and (ii) due to the wave. Theoretical values of the reflection and eclipticity effects were obtained using Merrill's (1970) equations. After removing the reflection and ellipticity terms from the derived Fourier coefficients for each year, the semiamplitudes $K_1=(A_1^2(\text{wave})+B_1^2)^{\frac{1}{2}}$ and $K_2=(A_2^2(\text{wave})+B_2^2)^{\frac{1}{2}}$ and the minima ($\theta^{I}_{min}$ and $\theta^{II}_{min}$) of the distortion wave were derived. It was found that $A_3$(wave), $A_4$(wave), $B_3$ and $B_4$ do not make any significant contribution to the distortion wave. The semiamplitudes $K_1$ and $K_2$ of the distortion wave are found to vary from year to year (range 0.002-



*P. B. Byrne and M. Rodonò (eds.), Activity in Red-Dwarf Stars, 399–400.*

0.021) and are found to be independent of wavelength. From a plot of $\theta^{I}_{min}$ and $\theta^{II}_{min}$ versus mean HJD it was found that the distortion wave in WY Cancri completes one cycle in a period of about 3 years and appears to migrate towards increasing orbital phase which is in contradiction to the conclusions of Oliver (1974). Taking the values of $L_h$ and $L_c$ from Awadulla and Budding (1979) the percentages of light contributed by the wave were determined. From these percentages it is clear that the hotter component is responsible for the distortion wave in WY Cancri. In order to find which component is responsible for the overall light variation outside the eclipses, we have proceded as follows. Taking the average light level outside the eclipse for the 1973-'74 light curve as standard, the difference in light was determined for the light curves obtained by us, Chambliss (1965) and Oliver (1974). The percentage of light contributed by these variations with respect to $L_h$ and $L_c$ were obtained. Here once again we found that the hotter component is responsible for the overall light variations outside the eclipses. To determine the periodicity of these average light variations we made a plot of the average light outside the eclipse versus the year of observation taking the 1973-74 curve as standard. From this plot we found that the period of the light variations outside eclipse is of the order of 30 years or more.

A more detailed account of this paper will be submitted to the Journal of Astrophysics and Astronomy in 1983.

REFERENCES


Awadulla, N.S. and Budding, E.: 1979, Astrophys. Space Sci. 63, 319.
Chambliss, C.R.: 1965, Astron. J. 70, 741.
Hall, D.S.: 1976, in W.A. Fitch (ed.) IAU Colloq. 29, p. 287.
Merrill, J.E.: 1970, Vistas in Astronomy 12, 47.
Oliver, J.P.: 1974, Ph. D. Thesis, Univ. of California.
Sarma, M.B.K.: 1976, Bull. Astron. Inst. Czech. 27, 335.


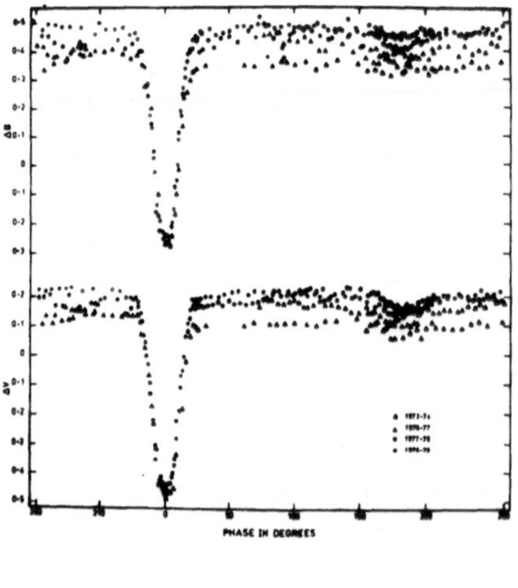


Figure 1. Blue and Yellow Light Curves of WY Cancri

# RECENT PHOTOMETRY OF AR LACERTAE

Young Woon Kang and Frank Bradshaw Wood
Rosemary Hill Observatory, University of Florida
Gainesville, Florida, USA

AR Lac is an eclipsing binary of the RS CVn type. The intrinsic variability of one of the components was first announced by Wood (1946) and the presence of sharp H and K emission lines was noted by Wyse (1934). More recent work by Nha and Kang (1982) shows clearly that the light curve varies from year to year. The period is variable.

The detailed study of the variations is made difficult by the period of almost exactly two days. Thus the present study is a co-operative effort by observatories in widely different longitudes. In Korea, the observations were made at the Yonsei University Observatory and at the Kongju Teachers College Observatory while in the United States two observatories (Rosemary Hill and Fernbank) at approximately the same longitude participated. The geographical location of these observatories are given in Table I with the telescopes used and the number of observations obtained. During the 1980 observing season, complete coverage was achieved and an excellent light curve was obtained; the 1981 observing season was not nearly as good at Rosemary Hill, but even so the light curve, as plotted and shown in Figure 1, was obtained. Seasonal difference, especially in the depths of the minima, are clearly shown and will be discussed in more detail in the forthcoming publication. The accompanying plot of comparison-check star measures shows clearly that these variations are in the AR Lac system.

Table I. Location of the observatories

| | Telescope | $\lambda/\phi$ | No. of observations V | B | U |
|---|---|---|---|---|---|
| Yonsei University Observatory | 40 cm L<br>61 cm L | 126° 46!0 E<br>37° 41!2 N | 504 | 482 | --- |
| Kongju Teachers College Observatory | 40 cm L | 127° 08!4 E<br>36° 28!0 N | 104 | 108 | 82 |
| Rosemary Hill Observatory | 76 cm L<br>46 cm L | 82° 35!2 W<br>29° 24!0 N | 88 | 88 | 88 |
| Fernbank Observatory | 91 cm L | 84° 19!1 W<br>33° 46.7 N | 154 | 155 | 156 |
| Total | | | 850 | 833 | 326 |



*P. B. Byrne and M. Rodonò (eds.), Activity in Red-Dwarf Stars, 401–402.*

We intend to continue observation, probably with collaboration from other observatories, to try to achieve a better understanding of these variations.

References
Nha, I.-S. and Kang, Y. W. (1982), P.A.S.P. (in press).
Wood, F. B. (1946), Princeton Contr. No. 21, 10.
Wyse, A. B. (1934) L. O. B. 17, 37 (464).

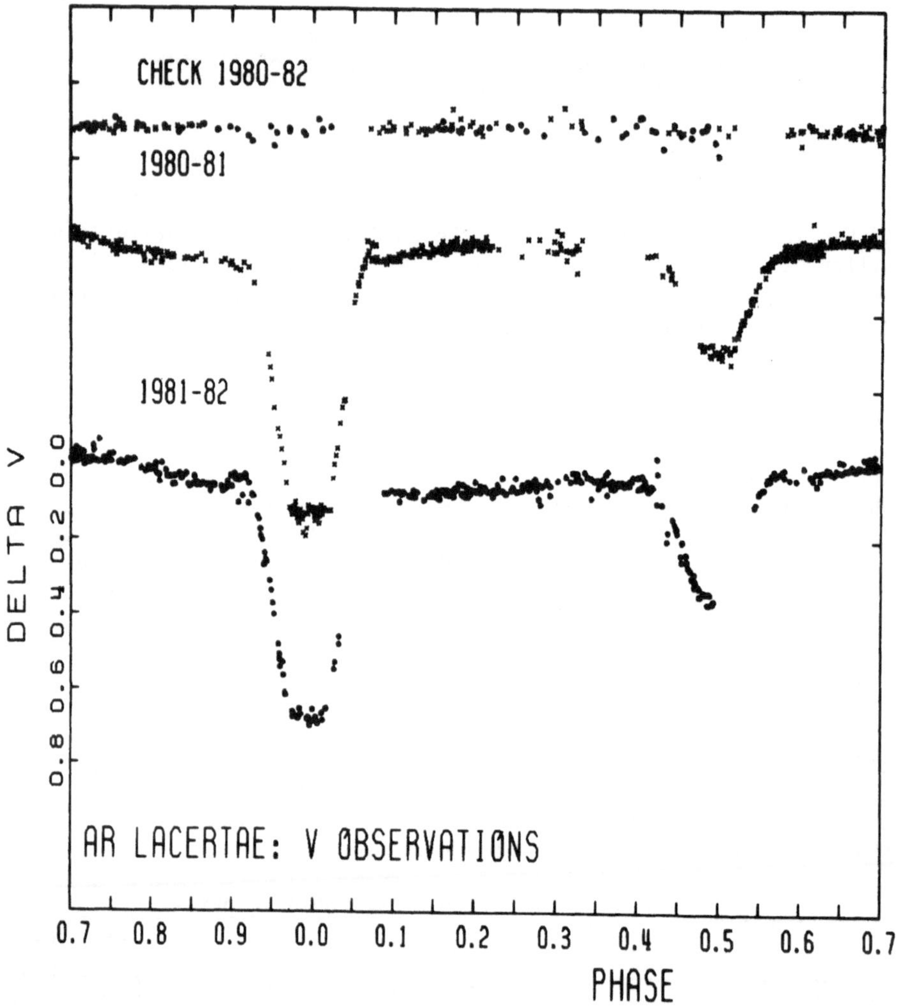


Fig. 1. Yellow light curves of AR Lac and check star, BD+44°4041. Symbols used are xs for 1980-81 and dots for 1981-82.

# HK LACERTAE

Katalin Olàh
Konkoly Observatory, Budapest, Hungary

HK Lac was observed at Konkoly Observatory between 1976-1981, the details of its story have been described in a previous paper (Olàh, 1979).

The aim of this investigation is to give a possible description of the strong variations in the light curve of the star.

Searching for a more accurate period, it became obvious that the same period was not applicable to all the data. Therefore, a new period was derived for each different cycle by the Lafler-Kinman method. Unfortunately, due to the small number of data in the data sets these newly derived periods have an error of about 0.1 day.

For the starspot modelling the method of Budding (1977) was used with some modifications. Since none of the observed light curves were symmetrical, the existence of two spots had to be assumed.

A serious complication of the modelling was the fact that the median brightness of the star was strongly varying (Olàh, 1979), which is clearly seen in Figure 1. As in the case of II Peg (Bopp and Noah, 1980) it was necessary to suppose a generally spotted band around the star which was

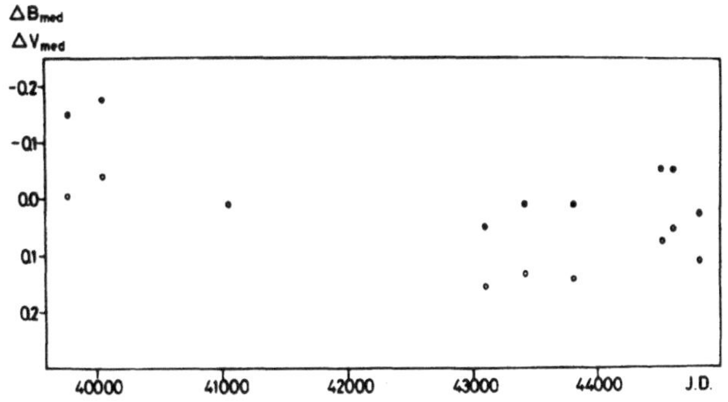


Figure 1. Median brightness variation of HK Lac



*P. B. Byrne and M. Rodonò (eds.), Activity in Red-Dwarf Stars, 403–405.*

responsible for the general dimming of the star. For this reason the 1.0 (maximum) level of the intensity curve was fixed to the individual maxima in each cycle.

The orbital inclination of the system was chosen as $60^\circ$: as the unseen companion is not visible in the spectra at all (Vogt, 1981), therefore, it is at least $2^m$ fainter than the primary component and, knowing the mass function (Gorza and Heard, 1971) the lower limit of the inclination can be calculated. On the other hand, there is no eclipse in the system and this fact gives an upper limit of about $80^\circ$ for i.

The limb darkening coefficient was taken from the table of Manduca et al. (1977) as 0.75 and 0.85 in V and B colour, respectively.

The flux ratio 0.4 in V and 0.36 in B was used supposing that the temperature of the spots were $2000^\circ$ cooler than the surface temperature of the star. This supposition does not contradict the result of Vogt (1981

The calculations of the spots' sizes and locations were performed in V and B colours separately. As final results, the weighted mean values (V:B = 2:1) of the two runs were accepted. (This ratio is about equal to the ratio of the standard deviations in the V and B observations.)

Because of the relatively small number of observations and the different scatter in B and V, the results of the two series of calculations are slightly different. The small differences in the total $\gamma$ values (angular extents of the spots which basically define the light amplitude) determined from the V and B light curves can be explained to some extent if we assume that the temperature of the spots is not always the same, on the contrary it is varying from cycle to cycle.

The strong variations in the observed light curves and in the periods show that the locations and the sizes of the spots are considerably different in consequent cycles, as it is also seen from the calculated values of $\lambda$, $\beta$ and $\gamma$ (Table 1). It is important to note that not only the

Table 1

| P | $\lambda_1$ | $\lambda_2$ | $\beta_1$ | $\beta_2$ | $\gamma_1$ | $\gamma_2$ | J.D. |
|---|---|---|---|---|---|---|---|
| $25^d.3$ | $39^\circ$ | $174^\circ$ | $41^\circ$ | $42^\circ$ | $17^\circ.5$ | $19^\circ$ | 39751-819* |
| 25.3 | 201 | 264 | 69 | 0 | 23.5 | 10 | 40004-099* |
| 25.0 | 310: | 316: | 9: | 24: | 19: | 6.5: | 43045-108 |
| 24.8 | 267 | 42 | 37 | 18 | 20.5 | 19.5 | 43355-482 |
| 25.0 | 351 | 79 | 59 | - 8 | 33 | 11 | 43713-881 |
| 25.3 | 298 | 35 | 41 | 31 | 17.5 | 14 | 44502-561 |
| 24.0 | 256 | 17 | 21 | 29 | 18 | 23.5 | 44605-638 |
| 24.4 | 332 | 1 | 34 | 42 | 20.5 | 7.5 | 44783-873 |

*Blanco, Catalano (1970) (their calculated period was used)

motion of the spots in latitude (β) can shorten or lengthen the period (because of the differential rotation) but if the motion of the spots is fast enough in the longitudinal direction (λ) it can give rise to changes of the period, too.

Figures 2 and 3 display two examples of the light curve fitting.

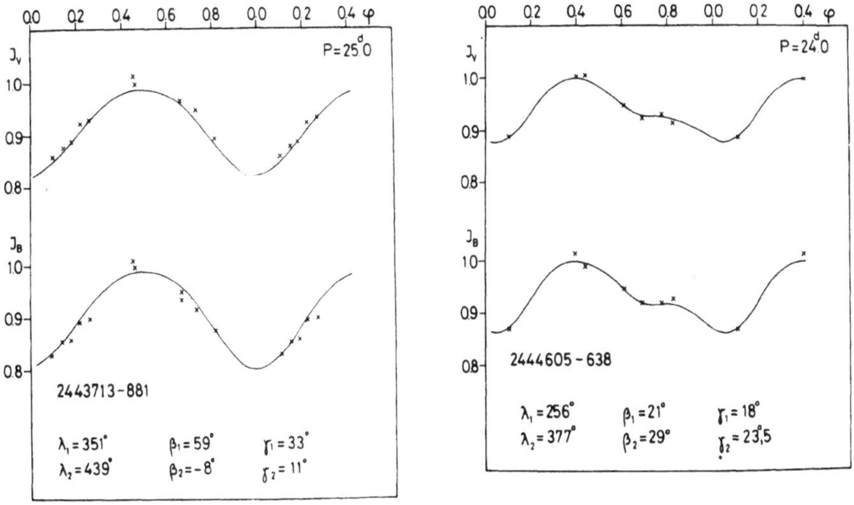


Figures 2 and 3. Light curve fitting of HK Lac

My thanks are due to Dr. B. Szeidl for many helpful discussions.

## REFERENCES

Blanco, C., Catalano, S.: 1970, Astron. Astrophys. 4, 482
Bopp, B.W., Noah, P.V.: 1980, Publ. Astr. Soc. Pacific 92, 717
Budding, E.: 1977, Astrophys. Space Sci. 48, 207
Gorza, W.L., Heard, J.F.: 1971, Publ. David Dunlap Obs. 3, 107
Herbst, W.: 1973, Astron. Astrophys. 26, 137
Manduca, A., Bell, R.A., Gustaffson, B.: 1977, Astron. Astrophys. 61, 809
Olåh, K.: 1979, Inf. Bull. Var. Stars No. 1717
Vogt, S.S.: 1981, Astrophys. J. 250, 327

# THE SOLAR-TYPE ECLIPSING BINARY SYSTEM AI PHOENICIS

E.F. Milone and B.J. Hrivnak
Rothney Astrophysical Observatory, The Univ. of Calgary

## INTRODUCTION

AI Phe was discovered to be variable and identified as an EA binary by Strohmeier (1972). Reipurth (1978) subsequently carried out uvby photometry, determined the period (24.5923 d), and noted the lengthy totality of primary minimum and the displacement of the secondary minimum. Imbert (1978) obtained radial velocity curves and determined spectroscopic orbital elements. Imbert also gives a spectroscopic classification of G2V for the primary (hotter) component and approximately G5 for the secondary. AI Phe thus appeared to offer a unique opportunity to study the limb darkening of a non-interacting solar-type star. As an extension of a solar UV limb darkening study (Kjeldseth Moe and Milone 1978), ten IUE spectra were obtained on Aug. 12 in 2 successive shifts (cf Milone et al 1981). In conjunction with this, ground-based 5-colour photometry and spectroscopy were carried out at CTIO and at UTLCO.

## PRESENT WORK

UBVRI photometry was carried out in September 1981 by Hrivnak at CTIO, and over the season, by I. Sheldon at Las Campanas. Figure 1 displays the differential V light curve relative to comparison star HD 6236. Standardization to the Johnson system permitted the determination of magnitude and colours of the two components, since the secondary is seen alone at primary minimum.

| Star | V | U - V | B - V | V - R | V - I |
|---|---|---|---|---|---|
| S | 9.326 | 1.35 | 0.85 | 0.70 | 1.15 |
| P | 9.335 | 0.47 | 0.49 | 0.47 | 0.70 |

Spectra taken in eclipse beginning at phase $0^{p}.0048$ reveal a composite spectral type of about G5. From this phase where the primary contributed 24% of the B light to the last exposure, at $0^{p}.009$, when the contributions were equal, the strength of the Balmer lines increased to suggest a composite spectral type no later than early G. Spectra at maximum, where the primary contributes just 58% (in B), appear still



*P. B. Byrne and M. Rodonò (eds.), Activity in Red-Dwarf Stars, 407–408.*

earlier, but later than ~G0.

We conclude that colours and spectra are consistent with a system containing an F6-7 main sequence primary and a mid-G, probably subgiant, secondary, although the latter's colours are somewhat peculiar. The colour problem created difficulties in modelling the system. Analysis initiated with the Wilson-Devinney code (modified by Wilson to treat orbital eccentricity) failed to give satisfactory fits to the primary minimum in all wavelengths. However, we have obtained a provisional set of elements.

$q = 1.044$ (Imbert), $i = 89^\circ\!.2$, $e = .14$, $\omega = 115^\circ$

| | Primary | Secondary |
|---|---|---|
| T | 6400 (assumed) | 4960 |
| M (☉) | 1.10 (Imbert) | 1.15 (Imbert) |
| r/a | 0.030 | 0.055 |
| R(☉) | 1.4 | 2.6 |
| ρ(☉) | 0.39 | 0.15 |

$L_p$ = 0.65 (U), 0.58 (B), 0.51 (V), 0.44 (R), 0.40 (I)

Each component is well within its Roche Lobe. The results suggest that the secondary star is slightly evolved off the main sequence, a result consistent with its slightly higher mass (Imbert 1978). A fuller discussion will appear in the Astronomical/Astrophysical Journal.

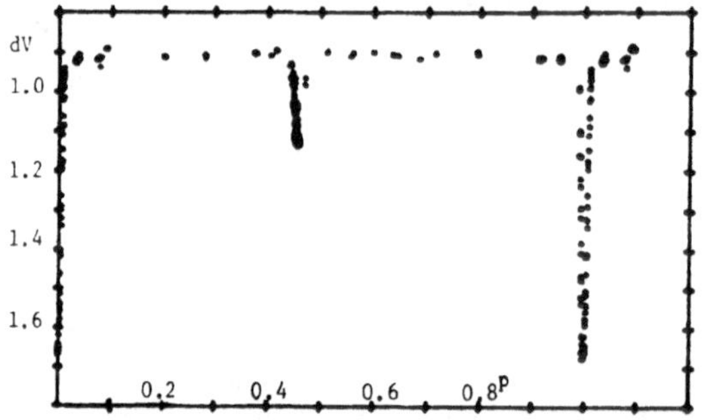


Figure 1 - Light curve of AI Phe from a single season by Hrivnak and Shelton. Note the displaced secondary.

REFERENCES

Imbert, M. 1979, Astron. Astrophys. Suppl. 36, 453.
Kjeldseth Moe, O. and Milone, E.F. 1978, Astrophys. J. 225, 301.
Milone, E.F., Hrivnak, B.J., Clark, T.A., Kjeldseth Moe, O., Blades, J.C. and Shelton, I. 1981, Info. Bull. Var. Stars 2060.
Reipurth, B. 1978, Info. Bull. Var. Stars 1419.
Strohmeier, W. 1972, Info. Bull. Var. Stars 665.

# LIGHT CURVES AND CA II EMISSIONS OF V711 TAURI DURING 1981-82

Il-Seong Nha and Jun Young Oh

Yonsei University Observatory
Seoul, Korea

The light variation of V711 Tauri was first discovered by Cousins, who suspected the variation with an amplitude of $\Delta \underline{V} = 0\overset{m}{.}11$. This light variation was confirmed by Landis and Hall (1976). This star has been identified as the brightest RS CVn-type star by Bopp and Fekel (1976). Bopp et al. (1977), using Cousins' old data and their data made nearly 13 years later than those of Cousins, found that the observations show the same light curve shape and amplitude and the minimum light falls very nearly at $0\overset{p}{.}0$ computed with their ephemeris.

A series of intensive photoelectric observations in UBV has been reported at a number of observatories after the Oct 1977 campaign by Weiler (1977). All light curves agree, in general, for the light curve shape, and they all seem to have a migration wave towards increasing phase. A dramatic change in the light curve of V711 Tau, however, was reported by Blanco et al. (1981). Since the light variations were so unusual , this discovery has changed our present understanding about the wave migration and the light curve evolution entirely.

Observations made during Nov 1981 - Mar 1982 at Yonsei University Observatory confirm, in part, the dramatic light change of V711 Tau reported by Blanco et al. for 1980-81 season. However, the lower peak at phase about $0\overset{p}{.}3$ of the double-peaked light curves of the 1980-81 appeared to be absent in our $\underline{V}$-light of the 1981-82 season. Only a single peak of amplitude $0\overset{m}{.}09$ remained in the light curve, and the peak is shifted towards decreasing phase.

Nine yellow light curves available up to date are given in a chronological order in Figure 1. Ephemeris used for the phase is

$$\text{JD (hel.)} = 2442766.069 + 2\overset{d}{.}83782\underline{E}.$$

There seem to exist two discontinuities in the light curve evolution. The first of these happened right after the radio outburst in 1978, which caused a sudden increase of the amplitude of the light variation from about $0\overset{m}{.}08$ to about $0\overset{m}{.}21$ in 1978-79. The second one is clearly



*P. B. Byrne and M. Rodonò (eds.), Activity in Red-Dwarf Stars, 409–410.*

present sometime after the observations of Sarma and Ausekar (1981) in 1979-80 season but before those of Blanco et al. in Jan 1981. None of these sudden light variations of V711 Tau can, easily, be understood with the present knowledge.

Simultaneous spectroscopic and photometric observations were made on 1982 Feb $10^{d}.5$ UT, the former with the 188-cm coude of Okayama Astrophysical Observatory, Japan, and the latter with the 40-cm and 61-cm reflectors of Yonsei University Observatory. Two spectrograms with a dispersion of 36 A/mm are obtained in the Ca II H and K line regions. A pronounced emissions of H and K are recorded with a slight wavelength shift in the K line as shown in Figure 2 in a logarithmic intensity scale.

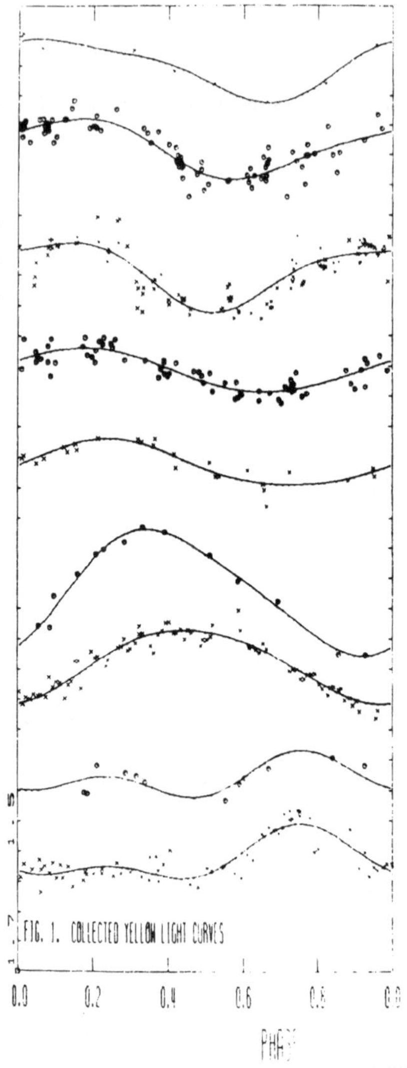


FIG. 1. COLLECTED YELLOW LIGHT CURVES

References:
Blanco, C., Catalano, S., Marilli, E., Rodono, M., and Scaltriti, F., 1981. IBVS 2000.
Bopp, B. W., and Fekel, F., 1976, A. J., 81, 771.
Bopp, B. W., Espenak, F., Hall, D. S., Landis, H. J., Lovell, L. P., and Reucroft, S., 1977, A. J., 82, 47.
Landis, H. J., and Hall, D. S., 1976, IBVS, 1113.
Sarma, M. B. K., and Ausekar, B. D., 1981, A. A., 31, 103.
Weiler, E. J., 1976, IAU Circ. No. 3089.

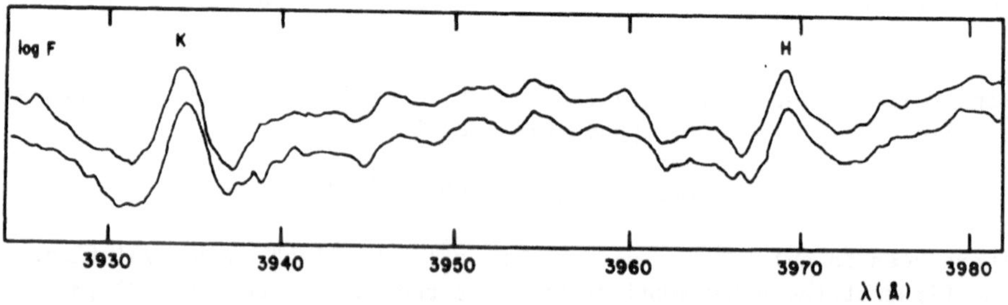


Fig. 2. Ca II H and K in the spectra of V711 Tau on 1982 Feb $10^{d}.5$ UT.

# A FLARE-LIKE EVENT FROM THE SHORT-PERIOD SYSTEM XY UMa

M. Zeilik, R. Elston, G. Henson and P. Smith
Capilla Peak Observatory, The University of
New Mexico, Albuquerque, New Mexico USA

XY UMa (+55°1317, SA027143) is a short-period (P 0.48 d) cousin of the RS CVn stars. The primary star is G2-G5V; the secondary K5 (Geyer, quoted by Lorenzi and Scaltriti, 1977). Geyer (1977) has done the bulk of the observational work to date, including the first photoelectric observations. The rapid, annual changes in XY UMa's light curve, and the fact that the last published light curves was from the 1977 season convinced us to reobserve this active system.

We did all observations with the 61-cm telescope at the University of New Mexico's Capilla Peak Observatory, located at an altitude of 2.84 km. The telescope has a single-channel, photon-counting photometer with a cooled (-20°C) EMR 641A phototube. The statistical error in each observation is no more than 0.01 mag. The star +54°1278 (F5) was used as the comparison star for all observations. The magnitudes of the comparison star +54°1278 are: U=9.90, B=9.94, V=9.56. Phases were calculated from HJD = 2435216.5086 + 0.478995E (Wood et al., 1980) and range from 0.41 to 0.20 on 31 January 1982 (UT) and 0.12 to 0.62 on 20 February 1982 (UT).

We detected a flare-like episode on 31 January 1982 (UT), which began at 5:14 UT at phase 0.54 (Figures 1-3). The flare peaked at 5:29 UT at phase 0.57; its total duration was about 30 minutes. At peak, the system magnitude was U=10.59, B=10.33, V=9.55, and R=8.88. Relative to the average light curve of the system, the apparent magnitude increase was 0.33 at U, 0.13 at B, and 0.09 at V.

To look at the event in its intensity profile, we converted magnitudes to fluxes and integrated over its duration (Table 1). From the spectral types of the stars, we estimated a distance of 84 pc to the system and calculated the mean and integrated energies (Table 1). Note that, although the flare was strongest relative to the system at U, most of the energy was released at B and V.

Similar flare-like episodes have been reported by Patkós (1981) from the very similar system SV Cam (period = 0.59 d, spectral types



*P. B. Byrne and M. Rodonò (eds.), Activity in Red-Dwarf Stars, 411–413.*

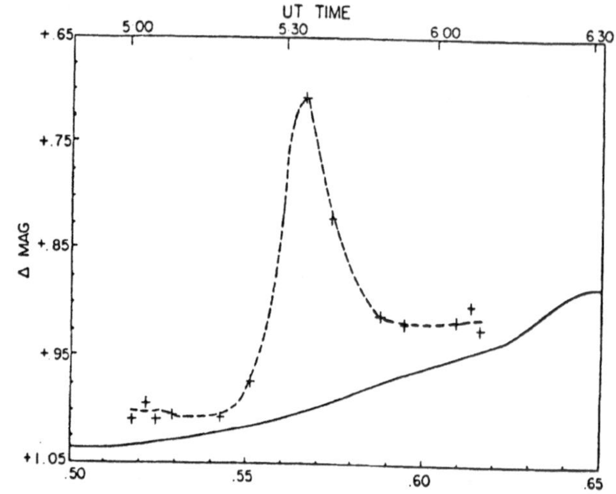


Fig. 1
XY Uma Flare like event on 1/31/82. In U band crosses are actual data points; dashed line is an interpolated fit to the data. The solid line is the average system brightness as observed on 2/20/82.

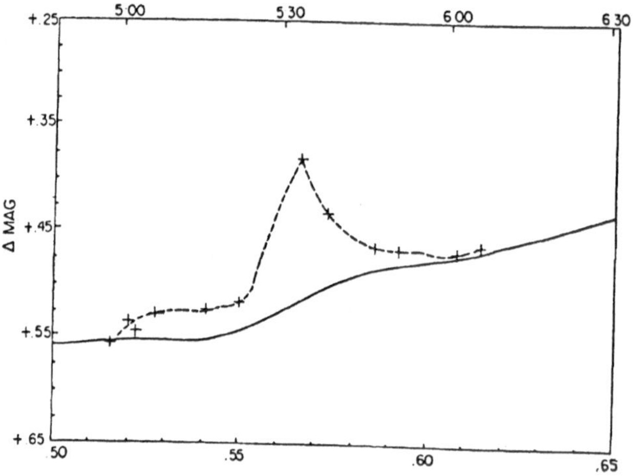


Fig. 2
Same as Figure 1, but for B band, note that UT runs along the top, phase along the bottom.

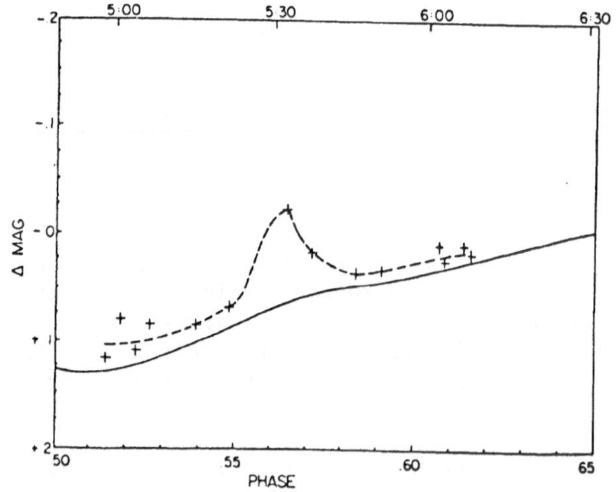


Fig. 3
Same as Figure 2, but for V band.

G3 and K4). The strongest of these (Flare 1 on 8/9 December 1980) took place at phase 0.61; its duration was 43 minutes, its peak reached 0.12 mag. above the system at U (or about 5%) of system's total intensity at U). Patkós positions the flare-active region as a sector 40° wide on the spotted half of the secondary star. If the primary star really is the active one in XY UMa, then our event also occured on the spotted hemisphere.

We can also compare our event to those of flare stars (Kunkel, 1975), which are typically spectral type dMe or dKe. Some of those, such as BY Dra, show distortion waves that can be attributed to starspot activity (Bopp and Evans, 1973). Flares from UV Ceti have B-band powers of $10^{29}$ to $10^{31}$ ergs/sec, whereas those from BY Dra cluster near $10^{33}$ ergs/sec. Solar white-light flares are typically a few times $10^{27}$ ergs/sec at B. Our event was not only more energetic than these, but also redder in color; our U-B ≅ -.66 in contrast to U-B ≅ -1 for solar and dMe flares

Geyer (1982) has reported brightness spikes from XY UMa, similar to, be less intense than our event. In February, March, and April 1982, Geyer (1982) observed a brightness spike at phase 0.63 with magnitude increases about 1/3 the amplitude of ours. Rather than an actual flare our event may be the appearance of an active, spot-enhanced chromosphere of the primary star.

We thank the Research Corporation for providing partial funding for the computer control system at Capilla Peak.

Table 1. Flux Densities and Integrated Energies

| Filter | Integrated Flux (mJy.sec) | Flare/System | Mean Energy (erg/Hz) | Integrated Energy (ergs) |
|---|---|---|---|---|
| U | $2.4 \times 10^4$ | 0.10 | $2.0 \times 10^{20}$ | $3.2 \times 10^{34}$ |
| B | $4.6 \times 10^4$ | 0.05 | $3.8 \times 10^{20}$ | $5.9 \times 10^{34}$ |
| V | $8.0 \times 10^4$ | 0.05 | $6.7 \times 10^{20}$ | $5.9 \times 10^{34}$ |

REFERENCES

Bopp, B. W. and Evans D. S. (1973), Mon. Not. Roy. Astro. Soc., 164, 343

Geyer, E. H. (1977), In IAU Colloquium No. 42, p. 292; (1982), private communication.

Kunkel, W. E. (1975), In Variable Stars and Stellar Evolution, edited by Sherwood and Plant, p. 15.

Lorenzi, L. and Scaltriti, F. (1977), Acta Astronomica, 27, 273.

Patkós, L. (1981), Astrophys. Lett., 22, 131.

Wood, F. B., Oliver, J. P. Florkowski, D. R. and Koch, R. H. (1980), A Finding List for Observers of Interesting Binary Stars (Univ. of Pennsylvania Press).

# RS CVn SYSTEMS : THE HIGH ENERGY PICTURE

P. A. Charles
Department of Astrophysics, University of Oxford

## ABSTRACT

The discovery of X-ray emission from RS CVn systems by HEAO-1 and subsequent surveys by the Einstein Observatory have shown that these close binaries exhibit greatly enhanced coronal activity. Here we review the 3 main observational areas: (1) results of the X-ray surveys of RS CVn systems and other late-type stars which indicate how the X-ray luminosity is correlated with the binary period (and hence stellar rotation) and other coronal activity indicators. This will be discussed in the context of scaled models of the solar corona; (2) X-ray spectroscopy of the most active systems which show multi-temperature spectra and line emission consistent with solar abundances of the heavy elements; (3) observations of X-ray "flare-type" activity that has been associated with several RS CVn systems.

## 1. INTRODUCTION

Following the discovery by HEAO-1 (Walter, Charles and Bowyer, 1978) of strong soft X-ray emission from the RS CVn binary systems these objects have been the subject of intense scrutiny at all wavelengths. A "basic" RS CVn system consists of a KO IV star synchronously rotating (period $\sim$ 1-14 days) about a F-G main sequence star (Hall, 1976). Although of short period the system is detached and non-interacting. The photometric wave observed in the light curve (usually $\sim 0.1 - 0.2^m$) is almost certainly due to intense starspot activity on the cooler component (Hall, 1972), a model which explains many of the intriguing features of their activity. The main sequence requirement for the hotter component is relaxed for the longer period systems such as Capella.

Since neither component in the binary is degenerate the RS CVn systems likely represent the extreme end of the luminosity range ($\sim 10^{30-32}$ erg $s^{-1}$) for late-type stellar coronal X-ray emission. They are therefore of great importance for theoretical studies of coronal



*P. B. Byrne and M. Rodonò (eds.), Activity in Red-Dwarf Stars, 415–427.*

heating mechanisms assuming that the same process occurs in all stellar coronae. RS CVn's have been surveyed most extensively now by the Einstein Observatory (Walter and Bowyer, 1981 hereafter WB) as well as high quality X-ray spectroscopy being available for the brighter sources. In this review of the X-ray picture of RS CVn systems I will concentrate on; (1) the results of this survey and compare it with similar surveys that have been undertaken of less active systems. The X-ray activity has been correlated with other indicators of coronal/chromospheric emission; (2) the X-ray spectra of RS CVn and related systems which are interpreted as thermal emission from hot ($\gtrsim 10^7$K plasma; and finally (3) the (necessarily) more scarce observations of X-ray flares from these systems.

## 2. THE HEAO SURVEYS AND THE ROTATIONAL-LUMINOSITY RELATION

All RS CVn systems were observed as part of the soft X-ray (0.1 - 3 keV) all-sky survey of the HEAO-1 A2 experiment (Rothschild et al, 1979). However, because of the limited sensitivity of this non-imaging satellite only about 15 of the nearest or brightest RS CVn and related systems were detected (Walter et al, 1980) with soft X-ray luminosities ranging from $\sim 10^{30}$ to $\gtrsim 10^{31}$erg s$^{-1}$. The proportional counter X-ray spectra were consistent with thermal emission from plasma at $T \sim 10^7$K.

A more systematic survey of RS CVn systems was then undertaken by WB with the dramatically greater sensitivity of the Einstein observatory. Table 1 gives the observed properties of the stars in the HEAO-1 and Einstein surveys (from WB, Walter et al, 1980, Charles, Walter and Bowyer, 1979, Hoffleit and Jaschek, 1982 and Hall, 1976). From these data WB show that the luminosity function is the same for giants and dwarfs, and hence that they are equally luminous. However, there is a correlation of the form $L_x/L_{bol} \propto \Omega^{1.2}$ which is displayed in figure 1. The advantage of using $L_x/L_{bol}$ is that, for similar effective temperatures, it is proportional to the surface X-ray flux and is independent of stellar radius and distance. The correlation is slightly less marked for the long and short period systems with power law indices of 0.8 and 0.6 respectively. WB interpret this relation as due to magnetic effects increasing with $\Omega$ and hence causing increased activity. Walter (1981) extends this work to rapidly rotating G and K stars with similar results. However, the G stars are about 10 times lower in $L_x/L_{bol}$ than the K stars, thus indicating that the hotter (earlier) component in RS CVn's contributes only $\sim$10% of the X-ray flux.

Pallavicini et al (1981, hereafter PAL) have investigated the rotation - luminosity relation for all stellar types using mostly Einstein data. Their results for F7 - M5 stars are shown in figure 2 and gives $L_x = 10^{27}$ (v sin i)$^2$ erg s$^{-1}$ which they find to be valid for giants and dwarfs. PAL cover a large range in $L_x$ and the RS CVn's as a group are consistent with this relation, again

Table 1. : OBSERVED PROPERTIES OF RS CVn SYSTEMS

| System | $V_{max}$ | log $L_x$ (erg $s^{-1}$) | d(pc) | $P_{rot}$(days) | Sp | Notes |
|---|---|---|---|---|---|---|
| $\zeta$ And | 4.1 | 30.14 | 31 | 38.9 | K1IIe/? | |
| $\lambda$ And | 3.8 | 30.56 | 25 | 53.7 | G8III-IV/? | |
| UX Ari | 6.5 | 31.32 | 50 | 6.5 | G5V/K0IV | HEAO-1 |
| CQ Aur | 9.6 | 30.94 | 220 | 10.7 | G0/? | |
| SS Boo | 10.2 | 30.80 | 220 | 7.6 | dG5/dG8 | |
| SS Cam | 10.1 | 30.89 | 255 | 4.8 | dF5/gG1 | |
| 12 Cam | 6.2 | 31.18 | 180 | 79.4 | K0III/? | |
| RZ Cnc | 9.8 | 31.54 | 340 | 21.9 | K1III/K4III | |
| AD Cap | 9.3 | 31.28 | 250 | 3.0 | G5/? | |
| RS CVn | 8.4 | 31.27 | 150 | 4.8 | F4III/K0IV | |
| 39 Cet | 5.4 | 31.18 | 59 | - | gG5/? | |
| UX Com | 10.8 | 31.51 | 350 | 3.6 | G5-9/? | |
| RT CrB | 10.2 | 30.74 | 360 | 5.1 | G0/? | |
| WW Dra | 8.8 | 30.84 | 180 | 4.7 | sgG2/sgK0 | |
| AS Dra | 8.0 | 29.39 | 31 | 5.4 | G/? | |
| RZ Eri | 8.4 | 31.33 | 275 | 38.9 | A5-F5V/sgG8 | |
| $\sigma$ Gem | 4.3 | 31.31 | 59 | 19.5 | K1III/? | HEAO-1 |
| Z Her | 7.3 | 30.3 | 85 | 4.0 | F4V-IV/K0IV | |
| AW Her | 9.5 | 30.94 | 315 | 8.9 | G2IV/sgK2 | |
| MM Her | 9.8 | 30.76 | 190 | 7.9 | G8IV/? | |
| PW Her | 10.7 | 31.01 | 285 | 2.9 | G0/? | |
| GK Hya | 9.1 | 30.90 | 220 | 3.6 | G4/? | |
| RT Lac | 10.0 | 31.46 | 210 | 5.1 | sgG9/sgK1 | SSS* |
| AR Lac | 6.9 | 31.18 | 40 | 2.0 | G2IV/K0IV | HEAO-1 |
| HK Lac | 6.5 | 31.14 | 139 | 25.1 | FIV/K0III | |
| RV Lib | 9.8 | 31.06 | 276 | 10.7 | G5/K5 | |
| VV Mon | 9.6 | 30.79 | 260 | 6.0 | G0/? | |
| AR Mon | 10.1 | 31.22 | 426 | 21.4 | F-G/K0II | |
| II Peg | 7.2 | 30.30 | 26 | 6.7 | K2-3 IV-V/? | = HD224085 |
| LX Per | 8.1 | 30.80 | 145 | 8.1 | G0V/K0IV | |
| SZ Psc | 8.0 | 31.40 | 100 | 4.0 | F8V/K1IV- V | |
| RW UMa | 10.2 | 30.64 | 150 | 7.4 | dF9/K1IV | |
| $\epsilon$ UMi | 4.2 | 30.56 | 71 | 39.8 | dA8-dF0/G5 III | |
| RS UMi | 10.6 | 30.95 | 350 | 6.2 | - | |
| ER Vul | 7.3 | 30.19 | 36 | 0.7 | G0V/G5V | |
| HR 1099 | 5.7 | 31.41 | 36 | 2.8 | G5V/K0V | |
| HR 4665 | 6.3 | 31.41 | 130 | 64.6 | K0III/? | DK Dra |
| HR 5110 | 5.0 | 31.11 | 53 | 2.6 | F2IV/? | BH CVn |
| HR 7275 | 5.8 | 30.56 | 48 | 28.8 | K1IV/? | |
| HR 7428 | 6.4 | 31.16 | 302 | 109.6 | A0V/K2III-IIe | |
| HR 8575 | 6.4 | 30.23 | 69 | 17.8 | K2III/? | V350 Lac |
| HR 8703 | 5.6 | 30.64 | 51 | 25.1 | K1IV-IIIp/? | IM Peg |
| HD 5303 | 7.8 | 30.37 | 66 | 1.8 | G0/? | |
| HD 155555 | 6.8 | 30.70 | 17 | 1.7 | G5/? | |
| BD+61°1211 | 9.4 | 31.24 | 130 | 7.6 | K0-1III-IV/? | |

* both stars active

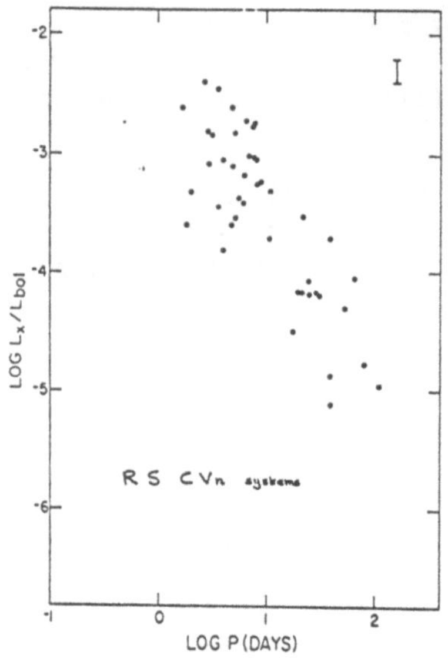


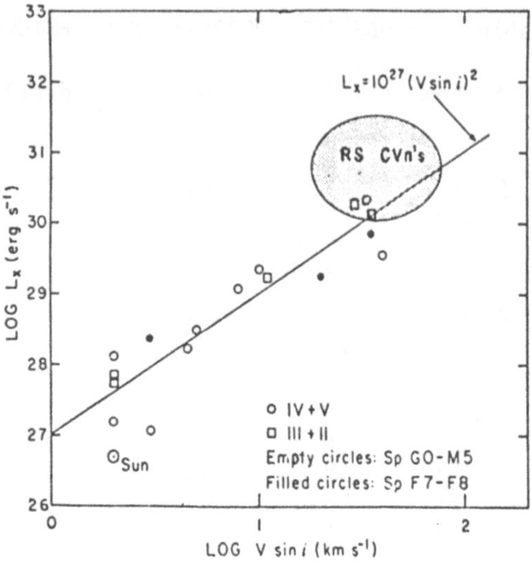


Fig. 1 : The WB plot of $L_x/L_{bol}$ against rotational period.

Fig. 2 : The PAL correlation of $L_x$ against rotational velocity for F7-M5 stars, including the RS CVn's as a group from fig.1.

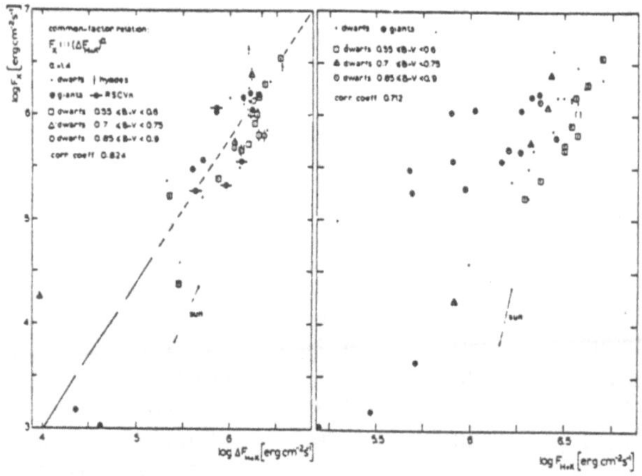

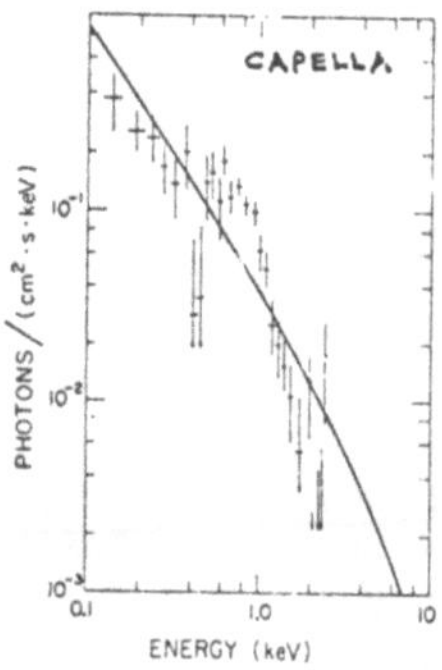


Fig. 3 : The SMZ correlation of surface X-ray flux with excess calcium flux is clearly shown in the left diagram compared to the much greater scatter of $F_x$ against $F_{H+K}$ on the right.

Fig. 4 : The original HEAO-1 LED spectrum of Capella (adopted from Cash et al, 1978) showing the Fe L emission blend at 0.85 keV.

implying that enforced rotation leads to enhanced emission. However, the RS CVn's alone give only a low power correlation with v sin i.

Finally, Walter (1982) points out that a double power law gives a much better fit to the earlier (F8 - G2) stars $\Omega - L_x/L_{bol}$ relation with a break occurring near P = 10 days. The fall-off of $L_x/L_{bol}$ at longer periods is dramatic. This period corresponds (Skumanich, 1972) to an age $\sim 10^9$ years and indicates that single stars remain active until that time.

## 3. THE X-RAY / CALCIUM II CORRELATION

Mewe, Schrijver and Zwaan (1981) used the Einstein Observatory to demonstrate a correlation between the X-ray surface flux, $F_x$, and the emission core flux in the Ca II H + K lines, $F_{H+K}$, for a small sample of cool stars (luminosity class III - V). $F_{H+K}$ was used because, from observations of the Sun, it is found that chromospheric line emission originates at sites of strong magnetic fields. This work has been extended by Schrijver, Mewe and Zwaan (1982, hereafter SMZ) who use a larger sample (52) of stars covering a 4 dex range of $F_x$ and which includes several RS CVn systems. SMZ find that the correlation between $F_x$ and the calcium flux is greatly improved by plotting $F_x$ against $\Delta F_{H+K}$ as shown in figure 3. $\Delta F_{H+K}$ is the difference between the observed flux and a lower limit to $F_{H+K}$ that depends on spectral type (SMZ find that this lower limit flux is uncorrelated with the stellar X-ray emission). All the stars in their analysis (single giants, dwarfs, binaries and RS CVn's) follow the relation

$$F_x = 2.4 \times 10^{-3} \, \Delta F_{H+K}^{1.4} \quad \text{erg cm}^{-2} \text{s}^{-1}$$

SMZ find no correlation of magnetic activity with the structure of the star (i.e. M or R) but speculate that it may be related to rotation.

## 4. X-RAY SPECTRA : HOT CORONAL PLASMAS

Only a handful of these systems are bright enough for the HEAO-1 proportional counters to register spectral information. The first of these to be observed was Capella ($\alpha$ Aur) by Cash et al (1978). Although not strictly an RS CVn system (see Popper, 1980) it certainly displays the RS CVn phenomenon as does Algol ($\beta$ Per; see White et al 1980). The HEAO-1 spectrum of Capella (figure 4) clearly shows the intense iron L emission blend at 0.85 keV which enables the spectrum to be described by an $\sim 10^7$K solar abundance coronal plasma. The other bright RS CVn system, UX Ari, was also consistent with a temperature of $\sim 10^7$K (Walter et al, 1978) although, because of the simplicity of the model, it was thought that iron was under-abundant.

## 4.1 Einstein SSS Observations : Medium Resolution Spectra

The Solid State Spectrometer (SSS; Joyce et al, 1978) onboard the Einstein Observatory was the first medium sensitivity detector to be employed for cosmic X-ray spectroscopy with a resolution at low energies substantially better than conventional proportional counters. The RS CVn and related systems were thus ideal targets for the SSS to investigate the relative line and continuum contributions. The SSS spectra of Capella, UX Ari, $\sigma$ CrB, AR Lac and Algol (from Swank and White, 1980, Agrawal et al, 1981 and White et al, 1980) are shown in figure 5 together with the best-fitting thermal models. In all cases these models require 2 temperature components (assuming each to be in collisional ionisation equilibrium), the parameters of which are given in Table 2 (from Swank et al, 1981), and abundances within a factor 2 of solar. Figure 6 is a plot of the emission integral of each component versus temperature for all 8 RS CVn's observed. This shows that:

(i) the low temperature components are between 4 and 8 x $10^6$K and have a narrow range of emission integrals ($\sim$1 dex) near $10^{53}$ $cm^{-3}$;

(ii) the high temperature components have a much larger scatter in both temperature (26 - 93 x $10^6$K) and emission integral ($\sim$3 dex);

(iii) Capella, UX Ari and AR Lac all show <u>independent</u> variations of the cool and hot components.

The prominent emission lines (due mainly to Fe, Mg, Si and S) are provided by the low temperature components, as most of these elements are fully ionised at the higher temperatures.

## 4.2 Einstein OGS Observations : High Resolution Spectra

The Objective Grating Spectrometer (OGS; Gronenschild et al, 1981) onboard the Einstein Observatory can give much higher resolution soft X-ray spectra than the SSS but has a much <u>lower</u> sensitivity. for this reason it has only been able to acquire a spectrum of Capella, and this is shown in figure 7. Gronenschild et al fit 2 temperature components to this spectrum of 5 and 10 x $10^6$K. They are insensitive to X-rays of energy $>$1 keV and thus cannot detect the very high temperature components observed by the SSS.

## 5. STELLAR FLARES AND X-RAY TRANSIENTS

Long before RS CVn's were suspected of being X-ray sources they were known to exhibit large-scale radio flaring (Gibson et al, 1975). Indeed, a simultaneous radio, H$\alpha$, Ly$\alpha$ and X-ray outburst from HR 1099 was observed in 1976 (White, Sanford and Weiler, 1978) and a much more powerful and extended period of radio outbursts from HR 1099 was seen in 1978 (Feldman et al, 1978) but no X-ray observations could be made then. The HEAO-1 satellite enabled both the identification (Schwartz

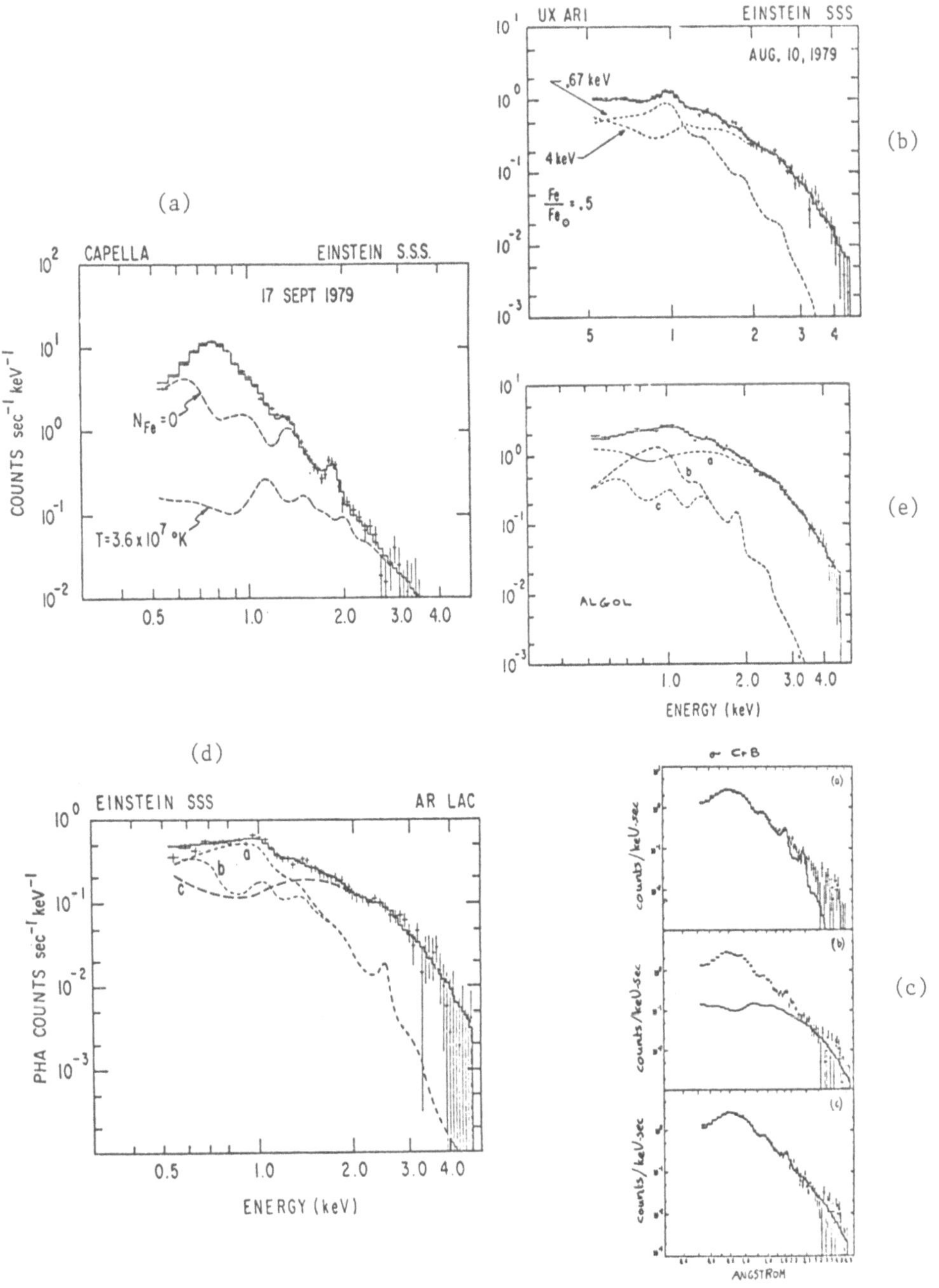


Fig. 5 : SSS spectra of (a) Capella,, (b) UX Ari, (c) σ CrB, (d) AR Lac and (e) Algol (adopted from Swank and White, 1980, Agrawal et al, 1981 and White et al, 1980). The individual temperature components of the best fit spectra are also plotted. The contribution of everything but Fe are shown for Capella, AR Lac and Algol.

Table 2 : X-RAY SPECTRAL PARAMETERS (from Swank et al, 1981)

| Source | $T_1$ ($10^6$K) | $T_2$ ($10^6$K) | log $EM_1$ ($cm^{-3}$) | log $EM_2$ ($cm^{-3}$) |
|---|---|---|---|---|
| σ CrB | 5.8 | 46 | 53.3 | 52.9 |
| AR Lac | 6.7-7.4 | 93-43 | 53.5 | 53.5-53.9 |
| HR1099 | 6.7 | 48 | 53.4 | 53.7 |
| Algol | 7.0-8.2 | 33-48 | 53.1 | 53.5-53.9 |
| RS CVn | 7.0 | 96 | 53.4 | 54.0 |
| UX Ari | 7.5-7.9 | 31-65 | 53.8 | 54.0-54.5 |
| λ And | 6.8 | 36 | 53.3 | 53.3 |
| Capella | 4.0-6.0 | 30 | 53.0 | 52.3 |

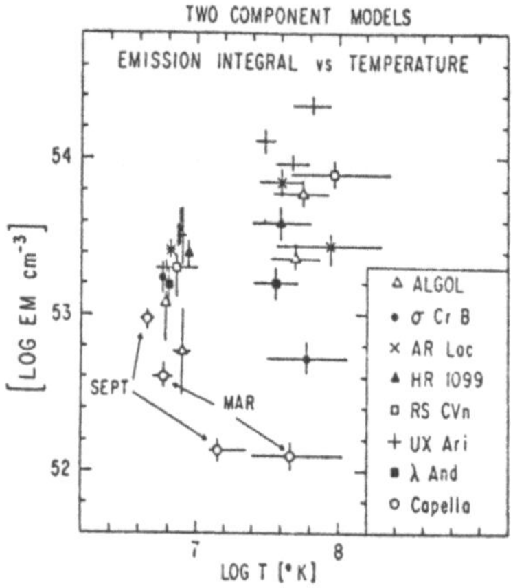


Fig. 6 : The Swank et al (1981) plot of emission integral against temperature for the 8 systems observed by the SSS. The low and high T components are clearly segregated (some sources were observed several times).

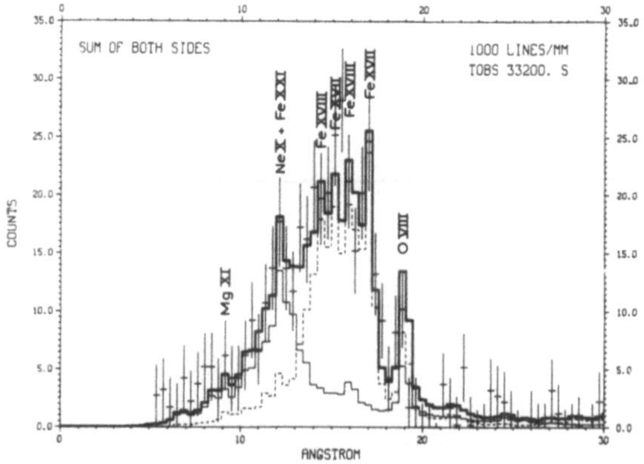


Fig. 7 : The Einstein OGS high resolution spectrum of Capella (adopted from Gronenschild et al, 1982). The model fit is the sum of 4 and 8 x $10^6$K components and clearly shows individual ionic species.

et al, 1979) and subsequent soft X-ray and optical study (Charles, Walter and Bowyer, 1979; Kimble, Kahn and Bowyer, 1981) of 2A1052 + 606 ( = SAO 015338 = BD + 61°1211 = DM UMa), another X-ray source to exhibit flaring behaviour. A summary of all 6 such objects is given in Table 3. This flaring or transient-like activity may represent yet another extrapolation of the solar analogy. Of the stars in Table 3 it should be noted that:

(i) HR1099 has the strongest Hα emission of the objects in Hall's (1976) list;

(ii) the 3A1431-409 counterpart has much stronger (and steady) Hα emission than HR1099 (Booth & Charles, 1982);

(iii) BD+61°1211 and HD 224085 have much larger amplitude photometric waves ($\Delta V \sim 0.3$; Kimble et al, 1981; Rucinski, 1977) than normal RS CVn's ($\sim 0.1 - 0.2$ mag) and both are asymmetric (see figure 8).

## 6. THE CORONAL AND STARSPOT MODELS

It was recognised soon after the discovery of X-rays from RS CVn's that some containment mechanism was required for the $10^7$K and hotter gas (Cash et al, 1978; Walter et al, 1978). As pointed out by Swank and White (1980) the ratio of pressure scale height to stellar radius (assuming approximate hydrostatic equilibrium) is $H/R = 0.7\ (T/10^7)\ (R/R_\odot)\ (M_\odot/M)$ which is $\sim 1$ for Capella. Hence the gas is not controlled by the gravitational field. Instead Walter et al (1980) and Swank et al (1981) use the coronal loop model of Rosner, Tucker and Vaiana (1978 ; hereafter RTV; see also these proceedings) to interpret their data. In this model the X-ray emitting gas is confined at constant pressure to loops by magnetic fields. It is possible to use the RTV scaling law for loops of length $L = 5(T/10^7)^3\ p^{-1}$ to estimate the number of loops and area of the star covered by them given the value of pressure p. Since this is $\sim 0.1 - 10$ dyne cm$^{-2}$ (for Capella, λ And and UX Ari) then L can be comparable to the stellar radius for the low T component but is comparable to the binary separation for the high T component unless the pressure is much higher.

It must be noted that if the X-ray emitting gas is confined to loops close to the surface of the active star then you would expect:

(i) an X-ray modulation of the light curve synchronised with the phometric wave;

(ii) X-ray eclipses.

The most extensive search for evidence of an eclipse has been made in AR Lac and none was seen (Swank & White, 1980). Nor have any phase-dependent effects been seen. The variations observed are thus presumed to be due to flares and these would arise from hot gas in the radio emitting region between the 2 stars (see Simon & Linsky, 1980).

To fit the observed photometric waves the starspots must cover between 6 - 18% of the surface of the active component (Eaton and Hall,

Table 3 : RS CVn / X-RAY TRANSIENT AND FLARING SYSTEMS

| X-ray source | Optical Star | $L_x$ (max) erg $s^{-1}$ | $V_{max}$ | ref. |
|---|---|---|---|---|
| 4U0336+01 | HR1099 | $4 \times 10^{32}$ | 5.9 | White, Sanford & Weiler, 1978 |
| 2A1052+606 | BD+61°1211 | $10^{32}$ | 9.4 | Charles, Walter & Bowyer, 1979 |
| H0123+075 | HD8357 | $10^{33}$ | 7.2 | Garcia et al 1980; Hall et al, 1982 |
| 4U1137-65 | HD101379 | $4 \times 10^{31}$ | 5.1 | Garcia et al 1980 |
| A0000+28 | HD224085 | $10^{32}$ | 7.2 | Schwarz et al, 1981 (= II Peg) |
| 3A1431-409 | "V532 Cen" | $\lesssim 8 \times 10^{33}$ | 12.4 | Booth & Charles, 1982; McHardy et al, 1982 |

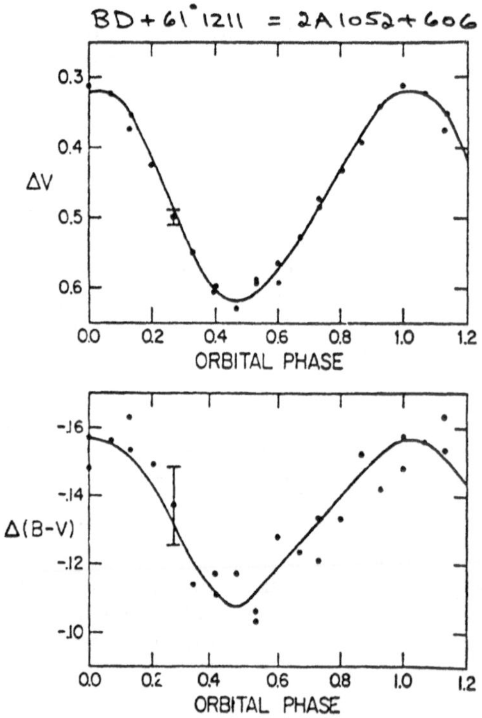


Fig. 8 : The large amplitude photometric wave and colour of BD+61°1211 (adopted from Kimble et al, 1981).

1979; Dorren et al, 1980). Kimble et al (1981) show that this fraction must increase to ~1/3 in the case of BD + 61°1211 (with a lower limit of 16%) but again the starspot model does fit the observations very well (figure 8).

## 7. FUTURE OBSERVATIONS

The work I have tried to summarise above clearly shows the tremendous resurgence of observational interest in these very active cool stars at all wavelengths. As these proceedings show this has been matched by theoretical developments in coronal physics. At X-ray wavelengths the immediate future holds the bright prospect of the EXOSAT mission (due for launch Nov. 1982) which should be able to capitalise on the Einstein work shown here. The major areas of study include:

(i) more sensitive and longer observations of RS CVn's at high spectral resolution;

(ii) further study of RS CVn's with imaging systems to obtain crude spectra and variability information;

(iii) more extensive simultaneous observations with IUE and EXOSAT as well as ground-based optical and radio coverage.

## 8. REFERENCES

Agrawal, P., Riegler, G.R. & White, N.E.: 1981, M.N.R.A.S., 196, 73P.
Booth, L. & Charles, P.A.: 1982, M.N.R.A.S., (in press).
Cash, W., Bowyer, S., Charles, P., Lampton, M., Garmire, G. & Riegler, G.: 1978, Ap.J., 223, L21.
Charles, P., Walter, F. & Bowyer, S.: 1979, Nature, 282, 691.
Dorren, J.D., Siah, M.J., Guinan, E.F. & McCook, G.P.: 1981, A.J., 86, 572.
Eaton, J.A. & Hall, D.S.: 1979, Ap.J., 227, 907.
Feldman, P.A., Taylor, A.R., Gregory, P.C., Seaquist, E.R., Balouek, T.J. & Cohen, N.L.: 1978, A.J. 83, 1471.
Garcia, M., Baliunas, S.L., Conroy, M., Johnston, M.D., Ralph, E., Roberts, W., Schwartz, D.A. & Tonry, J.: 1980, Ap.J. 240, L107.
Gibson, D.M., Hjellming, R.M. & Owen, F.N.: 1975, Ap.J., 200, L99.
Gronenschild, E.H.B.M., Mewe, R., Westergaard, N.J., Heise, J., Seward, F.D., Chlebowski, T., Kuin, N.P.M., Brinkman, A.C., Dijkstra, J.H. & Schnopper, H.W.: Sp.Sci.Rev. 30, 185.
Hall, D.S.: 1972, P.A.S.P., 84, 323.
Hall, D.S.: 1976, IAU Colloquium# 29, 287, in Multiple Periodic Variable Stars, ed. by W.S. Fitch (Holland-Reidel).
Hall, D.S., Henry, G.W. & Louth, H.: 1982, Ap.J., 257, L91.
Hoffleit, D. & Jaschek, C.: 1982, Bright Star Catalogue, 4th Edition, Yale University Observatory.
Joyce, R.M., Becker, R.H., Birsa, F.B., Holt, S.S. & Noordzy, M.P.: 1978, IEEE Trans.Nucl.Sci. 25, 453.

Kimble, R.A., Kahn, S.M. & Bowyer, S.: 1981, Ap.J., 251, 585.
McHardy, I.M., Pye, J.P., Fairall, A.P., Caldwell, J. & Spencer-Jones, J.: 1982, M.N.R.A.S. (submitted)
Mewe, R., Schrijver, C.J. & Zwaan, C.: 1981, Sp.Sci.Rev. 30, 191.
Pallavicini, R., Golub, L., Rosner, R., Vaiana, G.S., Ayres, T. & Linsky, J.L.: 1981, Ap.J., 248, 279 (PAL).
Popper, D.: 1980, in Close Binary Stars: Observations and Interpretation ed. Plavec et al (Boston: Reidel) p.387.
Rosner, R., Tucker, W.H. & Vaiana, G.S.: 1978, Ap.J., 220, 643 (RTV).
Rothschild, R., Boldt, E., Holt, S., Serlemitsos, P., Garmire, G., Agrawal, P., Riegler, G., Bowyer, S. & Lampton, M.: 1979, Sp.Sci.Instr. 4, 269.
Rucinski, S.M.: 1977, PASP 89, 280.
Schrijver, C.J., Mewe, R. & Zwaan, C.: 1982, Astron.Astrophys. (submitted) (SMZ).
Schwartz, D.A., Bradt, H., Briel, U., Doxsey, R.E., Fabbiano, G., Griffiths, R.E., Johnston, M.D. & Margon, B., 1979, A.J. 84, 1560.
Schwartz, D.A., Garcia, M., Ralph, E., Doxsey, R.E., Johnston, M.D., Lawrence, A., McHardy, I.M. & Pye, J.P.: 1981, M.N.R.A.S. 196, 95.
Skumanich, A.: 1972, Ap.J., 171, 565.
Simon, T. & Linsky, J.L.: 1980, Ap.J., 241, 759.
Swank, J.H. & White, N.E.: 1980, in Cool Stars, Stellar Systems and the Sun ed. A.K. Dupree, SAO Sp.Rep. 389, 47.
Swank, J.H., White, N.E., Holt, S.S. & Becker, R.H.: 1981, Ap.J., 246, 208.
Walter, F.M.: 1981, Ap.J., 245, 677.
Walter, F.M.: 1982, Ap.J., 253, 745.
Walter, F.M. & Bowyer, S., 1981, Ap.J., 245, 671 (WB).
Walter, F.M., Cash, W., Charles, P.A. & Bowyer, S.: 1980, Ap.J. 236, 212.
Walter, F., Charles, P. & Bowyer, S.: 1978, Ap.J.Letters 225, L119.
White, N.E., Holt, S.S., Becker, R.H., Boldt, E.A. & Serlemitsos, P.J.: 1980, Ap.J., 239, L69.
White, N.E., Sanford, P.W. & Weiler, E.J.: 1978, Nature, 274, 569.

## DISCUSSION

Budding: Have you looked at the Durney et al dynamo number parameter which will run together quantities like colour and rotation period. One has the impression that it may take out some of the scatter from some of your correlations. Secondly I question your lumping together of the RS CVn and Algol systems. I think that these stars are likely to be structurally rather different. They may be superficially similar in some respects but the Algol subgiants have been massive stars and are now in the thin shell-burning phase with highly collapsed cores. The RS CVn stars have just left the main sequence and are in the thick shell-burning phase.

Charles: I think you are probably right there. To take the second point first, I have not tried to say the Algols are RS CVn's. I rather want to say that the Algols exhibit the RS CVn phenomenon and that they fall below the RS CVn's activity relations. They are related but one does not know exactly how. So I present as much of the observational material as possible so that people can make up their own minds. I agree with you that they are different but on the other hand there are many similarities such as when one compares the X-ray spectrum of Algol with any of the RS CVn's. With regard to your first point, I did not actually do any of that work so I cannot comment fully. However I thought that something along the lines of your suggestion had been mentioned earlier on. Perhaps Fred Walter or Dr. Pallavicini would comment?

Catalano: I agree with you about the classification of secondary components on Algol-type stars but there is evidence that they show RS CVn features, at least from a photometric point of view. There are changes in the depth of eclipse in Algols, a time when we see the secondary, and these may be related to the out-of-eclipse variability in RS CVn's. We cannot see the wave variation outside of eclipse in the Algol because the primary is so bright. When one makes a Fourier-type analysis of the outside-of-eclipse light in Algols however there is some evidence of RS CVn-type variability also.

# PRELIMINARY RESULTS OF A FIVE-YEAR SURVEY OF RADIO EMISSION FROM RS CVn AND SIMILAR BINARIES

P.A. Feldman
Herzberg Institute of Astrophysics,
National Research Council of Canada, Ottawa, Canada

This is a summary of the systematic program of $\lambda$3cm radio observations of RS CVn and similar binaries which has been carried out since 1977 at the Algonquin Radio Observatory*. The observations were made at X-band frequencies (10-11 GHz) with the 46m telescope using the same observing procedures described in Feldman et al. (1978). The detection limit is generally 25 mJy, although it is possible to do somewhat better than this if the star is located well off the galactic plane and if it maintains an elevated flux level (i.e. $\gtrsim$ 15 mJy) for several hours. The observing list has grown roughly four-fold since the start of the program in 1977. It now encompasses 65 of the 69 "bona fide" RS CVn binaries compiled by Hall (1982); the rest are suspected RS CVn binaries, BY Dra variables, Algol-type binaries, and W UMa binaries. To date, a total of some 85 observing days, representing more than 1000 hours of telescope time, has been devoted to this program.

When the program was first instituted, there were three main objectives: (1) to detect new radio binaries; (2) to study the physics of the radio sources by observing the complete rise and decay phases of at least several flares; and (3) to detect major flaring events early enough to enable useful collaboration with observers at other wavebands.

(1) Five new RS CVn binaries were detected at $\lambda$3 cm, adding substantially to the number of these systems known to emit radio waves. These are SZ Psc (detected independently by Owen and Gibson (1978); HR 5110 = BH CVn; HK Lac; HR 8575 = HD213389 = V350 Lac; and HR 9024, which is not a "bona fide" RS CVn binary

* The Algonquin Radio Observatory is operated by the National Research Council of Canada, Ottawa, as a national radio astronomy facility.



*P. B. Byrne and M. Rodonò (eds.), Activity in Red-Dwarf Stars, 429–437.*

according to Hall's criteria but which may be either such a system or an FK Comae star seen pole-on (Feldman and Fraquelli 1983). All the new detections reported here are restricted to sources with flux densities in excess of 25 mJy and for which extensive enough observing was done at different times and parallactic angles to give reasonable confidence in the results. In addition, several possible new detections have been made at the $\approx$ 15 to 25 mJy level, but these require further confirmation.

(2) The complete rise and decay profiles of individual RS CVn radio flares have now been observed many times (see, e.g., Feldman et al. 1978, Hjellming and Gibson 1980). The basic radiation mechanism at short cm wavelengths now seems firmly established as incoherent gyro-synchrotron radiation of electrons with Lorentz factors $\gamma \lesssim 10$ in magnetic fields $B \approx 30$-100 G. Coherent gyro-synchrotron or possibly plasma wave emission processes may be operative at longer radio wavelengths ($\lambda \approx 10$ to 20 cm) as evidenced by the high degree and rapid variability of the circularly polarized flux which are sometimes observed. However, these effects have not yet been found at the relatively short ($\lambda$3cm) wavelengths used in this study.

(3) A major goal and success of this program has been the early detection of major flaring events. This has enabled valuable "target-of-opportunity" collaboration with observers at other wavebands. The prime example of such joint efforts occurred during the 1978 February-March outburst of HR 1099 = V711 Tau (Feldman et al. 1978; see special issue of Astron. J., December 1978). However, several other RS CVn superflares have been observed extensively, especially the 1978 Dec. - 1979 Jan. flare of UX Ari and both major events of HR 5110 (see Table 1). In general, it has proved feasible to obtain timely, useful collaboration from ground-based optical spectroscopists and from IUE observers (under a pre-arranged "target-of-opportunity" program). However, it has not been possible to obtain radio "back-up" at short notice, especially after the NRAO interferometer ceased operation as a national facility several years ago. On rare occasions, unscheduled VLBI observations of major radio flares were arranged with difficulty, but the most useful VLBI measurements to date have resulted from a fortuitously pre-scheduled program (Cohen et al. 1983). For operational reasons "target-of-opportunity" X-ray observations could never be scheduled during the Einstein era.

TABLE 1. Major Radio Flares of RS CVn Binaries observed at Algonquin Radio Observatory (1977-1981)

| Star | Distance (pc) | Date | Peak Flux Density (mJy) | Observing Frequency (GHz) | Peak Radio Luminosity (Lu)* |
|---|---|---|---|---|---|
| UX Ari | 55 | 1978 Dec.-<br>1979 Jan. | 255 | 10.65 | 920 |
| | | 1979 Dec. | 140 | 9.90 | 500 |
| HR 1099 = V711 Tau | 33 | 1978 Feb.-Mar. | 960 | 10.52 | 1300 |
| | | 1979 July | 1210 | 10.76 | 1600 |
| HR 5110 = BH CVn | 52 | 1979 May-June | 460 | 10.52 | 1500 |
| | | 1981 April | 410 | 10.46 | 1300 |
| AR Lac | 47 | 1977 May | 550 | 10.48 | 1500 |
| HD216489 = HR 8703 | 200 | 1980 June | 40 | 10.48 | 2000 |
| SZ Psc | 100 | 1977 May | 65 | 10.48 | 780 |
| | | 1977 Sept. | 60 | 10.29 | 720 |
| | | 1978 April | 85 | 10.52 | 1000 |
| | | 1979 Nov.-Dec. | 110 | 10.82 | 1300 |
| II Peg = HD224085 | 29 | 1979 Nov.-Dec. | 255 | 10.82,9.90 | 260 |
| | | 1980 Sept. | 170 | 10.45 | 170 |

* 1 Lu $\equiv 10^{15}$ erg s$^{-1}$ Hz$^{-1}$

The (daily-averaged) absolute radio luminosity distribution for RS CVn binaries was first determined at 5 GHz for five comparable systems (AR Lac, UX Ari, HR 1099, SZ Psc, and II Peg) by Owen and Gibson (1978). This survey was done with sensitive instruments and thus probably provides a good estimate of the low-luminosity (≲ 250 Lu) portion of the radio luminosity function during 1974-1977. Owen and Gibson

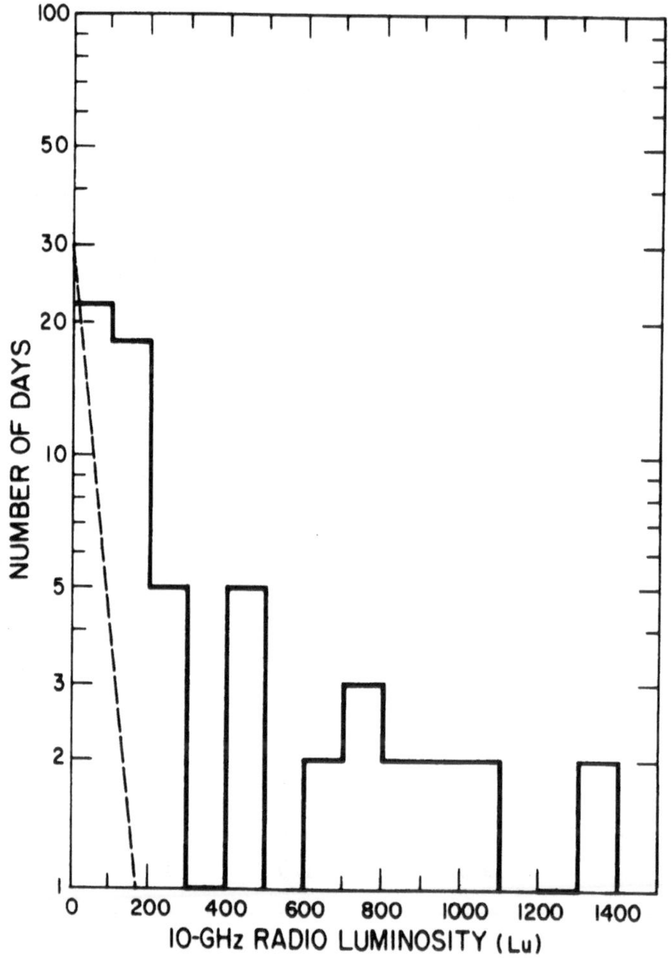


Figure 1. Cumulative differential 10-GHz radio luminosity distribution of six RS CVn binaries (UX Ari, HR 1099, HR 5110, AR Lac, SZ Psc, and II Peg). The ordinate is the number of independent days that these systems were detected with a daily-averaged radio luminosity given as the abscissa (in Lu, where 1 Lu ≡ $10^{15}$ erg $s^{-1}$ $Hz^{-1}$). The dashed line is the cumulative distribution obtained for five of these binaries (excluding HR 5110) at 5 GHz by Owen and Gibson (1978).

approximated the distribution of radio luminosities above the median luminosity (37 Lu) as a simple exponential function with an e-folding luminosity of 50 Lu. Therefore, exceedingly long times would seem to be required before an observer can expect to detect a radio superflare (>250 Lu). For example, 1.3 million years of observing is formally estimated as needed before a 1000-Lu flare will be detected from one of the five RS CVn binaries considered! However, even before the Algonquin survey it was clear that RT Lac was substantially over-luminous with respect to Owen and Gibson's luminosity function. Thus it is perhaps no great surprise that many radio superflares have subsequently been detected from these (and other) systems.

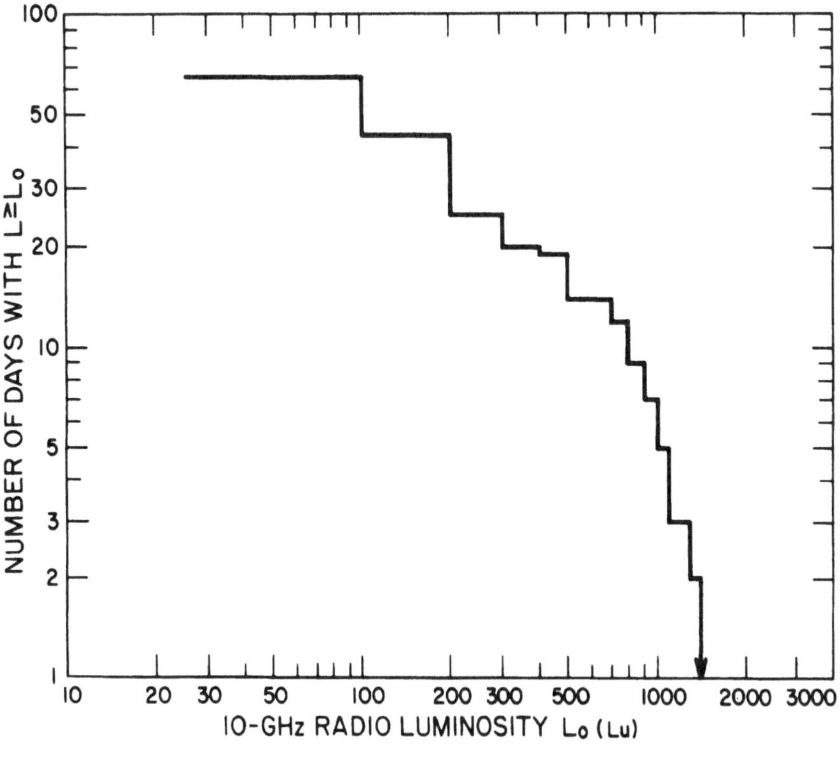


Figure 2. Cumulative integrated 10-GHz radio luminosity distribution of six RS CVn binaries (UX Ari, HR 1099, HR 5110, AR Lac, SZ Psc, and II Peg). The ordinate is the number of independent days that the daily-averaged radio luminosity equaled or exceeded the value given as the abscissa (in Lu, where 1 Lu ≡ $10^{15}$ erg $s^{-1}$ $Hz^{-1}$). The arrow indicates that no (daily) events exceeding 1400 Lu have been recorded.

Figure 1 shows the cumulative differential radio luminosity distribution measured at λ3cm for the five binaries considered by Owen and Gibson plus HR 5110 which is a similar system. Daily averages were used to make the data more directly comparable with the earlier work. A conscious attempt was made to eliminate carefully measurements performed because earlier observations had shown certain stars to be flaring. As an example, in the case of HR 1099 the period 1978 Feb.-March is represented by only two data, corresponding to when this star would normally have been surveyed. It is obvious from the figure that the exponential distribution suggested by Owen and Gibson (1978) is inappropriate to describe the frequency of occurrence of major radio flares in RS CVn systems. Evidence is found for a cutoff in the absolute radio luminosity, at ≈ 2000 Lu. This is seen even more clearly in the cumulative integrated radio luminosity distribution, given in Figure 2, and in the peak radio luminosity results given in Table 1. The cutoff is probably due to the brightness-temperature limit imposed by the $\gamma \lesssim 10$ electrons themselves for radio sources of stellar dimensions (Feldman 1983). In plotting the radio luminosity distributions (Figs. 1 and 2) upper limits have been disregarded. These affect only the very low end (< 300 Lu) of our observed range of values and hence are insignificant to the basic conclusion that radio superflares are unexpectedly commonplace. A more complete version of these results will be published elsewhere (Feldman 1983).

Finally, the 1981 April outburst of HR 5110 is shown in Figure 3, as an example of a RS CVn radio superflare. A routine survey observation detected the binary in outburst at ≃ 240 mJy on 1981 April 4, and Mark III λ6cm VLBI observations were made on the two subsequent days (Feldman 1981). Unfortunately, much of the data was accidentally erased from our magnetic tapes before it could be processed, but some useful data survived for the April 6 observations. Preliminary reduction of these data (a one-baseline VLBI "snapshot") indicates that the fringe-visibility amplitude was approximately 0.5 with a fringe spacing of 3.7 milliarcsec. This implies that, during the later (plateau) portion of the outburst, half the radio flux was emitted from within a volume whose size scale was ≈ $3\times10^{12}$ cm, or several times the binary star separation, with a brightness temperature $T_B > 3\times10^8$ K. This is in itself not surprising (cf. Feldman et al. (1978)). However, it also means that fully half the radio flux at 6cm was emitted over a size scale greater than this dimension. If we adopt the picture of interacting magnetic flux tubes in a binary system as proposed by Simon et al. (1980), the loops containing radiating MeV electrons would have to be disrupted on a very large scale indeed in the later stages of the flare. This is reminiscent of the behaviour of giant moving Type IV solar radio bursts, although the radio frequencies which are produced are, of course, quite different. IUE spectra taken coincident with the radio flaring (see Fig. 3) show evidence for hot gas with large line-of-sight motions, possibly indicating the infall of plasma onto one or both stars of the binary system (Linsky 1981). Similar behaviour was seen in the case of a superflare of UX Ari (Simon _et al._ 1980).

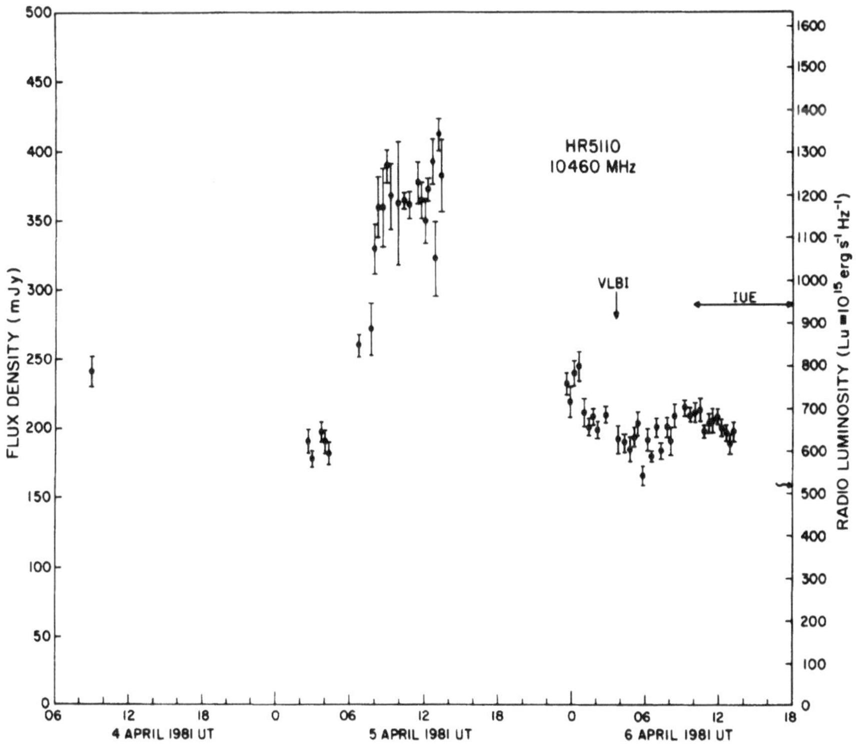


Figure 3. Radio "light curve" of HR 5110 during the 1981 April flare. The ordinates are the 10460-MHz flux density in mJy (1 mJy ≡ $10^{-26}$ erg $s^{-1}$ $Hz^{-1}$ $cm^{-2}$) and radio luminosity in Lu (1 Lu ≡ $10^{15}$ erg $s^{-1}$ $Hz^{-1}$) measured at the Algonquin Radio Observatory as a function of UT. The vertical arrow indicates the time of the VLBI measurement; the horizontal straight arrows indicate the duration of the IUE observations. The wiggly arrow indicates the flux-density/radio-luminosity levels maintained for several days after 6 April.

## References

Cohen, N.L., Feldman, P.A., Crane, P.C., and Geldzahler, B.J., 1983, in preparation
Feldman, P.A., 1981, I.A.U. Circ. No. 3591
Feldman, P.A., 1983, in preparation
Feldman, P.A., and Fraquelli, D.A., 1983, in preparation
Feldman, P.A., Taylor, A.R., Gregory, P.C., Seaquist, E.R., Balonek, T.J., and Cohen, N.L., 1978, Astron. J. 83, 1471
Hall, D.S., 1982, private communication and Space Science Reviews (in preparation)
Hjellming, R.M., and Gibson, D.M., 1980, in Proceedings of IAU Symp. 86, "Radio Physics of the Sun", eds. M.R. Kundu and T.E. Gergely (D. Reidel Publ. Co., Dordrecht: Holland, 1980), p. 209
Linsky, J.L., 1981, I.A.U. Circ. No. 3591
Owen, F.N., and Gibson, D.M., 1978, Astron. J. 83, 1488
Simon, T., Linsky, J.L., and Schiffer, F.H., 1980, Astrophys. J. 239, 911

## DISCUSSION

Kuijpers: Could you follow the time development of this flare with VLBI?

Feldman: The VLBI was unfortunately just a "snapshot" study. We had multiple baselines for only 15 minutes. However, verbal reports of unpublished work on such observations indicates that flares remain unresolved on an angular scale of 1 milli-arc sec in the early phases of the flare but may be partially resolved in the later phases. People all think of moving Type IV bursts in this context. I would anticipate that progress in this field will begin to show expansions during these events.

Venugopal: What is the lowest frequency at which the RS CVn stars have so far been detected?

Feldman: Dave Gibson would probably know this better than I would. However as far as I know the answer is about 21 cm. Would you like to comment, Dr. Gibson?

Gibson: Yes, 21 cm seems to be the longest wavelength at which RS CVn flares have been detected so far. I would like to add a comment in defence of your jabs at us that we must have made a mistake in our figures because we have observed a 1500 Luminosity units (Lu's) flare on RT Lac. That would bring us right to the top of your distributions (Figs. 1,2).

Feldman: But you did not include RT Lac in the figure which I referenced.

Gibson: No, I may not have done but it should have been there. So we would not have concluded the same thing either.

Catalano: Have you any idea of the size of the coronal region in the BY Dra or short period RS CVn binaries?

Feldman: Not me because BY Dra binaries have not been observed in the radio to my knowledge.

Catalano: As I showed this morning there are differences in that one sees optical flares in BY Dra and short-period binaries but not in long-period ones. So are the radio flares you observe related to the magnetosphere outside the system or at least that which envelopes the entire system? Or could they arise from the photosphere in a way different from what is observed in the short-period systems or flare stars?

# ACTIVITY CORRELATIONS IN CLOSE BINARY SYSTEMS

Gibor Basri[1] and Robert Laurent
University of California, Berkeley
[1]Guest Observer, International Ultraviolet Satellite

The best way to learn about the detailed physical conditions associated with stellar activity is to construct detailed self-consistent model stellar atmospheres based on the wide range of diagnostics now available. A quicker rough analysis of the activity can be accomplished by studying the interrelations of the various diagnostics. Tight correlations between diagnostics suggest physical association, and the power law indices of the correlation can help to elucidate the general physical relation the diagnostics have with each other. Our aim in this study was to observe with the IUE satellite a homogeneous sample of stars for which there is extensive coronal data. The X-ray studies of close binaries by Walter and Bowyer (1981) and Walter (1982) provided such a sample; these stars are also accessible to the IUE because of their tendency to strong activity. Because of intrinsic stellar variability, this type of study is best carried out when all diagnostics are measured simultaneously. We have come fairly close to this ideal; only the coronal data is non-contemporaneous for most of the sample. Our reduction of the IUE data is quite similar to that of Ayres, Marstad and Linsky (1981, hereafter AML) and we have used essentially the same diagnostics.

In order to study the relations among the diagnostics we have constructed correlation plots in which the normalized flux in each diagnostic is plotted in a log-log plane against various canonical diagnostics, normalized to the bolometric flux measured at the earth (corrected for inactive components in the binary system). Thus we are studying a measure of the activity per unit surface area, free of errors in the stellar radius or distance. We chose three canonical diagnostics for which to make the correlations: MgII for the chromosphere, CIV for the transition region, and X-rays for the corona. In order to make a good quantitative assessment of the correlations and their significance, we have employed a numerical least-squares linear fitting program which minimizes the perpendicular distance to the data points from the fitted line. We have examined the relations both for the binary sample as a whole, which includes active K and G subgiant RS CVn stars and G main sequence components, and for the K subgiants separately. These latter



*P. B. Byrne and M. Rodonò (eds.), Activity in Red-Dwarf Stars, 439–442.*

comprise an active subclass for which the fundamental stellar parameters are very similar and which have all just evolved off the main sequence. Due to tidal synchronization with their companions, these stars tend to have much higher rotational velocities than their field counterparts, which may account for their increased activity (see also Basri, Laurent and Walter 1982, hereafter BLW).

Our results for diagnostics compared to MgII correspond to the data presented in AML; we also see fairly good correlations between most of the diagnostics (see Table 1). The fits are uniformly much poorer for the full sample of stars than for the K subgiants. Examination of the data in detail reveals that the different subgroups of stars tend to have different general levels of activity. The full sample slopes are therefore somewhat misleading, so we concentrate on the results for the K star subgroups. For these, our correlation lines lie above the ALM lines and are steeper. The less active binaries show intermediate behavior. Thus it appears that not only do higher temperature lines increase more rapidly with activity, but the amount of increase itself also becomes greater with increasing activity. The lines fall into three classes of behavior: OI (with large scatter), CI and NV with slope $\sim$ 2.0, HeII, SiII, CII, CIV with slopes 2.5-3.0, and SiIV and X-rays with slopes $>$ 4.0. The formal errors in our slope determinations are less than 0.1. Note the puzzling behavior of NV, which should be formed at $10^5$ K and might be expected to behave most like a coronal line.

The slopes of the relations are naturally much shallower when viewed relative to the transition region line CIV (see Table 1). In fact most of the lines have unit slope relative to this line for the K subgiants, and generally display the smallest scatter about the correlation line. Some of the relations are quite tight even with the full sample of stars, particularly CII. A lot of the other diagnostics behave as though they were also formed in the transition region for the K subgiants, including CI, OI, SiII, HeII, CII, and NV (with larger scatter). Only MgII shows significantly less than unit slope, while SiIV and X-rays show increasingly greater slope; these relations all have larger scatter. The different groupings relative to MgII may therefore partly just reflect the greater physical disconnection of this diagnostic in both temperature and optical thickness from the other lines, and partly the greater sensitivity of steeper slopes to small fluctuations in the data.

The X-ray relations show the most scatter; enough to make them unreliable as predictors of activity. This may reflect intrinsic stellar variability coupled with non-similtaneity of the X-ray data, and the fact that the corona is more sensitive to changes in activity levels than the cooler diagnostics. The most interesting aspect of these relations is that the line which shows most nearly unit slope is SiIV (not either HeII or NV). This is consistent with our other results and those of BLW that SiIV has the behavior most like coronal, and NV actually looks almost like a chromospheric line. Hartmann, Dupree, and Raymond (1980) have proposed that earlier studies of the formation of the HeII line at λ1640 which suggested a strong dependence on coronal X-ray fluxes are corroborated by a linear dependence of HeII and X-rays in their data. We find in fact that the slope of the K star relation is about 2/3 (slightly more for the full sample), but there is a very definite connection with X-rays in the sense that this relation shows much lower scatter than for any other diagnostic.

We have not yet carried out a detailed analysis to explain our results, so we offer here only a few possibilities. The behavior of the diagnostics would best be explained by an increasingly greater pressure at the transition region (and thus also in the corona) with increasing activity. One might expect something like a $P^2$ dependence of diagnostics in this regions but not the mid-chromosphere due to both the increasing populations of ions and increasing collisional excitation of the resonance lines. As the transition region moves low enough, even low ionization stages like CI and SiII are able to form in the lower transition region where collisional excitation is increasingly important. In the less active single stars some of these lines are more chromospheric, leading to the ALM results. Studies of the period-activity relations for these close binaries (BLW) leads to the suggestion that increasing total magnetic flux may largely be responsible for the increasing activity. The additional sensitivity of the hottest diagnostics to activity may derive from a more purely magnetic dependence of the heating and temperature in the corona compared to the chromosphere.

Table 1: Activity Correlations

| | vs. $MgII/\ell_{bol}$ | | vs. $CIV/\ell_{bol}$ | |
|---|---|---|---|---|
| | Slope | $\chi^2[\langle\sigma\rangle\sim .17]$ | Slope | $\chi^2[\langle\sigma\rangle\sim .17]$ |
| MgII 13(23)[1] | - | - | .4(.6) | 1.2(1.9) |
| OI 12(24) | 1.9(1.4) | .9(1.0) | .9(.7) | .5(1.2) |
| CI 10(21) | 2.0(1.4) | .4(1.3) | .9(.8) | .4(1.6) |
| SiII 12(15) | 2.9(2.1) | .9(2.0) | .9(.7) | .5(1.2) |
| HeII 11(20) | 2.6(2.0) | .6(.8) | .9(1.2) | .5(1.3) |
| CII 12(25) | 3.0(1.75) | .7(1.3) | 1.0(.9) | .7(.6) |
| CIV 13(23) | 2.8(1.7) | .7(1.6) | - | - |
| SiIV 11(19) | 4.3(2.6) | .4(1.6) | 1.25(1.1) | .8(1.5) |
| NV 11(20) | 2.3(1.8) | .4(.1) | 1.0(1.0) | 2.0(2.6) |
| X-ray 17(30) | 4.8(2.7) | .6(1.2) | 1.5(1.6) | .95(.7) |

[1]number of stars used: K RSCVn (all binaries)

Ayres, T., Marstad, N., and Linsky, J.: 1981, Ap.J. 247, p.545.
Basri, G., Laurent, R., and Walter, F.: 1983, IAU Symp. 102, J.Steflo(ed.)
Hartmann, L., Dupree, A.K., and Raymond, J.C.: 1982, Ap.J., in press.
Walter, F.: 1982, Ph.D. Thesis.
Walter, F. and Bowyer, S.: 1981, Ap.J. 247, p.545.

DISCUSSION

Walter: Is not surprising that you get such a tight correlation for He since the observations are non-contemporaneous and it varies all over the place by factors of up to an order of magnitude?

Basri: That's true. In fact I find the tightness of many of these relations surprising. Even when the data is non-simultaneous the correlations are better than you might expect.

Dupree: Do we actually know how much the X ray flux varies in RS CVn stars?

Basri: Yes, the X ray flux in a few of them has been observed to vary by as much as 0.5 dex and also flares have been seen. The rotation-activity correlations in fact show a scatter of about ± 0.3 dex. So from that alone one would expect them to be variable.

Bromage: I notice that there are no error bars on any of your points. Have you worked out correlation coefficients allowing for these errors and if so does the He II relation still appear tighter than the others?

Basri: We have done a fairly careful error analysis of this data. The table in fact gives both standard deviations and chi-squared values. I should also mention that the fits are not standard least squares fits in that the X and Y axes are dependent variables. We allow for errors in both quantities and we fit the points perpendicular to the line. These are the chi squares which result from that procedure. The measurement and intrinsic errors are reflected in the standard deviation. The standard error is in the log.

Bromage: So these are maximum likelihood fits. Are they straight line fits?

Basri: Yes. They are all straight line fits. Given the number of stars observed a more complicated fit is not warranted. The best case is C II vs C IV where the scatter is surprisingly low. So I believe that these relationship are physically significant and it is up to us to figure how these translate into the physics of the stellar atmosphere.

## IUE SPECTRA OF RS CVn STARS

A.D.Andrews[1], P.B.Byrne[1], C.J.Butler[1], J.L.Linsky[2], T.Simon[2], N.Marstad[2], M.Rodonò[3], C.Blanco[3], S.Catalano[3], E.Marilli[3] and V.Pazzani[3]

[1] Armagh Observatory, Armagh, N.Ireland
[2] JILA/NBS, University of Colorado, Boulder, USA
[3] Osservatorio Astrofisico and Università di Catania, Italy

### INTRODUCTION

The characteristic optical light curves of RS CVn - and BY Dra-type variables are believed to represent non-uniform distribution of dark spots akin to sunspots which are revealed by rotation. By analogy with the Sun, strong magnetic fields probably underlie this phenomenon, extending upwards into the chromosphere and corona, enhancing the emission from regions that overlie the spots. Previous work on the BY Dra variable AU Mic (Ref.1) did not clarify whether the fluxes from chromospheric and transition region lines were rotationally modulated in the sense that the phase of maximum emission was in register with spot visibility or minimum light. This important question prompted the need for further collaborative IUE, optical and radio work.

### OBSERVATIONS

Three RS CVn stars were observed during contiguous ESA-NASA-SRC shifts on 1-7 October 1981. Eight spectra of II Peg, ten of HR1099 and four of AR Lac, in each of IUE's long-(LW) and short-wavelength (SW) intervals were obtained at high LW dispersion and low SW dispersion (except for one high SW spectrum of HR1099). Observations were well distributed with respect to phase of the optical light curves. Simultaneous ground-based photometry was obtained in addition to data on the mean optical light curve prior to IUE observations. Also, the fine error sensor (FES) visual magnitudes reduced using recent calibrations (Ref.2) allow a fairly

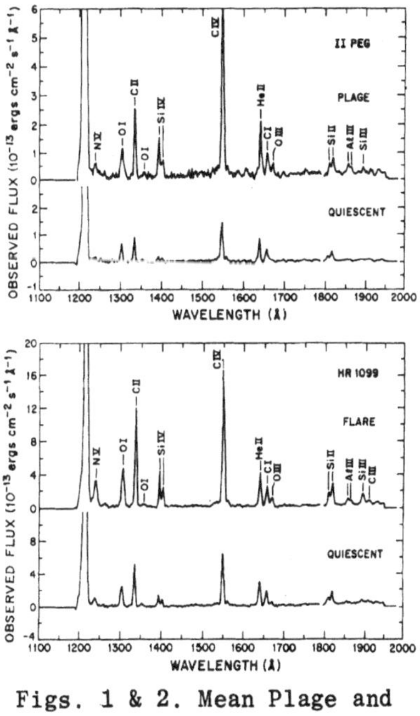


Figs. 1 & 2. Mean Plage and Quiescent Spectra & Flare Spectrum of II Peg & HR1099.



*P. B. Byrne and M. Rodonò (eds.), Activity in Red-Dwarf Stars, 443–444.*

accurate check on the behaviour of the stars adjacent to the spectral exposures. Radio coverage revealed flares on II Peg (60 mJy peak) on 2 October at 0200 UT and on HR1099 (180 mJy peak) extending over 24 hrs on 2-3 October.

## RESULTS AND DISCUSSION

Figs.1 & 2 show our summarized results for quiescent, plage and flare spectra from the data along with line identifications. Typical chromospheric transition region lines are seen ranging from OI to CIV and NV. Ground-based data show that all three stars had well established wavelike variations e.g. see top of Figs.3 & 4. Integrated line fluxes for the strongest chromospheric features are plotted below the photometric data as a function of phase in Figs.3 & 4. II Peg (Fig.3) clearly demonstrates that the maximum of the emission-line fluxes coincides with the minimum of the established mean light curve around phase 0.7-0.8, in agreement with the solar-type spot model. The variations at phase 0.32 and 0.79 illustrate variations in chromospheric emissions at small timescales. For HR1099 (Fig.4) the radio flare coincides with the emission maximum at phase 0.72 (shown by dotted line), and removing this peak, the strongest lines confirm the anti-phase correlation found in II Peg. Similar results are found in the sparser data available for AR Lac (not shown).

## CONCLUSIONS

We present clear evidence that in RS CVn stars the UV chromospheric and transition region line fluxes vary such that maximum line flux corresponds to optical minimum or maximum spot visibility, consistent with a solar-type dark spot model.

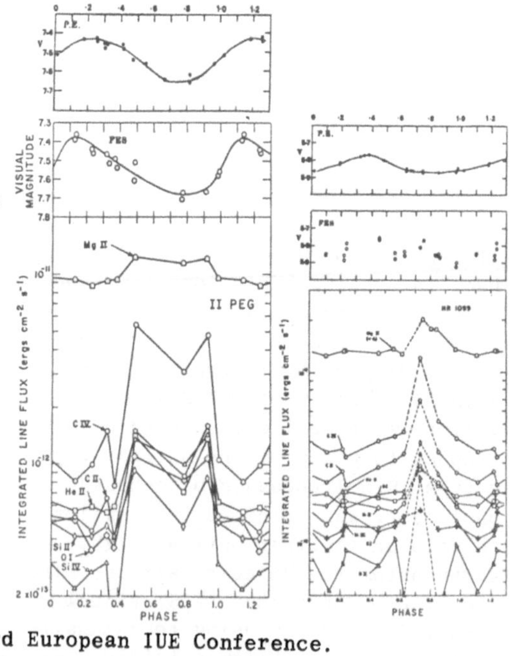


Figs. 3 & 4. Mean wave, FES mags & line flux against phase for II Peg & HR1099.

## REFERENCES

1. Linsky,J.L. et al. 1982. Third European IUE Conference.
2. ESA Memorandum 1981. FES Calibration at Vilspa by Holm & Rice.

# CORONAL AND CHROMOSPHERIC STRUCTURE IN AR LAC. II. PHYSICAL CHARACTERISTICS OF THE ATMOSPHERE

F. M. Walter,[1] D. M. Gibson,[2] and G. S. Basri[3]
[1]JILA, Univ. of Colo. and NBS, Boulder, CO 80309
[2]Dept. of Physics, New Mexico Tech., Socorro, NM 87801
[3]Dept. of Astronomy, Univ. of Calif., Berkeley, CA 94720

In this series of papers we report on an effort to deduce the morphology of the chromosphere and corona of AR Lacertae, an eclipsing RS CVn system (Hall 1976), via observations at UV, radio, and X-ray wavelengths during a complete orbital cycle. In another paper (Gibson, Walter and Basri 1982, Paper I), we present the observational data and general constraints on the morphology of the coronae. The full study will be published elsewhere (Walter, Gibson, and Basri 1982).

In modeling the corona of AR Lac, we have tried to keep the model simple, yet realistic. We have drawn upon solar analogy except that, in light of the much larger observed coronal emission measures, we model the corona as a volume of uniform volume emissivity rather than as individual loops. Our model is symmetric about the equator, and confines the emission to within a band in latitude.

Because of the rapid X-ray egress from primary minimum the G star corona must be small, with a slab height of $\lesssim 0.02\ R_*$. Conversely, the corona about the K star must be extended. Both must be asymmetric in longitude, and the extended component of the K star corona must be highly confined in latitude. There is also a small, poorly constrained coronal component ($R_c \sim 0.01\ R_*$) on the K star. Figure 1 shows the projected X-ray surface brightness contours as observed for the two stars.

From the observed emission measure $N_e^2V$, the coronal temperatures (Paper I), and the emitting volume V inferred from the model, we can compute mean densities $N_e$ and pressures $P = 2N_ekT$; these are given in Table 1 and are not model dependent. The derived pressure scale heights are much larger than the observed extent of the corona, implying the necessity of active confinement of the gas. The deduced pressures and densities are large, but are comparable to that of a hypothetical Sun covered with small flares (Vaiana and Rosner 1978). Transition region pressures appear to be significantly lower.

The simultaneous Einstein MPC data strongly suggest that during egress from primary eclipse we observed either a small flare somewhere in the system or the emergence from eclipse of a flaring region. If the latter, we can compute an upper limit to the size of the region ($<2 \times 10^{10}$ cm), and a lower limit of $\sim 10^{11}$ cm$^{-3}$ on the particle density. At this density the plasma frequency is ~18 GHz, which may explain the lack of a radio flare.

The asymmetries in the X-ray and UV light curves require that the coronae be highly asymmetric. Figure 2 is a sketch to scale of the



*P. B. Byrne and M. Rodonò (eds.), Activity in Red-Dwarf Stars, 445–446.*

system at this epoch. The leading (spotted) hemisphere of the G star contributes most of the X-ray and UV luminosity. The inner (substellar) trailing hemisphere of the K star is bright in the UV, while the inner leading quadrant is brightest in X-ray. We infer surface flux differences of more than a factor of three in the chromosphere and suggest that chromospheric models based upon homogeneous atmospheres may be unacceptable for highly active stars.

Table 1. AR Lac Coronal Parameters

| Component | G | $K_{extended}$ | $K_{inner}$ | Flaring Sun |
|---|---|---|---|---|
| Coronal radius ($R_*$) | 1.0-1.02 | 1.5-2.0 | 1.0-1.01 | --- |
| $\overline{N}e_{corona}$ ($cm^{-3}$) | $5.9\times10^{10}$ | $2.9\times10^{9}$ | $7.1\times10^{10}$ | $5\times10^{10}$ |
| $\overline{P}_{corona}$ (dynes $cm^{-2}$) | 96 | 24 | 67 | 140 |
| $\overline{N}e_{Tr}$ ($cm^{-3}$) | $>2\times10^{11}$ | --- | $10^{11}$ | |
| $P_{Tr}$ (dynes $cm^{-2}$) | >3 | --- | 1.7 | |

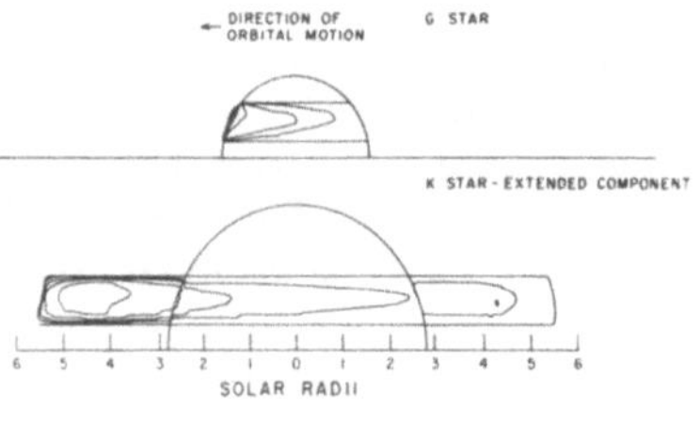


Figure 1. Contour plot of the projected X-ray surface flux at Earth. The contour levels are 0, 0.25, 0.5, 1.0, 2.0, 4.0, and 8.0×1.4E-33 ergs $cm^{-2}$ $s^{-1}$ for the G star, and 0, 0.5, 1.0, 2.0, 4.0×1.1E-33 ergs $cm^{-2}$ $s^{-1}$ for the K star.

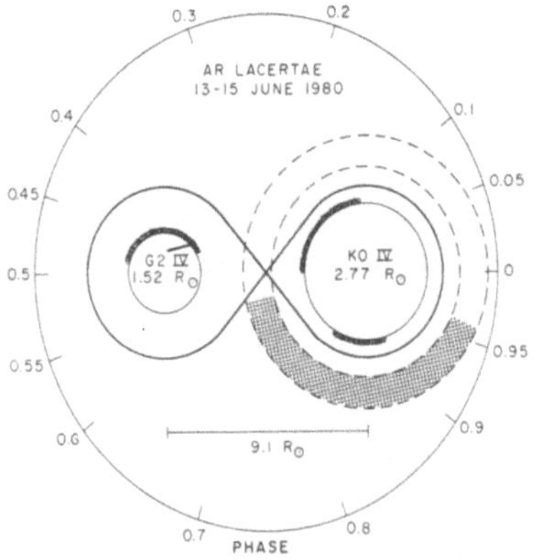


Figure 2. The state of the coronae of AR Lac. The dark highlighting on the stellar surface indicates bright coronal and chromospheric emission. The cross-hatched region indicates the extended component of the K star's corona. Orbital phases are indicated.

REFERENCES

Gibson, D., Walter, F. and Basri, G. 1983, IAU Symp. 102 (Paper I).
Hall, D. S. 1976, in IAU Colloquium No. 29 (Dordrecht: Reidel).
Vaiana, G. S., and Rosner, R. 1978, ARAA 16, 393.
Walter, F., Gibson, D. and Basri, G. 1982, Ap. J. (submitted).

# CONTACT BINARY STARS

A. K. Dupree
Harvard-Smithsonian Center for Astrophysics
Cambridge, MA 02138 U.S.A.

Abstract: Ultraviolet and X-ray surveys of the W Ursae Majoris type stars are reviewed. These systems exhibit extended coronas and transition regions that are confined close to the optically determined surfaces. Correlations of X-ray activity with period or rotational velocity indicate a turn-over or saturation of emission at the short periods or high velocities found in the W UMa-type systems. For a number of systems, ultraviolet emission appears to be anti-correlated with the strength of X-ray emission. These observations are discussed in terms of solar structures, activity, and evolution.

## 1. INTRODUCTION

Contact binary stars of the W Ursae Majoris type are eclipsing systems composed of cool dwarf stars, with orbital periods of less than one day. The two components are spectroscopically similar in the optical spectral region although their masses are not equal. Mass ratios, $M_2/M_1$ can range from 0.5 to 0.1. It is generally thought that both components are contained in a common envelope, and the more massive component generates most of the luminosity of the system which is then transferred to the outer envelope of the secondary.

Some of these systems show period changes either of an abrupt or gradual nature, and the light curves can change in character. It is possible that some form of surface activity is present (Hall 1976; Guinan 1982).

These systems are believed to evolve toward lower mass ratios, higher effective temperatures and longer orbital periods. There may be loss of angular momentum from a system due to braking by a stellar wind. It has been suggested from theoretical calculations (Webbink 1976) that the final state of W UMa type binaries is a rapidly rotating single giant star as the more massive star engulfs its companion. Stars of the FK Comae class may represent such a final state (Bopp and Rucinski 1981; Bopp and Stencel 1981).



*P. B. Byrne and M. Rodonò (eds.), Activity in Red-Dwarf Stars, 447–461.*

The discovery that W UMa stars are a rich source of X-ray (Carroll et al. 1980) and ultraviolet emission (Dupree et al. 1979, 1980) enables the study of stellar activity under extremes of physical conditions and the definition of the influence of rotation on radiative emission levels. It is also of interest to investigate the extent of validity of analogies to solar atmospheric structures. This review will highlight the ultraviolet spectroscopic survey of W UMa systems made with IUE (Dupree and Dussault 1982; Rucinski and Vilhu 1982; Eaton 1982), and the X-ray survey undertaken with the HEAO-2 ("Einstein") satellite by Cruddace and Dupree (1982).

## 2. ULTRAVIOLET OBSERVATIONS OF W UMA-TYPE SYSTEMS

Ultraviolet observations of the short period binary systems VW Cep and 44 Boo were undertaken with the International Ultraviolet Explorer (IUE) satellite (see Dupree et al. 1979) immediately after the discovery by Carroll et al. (1980) that VW Cep was a source of X-rays.

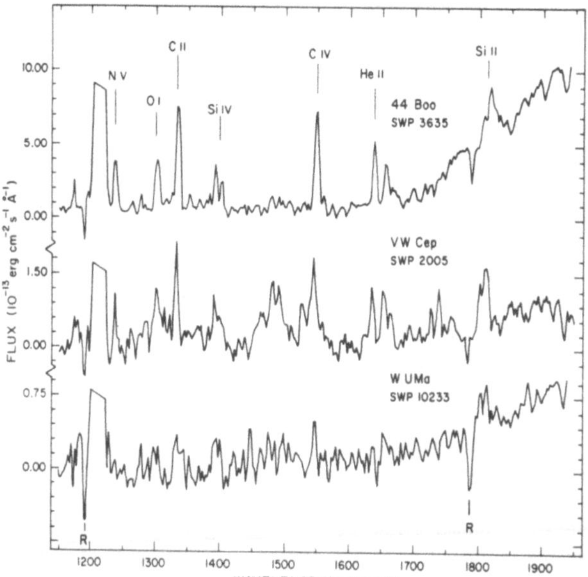


Figure 1. IUE spectra of three W UMa type systems: 44 Boo (P = 0.27 days); VW Cep (P = 0.28 days); and W UMa (P = 0.33 days). R denotes a fiducial mark on the camera faceplate (reseau). The emission feature at 1216 A is due principally to geocoronal Lyman-$\alpha$ and has been truncated. 44 Boo shows an increase in flux at long wavelengths. This emission arises from the companion star in the triple system. (From Dupree and Dussault 1982).

Subsequent to these initial observations, a number of systems have been measured with IUE by various authors (Dupree 1980; Rucinski and Vilhu 1982; Eaton 1982).

Short wavelength IUE spectra cover the range 1150 - 1950 A and sample spectra are shown in Figures 1 and 2.

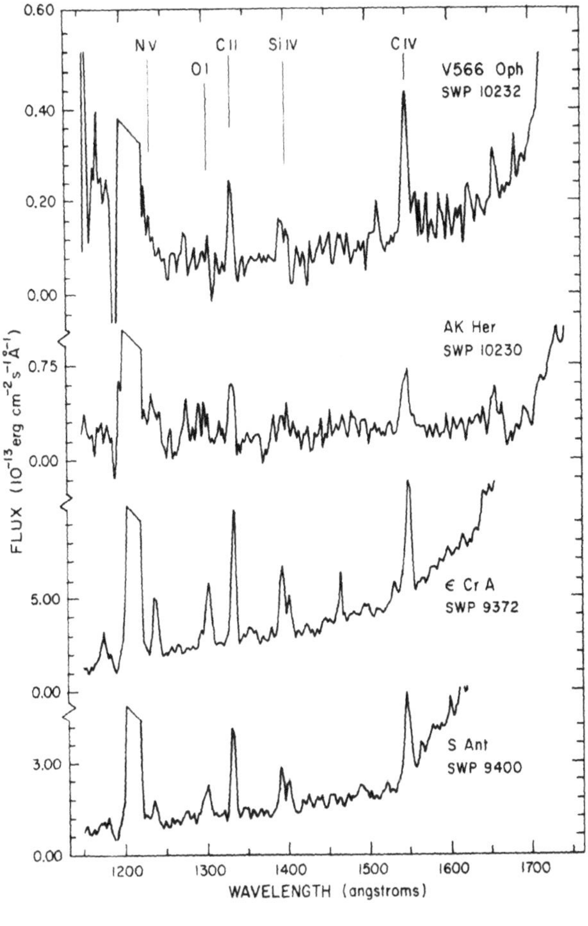


Figure 2. IUE spectra of three long-period W UMa systems: V566 Oph (P = 0.41 days); AK Her (P = 0.42 days); ε CrA (P = 0.59 days); and S Ant (P = 0.65 days). The systems of longest period, ε CrA and S Ant show the continuum increase longwards of 1500 A that is typical of hotter stars.

All of the systems show the typical strong series of emission lines, C II, Si IV, C IV, and N V that are present in single dwarf stars. A comparison of IUE spectra of cool stars in general can be found in

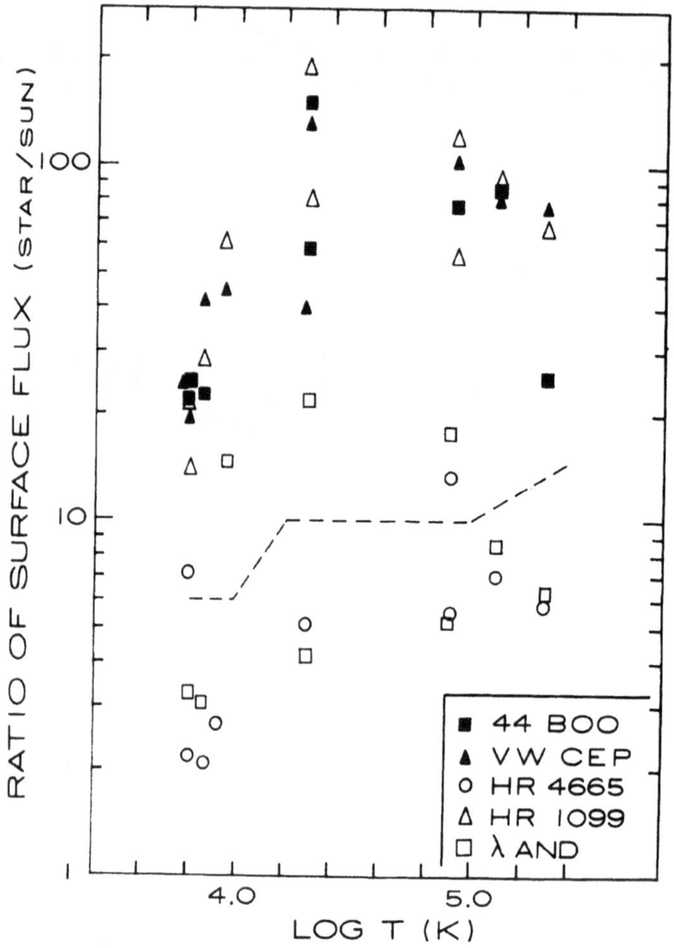


Figure 3. Homogeneous surface flux, in units of the quiet solar flux, for strong ultraviolet emission lines in five binary systems: 2 of W UMa type (44 Boo and VW Cep); 3 of RS CVn type (HR 4665; HR 1099; λ And). The emission lines are assigned to a typical temperature based on their formation in a collisionally dominated plasma. The broken line indicates the behavior of lines in a typical solar active region. Figure from Dupree et al. (1979).

several publications (Dupree 1980; Hartmann, Dupree and Raymond 1982). The relative intensities of the ultraviolet lines are similar to, but not precisely the same as, their relative intensities in a solar spectrum. There is clear indication of the presence of plasma having temperatures up to and including $2 \times 10^5$ K. The He II line at 1640 A is strong in the shortest period systems, 44 Boo and VW Cep. This is of interest because the line is thought to be produced in part by recombination following photoionization of helium by an X-ray continuum (Hartmann et al. 1979,1982); these two systems have the strongest X-ray flux. The longest period systems, although thought to be evolved, do not exhibit the characteristic strong O I line typical of giant stars with extended atmospheres where the O I is believed to be strengthened by fluorescence with Lyman-$\beta$.

It is particularly interesting to evaluate the surface flux of the emission lines (Figure 3).

For the two short period systems in Figure 3, the behavior of their emissions is similar in character to that found in solar active regions and RS CVn stars, namely an increasingly enhanced surface flux with increasing temperature of formation. It is the helium transition at 1640 A that provides the high points at a temperature of 4.5 dex K.

Most noteworthy is the enhancement of the surface fluxes in the short-period systems relative to those at longer period (Dupree et al. 1979). The RS CVn binaries have periods of 64 days (HR 4665) and 54 days ($\lambda$ And, photometric period). It is only the short-period member of the RS CVn class HR 1099 (2.8 days) that possesses fluxes of commensurate value to the W UMa binaries. Rotation is a clear determinant of radiative losses.

Some of the brightest systems can be measured spectroscopically in the ultraviolet in a time short as compared to the orbital period. The flux variation in a number of lines has been measured for four systems in the low resolution mode of IUE. Two of the systems are bright enough to allow high dispersion spectra to be obtained at various phases. A typical light curve for $\varepsilon$ CrA is shown in Figure 4.

Ultraviolet line emission is visible at all phases, and the variation of the emission is similar to the optical variation. The extension of the atmosphere is then similar to the optically determined surfaces. The depth of the ultraviolet primary minimum is comparable to that in the optical, and it appears that the depth of the minima are more nearly equal in the ultraviolet. High resolution ultraviolet spectra of the Mg II emission have been obtained from two systems with enough time resolution to separate the component stars. Both 44 Boo (Dupree and Preston 1980) and $\varepsilon$ CrA (Dupree and Dussault 1981), when observed at elongations show two distinct Mg II features which are attributable to the individual stellar components of the binary. Moreover, their fluxes are in the ratio of apparent surface areas of the two components and the individual line widths are broadened consistent with that expected from synchronous rotation.

Thus the general behavior of the ultraviolet emissions is consistent with the optically determined physical parameters of the stars. The level of radiative losses or activity is substantially above that found in single stars of similar effective temperature and gravity.

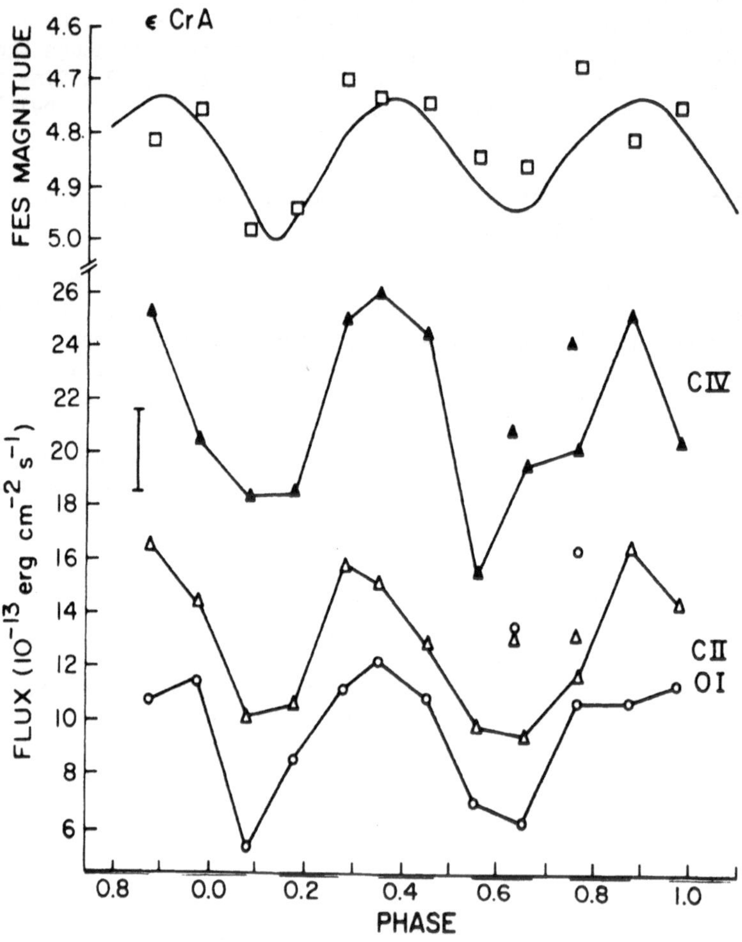


Figure 4. The light curve of ε CrA in a number of ultraviolet emission lines during a single epoch. The fine error sensor (FES) of IUE was used to determine the visual magnitude of the system immediately prior to each exposure. Symbols unconnected by lines represent spectra taken at other epochs. (From Dupree and Dussault 1981).

## 3. X-RAY OBSERVATIONS

A survey of 17 contact binary systems was carried out by Cruddace and Dupree (1982) using the Imaging Proportional Counter (IPC) of the HEAO-2 ("Einstein") Observatory as Guest Investigators. The stars in the survey represented a variety of short and long period systems. A total of 14 stars was detected. The X-ray luminosities of these systems (see Figure 5) span $\sim 10^{29}$ to $10^{30.2}$ erg $cm^{-2}$ $s^{-1}$.

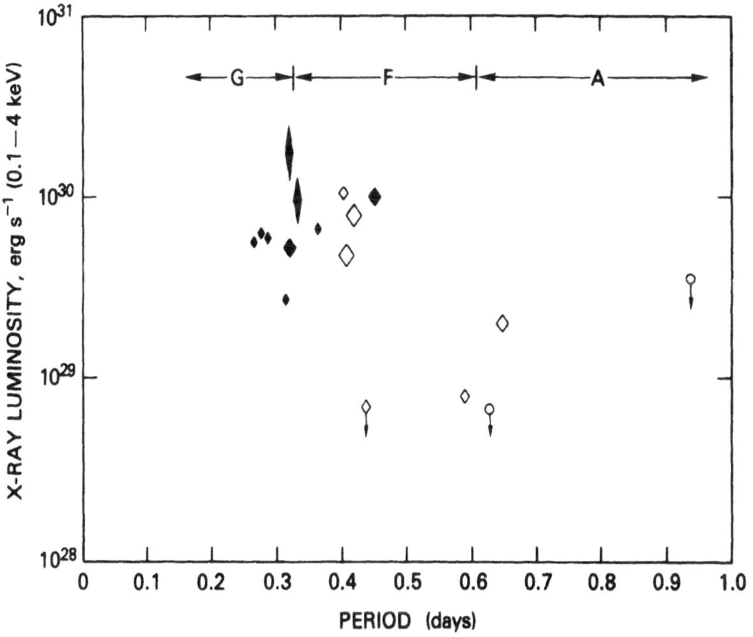


Figure 5. The X-ray luminosity of contact systems as a function of orbital period. The filled symbols represent the W-type systems; the open triangles denote the A-type binaries, and the open circles are the hot early type contact systems. The vertical extent of the symbols results from the uncertainty in distance (From Cruddace and Dupree 1982).

These values are in excess of the X-ray luminosities of single stars of similar effective temperatures as found in the Center for Astrophysics survey (Vaiana et al. 1981), and also exceed the X-ray luminosity of dwarfs in the Hyades (Stern et al. 1981) at the same spectral type. There is a clear dependence of X-ray luminosity on orbital period; the short period systems exhibit a generally higher luminosity than those of longer period. The data are insufficient to distinguish between a continuous or bimodel distribution of X-ray luminosity with period.

A quantity of interest is $L_X/L_{bol}$ as a function of orbital period (see Figure 6) for it eliminates the uncertainty in distance.

The contact W UMa systems do not show a continuous extension of the $L_X/L_{bol}$ relation from the RS CVn stars. As the orbital period shortens, there is a turnover of the $L_X/L_{bol}$ dependence upon period - as if a saturation of flux occurs towards short periods. The position of the turnover may be associated with spectral type, but the data are too scanty to define this (Rucinski 1983; Cruddace and Dupree 1982).

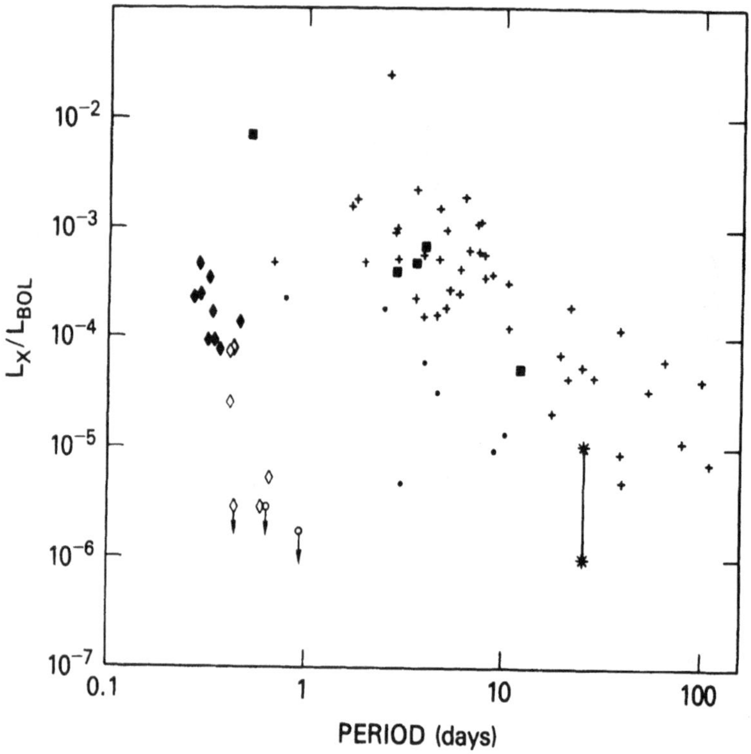


Figure 6. The quantity $L_X/L_{bol}$ as a function of orbital period for the contact binaries (filled and open diamonds; open circles), RS CVn stars (plus marks) from Walter and Bowyer (1981), single F and G dwarfs (filled squares and dots respectively) from Walter (1981), and the Sun in an active and quiet state (stars).

Long and continuous measurements of the X-ray flux allowed a complete orbital period to be monitored in VW Cep (Dupree and Cruddace 1982). This demonstrated (see Figure 7) that X-rays are present at all phases suggesting a global distribution of hot plasma. There is not any well-defined enhancement at a repeatable phase which might be identified with long lived star spots or a splash-down region where mass transfer is occurring. The X-ray modulation clearly does not follow the optical light variation indicating that the corona is much larger than the scale of the optically determined system. The enhancements appear to result from short-term enhancements having a time scale on the order of the orbital period. The behavior at orbital phase 0.3 exhibits this short-term variation with a maximum enhancement on the order of a factor of two.

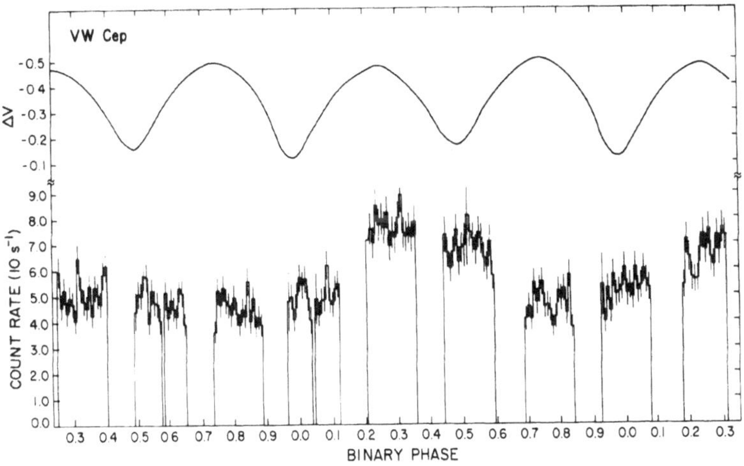


Figure 7. A continuous measurement of VW Cep through two orbital periods. The optical light variation is indicated by the top curve; the IPC count rates are given below where the lack of data corresponds to Earth occultations or data dropouts. (From Dupree and Cruddace 1982).

## 4. ATMOSPHERIC STRUCTURE

We can investigate whether there are any similarities between the structures suggested by these ultraviolet and X-ray observations and those familiar to us from the Sun. A useful quantity is the relative amount of hot plasma over the decade in temperature $2 \times 10^5$ to $2 \times 10^6$ K. This can be evaluated observationally using the N V and X-ray emission; simple loop models also predict this ratio. A summary of observations of various stars is shown in Figure 8.

Dwarf stars, the short period W UMa systems (the W-types), and RS CVn systems lie on the same theoretical relation as the Sun, and these three have a generally increasing scale of emitting material as expected from their luminosity class. Emitting structures of the short period W UMa binaries appear to be similar to coronal structures found on the solar surface.

Evolved stars such as field giants, the Hyades giants, and the hybrid supergiants show a deficit of high temperature material as do the long-period (A-type) W UMa stars. This suggests that the atmospheric structure of the long-period contact binaries is similar to that of more luminous stars, even though the contact binaries are relatively close to the main sequence. The response and evolution of an atmosphere may proceed more rapidly for a star in a binary system. Stellar winds are likely to be enhanced, a phenomenon known to cause a depletion of high temperature material, as found in solar coronal holes.

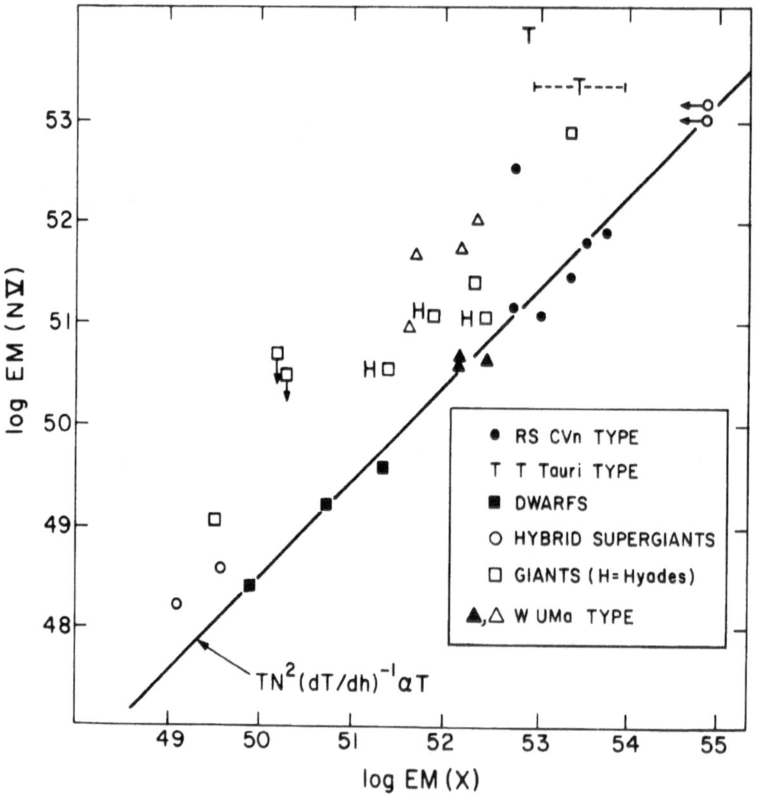


Figure 8. Emission measures as defined by N V and X-ray fluxes for stars of various classes. W-type contact binaries are noted by filled triangles; the A-type binaries are marked by open triangles. The Sun in an active and quiet state is indicated by the open circles. The straight line is the theoretical relation predicted by standard loop models. (From Hartmann, Dupree, and Raymond 1982, and Dupree 1982).

Another indication of atmospheric structure may be found in the behavior of the C IV and X-ray luminosity (Figure 9) which is anticorrelated for seven systems studied both with IUE and in X-rays.

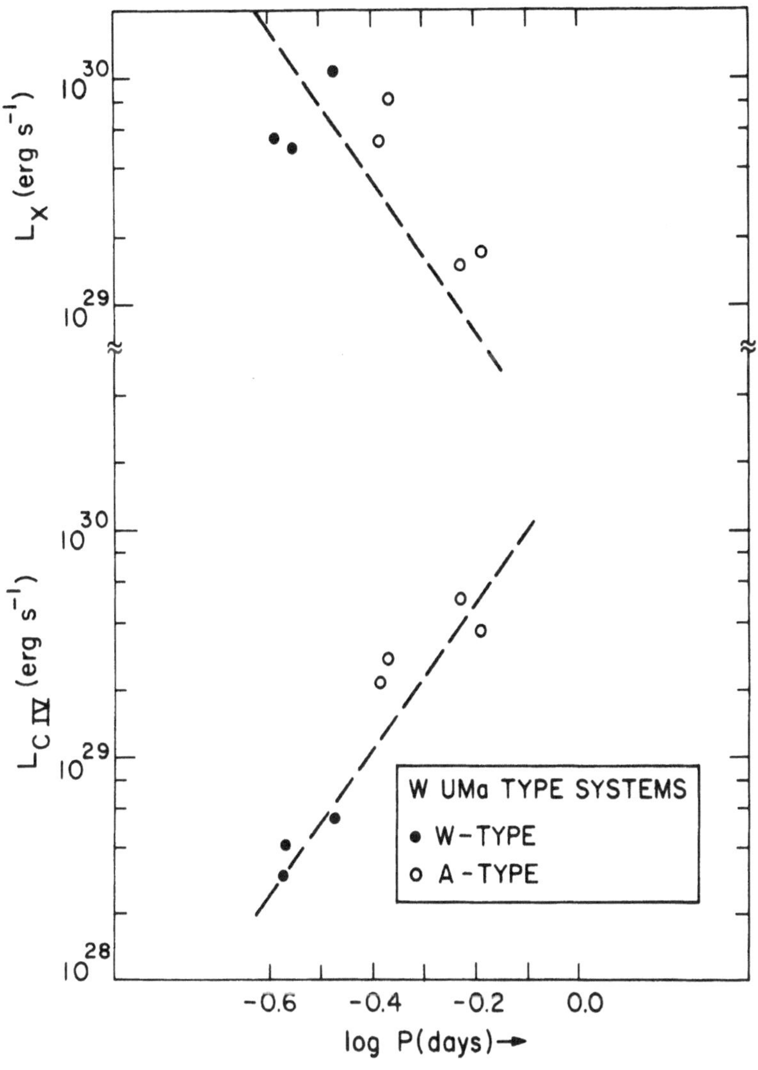


Figure 9. Total C IV and X-ray luminosity (erg $s^{-1}$) as a function of period for seven W UMa-type systems (Dupree 1982).

The X-ray luminosity decreases with increasing period, but the C IV luminosity does not follow suit, and increases with increasing period of rotation. This can be interpreted as a different heating mechanism for the corona as distinct from the atmosphere at lower temperatures. The results in Figure 9 may also reflect the response of an atmosphere to different magnetic field configurations, and the same total amount of energy might be required. The long-period systems exhibit a brightening in transition region lines, and relatively more material is at temperatures close to $10^5$ K. However, the total radiative flux (here taken to be $L_X + L_{C\ IV}$) in the transition region and corona remains approximately constant.

5. DISCUSSION AND COMMENTS

Studies of these W UMa type systems reveal a high level of chromospheric and coronal activity from the majority of the objects. Stars in contact configurations emit enhanced X-ray and ultraviolet emission at all spectral types (here, generally late A through early K), and at all phases of their orbits.

Measurements of these binaries define the short period region of stellar activity and reveal that the extrapolations made from study of binaries not in contact do not predict the emission levels or behavior of the contact systems. Whereas rotation clearly enhances emission from stellar atmospheres - a fact long known from optical studies, and in general consistent with our understanding of the operation of the dynamo mechanism - the emission level does not continue to rise as the orbital period decreases below one day. Somehow, the dynamo "saturates". An explanation may reside in the structure of the subsurface layers, and the extent to which differential rotation is maintained.

Many other measurements need to be acquired. Very few observations have been made of these systems for any extended period of time so as to separate the transient behavior from the steady level, and to identify the extent of homogeneity of the surface emissions. A fundamental datum - the time of eclipse of the systems and the period - has not been recently measured for most of even the brightest systems. In one case, that of ε CrA, it was necessary to advance the published phase by 0.25 in order to force agreement with the IUE optical photometer.

The behavior of X-ray and ultraviolet emissions offers an opportunity to evaluate the effects of evolution upon the structure of a stellar atmosphere, and, due to the contact configuration, an accelerated form of evolution may occur.

This work was supported in part by NASA Grant NAG 5-87 to the Smithsonian Astrophysical Observatory and NASA Defense Purchase Requests H-43050B, H-43055B, H-46079B, and H-57323B with the Naval Research Laboratory. I am grateful to the Local Organizing Committee of this Colloquium for hospitality in Catania.

## REFERENCES

Bopp, B. W., and Rucinski, S. M. 1981, in "Fundamental Problems in the Theory of Stellar Evolution", IAU Symposium No. 93, (D. Sugimoto, D. Q. Lamb, and D. N. Schramm, eds.), p. 177.
Bopp, B. W. and Stencel, R. 1981, Ap. J., 247, L131.
Carroll, R. W., Cruddace, R. G., Friedman, H., Byram, E. T., Wood, K. Meekins, J., Yentis, D., Share, G. H., and Chubb, T. A. 1980, Ap. J., 235, L77.
Cruddace, R. G., and Dupree, A. K. 1982, Ap. J., submitted.
Dupree, A. K., Black, J. H., Davis, R. J., Hartmann, L., and Raymond, J. C. 1979, in The First Year of IUE, ed. A. J. Willis, University College London, p. 217.
Dupree, A. K. 1980, in Highlights of Astronomy, P. Wayman (ed.), 5, 263.
Dupree, A. K. 1982, in Second Cambridge Workshop on Cool Stars, Stellar Systems, and the Sun, M. S. Giampapa and L. Golub, eds., SAO Special Report 392, Vol. II, p. 3.
Dupree, A. K., and Dussault, M. 1981, Bull. A. A. S., 13, 873.
Dupree, A. K., and Dussault, M. 1982, in preparation.
Dupree, A. K., and Preston, S. 1980, in The Universe at Ultraviolet Wavelengths - The First Two Years of IUE, NASA Conf. Pub. 2171, p. 333.
Eaton, J. 1982, preprint.
Guinan, E. 1982, personal communication.
Hall, D. S. 1976, IAU Colloq. No. 29, "Multiple Periodic Variable Stars", (Reidel: Dordrecht), p. 287.
Hartmann, L. W., Davis, R., Dupree, A. K., Raymond, J., Schmidtke, P. C., and Wing, R. F. 1979, Ap. J., 233, L69.
Hartmann, L., Dupree, A. K., and Raymond, J. C. 1982, Ap. J., 252, 214.
Rucinski, S. 1983, this volume.
Rucinski, S., and Vilhu, O. 1982, preprint.
Stern, R. A., Zolcinski, M. C., Antiochas, S. K., and Underwood, J. H. 1981, Ap. J., 249, 64.
Vaiana, G. S., et al. 1981, Ap. J., 245, 163.
Walter, F. M. 1981, Ap. J., 245, 677.
Walter, F. M., and Bowyer, C. S. 1981, Ap. J., 245, 671.
Webbink, R. F. 1976, Ap. J., 209, 829.

## DISCUSSION

Walter: I have two quick comments. The first is that you made some deal about the difference between the F and G stars when you plotted $L_X$ vs $L_{Bol}$. It is very important to realize that you expect a difference because $L_{Bol}$ has a $T^4$ term in it. So you have to be careful of what you are talking about. If you compare X-ray surface fluxes they may be the same whereas when you normalize to $L_{Bol}$ they may be different.

Dupree: I concur but it is difficult to decide what to plot since surface areas, for instance, are not well determined.

Walter: With regard to the discontinuity at a period of 1 day, there is an interesting system, V471 Tau, which is not an RS CVn but has similar properties. It comprised a K dwarf and a white dwarf with an orbital period of 12 hours. So it is a very rapid rotator and in X-rays it looks just like and RS CVn star, falling on an extrapolation of the locus of the RS CVn's. Its period it is commensurate with of the WUMa systems. So this discontinuity at 1 day may not be a discontinuity but rather a change in the nature of the objects' convective zone or something. When dealing with non-single stars we may have very peculiar convective patterns.

Dupree: I concur. Nevertheless it is good to place them in perspective. But I concur, since these are contact systems we do not know much about their internal structures or atmospheres. So I agree, we may well be comparing "apples and oranges".

Linnell: From an evolutionary point of view it has become popular to think of evolution from the W subclass. It is important to note that Hoffman has identified a subclass whose light curve switches W-type to A-type. So a simple evolution from W to A subclass is certainly not entire picture. Webbink has pointed to a few systems in which the masses are large enough that you would never have expected them to pass through the W subclass at all and to others where the mass is so small you would never have expected them to have reached the A subclass.

Dupree: I concur. In every class of objects there are always some which are "schizophrenic". I was trying to term it in terms of physical quantities i.e. the mass and things like that. Let me make a point now which I had overlooked but which is of interest here. In evolutionary terms, it has been suggested that there is cannibalism among the dwarfs. If there is evolution from the short- to long-period, systems and then to the coalesced giants, then the position of FK Comae suggests that a substantial restructing and enhancement of the X-ray luminosity may be necessary to agree with Webbink's evolutionary calculations.

Rucinski: I have one comment and one quetsion. The comment is that the discontinuity at 1 day is real. Some of these contact binaries arise from main sequence stars which come closer because of the constant loss of angular momentum. It is easy to show that this loss of angular momentum should speed up greatly as the stars come into contact. Once they are actually in contact the rate of loss of angular momentum should slow considerably. So it is probably quite natural that we have this break. My question is since contact binaries are under-active in X-rays, that is they lie lower than the extention of the RS CVn period relationships, do you think that this could be explained by the matter lost to the systems. This matter could be floating about and decrease the X rays.

Dupree: I have two comments. Firstly, the mass column density to all of these systems, determined spectroscopically, is of the order of $10^{18}$ $cm^{-2}$ which is low. Secondly there is the question of whether we are really seeing the X-rays. In many of these systems we do not see the HeII (λ1640) line. We believe that this line is formed by recombination following ionization by X-rays. So this would be another indirect point against there being undetected X-rays. We also have looked at the Mg II line in high-resolution for signs of circumstellar absorption and not found any. There is... (part of comment lost)... in CrA at conjunction where you see an asymmetric profile which can be interpreted as indicating outflow.

Linsky: If one takes a limited sample of X-ray and optical data one could investigate whether the X-rays vary in phase or out-of-phase with the optical. Those who allocate X ray and UV telescope time ought to recognize that fact. Secondly, this is the only group of stars I know of in which the luminosity of the C IV line normalized to $L_{Bol}$ decreases with decreasing period. Would you care to make a comment as to why this is?

Dupree: I would prefer to address this via the dependence of $L_{CIV}$ on $L_X$. The ratio of $L_X/L_{CIV}$ increases with decreasing period. It appears to be consistent with the evolution of an atmosphere in the sense that the CIV relative to X-rays gets larger. That is what we see in the evolved stars.

# THE EVOLUTIONARY STATUS OF SHORT-PERIOD RS CVn AND RELATED W UMA ECLIPSING BINARIES

Russo, G., Milano, L., Mancuso, S.
Capodimonte Astronomical Observatory

Among the RS CVn stars showing solar-type activity, with spectral types ranging from F to K and total masses up to 4 $M_\odot$, there are two peculiar groups with period less than one day: a) a group with components well inside their Roche lobes (Short-Period-Group, hereinafter SPG) and b) a group with their components in a thin or marginal degree of contact, with lightcurves of W UMa-W type (hereinafter WWG).

We hypothized (Milano, 1981, Milano, Russo, Mancuso, 1981) a possible linkage between SPG and WWG in the sense of an evolution SPG WWG taking into account the following observed properties:

a) in the period-colour diagram the representative points of the two groups lie on two parallel bands, showing a reddening of SPG with increasing period;
b) high variability of the observed lightcurves with pseudo-regular migrating waves for SPG, and less regular variations for WWG;
c) a possible correlation between orbital periods and mass ratios;
d) the angular orbital momentum as a function of the total mass of the systems considering empirical relations for detached, semidetached and contact systems (Chaubey, 1979) for SPG group (d systems) is typically of sd systems, whilst for WWG group is typical of contact systems;
e) Ca II and Mg II emission lines are present in both groups.

Among the observed properties there is the log $H_{orb}$ vs log $M_{tot}$ diagram that could be a strong indication of evolution with angular orbital momentum loss (AML) and this fact can support the current theory by Vilhu (1982), Rahunen (1981), van't Veer (1979). We computed the orbital angular momentum of a set of about 450 close binary systems with spectral types ranging from A8 to M, and total masses up to 6$M_\odot$ using the data by Brancewicz and Dworak (1980), and there is a spread of points between the d, sd and c lines already mentioned above. This could imply that the AML process is slow. We think, notwithstanding these evidences, that it is necessary to try to estabilish an age scale, using the anti-correlation between Ca II emission strength and stellar age (Wilson and Skumanic, 1964, Skumanic, 1972). Moreover, the primary density vs primary colour diagram (Mochnacki, 1981) which should contain the same



*P. B. Byrne and M. Rodonò (eds.), Activity in Red-Dwarf Stars, 463–464.*

informations of the period-colour diagram, but with the effects of different mass-ratios and fill-out removed does not seem a good indicator of the evolutionary status, owing to the lack of knowledge of the behaviour of the radius in the case of AML evolution.

To throw light on the problems outlined above, we have begun a systematic analysis both of photoelectric lightcurves and of optical and ultraviolet spectra.

## REFERENCES

Brancewicz, H.K. and Dworak, T.Z.: 1980, Acta Astronomica 30, 501.
Chaubey, U.S.: 1979, Astrophys. Space Sci. 64, 177.
Milano, L.: 1981, in Photometric ans Spectroscopic Binary Systems, Proceeding of the NATO ASI (Maratea), ed. E.B. Carling and Z. Kopal, (Dordrecht, Reidel), p. 331.
Milano, L., Russo, G., Mancuso, S.: 1982 in Binary and Multiple Stars as Tracers of Stellar Evolution, Proceedings of the IAU Coll. 69 (Bamberg, FRG), ed. Z. Kopal and J. Rahe, (Reidel).
Mochnacki, S.W.: 1981, Astrophys. J. 245, 650.
Rahunen, T.: 1981, Astron. Astrophys. 102, 81.
Skumanic, A.: 1972, Astrophys. J. 171, 565.
Vilhu, O.: 1982, Astron. Astrophys., in press.
Wilson, O.C. and Skumanic, A.: 1964, Astrophys. J. 140, 1401.

## DISCUSSION

Dupree: Could you comment on what you see as the evolutionary status of the contact systems?

Russo: We know that the W UMa binaries are divided into A-type and W-type and that the A-type are evolved systems. In this view the W-type should be less evolved than the A-type. I do not think however that the next stage of evolution of the W-type will be the A-type. I believe that the process of formation of contact systems is not unique and we can have a different kind of life for the different systems.

Dupree: You do not then see a continuity in evolution form the short- to the long-period systems i.e. from the W's to the A's.

Russo: I think that there is no continuity. I believe that the A-type are evolved systems which originate as contact systems, while the W-type are contact systems which do not originate as contacts but detached systems. These latter are young because the process of going from detached to contact should last about $10^8$ years according to some calculations of magnetic braking, if there is enough magnetic field i.e. about 1000 G.

# ANGULAR MOMENTUM LOSS AND THE FORMATION OF W UMa SYSTEMS

E. Budding

Department of Astronomy
University of Manchester
Manchester M13 9PL
England

Angular momentum loss (AML) may be introduced as part of the explanation of the peculiar overabundance of W UMa-type contact binaries (W UMS) (for a recent review, see Vilhu, 1981). Among the various possible mechanisms to achieve this are (i) overflow through the outer ($L_2$) Lagrangian point in a deep contact phase (Kuiper, 1941; Nariai, 1979), (ii) magnetic braking, where magnetic lines of force "stiffen" and thus enhance the efficiency of angular momentum loss associated with a stellar wind (Huang, 1966; Mestel, 1968). Such subjects have been investigated in a number of more recent detailed studies.

The existence of a common envelope for W UMS entitles us to regard them as being involved in some kind of mass transfer episode of their close binary evolution, since, in general, stars could not be both simultaneously filling adjoining Roche lobes and remain in thermal equilibrium (Kuiper, 1941). In the usual language, we may speak of Case A, in which the cores of both stars are still similar to those of the Main Sequence, or Case B, in which core hydrogen burning has already ceased prior to Roche lobe overflow (RLOF). Considerations of AML in Case A and Case B can be tested by observing "protomorph systems" (PS).

Even with a generous estimate of candidate systems, however, there still appear insufficient PS to account for W UMS, considering only Case A without AML on approach (Budding, 1982a). AML through magnetic braking may ease the pressure for a high PS supply under certain supposed conditions. This process should have a determinable effect on observed binary incidence at low period.

In Figure 1, Mestel's (1968) rather thorough description of AML by magnetic braking has been used to illustrate the variation of a characteristic braking time $T_J$ (= $J/\dot{J}$) with P for some feasible range of PS parameters. Three basic parameters were involved; $\kappa$ , the most significant one for the present context, denoting a normalized centrifugal force; $\ell$ , relating to the thermal energy of the source corona, and $\zeta$ , relating to the magnetic energy of the source. Okamoto (1974) generalized this discussion to allow for radial field structure in terms



*P. B. Byrne and M. Rodonò (eds.), Activity in Red-Dwarf Stars, 465–468.*

of an additional parameter λ. The existence of a power-law rise and "cut-off", in the response of AML to period (κ) variation, may have some bearing on the AML discussion of Ruciński and Vilhu (1982).

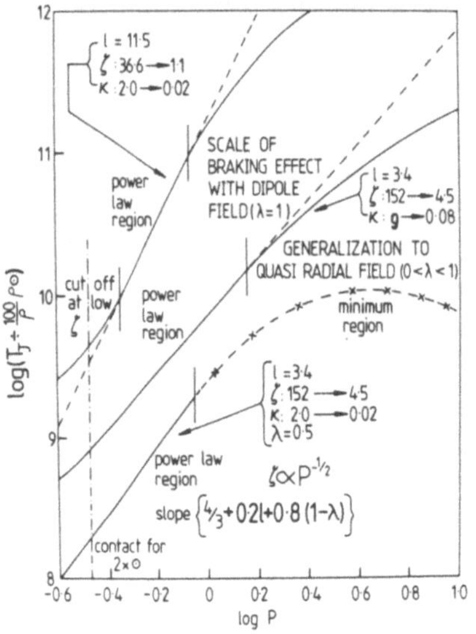


Figure 1

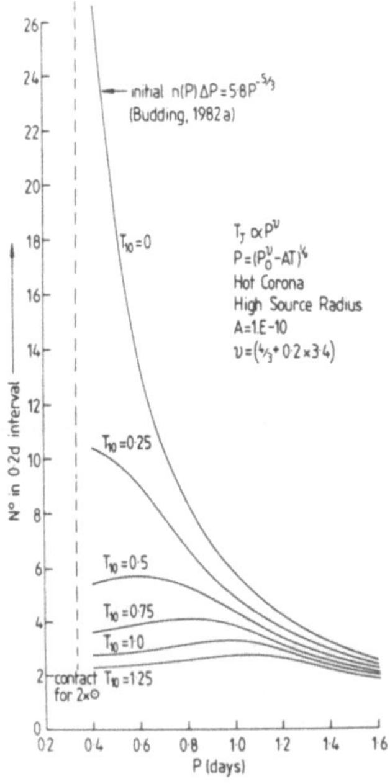


Figure 2

The power-law form for AML has been incorporated into a study of the evolution of n(P), with an initial form deduced from observations of detached close binaries of different spectral types (Budding, 1982a) and a constant rate of binary formation. Some results are shown in Figure 2. It is very difficult to understand the "bunching" effect of cool detached binaries on this basis (Budding, 1981), let alone the pronounced peak which occurs at contact.

Other problems with the Case A scenario can be noted as follows: (i) A number of W UMS are observed actually to exhibit period increases (Yamasaki, 1975). (ii) With Case A brought about by AML it is the more massive star which should first undergo RLOF. This should then tend towards equalising the masses, in contrast to the well-known inequality

An alternative, Case B, picture will be presented in more detail elsewhere (Budding, 1982b). The basic premise is that PS are supposed to come from a large number of order $2M_\odot$ primaries accompanied by low mass secondaries ($q \lesssim 0.2$) at initial separations of order 10 $R_\odot$. There is evidence that such systems might indeed be very common (Trimble, 1978; Lucy and Ricco, 1979), despite strong selection effects against their discovery. Such PS could result from an initial fission process (Lucy, 1981). The contraction of the system, after RLOF starts, ensures overfilling of the outer critical Roche surface (whose volume is about twice that of the inner contact volume). AML by process (i), above, then appears as a necessary feature of this mode of binary evolution.

Secondary overluminosity and period increases on a less than core nuclear timescale can both be quite naturally understood within this framework, while it is also possible to make the long sought connection with U Gem systems. A number ($\sim$ 10%) of semi-detached binaries of low mass, currently having extremely low mass ratio (e.g. AS Eri), may be noted here. It seems very likely that such systems have been in over-contact in the past (Refsdal, et al., 1974).

In this picture magnetic braking plays a secondary role up until the time when it might delay binary separation at low q. In Case A, magnetic braking is generally required to slow down once contact has been reached (Ruciński, et al., 1982). The depth of the secondary's convection zone may also be relatively large, particularly at higher mass ratio, though light curve asymmetries should still tend to have, as with classical Algols, a preferred sense.

In spite of such differences, there are sufficient encouraging points in the comparisons with observations to cause the Case B scenario to be taken seriously, at least for some W UMS.

REFERENCES

Budding, E.: 1981, in Investigating the Universe (ed. F. D. Kahn) Reidel, p.271.
Budding, E.: 1982a, in Binary and Multiple Stars as Tracers of Stellar Evolution, IAU Coll.No.69, Bamberg, eds. Z. Kopal and J. Rahe, p. 351.
Budding, E.: 1982b, in preparation.
Huang, S. S.: 1966, Ann. d'Astrophys., 29, 3.
Kuiper, G. P.: 1941, Astrophys. J., 93, 133.
Lucy, L. B.: 1981, in Fundamental Problems in the Theory of Stellar Evolution, IAU Symp.No.93, Kyoto, eds. D. Sugimoto, D. Q. Lamb and D. N. Schramm.
Lucy, L. B. and Ricco, E.: 1971, Astron. J., 84, 401.
Mestel, L.: 1968, Mon. Not. R. astr. Soc., 138, 559.
Nariai, K.: 1979, Publ. Astron. Soc. Japan, 31, 299.
Okamoto, I.: 1974, Mon. Not. R. astr. Soc., 166, 683.
Refsdal, S., Roth, M. L. and Weigert, A.: 1974, Astron. Astrophys., 36, 113.
Ruciński, S. M. and Vilhu, O.: 1982, Mon. Not. R. astr. Soc., (preprint).
Ruciński, S. M., Pringle, J. E. and Whelan, J. A. J.: 1982, in Binary and Multiple Stars as Tracers of Stellar Evolution, IAU Coll. No.69, Bamberg, eds. Z. Kopal and J. Rahe, p. 309.
Trimble, V.: 1978, The Observatory, 98, 163.
Vilhu, O.: 1981, Astrophys. Space Sci., 78, 401.
Yamasaki, A.: 1975, Astrophys. Space Sci., 34, 413.

DISCUSSION

Rucinski: I just wanted to contest your initial statement about the frequency of these stars, that is one W UMa-type star per thousand stars seems to me to be a small number since almost every star is a binary. It is true, of course, that many stars are in wide binaries and you might have trouble with photomorphs.

Budding: Maybe we have to agree to differ about that. To me when one compares with other, detached close binaries, one in a thousand is about 5-10 times as much as one would expect. I am not the only one to have commented on this. The idea goes back to Shapley.

Rucinski: W UMa-type binaries are very easy to discover so they are over-represented with respect to other eclipsing binaries. If we consider all of the stars in the sky, most of them are in binaries. Among these W UMa stars are very rare although very interesting.

# ACTIVITY OF CONTACT BINARIES

S.M. Rucinski
Max Planck Institute for Astrophysics, Munich
(on leave from Warsaw University Observatory)

O. Vilhu
Observatory and Astrophysics Laboratory, Helsinki

J. Kaluzny
Warsaw University Observatory

We discuss observational results for indicators of elevated activity of contact (W UMa-type) binaries. References are made to results for fast rotating single and detached binary (MS) stars: they are discussed in the accompanying paper by Vilhu and Rucinski.

$\delta$(u-b): The ultraviolet excesses $\delta$(U-B) were discovered by Eggen (1967) who explained them by the lowered metallicity in older among W UMa systems. However, because of the properties of the U-band, it was not possible to disentangle effects of metallicity from any Balmer-continuum emission. The four-colour Strömgren photometry for 36 W UMa systems (Rucinski and Kaluzny 1981) which will be soon supplemented by results for 17 southern systems enabled to obtain the following results: The interstellar extinction corrected data (Fig. 1) show concentration of genuine ultraviolet (u-band) excess at the short period border of the period-colour relation. This can be interpreted by the Balmer continuum emission related to the period-activity connection; the latter must, however,

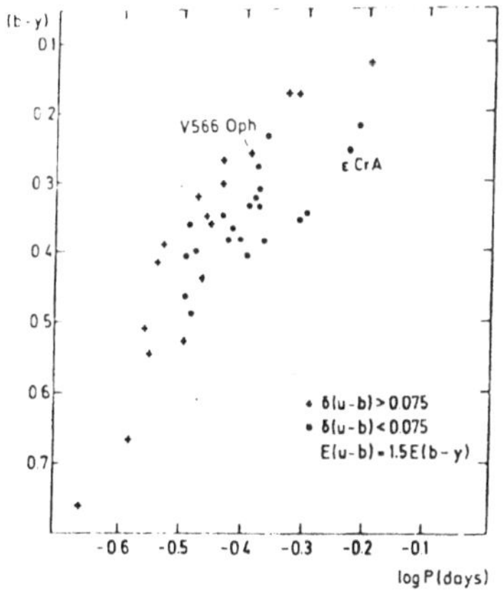


Fig. 1 Period-colour relation in the four-colour photometry



*P. B. Byrne and M. Rodonò (eds.), Activity in Red-Dwarf Stars, 469–473.*

differentiate between systems differing in periods by less than $\Delta \log P = 0.2$. Such a strong dependence is quite unexpected. Two systems, V566 Oph and ε CrA (mentioned further) having similar (b-y) but differing in $\delta$(u-b) are marked in Fig. 1 for visualization of the ultraviolet excess effect.

The concentration of systems with ultraviolet excesses at the short period border in Fig. 1 can result to some extent from an improper removal of the interstellar extinction, especially if circumstellar extinction of unknown properties were to be present. A downward shift by E(b-y) should produce a 1.5 times larger (and positive) effect on (u-b) resulting in a decrease of $\delta$(u-b). The "standard" value of $\delta$(u-b) for contact binaries would correspond then to an excess > 0.075. It would be impossible to explain the whole morphology of the period-colour relation by effects of reddening but this possibility should be kept in mind, especially in view of the systematically too positive reddening-corrected (b-y) colours for earlier spectral-type systems.

$f_{TR}/f_{BOL}$. The summed fluxes for the transition-region (TR) lines observed with the IUE in SWP and normalized to the bolometric fluxes are surprisingly little dependent on the spectral type or period for the sample of 9 contact binaries (Rucinski and Vilhu 1982). Very rare fast-rotating giants seem to be even more active but for MS binaries there seems to exist a plateau in the TR activity for $P < 3$ days which continues into the contact binary domain (cf. Fig. 1 in the paper by Vilhu and Rucinski). We call this effect "saturation" of the TR activity. The origin is unknown but is related only to the period and it can be shown that it is not due to the transfer-of-radiation saturation effects in resonance lines.

$f_{COR}/f_{BOL}$. The normalized coronal (COR) fluxes, as observed by Dupree et al. (cf. Dupree 1981 for references) in soft X-rays with the Einstein satellite show strong dependences on period and spectral type (Fig. 2). The spectral type dependence is in the same sense as for detached binaries and single stars: more active are stars of later spectral types having thicker convection zones. However, the period dependence is quite unexpected showing decrease of coronal activity for shorter periods. It should be noted that the almost orthogonal dependence on period for contact binaries by Dupree (1981) resulted from overlooking the period-spectral type correlation; therefore, the contact systems must be split into spectral groups, as in Fig. 2, and then it is clearly seen that the strongest COR activity is observed for systems located close to the border between contact and detached binaries. This result has no obvious interpretation at present but a systematic decrease in loop sizes for very short-period systems seems to be a plausible interpretation. Notice in Fig. 2 locations of systems V566 Oph and ε CrA with identical (b-y) but with different $\delta$(u-b).

$f_{MgII}/f_{BOL}$. The data for contact binaries are very fragmentary (Rucinski et al. 1982, Vilhu and Rucinski 1982) but suggest reduced MgII fluxes when compared with fast rotating stars, single and in binaries: W UMa itself seems to have similar MgII fluxes as that observed for the Sun. The total radiative losses (CHR, TR, COR) are difficult to estimate but might be roughly consistent with the $(\text{Period})^{-1}$ dependence when comparison is made relative to the Sun.

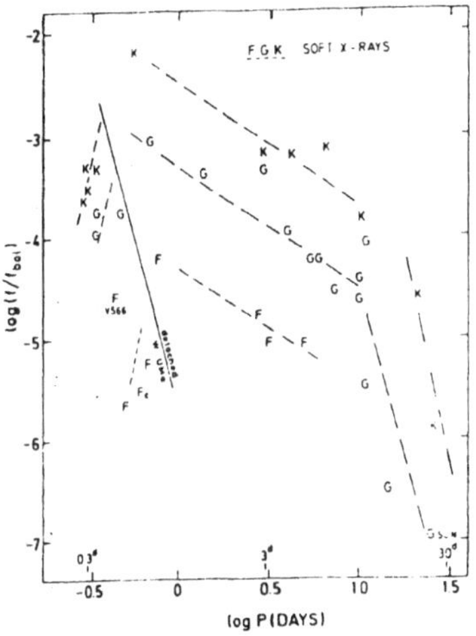


Fig. 2 Coronal activity - rotational period relation

Spots. The presence of spots on more massive components seems to be the best explanation of the W-subtype syndrome (Eaton et al. 1980, Hoffmann 1978, Stepien 1980). The spots must cover up to 20% of the primary's surface and be relatively smoothly distributed (e.g. small spots). Preference for more massive components can be plausibly accommodated into all presently considered theories of contact binaries: in the thermal-relaxation-oscillation and angular-momentum-loss theories, it would be due to the thicker convective zone of the primary component; in the contact-discontinuity theory, it would be due to difficulties for the magnetic field to penetrate the discontinuity. However, the W-subtype light curves can be obtained by assuming existence of belts of darkening extending to similar latitudes in both components. This non-magnetic darkening would result from the reduced efficiency of convection in equatorial regions and from different convection modes at different latitutes (like on Jupiter). This possibility, however, is difficult to incorporate into the light-curve-synthesis modelling because of unknown properties of such belts.

References

Dupree, A.K.: 1981, in Solar Phenomena in Stars and Stellar Systems, NATO Adv. Study Inst., eds. R. Bonnet and A.K. Dupree, p. 407(Reidel)
Eaton, J.A., Wu, C.-C., Rucinski, S.M.: 1980, Ap.J., 239, 919.
Eggen, O.C.: 1967, Mem. Roy. Astr. Soc., 70, 111.
Hoffmann, M.: 1978, Astr. and Astroph. Suppl., 33, 63.
Rucinski, S.M., Kaluzny, J.: 1981, Acta Astr., 31, in press.
Rucinski, S.M., Pringle, J.E., Whelan, J.A.J.: 1982, in Binary and Multiple Stars as Tracers of Stellar Evolution, IAU Coll. 69, Bamberg, eds. Z. Kopal and J. Rahe, (Reidel)
Rucinski, S.M., Vilhu, O.: 1982, Mon. Not. Roy. Astr. Soc., in press.
Stepien, K.: 1980, Acta Astr., 30, 315.
Vilhu, O., Rucinski, S.M.: 1982, in preparation.

DISCUSSION

Linnell: The model which you use to explain the W-type phenomenon using Mullan's suggestion of starspots, usually on the more massive component, seems to me to suffer from a fatal error. This is because it fails to account for the anomalous mass-luminosity relationship for the less massive components. How do you get energy transferred to the less massive component?

Rucinski: Can you explain your point a little further?

Linnell: It is this. You explain the occultation eclipse being the deeper one by reducing the surface brightness of the more massive star by putting dark spots on it. That does not provide an explanation of why the secondary, less-massive component is over-luminous for its mass.

Rucinski: No, but that is quite a separate problem.

Worden: The starspot models for W UMa systems are quite interesting. I have a comment and a question. It would seem that dynamo models for the generation of magnetic fields would be very much complicated by the dumb-bell-shaped convection zone. Are you aware of any models which predict what kind of convection zones could be expected in these stars?

Rucinski: No, I have no idea. My personal feeling is that there is no differential rotation because of the existence of something similar to meridional circulation. We might then have turbulence in very small cells and therefore more rigid rotation. Buth that is just a "handwaving" argument.

Budding: I have a comment about the A- and W-type systems which a lot has been made of. To my mind non convincing case has been made to show that all of the A-type systems are in contact. There are cases, for instance, RR Cen and ε CrA which may be close but detached systems. This is true for W-type systems although among those low-mass, cooler systems one cannot easily distinguish between W- and A-type and some change between the two types. So I wonder about the reality of this sub-division.

Rucinski: The answer to the first point would be that there is no way of transferring sufficient energy from one component to the other without contact. We observe that both stars have the same $T_{eff}$ but different masses.

Budding: I do not think that is necessary. We observe different luminosities but there are not good spectroscopically determined mass ratios for many. We are using photometric mass ratios. RR Cen is an example where there is no spectroscopic mass ratio. The luminosities are different but because one star is quite a bit bigger than the other the $T_{eff}$'s work out comparable.

Rucinski: It was very nice to see this splitting in the case of ε CrA, which works out as predicted, in spite of the fact that no spectroscopic mass ratio was available. Tapi and Whelan just assumed something about the primary component.

Dupree: Yes and we saw the velocity separation giving the predicted massa ratio of 10:1.

Rucinski: Even in the case of AW UMa which has a very extreme mass ratio, the secondary component was recently observed and shown to have a very small mass.

Budding: In the case of ε CrA you have a mass ratio of 0.8 which is a long way from indicating a low-mass secondary.

Rucinski: Anyway to me the fact that they have equal temperatures and differ very much in the masses tells us that they must transfer immense amounts of energy from one component to the other and so that spots are a minor consideration. 20% spot coverage on one component or the other does not matter. Concerning your question about the division into A- and W-types, that it if one considers cooler stars they will all be W-type and if one considers hotter stars they will be A-type. In between there is certainly a group which switches from one type to the other.

more comparable.

[illegible]

# ROTATION-ACTIVITY CONNECTIONS IN MAIN SEQUENCE BINARIES

O. Vilhu
Observatory and Astrophysics Laboratory
Univ. of Helsinki, Finland

S.M. Rucinski
Max-Planck-Institut für Astrophysik
Garching b. München, W. Germany

We have studied transition-region (NV + SiIV + CIV) and chromospheric (MgII) emission observed with IUE (Vilhu and Rucinski, 1982; Rucinski and Vilhu, 1982; Ayres et al., 1981) together with coronal (soft X-ray) emission observed with the Einstein satellite (Cruddace and Dupree, 1982; Walter, 1982) for contact (W UMa) and detached main sequence binaries. The components are main sequence stars (or near ms) and rapid rotators due to spin-orbit coupling. They can thus be expected to give information of dynamo-processes in rapid rotators, although the binary effects clearly produce extra complications in this discussion.

We note that the transition region (NV + SiIV + CIV, $10^5$ °K) fractional fluxes ($f/f_{bol}$) become "saturated", i.e. unsensitive to the period and spectral type, below P ≈ 3 days; at the same time soft X-rays still are quite sensitive on both quantities, as shown in Fig. 1. Notice also a clear distinction in coronal emission between detached and contact binaries.

Separate wavelength-bands (excitation levels) have quite different rotational-period dependences so that estimates of the total radiative losses are rather difficult to do. We find that the magnetically-confined coronal-loop models are a convenient way to try to estimate these losses. In the simplest models (see e.g. Rosner et al., 1978) the loop pressure p, multiplied by the filling factor ξ, is proportional to the transition region surface flux, whereas the maximum loop-temperatures $T_{max}$ can be estimated using the X-ray data (for details see Vilhu and Rucinski, 1982). The total radiation from loops (mainly in EUV) can then be computed from

$$f_{loops} \propto \xi p \cdot T_{max}^{1/2} \text{ erg cm}^{-2} \text{ s}^{-1}$$

Dependence of this radiation ($f_{loops}/f_{bol}$) on period is shown in Fig. 2 for stars in our sample for which both IUE- and Einstein-data are available. The dashed line shows the dependence as estimated from the Ayres et al. (1981) scaling laws for fluxes (X ∿ $(NV + SiIV + CIV)^2$) used



*P. B. Byrne and M. Rodonò (eds.), Activity in Red-Dwarf Stars, 475–477.*

together with the Walter's (1982) X(P)-relation for G stars (see Fig. 1). Comparing the Sun and the most rapidly rotating stars ($P \sim 10^{-2} P_{sun}$) we find $f_{loops}/f_{bol}$ on average to increase with decreasing period slightly more steeply than the linear $P^{-1}$ - relation; this rise may not, however, be necessarily linear. Note that the short-period RS CVn-stars HR 1099 and UX Ari (1099 and UX in the figure) are not strictly main sequence stars.

The chromospheric radiative losses are difficult to model, but in the Sun MgII 2800 resonance lines (h and k) may represent roughly 30 % of the identifiable emission-line cooling of the chromosphere. If a similar picture holds also in the rapidly rotating active stars (i.e. losses equal to MgII/30 % + loops), we find that in the Sun chromospheric radiation dominates while in rapid rotators most of the radiation comes out from the loops. In this way, if the total (chromo + TR + coro ≡ chromo + loops) radiation is considered, only the Sun in the Fig. 2 will be considerably shifted upwards by approximately 1.0 dex. In this case the rapid rotators ER Vul and $\sigma^2$CrB would follow quite closely the $P^{-1}$ - relation when compared with the Sun. With still decreasing the period (but then a physical contact sets in) the total radiative losses "freeze in" or even slightly diminish (see Fig. 2).

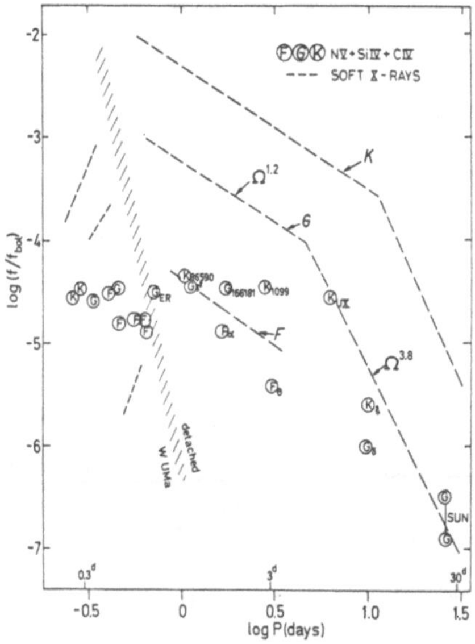

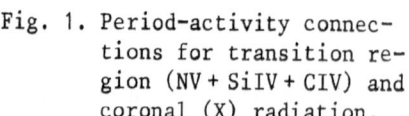

Fig. 1. Period-activity connections for transition region (NV + SiIV + CIV) and coronal (X) radiation.

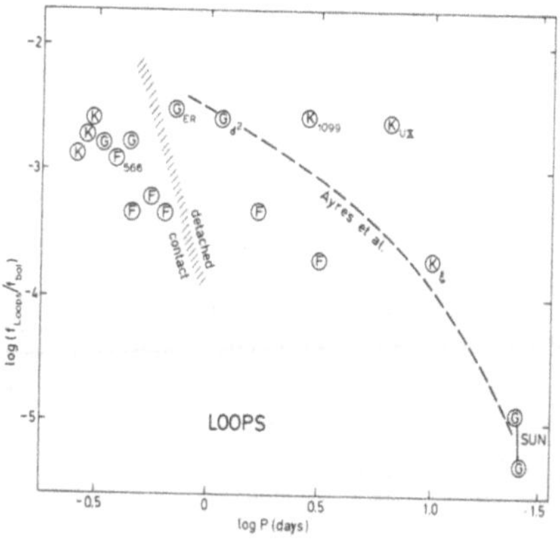


Fig. 2. Period-dependence of the total radiation from loops.

## References

Ayres,T.R., Marstad,N.C. & Linsky,J.L., 1981, Astrophys. J. 247, 545.
Cruddace,R.G. & Dupree,A.K., 1981, preprint.
Rosner,R., Tucker,W.H. & Vaiana,G.S., 1978, Astrophys. J. 220, 643.
Rucinski,S.M. & Vilhu,O., 1982, MNRAS, in press.
Vilhu,O. & Rucinski,S.M., 1982, in prep.
Walter,F., 1982, Astrophys. J. 253, 745.

# THE VW CEPHEI SYSTEM

Albert P. Linnell
Department of Physics and Astronomy, Michigan State University, East Lansing, MI 48824, USA

It is well known that the original Lucy CCE model (Lucy, 1968a, 1968b) was unable to represent W-subclass light curves of W Ursae Majoris stars. Lucy's zero order barotropic model, with gray atmosphere limb darkening produces a deeper transit eclipse and zero color variation with phase. These predictions disagree with observation. Changing the limb darkening coefficient alone, within acceptable limits, cannot deepen occultation eclipse enough to make it deeper than the transit eclipse, as observations require.

Changing the bolometric albedo, A, and the gravity brightening coefficient, b, from their barotropic values of 0.0 produces a baroclinic photosphere. The hydrodynamic consequences have not been investigated. Part of the controversy between Lucy and Shu (Lucy, 1976; Lucy and Wilson, 1979; Shu, 1980; Shu and Lubow, 1981) involves a disagreement over the proper theoretical values of these coefficients. Lucy asserts that $A = 0.5$ and $b = 0.08$ (Lucy, 1967, 1973). Shu and associates (Anderson and Shu, 1977, 1978) maintain that $A = 1.0$ and $b = 0.0$.

Light synthesis studies of VW Cephei show that values of A and b within these ranges cannot reproduce the observed light curves, assuming nominal limb darkening of 0.6 and an orbital inclination of $i = 66.7^{\circ}$, adopted from Lucy's original solution (Linnell, 1982b). Shu and associates suggest a reduction in limb darkening to zero. This would deepen the occultation eclipse relative to the transit eclipse, but zero limb darkening is objectionable on physical grounds.

Wilson and Biermann (1976) have generated W-type light curves by assuming von Zeipel values, $b = 0.25$ or larger, together with large positive A values, or negative A values in some cases. Since von Zeipel gravity brightening violates zero radiative flux divergence on equipotential surfaces, and leads to Eddington-Sweet circulation currents which tend toward isothermal equipotentials, b values approaching 0.25 lack theoretical justification, even for radiative envelopes.

The canonical explanation of W-subclass light curves, using the postulate of a hot secondary, by Rucinski (1973,1974), meets difficulty in a uv study of W UMa, as Eaton, Wu, and Rucinski (1980) have shown. Light synthesis simulation demonstrates that a hot secondary model produces a color change at secondary minimum opposite to that observed



*P. B. Byrne and M. Rodonò (eds.), Activity in Red-Dwarf Stars, 479–480.*

(Linnell, 1982b). The hot secondary model, in its simplest form, therefore is inadmissible.

A possible model adopts elevated temperatures on the facing hemispheres. This model produces acceptable color curves and light curves (Linnell, 1982b). Two physical interpretations of this model are possible. One attributes the observational effects to starspots on the more massive component, in accordance with a theoretical model by Mullan (1975). There is separate evidence for starspot activity in this star (Linnell, 1982a). However, an exclusive interpretation in terms of starspots would require an extremely smooth underlying starspot distribution, with starspots at all longitudes but concentrated on the opposed hemisphere of the primary, together with localized starspot development in time intervals of a few days. The starspot concentration is opposite to that expected theoretically. The exclusive starspot hypothesis would leave no room for thermal effects of a mass circulation model, such as that proposed by Webbink (1977). Some form of mass circulation model is necessary to explain the Lucy paradox. The second physical model adopts localized starspot activity to explain cycle-to-cycle light curve changes and the O'Connell effect. The temporally stable temperature excesses on facing hemispheres, and the observationally-indicated transverse temperature gradient then prospectively become the observational signature of the energy transfer process.

This research has been supported by grant AST-80-02116 from the National Science Foundation.

REFERENCES

Anderson, L. and Shu, F.H.: 1977, Astrophys.J. 214, 798.
Anderson, L.: 1978, Ibid., 221, 926.
Eaton J.A., Wu, C.C. and Rucinski, S.M.: 1980, Astrophys.J. 239, 919.
Linnell, A.P.: 1982a, Astrophys. J. Suppl., in press.
Linnell, A.P.: 1982b, preprint.
Lucy, L.B.: 1967, Zeit. f. Astrophyz. 65, 89.
Lucy, L.B.: 1968a, Astrophys. J., 151, 1123.
Lucy, L.B.: 1968b, Ibid. 153, 877.
Lucy, L.B.: 1973, Astrophys. Space Sci. 22, 381.
Lucy, L.B.: 1976, Astrophys. J. 205, 208.
Lucy, L.B. and Wilson, R.E.: 1979, Astrophys. J. 231, 502.
Mullan, D.J.: 1975, Astrophys.J. 198, 563.
Rucinski, S.M.: 1973, Acta Astron. 23, 79.
Rucinski, S.M.: 1974, Ibid. 24, 119.
Shu, F.H.: 1980 in IAU Symposium 88;"Close Binary Stars: Observations and Interpolation", ed.M.J. Plavec, D.M. Popper and R.K. Ulrich (Dordrecht: Reidel), p.477.
Shu, F.H. and Lubow, S.H.: 1981, Ann.Rev. Astron. Astrophys. 19, 277.
Webbink, R.F.: 1977, Astrophys. J. 215, 851.
Wilson, R.E. and Biermann, P.: 1976, Astron. Astrophys. 48, 349.

# A BRIGHT SPOT AND A SERENDIPITOUS STELLAR FLARE ON THE CONTACT-BINARY VW Cep.

K. E. Egge and B. R. Pettersen
Institute of Mathematical and Physical Sciences
University of Tromsø, N-9001 Tromsø, Norway.

## INTRODUCTION

Almost twenty years ago a large flare event was observed on the prototype contact binary W UMa by Kuhi (1964). Similar events have been reported on 44 i Boo (Eggen 1948) and U Peg (Huruhata 1952). In this paper we present photoelectric observations at three wavelengths of a flare on VW Cep. This is the first event of this kind to be reported for this star. VW Cep is a triple system. The main contributor to the visual flux is the eclipsing binary, consisting of a K1 primary and a G6 secondary (Kopal 1978), classified to be in contact. The orbital period is $6^h41^m$. Seven per cent of the total flux in the visual filter is due to the third component, a late K type dwarf at a distance of 12 AU from the eclipsing system (Hershey 1975).

## OBSERVATIONS

The photoelectric observations were made with the 50 cm Ritchey-Chrétien telescope at Skibotn Astrophysical Observatory using a chopping photometer which alternates between star and sky background at a frequency of 5 Hz. Data were taken through three filters approximating the UBV system at a time resolution of about two minutes per filter. A comparison star was measured at regular intervals.

The observations were made on the night 27-28 December 1978 UT, and photometric conditions prevailed for ten hours. Whereas the primary eclipse is perfectly symmetrical, the light curves reveal that the secondary eclipse is asymmetrical. Relative to mid-eclipse at phase 0.5, each point on the descending branch (0.25<phase<0.5) has a systematically higher flux than the corresponding points on the rising branch. In fact, the excess flux prevails until phase 0.52. The observations cover two succesive secondary eclipses. The excess flux is detected from about phase 0.25 in both cycles. Figure 1 shows portions of the light curves, corrected for the effects of eclipses. The excess flux is present in all filters, but has diminished in amplitude by 0.02 mag from the first



*P. B. Byrne and M. Rodonò (eds.), Activity in Red-Dwarf Stars, 481–483.*

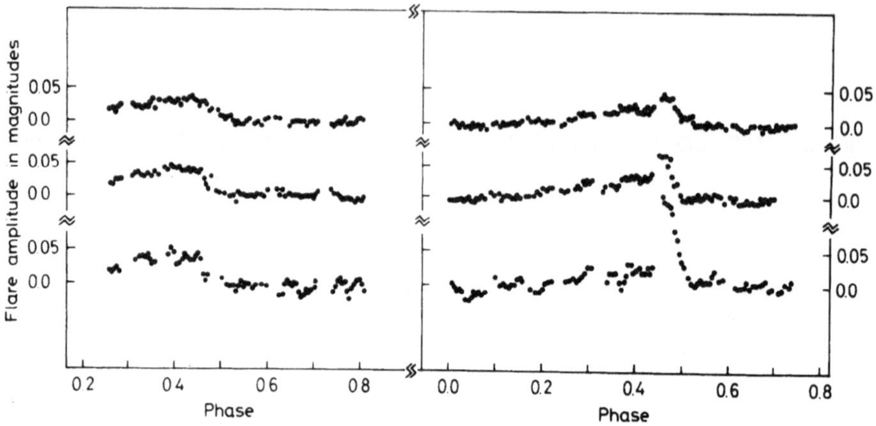


Fig. 1. Two succesive cycles on VW Cep displaying excess flux and a flare

cycle to the second. During the later part of the excess plateau in the second cycle a flare event occurred. Correcting for the shape of the plateau we obtain the light curve of the flare itself, as shown in Fig. 2. The flare decays to one half of the peak value in 13 minutes, and the total duration is less than 35 minutes. We have no time resolved observation of the rise phase, and can only set an upper limit of 5 minutes to this parameter. Adopting Johnson's(1966) calibration of the UBV system, the peak fluxes of the flare is $7 \cdot 10^{30}$ erg/s(U), $6 \cdot 10^{30}$ erg/s(B), and $5 \cdot 10^{30}$ erg/s(V). Integrating numerically under the light curves we obtain for the total energy during the flare $4 \cdot 10^{33}$ erg(U), $6 \cdot 10^{33}$ erg(B), and $4 \cdot 10^{33}$ erg(V).

## DISCUSSION

Because of the perfect symmetry of the primary eclipses and the consistenc obtained in the results of the light curve analyses of each of the three filters, we assume here that the phase interval from about 0.5 to 1.25 represents the unperturbed system. Thus the interpretation of the asymmetry of the secondary eclipses is that the excess flux on the descending branch is due to a bright spot. This spot rotates into view near phase 0.25 when the two stars of the binary have their largest separation as seen by the observer. If the stars rotate in synchronism with the orbital revolution, the spot is near the disk meridian at phase 0.5 and is being eclipsed. Consequently the spot is situated on the larger star in the system. The observed close relationship in time between the spot and the flare is indicative of a close physical relationship. We suspect that the flare occurred on the leading edge of the spot and that it was actually eclipsed during the decay phase. The flare light curve has no slow tail and falls of rather abruptly. As the active region is facing the other component, it is possible that the activity observed is related to mass transfer phenomena rather than solar-like flare activity.

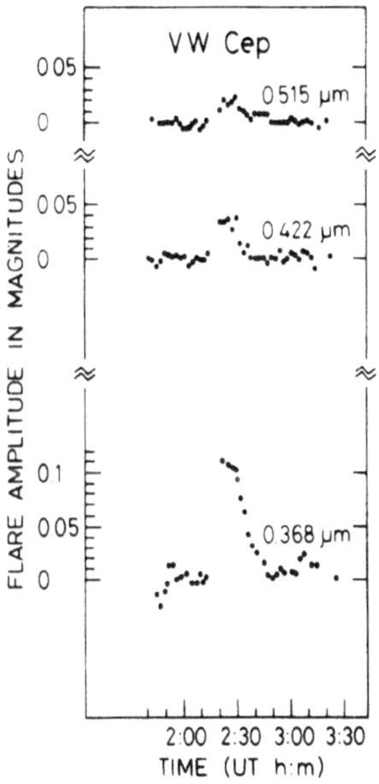


Fig. 2. A flare on VW Cep at three wavelentghs.

References
Eggen, O.J., 1948, Ap.J. 108, 15.
Hershey, J.L., 1975, AJ 80, 662.
Huruhata, M., 1952, Publ.Astron.Soc.Pacific 64, 200.
Johnson, H.L., 1966, Ann.Rev.Astron.Astrophys. 4, 193.
Kopal, Z., 1978, "Dynamics of Close Binary Systems", D. Reidel Publ. Co., Holland, p. 433.
Kuhi, L.V., 1964, Publ.Astron.Soc.Pacific 76, 430.

NUMERICAL ANALYSIS OF ORBITAL PERIOD VARIATIONS AND A MECHANISM FOR CHANGES IN THE LIGHT CURVE OF VW CEP

L. Xuefu and L. Chengzhong
Beijing Normal University, People's Republic of China

## NEW INTERPRETATION OF THE KWEE EFFECT

The WUMa contact binary VW Cep has a period of 0.2783 days, spectral type G5+K1 and mass 1.1+0.4 $M_{\odot}$. Our three-colour photoelectric measurements made in Sept. and Oct. 1964 confirmed the existence of the Kwee effect in this system. We believe that if the Kwee effect is to be explained in terms of moving gas streams the influence of magnetic fields must be considered. It is assumed that there is a deep convection zone on the primary accompanied by a dipole magnetic field. This field is distorted by the moving gas streams to form a ring-like configuration with an enhanced field strength close to the inner Lagrangian point, $L_1$ (see figure 1), where large groups of spots form. We suppose the magnetic field in the spots to be 5000G and the general field to be ~10-20 G. The cycle of magnetic activity can be calculated using the theory of magnetic flux oscillation. The period, $T= 4\pi^{3/2}(\rho Hd/B_pB_t)$ we calculate as 747 days, close to the observed period of 718 days. Hot gas streams from the secondary could gradually develop into a shock and form a heat flare or local emission region on the primary. Such flare activity would also be seen on the light curve but would not affect the basic period. In short the Kwee effect is the result of spot activity and a local emission region on the primary. Figure 1

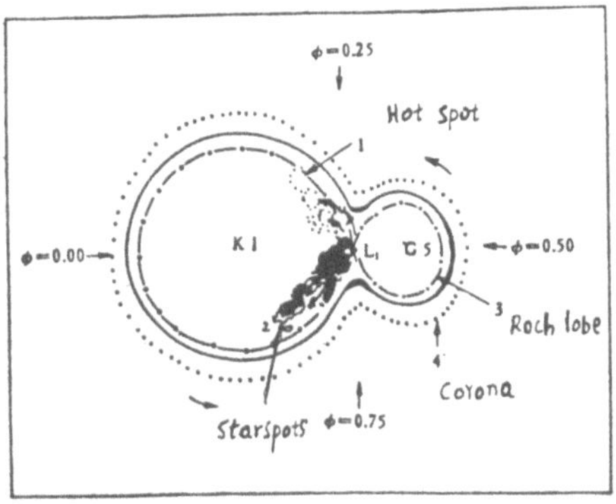


Figure 1. A model of VW Cep.



*P. B. Byrne and M. Rodonò (eds.), Activity in Red-Dwarf Stars, 485–486.*

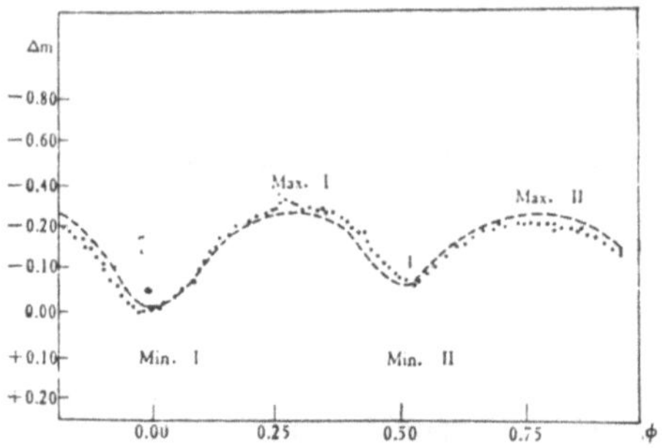


Figure 2. Light curve of VW Cep distorted by additional or missing light due to hot spots or starspots respectively.

shows the model and Figure 2 shows the Kwee effect on the light curve. Note that the light curve has two maxima at phases 0.25 and 0.75 respectively.

NUMERICAL ANALYSIS OF THE ORBITAL PERIOD CHANGES.

The change in the orbital period (O-C) referred to the minimum at phase zero in Figure 2 is given by JD(hel)2424658.758-0.2783199E based on data from 1926 to 1977. The resulting (O-C) curve is given in Figure 3. Let (O-C)=f(E). f(E) oscillates about a mean relation $f(E)=0.047-2.029\times10^{-6}E$. A Fourier method was applied to the oscillations of the form.

$$f(E)=a+bE+\sum_{k=1}^{m} A_k \sin(kwE+\phi_k)$$

where $w= 2\pi/T$. Altogether 35 solutions for $A_k$, $\phi_k$ and $T_k$ were obtained. We believe that the variation of the orbital period is not caused by a third body alone. There are also effects due to gas streams, tidal action, stellar wind, mass loss, etc.

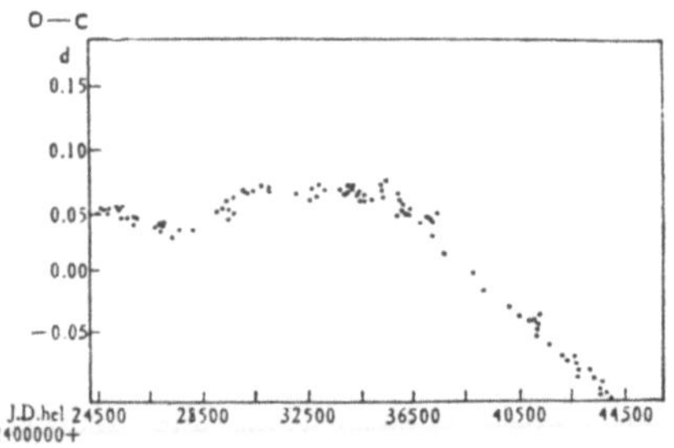


Figure 3. Period changes in VW Cep (1926-1977).

# ON ACTIVITIES OF UV CET-TYPE FLARE STARS AND OF T TAU-TYPE STARS

R.E. Gershberg
Crimean Astrophysical Observatory, USSR

Comparisons of the activities of UV Cet-type flare stars and of T Tau-type stars with solar activity permits the conclusion that non-stationary processes in the UV Cet-type stars and in the Sun are of an indentical physical nature but that they differ qualitatively from active events in the T Tau-type stars. The identity of the activity in flare stars and the Sun makes it possible to study successfully stellar activity with the help of known models of various solar events and, on the other hand, to have a more general approach to the physics and evolution of solar activity on the basis of established features of numerous flare stars of different ages and masses. The hypothesis on hydromagnetic activity of the T Tau-type stars is sketched; within this framework, one supposes that the main feature of such stars is an occurrence at every point of the stellar surface, of conditions necessary for the existence of dark spots, i.e. a lowering of the photospheric brightness due to strong local magnetic fields. It is noted that the total energy losses from non photospheric radiation of flare stars are comparable in power to the total radiation deficit of dark spots on such stars. This fact confirms the possibility of transmitting effectively a significant part of a subphotospheric layers energy into higher regions of the stellar atmosphere by hydromagnetic waves and is an important argument in favour of the proposed hypothesis on the nature of T Tau-type stars.

As is known, the T Tau-type and the UV Cet-type variables were discovered by Joy and by Luyten practically simultaneously, in the middle of the forties, and are often considered as related objects physically and generically. Indeed, these eruptive stars have many common features: low luminosities, non-periodic brightness variations with amplitudes up to several magnitudes, advanced spectral classes at brightness minima and strong emission spectra that are seen even in the quiet state and correspond to a rather low, chromospheric, degree of excitation. It is



*P. B. Byrne and M. Rodonò (eds.), Activity in Red-Dwarf Stars, 487–495.*

a widespread opinion that young T Tau-type stars are immediate predecessors of older flare stars of the UV Cet type. This opinion is based on series of observable facts: flare stars do not show a relation with diffuse matter, but such a relation is very characteristic for T Tau-stars and is an important argument in favour of their youth; many flare stars have been found in stellar clusters but T Tau-stars are no longer there; flare and chromospheric activity in UV Cet-stars involves processes that are of several magnitudes lower in energy than similar processes in T Tau-stars, and the abnormally high abundance of Li that is inherent in T Tau-stars is found only in one flare star. However, there are considerations that cast doubt on a thesis that observable T Tau-stars are immediate predecessors of typical flare stars. Thus, T Tau-stars belong to the flattest component of the Galaxy but among flare stars we have objects of the galactic disc and even of the galactic halo. Then, masses of T Tau-stars are evaluated to be about 1-3 solar masses but masses of flare stars are an order of magnitude lower and observations do not show such a strong outflow which could diminish a stellar mass so much in $10^6$ - $10^7$ years, the time scale of T Tau phase of stellar evolution. Therefore, the problem of the generic relation between eruptive variables of T Tau-type and of UV Cet-type requires more careful study. However, problems of stellar evolution are outside the scope of this Colloquium. So now let me turn your attention to a comparative consideration of the observable activities of these stars.

In the very beginning of investigations of T Tau and UV Cet stars there were noted some common features in their activities and those of the Sun: emission spectra in the quiet state of variables of both types lead to ideas about strong stellar chromospheres and short-lived flares suggest an analogue with the solar chromospheric flares. However, these original general considerations evolved into very different model conceptions.

1. After a lot of observational data on the UV Cet star flares and on the features of photospheres, chromospheres and coronae of such stars had been accumulated, the identity of the physical nature of activities of these variables and of the Sun was proposed (Gershberg, 1975). Direct discoveries of the transition region between chromospheres and coronae (Hartmann et al., 1979; Haisch and Linsky, 1980), of thermal radiation from stellar coronae (Johnson, 1981; Vaiana et al., 1981), recordings of stellar flares in the EUV range (Butler et al., 1981) and in the X-ray range (Kahn et al., 1979; Haisch et al., 1980) converted this suggestion into an obvious statement. A useful aspect of this suggestion should be stressed.

Firstly, the identity of the physical nature of active processes in the Sun and in flare stars permits us to use models of solar events to understand the nature of events observed in red dwarfs. An excellent example of a successful application of solar physics to the study of flare stars is the recent paper by Katsova et al. (1980). Developing the Pickel'ner-Kostyuk model for the solar chromospheric flares, they have carried out refined calculations of a strong hydrodynamical disturbance in the solar atmosphere subjected to impulsive heating with a fast electron flux. Calculations showed that in such a situation a strong shock propagation downward to the photosphere is formed, and intensive radiation of hydrogen must originate in a thin layer behind the shock front. In extreme cases the optical thickness in the continuum of this layer can approach unity, and such a layer can be responsible for a white light flare. Apparently, an application of this scheme to flare stars gives a solution of a problem of the short-lived continuum emission in stellar flares. According to calculations, in the denser atmospheres of K-M stars the downward propagating shock practically always approaches the conditions where the optical thickness of the compressed gas is of order of unity, i.e. white light flares must be rather ordinary events in flare stars and not extraordinary events as they are in the Sun. It should be noted that calculation of the chromospheric response to the initial temperature disturbance, including the formation and propagation of a shock within an essentially heterogeneous medium with variable radiation losses, is rather a complex subject. For instance, it is hard to evaluate flare color indices or to compute a set of theoretical light curves. All such photometric characteristics can be found only with a numerical integration over a very non-homogeneous disturbed volume. However, the physical clarity of this scheme makes it much more attractive than numerous homogeneous models (e.g. a hot photospheric spot, a hot gas at chromospheric or subcoronal temperature, relativistic or subrelativistic particles) or simple combinations of such models, that have been used for a long time to represent observable flare color indices, to compute theoretical light curves and to understand the short-lived continuum radiation from the UV Cet star flares (see Gershberg, 1978). It is not possible to uniquely pick out the best homogeneous model or simple combination of such models to represent this continuum emission, but the two best models (optically thin gas of subcoronal temperature (Gershberg, 1975; Mullan, 1976; Kodaira, 1977) and gas of chromospheric temperature and of a significant optical thickness in the Balmer continuum (Grinin and Sobolev, 1977)) are natural components of the non-homogeneous model by Katsova et al.

Secondly, the identity of the physical nature of the activity of flare stars and of the Sun does not mean that problems of the physics of these eruptive stars become problems of the solar physics. In reality, we have here a two-way fruitful interaction. For instance, it is obvious that the observations of a number of flare stars of different masses, sizes and ages give a lot of data for a more general approach to the solar activity phenomenon, e.g. finding new regularities in this activity or possible evolution of the activity, in total. The total stellar flare radiation energy can be obtained with a rather simple observation and therefore flare energy spectra in the UV Cet-type stars have been investigated in much more detail than in the Sun. In Shakhovskaya's (1979) paper updated for this Colloquium, the most complete description of the stellar flare energetics is given based on such a spectral-energetic approach. A similar analysis of the solar data would be valuable. Then, if flares on the Sun and on red dwarfs are physically identical, only a theory of solar flares may be correct, which can represent successfully, by varying the magnitudes of essential parameters, the much more powerful and faster flares of the UV Cet-type stars. It is a rather strict criterion for correctness of a solar flare theory, but it seems to me that solar physicists have not realized in full the importance of this criterion yet. Only when it was found that thermal radiation from flare star coronae contributes several percent of the stellar luminosity, was the necessity to revise the tradition concept of stellar (and solar) coronae realized.

2. Relations between the T Tau-type stars and the Sun are of quite another character. As we know more about these variables we find less place for a direct analogue between the activities of these stars and of flare stars and the Sun. Actually, emission spectra of typical T Tau-stars are similar to the solar chromospheric spectrum in excitation but contain a noticeable amount of stellar optical radiation while the solar chromospheric contribution does not exceed $10^{-4}$ of the total luminosity of the Sun. Then, time-averaged variable components of T Tau-star luminosities are comparable in power to their constant component while flares in the Sun and in flare star give negligible contributions to their total luminosities. These quantitative differences are so large that we have no practical hope of describing T Tau-star activity with known schemes of the solar activity, using variations of the magnitudes of essential parameters. The most important fact is that in T Tau-stars we observe events that give evidence of qualitative differences between these variables and the Sun. In some T Tau-stars, infrared emission exceeds noticeably the photospheric radiation in power while no infrared excess has been found either in the Sun, or in flare stars. During the FU Ori-type flares that are inherent in T Tau-stars stellar luminosities increase up to two orders of magnitudes and remain at such a high level for decades. Neither the Sun, nor flare stars show such powerful processes.

As is known, in spite of a large volume of accumulated data on the T Tau-type stars, we have no common model for the activity of these stars yet. We do not know a physical mechanism that may be certainly regarded as responsible for irregular variations of stellar brightness. There is no commonly accepted interpretation for the emission line spectra of these variables. There exist different hypotheses on excitation sources, on structure and kinematics of irradiated gas. The nature of the blue continuum emission is not clear. There exist many qualitative schemes to explain separate features of T Tau-stars. Recently Lynden-Bell and Pringle (1974) have proposed the quantitative model of disc accretion and within this framework they have represented some photometric features and infrared radiation of these variables: Larson (1980) has worked out quantitative model of rotational non-stability to explain the FU Ori-type flares. The close relation of these models with current ideas on the youth of T Tau-stars and their quantitative nature make them very attractive. They are not however sufficient to represent all the main observable features of the T Tau-type stars and are unlikely to be mutually compatible.

Dr Petrov and myself (Gershberg and Petrov, 1976) have attempted to understand the whole variety of observable features of these stars within the framework of a common scheme based on a decisive role of stellar magnetism. In the most complete version this hypothesis on hydromagnetic activity of the T Tau-type stars is published in the Astronomische Nachrichten N 4, 1982. There are two suggestions that are the basis of this hypothesis:

a) at the phase of minimum brightness, at practically every point of a stellar surface conditions similar to those in Sunspots are found where the presence of a strong magnetic field leads to a significant suppression of the photospheric brightness; the whole stellar surface is supposed to be a mosaic consisting of magnetic spots.

b) the degree of suppression of optical radiation in these stellar spots exceeds that in the Sunspots and approaches 100%.

In spite of the exotic character of these suggestions, apparently, they do not contradict to observations and known laws of physics (Gershberg, 1982), and in the framework of current ideas on stellar evolution an essential role for magnetic fields at early phases of stellar life is very probable (Mestel, 1977; Lynden-Bell, 1977). In the framework of the proposed hypothesis, observed irregular light curves of the T Tau-stars should be considered a result of numerous quasi-independent processes "dark spot-normal bright photosphere", which take place due to magnetic field strength variations in different areas of the stellar surface. The FU Ori-type flare corresponds to a rare phase when practically the whole stellar surface is free from dark spots. Infrared excess and intense chromospheric emission spectra are related in this scheme to strong fluxes of non-radiative energy that supply most of the stellar energy losses at the minimum brightness phase. Recently discovered very fast brightness variation in T Tau-stars that are similar to the UV Cet-type flares arise in a natural manner from the proposed hydromagnetic situation at the T Tau-star surfaces. However, these stars flares which are much more powerful and much slower than the UV Cet-type flares do not permit us to reduce them to superpositions of short-lived flares of red dwarf type.

The proposed hypothesis on hydromagnetic activity of the T Tau-type stars is a development of Danielson's (1965) idea on the outflow of the sunspot radiation deficit beyond the photosphere by hydromagnetic waves. As is known, this idea had been developed by Mullan (1974a) and by Parker (1974) and then applied to flare star dark spots by Mullan (1974b). This model is not commonly accepted and the problem of sunspot radiation deficit is being hotly debated (Cowling, 1976; Parker, 1977; Sreenisavan, 1977; Obridko and Teplitskaya, 1978). However, recent observations by Willson et al. (1981) give an important argument in favour of this model. Our hypothesis is supported as well by Bray's (1981) results: he has found that in one sunspot center the surface brightness is equal to 14% of the quiescent photospheric value, much less than previously found.

3. Thus, as the observations are accumulated, it becomes more clear that the UV Cet-type flare star activity is identical to the solar activity but has essential qualitative differences from the T Tau-type star activity. In such case a question arises: why are we actually discussing the T Tau-stars at the Colloquium on the UV Cet-stars?

In a paper published in the Memorie della Società Astronomica Italiana (49, 781, 1978) I have suggested that stellar magnetism is a decisive factor in the evolution and physics of activities of variables of both types and, in particular, that this is a distinguishing factor for flare and non-flare stars. Powerful coronae of flare stars have since been found by the Einstein Observatory which do not fit the traditional models of stellar coronae supported with acoustic waves are the most important argument in favour of such a suggestion. Correlation between the BY Dra syndrome and axial rotation rates (Bopp and Espenak, 1977) and between the X-ray luminosity and axial rotation rates (Katsova, 1981) suggest that magnetic fields in flare stars are of a secondary nature. However, independent of the relation between fields and stellar rotation, it is clear that magnetic loops carrying stellar coronae are supported on extended active regions in photospheres. The problem of dark spots in active regions of flare stars seems to be the main un-resolved problem in flare star physics. Just in this point - in the problem of dark stellar spots connected with strong local magnetic fields - there is a deep physical relation between flare stars and T Tau-stars. It should be noted that the origin of magnetic fields of these variables may be quite different. In T Tau-type stars, that are young objects, the relic fields can be dominant while in older UV Cet-type stars the fields originated during the stellar evolution can be dominant. Since similar local events in these stars - dark spots, short-lived flares, chromospheric emission - are defined mainly by the strengths of local magnetic fields, the noted similarity in observed events does not exclude essential differences in the global structure and evolution of magnetic fields of these stars.

According to observation, the total radiation deficit of flare star spots approaches 10% of the stellar luminosities. The problem of so large a deficit has been discussed in details by Hartmann and Rosner (1979). They have concluded that neither the traditional Biermann-Cowling conception with a space redistribution of the sunspot radiation deficit, nor the Danielson-Mullan-Parker scheme with a transformation of the electromagnetic into hydromagnetic radiation gives a successful solution for the problem of the large radiation deficit. They proposed the idea of a temporal redistribution of the stellar luminosity due to a possible effect of the magnetic field on the convective transfer efficiency. In connection with such an idea it should be noted that the current theory of convection accounting for magnetic fields does not predict the existence of structures where the energy could be accumulated and which, in the end, could be responsible for temporal variations of stellar luminosities. However, since this theory is a phenomenological one only, we cannot reject the possible appearance of such structures in a future complete physical theory of stellar convection that might be constructed on the basis of the principles of the thermodynamics of open systems. It is the only possibility that gives hope of the correctness of the Hartmann-Rosner hypothesis. On the other hand, as Shakhovskaya and myself find, the sum of the energy losses of a flare star for short-lived flares to occur and for maintenance the stationary stellar chromosphere, corona and stellar wind turns out to be compatible in order of magnitude with the total radiation deficit of dark stellar spots. This circumstance removes the main objections by Hartmann and Rosner against the Danielson-Mullan-Parker scheme, since similar values of energy losses mentioned and total radiation deficits of stellar spots can be interpreted as observable evidence of a transmission of a significant part of the subphotospheric layers' energy into higher regions of the stellar atmosphere by hydromagnetic waves; although a mechanism for such a transmission is not clear yet. Therefore, it is very desiderable to carry out X-ray observations for different degrees of spottedness of flare stars and long-term bolometric monitoring for several flare stars and T Tau-type stars similar to the solar experiment ACRIM (Willson, 1981).

REFERENCES

Bopp B.W. and Espenak F. 1977, Astron.J. 82, 916.
Bray R.J. 1981, Solar Phys. 69, 3.
Butler C.J., Byrne P.B., Andrews A.D. and Doyle J.G. 1981, Monthly Notices Roy.Astron.Soc. 197, 815.
Cowling T.G. 1976, Monthly Notices Roy.Astron.Soc. 177, 409.
Danielson R.E. 1965, in "Stellar and Solar Magnetic Fields" Ed. R.Lüst, North Holland Publ.Co., Amsterdam, p. 314.
Gershberg R.E. 1975, in "Variable Stars and Stellar Evolution" Eds. V. Sherwood and L.Plaut, D.Reidel Publ.Co., Dordrecht, p. 47.
Gershberg R.E. 1978, "Low-mass Flare Stars", Nauka, Moscow.
Gershberg R.E. 1982, Astron.Nachr. 303, 251.
Gershberg R.E. and Petrov P.P. 1976, Soviet Astron.Letters 2, 499.
Grinin V.P. and Sobolev V.V. 1977, Astrofizika 13, 587.
Haisch B.M. and Linsky J.L. 1980, Astrophys.J.Letters 236, 33.
Haisch B.M. et al. 1981, Astrophys.J. 245, 1009.
Hartmann L. and Rosner R. 1979, Astrophys.J. 230, 802.
Hartmann L. et al. 1979, Astrophys.J.Letters 233, 69.
Johnson H.M. 1981, Astrophys.J. 243, 234.
Kahn S.M. et al. 1979, Astrophys.J.Letters 234, 107.
Katsova M.M. 1981, Astron.Circ.USSR No. 1154.
Katsova M.M., Kosovichev A.G. and Livshits M.A. 1980, Soviet Astron. Letters 6, 498.
Kodaira K. 1977, Astron.Astrophys. 61, 625.
Larson R.B. 1980, Monthly Notices Roy.Astron.Soc. 190, 321.
Lynden-Bell D. 1977, in "Star Formation" Eds. T. De Jong and A. Maeder, D.Reidel Publ.Co., Dordrecht, p. 291.
Lynden-Bell D. and Pringle J.E. 1974, Monthly Notices Roy.Astron.Soc. 168, 603.
Mestel L. 1977, in "Star Formation" Eds. T. De Jong and A. Maeder, D.Reidel Publ.Co., Dordrecht, p. 213.
Mullan D.J. 1974a, Astrophys.J. 187, 621.
Mullan D.J. 1974b, Astrophys.J. 192, 149.
Mullan D.J. 1976, Astrophys.J. 210, 702.
Obridko V.N. and Teplitskaya R.B. 1978, in "Results of Science and Technique", Astronomy vol. 14, Moscow.
Parker E.N. 1974, Solar Phys. 36, 249.
Parker E.N. 1977, Monthly Notices Roy.Astron.Soc. 179, 93P.
Shakhovskaya N.I. 1979, Izv.Krymsk.Astrofiz. Obs. 60, 14.

Sreenisavan S.R. 1977, Astrophys.Space Sci. 51, 341.
Vaiana G.S. et al. 1981, Astrophys.J. 245, 163.
Willson R.C. et al. 1981, Science 211, 700.

DISCUSSION

Vaiana: I did not quite understand where the luminosity of these stars would appear in your picture. Would it be in X-rays, in the optical, in the infra-red or where?

Gershberg: I have not time to go into this fully. However it is possible to explain in this hydromagnetic theory the problems of the brightness variations, of the chromospheres, the darkness and FU Ori. In the minimum brightness state the surface is completely covered in spots. When these separate into several individual spots the magnitude increases to some critical level at which one sees a normal photosphere. When the star is free from any spots we have the FU Ori-type stars. So the normal state of the star is that of maximum brightness.

Rodonò: In your slide showing flare frequency spectrum as a function of luminosity the anomalous behaviour of EQ Peg is due to the fact that both stars flare. This applies also to UV Ceti. I also have almost the same question as Dr.Vaiana, that is, where is the energy flux missing from the spots deposited? We can for instance have deposition temporally, leading to variability or we can have deposition into another spectral band. Which do you think occurs?

Gershberg: In my model there is no energy storage. An any moment the total energy flux is constant. However depending on the total strength of the magnetic field it acts as a regulator. If the field is weak we have a normal photosphere. If the field is strong we have the spotted condition and the energy flux is carried in hydromagnetic waves.

Feldman: Do you have an explanation for the strong bipolar mass flows that are observed in the CO line in many T Tauri stars? There is a lot of energy in those flows.

Gershberg: May I discuss this with you later?

# THE VARIABILITY OF THE T TAURI STAR DI CEPHEI

G.F. Gahm[1], P.P. Petrov[2]

1. Stockholm Observatory, S-133 00 Saltsjöbaden, Sweden
2. Crimean Astrophysical Observatory, P/O Nauchny, Crimea, USSR

Abstract: It is demonstrated that the light variations of DI Cephei can be explained as due to the combined effect of variable completely grey circumstellar extinction and variable continuous and line emission.

Several attempts have been made to explain the light variability of T Tauri stars. The variations have been explanied as due to variable circumstellar extinction, variable physical properties in a stellar wind or accretion envelope, activity at the stellar surface (spots, plages, flares, convection cells) or geometrical changes in the stellar body (see e.g. Gahm, 1978 for a review).

The following is a brief summary of some of the results that we have obtained for DI Cephei, a T Tauri star with a fairly strong continuous and line emission spectrum superimposed on an absorption line spectrum of late spectral type. This study is based on spectrograms taken at the Crimean Astrophysical Observatory and covering the blue and visual spectral regions and on photometric UBV data kindly provided by Drs. E. Kolotilov and T.S. Beljakina. In addition we have made use of all published UBV data (Cohen and Schwartz, 1976; Gahm et al., 1977; Grinin et al., 1980).

Our spectrograms yield a spectral class of G8 IV for DI Cep. The relative contribution of light from the star and from the emission region was evaluated for three spectrogram taken simultaneouly with photometry. The details of this procedure will be published elsewhere. The results are given in Table 1.

Table 1. Spectroscopic results

| Date (1980): | June 19 | June 21 | June 22 |
|---|---|---|---|
| V | 11.47 | 11.48 | 11.66 |
| ΔB | 0.12 | 0.12 | 0.43 |
| E(B-V) | 0.12 | 0.11 | 0.07 |

It is seen that the contribution from the emission region in B (ΔB) is variable and that the extinction is low. The contribution from emission in V is negligible. It is obvious that variable emission activity influences the brightness level of DI Cep in the B band.

In Figure 1 (top) we have indicated the number of observations when DI Cep was found within boxes of 0.1x0.1 mag. in the V/B plane (left) and V/U plane (right). The question mark refers to 2 uncertain observations which we ignore in the following discussion. Kholopov (1952) presented photographic magnitudes that indicate that the star on occasion can reach a minimum in B of around 13.0. It is seen that a gross pattern exists in the sense that increasing V is followed by increasing B and U. A con-



*P. B. Byrne and M. Rodonò (eds.), Activity in Red-Dwarf Stars, 497–500.*

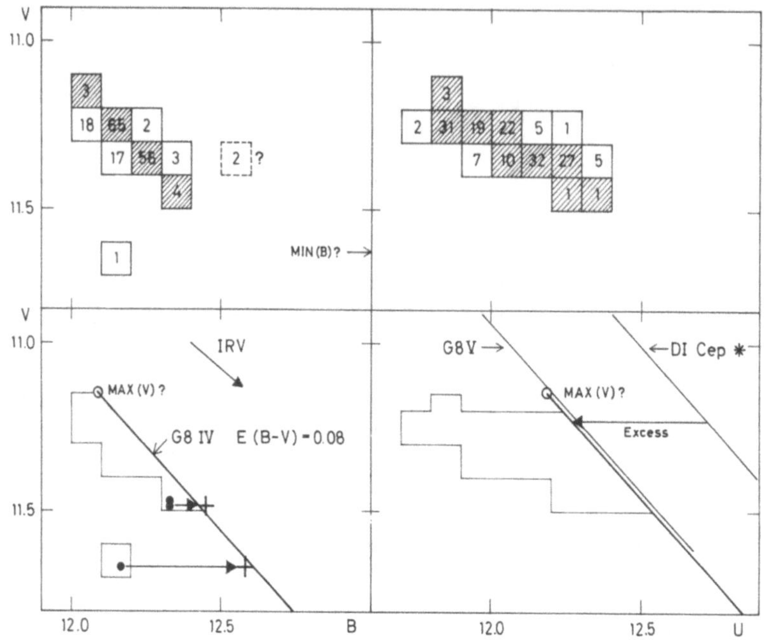


Fig. 1. Variability pattern of DI Cep in U, B and V. Top: Number of observations in squares of 0.1 mag. Preferred areas hatched. Bottom: The thick lines are loci for the most quiescent phase of the star under various amount of grey extinction. IRV = interstellar reddening vector. MAX(V)? = possible maximum in visual brightness under quiescent phase. Loci for a G8 V (unreddened) and G8 IV (present solution) are given and the remaining ultraviolet excess indicated (right).

siderable scatter is present, especially in U, that permits the star to become both redder and bluer with decreasing visual brightness.

The gross pattern is inconsistent with a model where variable emission activity is the only cause of the light variability. Instead, Fig. 1 indicates that there are two causes of the variations. One, producing the gross pattern and another causing the spread in this relation. We have attempted two models to explain the variations. 1. Changing spectral class producing the gross pattern and emission activity causing the spread. 2. Variable circumstellar extinction producing the gross pattern and emission activity causing the spread. On the assumption that the colours of the star is similar to those of normal stars we can exclude model 1.

If the variations follow model 2 we expect all observed points in Fig. 1 (top) to lie to the left of a reddening line which defines the most quiescent state of the emission activity. If the contribution of emission activity is $-\Delta B$ and $-\Delta U$ we expect a relation $\Delta B = \text{const}\ \Delta U$. Several laws of extinction by dust were tried. The best solution was obtained for a completely grey extinction law which yields (see Fig. 1, bottom):

$B = V + 0.94 - \Delta B$

$U = V + 1.04 - \Delta U \qquad \Delta U = 2.0\ \Delta B$

From these relations we can express U in B and V as:

$U = - V + 2B - 0.91$

From all available information on UBV for DI Cep we conclude that with given values of V and B this formula predicts U to within a standard deviation of 0.08 mag.

In Fig. 1 (bottom, left) we can see that the corrections for blue emission as derived from the spectroscopic analysis are fully consistent with the photometric analysis. Furthermore the derived extinction E(B-V) is small even at V = 11.66. However, the locus of the quiescent state in the V/U plane (Fig. 1, bottom, right) is ~ 0.4 magnitudes in excess of the expected colours for a G8 IV star with E(B-V) = 0.08. A change in spectral type or luminosity class of one step in any direction leads to larger discrepancies. Hence, the present model implies that when the star is in its most quiescent phase there is a remaining ultraviolet excess but no, or very little, veiling.

## References

Cohen, M., Schwartz, R.D.: 1976, Mon.Not.Roy.Astron.Soc. 174, 137.

Gahm, G.F.: 1978, Stockholm Obs. Report, No 14.

Gahm, G.F., Gehrsberg, R.E., Petrov, P.P., Shcherbakov, A.G., Kolotilov, E.A., Zaitseva, G.V., Shanin, G.I.: 1977, Variable Stars 20, 381.

Grinin, V.P., Efimov, Ju. S., Krasnobatsev, V.I., Shachovskaja, N.I., Shcherbakov, A.G., Zaitseva, G.V., Kolotilov, E.A., Shanin, G.I., Kiselev, N.N., Gjulaliev, Ch. G., Salmanov, I.R.: Variable Stars 21, 247.

Kholopov, P.N.: 1952, Variable Stars 9, 157.

## DISCUSSION

Worden: I would like to caution people about making too much of changes in colour. A few years ago we made some high-speed photometry and especially in the U-band the T-Tauri stars appear to be continuously flaring, or at least they look as though they are flaring. One should be careful to ensure that colours are measured simultaneously. One can get spurious colour changes due to the fact that either the star was flaring in one colour or the other or the flare light is blue.

Gahm: You are saying that the colour will be different from flare to flare. I agree.

Worden: I am saying that any observation of variable colour may be due to flaring (Part lost).

Gahm: Yes, but the changes I am discussing are slow. I can believe that there are flare-like peaks on a short timescale. Were your flares on a short timescale or were they on a timescale of hours?

Basri: I would just like to muck up the situation a little further. Models of T-Tauri chromospheres that Dr.Calvet and I have calculated indicate that it is relatively easy to produce changes in the visual continuum as well as the blue due to the chromospheric activity which takes place. So again one has to be careful in interpreting colour changes in these stars.

Gahm: Yes, that is true. But these colour changes are tied up with spectroscopic observations. Since the star varies so much, it goes down to B = 13, it is very difficult to imagine that when you leave out all the excess continuum and line emission you will find a normal star there. I believe it is impossible. So either you invoke a star completely covered in spots which will be completely black or it must be something circumstellar such as extinction.

Hartmann: If one looks with good resolution at the optical spectrum of this star the photospheric lines are seen to be filled-in with emission and the Mg I B line sticks up in emission.I don't see how you can make a simple veiling correction given the complicated nature of the filling-in.

Gahm: (First part of answer lost) ... over the red spectral region. What we have is a set of standard stars and for the red spectrogrammes you can find that the amount of veiling and emission is very low indeed. Of course you can find lines like the Na-D lines and the Mg lines and Hα, of course, but in the V band and in the red it is a typical G8IV star.

# OPTICAL MONITORING OF T TAURI STARS

William Herbst
Van Vleck Observatory, Wesleyan University

At the Van Vleck Observatory of Wesleyan University, Middletown, Connecticut, we have initiated a program of monitoring T Tauri stars with the 24-inch reflector (Perkin Telescope). During the first observing season, we have obtained UBVRI and Hα (30Å and 150Å) data for five stars: T Tau, RY Tau, SU Aur, RW Aur, and CO Ori. The data will be presented elsewhere (Herbst, Holtzman and Phelps, 1982). Here we report some of our conclusions.

1. Many T Tauri stars exhibit "flare-like" activity which manifests itself as short time scale variations in Hα flux correlated with variations in the U bandpass (and, in extreme cases, B bandpass). This activity was seen in all the stars observed except SU Aur, which has the weakest Hα emission. The flare-like activity appears to be independent of the variations at longer wavelengths and, in fact, is easiest to see in those stars (e.g. T Tau and RY Tau) which did not change brightness in B, V, R, or I. The time behaviour of this component is discussed by Kuan (1976) and Worden *et al.* (1981) who monitored T Tauri Stars in U.

2. At longer wavelengths (V, R and I), all T Tauri stars studied behave in the same manner; they get redder when fainter. The stars with weaker emission lines also get redder in (B-V) and (U-B) when fainter. Stronger emission line stars tend to mask this behaviour with the independent "flaring activity" discussed in point 1. If such stars are studied only in U, B and V, or (worse yet) only in $m_{pg}$, the combination of two components of variation can result in very complex and confusing behaviour. Future monitoring of T Tauri stars should always include bandpasses redward of V.

3. Stars with moderate or small Hα equivalent widths and absorption line spectra in the optical region (SU Aur, CO Ori, BM And) show an increase of EW(Hα) as the star fades, while the flux in the Hα line remains constant or increases slightly. This shows that, if circumstellar dust is responsible for the variations at longer wavelengths, it must be located closer to the photosphere of the star than is the region responsible for the Hα emission. If the Hα emission originates



*P. B. Byrne and M. Rodonò (eds.), Activity in Red-Dwarf Stars, 501–502.*

in a chromosphere-like region of higher temperature above the photosphere, this means that the dust would have to be between the photosphere and chromosphere - an unlikely circumstance! At present, the more appealing alternative is that the variations of these stars come primarily from photospheric changes - presumably changes in the global effective temperature, perhaps related to magnetic activity such as that which produces sunspots.

4. In one star, RW Aur, a strong emission line object, the variations at all wavelengths from U to I are well correlated with Hα flux. The magnitude-color relations, however, show that in U (and to a lesser extent B), there are two components to the variability - one coming from the flare-like activity described in point 1, and one associated with the variations at longer wavelengths. The spectrum of RW Aur is dominated by emission lines and a largely featureless continuum over the entire optical range from U to I; however, high signal-to-noise data do show the presence of photospheric absorption lines at least occasionally (Mundt and Giampapa, 1982). Hence, it is not clear from the spectrum alone whether the variations at longer wavelengths could be "photospheric" or "chromospheric" in nature, although excellent correlation with Hα flux argues that they are chromospheric.

5. Since virtually all T Tauri stars, a class defined solely by spectroscopic criteria, are irregular variable stars (Herbig 1962), it is reasonable to suppose that there is a single underlying cause of that variability, however complex its manifestations may be. This statement is in contrast to some discussions in the literature which attempt to attribute variations in one star to one cause and others to other causes. If there were really several variability mechanisms that selectively operated in some stars and not in others, then one might expect to find a group of stars in which none of the mechanisms operated. Yet, non-variable T Tauri stars do not appear to exist (although the author is admittedly unaware of a systematic study which firmly establishes this statement as a fact). Our impression from this study and the literature is that, if a single underlying cause of the irregular variability of T Tauri stars does exist, it is related to surface changes (e.g. localized flare events and perhaps spotting activity) presumably associated with magnetic fields, in the photosphere and/or chromosphere, and not to circulating protoplanets or other forms of dust. A test of the "photospheric" hypothesis should be whether or not stars like CO Ori change their spectral type when undergoing large changes in brightness. We intend to monitor that star as well as SU Aur both photometrically and spectroscopically during the next observing season to this hypothesis.
The author thanks the Perkin Fund, Research Corp. and NSF for support.

References

Herbig, G.H. 1962, Adv. Astron. Astrophys. 1, 47.
Herbst, W., Holtzman, J.A., and Phelps, B.E. 1982, A.J., submitted.
Kuan, P. 1976, Astrophys. J. 210, 129.
Mundt, R. and Giampapa, M.S. 1982, Astrophys. J. 256, 156.
Worden, S.P., Schneeberger, T.J., Kuhn, J.R., and Africano, J.L. 1981, Astrophys. J. 244, 520.

PERIODIC LIGHT VARIABILITY IN FOUR LATE-TYPE PRE-MAIN-SEQUENCE STARS

F.J. Vrba
U.S. Naval Observatory, Flagstaff Station, Flagstaff, Arizona

A.E. Rydgren, J.T. Schmelz
Physics Dept., Rensselaer Polytechnic Institute, Troy, New York

While the T Tauri stars are the best known of the late-type pre-main-sequence (PMS) stars, there are also some late-type PMS stars with only weak line emission in their visible spectra. Several years ago we noted that the weak-emission PMS stars have B-V colors too blue for their V-I colors and suggested that their surfaces might have regions of differing temperature. During October 1981 we used the USNO 40-inch and Kitt Peak National Observatory No. 4 16-inch telescopes to monitor, over a 7 night interval with UBVRI photometry, four of these weak-emission PMS stars: HD 283447, V410 Tau, and X-ray stars 1 and 2 of Feigelson and Kriss (1981). The PMS nature of these stars is established from (1) their membership in the Taurus dark cloud T-association and (2) their location within the T Tauri band region of the H-R diagram.

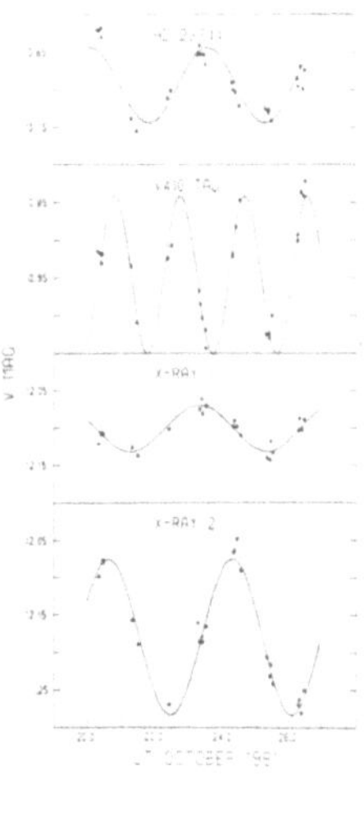

Figure 1

Figure 1 shows the V magnitudes of these stars as a function of time along with the best-fit sine curves to the data. All four stars show quasi-sinusoidal light variations with timescales ranging from 1.9 to 4.0 days. In Figure 2 we show the amplitude of the light curves as a function of wavelength for the three stars with largest amplitude along with the best fits to the data of a simple, geometrical starspot model. This over-simplified model indicates starspots with temperatures typically 600° K cooler than the photosphere and with fractional disk areas covering from 0.17 to 0.27 of the visible hemisphere.

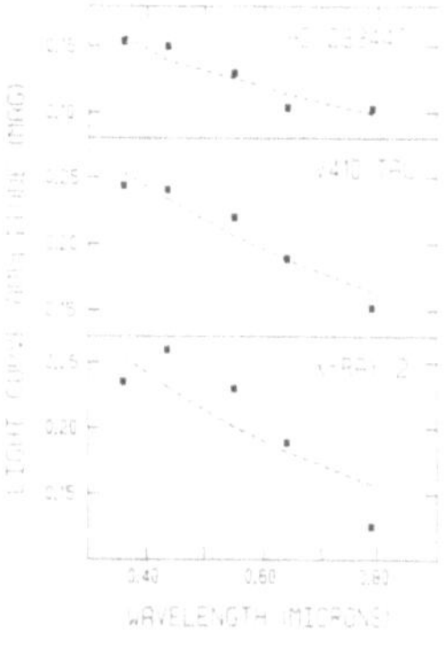

Figure 2

During the course of our monitoring we detected two U flares in X-ray 2 and one in V410 Tau. The apparent peak of one of the X-ray 2 flares



*P. B. Byrne and M. Rodonò (eds.), Activity in Red-Dwarf Stars, 503–504.*

was observed for which we estimate total amplitudes of about 0.54 mag in U and 0.07 mag in B.

Our observations of the weak-emission PMS stars have shown: (1) periodic light variations which are approximately sinusoidal in form, (2) decreased light-curve amplitudes toward longer wavelength which are compatible with reasonable starspot parameters, and (3) UV flares in half of our sample of stars during a limited span of observation. From these results, along with the previously known weak line emission, we conclude that the program stars exhibit magnetic surface activity similar to that which is seen in BY Dra and RS CVn stars.

How do the weak-emission PMS stars relate to the T Tauri stars? Using the computed positions of our program stars in the log $(L/L_{\odot})$, $\log(T_{eff})$ plane along with conventional evolutionary tracks we derive masses (0.8-2.0 $M_{\odot}$) and ages ($\simeq 10^6$yrs) which are comparable to those of the T Tauri stars. These ages do not clearly indicate that these stars should be considered "post-T Tauri" stars.

The program stars' rotational velocities have been calculated from their light curve periods and from radii deduced from the observed luminosities and effective temperatures. The computed values range from about 20 km/s for X-ray 1 to 75 km/s for V 410 Tau. V sin i measurements of T Tauri stars by Vogel and Kuhi (1981) indicate rotational velocities of 50-100 km/s for stars of several $M_{\odot}$ to upper limits of 25-35 km/s for stars of 1 $M_{\odot}$ or less. Their measurement of V sin i $\simeq$ 76 km/s for one of our program stars, V 410 Tau, is in excellent agreement with our calculated results. We thus conclude that our weak-emission line program stars rotate at least as rapidly as typical T Tauri stars of similar mass.

In summary, the fundamental distinction between the weak-emission line PMS stars, such as we have observed, and the true T Tauri stars does not appear to be due to age, mass, or rotational velocity differences. Thus, it is by no means obvious that the weak-emission PMS stars are in a "post-T Tauri" phase of evolution. Additionally, the combination of relatively rapid rotation and weak line emission in these stars raises doubts that rapid rotation in a deeply convective PMS star is the fundamental cause of the T Tauri phenomenon via dynamo-generated intense magnetic fields.

A full report of our investigation has been submitted to The Astrophysical Journal for publication. Part of our observations were obtained at Kitt Peak National Observatory, operated by AURA, Inc. This work is partially supported by NSF grant AST80-18229.

REFERENCES

Feigelson, E.D., and Kriss, G.A., 1981, Ap.J.(Letters), 248, L35.
Vogel, S.N., and Kuhi, L.V., 1981, Ap.J., 245, 960.

# NEW RESULTS ON THE BINARY COMPANION OF T TAURI

Theodore Simon[1], P. R. Schwartz[2], H. M. Dyck[1], B. Zuckerman[3]
[1]Institute for Astronomy, University of Hawaii
[2]Naval Research Laboratory, Washington, D. C.
[3]Astronomy Program, University of Maryland

We have recently reported the discovery of a cool (650-800 K) low-luminosity companion to the pre-main-sequence star, T Tauri (Dyck et al. 1982). We proposed that the optical star and its infrared companion form a physical pair with a N-S separation of 100 a.u. However, there remained in our 2-5 μm speckle interferometry an ambiguity of 180° in the position angle of the secondary. In addition, Cohen et al. (1982) noted an 800 milliarcsec (mas) offset between the visual and 6 cm radio positions at T Tau. Both of these positional discrepancies have now been clarified by accurate visual and radio astrometry of T Tau, and by further near-IR speckle interferometry.

We have made additional speckle observations of T Tau at the IRTF on Mauna Kea, Hawaii, which confirm our earlier measurements at the University of Hawaii's 2.2 m telescope. The co-added speckle data at 3.8 μm are presented in Figure 1 in the form of a visibility curve, along with a computation for a binary model with separation 590 mas and magnitude difference $\Delta m = 1.47$. The slight high frequency roll-off of the visibility curve at its second maximum indicates that one (or both) of the binary components may be spatially resolved at ~100 mas (16 a.u.), but more observations are needed to confirm this. The binarity of T Tau is also quite clearly evident in N-S scans formed by shifting and stacking individual rapid scans obtained on IRTF.

New radio observations have been made with the VLA at 1.3, 2, 6 and 20 cm. Our preliminary reductions of the 6 cm observations (representing almost 180 min. of observing) are shown in Figure 2. The coordinates are for equinox 1950.0 (epoch 1982.482) and have been determined with reference to nearby quasars, for which K. J. Johnston has established accurate offsets from the nearby FK4 star, Algol. At 6 cm T Tau is clearly seen to be a radio binary, with a N-S separation of 540 mas. The two radio sources are unresolved at our resolution of 250 mas (FWHP). The 6 cm flux of the strong southern peak is 4.3 ± 0.5 mJy; the northern source is nearly 10 times fainter, although significantly above our peak-to-peak noise level of 0.1 mJy per beam.



*P. B. Byrne and M. Rodonò (eds.), Activity in Red-Dwarf Stars, 505–507.*

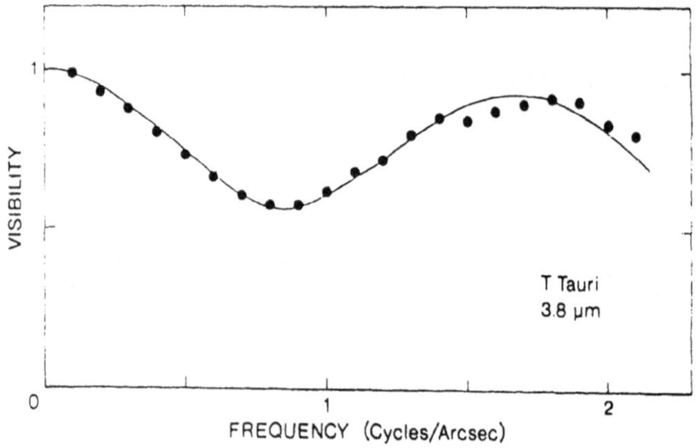


Figure 1. North-South Visibility of T Tau at 3.8 μm. The solid curve is a model with a binary separation of 590 mas and $\Delta m = 1.47$.

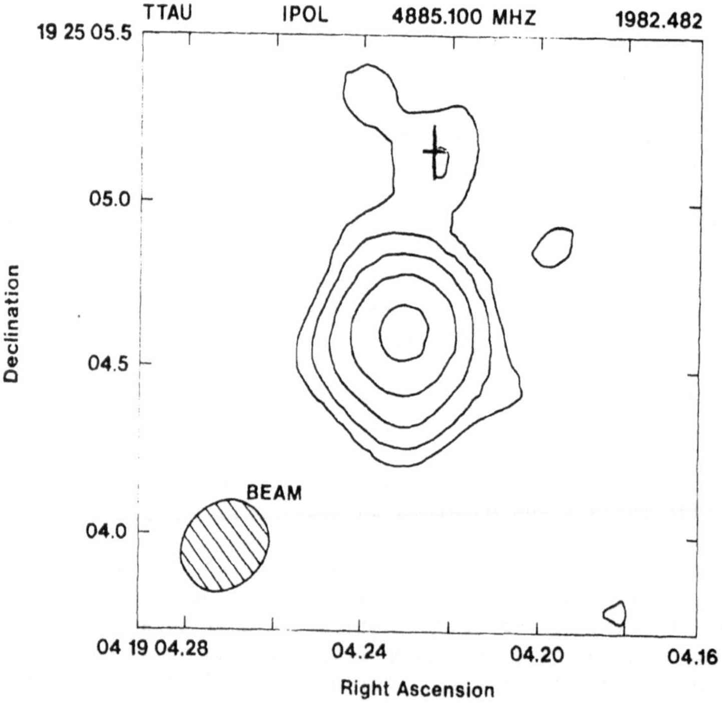


Figure 2. VLA 6-cm Map of T Tau. Coordinates are equinox 1950.0. The cross denotes the optical position and its uncertainty. The absolute radio positions have an uncertainty of 100 mas, or less, in each coordinate. Contour levels are given at 5, 10, 20, 40 and 80% of the peak flux, 3 mJy per beam.

Also shown for comparison is a new visual astrometric position for T Tau, corrected for proper motion to epoch 1982.482, which was kindly communicated by B. Jones and R. Hanson (1982). The exact coincidence of the optical star and northern radio peak confirms the positional discrepancy noted by Cohen et al. (1982) and appears to settle the question of the orientation of the binary. We infer that the IR companion lies to the S at the strong radio peak, is fainter at all wavelengths shortward of 5 μm than the primary, and has a near-infrared color temperature of 800 K. If the detected radio emission arises in a stellar wind, then the high mass loss rate previously derived by Cohen et al., $\dot{M} = 4 \times 10^{-7}\ M_\odot\ yr^{-1}$, must be associated not with the optical star, but the IR companion. A lower $\dot{M}$ for T Tau(opt.) is consistent with the Alfven wave driven wind model of Hartmann et al. (1982), but this theory leaves unexplained the potency of the wind from the ~1.5 $L_\odot$ IR companion of T Tau.

A complete version of this paper is to be submitted to the *Astrophysical Journal.*

REFERENCES

Cohen, M., Bieging, J. H., and Schwartz, P. R.: 1982, Astrophys. J. 253, p. 707.
Dyck, H. M., Simon, T., and Zuckerman, B.: 1982, Astrophys. J. (Letters) 255, p. L103.
Hartmann, L., Edwards, S., and Avrett, E.: 1982, Astrophys. J. in press.
Jones, B. F., and Hanson, R. B.: 1982, private communication.

# CHROMOSPHERIC PROPERTIES OF T TAURI STARS DETERMINED FROM EUV SPECTRA

A. Brown
Department of Physics,
Queen Mary College,
London, England

C. Jordan
Department of Theoretical Physics,
University of Oxford,
Oxford, England

IUE spectra of six pre-main-sequence (PMS) stars are analysed and the resultant emission measure distributions compared with that of T Tau for which a chromospheric model has been calculated. The general shape and absolute level of the mean emission measure distributions are remarkably similar, indicating the relevance of the T Tau chromospheric model to other PMS stars. Evidence for the influence of large scale motions and/or stellar winds on the transition region and coronal emission measures is found. The relative importance of different energy balance terms is discussed.

## 1. INTRODUCTION

T Tauri stars are among the most chromospherically active stars known, showing EUV emission lines with surface fluxes $\sim 10^4 - 10^5$ times those of the quiet sun. However they are sufficiently distant that for most stars only low resolution spectroscopy can be attempted with the IUE satellite in the important 1200 - 2000 A wavelength range (see Gahm (1980) and Imhoff and Giampapa (1980) for general discussion of early IUE results). One of the few stars for which detailed investigation is possible is T Tau and Brown, Ferraz and Jordan (1982) have constructed chromospheric models for this star using high and low dispersion IUE spectra. In this paper we analyse low resolution spectra from the IUE data archive and compare the results with those for T Tau.

## 2. ANALYSIS

The stars studied are RW Aur (SWP 4838, 420 min), RU Lup (SWP 5548, 240 min; SWP 5569, 180 min), RY Tau (SWP 7034, 180 min), SCrA (SWP 1755, 200 min), DR Tau (SWP 7189, 292 min) and CoD-35$^{\circ}$ 10525 (SWP 5436, 310 min). Emission measure ($E_m$) distributions have been calculated, including corrections for interstellar and circumstellar absorption, and compared with that of T Tau. The methods used to calculate these distributions are as described by Brown, Ferraz and Jordan (1982). Stellar parameters were taken from Giampapa et al. (1981) and Cohen and Kuhi (1979).



*P. B. Byrne and M. Rodonò (eds.), Activity in Red-Dwarf Stars, 509–511.*

The absolute level of the $E_m$ distributions up to log $T_e \sim 4.8$ vary remarkably little from star to star, with values between one and ten times those of T Tau, even though the sample contains representatives of the main PMS subgroups. SCrA, DR Tau and CoD-35$^o$ 10525, which all show the YY Orionis infall phenomenon to some extent, have enhanced MgII emission but roughly similar CIV emission to the other stars. The CIII 1908 intersystem line is probably near the low density limit as found in T Tau, but the SiIII 1892/CIII 1908 ratio suggests higher densities. These similarities suggest that the detailed modelling of T Tau has relevance to PMS stars in general and many of the results, particularly the use of two-component models, should hold for other stars.

The available x-ray data indicates that the $E_m$ distribution above $10^6$ K is much lower than expected from observations of the sun and other late-type dwarfs (Walter and Kuhi (1981)). Certainly a $^3/_2$ power law does not seem applicable between the transition region and corona. In our sample of stars a related effect is that the CIV and NV emission measures show a downturn in the mean $E_m$ distribution, which becomes more pronounced with increasing SiIII 1892/CIII 1908 ratio. Our interpretation is that the declining $E_m$ distribution is associated with an increasing importance of large scale motions and/or outflow in the chromospheric energy balance. Since the energy input for these stars seems roughly the same and conduction is relatively unimportant, an increase in turbulent and flow energy leads to smaller radiative losses. Finally it should be noted that the fluorescent $H_2$ emission lines seen in the spectrum of T Tau (Brown et al., 1981) are also present in RU Lup and probably in RW Aur, indicating the likely presence of shocked $H_2$ molecules around these stars.

A full version of this work, hopefully including more stars, will be submitted to Mon.Not.R.astr.Soc..

REFERENCES

Brown, A., Ferraz, M. de M. and Jordan, C.: 1982, in preparation.

Brown, A., Jordan, C., Millar, T.J., Gondhalekar, P. and Wilson, R.: 1981, Nature, 290, pp.34-36.

Cohen, M. and Kuhi, L.V.: 1979, Astrophys.J.Suppl., 41, pp.743-843.

Gahm, G.F.: 1980, Proc.Symp. "The Universe at Ultraviolet Wavelengths: The First Two Years of IUE", Goddard Space Flight Centre, NASA-CP2171, pp.105-114.

Giampapa, M.S., Calvet, N., Imhoff, C.L. and Kuhi, L.V.: 1981, Astrophys.J., 251, pp.113-125.

Imhoff, C.L. and Giampapa, M.S.: 1980, Proc.Symp. "The Universe at Ultraviolet Wavelengths: The First Two Years of IUE", Goddard Space Flight Center, NASA-CP2127, pp.185-191.

Walter, F.M. and Kuhi, L.V.: 1981, Astrophys.J., 250, pp.254-261.

DISCUSSION

Walter: If one observes in X-rays most T-Tauri stars are emitters. It is only the most extreme cases like RU Lup and RW Aur that are not. There appears to be an anti-correlation with optical emission. This is something that most people do not appear to realize i.e. most T-Tauri stars do not have strong emission lines. If one looks at those in Taurus, for instance, about 80% are X-ray sources. They look just like RSCVn stars or any other kind of late-type rapidly rotating stars.

Brown: Are they at a level of $10^4$-$10^5$ times the solar output?

Walter: They are at a few times $10^{30}$ ergs $s^{-1}$ i.e. $10^3$-$10^4$ times solar.

Brown: This is certainly not like the chromospheric-transition region-coronal emission measure distribution that one gets in stars like the Sun. Would you agree?

Walter: I certainly don't think they are distinguishable from RSCVn stars. The luminosities are similar and the X-ray emission measures are not an order of magnitude lower. They may be lower by a factor of two but they have a scatter of a factor of three to five. So they are really comparable.

Dupree: Does RW Aur have a He II λ1640 line? If so that could be used to indicate the presence of X-rays if, for instance, the X-rays were being absorbed in a massive shell.

Brown: I think it does but it is relatively weak.

Dupree: That might be worth looking into.

Brown: There might be other things there too since our data is low resolution.

Simon: What correction did you make for the UV lines before you did your emission measure analysis?

Brown: I took data from Giampapa et al and from Cohen and Kuhi and did a simple correction using a normal extinction law.

Gahm: I recall having seen N V in emission in RU Lupi. Would you like to comment on that?

Brown: Yes, it is there, but it is very weak.

# QUASI-PERIODIC VARIATIONS OF BALMER LINE PROFILES IN SPECTRA OF THE T TAU-TYPE STAR RW AURIGAE

V.P. Grinin, P.P. Petrov, and N.I. Shakhovskaya
Crimean Astrophysical Observatory, USSR

RW Aur is one of the most active T Tau-type stars with essentially variable emission line spectrum. In order to study the nature of this variability,patrol spectroscopic observations of the star were carried outfrom November 1979 to March 1982 at the 2.6-meter telescope of the Crimean Astrophysical Observatory. The spectral region λλ3800-6700 A has been observed with a dispersion of about 40 A/mm. All nights were supplied with simultaneous spectrophotometry of the star.

The Hβ and Hα lines in the spectra of RW Aur have a constant double peak structure with variable ratio of the blue to red component intensities. To define the shape of the lines we take the index $j=(V-R)/(V+R)$ where V and R are, respectively, the blue and the red component intensities, expressed in units of the adjacent continuum.

We have searched for periodicity in the variations of the j index of the Hβ line (jHβ) applying the Jurchevich method, in the period interval 4 to 15 days. The dispersogram computed on the base of all 54 nights data is in Fig. 1a, showing the most probable period P=5.39 days with the confidence level ≈0.9 according to the Jurkevich criterium. Using a somewhat shorter run of the data (37 nights) we have localized the time interval when the periodic variations of the line profiles were more pronounced. The dispersogram on Fig.1b corresponds to this interval of time (October 1980 - January 1982) and shows the prominent peak of the period P=5.39 days with the confidence level of 0.96. The variations of the jHβ index with the period P are shown on Fig.2. Apparently, the relatively stable periodic variations of the Hβ line profile last for about a year.

The jHβ index varies because of the shift of the central absorption by Hβ over the radial velocity interval -70 to +45 km/s (heliocentric). The Hα line profile reveals similar changes with the same period. We have not found any relation between the jHβ index and the brightness of the star.



*P. B. Byrne and M. Rodonò (eds.), Activity in Red-Dwarf Stars, 513–514.*

The analysis of the data leads us to the assumption that local magnetic fields (about 1000 gs) exist on the surface of RW Aur and affect the gas envelope structure around the star. In this case the rotation of RW Aur would cause the effect of modulation of the Balmer lines profiles.

The complete paper will be published in the Soviet Astronomical Letters.

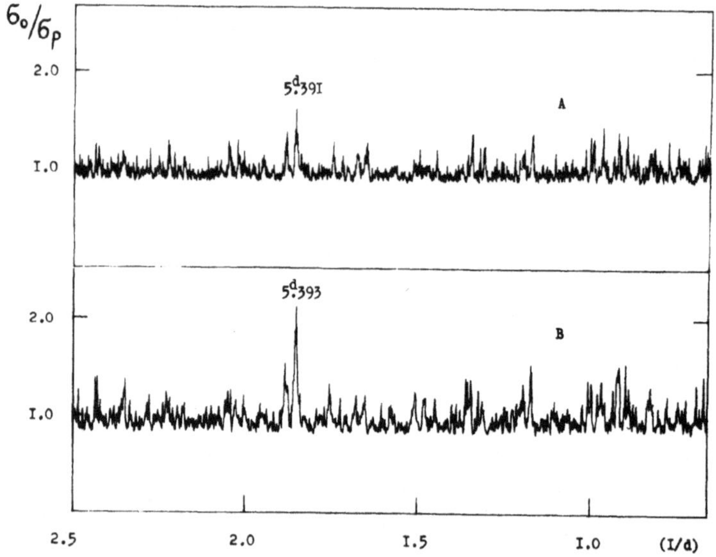


Figure 1. The dispersograms according to the Jurkevich method:
a) J.D. 2444180.65 - 2445045.28,
b) J.D. 2444526.49 - 2444994.37.

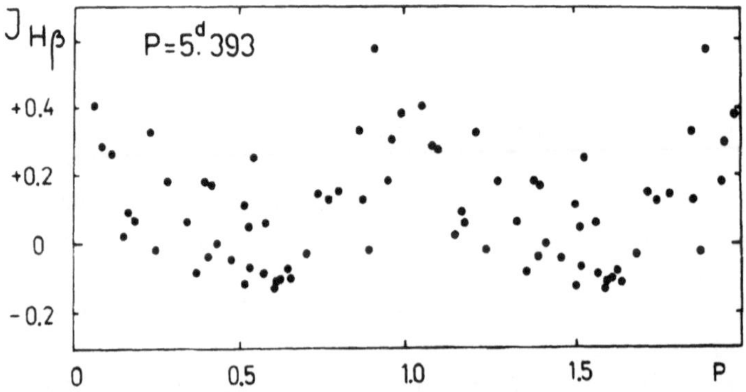


Figure 2. The periodic variations of the jHβ index,
J.D. 2444526.49 - 2444994.37

# ERUPTIVE PHENOMENA AS A SOURCE OF THE OBSERVED VARIATIONS OF INTRINSIC POLARIZATION IN LONG-PERIOD AND T TAURI VARIABLES

M. Solc, J. Svatos
Department of Astronomy and Astrophysics, Charles University,
Prague, Czechoslovakia

## ABSTRACT

The correlation between the rapid rise of intrinsic polarization, increasing U-B and enhanced equivalent width of Balmer, H and K emission lines is interpreted as a consequence of the activation of colour centres in circumstellar silicate dust by X/UV photons from flares. The temperature dependence of bleaching is discussed and the number of centres producing polarization changes is estimated.

Flare activity of some T Tauri stars with individual events being $10^3$ times more energetic than large solar flares has been recently reported by Worden et al. (1981). The energies released at X/UV and optical wavelengths by a flare have characteristic values L(X/UV, flare) ≈0.05 L(X/UV, star) and L(opt, flare)≈2.4x 0.05 L(X/UV, star). Infrared studies by Cohen 41981), Rydgren et al.(1982) and others have confirmed the presence of circumstellar silicate dust around many T Tauri stars. Dust is considered to cause the infrared excess observed in these stars and perhaps the intrinsic polarization as well. A strong interaction of flare X/UV photons and stellar wind particles with dust is thus expected. We suggest that this interaction can relate to the variations of polarization degree.

Simultaneous observations referring to brightness (UBVR, $H_\alpha$) and polarization ($P_V$) of T Tauri itself, carried out by Redkina et al. (1980) with a 70-cm telescope at Gissara Observatory show clearly the correlation mentioned above (Fig. 1). The mean error in the U-band is 0.02-0.05 mag, in B, V is 0.01-0.03 mag and in $H_\alpha$ is 0.02-0.04 mag. The polarization degree is accurate to 0.1-0.14%. The joint observational program of T Tauri star DI Cep has been reported by Grinin et al. (1980). In Fig. 2 can be seen some strong UV bursts of long duration, such as at JD 2442640. The simultaneous rise of polarization degree and U-B was first mentioned by Shawl (1975) for Mira at phase 0.8.



*P. B. Byrne and M. Rodonò (eds.), Activity in Red-Dwarf Stars, 515–518.*

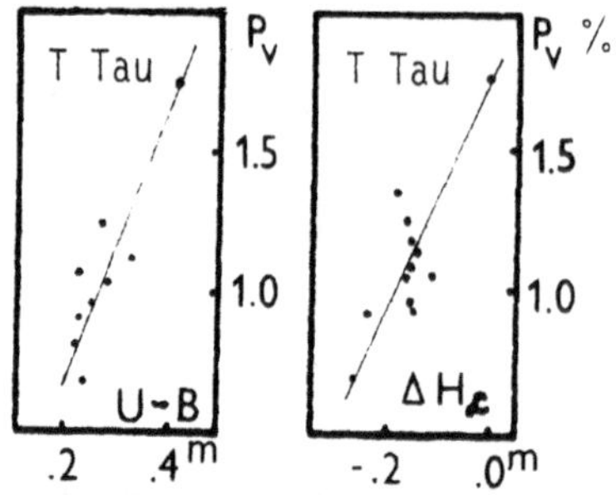


Fig. 1 (by Redkina et al.,1981)

Flaring processes in Mira variables can be created by a shock wave propagating through the atmosphere just at this phase (Wing, 1979). If the following interpretation of this phenomena proves to be valid, the underlying process causing polarization variability should be quite common among the stars with circumstellar dust.

Let us consider circumstellar dust as the only agent responsible for the varying polarization. Several theoretical models of dust envelopes have been suggested, such as: polarization due to scattered starlight on moving circumstellar dust clouds; sudden condensation/evaporation of grains by decreased/increased temperature; or enhanced alignment of nonspherical dust grains by increasing magnetic field due to the Davis-Greenstein mechanism (for review see Greenberg, 1976). However, a detailed discussion of these models shows that none of them proceed so quickly as to explain the simultaneous rise of P and other observed characteristic unless unlikely physical conditions in the envelope are assumed. This was the very reason for seeking models with a more flexible response to flare events.

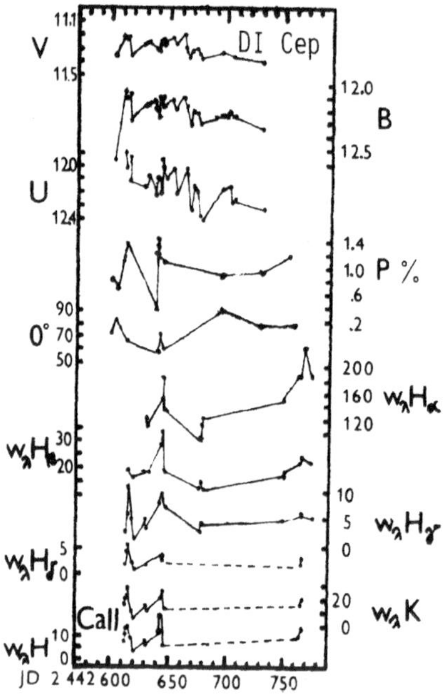


Fig. 2 (by Grinin et al.,1980)

As suggested by Svatos (1980) the X/UV activation of color centers (CC) in silicates could explain the polarization rise. Short wavelength irradiation leads to an increase of the imaginary part (k) of the refractive index for a wide variety of silicates and glasses, as has been shown in the comprehensive laboratory study by Kats and Stevels (1956). Let us now consider a model of a circumstellar shell with as few free parameters (and assumptions) as possible, which can account both for T Tauri stars and Miras:

- the dust grains are needle-shaped of radii 0.1-1.0μm (with no distribution function), being perfectly aligned, consisting mainly of silicates with complex refractive index m=1.7-ik;
- the only particles projected on the stellar disk are assumed to

polarize starlight (polarization is seen in the transmitted light); $\tau \simeq 0.1$.

Simple formulae by Hulst (1957) giving efficiency factors of cross-section for very long thin cylinders can be applied:

$$Q_{||} = -\pi^2 a \lambda^{-1}\, I_m(m^2-1),\quad Q_{\perp} = -2\pi^2 a\lambda^{-1}\, I_m(m^2-1)/(m^2+1)\ ;$$

where subscript $||$, $\perp$ refer to the light wave oscillating in parallel and perpendicular direction with respect to the axis of the cylinder.

By definition, the polarization degree is

$$P = |I\ \exp(-\tau_{\perp}) - I\ \exp(-\tau_{||})| / |I\ \exp(-\tau_{\perp}) + I\ \exp(-\tau_{||})|$$

where I is the luminosity of the star. Substituing $\tau = N\, \ell\, Q\, C$, where $N\, \ell$ is the column density and C the geometrical cross-section of one grain, a linear relation between P and k is found. As a consequence of this we obtain (replacing P by $\Delta P$ and k by $\Delta k$)

$$\Delta P = 42.7\ N\, \ell\, a^3\, \lambda^{-1}\, \Delta k.$$

If the concentration ($\underline{n}$) of CC in circumstellar dust is comparable with that in common glasses, i.e. at least $10^{18}$ per gram, the column density of CC per $cm^2$ required for an increase $\Delta P$ is $\simeq 10^{13}\ \Delta P\ \lambda(\mu m) \Delta k^{-1}$. If $\Delta P$=1%, number of X/UV photons per 1 CC at a distance $\simeq$ 50 R ($\simeq 150 R_{\odot}$) from a T Tauri star or at a distance 4-5 R ($\simeq$700-900 $R_{\odot}$) from a Mira star is of order 10-100 per sec. Supposing a 10% efficiency of CC's production, this rate is sufficient to maintan CCs completely activated in the case of absent thermal bleaching.

In the most simplified model the thermal bleaching is a two-body reaction inverse to the activation of CC. The temperature dependence of CC's bulk concentration (n) is given by the equation

$$-\frac{dn}{dt} = K\ \exp\ (E/kT) n^2\ ,$$

where E is the activation energy and K is the constant characteristic for the given kind of CC. During the flare, the dust particles are being heated and hence $T \sim t$ can be substituted in the equation above. The solution n(T) rapidly fades at temperature $T_a \simeq 700$ K depending on the kind of CC due to progressive annealing of CC.

In conclusion, the CC's production in dust during steady continuous flaring or during quiet phases is balanced by thermal bleaching. Following immediately the X/UV burst, the production of CC exceeds their decay, and enhanced polarization should be observable until the dust temperature reaches $T_a$, which can still occur during the flare (Fig.2). The stream of particles from flare reaches the dust region with a delay, so that we can neglect it in this simple model.

Thus, simultaneous polarimetry and spectroscopy with convenient time resolution, specifically for T Tauri stars rather than for long-period variables, seems to be fundamental for understanding the intrinsic polarization. The crucial point for theoretical models is a good timing of the polarization rise and enhanced line emission and the subsequent polarization fall and the rise of infrared thermal flux from circumstellar dust.

REFERENCES

Greenberg J.M. 1976, "Interstellar Dust", in Cosmic Dust, ed. McDonnell.

Grinin V.P., Efimov J.S., Krasnobabtsev V.I., Shachovskaya N.I., Shachovskoy N.M., Shcherbakov A.G., Zaitseva G.V., Kolotolov E.A., Shanin G.I., Kiselev N.N., Gyulaliev C.G. and Salmanov I.R. 1980, Peremennye Zvezdy 21, 247 (in russian).

Hulst H.C. 1957, "Light Scattering by Small Particles", Chapman and Hall.

Kats A. and Stevels J. 1956, Philips Res.Rep. 11, 115.

Redkina N.P., Zubarev A.V. and Chernova G.P. 1981, Astron.Circ. USSR No. 1149 (in russian).

Rydgren A.E., Schmelz J.T. and Vrba F.J. 1982, Astrophys.J. 256, 168.

Shawl S.J. 1975, Astron.J. 80, 602.

Svatos J. 1980, Bull.Astron.Inst.Czech. 31, 302.

Wing R.F. 1979, "Mira Variables, an Informal Review", Contr.Perkins Obs. Ser.II No. 80.

Worden S.P., Schneeberger T.J., Kuhn J.R. and Africano J.L. 1981, Astrophys.J. 244, 520.

# THE INTERACTION OF FAST PARTICLES WITH FROZEN GASES IN T TAU NEBULAE: THE PHYSICAL BACKGROUND

V. Pirronello[1,2], G. Strazzulla[2] and G. Foti[3]
[1]Istituto di Fisica, Facoltà di Ingegneria, Università di Catania, Catania, Italy
[2]Osservatorio Astrofisico di Catania, Italy
[3]Istituto di Struttura della Materia, Università di Catania

The interaction of energetic particles with frozen gas layers plays a relevant role in a variety of astrophysical scenarii and also in T Tau nebulae because of the large fluxes of particles ejected by these stars during their continuous flaring activity and the observed presence of ice grains at least around the very young star KL Tau (Cohen, 1975).

In order to better understand the problem we will now describe briefly the most important physical processes occurring during ion bombardment of frozen gas layers (mantles).

When an ion ($H^+$, $He^+$,....) impinges on a mantle it loses energy through subsequent collisions with atoms and molecules mainly ionizing and exciting them.

The collision time "$\tau_c$" of a proton in the range of energy between 100 keV and 1 MeV is $10^{-17} \div 10^{-18}$ sec. and the mean time between collisions "$\tau$" that excite or ionize is about $10^{-16}$ sec. Secondary electrons produced by primary interactions can lose their energy on a temporal scale about ten times longer than "$\tau$" through a cascade of collisions till when they remain with an amount of energy below the threshold for further ionizations or excitations.

The transfer of this energy which is still in the electronic system, to atoms and molecules may happen by means of the generation of phonons that may establish transient local quasi-thermodynamic equilibrium in a cylindrical region 50 Å wide around the track of the impinging ion if energy remains confined for at least $10^{-12}$ sec. The occurrence of such an equilibrium can give rise to interesting consequences; two main effects of the energy deposition of fast ions in a frozen gas layer are:

1) ejection of atoms or molecules, and
2) chemical modification of the layer itself.



*P. B. Byrne and M. Rodonò (eds.), Activity in Red-Dwarf Stars, 519–521.*

The first one takes place because when the transient local quasi-thermodynamic equilibrium is reached the most energetic particles, which are near the surface and belong to the quasi-Maxwellian distribution, may overcome the surface barrier and escape from the solid evaporating (Fig. 1). This effect plays a relevant role in the stability problem of ice grains in T Tau nebulae.

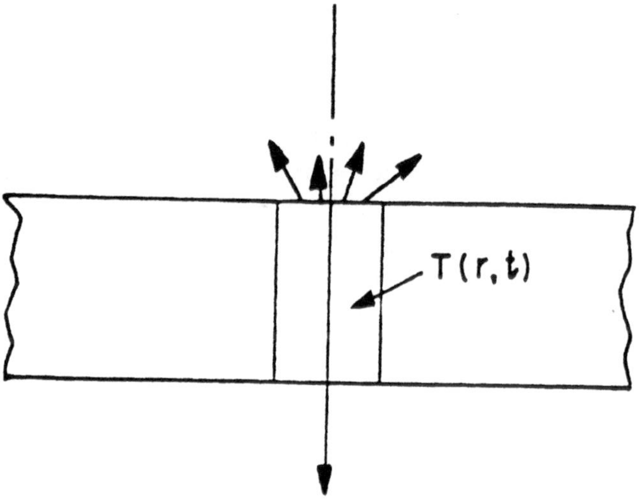


Figure 1. Schematic view of the superficial evaporation (erosion) of the frozen gas layer induced by an impinging ion.

The second effect is due to the breaking of molecular bonds in the target by the fast impinging ion. The cracking fragments, being mobile because in a high temperature region, have a finite chance to recombine forming "new" molecules.

A consequence of such a formation mechanism is that when these molecules are released in gas phase enrich the surrounding medium. This could be an interesting channel to form molecules, observed in space but difficult to obtain by gas phase reactions.

Experimental simulations (Brown et al 1978, Pirronello et al 1981, Brown et al 1982, Ciavola et al 1982) of the interaction between ions and frozen mantles have been performed bombarding thin ice layers (3000-4000Å thick), deposited by vapour on a silicon substrate cooled at 4 K or 77 K, with protons or helium ions accelerated to keV-MeV energies by an ion inplanter or a Van de Graaff.

Erosion yields have been evaluated measuring the thickness of the frozen layer before and after a known dose of projectiles reached the target. The number of removed atoms or molecules versus the projectile dose gives the erosion yield coefficient "Y" (number of atoms or molecules ejected per impinging ion).

The thickness of the layer is obtained measuring the energy lost by 1.5 MeV $He^+$ ions in the layer itself before and after being backscattered by a thin (250Å) gold marker previously deposited on the substrate.

Huge Y's have been obtained for $H_2O$, $N_2$, and other frozen volatiles. Y is regularly temperature independent at low values of the substrate temperature T and strongly dependent from it at higher T, when diffusion processes become relevant.

Chemical modifications of the target due to particles irradiation have been investigated analyzing "in situ" ejected species by a quadrupole mass spectrometer. In particular ice mantles release in a strongly temperature dependent way $H_2$ and $O_2$ together with $H_2O$; furthermore a mixture of equal parts of $H_2O$ and $CO_2$ prepared in gas phase and deposited on the substrate at 9K has shown, when bombarded by $He^+$ ions, an high rate of formation of formaldehyde molecules.

REFERENCES

Brown, W.L., Lanzerotti, L.J., Poate, J.M., Augustyniak, W.M.: 1978, Phys. Rev.Lett. 40, 1027.

Brown, W.L., Augustyniak, W.M., Simmons, E., Marcantonio, K.J., Lanzerotti, L.J., Johnson, R.E., Boring, J.W., Reimann, C.T., Foti, G., Pirronello, V.: 1982, Nucl. Instr. Methods. 198, 1.

Ciavola, G., Foti, G., Torrisi, L., Pirronello, V., Strazzulla, G.: 1982, Radiation Effects (in press).

Cohcn, M.: 1975, M.N.R.A.S. 173, 279.

Pirronello, V., Strazzulla, G., Foti G., Rimini, E.: 1981, Astron. Astrophys. 96, 267.

# ON THE ORIGIN OF $H_2$ IN T TAU STARS

G. Strazzulla[1], V. Pirronello[1,2], G. Foti[3]
[1]Osservatorio Astrofisico,V.le A. Doria,6,I-95125 Catania, Italy
[2]Istituto di Fisica, Facoltà di Ingegneria, Catania
[3]Istituto di Struttura della Materia, Catania

Narrow-band spectrophotometric observations (Cohen, 1975) showed that the extremely young star HL Tau has an absorption feature at 3.1 μm well matched by extinction from pure ice solid grains (radius $a \approx 0.3 \mu m$, $m_{ice} \approx 10^{-4}$ gr $cm^{-2}$). This feature, not present in others (more evolved) T Tau stars, suggests that ice may be only found around the very youngest T Tau stars and progressively disappears as stars become increasingly visible through their circumstellar shells (Cohen, 1975).

Here we suggest that the mechanism responsable for ice grains destruction in these stars is erosion by energetic protons (≈1 MeV) copiously produced by stellar flare activity. During erosion the production of a large quantity of $H_2$ molecules occurs. The proposed mechanism is based on the huge experimentally measured erosion yields of frozen gas bombarded by energetic particles (Brown et al. 1978, Pirronello et al. 1981 and references therein) and on the molecular character (mainly $H_2$, $O_2$ and $H_2O$) of the products eroded from frozen $H_2O$ and observed by mass spectrometry (Ciavola et al. 1982, Brown et al. 1982, Pirronello et al. 1983).

Worden et al. (1981) on the basis of Johnson U band observations derived $L_{Tau\ Flares}/L_{T\ Tau}$ and assumed a similar flare spectrum to UV Ceti flares. They also assumed that the $L_{proton}/L_{Flare}$ ratios for the Sun and T Tau star were the same and found a flux of protons $J$ ($\gtrsim$ 10 MeV)$\approx 10^8$ $cm^{-2}$ $sec^{-1}$ at 1 AU. Assuming a law of type $dJ/dE \sim E^{-\gamma}$ with mean $\gamma = 3$ we find $J$ ($\gtrsim$ 1 MeV)$\approx 10^{10}$ $cm^{-2}$ $sec^{-1}$ at 1 AU. The lifetime against erosion by protons of a typical grain ranges then from $t_g \approx 2 \times 10^4$ yr at a distance from the star $R \approx 2 \times 10^{15}$ cm, to $t_g \approx 3 \times 10^6$ yr, at $R \approx 2 \times 10^{16}$ cm that according to Scwartz (1974) is the radius of the T Tau nebula. Thus lifetime always result shorter than $3 \times 10^6$ yr, the typical age of a T Tau star. For $T_g \lesssim 100$ K ($R \gtrsim 4 \times 10^{15}$ cm) these lifetimes are orders of magnitude lower than those expected against sublimation that



*P. B. Byrne and M. Rodonò (eds.), Activity in Red-Dwarf Stars, 523–524.*

can not explain the observed lack of ice grains in evolved T Tauri stars.

A consequence of ice erosion is the restore of molecules in the gas phase. In particular as experimentally shown (Ciavola et al. 1982, Brown et al.1982), the $H_2$ production increases with the temperature (i.e.decreases with the distance from the star) of the ice. Thus we find an integral $H_2$ production ranging between $n(H_2)\approx 10$ ($T_g \lesssim 100$ K) and $n(H_2)\approx 100$ ($T_g \gtrsim 100$ K). This should be only a marginal contribution to the total in the region where $T_g \lesssim T_{cr}$ (100 K or 40 or 30 or?) is the maximum temperature for which H-H recombination on grains effectively works to produce $H_2$ molecules (Hollenbach and Mc Kee 1979, Lequeux 1981). In the regions with $T_g > T_{cr}$ (if grains do not sublime too quickly) the present mechanism should be the only one that produces a noticeable amount of $H_2$.

## REFERENCES

Brown, W.L., Lanzerotti, L.J., Poate, J.M. and Augustyniak, W.M.: 1978, Phys. Rev. Lett., 40, 1027.

Brown, W.L., Augustyniak, W.M., Simmons, E., Marcantonio, K.J., Lanzerotti, L.J., Johnson, R.E., Boring, J.W., Reimann, C.T., Foti, G. and Pirronello, V.: 1982, Nucl. Instr. Meth. (in press).

Ciavola, G., Foti, G., Torrisi, L., Pirronello, V. and Strazzulla, G.: 1982, Radiation Effects (in press).

Cohen, M.: 1975, MNRAS, 173, 279 .

Hollenbach, D. and Mc Kee, C.F.: 1979, Astrophys. J., 41, 555.

Lequeux, J.: 1981, Comments on Astrophys., 9, 117.

Pirronello, V., Strazzulla, G., Foti, G. and Rimini, E.: 1981, Astr. Astrophys., 96, 267.

Pirronello, V., Strazzulla, G. and Foti, G.: 1983, (This volume)

Scwartz, R.D.: 1974, Astrophys.J., 191, 419.

Worden, S.P., Schneeberger, T.J., Kuhn, J.R. and Africano, J.L.: 1981, Astrophys. J., 244, 520.

# SESSION IV

## THEORETICAL ASPECTS

MODELS OF SPOTS AND FLARES

D.J. Mullan
Bartol Research Foundation, University of Delaware
Newark, Delaware 19711 USA

## 1. INTRODUCTION

MHD effects in stars are seen in their most spectacular form in the processes which are typical of flares. At first sight, it appears that the phenomena of dark spots (whose long lifetimes give an impression of quasi-equilibrium) are inevitably less interesting. However, this is not necessarily true. Laboratory experiments in recent years have shown that there are many more ways to drive a plasma out of equilibrium than to preserve equilibrium. In that sense, then, it is perhaps "easier to understand" why flares should occur in a stellar atmosphere (where convective jostling of field lines creates potential for driving a large number of instabilities) than why a long-lived feature such as a dark spot should persist. Various instabilities which may contribute to flares are discussed by Priest and Spicer (this volume). Here, we summarize work on the equilibrium structure of cool spots in the sun and stars. Since spots involve complex interactions between convective flows and magnetic fields, we need to refer to observations for help in identifying the dominant processes which should enter into the modelling. This summary therefore begins by discussing certain relevant properties of spots in the solar atmosphere.

## 2. OBSERVED CHARACTERISTICS OF SUNSPOTS

A sunspot is a dark magnetized area of photosphere, with dimensions ranging from $\sim 10^3$ km up to several times $10^4$ km. Small spots (pores) are uniformly dark, but larger spots are composed of umbra and penumbra. The former is uniformly dark in many cases, but the penumbra is in all cases highly structured, typically filamentary. Penumbral filaments are bright and dark, and they radiate from the umbra towards the undisturbed photosphere. Umbral fields are mainly vertical, but fields are almost horizontal at the outer edge of the penumbra, with a smooth transition in between. Umbral fields are typically 2-4 kG, essentially independent of area.

The visible radiation from an umbra has effective temperature $\sim$4000 K,



*P. B. Byrne and M. Rodonò (eds.), Activity in Red-Dwarf Stars, 527–543.*

with surprisingly small deviations from this mean value for spots of widely different areas. Only a small fraction (2%) of the umbral thermal flux is carried by convection at $\tau_i$=1. (Subscript i refers to quantities inside the spot.) This represents a striking reduction in the efficiency of convective transport relative to the photospheric convection, where at $\tau_e$=1, convection carries $\sim$15% of the total flux (Beckers, 1977; Stellmacher and Wiehr, 1976). (Subscript e denotes quantities external to the flux tube.) The missing flux, in the case of a spot occupying an area of 500 millionths of the hemisphere ($A=10^{19}$ $cm^2$) is about $5\times10^{29}$ ergs/s, i.e. $\sim$0.01% L(sun). The sun apparently stores this missing flux, or converts it into non-thermal form, in the case of spots which live for more than about 7 days (Willson et al., 1981). Although this missing flux is small compared to the solar luminosity, it is very large compared to the non-thermal energy flux which is responsible for heating the chromosphere/corona over the spot. Hence, if even a small fraction of the missing flux could be tapped, large flares could be energized above a spot.

A spot appears initially as an enlarged darkening of intergranular material. When the scale size becomes comparable to one granule diameter ($\sim10^3$) km, the darkening becomes definitely pronounced: this is a "pore". Most pores survive for only a short time ($\sim$1 hr), and then fade away. Those which survive somewhat longer than this appear to grow by coalescing with other pores. However, even larger pores do not survive long unless they develop a penumbra. Spots which survive for periods of days or weeks in all cases possess penumbrae.

Groups of spots are formed when a large magnetic flux rope erupts through the surface from below. The overall structure is bipolar, but there is usually a pronounced asymmetry between leading and following parts of the bipolar pair: leader spots are usually large and single, whereas the following flux is distributed over an irregular group of smaller spots and pores. The longest-lived member of a group is usually the leader. Spots which are especially long-lived appear to be surrounded by an organized flow pattern which emerges from the edge of the penumbra and radiates to distances of $\gtrsim10^4$ km. This flow, which sweeps the surrounding area clean of magnetic flux for a time, is called a "moat" (Pardon et al., 1979). Spots decay when separate strands of flux break away (or are torn away) from the main flux rope, and migrate elsewhere. The lifetime of individual flux strands is difficult to estimate empirically, partly because when the strands become small enough to be filled in by radiation, they are difficult to detect. The lifetimes might be as long as a year or more, without contradicting empirical data (Zwaan, 1978). The question of lifetime is an important one in deciding whether the birth of a new active region necessarily indicates the creation of new flux, or whether it may simply represent a re-concentration of already existing (but dispersed) flux.

Flux tubes exist in a variety of forms ranging from large dark spots down to small (sub-arcsecond) bright points in faculae. These forms appear to belong to a single-parameter family. The parameter, magnetic flux $\Phi$, is smaller than $\sim 10^{19}$ maxwells in small flux tubes which appear bright at all altitudes in the photosphere, whereas for $\Phi \gtrsim 10^{19}$ maxwells, the flux tube appears dark, at least in the photosphere (Frazier, 1977). We note that, since photospheric fields are in almost all cases of order 1-2 kG (Stenflo, 1973), a critical flux of $10^{19}$ maxwells corresponds to a critical length scale of order $10^3$ km, i.e. about one granule diameter. Thus, interference between magnetic fields and convection is central to the phenomenon of dark spots on the sun.

The flux which emerges through an umbra is only part of the total flux in an active region. Surrounding the spots, magnetic fields exist throughout the active region, and these cause not a darkening, but a brightening in the chromosphere. (We refer to this generically as "plage".) In plage, the fields are not organized into clearly identifiable large-scale entities: "cell"-structure is very irregular, apparently indicating that the fields in plage are generally more random in direction than in spots. (Randomness of the fields contributes to chromospheric heating, whether one considers heating due to wave dissipation or reconnection.) This suggests that we might usefully consider the following definition for the distinction between spotted and non-spotted (plage) areas of an active region: in a spot, the field is (somehow) organized into an entity of vertical flux which is coherent over a horizontal scale in excess of one granule diameter. Plage represents "everything else" in the active region. The ratio of spotted area to plage area in an active region is difficult to extract from the literature. It clearly depends on the age of the active region: in an old active region, plage persists even after spots have disappeared, whereas in the young phase, spots are the pre-eminent characteristic of the active region at the photospheric level. In the context of the proposed definition, the penumbra should be classified as "plage". Thus, a pore represents the limit of large spot/plage ratio.

Although the photospheric level of an umbra remains darker than normal for prolonged times, this does not mean that time-dependent phenomena are entirely absent from the umbra. On the contrary, umbrae and penumbrae are observed to be sites of a rich variety of dynamic phenomena, although the energy transport associated with these is only a small fraction of the missing flux. When an umbra is observed in chromospheric radiation, transient brightenings, waves, and various oscillations (with periods ranging from 100-500 seconds and longer) are detected (Moore, 1981). The ultimate manifestation of energetic processes in an umbral flux tube occurs during certain very large flares: in "proton flares", the umbral flux tube emits enhanced chromospheric radiation, suggesting that the umbral chromosphere has been forced down to lower altitudes. De Jager (1968) suggested that when large flares occur close together in time in a particular active region, the most likely source of energy is the missing flux in spots in that active region: no other source of energization seems adequate. However, De Jager did not specify how the missing flux might be tapped in order to power the large flares.

## 3. THEORETICAL STUDIES OF SPOTS: SUMMARY

Of the several topics which arise in connection with theoretical studies of spots on the sun and other stars, we select four for comment in the present paper: field concentration, stability, cooling and missing flux, and relationship with flares. In each case, we will discuss the topic initially with reference to the sun, and then go on to consider implications for spots on cool dwarfs. In the case of starspots, the primary aim of theoretical work at the present time (in view of the available observational data) should be the prediction of the areas of starspots on stars of various spectral types, the effective temperatures of the starspots, and the ratio of spot area to plage area in a stellar active region. Unfortunately, as will become clear in the discussion, none of these quantities can be predicted with great confidence at the current level of our knowledge. Thus, it appears likely that observations will lead theory in the starspot problem for some years to come.

## 4. CONCENTRATION OF MAGNETIC FIELDS

During the growth phase of large sunspots, the magnetic flux in the spot grows so rapidly that it cannot be the result of concentration of weak flux already in the photosphere by supergranule velocities (Zwaan, 1978): in order for this to be the case, some 35 supergranules would have to be "swept clean" and concentrated into one region, and the velocity of the flow would need to be 1 km/s. The velocity is too high to be consistent with observed supergranule velocities, and it seems improbable that a coherent convergent flow would arise embracing 35 supergranules at one time. Hence, the flux rope which emerges in the form of a sunspot must have already been concentrated at depth, prior to eruption (Piddington, 1976). It is not possible at present to estimate how strong the magnetic field might be at great depth in the convection zone. However, Zwaan (1978) suggested that equipartition with turbulent velocities might be appropriate: this leads to about 10 kG at depths of order $10^5$ km. Near the surface, the equipartition field strength is found to be about 600 G. Although there is some evidence that flux does emerge at the solar surface with fields of this order, subsequent strengthening of the field apparently occurs, since the flux tubes which are most common in the photosphere at any one time appear to have fields appreciably stronger than the above value: typical strengths are 1.5-2 kG. The cause of the extra strengthening is probably related to surface cooling of the flux tube: with cool gas inside the flux tube, the pressure scale height becomes less than that outside, and the internal gas slumps to lower levels than the external gas. At any particular geometric depth, then, the external pressure exceeds the internal pressure, and the effect is to compress the flux tube. Thus, field concentration may have two separate components, one operating at depth, the other operating near the surface where cooling is permitted.

According to this view, pores and individual flux tubes in an active region on the sun's surface can be considered as separate strands of flux belonging to the main flux rope which creates the active region as a whole. The strands are thought to be attached to the main rope somewhere beneath the surface. This view explains why pores appear to coalesce into a single spot on the surface: what one is watching on the surface is a two-dimensional "cut" across a three-dimensional "rising tree" of flux (Vrabec, 1974), in which separate strands of flux correspond to the "branches", while the emergence of the main "trunk" of the tree creates a large umbra with the appearance that all of the branches have been "swept" into the large umbra. The sweeping of pores into a single umbra is therefore only a projection effect, rather than being evidence for surface motions acting to push magnetic strands together.

The creation of magnetic flux at great depths in the convection zone of a star (or below the convection zone) is perhaps due to dynamo action (Belvedere, this volume). Hence, if we wish to predict fluxes in active regions in cool dwarfs, it is necessary to know the physical conditions at great depth. In the limiting case of low mass stars, this means, in effect, knowing the physical conditions essentially all the way to the center of the star. Unfortunately, there are at present serious uncertainties in such basic quantities as the depth of the convection zone, the convective efficiency, the turbulence velocities etc., in models of low mass stars (Cox et al., 1981). Thus, by changing the opacities near the surface to include previously neglected contributors, Cox et al. (1981) found that a star which used to be believed to be fully convective may instead possess an extensive radiative core. Hence, application of various scaling laws to the various internal physical variables along the lower main sequence (in order to predict dynamo properties) must be considered as quite uncertain at present.

The area $A_s$ of a flux rope on the surface of a star is determined by the total flux, $\Phi$, and the surface field strength, $B_s$: $A_s=\Phi/B_s$. The value of $B_s$ is observed to be remarkably uniform in the sun: from the largest spot down to the smallest detectable facular bright point, the value of $B_s$ is within a factor of 2 of 2 kG. Within the framework of the ideas mentioned above concerning the slumping of gas within a flux tube, we denote by D the depression of the layer $\tau_i=1$ (i.e. optical depth unity inside the flux tube, where pressure is $p_i$) below the layer $\tau_e=1$ (i.e. optical depth unity in the photosphere). At depth D below $\tau_e=1$, we can write $p_e(D)=p_i+B^2/8\pi$, if the flux tube is assumed to be in magnetohydrostatic equilibrium (MHSE). In the limit of total evacuation of the flux tube, the value of $B_s$ can be approximated by $(8\pi p_e(D))^{1/2}$. For example, to reproduce a surface field of, say, 4kG, as is observed in some large sunspots, this relation requires us to pick $D\sim400$ km (according to the convection zone model of Spruit, 1977). Finite $p_i$

increases D. Weaker fields require smaller values of D, the Wilson depression. The fact that flux tubes at different parts of the sun are observed to have more or less the same values of $B_s$ reflects the condition that on a large scale, the solar photosphere is spherically symmetric.

If this is a correct interpretation of the properties of flux tubes near a stellar surface, then evaluation of $B_s$ in red dwarfs will require a knowledge of the depth-dependent pressure, as well as an estimate of the appropriate Wilson depression. When we recall that in a star such as UV Ceti, where photospheric temperatures are of order 2600 K, and gravity is of order $10^5$ cm$^2$/s, the pressure scale height is only about 10 km, it is clear that a small error in estimating D will result in a large uncertainty in $p_e(D)$, and therefore in estimating $B_s$ in such a star. The serious uncertainties associated with choice of opacity in the outer layers of cool dwarfs, as well as with choice of the adiabatic gradient (due to molecule formation) have been stressed by Bohn (this volume). Even if both of these variables were well known, however, there remains the long standing uncertainty in the treatment of stellar convection, particularly in the near-surface layers where the degree of super-adiabaticity is most pronounced.

Thus, prediction of starspot area $A_s$ from first principles requires knowledge of physical conditions not only near the base of the convection zone but also near the top. In view of these difficulties, it is not surprising that simple prescriptions have occasionally been used to "predict" spot sizes. Supergranules were considered important elements in the prescriptions proposed by Mullan (1973) and Rucinski (1979). However, as the importance ascribed to supergranule flows in concentrating spot fields has diminished over the years (Zwaan, 1978), the arguments based on supergranule properties seem to loose much of their forcefulness. In fact, it now seems more likely that supergranule sizes and flow patterns may be controlled by the magnetic field, rather than the reverse, at least in the regions of primary interest to us here (i.e. regions of strong B)(cf. Zwaan, 1978). For example, the only place where the supergranule flow is "permitted" to organize itself into a large annular cell appears to be in vicinity of a well developed sunspot; and in an active region, the supergranule cellular pattern is not "permitted" to be established, and the network is quite chaotic in character in an active region.

## 5. STABILITY OF SPOTS

The one-parameter family of magnetic flux elements has been interpreted by Spruit (1977) in terms of a model of flux tubes in MHSE. In this model, a vertical flux tube is supposed to be the site of a reduced upward flux of thermal energy ($F_i < F_e$) relative to the external undisturbed photosphere, where the upward flux is $F_e = \sigma T_e^4$ ($T_e$ is the solar effective temperature). The reduction in flux inside the flux tube is presumably related to the reduction of convective efficiency by the vertical magnetic field, although the details of the origin of $F_i<F_e$ are not crucial for Spruit's models. The flux tube is also characterized by lower temperatures, $T_i<T_e$. Again, the specification of the cooling agent is not crucial (cf. also Low, 1980). Influx of radiation from the walls of the tube reduces the difference $T_e-T_i$ in the surface layers, at least in the case of small tubes (with diameters less than about $10^3$ km). Larger tubes, such as spots, do suffer somewhat from lateral influx of radiation, but not enough to compensate for the darkness of the spot due to reduced $F_i$. Spruit showed that as the flux tube diameter increases from values much less than $10^3$ km to larger than that, there is a smooth progression from bright points (such as faculae) to dark spots, in good agreement with the one-parameter family discovered by Frazier (1977) in the sun.

The question which arises in connection with equilibrium models of flux tubes now is the following: are the equilibria stable? In general, the answer is no. A vertical flux tube is a region in which the magnetic field is attempting to exclude the external, higher pressure gas from entering. This situation, which arises in plasma fusion devices such as bottles and various pinches, is known to be unstable to interchange and flute instabilities (Spitzer, 1962; Priest, this volume). Thus, in the case of a bundle of vertical flux strands surrounded by field free gas, if one interchanges one of the flux strands with a column of field free gas from outside the bundle, the total potential energy of the system can be reduced. As a result, a bundle such as a spot should shred into separate strands in a time scale which is essentially the Alfven wave crossing time, $R/v_A \sim 1$ hour in the case of spots on the sun. The fact that many spots survive for periods of time which exceed this disruption time-scale by several orders of magnitude indicates that powerful stabilizing forces of some kind are at work to hold the spot together. We can discuss the stabilizing forces most conveniently by artificially separating the discussion of surface layers from the discussion of the layers at great depth below the surface.

To stabilize the surface layers, Meyer et al. (1977) proposed to use buoyancy forces. Suppose the flux tube, instead of being vertical, fanned out in some region so that the strands of flux were inclined to the vertical, and extended out from the main flux rope over the top of the non-magnetic gas outside the rope. Because of MHSE, the gas inside the strands is of lower density than the field-free gas. Now consider the process of attempting to interchange a flux strand inside the rope with some field-free gas from outside. Since the field-free gas lies deeper than the overriding strands which fan out from the rope, work must be done to raise the field free gas up to the level where the strand of flux is. And since the density inside is less than density outside, this requires more work to be done against gravity than is released when the flux strand completes the interchange by sinking into the field-free gas. As a result, buoyancy can convert the equilibrium into a stable one, provided that the inclination of the fanning-outflux strands to the vertical is larger than a critical value. Now fanning out is precisely the characteristic which most closely describes penumbral structure, and it is interesting that all spots which survive for periods of time of, say, a day or more possess penumbrae. It seems that the formation of a penumbra signals the spot's attempt to stabilize itself, at least in the surface layers.

The requirement of a critical amount of fanning out of field lines is necessary in order that buoyancy forces can operate during an attempted interchange. Now, in MHSE, the cause of fanning is a rapid drop in external pressure. Suppose, however, that a spot is so small that its radius R is less than the pressure scale height, $H_p$, in the external gas. Then over a height range of order R, the pressure does not vary significantly outside the tube, and hence the tube has little incentive to fan out. As a result, Meyer et al. (1977) concluded that flux tubes could not be stable if they were too small, i.e. if the total amount of flux fell below a certain limit. Using empirical solar models, Meyer et al. found that the critical flux which is required for surface stability is $\sim 10^{19}$ maxwells. This is in remarkably good agreement with the empirical boundary between dark and bright spots. (Meyer et al. did not consider lateral influx of radiation, and therefore did not deal with bright flux tubes, as Spruit (1977) did.)

To stabilize a spot at great depth requires a different mechanism. There, the flux tube is expected to become essentially vertical, or perhaps even fan out again. In either case, buoyancy forces cannot be called upon to stabilize the structure at depth. What then can combat instability? Meyer et al. (1974) proposed that converging flow due to reversed supergranule flow at depths $d_c \sim 2R$ might serve as a "collar" to hold the flux rope together. They proposed that the convergent flow, when it encounters the wall of the tube, flows upwards, and eventually creates the outward flow around the surface layers of the spot which is observed as the "moat" around long-lived spots. This is consistent, therefore, with the observation that spots with moats are the most stable spots, in general. Since a moat is a three-dimensional annular flow which is organized coherently over scale sizes of order $d_c$ (i.e. several times $10^4$ km, in the case of the largest spots), the flux tube beneath the surface of the spot must remain more or less vertical over depths of order $d_c$ if moat formation is to be permitted. A non-vertical flux tube is not conducive to setting up and preserving the 3-D annular flow pattern of a moat. This has an important implication for differences between leading and following spots in the sun, where the angular velocity of rotation, $\Omega$, increases inwards (as various data suggest). Meyer et al. (1977) pointed out that because $d\Omega/dr<0$, the flux rope beneath a leader spot is more likely to be vertical than the flux rope beneath a follower spot in the sun. Thus, moat formation would be easier around the leader, and this could explain why leaders usually are the longest-lived spots. Moreover, in the case of the follower spot, not only is the flux tube non-vertical, but the field lines are strongly curved in the surface layers on the side of the follower spot which lies closest to the leader. As a result, curvature forces are more severe on the field lines in the follower spot, and flux tends to shred away from the follower. This would explain why follower flux tends to be distributed among many small spots and pores.

What are the implications of these ideas for starspots? The most prominent characteristic of active stars is fast rotation. This is especially true of stars which are observed to have cool spots on their surface: stars which rotate with periods longer than about 7 days tend to show strong emission in, say, Ca K, i.e. they are "plage" stars, whereas stars which rotate with periods shorter than 7 days are spotted stars (Vogt, this volume). If we accept the definition proposed above for the distinction between spotted and plage areas of an active region, then we conclude that fast rotation apparently encourages the appearance of flux ropes which are primarily vertical at the stellar surface and coherent over length scales of at least one granule diameter. Slower rotation may also create magnetic flux, but apparently it intersects the surface in a more chaotic manner, or else the flux in individual tubes is simply not large enough to engulf a single granule (given the constraints of surface field strength imposed by the photospheric pressure distribution). However, given the fact that faster rotation favors creation of spots, what does the faster rotation do to the stability of those spots? It seems plausible that as $\Omega$ increases, $|d\Omega/dr|$ should also increase. If this is true (and it must be admitted that high turbulent viscosity in convection zones would generally be expected to keep $|d\Omega/dr|$ small: quantitative evaluation will ultimately be required), it leads us to the following conclusion: when a flux rope erupts to the surface of a spotted star, the rope will be sheared beneath the surface more seriously than in the solar case. Therefore, even the leader spot in a bipolar pair might have its flux tube tilted seriously away from the vertical beneath the surface (and the follower more so). As a result, a moat could not form even around the leader spot. This argument leads us to the conclusion that large stable spots may not exist on the surfaces of fast rotating active stars. Instead, the flux rope is expected to break up into many small spots, each of which is in a stage of transient evolution, breaking away from the main flux rope. Thus, spotted stars would not possess analogs of what is perhaps the most prominent hall-mark of solar activity: large, single long-lived more-or-less symmetric spots. Ideally, the presence of multiple spots on stellar surface can be proven by careful photometry of an eclipsing system during ingress or egress. However, the expected amplitudes of fluctuations from individual spots may be too small to be detectable, even with the best available photometers (Caton, 1981).

A further conclusion can be drawn from the necessity of many small spots (rather than a single large spot). Small spots are, as we have mentioned, less likely to form penumbrae. Spot groups on active stars may therefore be characterized by larger numbers of umbrae, without as much penumbral coverage as in the sun. In such a case, the spot/plage ratio would be larger than in the solar atmosphere. Of course, these arguments will remain speculative until more is known about $d\Omega/dr$ in cool dwarfs.

So far, we have mentioned the proposal of Meyer et al. (1974) to stabilize spots at depth by means of moat flow. Now, most spots are not observed to be surrounded by moats. In the absence of any other stabilizing influence (such as strong twists), most spots may not have an appropriate "collar" at depth, and should therefore not be stable. This led Parker (1979) to propose that indeed a spot flux rope succumbs to instability beneath the surface (below 1-2 thousand km), and filaments into separate flux strands, with field-free gas surrounding each separate strand. In Parker's view, a spot is a dynamical clustering of separate tubes, and the spot will disperse as soon as the clustering force disappears. As a source for the clustering force, Parker proposed a converging flow (as Meyer et al. did), but instead of the flow turning upwards along the walls of the tube, Parker proposed that the flow went downwards. (This helps then not only with the stability problem, but also with the cooling problem: see below). Evidence for large-scale organized sub-surface flows persisting for days or weeks around spots is weak at present. However, Parker's suggestion explains why spots of very different areas can have properties (such as effective temperature and magnetic field strengths) which are remarkably uniform: these properties are determined by the elementary flux tubes themselves, and different spots differ only in total numbers of elementary flux tubes which they contain.

## 6. SPOT COOLING AND MISSING FLUX

So far, we have said nothing about thermal equilibrium of flux tubes. Models of flux tubes in MHSE have simply assumed that the thermal flux inside the tube is reduced somehow, without discussing the energy equation. In fact, it is by no means obvious that the constraints of thermal equilibrium can be satisfied within the context of some of the MHSE models which have been constructed (Low 1980). Zwaan (1978) has summarized the topic of spot cooling succinctly: "the heat balance is the ultimate problem in long-lasting flux tubes." The missing flux is certainly not present in thermal form (Willson et al., 1981): when spots are present on the surface of the sun for at least 7 days, the solar power output is reduced by essentially the spot missing flux. It is true that some compensation for the spots may be provided by facular emission, but this compensation is not total. The missing flux may be emerging from the sun in non-thermal form (thereby escaping detection by the radiometer used by Willson et al.), or else it may be stored inside the sun. A recent theoretical discussion by Spruit (1982) on the time-dependent heat transfer problem which arises when a cool spot appears on a stellar surface has shown that almost all of the missing flux can be stored inside the sun for long time scales, if necessary (longer than a year).

The question of storage of missing flux of starspots inside the star is in principle susceptible to observational testing: what is necessary is bolometric luminosities at times when a spot is present on the surface and when it is absent. In view of the low effective temperatures of red dwarfs, this requires photometry out to wavelengths of at least several microns. As far as I know, such measurements have not been made for any red dwarfs. The simpler problem of searching for a "bright ring" around starspots has, to the best of my knowledge, also not been attacked. Historically, prior to the data of Willson et al., it was always possible to say that the sunspot missing flux was simply re-distributed over a large area of the surrounding photosphere, so large that the enhancement relative to the normal photosphere would be lost in the noise of granulation fluctuations. In fact, Spruit (1977) showed how easy it is for the sun to "hide" the missing flux in this way. However, in a red dwarf, where the missing flux may amount to some tens of percent of the total luminosity, the brightness of the surrounding "undisturbed" photosphere would certainly be enhanced by detectable amounts even if the missing flux were spread out over the entire area of the unspotted photosphere. There are now less pressing reasons to search for bright rings around starspots, since solar data suggest that temporal redistribution of the missing flux is at work, rather than spatial redistribution (see also Hartmann and Rosner, 1979).

Very recently, however, the possibility that the missing flux in starspots is actually emerging into the corona as hydromagnetic waves, and dissipating there, has been proposed by Gershberg (this volume). He has pointed out that the X-ray power emitted by flare stars in their quiescent condition is comparable to the missing power in the dark spots on the stars concerned. He interprets this as evidence that a significant fraction of the energy trapped in subphotospheric gas can be transferred effectively into higher regions of a stellar atmosphere by hydromagnetic waves. Our solar prejudices force us to consider this suggestion as a daring one, indeed. It is well known that the corona above an active region emits in X-rays at a level which is orders of magnitude below the level of missing power due to spots in that region (Evans et al., 1977). And there are theoretical arguments which suggest that Alfven waves coming up from sub-photospheric gas along vertical magnetic field lines are strongly reflected (Thomas, 1978). This has led to the conclusion that if Alfven waves are indeed carrying the missing flux in spots, only a small fraction (of order $10^{-5}$) would normally be able to leak upwards into the corona (Mullan, 1981). On the other hand, other studies of the propagation of Alfven waves in the solar atmosphere have shown the possibility of resonances, such that waves of certain preferred periods may encounter much less serious reflection at the temperature minimum. In particular, the propagation characteristics along a coronal arch may be such as to raise the transmission coefficient of certain Alfven waves to the corona by orders of magnitude, thereby allowing those waves to heat the corona strongly along the arch (Zhugzhda, 1982). Loops which emerge from sunspot umbrae are often rather cool (Orrall 1981). Hence, the resonance behavior of loops apparently does not help to improve the transmission of Alfven waves from umbrae to corona in the sun (assuming Alfven waves are present in umbrae).

If Gershberg's proposal can be validated, it may indicate that conditions in coronal loops which emerge from umbrae of starspots may be more conducive to transmission of MHD waves from beneath the surface of the umbra into the corona. For example, a resonance condition of the kind $P=L/v_A$ (where P is wave period, L is loop length, $v_A$ = Alfven speed) is probably not satisfied in the solar loops, where $L \sim 10^9$ cm, $v_A \sim 10^8$ cm/s. In such loops waves would need to have periods of order 10 s to resonate. However, MHD waves in sunspots are expected to have periods of a convective turnover time, or times of order $10^2$-$10^3$ sec (Mullan, 1981). Application of the same arguments to red dwarfs, however, suggest that since loop lengths are in all likelihood longer in such stars (perhaps as large as a stellar radius; Kodaira, this volume), whereas $v_A$ may not be very different from solar values (Mullan, 1975), the resonant periods may increase to greater than $10^2$ seconds. Such periods would lie within the spectrum of periods predicted by convective turnover time-scales (the latter can be extracted from the models of Mullan, 1974b), and it is therefore not excluded that resonant transfer is considerably more efficient in red dwarfs than in the solar case.

The possibility that the missing flux from spots may be used for coronal heating in red dwarfs needs to be explored in some detail. It is characteristic of Alfven waves that they have great dissipation lengths. For this reason, they have traditionally been considered as the favorite wave mode for heating of the corona of the sun (since other modes would have dissipated low down in the chromosphere). However, in the absence of evidence for Alfven waves passing up through the chromosphere, the role of Alfven waves in heating the solar corona has diminished in favor of other mechanisms. Gershberg's suggestion indicates that red dwarf coronae may be tapping a reservoir of mechanical energy which is unimportant in the sun, namely, MHD waves in spots. In view of the long dissipation lengths, tapping of such a reservoir would be expected to enhance coronal heating more than chromospheric heating. This would be a very useful feature, since it has been found empirically that red dwarfs are more efficient at heating their coronae than at heating their chromospheres (Mullan 1979), whereas in the sun, the reverse is true.

How, then, can we estimate how cool a spot should be, and how much missing flux should there be in a starspot? Attempts to estimate spot cooling by prescribing velocity fields are valuable in the case of the sun, where the velocity fields can be observed (Schatten, 1981). However, this approach cannot be carried over to red dwarfs. The question of spot cooling can be discussed consistently only by means of a model for stellar convection. In this regard, discussions of the onset of convection (i.e. in terms of critical Rayleigh numbers) seem to be of rather limited applicability. For example, the criteria derived from linear theories for the magnetic inhibition of convection are too strong to apply to the solar case (Staude, 1978). Even in the presence of a magnetic field of 3-4 kG, solar convection cannot be suppressed completely because the electrical conductivity is far from infinite. Thus, from the point of view of modelling, it is natural to consider convection in a spot in terms of the same model as one uses for the undisturbed convection zone, with some modification for the presence of the vertical field. In particular, in a compressible medium, the convective heat flux depends on the local temperature gradient in an essentially non-linear way (including threshold behavior). Failure to appreciate this has led to conclusions about spots which were seriously in error (cf. Spruit, 1977). Moreover, the convection model must include horizontal motions: in a convection model where only vertical flows are calculated, there is no physically meaningful way to parameterize the reduction of convective efficiency in the presence of a vertical field. Furthermore, the combination of horizontal flows with vertical fields allows the model to include conversion of convective energy to MHD waves mechanically. Thus, the severe thermodynamic and conceptual problems which arise in attempting to discuss a cool spot in terms of a heat engine which emits Alfven waves by means of overstable oscillations (Cowling, 1977) do not arise. Reduction of convective efficiency and copious emission of MHD waves are not mutually exclusive in a sunspot.

A model which incorporates the above requirements was published several years ago (Mullan, 1974a). It led to sunspot effective temperatures of less than 3000 K in a 3 kG field, and effective temperatures of less than 2000 K in the case of a spot on a red dwarf (Mullan, 1974b). In the latter case, arguments were presented which suggested that the surface fields in starspots should be larger than in the solar case by an order of magnitude. As far as I know, there are no other predictions of effective temperatures of starspots. Staude (1978) applied a cellular model of convection to the sunspot problem, but no application to starspots was made. In my models, the missing flux was carried by Alfven waves. In the case of sunspots, these are expected to be reflected inside the sun. They will eventually dissipate and return their energy to the thermal pool inside the sun. However, as far as the surface of the sun is concerned, the missing flux will essentially be stored as long as the spots are visible. We can estimate the time-scales for Alfven wave dissipation inside the sun due to ohmic effects, if we select appropriate parameters, e.g. fields of $10^4$ G at depths of $10^5$ km, where $\rho \sim 0.2$ gm/cm$^3$, $T \sim 2 \times 10^6$ K. Then for waves of periods $P=10^2-10^3$ sec (typical of sunspots, according to the models), we find dissipation times of 6-600 days. The solar radiometer experiment (Willson et al., 1981) shows that the solar output shows no fluctuations on time-scales shorter than about 7 days. It appears that storage of sunspot missing flux in the form of Alfven waves inside the sun is not incompatible with the solar radiometer data.

## 7. RELATIONSHIP BETWEEN SPOTS AND FLARES

Most (99%) solar flares occur in the vicinity of spots. Is the spot-flare relation purely passive (in the sense that both require strong fields), or is it active? In the former case, the major role to be played by spots would be in defining the separatrices across which flux transfer would occur during flares (Baum and Bratenahl, 1980). In the latter case, the spot would contribute to flare energization, particularly in flares where storage of energy in the corona is inadequate (Piddington, 1973; Spicer, this volume). The proposal that Alfven waves carry the missing flux of spots, and that these waves are normally trapped beneath the surface, provides a natural context for exploring the possibility of an active spot-flare connection. During large proton flares, the chromosphere/corona is forced downwards over umbrae, thereby enhancing the transmission of Alfven waves (non-resonantly) into the corona. Even if only 1% of the missing flux reaches the corona, a large fraction of the flare energy budget can be supplied (Mullan, 1981). This is therefore a specific mechanism to implement De Jager's (1968) suggestion that large solar flares may be powered by the missing flux in spots. In the context of stellar flares, the proposal that flares are powered by the missing flux of spots (or a fraction thereof) can be used to predict flare frequency as a function of stellar luminosity with some success over a range of some 10 magnitudes (Mullan, 1975b).

To the extent that there is an active relationship between spots and flares, spots should be viewed therefore not simply as cool areas which are dull, compared to the more interesting behavior exhibited by flares: rather, spots should be viewed as engines which do the work of converting the energy of convective flows into flare-compatible form.

## REFERENCES

Baum, P.J. and Bratenahl, A.: 1980, Solar Phys. 67, 245.
Beckers, J.: 1977, Ap. J. 213, 900.
Caton, D.B.: 1981, Publ. Obs. Appalachian St. Univ., Vol. I, part 2.
Cowling, T.G.: 1976, Mon. Not. Roy. Astron. Soc. 177, 409.
Cox, A.N. et al.: 1981, Ap. J. Letters, 245, L37.
DeJager, C.: 1968, in Structure and Evolution of Active Regions, ed. K. Kiepenheuer (Dordrecht: Reidel), p. 480.
Frazier, E.N.: 1977, Astron. Ap. 64, 355.
Hartmann, L. and Rosner, R.: 1979, Ap. J. 230, 802.
Low, B.C.: 1980, Solar Phys. 67, 57.
Meyer, F. et al.: 1974, Mon. Not. Roy. Astron. Soc. 169, 35.
Meyer, F. et al.: 1977, Mon. Not. Roy. Astron. Soc. 179, 741.
Moore, R.L.: 1981, Space Sci. Rev. 28, 387.
Mullan, D.J.: 1973, Ap. J. 186, 1059.
Mullan, D.J.: 1974a, Ap. J. 187, 621.
Mullan, D.J.: 1974b, Ap. J. 192, 149.
Mullan, D.J.: 1975a, Astron. Ap. 40, 41.
Mullan, D.J.: 1975b, Ap. J. 200, 641.
Mullan, D.J.: 1979, Irish AJ 14, 73.
Mullan, D.J.: 1981, Solar Phys. 70, 381
Orrall, F.Q.: 1981, Space Sci. Rev. 28, 423.
Pardon, L. et al.: 1979, Sol. Phys. 63, 247.
Parker, E.N.: 1979, Ap. J. 230, 905.
Piddington, J.H.: 1973, Solar Phys. 31, 229.

Piddington, J.H.: 1976, Ap. Space Sci. 40, 73.
Rucinski, S.: 1979, Acta Astron. 29, 203.
Schatten, K.H.: 1981, Ap. J. 247, L139.
Spitzer, L.: 1962, Physics of Fully Ionized Gases (Interscience).
Spruit, H.C.: 1977, dissertation, Utrecht.
Spruit, H.C.: 1982, Astron. Ap. 108, 348.
Staude, J.: 1978, Bull. Aston. Inst. Czechoslov. 29, 71.
Stellmacher, A. and Weihr, E.: 1976, Astron. Ap. 47, 479.
Stenflo, J.O.: 1973, Solar Phys. 32, 41.
Thomas, J.H.: 1978, Ap. J. 225, 275.
Vrabec, D.: 1974, in IAU Symo. 56, 201.
Willson, R.C. et al.: 1981, Science 211, 700.
Zhugzhda, Y.: 1982, Solar Phys. 81, 245.
Zwaan, C.: 1978, Solar Phys. 60, 213.

DISCUSSION

Uchida: The idea of storing energy in the form of Alfvèn waves is interisting but it is possible to store the waves without dissipation down in the high β region?

Mullan: Estimates of the time for dissipation of an Alfvèn wave depend critically on the field strength and the density i.e. on the Alvèn speed to a high power. I do not believe that our models of the deep convection zone are yet good enough to provide reasonable estimates of what the value might be. So I cannot estimate the lifetime of an Alfvèn wave below the surface at this moment.

Gibson: I would like to offer some independent evidence for the leading spot hypothesis. When you look at the non-eclipsing RS CVn binaries, i. e. the ones whose orbital planes are apparently titled with respect to our line-of-sight showing more of the pole, you expect on the leading spot hypothesis to see more intense magnetic field either coming toward you or going away from you. In the radio data for these stars one sees a particular handedness of circular polarization, e.g. for HR 1099 we consistently see 20% circular polarization. In the eclipsing systems, where presumably you see equal contributions from the Northern and Southern hemispheres, we see no circular polarization especially at the quiescent times.

Mullan: I wonder whether you can separate, in that case, the effects of local from those of global fields. If you happen to be looking at a non-eclipsing system then one of the poles is tilted toward you. So are you sure that you can distinguish this effect?

Gibson: No, I cannot do that. Buth what I am saying is that the leading spot would presumably have a larger field than the trailing field and so, since the non-thermal emission varies as $B^2$, the bulk of that emission would be coming from over the leading spot. That is the reason for my comment.

Mullan: I do not think that I said that the field of a leading spot would be larger than that of a following spot. At least, I did not mean to say it. What I meant to say was that the leading spots on the Sun are more stable than those following.

Weiss: It is true that if the following spot broke up into separate flux tubes their field strengths would be small up in the corona.

Spicer: You did not say anything about the possibility of parallel currents flowing in these flux tubes attracting one another. Suppose, for instance, one had a lot of pores and they had parellel currents then they would attract and coalesce. So parallel currents give rise to coalescence instability, to filamentation effects and also to an energy transport mechanism. They also help to stabilize. Neither Spruits' nor Parker's calculations, as well as elsewhere in the literature, make allowance for these currents.

Mullan: Well, the coalescence of pores has been observed but one is not forced to conclude that they are being swept together by the velocity fields. There could be other interpretations. If your suggestion is that parallel currents can sweep pores together then that strengthens the point I was making. That point was that one cannot rely on surface velocity fields to do the coalescence. The problem of currents in spots has not been treated except in one case by Gopelli (?) and Zwaan who believed they could help to stabilize the spot by having a current sheet around the flux tube. There are however some reasons why the numerical values they used might be difficult to justify.

Weiss: I would like to make two quick comment on that. First of all, Gene Parker has studied in some detail the interaction between small flux tubes in a whole series of papers. Secondly, some calculations on magneto-convection suggest, although in a rather uncertain way, that in a system of magnetic fields in the presence of convection the favoured configuration is one where the smaller flux tubes are amalgamated into larger fields rather than being disseminated.

Spicer: I have examined Parker's publications and Spruit and I have discussed this and parallel currents are not considered.

Weiss: Well, I think that parallel currents are irrelevant myself.

Spicer: But they are attractive. They are the reason for a coalescence instability.

Weiss: They are irrelevant in the atmosphere, I believe, but not in the interior.

Venugopal: How does one explain the Evershed Effect i.e. the outflow of gases.

Mullan: The Evershed Effect is a surface cellular effect in which outflow occurs at the photospheric level and inflow occurs at the chromospheric level. That probably involves Lorentz forces which are going on in the higher levels of the atmosphere and which I have not considered here. The calculations of the convective models which I have done begin in the convective zone. I believe that there is an array of surface dynamic phenomena which are occuring in the surface layers and which include the Evershed Effect which I have not dealt with here. If you want a very important example I can only refer to the very important survey which Ron Moore has carried out recently. But I agree that the surfaces of spots are the sites of very interesting dynamic phenomena.

Rodonò: I think we have very clear observational evidence in the stellar case of the connection between spots and flares. In fact the most energetic flares are, on average, observed in the more luminous stars i.e. the K stars and these are also those showing spots or at least exhibiting phenomena which can be interpreted as spots.

van Leeuwen: I would object to that. We have observations too of those stars which got ... (rest of comment lost) ... changes in brightness and colour that the total integrated brightness remained constant. So we have an indication that for these stars we have a redistribution of flux rather than flux storage with later release.

Rodonò: I would like to refer to the IAU Colloquium in 1974 in Moscow where a plot was presented which clearly shows what I mean.

# MAGNETIC INSTABILITIES IN STELLAR ATMOSPHERES

E.R.Priest

Applied Mathematics Department
The University, St.Andrews, Scotland.

ABSTRACT. The extensive theory for magnetohydrodynamic instability of a flux tube is briefly reviewed, together with its application to tokamaks and solar flares. In a star a single coronal loop whose foot-prints are anchored in the dense photosphere may become unstable to the kink instability when it is twisted too much. Magnetic arcades may also be subject to an eruptive instability when they are sheared too much. After the eruption the magnetic field closes back down by reconnection and continues to heat the plasma long after the impulsive phase. Global instability of a large part of the coronal magnetic field is also possible when the stored energy is too great.

## 1. INTRODUCTION

My aim in this review is to tell you some of the interesting lessons we have learnt from studying magnetic instabilities in tokamaks and solar flares. Hopefully, some of them may also be relevant to stellar flares. Twenty years ago it was recognised that typical magnetic structures on the Sun (with size 100,000 km and magnetic field a few hundred Gauss) contain enough energy ($3 \times 10^{25}$ J) to provide a flare. The main problem for theorists was to explain how to release the energy fast enough (over $10^3$ sec). The time-scale for such release by ohmic dissipation is the diffusion time

$$\tau_d = \ell^2/\eta \quad , \tag{1.1}$$

where $\eta$ is the magnetic diffusivity. With a typical length-scale ($\ell$) of 10,000 km this gives times that are much too long ($10^{11}$ sec), and so it was realised that a current sheet is needed with a width of only 1 km or less. This led to the development of theories for fast magnetic reconnection at such sheets. Today we know much more about the solar flare, especially since the spectacular observations from Skylab and the Solar Maximum Mission. There are now many more constraints on the imagination of theorists, and the emphasis has shifted to trying to explain the basic magnetic instability that produces a flare. The observations have been reviewed in the books by Svestka (1976) and Sturrock (1980), while summaries of the theories can be found in, for example, Priest (1981,1982)



*P. B. Byrne and M. Rodonò (eds.), Activity in Red-Dwarf Stars, 545–558.*

or Spicer and Brown (1981).

A typical large solar flare has three stages to its development. During the preflare phase (for ½ hr) one sees a slow rise of a large magnetic flux tube (a filament or prominence), together with a soft X-ray brightening. At the rise phase (lasting for 5 mins - 1 hr) the flux tube suddenly erupts much more rapidly. There is a steep rise in Hα emission from the chromosphere and in soft X-rays from the overlying corona, sometimes accompanied by a hard X-ray burst. The Hα comes from two ribbons. During the main phase energy continues to be released, but the intensity declines slowly over an hour or a day. At the same time the Hα ribbons separate and are joined by a rising arcade of hot loops up to 100,000 km high. The rise speed of the loops is at least 20 $kms^{-1}$ early in the event and only 0.5 $kms^{-1}$ later on. Their density and temperature are typically $10^{17} m^{-3}$ and $2 \times 10^7$K at first, falling to $10^{16} m^{-3}$ and $5 \times 10^6$K after a few hours. In regions where the magnetic field is weak (10G) a flare near a quiescent filament tends to be slow, long-lived and thermal, often with no Hα emission at all. When the field is strong (500G) and complex near a plage (or active-region) filament the flare is violent, fast and non-thermal.

## 2. BASIC MHD INSTABILITIES OF A FLUX TUBE

### 2.1. Ideal Modes.

The theory of magnetic instabilities is now highly developed and has been clearly summarised by Bateman (1978) and Wesson (1978,1981), whom I shall follow here. The ideal modes grow fastest and have the magnetic field frozen to the plasma, whereas the resistive modes have lower thresholds for instability and allow the magnetic field to slip through the plasma in a narrow layer around a so-called resonant surface.

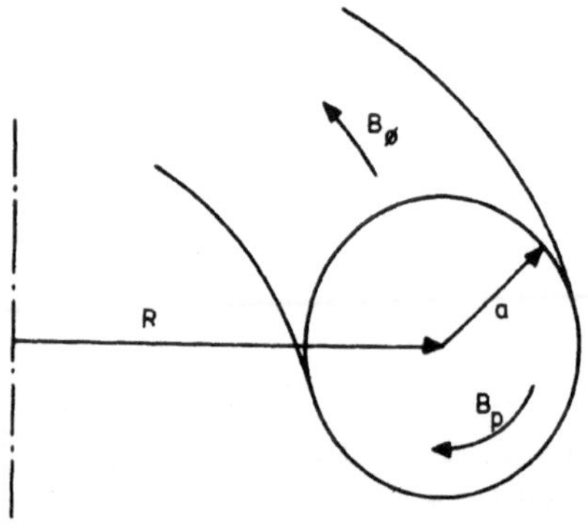


Figure 1. Notation for a curved flux tube.

Consider a magnetic flux tube of major radius R and minor radius a (both constant) with field components $B_p(r)$ (poloidal) and $B_\phi(r)$(toroidal) that depend on the distance (r) from the magnetic axis. Several useful quantities may be defined as follows. The *plasma beta* $\beta = 2\mu p/B^2$ is the ratio of plasma to magnetic pressure, and $\beta_p = 2\mu p/B_p^2$ is the corresponding ratio with the poloidal magnetic pressure. For a semicircular flux tube the amount by which a field line is twisted about the axis in going from one end of the tube to the other is $\Phi(r) = \pi R\, B_p/(rB_\phi)$. A related quantity is the *safety factor* $q(r) = r\, B_\phi/(R\, B_p) = \pi/\Phi$, which for a whole torus is the ratio of the wavelength of a field line to the major circumference $(2\pi R)$ or, in other words, the number of turns that a field line makes around the major axis during one turn around the minor axis. Thus $q = 1$ would mean the field line twists once around the minor axis of a torus, whereas $q = 2$ would mean it undergoes half a twist. The inverse *aspect ratio* $\varepsilon = a/R$ will be assumed much smaller than unity with $q \sim 1$ and $\beta \sim \varepsilon^2$ (i.e. $\beta_p \sim 1$). The *shear* is $d/dr(q^{-1})$ and is a measure of the way the twist varies with radius. The electric current density along the flux tube is $j_\phi(r) \sim r^{-1}\partial/\partial r(rB_p)$, which takes the value $2B_\phi/(Rq)$ on the axis $(r = 0)$. We shall consider a typical flux tube in which $B_p$ increases with r from the axis while $B_\phi$ is roughly constant, $j_\phi$ decreases from a maximum and q increases from a minimum. As the tube is twisted up more, so $j_\phi$ increases and q falls in value.

A radial perturbation $\xi$ proportional to $e^{i(m\theta - n\phi)}$ produces a shape like a single helix if $m = 1$ or a double helix if $m = 2$. The radius $(r_s)$ where $q(r) = m/n$ is called a *resonant surface*, and is such that the orientation of the perturbation matches that of the field so that the crests and troughs of the helix follow the field lines.

To second order in $\varepsilon$ there are no toroidal effects and the change in potential energy produced by the perturbation (assuming a vacuum outside the tube) is

$$\delta W_2 = \frac{\pi^2 B_\phi^{\,2}}{R}\left\{\int_0^a \left[\left(r\frac{d\xi}{dr}\right)^2 + (m^2-1)\xi^2\right]\left(\frac{n}{m} - \frac{1}{q}\right)^2 r\, dr + S_a\right\}, \qquad (2.1)$$

or

$$\delta W_2 = \pi^2 R\int_0^\infty \left[B_1^2 + B_\theta(1 - nq/m)\frac{dj_\phi}{dr}\,\xi^2\right] r\, dr\,, \qquad (2.2)$$

where

$$S_a = \left[\frac{2}{q_a}\left(\frac{n}{m} - \frac{1}{q_a}\right) + (1+m)\left(\frac{n}{m} - \frac{1}{q_a}\right)^2\right] a^2\xi_a^2 \;.$$

Here $\xi_a$ is the surface perturbation and $\underset{\sim}{B}_1 = \underset{\sim}{\nabla} \times (\underset{\sim}{\xi} \times \underset{\sim}{B}_0)$ is the magnetic field perturbation. When $\delta W < 0$ the equilibrium is unstable and otherwise it is stable.

*Kink modes* are *driven by the current gradient* and are *surface* modes in the sense that they distort the surface of the tube. They are the instabilities that arise at second order in $\varepsilon$ and are potentially the strongest. It can be seen from (2.1) that they need $q_a < m/n$, so that the resonant surface is outside the tube. Also, the second term in (2.2) shows that it is the torque arising from the current gradient $dj_\phi/dr$ that drives the instability and that the destabilising region is inside the resonant surface (i.e. $q(r) < m/n$). Wesson (1978) has considered the

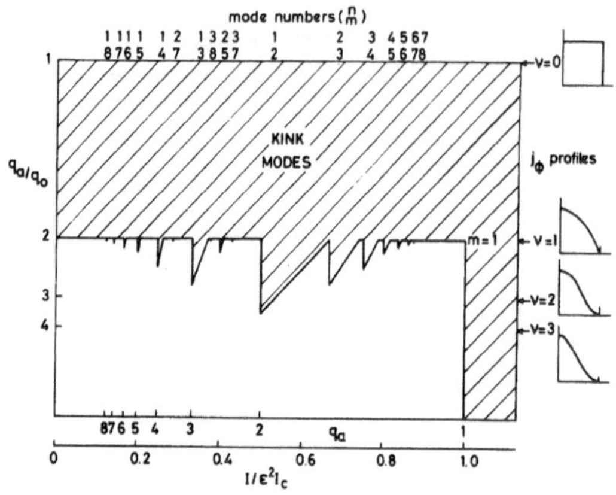


Figure 2. Kink instability diagram (Wesson, 1978).

current profile $j_\phi = j_{\phi o}(1-r^2/a^2)^\nu$ for which the total current is $I = \pi a^2 j_{\phi o}/(\nu+1)$ and the ratio of the q-values at the edge and axis of the tube is $q_a/q_o = \nu + 1$. Figure 2 shows that when there is no shear ($q_a = q_o$) the tube is always kink unstable. At some value of ν between 1 and 2.5 (depending on $q_a$) so that the current is sufficiently peaked, the mode becomes *stabilized by shear*. However, when $q_a < 1$ (the Kruskal-Shafranov boundary) the mode is always unstable. The effect of a potential or force-free plasma surrounding the tube is to provide some extra stability.

*Internal (interchange) modes* are *driven by a pressure gradient* and do not require a surface perturbation (i.e. $\xi_a = 0$). A resonant surface now lies inside the tube and the potential energy is of fourth order in ε with growth-rates smaller than those of the kinks by a factor ε. The modes with m > 1 are *localized around* $r_s$ (i.e. ξ = 0 except near q=m/n) so that $\delta W_2 \approx 0$. In a cylindrical plasma they are unstable if

$$p' + r\,B_z^2/(8\mu)(q'/q)^2 < 0 \qquad \text{(Suydam's criterion)}$$

The first term is destabilizing when $p' < 0$ and the second term represents the *stabilizing effect of shear*. In a torus the curvature provides extra stability by multiplying $p'$ by $(1-q^2)$ (Mercier's criterion), so that a negative pressure gradient is only destabilizing when $q_o < 1$. Thus the internal modes occur below a diagonal line $q_o = 1$ in Figure 1. For sufficiently high β these modes *balloon* (i.e. have a large variation along $\underset{\sim}{B}$) on the outer surface of a curved tube where the curvature is unfavourable.

2.2. Resistive modes.

The inclusion of resistivity removes a constraint by allowing field-lines to break and rejoin in narrow layers around the resonant surfaces. The growth-times for the resulting instabilities lie between the diffusion time ($\tau_d = a^2/\eta$) and the Alfvén time ($\tau_A = a/v_A$), where $\tau_d \gg \tau_A$.

The resistive form of the kink mode is called a (surface) *tearing mode*. It is driven by the current gradient but now occurs when $q_a > m$ so that the resonant surface lies inside the tube. The Euler-Lagrange equation for (2.1) when $\eta = 0$ is

$$\frac{d}{dr}\left(r\frac{d}{dr}(r\,B_{r1})\right) - m^2 B_{r1} - \frac{dj_\phi/dr}{(B_p/mr^2)(m-nq)}\,B_{r1} = 0\ , \qquad (2.3)$$

where $B_{r1} = iB_p(m-nq)\xi/r$. This is also the equilibrium equation $\nabla \times (j \times B)_1 = 0$ since the smallness of the growth-rate makes inertia negligible. The solutions to (2.3) starting at the axis and at infinity become singular at $r = r_s$, and so they need to be matched with those in the resistive layer. The result is that the mode is unstable when

$$\Delta' = \left[\frac{1}{B_{r1}}\,\frac{dB_{r1}}{dr}\right]_{r_s-\varepsilon}^{r_s+\varepsilon} > 0, \qquad (2.4)$$

and the growth-time behaves like $\tau_d^{3/5}\,\tau_A^{2/5}$. For m = 2 the effect of twisting up a flux tube is to move to the right in Figure 3 and so cross the threshold first for tearing ($q_o = 2$) and then, as the resonant surface crosses r = a, for kinking ($q_a = 2$). (One may be tempted to equate the crossings of such thresholds with the onset of the preflare phase and rise phase of a flare, but we shall see below that tokamak phenomena are not so simple and require nonlinear theory for their interpretation.) The lower boundary in Figure 3 appears because of shear stabilization when $q_a/q_o$ is large enough. A similar figure is obtained for m = 3, but modes with m > 3 are stabilized because of the tension term ($-m^2 B_{r1}$) in (2.3).

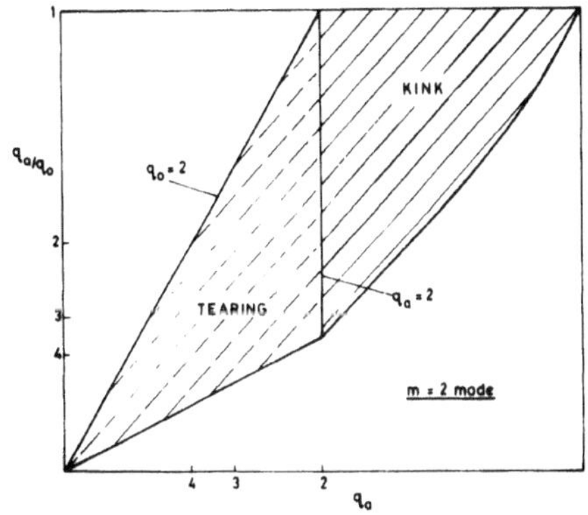


Figure 3. Instability of m = 2 mode (Wesson, 1978).

The *resistive interchange* (or resistive g) modes are the resistive form of the internal modes with m > 1 and have growth times of order $\tau_d^{1/3}\,\tau_A^{2/3}$. The effect of increasing the shear is to localize the modes and reduce their growth-rate. They are unstable if

$$(-p')\left\{q^2 - 1 + \frac{q^3 q'}{r^3}\int_0^r \frac{r^3}{q^2} + \frac{2R^2 r^2}{B_\phi^2}(-p')dr\right\} < 0\ ,$$

and so under normal conditions ($p'<0$, $q'>0$) they require $q_o<1$. Increasing $\beta$ provides more stability by changing the solution in the resistive layer and modifying the instability criterion to $\Delta' > \Delta'_c$, where $\Delta'_c$ increases with $\beta$. The threshold for stability of m = 2 is moved to the right from $q_o = 2$ in Figure 3 and the mode is completely stabilized when $\Lambda \gtrsim 60$ ($\Lambda \gtrsim 10$ for m = 3), where $\Lambda = \beta_p^{5/6}\, \varepsilon^2\, (\tau_d/\tau_A)^{1/3}$. However, resistive ballooning modes appear. The *m = 1 internal resistive kink mode* becomes unstable when $q_o < 1$. As the twist is increased so the flux tube becomes tearing mode unstable first with $i\omega \sim \eta^{3/5}$; then it passes through a region where $i\omega \sim \eta^{1/3}$ and finally it becomes unstable to the ideal mode. The importance of other resistive effects in enhancing the dissipation in flares has been debated by Spicer(1977) and Van Hoven(1981).

## 3. APPLICATION TO TOKAMAKS.

A standard tokamak is a torus with $q \sim 1$, $\varepsilon << 1$ and $\beta \sim \varepsilon^2$ so that $a << R$, $B_p << B_\phi$ and $p \sim B_p^2/(2\mu)$. The aim is to confine plasma at $T \sim 15$keV ($10^8$K), $n \sim 10^{20}$ m$^{-3}$, $B \sim 20$-100 kG, $\beta \sim 5\%$ for a time $\tau \sim 1$sec. At present the values reached are $\beta \sim 2\%$ and $\tau \sim 0.1$sec. In contrast, a reversed field pinch has $B_p \sim B_\phi$ with a higher $\beta$(10%) but (so far) a lower confinement time ($10^{-2}$ sec). Three phenomena appear to be produced by MHD instabilities in tokamaks as follows.

*Mirnov oscillations* show up with magnetic pickups as a small regular vibration at several values of m > 1. They are due to resistive modes near the q = m surfaces that have saturated nonlinearly to a steady state containing islands. Finite Larmor radius effects cause the steady helical structure to rotate around the torus.

*Sawtooth oscillations* with a period of a few millisecs appear in the soft X-ray emission from the core of the plasma. They are relaxation oscillations due to nonlinear effects of the internal resistive m = 1 mode (Figure 4). The ohmic heating ($\sigma E^2$) driven by the applied electric field (E) slowly concentrates the current towards the axis, because the heating raises the temperature, the conductivity ($\sigma \sim T^{3/2}$) and therefore the current ($j \sim \sigma E$). When $q_o$ falls below 1 the m = 1 resistive mode occurs rapidly and creates a magnetic island with q > 1. This grows by reconnection and eventually displaces the old island creating a new stable state. The true $B_p$ does not reverse sign, but Figure 4 refers to the magnetic field you would see looking at angle such that the line-of-sight field vanishes at the radius where q = 1.

*Disruptive instability* is the most dramatic event in tokamaks. When the current is so large that $q_a$ falls below about 2.5 the m = 2 oscillations grow rapidly over a few millisec, hard X-rays are emitted, and the soft X-ray emission falls, followed by a rapid collapse of the current. The explanation is rather controversial (like that of flares!). One possibility is the nonlinear destabilization of the m = 3, n = 2 mode by the m = 2 mode (Waddell et al, 1979), producing many magnetic islands with ergodic fields and a rapid escape of heat from the core. Alternatively, it may be due to a magnetic catastrophe. Usually, the tearing mode

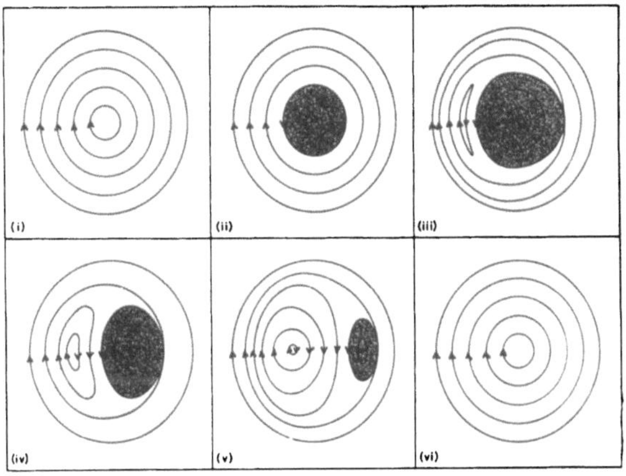


Fig. 4. Model for sawtooth oscillations (Wesson, 1980).

is self-stabilizing with a saturated state that reduces the current gradient ($j_\phi'$), but if the current is too large Sykes and Wesson(1981) and Wesson et al(1982) have demonstrated that the saturated equilibrium no longer exists. The current profile $j_\phi(r)$ is modified by the m = 1 and m = 2 modes, whose islands flatten it near the resonant surfaces at q(r) = 1 and q(r) = 2. The widths ($\omega$) of the islands are determined by $\Delta'(\omega) = 0$, with $\Delta'$ given by (2.4), $B_{r1}$ given by (2.3) and $j'_\phi$ in (2.3) modified by the presence of the islands. As the current grows so the island widths grow until, at some critical width, $\Delta'(\omega) = 0$ no longer has a solution.

Lessons that may be learnt for astrophysics from tokamak studies include the following. The details of the magnetic structure ($B_\phi/r$) and $j_\phi(r)$ are important for determining the relevant instability. The resistive modes occur at lower thresholds than the ideal ones, but they are not necessarily destructive since they may instead produce gentle oscillations (Mirnov or sawtooth). It is crucial to study the nonlinear development of any instability to see whether it saturates (sawtooth) or grows explosively (disruptions).

## 4. SOLAR FLARES.

### 4.1. Loop configuration.

The ideal kink instability of a force-free loop has been studied by several authors (Raadu, 1972 ; Hood and Priest, 1979 ; Van Hoven etal, 1981 ; Einaudi and Van Hoven, 1981) using different forms for $\xi$ and different equilibrium fields. We saw in §2.1 that a tube of uniform twist is always unstable, but in the Sun there is a most important extra stabilizing influence, namely *line-tying*, since the feet of the loop are anchored in the dense photosphere. Such a line-tied perturbation takes the form

$$\boldsymbol{\xi} = (\xi_r, -B_{oz}\xi_o/B_o, B_{o\theta}\xi_o/B_o)e^{i(\phi+\omega t)}$$

in a cylindrical geometry such that $\boldsymbol{\xi}\cdot\mathbf{B}_o = 0$. The equation of motion is

$$\rho_o \partial^2\boldsymbol{\xi}/\partial t^2 = \mathbf{j}_1 \times \mathbf{B}_o + \mathbf{j}_o \times \mathbf{B}_1 ,$$

where $\mathbf{j} = \nabla \times \mathbf{B}/\mu$ and $\mathbf{B}_1 = \nabla \times (\boldsymbol{\xi} \times \mathbf{B}_o)$, subject to the conditions that $\boldsymbol{\xi}$ vanish as $r \to \infty$ and at the ends ($z = \pm L$) of the loop. This gives a pair of partial differential equations for $\xi_r(r,z)$ and $\xi_o(r,z)$, whose numerical solution (Hood and Priest, 1981) gives stability when the tube is slightly twisted and instability when the twist ($\Phi$) exceeds $2.5\pi$ (i.e. $q < 0.4$). The effect of pressure gradients and line-tying has been included by Hood and Priest (1979) and Hood, Priest and Einaudi (1982), who obtain analytical, line-tied solutions for a magnetohydrostatic loop with a sharp boundary. They find that stretching or twisting it eventually makes it unstable.

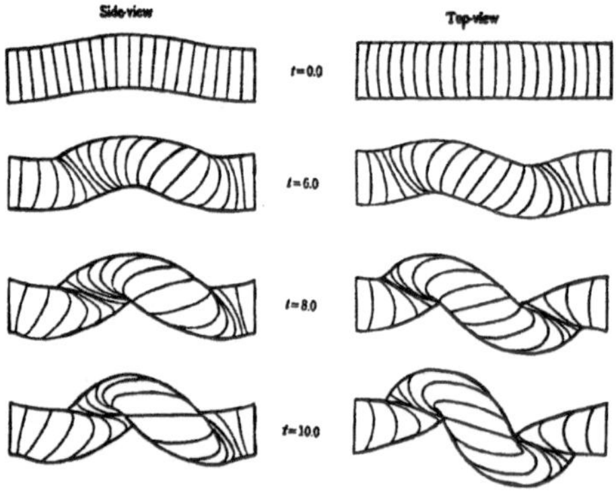


Figure 5. Kinking of a flux tube (Sakurai,1976).

In future, it is important to study several other effects due to the resistive modes, shear stabilization and the curvature and non-uniform cross-section of a loop. A start has been made on following the non-linear development by Sakurai (1976). He considers an infinitely long cylinder of uniform current or a force-free field and simulates line-tying roughly by requiring that the axial wavenumber ($k$) be $\pi/L$. His numerical solutions (Figure 5) show how the flux tube rises and twists up. Tubes with a weak twist develop strong helical kinks and rise less than those with strong twist because of the stabilizing tension force from the axial field.

## 4.2. Magnetic Arcades.

A coronal magnetic field evolves passively through a series of (largely) force-free equilibria and stores more and more energy in response to the motions of its photospheric footpoints until it becomes unstable. This point of instability may be found by seeking multiple equilibrium solutions to the same photospheric boundary conditions on the plane $y = 0$ (Low,1977 ; Birn et al, 1978; Priest and Milne, 1980).

For a magnetic field of the form

$$\underset{\sim}{B} = (\partial A/\partial y,\ -\partial A/\partial x,\ B_z(A)) \qquad (4.1)$$

the force-free field equation ($\underset{\sim}{j} \times \underset{\sim}{B} = \underset{\sim}{0}$) reduces to

$$\nabla^2 A = -\ B_z\ dB_z/dA\ ,$$

where the horizontal displacement of the footpoints from the x-axis in the z-direction is $d(x) = B_z \int dx/B_x$ and $B_y(x,0)$ is prescribed. Good progress has been made so far, but mainly with the case when $B_z$ rather than the footpoint displacement (d) is prescribed. In particular, Heyvaerts et al (1980) have discovered at least three solutions and therefore the possibility of jumping violently from one to another.

Magnetic stability of arcade equilibria has been tested directly by Van Tend and Kuperus (1978), Hood and Priest (1980) and Birn and Schindler (1981). Hood and Priest included line-tying but were unable to find a perturbation that destabilizes a simple arcade with a field of the form (4.1) and no magnetic island above the photosphere. This led them to consider arcade fields with the axis (an O-type neutral point in the x-y plane) a distance d above the photosphere (y=0) and the field lines twisted about this axis by an amount $\Phi$ as they ascend from the photosphere and return to it again. Such arcades are stable when d and $\Phi$ are small but become unstable when $\Phi$ or d exceed critical values. Thus, the eruption of an arcade may be due to a spontaneous MHD instability when $\Phi$ or d are too big or due to a resistive instability below the filament when d is too large. Alternatively, it may be triggered by some extra effect such as : emerging flux, which can push the filament up and tear away some of the overlying field lines or lower the tearing mode time by creating a small region of enhanced resistivity and initiating a large-scale reconnection (Heyvaerts et al, 1977); a thermal instability in the filament, which causes the plasma to expand and makes the tube rise until the critical d is reached; a fast magnetoacoustic wave, which can couple to and trigger tearing with a faster growth-rate than normal (Sakai and Washimi, 1982).

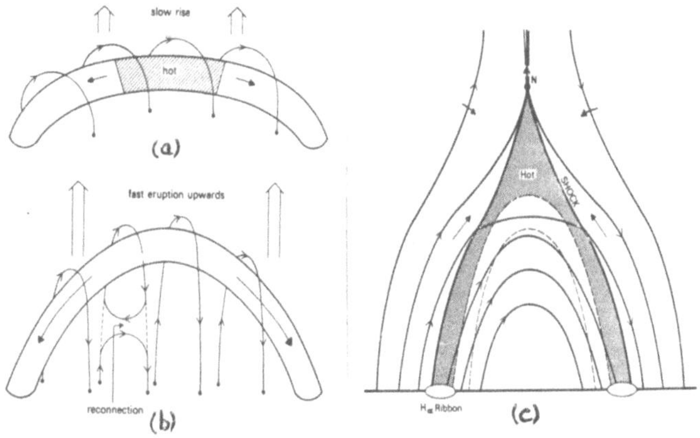


Figure 6. Overall behaviour of a flare.

4.3. Main phase.

The overall behaviour of a solar flare is indicated in Figure 6. First of all a magnetic flux tube rises slowly and stretches out the field lines of the overlying arcade. Then the rapid eruption is triggered and the field below the flux tube starts to tear. Later on (Figure 6c in a cross-section across the arcade), the field lines that have been dragged open by the eruption continue to close down and reconnect at the neutral point N. This rises and trails behind it a pair of slow MHD shock waves, which heat the plasma and create hot loops (Cargill and Priest,1982). This process of line-tied reconnection has been simulated numerically by Forbes and Priest (1982) with typically $\beta = 0.1$ and $\tau_d/\tau_A = 10^3$. They find that the field lines that are being stretched out by the erupting flux tube tear first near the base. In the nonlinear phase the flows build up to a large fraction of the Alfvén speed and a quasi-steady state of Petschek-like reconnection develops with the X-type neutral point (N) rising and a region of closed loops being created below N. Above N a plasmoid is ejected and the sheet thins. Eventually, it tears again creating a pair of O and X neutral points (Figure 7). Reconnection at the upper X dominates and the O is rapidly shot down to coalesce with the lower X. This process can repeat and may be an efficient means of accelerating particles.

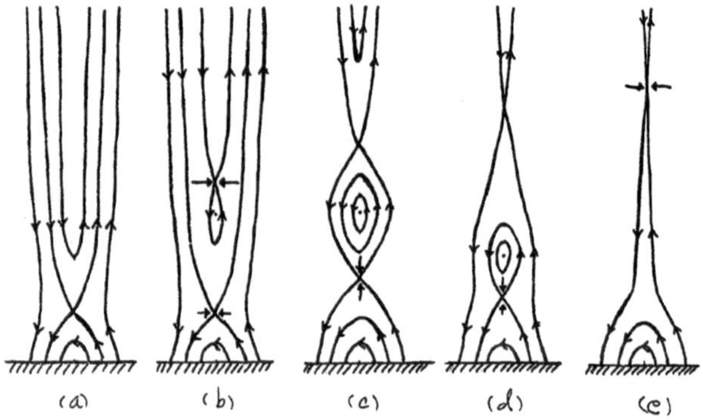


Figure 7. Creation and annihilation of neutral point pairs during the main phase of a flare (Forbes & Priest, 1982).

5.STELLAR FLARES.

Stellar flares may be caused by similar instabilities to the ones we have described for solar flares. The energy of a flare should scale like $W = L^3B^2/(2\mu)$ in terms of the local magnetic field (B) and the size (L) of the magnetic region, while the time-scale for energy release is just $\tau \approx L/(0.01V_A)$, where $V_A = B/(\mu\rho)^{\frac{1}{2}}$, and the pressure is $p \approx B^2/(2\mu)$, although on the Sun itself this scaling is only rough, since one has a wide range in W, $\tau$ and p. On a star one may find instability when the plasma pressure builds up so much that it cannot be contained by the magnetic field, or when the density builds up so much that it cannot be supported against gravity by a sheared field (Rayleigh-Taylor instability). Again, if a giant flux tube erupts by magnetic buoyancy in a violent manner it may produce the giant equivalent of the tiny X-ray bright points seen on the Sun.

A solar flare is caused by the magnetic instability and subsequent reconnection of a *local* magnetic region. But stellar flares may instead arise when *global* magnetic equilibrium breaks down. Several solutions have been presented for the global magnetic field in the corona of a star. Consider an axisymmetric field

$$\underset{\sim}{B} = \frac{1}{r\sin\theta}\left(\frac{\partial A}{r\partial\theta}, -\frac{\partial A}{\partial r}, B_\phi(A)\right)$$

with pressure $p = p_0(A)\exp - \int_{r_0}^{r} GM/(RTr^2)dr$. Then the force balance equation $(\underset{\sim}{J} \times \underset{\sim}{B} - \underset{\sim}{\nabla} p - \rho GMr^{-2}\hat{\underset{\sim}{r}} = \underset{\sim}{0})$ reduces to the basic equation

$$\frac{\partial^2 A}{\partial r^2} + \frac{\sin\theta}{r^2}\frac{\partial}{\partial\theta}\left(\frac{1}{\sin\theta}\frac{\partial A}{\partial\theta}\right) + B_\phi\frac{dB_\phi}{dA} + 4\pi r^2\sin^2\theta\frac{\partial p}{\partial A} = 0. \qquad (5.1)$$

Analytical solutions have been discovered by Uchida and Low (1981) with $B_\phi \equiv 0$ and $p = Q(r)A$, so that (5.1) reduces to a linear equation with solutions of the form $A = r^2 u(r)\sin^2\theta$. In particular, they discuss fields that are dipolar at the surface and uniform at large distances, and they consider the effect of different mass loadings.

When the pressure gradient and gravitational forces are negligible, (5.1) describes a *force-free* field. Raadu (1972) started with a quadrupole field and calculated the effect of increasing $B_\phi$ due to differential rotation of the footpoints. He found that the poloidal field lines expand outwards to provide an extra magnetic tension to contain the magnetic pressure of $B_\phi$. The magnetic energy increases by 25% in one rotation. Later, Milsom and Wright (1976) considered a particular form of $B_\phi(A)$ such that $B_\phi dB_\phi/dA = \alpha A^n$, with $n = 4$ for instance. They matched numerically a dipole surface field to a dipolar field at infinity of the form $A = A_\infty r^{-1}\sin^2\theta$, which is only possible when $n > 3$. As $\alpha$ increases so the toroidal field ($B_\phi$) increases and the field distorts until, at a critical value of $\alpha \approx 0.8$, $A_\infty$ becomes infinite, so that the field cannot be contained any longer and blows open. A similar feature should be present if the coronal pressure increases too much due to, for instance, too much heating. A similar effect has been demonstrated analytically by Browning and Priest (1973) by putting $B_\phi = B_0 A$ and $A = A_0(r)\sin^2\theta$, so that (5.1) reduces to

$$d^2A_0/dr^2 + (B_0^2 - 2r^{-2})A_0 = 0,$$

with solution

$$A_0 = C_1/(\lambda r)\cos(\lambda r + C_2) + C_1\sin(\lambda r + C_2).$$

She supposes that a stellar wind makes the field radial at some radius $s_0$ and imposes a dipole field at the surface. Starting with a potential field, the effect of differential rotation in twisting up the field is considered by increasing $\lambda$. The field lines expand up to a maximum $\lambda$

of 1.4 $s_0^{-1}$, above which a physically reasonable field no longer exists, since field lines detached from the star appear near the equator which would be pulled out by the stellar wind. The same effect is present with a quadrupolar field or an increasing pressure ($p = p_0 A$).

## 6. CONCLUSION.

Depending on the detailed magnetic structure, there is a rich variety of ways in which a magnetic field can go unstable due to either an ideal or a resistive mode. It is essential to study the nonlinear development of such instabilities : they need not always be fatal since they may easily be saturated and reach new equilibria rather than growing explosively. We have learnt much about solar flares from the wonderful observations of Skylab and SMM - what riches would be in store for us if we could view their stellar counterpart with as much clarity. Solar flares are due to an eruptive instability of a loop or arcade and the subsequent reconnection process as the magnetic field closes back down. Stellar flares may be due to a similar process or due to a lack of global equilibrium. However, there is a need to study such equilibria and their stability in much more detail.

## REFERENCES.


Bateman, G.: 1978, *MHD instabilities*, MIT Press.

Birn, J., Goldstein, H. and Schindler, K.: 1978, *Solar Phys.*57,81.

Birn,J. and Schindler, K.: 1981, Ch.6 of *Solar flare MHD* (ed.E.Priest), Gordon and Breach.

Browning, P. and Priest,E.,: 1983,preprint.

Cargill, P. and Priest, E. : 1982,*Solar Phys.*76, 357.

Einaudi, G. and Van Hoven, G.: 1981, *Phys.Fluids* 24, 1092.

Forbes,T. and Priest, E. : 1982, *J.Plasma Phys.* 27,157.

Heyvaerts, J., Lasry, J., Schatzman, M., Witomsky,G.: 1980, *Lecture Notes Math.*782, 160.

Heyvaerts, J., Priest, E. and Rust,D.: 1977, *Ap.J.*216,123.

Hood,A. and Priest, E. : 1979, *Solar Phys.* 64, 303.

Hood,A. and Priest, E. : 1980, *Solar Phys.* 66, 113.

Hood, A. and Priest, E. : 1981, *Geophys.Astrophys, Fluid Dynamics* 17,297.

Hood,A.,Priest,E. and Einaudi,G.:1982, *Geophys.Astroph.Fluid Dynamics*, 20,247.

Low,B.C.: 1977, *Ap.J.*212,234.

Milsom,F and Wright, G.: 1976,*Mon.Not.Roy.Astron.Soc.*174,307.

Priest,E. : 1981, *Solar Flare MHD*, Gordon and Breach.

Priest, E.: 1982, *Fund.Cosmic Phys*, 8.

Priest, E. and Milne, A.: 1980, *Solar Phys.*65,315.

Raadu,M. : 1971, *Astrophys.Space Sci.*14,464.

Raadu,M. : 1972, *Solar Phys.*22, 425.

Sakai, J. and Washimi, H.: 1982, *Ap.J.*, in press.

Sakurai, T. : 1976, *Pub.Astron.Soc.Japan* 28,177.

Spicer,D. : 1977, *Solar Phys.* 53,305.

Spicer,D. and Brown,J. : 1981, *The Sun as a star* (ed.S.Jordan),Ch.18.
Sturrock,P. : 1980, *Solar flares*, Colo.Ass.Univ.Press.
Svestka, Z. : 1976, *Solar flares*, D.Reidel.
Sykes, A. and Wesson,J. : 1981, *Plasma physics and controlled nuclear fusion research*, 1, 237.
Uchida, Y. and Low, B.C. : 1981, *J.Astrophys.Astron.* 2, 405.
Van Hoven, G. : 1981, Ch.4 of *Solar flare MHD* (ed.E.Priest), Gordon and Breach.
Van Hoven, G., Maa, S., Einaudi, G. : 1981, *Astron.Astroph.* 99,232.
Van Tend,W. and Kuperus, M. : 1978, *Solar Phys.*59, 115.
Waddell, B., Carreras, B.,Hicks, H., Holmes,J., Lee,D. : 1978, *Phys. Rev.Letters* 41, 1386.
Wesson,J. : 1978, *Nuc.Fusion* 18, 87.
Wesson,J., Sykes, A and Turner,M. : 1982, preprint.

ACKNOWLEDGEMENT.

The author is most grateful to J.Wesson for many helpful discussions and to the staff of Culham Laboratory for their hospitality during his stay.

DISCUSSION

Nordlund: I will concentrate my question on your description of two-ribbon flares. I think it is fairly well established that the velocities we see when the loops rise are to a large extent apparent because successive loops brighten rather than a single loop brithtening. A second point with regard to the instabilities you talked about, in the loop case we have found that the resistive instabilities came before the ideal ones. So don't you suspect that would be also the case in the arcade case? You just treated the ideal case.

Priest: I will take the first point first. Certainly it is believed that the rise of the loops as observed does not represent a bodily rise of the plasma. Rather it represents the fact that the neutral point at which reconnection takes place is rising. So, as the magnitude field closes back down, you are forming new loops one on top of the other. With regard to the second point, I believe that we are only beginning to take account of the experience of the laboratory plasma physicists and, as a result, so far people have only taken the ideal instabilities into account in giving rise to two-ribbon flares. As I mentioned I think we certainly must put the effects of resistivity into these calculations. That is by no means a trivial matter however and as I have pointed out, resistive modes do not always give rise to explosive behaviour. They just give rise to rather mild oscillations instead.

Stencel: A question concerning your numerical simulation of reconnection with Forbes. Have you considered the amount of mass ejected upward as a function of magnetic intensity?

Priest: No, we have not. These simulations have only been carried out fairly recently and we have not got as far as that. But it is something well worth doing.

Uchida: The numerical calculation which you made forming a magnetic island is very interesting. Sato and Hayshi have also shown numerically that forced reconnection is very efficient. In your calculation is mass sucked out of the region or is it pressed into the region? This is a boundary condition problem.

Priest: The Sato and Hayshi calculation does not start with equilibrium. Rather they impose a flow from the sides and then investigate the structure of the reconnection. Our simulation, which numerically is very similar to theirs, starts out with an equilibrium and then follows the development of the tearing mode into its non-linear phase. The important effect which we put in is the effect of line-tying on the base and this we regard as the most important aspect of our work.

# FLARING PROCESSES IN STELLAR ATMOSPHERES: THE IMPORTANCE OF ELECTROMAGNETIC COUPLING

D.S. Spicer
ETH Zentrum, Institute of Astronomy, Zurich, Switzerland

Paper not received.
(cf. D.S.Spicer in Activity and Outer Atmospheres of the Sun and Stars, 11th Adv.Course Suiss Soc.Astron.Astrophys., Observatoire de Geneve, p. 91, 1981).

DISCUSSION

Linsky: We do not know all of the energy released in the course of a dMe star flare. What we actually measure however is often $10^4$ to $10^5$ times larger than what is seen in large solar flares. I have two questions. Why do you think flares in dMe stars are so much more energetic than in the solar case? Secondly, why are solar flares so much less energetic than those in dMe stars?

Spicer: First of all I know nothing about stellar flares. I believe the most likely reason is that the gradients of the free magnetic field are much steeper in dMe stars than they are in the Sun. The magnitudes of the fields are much larger. So the magnetic energy release should be faster, by whatever method of release one chooses.

Linsky: In the previous talk we heard that the flare energy should be proportional to the magnetic field squared and some volume element. Just saying that the magnetic fields or the shears are larger, then to get four orders of magnitudes in the energy you need two orders of magnitude in the magnetic field.

Spicer: Well, you must take into account the volume element. There are a lot of free parameters in the system.

Evans: One of the problems here is the constraints provided by the observations. Nobody has yet demonstrated convinvingly the presence of high magnitude fields on the dM stars. It may be that there are very localized areas (of high field). In the case of the Sun we are discussing a stately minuet as compared with the violent disco of the dM stars. Account must be taken, therefore, of the non-detection of magnetic fields and the fact



*P. B. Byrne and M. Rodonò (eds.), Activity in Red-Dwarf Stars, 559–560.*

that the energy release is undoubtedly limited to a very small area of the stellar surface.

Spicer: This may be a matter of perspective. Even in the solar case $10^{32}$ ergs is a hell of a lot of energy.

Nordlund: I think that it is very important as you have done here to bring out the aspect of closing of a circuit. For example, with regard to the case of the dMe stars - in the case of the Sun we are talking about instabilities in the coronal plasma. Suppose that with some mechanical force we can shear the magnetic field in the interior of a very large spot on stars having larger fields, then it is possible to consider a circuit going up into the visible photosphere of chromosphere which has its driving energy input or the storage below the surface. Then both the fields involved and the energy output are much stronger.

Spicer: I had intended to point this out but felt that there not enough time during my talk. There are two possible scenarios for magnetic energy storage. Firstly, the total energy for the flare may be stored in an equivalent inductor. Secondly, a much larger energy may be stored and only a small fraction released in the flare. This is possible for certain types of circuit. One of the nice things about this is that it allows you to explain homologous flares. Such flares occur at intervals of hours; whereas large flares in an active region generally occur after 1 or 2 days of quiescence. So the 2 days may be the time necessary for sufficient energy storage. If, however, the stoarge time is of the order 2 days then it is difficult to explain homologous flares except by having more energy than is needed for the flare there to start with.

Mullan: I wonder if you would like to say anything about the connection between coronal heating and flares. Do you believe that there is a fundamental difference between the processes needed for quiescent coronal heating and for triggering a flare? Could such a difference be explained in terms of equivalent circuits?

Spicer: A lot of workers have proved that the two mechanisms are identical, including such processes as anomolous Joule heating and reconnection. I have my doubts about both these for coronal heating. To other suggestions for explaining coronal heating are Alfvèn wave heating or some variation on this. One can show mathematically rigorously that Alfvèn waves are always associated with any system of parallel currents. In talking about flares one generally speaks of very low frequency Alfvèn waves, while for coronal heating higher frequencies are invoked. Maybe this is the difference. I must admit that I have not studied the problem myself. I think however that the question is related to that of flare energy storage in the corona. If you cannot store the energy there then it must come from below. This would clearly distinguish flare mechanisms from coronal heating mechanisms.

# EMPIRICAL MODELS OF STELLAR FLARES : CONSTRAINTS ON FLARE THEORIES

K. Kodaira
Department of Astronomy, University of Tokyo

## ABSTRACT

A review is presented on empirical flare models of UV Cet type stars based upon optical, UV, X-ray, and radio observations. The observational constraints on the flare energetics, nature of radiation sources, and flare structures are discussed, with special attention to the geometrical dimension and the magnetic field of the flaring region. The hot-plasma model in the solar analogy is critically examined by comparing it with observations. Possible future observations are suggested to tighten the constraints on the stellar flare theories.

## INTRODUCTION

The purpose of this paper is to critically review the empirical models for flares of UV Cet-type stars and to propose possible improvement which may lead to tightening constraints on flare theories. Observed flare phenomena are very varied on the Sun as well as on stars. The primary energy sources are generated under various conditions instantaneously or continuously, and the subsequent dissipative processes are even more manifold than the primary processes, depending upon the physical configurations in the adjacent regions. Consequently it would be misleading to construct a model flare based upon data taken piecewise from different events. The most orthodox way is to construct a model for a given flare event using only the data proper to the specific event. For this purpose it is essential to make simultaneous observations over a range of wavelengths as wide as possible. I adopt as the most successful and as the 'reference flare' in this paper the 1979 Oct. 25 event of YZ CMi observed in X-ray, optical and radio regions by Kahler et al (1982). Although this flare showed fine structures composed of three peaks each separated by about 23 sec, its light curve was essentially of type I according to the classification



*P. B. Byrne and M. Rodonò (eds.), Activity in Red-Dwarf Stars, 561–578.*

of Oskanian (1969). Gershberg and Chugainov (1968) found this type to be the most frequent and the brightest among different types in their statistical study of optically detected stellar flares. I believe, therefore, that the Oct. 25 event is a good choice as a reference flare at this moment.

In contrast to the solar case, we cannot disentangle the aspect effects in the stellar case. Some may be beyond -the-limb events in which we can observe only the radiation coming from the upper part of the flaring region. The X-ray flare of Proxima Cen observed by Haisch et al (1981) on 1979 Mar 6 might have been a beyond-the-limb or close-to-the-limb case because no optical flare was detected for this event. The aspect affects the outcoming radiation when there are substructures in the form of threads or columns which are optically thick along the length but thin in cross-section or when the generation and propagation of radiation is directional. In this matter we are bound by statistical studies.

Solar flares have often been compared with stellar flares as guides in constructing models (Kahler and Shulman 1972, Mullan 1976, Kahn et al 1979, etc.). Since the solar white-light flares are most similar to stellar flares in emitting pronounced optical continuum, we exclusively refer to white-light flares in solar case when not otherwise stated. A general description of white-light flares is found in Svestka (1976), and detailed studies of individual cases are given in McIntosh and Donnelly (1972), Kane and Winckler (1969), Mastus and Stover (1967), Donnelly (1971), Machado and Rust (1974), and Dezso et al (1980). The white-light flares comprise a group of the strongest solar flares some of which are accompanied by gamma-ray radiation. The radiation characteristics of a white-light flare observed on 1967 May 23 according to Kane and Winckler (1969) and Hudson (1972) are compared with those of the reference flare in Table 1.

In section 2 the observed time-profiles of stellar flares are summarized and compared with those of solar flares. The source models for flare radiations in the optical, X-ray, and radio regions are reviewed in sections 3,4 and 5, respectively. Finally empirical models of stellar flares are discussed in section 6.

## LUMINOSITY PROFILES

Figure 1 shows schematized time-profiles of the reference flare, based upon figure 2 of Kahler et al (1982). In drawing this I have ignored fine fluctuations in order to make the global features clear. The optical continuum radiation shows an impulsive component and a succeeding gradual component which are typical for type-I flares. Balmer-line emission shows essentially a gradual component, although a small impulsive component may exist. Profiles of soft X-ray and dm

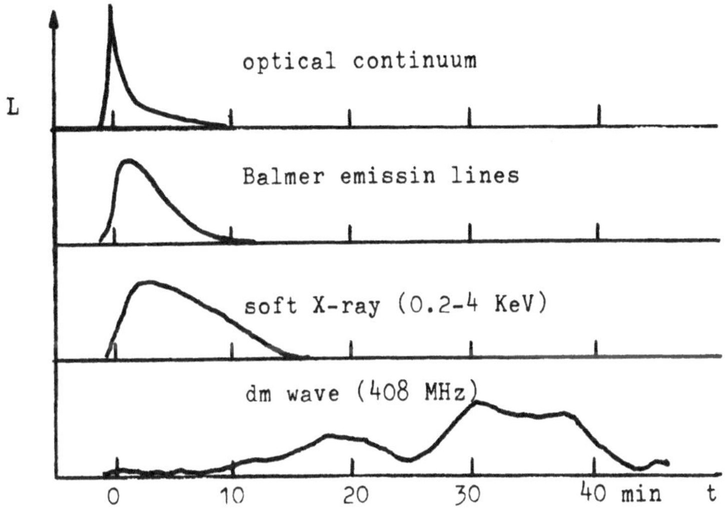


Figure 1. Time profiles (schematic) of 1979 Oct.25 flare of YZ CMi, according to Kahler et al (1982).

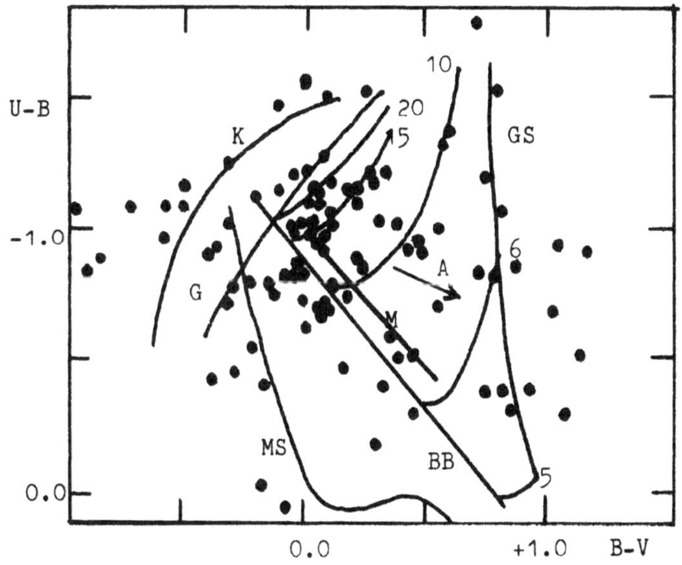


Figure 2. Color domains of light-source models. K and G: Kunkel's (1970) and Gershberg's (1967) nebular models, $2x10^4$K; upper left for cooler models.GS:Grinin and Sobolev's(1977) warmed photosphere model:Curves combining the thin and thick cases are labelled with T($x10^3$K).M:Mullan's(1976) hot-plasma model;$T_0=10^{7.9}-10^{4.2}$K. A:decay;BB:black body, and MS:main sequence. Dots: Observed flare colors.

radio-wave emissions consist of gradual components. It is important to notice that the maxima of the Balmer line emission and of the soft X-ray emission are delayed by one or two minutes after the peak of the optical continuum. This fact indicates that the radiation in the optical continuum of the impulsive component is not a secondary effect subsequent to the soft X-ray phenomena, but rather that the gradual components in Balmer line emission and soft X rays are the aftermath of the impulsive phenomena or that their energy is continuously supplied after the impulsive phase. The gradual component in the optical continuum, however, may be closely related to the gradual component in soft X-rays, as suggested by Kahler et al (1982).

By referring to McIntosh and Donnelly (1972) and Kane and Winckler (1969), we find strong similarities between stellar and solar flares, as far as the time-profiles of the optical continuum, Balmer line emission, and soft X-rays are concerned. In the case of solar flares, hard X-ray, EUV and microwave radio radiation, are detected to show a prominent impulsive component. Donnelly (1971) found that radio bursts below 500MHz are poorly correlated with these impulsive events for the solar case, just as we find for the stellar case (cf.Lovell 1971).

The luminosities for specific photon-energy ranges are summarized in Table 1. The data for the reference flare are adopted from Kahler et al (1982). The luminosity for the dm wave L(dm) was estimated by multiplying the flux at f=410MHz with an arbitrary range of about 400MHz. An upper limit is given for cm wave luminosity which is estimated from the minimum detectable signal of Algonquin dish, 20mjy.The upper part of the table shows the ratios of the maximum luminosities in individual passbands although the maximum occurred at different moments in different channels. The ratios of the luminosities at the moment of the maximum of the optical impulsive component are given in the lower part of the table. Note that passbands are slightly different between the stellar and the solar data, but the minor factors do not affect the following discussions. The solar data concerning HI, EUV, and soft X-ray at impulsive peak are adopted from McIntosh and Donnelly (1972) and Donnelly (1971). The value L (Hα)/L(opt)$\gtrsim$0.04 was estimated for the flare regions which were directly related to whitelight kernels. I could not find the ratio L (Hα)/L(soft X)specifically for solar white-light flares, so I referred to general statistics by Thomas and Teske (1971) which were cited by Kahler et al (1982). Kahler et al adopted a mean value 1.6, but the passband of soft X-rays was 8-12A. The corresponding value appears now as about 0.6 in Table 1, referred to a passband 2-12A. I have adopted also a value about 0.2 which was observed for the brightest flare of class 3 among Thomas and Teske's sample, because this flare may come closest to white-light flares in its magnitude.

Inspecting Table 1, I find that the luminosity profile of the

Table 1. Luminosity levels of flares.
Data are adopted from Kahler et al (1982), Kane and Winckler (1969), and Hudson(1972)

| | YZ CMi | Sun |
|---|---|---|
| | ΔU = 1.9 flare<br>1979 Oct. 25 | white-light flare<br>1967 May 23 |
| peak | | |
| L(soft X) erg/sec | $8\times10^{28}$<br>(0.2-4.0 KeV) | $2\times10^{27}$<br>(1 - 6 KeV) |
| peak ratio | | |
| L(opt)/L(soft X) | ∿ 2<br>(3500-6500 A) | ∿ 1<br>(3500-6500 A) |
| L(dm) /L(soft X) | $\sim 10^{-5}$<br>(200 -600MHz) | $\sim 10^{-7}$<br>(300 -900MHz) |
| L(hardX)/L(softX) | ------ | $10^{-5}$<br>( $\gtrsim$ 20 KeV) |
| L(cm) /L(soft X) | $< 6\times10^{-5}$<br>( 3 - 9 GHz) | $5\times10^{-6}$<br>( 3 - 9 GHz) |
| L(Hα)/L(soft X) | ∿ 0.2 | ∿ 0.2 1)<br>∿ 0.6 2) |
| L(EUV)/L(soft X) | ------ | ∿ 2 3)<br>(10-1030 A) |
| at the impulsive peak | | |
| L(Hα)/L(opt) | $\lesssim$ 0.1 | $\lesssim$ 0.04 4) |
| L(soft X)/L(opt) | ∿ 0.1 | ∿ 0.1 5) |

1) A importance 3 b flare, Thomas and Teske(1971).
2) Transformed from the value cited in Kahler et al.(1982).
3) Statistical value, McIntosh and Donnelly(1972).
4) Directly related with white-light kernels only, McIntosh and Donnelly(1972).
5) Donnelly(1971).

reference flare of YZ CMi resembles closely that of the solar white-light flare. When we assume the same luminosity ratios for the stellar and the solar flare, we expect a hard X-ray luminosity of about $8x10^{23}$erg/sec for the stellar event, that is a flux of about $2x10^{-16}$ erg/cm$^2$ sec. This flux corresponds to about $4x10^{-9}$ counts/cm$^2$ sec in 30keV photons. The expected cm wave luminosity is about $4x10^{23}$ erg/sec, and the corresponding flux is about 1.5 mJy, which is lower by one order of magnitude than the detection threshhold (20 mJy) of the radio telescopes used by Kahler et al (1982) (see also Moffett, Helmken, and Spangler 1978). The expected EUV luminosity is estimated at about $3x10^{29}$ erg/sec for the reference flare of YZ CMi. The luminosity for 2000-3000A may be estimated at L(UV)=1-$5x10^{28}$ erg/sec. Haisch et al (1981) cited a detection limit of about $1x10^{29}$ erg/sec in their IUE observation of Proxima Cen, that is, about $5x10^{29}$ erg/sec in the case of YZ CMi. In both cases the expected UV flux is lower than the IUE detection limit cited by Haisch et al.

We know another example of a type-I stellar flare which was detected simultaneously in soft X-ray and optical light ; Heise et al (1975) reported observation of peak luminosity L(0.2-0.28keV)=$6x10^{28}$ erg/sec for an optical flare of $\Delta$V=2.3 mag of UV Cet. Using the apparent visual magnitude of UV Cet V= 12.3 and the absolute calibration by Oke and Schild (1970), I estimate the V-band luminosity of the flare at $6.6x10^{29}$ erg/sec. When the empirical conversion equations among U,B,V, and total optical energy given by Lacy, Moffett, and Bopp (1976) are applied, the peak luminosity L(opt) of the flare is found to be $2.5x10^{30}$ erg/sec. The peak luminosity in another soft X-ray band L(1-7keV), which compares better with the data in Table 1, remained undetected. The upper limit was given as L(1-7keV)<$1.8x10^{30}$ erg/sec. Fortunately Heise et al. detected a flare of YZ CMi in both soft X-ray bands, but without optical coverage. Assuming that the energy spectrum in the soft X-ray region was similar between these two flares, I estimate the peak luminosity of the UV Cet flare in the 1-7 keV band at L(1-7keV)=$1x10^{30}$ erg/sec. Accordingly the resulting ratio is L(opt)/L(soft X)=2.5, which is compatible with corresponding values in Table 1. There were some disagreements among previous investigators about the optical/X-ray luminosity ratio of stellar flares and also of solar flares (cf.Kahler and Shulman 1972, Tsikoudi and Hudson 1975, Mullan 1976, Haisch et al 1977, Karpen et al 1977, Haisch et al 1978 and Kahn et al 1979). As for stellar flares, the disagreements were mainly caused by different ways of evaluating and defining the respective luminosities. As for solar flares, white-light flares were not always referred to. Another source of the discrepancy seems to be the lack of recognition that the time profiles are different between the optical and X-ray emissions.

There is no a priori reason to suppose that the luminosity ratios between different passbands be similar in different flares. If they turn out to be similar, however, one may assume a unit flare and regard all flares as a sum of many unit flares with more or less similar physical structure.

## 3. SOURCE MODEL FOR OPTICAL LIGHT

The physical nature of the light source has been investigated based upon the observed light curve, the broad-band colours at the peak and their later evolution, and the spectroscopic features (emission lines, Balmer jump). Studies of light curves were rewieved by Gershberg and Chugainov (1968), and the color studies were summarized by Cristaldi and Rodonò (1975). A good summary of spectroscopic observations is given in Schneeberger et al (1979).

Time resolved spectroscopic observations by Kunkel (1970), Moffett and Bopp (1971), Bopp and Moffett (1973), and recent multichannel photometry by Mochnacki and Zirin (1980) made it clear that the contribution of emission lines to the total optical luminosity is only about 5% at the impulsive peak and increases up to 40% in the late phase. No significant polarization has been reported for optical flare light (cf.Karpen et al 1977). Since the intense visible continuum distinguishes stellar flares from solar flares, its interpretation is of primary importance. In the last decade we were confronted with three categories of working hypotheses about the light source: fast-electron model, hot-plasma model, and nebular model.

The fast-electron model was proposed by Gurzadyan (1965, 1966,1972) to account for the observed continuum and was found to be successful in reproducing the broad-band colors by the inverse Compton effect of relativistic electrons (1-10MeV). This model, however requires too much energy input (about $10^{51}$ erg), and is due for revision since we now know that UV Cet-type stars have chromosphere-corona envelopes which are more rich in material than that of the Sun (Haisch et al 1980, Vaiana et al 1981, Giampapa, Worden and Linsky 1982), probably related to magnetic fields (Gray and Linsky 1981, Topka and Marsh 1982, Robinson, Worden and Harvey 1980). The large number of relativistic electrons in the Gurzadyan model would strongly interact with coronal plasma and the stellar surface (cf.Arutyunyan 1979). The resulting emitted radiation must dominate over the inverse Compton effect. Consequently I am inclined to retain only the essential feature of this model, namely, the non-thermal high-energy particles being a possible primary energy source.

The hot-plasma model was adopted by Andrews (1965) to explain the observed flat continuum (cf. Kodaira and Ichimura 1975) and the decay time-scale by bremsstrahlung of hot gas ($T \gtrsim 10^6$ K). As a single-layer model led to an increase of optical continuum with plasma cooling, the model was later refined by Mullan (1976, 1977) and Kodaira (1977), who included stratification effects and their time variation. In the refined model, the heat energy of hot plasma is partially radiated off as thermal bremsstrahlung but also conducted towards the stellar surface. The conducted energy flux is then efficiently radiated from denser and cooler layers. The flare soft X-rays are emitted from the hotter upper part ($T \gtrsim 10^6$K) while the optical light is predominantly from the cooler lower part ($10^5 \gtrsim 10^4$K). Mullan adopted the temperature structures of equilibrium models by Shmelva and Syrovatskii (1973) and showed that the model colors correspond well to the observed ones. Mullan and Tarter (1977) pointed out that a furter improvement of this model may be possible when the back-heating effect of the stellar surface by flare X-rays is taken into account. Cram (1982) showed that even at the quiescent phase, the back-heating effect of the coronal X-rays strongly affects the energy balance in the chromosphere of dMe stars. The hot-plasma model thus developed may explain the main features of flare soft X-rays, as will be discussed below, but leaves the details of the physical conditions of the dense lower part ($T \lesssim 10^4$ K) unspecified. This model is valid only for the gradual components and not for the rise of the impulsive optical light.

The nebular model was developed by Gershberg (1964, 1967a, 1967b) and Kunkel (1970) to represent the observed Balmer emission lines and Balmer jump. Recombination spectra are assumed to be emitted from a homogenous layer of electron density typical for the chromosphere, $N_e=10^{13}-10^{14}$ cm$^{-3}$. They found a temperature range $T_e=2-3\times10^4$K appropriate in order to reproduce the observed Balmer decrement ($H\gamma/H\beta=1.1-1.5$) and the Balmer jump ($I_-/I_+=4-5$), but failed to reproduce the observed broad-band colors.

The model colors are generally too blue and become bluer with decreasing temperature or density, contrary to the observations (see Figure 2). Thereupon Gershberg (1967b) suggested bremsstrahlung as an additional component, and Kunkel (1970) suspected admixture of slightly heated (about 100K) photospheric radiation.

Grinin and Sobolev (1977) proposed to shift the location of the "nebula" to a lower, cooler and denser layer ($T=0.5-1\times10^4$ K, $N_e=10^{15}-10^{17}$ cm$^{-3}$) than Gershberg and Kunkel had assumed. As the reason for this, they pointed out that the hot nebula of $T=2-3\times10^4$ K would give a rise not only to HI and HeI lines but also to strong lines of HeII and of other highly ionized atoms, which were not detected in stellar flare spectra (cf. Bopp and Moffett 1973, Schneeberger et al 1979). The high density they adopted corresponds to that of the bottom of the chromosphere or of the upper photosphere (cf. Giampapa, Worden, and Linsky 1982). In this model, when the temperature is low, the dominant $H^-$ opacity leads to a relatively flat continuum from the optically thin layer, and, when the temperature increases, sharply increasing HI opacity leads to a black-body-like radiation from an optically thick layer (see Figure 2). It is notable that Machado and Rust (1974) interpreted the continuum of white-light waves (but not the white-light kernels themselves) caused by a solar white-light flare as a free-bound radiation from a layer of 8500K, with an optical thickness at 5000A of 0.1. Grinin and Sobolev (1977) found that their "warmed photosphere" model of T= 8500K emits optical light as observed in stellar flares. Their picture may effectively be the same as that presented by Cram and Woods (1980), who increased the pressure at the bottom of the transition zone by 1-2 orders of magnitude and laid the photosphere-chromospheric temperature minimum deeper for the flaring relative to the quiescent region.

By assuming a black-body optical continuum, Mochnacki and Zirin (1980) found a peak color temperature of T=8500-9000K from their multichannel measurement of a flare of YZ CMi which had apparently a similar magnitude as the reference flare in Table 1. They obtained a projected area of about $10^{19}$ $cm^2$ for the blackbody emitter. Kahler et al showed that the reference flare also had a projected area of about $10^{19}$ $cm^2$ when a black body of 8500K was assumed as optical light source. In Mochnacki and Zirin's observation, the color temperature increased before the impulsive peak and then decreased down to 5000-6000K, while the projected area reached a maximum a few minutes later than the impulsive peak.

The spectroscopic observations by Kunkel (1970), Moffett and Bopp (1971), and Bopp and Moffett (1973) revealed that HeI emission lines appear sporadically or weak around the impulsive phase, and that CaII H and K lines decay slower than Balmer lines. These facts suggest that the temperature of the light-emitting region reaches the maximum around the impulsive phase and decreases gradually. Their observations as well as those by Gershberg and Chugainov (1967) also revealed that in some flares the half widths of emission lines increased up to about several tens km/sec during the impulsive phase of less than a few minutes and then rapidly returned to normal. In another case, they observed so-called "red asymmetry" of line wings with a corresponding velocity of 600-1100km/sec during the impulsive phase, which had been known for solar flares (cf. Svestka 1966). Although Schneeberger et al (1979, 1980) detected no significant line broadening for a $\Delta U$=0.5 flare of AD Leo and for $\Delta U$=1.5 flare of YZ CMi, it seems to me that dynamical phenomena are involved during the impulsive phase, which, however, may not be manifested clearly under certain aspect conditions. The motion has probably directivity. McIntosh and Donnelly (1972) reported that eruptive Hα structure of 600-1000km/sec are often observed for solar white-light flares seen near the limb.

A valuable study was done by Katsova et al (1980) and Livshits et al (1981) concerning the dynamical response of chromospheres to an impulsive energy injection at the top. Their one-dimensional study shows that the shock waves propagate both upwards and downwards. The post-shock layer of the downwards wave becomes in some cases optically thick, therefore, it may produce the continuum discussed above. (A one-dimensional dynamical heat propagation into the corona was discussed by Syrovatskii and Shmelva 1972). They assumed the duration of a unit injection of energy of flux $F=10^{12}$ erg/cm$^2$ sec to be about 10 sec, and that the impulsive phase of a flare consists of a group of such pulses. As a physical heat-source model for this, they assumed, as Hudson (1972) and Machado and Rust (1974) did for solar white-light flares, the "thick-target mechanism" for non-thermal high-energy particles.

## 4. SOURCE MODEL FOR SOFT X-RAYS

The termal nature of stellar flare radiation was clearly confirmed for the first time by Kahn et al (1979). Their data points (0.2-10keV) from the HEAO-1 A-2 experiment observation of a flare of AT Mic on 1977 Oct.25 fit well to the energy distribution from a thin thermal plasma of $T=3x10^7$ K. They believe they have identified the 6.7 keV Kα line of FeXXV which supports the hot thermal nature of the source. A pulse-height analysis of the soft X-ray data of a flare on Proxima Cen by HEAO-2's IPC experiment by Haisch et al (1980) yielded a maximum temperature of $T=1.7x10^7$ K during the rising phase of the soft X-ray flux, which then decreased to $1.2x10^7$ K at the beginning of the decay phase. The above authors derived emission measures EM= $1.4x10^{54}$ cm$^{-3}$ for AT Mic, $7.5x10^{50}$ cm$^{-3}$ for the rising phase of a Proxima Cen flare, and $1.2x10^{51}$ cm$^{-3}$ for the decay phase of the Proxima Cen flare, by referring to the radiative power calculated by Raymond, Cox, and Smith (1976). Based upon the thin-thermal-plasma interpretation, Kahler et al (1982) showed that the reference flare of YZ CMi reached a maximum temperature of $T=2x10^7$ K with an emission measure of $EM=4x10^{51}$ cm$^{-3}$ around the flux maximum of soft X-rays. Both temperature and emission measure gradually decreased in the decay phase, parallel with the flux. The increase of emission measure during the rising phase of soft X-ray flux strongly suggests that either the plasma density or the volume (or both) rapidly increased during the optical impulsive phase. As a comparison Kahn et al cited an extremely strong solar flare of 1972 Aug.4 (cf. Colgate 1978) which had a peak luminosity L(soft X)=$4x10^{27}$ erg/sec with $EM=6x10^{49}$ cm$^{-3}$, whose magnitude is thus comparable to that of our reference solar white-light flare in Table 1. Although the plasma temperatures are comparable in the solar and stellar flares, the maximum emission measure seems to be larger in strong stellar flares than in the strongest solar flares by a factor of up to $10^2$-$10^4$. By applying the principles of the hot-plasma model studied by Kodaira (1977) and Mullan (1977) to the decay phase of the reference flare, we find that the radiative cooling rate is comparable to the conductive cooling rate of the hot plasma. The observed decay time scale $t=10^3$ sec then suggests an electron density of $N_e=10^{11}$ $cm^{-3}$. The volume turns out to be about $10^{29}$ $cm^3$.

The lenght scale may be estimated from the expansion time scale of hot plasma, which may be regarded equal to the delay time of the soft X-ray maximum relative to the optical impulsive peak, t=100sec. The resulting length scale is $L= 10^{10}$ cm, which is comparable to the stellar radius, $R=0.5-2x10^{10}$ cm. Accordingly the dimension of the cross-section is estimated at $A=10^{19}$ $cm^2$. This corresponds to 0.5-3% of the stellar disk and is comparable to the area for impulsive white-light emission estimated by Kahler et al (1982), and by Mochnacki and Zirin (1980) (see section 3). This area and length estimate is also consistent with the conductive cooling rate required. The thermal energy contained in the hot plasma at the beginning of the decay phase was found to be $E= 4x10^{31}$ erg for this model of the reference flare. The total emission of the gradual and impulsive optical light are comparable to this. Accordingly the total flare energy amounts to more than $10^{32}$ erg.

The expansion phase of the soft X-ray plasma might be explained by the particle injection model of Livshits et al (1981). One of their models for an injection flux $F=3x10^{12}$ $erg/cm^2$ sec and a pulse duration t=10sec, that is, a total energy of $E= 3x10^{32}$ erg for an area of $A= 10^{19}$ $cm^2$, shows an expansion of the upper chromospheric material with a velocity of $10^3$ km/sec after it is heated to $T \gtrsim 10^7$ K. When pressure balance is assumed between the soft X-ray plasma and the magnetic field which probably confines the plasma, one finds a field strength of $B= 10^2$ gauss.

## 5. SOURCE MODEL FOR RADIO WAVE EMISSION

A review of radio wave observations of stellar flares in the early period was given by Lovell (1971). He reported observations of radio flares in the 200-400MHz range which were temporally correlated with optical flares, but found also numerous cases in which only radio or optical emission was detected. There seems to be no clear correlation between the observed luminosities of the two emissions. When they are temporally correlated, the time profiles of the dm-m wave emission are generally much broader than those of the optical impulsive peak, sometimes even broader than that of HI lines of soft X-ray gradual component. Radio emission rises much slower than the latter, but in some cases isolated sharp peaks stand out above the broad components. The radio maximum mostly appears a few to several minutes later than the optical peak but the rise often begins one to a few minutes before the start of the sharp optical rise. In some small flares, the radio maximum appears even before the optical peak. When a flare was observed in two pass bands, the flux in the low-frequency band (say 200 MHz) is found to be larger than that in the high-frequency band (say 400 MHz) by a half order of magnitude or more, indicating a rather steep spectrum.

This diverse behavior of dm-m wave flares was further confirmed by Lovell et al (1974), Spangler, Shawman, and Rankin (1974), Spangler and Moffett (1976), and by recent investigations coordinated with X-ray observations (Karpen et al 1977, Kahler et al 1982). Spangler and Moffett (1976) underlined that there are no systematic correlations in the morphology of time profiles between radio and optical flares. The above facts suggest that the radiation source of the dm-m wave is not directly related to those of optical and soft X-ray emissions. Since the electron density for the plasma frequency f=400MHz is about $N_e=2x10^9\ cm^{-3}$, the emitting region of dm-m wave may lie outside the dense soft X-ray plasma ($N_e=10^{11}\ cm^{-3}$).

The reference flare of YZ CMi follows the general trend of stellar radio flares. The delay time of this flare, however, was much longer than the values cited above as average; 20-60 minutes. Such a large delay was observed for an extremely strong flare of YZ CMi by Lovell (1971) on 1969 Jul. 19. The dm-m wave enhancement of this $\Delta V=1.8$ flare continued over 4 hours after the optical impulsive peak and the first broad maximum appeared 30-60 minutes after the flare start. Judging from the optical peak luminosity $L(opt)=4x10^{30}$ erg/sec, this flare was one and half orders of magnitude stronger than the reference flare. Its dm-m wave luminosity is estimated at $L(200\text{-}600MHz)=10^{26}$ erg/sec. Thus the ratio $L(dm)/L(opt)=2.5x10^{-5}$ turns out to be comparable with the value $L(dm)/L(opt)=5x10^{-6}$ in the reference flare. If the spectrum falls off according to $f^{-n}$, the index n is about -1.5 for the 200-400MHz region. By extrapolating this, one obtains a flux of about 8mJy at 1428MHz for the reference flare, which lies far below the detection limit (100mJy) of the observation by Kahler et al (1982).

Detections of cm wave flares were reported by Karpen et al (1977), Gibson and Fisher (1980), and Haisch et al (1981). Most of the flux levels detected with single-dish telescopes were close to the detection limit and the observed flare durations were less than one integration time (a few minutes), accordingly, these data must be applied with caution. Haisch et al (1981) found three cases correlated with optical flares of Proxima Cen. One of them occurred simultaneously with a $\Delta U=4.5$ optical flare, the others with a delay of 2-5 minutes. When the simultaneous one is scaled to the reference flare according to the U magnitude, one would expect a flux level of only 0.03 mJy. The event reported on YZ CMi by Karpen et al (1977) had no optical coverage. It emitted 1.2 Jy at 160 MHz and 78 mJy at 5 GHz. The latter level is anomalously high compared with levels less than about 10 mJy observed by Haisch et al (1981) in Proxima Cen (see also Moffett, Helmken, and Spangler 1978). Gibson and Fisher (1980) reported that their 4.9 GHz VLA observation of YZ CMi indicated a significantly lower flux level and a much longer duration than single dish observations had suggested. The detected cm wave emission, therefore, might be the high-frequency tail of the gradual dm-m wave component.

It should be noted that quiescent coronal emissions in cm waves were also detected using VLA by Gray and Linsky (1981) and by Topka and Marsh (1982). The former detected 1.6 mJy emission from UV Cet, and the latter 0.7 and 0.4 mJy emissions from the binary components of EQ Peg. They interpreted these by gyroresonance of thermal electrons in their coronae.

The most interesting feature of the radio emission related to stellar flares is the high degree of circular polarization (50-100%) observed by Spangler, Rankin and Shawman (1974) at 430 MHz and by Gibson and Fisher (1980) at 4.9GHz. This suggests gyroresonance or gyrosynchrotron radiation in magnetoactive plasma as an emission mechanism. The gyrofrequency, which should be less than the observation frequency, corresponds to a magnetic field less than about B=50 gauss. The luminosity L(dm) of the reference flare, however, is too high to be interpreted as of thermal origin because the brightness temperature becomes higher than $T= 10^{13}$ K even when an area comparable to the whole stellar disk is assumed. This situation, the long delay time, and the gradual nature of the dm-m emission of the reference flare suggests that any sort of non-thermal emission process was induced by a series of large-scale structure changes of coronal magnetic field which followed the initial localized event.

## 6. DISCUSSION

A framework of a stellar flare model might be deduced from the considerations in the preceeding sections. A flare event presumably starts with a structural change of magnetic field, which first begins locally and slowly, supplying energy to particles and plasma, and appears as a gradual rise of chromospheric line emission and soft X-ray emission, sometimes also of dm wave emission. At a certain phase of the structural change a series of intense impulsive energy releases occur within limited regions, which leads to a sudden heat-up of the chromosphere. This produces shock waves propagating downwards and rapid upwards expansion of the heated plasma. The post-shock dense layers emit impulsive optical light, while the expanded thin hot plasma emits soft X-rays whose luminosity increases with increasing matter supply and decays gradually due to radiative and conductive cooling. The downwards heat conduction and the back-heating of the soft X-rays supplies the energy for the gradual component of the optical emission. Weak, gradual energy release may continue for a substantially longer period than the impulsive phase, contributing to stimulation of the gradual soft X-ray and optical emissions. Partially as an aftereffect of the intense energy release,but more likely as a continuation of the general structural change of the magnetic field, non-thermal energy release occurs in the magneto-active corona and gives rise to the gradual dm-m wave emission.

So far as this rough scenario of a flare event is concerned, we see a broad similarity between the stellar flare and the solar white-light flare and feel almost justified in applying the same model to both. There is, however an outstanding difference in magnitude between them. Although the estimated area $A = 10^{19}$ cm$^2$ is comparable to the H$\alpha$ area of a solar flare of importance 3 (cf. Svestka 1976), the energy of the reference flare which was emitted in the observed radiations ($E \gtrsim 10^{32}$ erg) is two orders of magnitude larger than the strongest solar flares, and comparable to the total energy of the most powerful solar flare, most of which, however, is deposited in the shock wave. Flares of the same magnitude as the reference flare occur on YZ CMi every 10 hours on average (cf.Lacy, Moffett, and Evans 1976) and the mean luminosity radiated by flares of this class amounts up to $3\times10^{27}$ erg/sec, which is about $3\times10^{-6}$ of the stellar luminosity. In the solar case, flares of importance class $\gtrsim 3$ occur 20-30 times in a year around the sunspot maximum, thus the mean luminosity of these flares is about $10^{26}$ erg/sec, $2.5\times10^{-8}$ of the solar luminosity. A stellar flare theory must first of all account for this efficient energy accumulation and release in the form of flare radiation.

Since the solar white-light kernels occupy only about 1/100-1/30 of the total flaring area (cf.McIntosh and Donnelly 1972), one may presume either that the total flaring area in the solar sense was about $10^{21}$ cm$^2$ for the reference flare, that is, almost the whole hemisphere of YZ CMi, or that the flaring region of YZ CMi was covered almost exclusively by white-light kernels. In the former case, the physical structure of a stellar flare may be replaced by that of strongest solar flares, except for collective effects which certainly become prominent if the whole hemisphere should be covered by the strongest (solar) flares. The question of how the vast area can simultaneously become unstable, remains to be solved.

In the latter case, we may suspect differences in the physical structure of flares. When we fail to detect the impulsive microwave component from stellar flares in contrast to the solar case, it would indicate that the primary energy release generally occurs in a deep layer where the plasma frequency exceeds microwave frequencies; the density of the layer must be as high as $N_e = 10^4$ $cm^{-3}$ corresponding to the density of the transition zone to upper chromosphere. If this is the case, one may further assume that the efficiency in the thermalization process of any non-thermal primary energy input and in the radiation process is much higher under this conditions, what would conform to the empirical picture.

Assuming that the energy of a homogenous magnetic field is converted into heat, the minimum magnetic field required for the reference flare is $B=10^{4.5}/(LA)^{(0.5)}$ gauss, L and A being the height (in $10^3$ Km) and the area (in $10^{19}$ $cm^2$) of the volume in which magnetic energy is to be consumed. The area A can vary between 1 and 100. If L=1 for the chromospheric scale, B=3000-300 gauss, while, if $L=10^2$ for the coronal scale, B=300-30 gauss, which appears not unreasonable. This concept leads to a picture of a stellar flare in which a large quantity of magnetic energy ($E \gtrsim 10^{32}$ erg) is converted into heat in a dense layer ($N_e \gtrsim 10^{11}$ $cm^{-3}$) of an area of $A=10^{19}$ $cm^2$ The question remains how to convert it into heat!

REFERENCES


Andrews,A.D.:1965, Irish Astron.J.,7,p.20.
Artyunyan,G.A.:1979, Astrofizika,15,p.431.
Bopp,B.W.,and Moffett,T.J.:1973, Astrophys.J.,185,p.239.
Colgate,S.A.:1978, Astrophys.J.,221,p.1068.
Cram,L.E.:1982, Astrophys.J.,253,p.768.
Cram,L.,and Woods,D.T.:1980, Bul.Amer.Astron.Soc.,12,p.914.
Cristaldi,S.,and Rodono,M.:1975, Variable Stars and Stellar Evolution IAU Symp.No.67,ed. V.E.Sherwood and L.Plaut,(Reidel),p.75.
Dezso,L.,Gesztelyi,L., Kondas,L., Kovas,A.,and Rostas,S.:1980, Solar Phys.,67,p.317.
Donnelly,R.F.:1971, Solar Phys.,20,p.188.
Gershberg,R.E.:1964,Izv.Krim.astrofiz.Obs.,32,p.133.
Gershberg,R.E.:1967a, Izv.Krim.astrofiz.Obs.,36,p.216.
Gershberg,R.E.:1967b, Astrofizika,3,p.127.
Gershberg,R.E.,and Chugainov,P.F.:1967, Astr.Zh.,44,p.260.
Gershberg,R.E.,and Chugainov,P.F.:1968, Izv.Krym.astrofiz.Obs.,40,p.7
Giampapa,M.S., Worden,S.P., Linsky,J.L.:1982, Astrophys.J., in press.
Gibson,D.M.,and Fisher,P.L.:1980, Bul.Amer.Astron.Soc.,12,p.500.
Gray,D.E.,and Linsky,J.L.:1981, Astrophys.J.,250,p.284.
Grinin,V.P.,and Sobolev,V.V.:1977, Astrofizika,13,p.587.
Gurzadyan,G.A.:1965, Astrofizika,1,p.313.
Gurzadyan,G.A.:1966, Astrofizika,2,p.217.
Gurzadyan,G.A.:1972, Astron.Astrophys.,20,p.145.
Haisch,B.M., Linsky,J.L., Harnden,F.R. Jr., Rosner,R., Seward,F.D., and Vaiana,G.S.:1980, Astrophys.J.,242,L99.
Haisch,B.M., Linsky,J.L., Lampton,M.,Paresce,F., Margon,B., Stern,R.: 1977, Astrophys.J.,213,L119.
Haisch,B.M.,Linsky,J.L.,Slee,O.B.,Hearn,D.R.,Walker,A.R.,Rydgren,A.E. and Nicolson,G.D.:1978, Astrophys.J.,225,L35.
Haisch,B.M.,Linsky,J.L.,Slee,O.B.,Siegman,B.C.,Nikoloff,I.,Candy,M., Harwood,D.,Verveer,A.,Quinn,P.J.,Wilson,I.,Page,A.A., Higson,P.,and Seward,F.D.:1981, Astrophys.J.,245,p.1009.
Heise,J.,Brinkman,A.C.,Schrijver,J.,Mewe,R.,Gronenschild,E., den Boggende,A.,and Grindlay,J.:1975,Astrophys.J.,202,L73.
Hudson,H.S.:1972, Solar Phys.,24,p.414.
Kahler,s.,Golub,L.,Harden,F.R.Jr.,Liller,W.,Seward,F.,Vaiana,G., Lovell,B.,Davis,R.J.,Spencer,R.E.,Whitehouse,D.R.,Feldman,P.A Viner,M.R.,Leslie,B.,Kahn,S.M.,Mason,K.O.,Davis,M.M. Crannell,C.J.,Hobbs,R.W.,Schneeberger,T.J.,Worden,S.P., Schommer,R.A.,Vogt,S.S.,Pettersen,B.R.,Coleman,G.D.,Karpen,J. T.,Giampapa,M.S.,Hege,E.K.,Pazzani,V.,Rodono,M.,Romeo,G., and Chugainov,P.F.:1982, Astrophys.J.,252,p.239.
Kahler,S.,and Shulman,S.:1972, Nature Phys.Sci.,237,p.101.
Kahn,S.M.,Linsky,J.L.,Mason,K.O.,Haisch,B.M.,Bowyer,C.B.,White,N.E., and Pravdo,S.H.:1979, Astrophys.J.,234,L107.
Kane,S.R.,and Winckler,J.R.:1969, Solar Phys.,6,p.304.
Karpen,J.,Crannell,C.J.,Hobbs,R.W.,Maran,S.P.,Moffett,T.J.,Bardas,D., Clark,G.W.,Hearn,D.R.,Li,F.K.,Markert,T.H.,McClintock,J.E., Primini,F.A.,Richardson,J.A.,Cristaldi,S.,Rodono,M.,

Galasso,D.A.,Magun,A.,Nelson,G.J.,Slee,O.B.,Chugainov,P.F., Efimov,Yu.S.,Shakhovskoy,N.M.,Viner,M.R.,Venugopal,V.R., Spangler,S.R.,Kundu,M.R.,and Evans,D.S.:1977, Astrophys.J., 216,p.479.
Katsova,M.M.,Kosovichev,A.G.,and Livshits,M.A.:1980,Pis'ma Astron. Zh.,6,p.498. = Soviet Astron.Lett.,6.
Kodaira,K.:1977, Astron.Astrophys.,61,p.625.
Kodaira,K.,and Ichimura,K.:1976, Publ.Astron.Soc.Japan,28,p.665.
Kunkel,W.E.:1970, Astrophys.J.,161,p.503.
Lacy,C.H.,Moffett,Th.J.,and Evans,D.S.:1976, Astrophys.J.Suppl.,30,p85.
Livshits,M.A.,Badalyan,O.G.,Kosovichev,A.G.,and Katsova,M.M.:1981, Solar Phys.,73,p.269.
Lovell,B.Sir:1971, Q.J.Roy.Astron.Soc.,12,p.98.
Lovell,B.Sir,Mavridis,L.N.,and Contadakis,M.E.:1974,Nature,250,p.124.
Machado,M.E.,and Rust,D.M.:1974, Solar Phys.,38,p.49.
McIntosh,P.S.,and Donnelly,R.F.:1972, Solar Phys.,23,p.444.
Mochnacki,S.W.,and Zirin,H.:1980, Astrophys.J.,239,L29.
Moffett,T.J.,and Bopp,B.W.:1971, Astrophys.J.,168,L117.
Moffett,T.J.,Helmken,H.F.,and Spangler,S.R.:1978, Publ.Astron.Soc. Pacific,90,p.93.
Mullan,D.J.:1976a, Astrophys.J.,207,p.289.
Mullan,D.J.:1976b, Astrophys.J.,210,p.702.
Mullan,D.J.:1977, Astrophys.J.,212,p.171.
Mullan,D.J.,and Tarter,C.B.:1977, Astrophys.J.,212,p.179.
Oke,J.B.,and Schild,R.E.:1970, Astrophys.J.,161,p.1015.
Oskanian,V.:1969, Non-Periodic Phenomena in Variable Stars,IAU Colloq. No.4,(Reidel),p.131.
Raymond,J.C.,Cox,D.P.,and Smith,B.W.:1976, Astrophys.J.,204,p.290.
Robinson,R.D.,Worden,S.P.,and Harvey,J.W.:1980, Astrophys.J.,236,L155.
Schneeberger,T.J.,Linsky,J.L.,McClintock,W.,and Worden,S.P.:1979, Astrophys.J.,231,p.148.
Schneeberger,T.J.,Linsky,J.L.,Worden,S.P.:1979, Astrophys.J.,231,p.148.
Schneeberger,T.J.,Worden,S.P.,DeLuca,E.E.,Giampapa,M.S.,and Cram,L.E.: preprint, Center for Astrophys. No.1517=Bul.Amer.Astron. Soc.,11,628(1979)
Shmelva,O.P.,and Syrovatskii,S.I.:1973, Solar Phys.,33,p.341.
Spangler,S.R.,and Moffett,Th.J.:1976, Astrophys.J.,203,p.497.
Spangler,S.R.,Shawman,S.D.,and Rankin,J.M.:1974, Astrophys.J.,190,L129.
Svestka,Z.:1966, Space Sci.Rev.,5,p.388.
Svestka,Z.:1976, Solar Physics,(Reidel)
Syrovatskii, S.I.,and Shmelva,O.P.:1972, Astron.Zh.,49,p.334.
Thomas,R.J.,and Teske,R.G.:1971, Solar Phys.,16,p.431.
Topka,T.,and Marsh,K.A.:1982, Astrophys.J.,254,p.641.
Tsikoudi,V.,and Hudson,H.S.:1975, Astron.Astrophys.,44,p.273.
Vaiana,G.S.,Cassinelli,J.P.,Fabbiano,G.,Giacconi,R.,Golub,L., Gorenstein,P.,Haisch,B.M.,Harnden,F.R.,Jr.,Johnson,H.M., Linsky,J.L.,Maxson,C.W.,Mewe,R.,Rosner,R.,Seawrd,F., Topka,K.,and Zwaan,C.:1981, Astrophys.J.,244,p.163.

DISCUSSION

Gershberg: It seems to me that if you wish to constrain theory you should use not the mean flare energy but rather the extreme case. There are known at least three optical flares which are more energetic than the mean by two orders of magnitudes viz. the 1969 YZ CMi, your own flare (on EV Lac) and the flare observed some months ago in Soviet Middle Asia. In all these cases the optical energy only exceed $10^{34}$ ergs. This appears to me to be a real constraint for the theories.

Kodaira: My answer would be that since such cases are rare they do not contribute significantly to the average energy output.

Priest: Can I just give a "back-of-the postage-stamp" type of estimate in how you can ... (part of the recording lost) ... the energy should scale as $B^2$ times $l^3$. The timescale should scale as the length-scale over the Alfvèn speed squared. So if you eliminate l between these two there results an expression for W. Now assume that a stellar flare is $10^3$ times larger in energy than a solar flare. Assume the timescale is a factor of 10 shorter and that the density is a factor of 10 larger. Put these into this expression and you find that the magnetic field has to be a factor of 30 times larger than in the Sun. The magnetic field in the solar case is between 100 and 500 G; then in the stellar case we expect it to be 3 - 10kG, assuming of course that we are dealing with the same mechanism.

Weiss: This would be pushing things. One can image sunspot-strength fields extending over large regions but not fields of 10kG.

# DYNAMO THEORY IN THE SUN AND STARS

G. Belvedere
Istituto di Astronomia, Università di Catania
I-95125 Catania, Italy

ABSTRACT

Complementarity of solar and stellar observations should increase our knowledge of the basic facts and mechanisms of activity. The great development of stellar activity observations has enhanced theoretical study, most of which is in the framework of dynamo theory.

Even if dynamo theory seems plausible and successful in capturing the essential processes, several uncertainties and intrinsic limits do still exist and are discussed here together with alternative or complementary suggestions.

It is stressed the importance of magnetoconvection and flux tube studies to improve our understanding of both large scale and small scale interaction of rotation, turbulent convection and magnetic field.

Finally, recent models of stellar activity are critically reviewed. It is pointed out that the confront with the new stellar data should extend our comprehension of the dynamo operation modes, which probably depend on stellar structure, rotation and age.

## 1. THE SOLAR-STELLAR CONNECTION

In the last decade the concept of a unified sight of solar and stellar activity has revealed worthwhile under many aspects gaining the increasing favour of observers and theoreticians. The term solar-stellar connection has recently been introduced to indicate the complementarity of solar and stellar observations (see e.g.Hartmann 1981, Noyes 1981, Worden 1981) to increase our knowledge and understanding of the basic facts and mechanisms of activity. While solar observations offer a detailed study of the activity phenomena in a relatively close astrophysical laboratory and provide a guide to explore stellar activity on the basis of what is learned from the Sun, it is also clear that the complex of solar activity phenomena is to be regarded as " a limiting regime, for a star of solar parameters and its present state of evolution, of features which must be seen in a broader context "



*P. B. Byrne and M. Rodonò (eds.), Activity in Red-Dwarf Stars, 579–599.*

(Mihalas 1981). Thus, for instance, even if dynamo theory has revealed promising for the Sun and it seems valuable its extension to the more general stellar case, nevertheless possible differences in the dynamo operating modes are expected as a consequence of different stellar properties (structure, angular velocity of rotation $\omega$, depth of the convection zone (c.z.)) and evolutionary age. So that, observation of activity in a large sample of stars may show all the various aspects of stellar activity on a multiplicity of time scales and physical situations leading back to a better understanding of solar activity itself (see, for example, the Maunder minima).

In the most recent years the data collected by IUE and EINSTEIN have widened our knowledge of chromospheric and coronal activity and, in particular, X-ray observations have pointed out the basic relevance of magnetic fields in sustaining stellar coronae (Vaiana 1980, 1981; Rosner 1980; Hartmann 1981; Linsky 1981). Nevertheless, the more classical CaII emission flux observations have still confirmed to be a powerful method of investigation (Wilson 1978; Vaughan and Preston 1980; Vaughan 1980; Vaughan et al 1981; Skumanich and Eddy 1981) and indirect detection of activity regions from photometric variations in RS CVn and BY Dra type stars(+) seems to be more than a promising tool to give further insight on surface phenomena (Hall 1981, Rodonò 1981, 1982). Flares observed in lower main sequence stars, red dwarfs and pre-main sequence objects are of extreme importance to our understanding of the energetics of such violent magnetic field instabilities (Gershberg 1978). Magnetic field measurements are now basically improved with the new Robinson's method that allows to measure field intensities (larger than a kilogauss) and filling factors by comparison of magnetically sensitive and insensitive lines (Robinson 1980). The great development of stellar activity observations has enhanced the theoretical effort to predict, interpret and reproduce the observed features, giving rise to several models, most of which are in the framework of dynamo theory. Dynamo theory must therefore confront with the new observational data, which will surely stimulate improvement of the theoretical background and refinement of the methods of analysis, leading to more general and realistic studies.

In the following we review first the up to date status of dynamo theory together with alternative or complementary suggestions, then we discuss recent stellar activity models.

(+) In the Sun, variations of the solar constant on short and cyclic timescales (Foukal 1980, 1982) should give insight on the convection-magnetic field interaction.

## 2. CRITICAL REVIEW OF DYNAMO THEORY

Dynamo theory attempts to explain the generation and evolution of cosmic magnetic fields in terms of induction effects in conducting fluid masses and is, in the author's opinion, a plausible and somewhat successful theoretical framework to understand solar and stellar activity even if several uncertainties and intrinsic limits do still exist.

Since the famous anti-dynamo theorem of Cowling (1934) and the pioneristic work of Parker (1955) on the dynamo mechanism in the Sun, many efforts have been done to develop an internally consistent theory of Mean Field Electrodynamics (Steenbeck, Krause, Rädler 1966) in connection with the basic ideas of dynamo action in rotating convective bodies (see for the development of the main concepts and an exhaustive and general formulation: Parker 1979, Moffatt 1978, Krause and Rädler 1980).

Dynamos can be separated into two classes: kinematic (linear) dynamos in which the velocity field is assigned independently, without taking into account the feedback of the magnetic field on the motion; hydromagnetic (non-linear) dynamos in which the back-reaction of the magnetic field through the Lorentz force is considered and the whole system of the magnetohydrodynamic equations is solved simultaneously, assuring the internal consistency. The reliability of the kinematic approximation depends on the (magnetic energy density)/(kinetic energy density) ratio. If this ratio is small compared to the unity the kinematic approximation is reasonable, otherwise the hydrodynamic approach should be dealt with.

### 2.1. The α-ω Dynamo in the Mean Field Electrodynamics

In its kinematic formulation, dynamo problem reduces to an eigenvalue problem for the magnetic field $\underline{B}$, governed by the induction equation:

$$\partial \underline{B}/\partial t = \mathrm{curl}\ (\underline{u} \times \underline{B}) - \mathrm{curl}\ (\eta\ \mathrm{curl}\ \underline{B}) \qquad (1)$$

where $\underline{u}$ is the velocity field and $\eta=1/\mu\sigma$, with $\mu$= magnetic permeability and $\sigma$= electric conductivity, is the ohmic diffusivity. For a given $\underline{u}$, solutions of the form $B \sim \exp\,(a+ib)t$ are searched to determine the growth rate a and the oscillatory frequency b of the eigenmodes, depending on some physical condition for the dynamo maintenance of the field.

In the framework of the Mean Field Electrodynamics (MFE), the vector fields $\underline{u}$ and $\underline{B}$ are expressed as sums of mean (large scale-slowly varying in time) parts and fluctuating (small scale-rapidly varying in time) parts:

$$\underline{u} = \langle \underline{u} \rangle + \underline{u}'$$

$$\underline{B} = \langle\underline{B}\rangle + \underline{B}'$$

It can easily be shown that this leads to the following equations for the mean and the fluctuating magnetic fields:

$$\partial\langle\underline{B}\rangle/\partial t = \text{curl}\ (\langle\underline{u}\rangle \times \langle\underline{B}\rangle + \langle\underline{E}\rangle) - \text{curl}\ (\eta\ \ \text{curl}\ \langle\underline{B}\rangle\ ) \qquad (2)$$

$$\partial\underline{B}'/\partial t = \text{curl}\ (\langle\underline{u}\rangle \times \underline{B}' + \underline{u}' \times \langle\underline{B}\rangle + \underline{G}) - \text{curl}\ (\eta\ \text{curl}\ \underline{B}') \qquad (3)$$

where $\langle\underline{E}\rangle = \langle\ \underline{u}' \times \underline{B}'\rangle$ represents an additional mean electromotive force generated by the turbulent interaction between the fluctuating velocity and magnetic fields and $\underline{G}=\underline{u}'\times\underline{B}' - \langle\ \underline{u}'\times\ \underline{B}'\rangle$ .

If we consider homogenous turbulence ($\langle\underline{u}\rangle = 0$ in a proper reference system) and introduce the so called first order smoothing approximation (or quasi-linear approximation), which consists in neglecting $\underline{G}$ in (3), we are led to a simplified equation for the fluctuating magnetic field $\underline{B}'$:

$$\partial\underline{B}'/\partial t = \text{curl}\ (\underline{u}' \times \langle\underline{B}\rangle\ ) - \text{curl}\ (\eta\ \text{curl}\ \underline{B}') \qquad (4)$$

This implicitly means to assume $\underline{B}'$ small compared to $\underline{B}$ and is consistent with equation (3) in two cases:

(1) Magnetic Reynolds number $R_m=U\ell/\eta \ll 1$, where U is a typical turbulent velocity and $\ell$ a typical length scale;

(2) Stroughal number $U\tau/\ell \ll 1$, where $\tau$ is the correlation time.

These conditions correspond respectively to the high resistivity case (in which the advection term is balanced by the dissipative one) and to the rapid fluctuation case (in which the advection term is balanced by the variation in time of the fluctuating magnetic field).

Integrating equation (4) over $\tau$, which is a time interval short enough for $\underline{u}'$ and $\langle\underline{B}\rangle$ to be considered time-independent, linearity of $\underline{B}'$ in $\langle B\rangle$ and its space derivatives allows to express $\langle\underline{E}\rangle = \langle\ \underline{u}'\times\ \underline{B}'\rangle$, in the case of isotropic turbulence (+), as:

$$\langle\underline{E}\rangle = \alpha\ \langle\underline{B}\rangle - \beta\ \text{curl}\ \langle\underline{B}\rangle \qquad (5)$$

Substituting (5) into (2) and dropping the brackets, we get the MFE dynamo equation:

$$\partial\underline{B}/\partial t = \text{curl}\ (\underline{u} \times \underline{B} + \alpha\ \underline{B}) - \text{curl}\ \left[(\eta+\beta)\ \text{curl}\ \underline{B}\ \right] \qquad (6)$$

Here $\alpha\underline{B}$ represents an electromotive force, parallel to the mean field $\underline{B}$, generated by turbulence and $\beta$ is the turbulent magnetic diffusivity.

(+) for anisotropic $\alpha$-effect dynamos see Busse and Miin (1979)

Turbulent diffusion operates by mixing large scale mean magnetic fields, while it does not destroy the small-scale fields ultimately smoothed out by ohmic diffusion. It can be shown (Krause and Rädler, 1980) that the coefficient $\alpha$ is proportional to the mean helicity $<\underline{u}'.\mathrm{curl}\ \underline{u}'>$ of the turbulent motion $\underline{u}'$ and does not vanish if the turbulence lacks mirror symmetry. In the so called $\alpha$-$\omega$ dynamos the advection term $\underline{u} \times \underline{B}$ generates the toroidal field from the poloidal field by differential rotation ($\ddot{\omega}$-effect), while the $\alpha\underline{B}$ term regenerates the poloidal field from the toroidal field by cyclonic turbulence (Parker 1955, 1979), through the twisting action of the Coriolis force on magnetic field loops in the convective cells ($\alpha$-effect). The relative strength of the poloidal to the toroidal field is given by $(\alpha/\Delta\omega\ R)^{\frac{1}{2}}$ where $\Delta\omega$ is the differential rotation and R the stellar radius. In the case of the Sun this ratio is of the order of $10^{-2}$, depending however on the magnitude of $\alpha$, whose estimates seem to be in excess.

The dominant time scales involved in the dynamo process are the period of the oscillatory field $(R/\alpha\Delta\omega)^{\frac{1}{2}}$ and the turbulent diffusion time $R^2/\beta$. For marginal dynamo instability these two times are expected to be of comparable order of magnitude(+). In the case of the Sun, the probably too large $\alpha$-value leads to a theoretical period shorter by an order of magnitude. The mean field (dynamo wave) propagates along the surfaces of isorotation (Parker 1955, Yoshimura 1975) in the direction of $\alpha\nabla\omega \times \underline{i}_\phi$ (where $\underline{i}_\phi$ is the azimuthal unit vector). This implies, with $\alpha \gtrless 0$ respectively in the northern and southern emispheres (Stix 1976), $\partial\omega/\partial r < 0$, if the observational constraint of the butterfly diagram (propagation towards the equator) is taken into account. The parity of the mean field with respect to the equator can be even or odd depending on which modes are excited at lower $R_{\alpha c}$. For the Sun the question is open, since no apparent preference is shown for the observed odd parity modes (Belvedere et al 1980b).

## 2.2. Limits and possible improvements of $\alpha$-$\omega$ dynamo theory

Criticism against $\alpha$-$\omega$ dynamo theory in the mean field electrodynamics concerns essentially two points: the rather crude method of closure- strictly justified only if $\underline{B}'$ is small compared to $<\underline{B}>$ - involved in the first-order smoothing approximation (for the Sun, indeed, $R_m \gg 1$ and $U\tau/\ell \approx 1$) and the role of turbulent diffusion, according to which magnetic field diffusion occurs on time scales considerably smaller than

(+) Marginal dynamo instability arises when the dynamo number $R_\alpha R_\omega$ is slightly larger than a critical value. For fixed $R_\omega = \Delta\omega\ R^2/\beta$, this occurs when $R_\alpha = \alpha R/\beta$ is slightly larger than a critical value $R_{\alpha c}$.

the ohmic ones ($\beta >> \eta$), therefore comparable with the observed solar cycle period. Piddington (see e.g.: 1981, 1982 and earlier references therein) does not agree with the analogy of turbulent diffusion of the magnetic field with turbulent diffusion of a scalar field (see also Knobloch 1977) and claims that eddy diffusivity leads to shear amplification of the field within the eddies. In his opinion no merging of fields can be accomplished by turbulence. Comparison between the merging rate and the amplification rate leads to a non-vanishing field, whose growth is limited only by the equipartition value. He contests both the applicability of the Petschek mechanism within the eddies for rapid reconnection of the field lines (see e.g. Parker 1979, chapter 15) and the accumulation of magnetic field at the cell boundaries shown by numerical experiments of magnetoconvection (see e.g. Galloway and Weiss 1981; Knobloch 1981a and references therein).

This criticism is in part due to some misunderstanding of the role of turbulent diffusivity which applies only to the mean field, not to the fluctuating field. Moreover, even if there are several doubts that the Petschek mechanism can operate at magnetic pressure not comparable with gas pressure as in the deep convection zone (Cowling 1981), Piddington's argument, based on a simple estimate of the rates of accumulation and diffusion of field within the eddies is not very convincing, and seems less founded than the non-linear simulations of the interaction of turbulent convection and magnetic field carried out in magnetoconvection studies. Piddington emphasizes also the fact that convective motions and buoyancy would tend to transport upwards the "newly" generated poloidal field, so that it could not come down to the lower part of the convection zone, where it has to be operated by differential rotation. A possible reply is that also downwards motions are involved in the convective transport and that turbulent diffusion may well do the job (Cowling 1981). Furthermore, the magnetic buoyancy argument suggests the location of both the $\omega$-effect and the $\alpha$-effect in the same region, which, according to the present view (see e.g. Parker 1975, Rosner 1980 a,d other references quoted later), should be at the lower boundary of the c.z.

The alternative scenario proposed by Piddington consists of a primordial non-reversing dipole - buried in the radiative region to avoid turbulence - whose field lines oscillate in the meridian planes with a period of 22 years and are acted by the $\omega$-effect, generating toroidal fields of opposite signs in two consecutive solar cycles. However, no clear fundament is given to the energy source and the mechanism of this dipole field oscillation. Moreover there are several doubts that a fossil field would have survived the fully-convective Hayashi phase.

Further, Piddington's alternative is not supported by a satisfactory quantitative treatment. This may also apply to Layzer's et al (1979) paper, whose criticism to $\alpha$-$\omega$ dynamo in the MFE formulation is correct in some points concerning the quasi-linear approximation (even if some misunderstandings about the role of turbulent diffusion and the concept of mean field are still present (see Stix 1981)), but which does not offer a real alternative. In the opinion of those autors there are serious observational and theoretical difficulties against dynamo theory, namely the absence of a surface large scale poloidal field, the Maunder minima and some physical and mathematical inconsistencies in the MFE formulation and in particular in the significance of $\alpha$ and $\beta$.

For instance $\beta$ could be negative (Kraichnan 1976, Knobloch 1977), this being in favour of accumulation of field lines in spatially intermittent flux tubes. To this regard, we have to point out with Stix (1981) that the mean field is not to be intended as a smooth diffuse background field between the flux concentrations but as an <u>average</u> field, where the average includes the highly concentrated intermittent fields whose existence in the convection zone is inferred from the surface observations. In this view, we think that observational evidence of intermittent fields is of no obstacle to dynamo theory, although only non-linear calculations can describe the formation of flux concentrations.

The presence in the past (and in the future? )of Maunder minima does not even affect heavily the $\alpha$-$\omega$ theory, which is in principle able to maintain fields at arbitrary low levels (Stix 1981), so that the dynamo mechanism can well operate even if the fields are so weak to give no observational evidence. It remains to be explained how this mode change happens. Following Ruzmaikin (1980) non-linear dynamos can have solutions diverging from a bifurcation point; so that dynamo can operate in a bimodal way, switching from one mode to another in the so called strange attractor behaviour.

Coming back to Layzer et al (1979), the alternative scenario they propose consists of an original field generated by the Biermann mechanism (Biermann 1950), amplified by a sort of dynamo mechanism during the fully convective phase and giving rise ultimately to a large scale tangled field in the uniformly rotating radiative core. Differential rotation acting on the field in the overshooting layer generates a toroidal component which is wound and unwound alternatively. This torsional field oscillation would explain the solar cycle, whose exterior manifestations would be due to fields leaving the toroidal flux region and floating to the surface. Anyway, this alternative model seems too speculative inasmuch as no sufficiently developed physical description and formal treatment of the torsional oscillation are given.

Incidentally, we recall that the 11-years period torsional oscillation discovered by Howard and La Bonte (1980) has been proposed by the same authors (La Bonte and Howard 1982) to sustain the magnetic cycle. However, it is difficult to think that this rather weak oscillation can compete with the much stronger differential rotation or turbulent convection fields. It seems more reasonable that the torsional oscillations is driven by the longitudinal component of the Lorentz force of the dynamo waves which generate the solar-cycle itself (Yoshimura 1981).

The success of the $\alpha$-$\omega$ theory in reproducing the main characteristics of the solar cycle (see e.g.: Köhler 1973; Stix 1976 a,b; Yoshimura 1975, 1978a, 1978b; Parker 1979; Belvedere et al 1980b) seems to confirm the capability of dynamo equations of capturing in a simple way the essential mechanisms that maintain the solar cycle (Weiss 1981). Nevertheless, several questions remain open and should be investigated deeply in a more consistent and detailed non-linear theory:

- The weakness of the first order smoothing approximation still remains, since in the Sun $R_m \gg 1$ and $\ell \approx U \tau$ . A possibility of overcoming this difficulty is in Cowling's (1981) argument that "$\underline{B}'$ varying rapidly in space is no longer large compared with $\langle \underline{B} \rangle$, being rapidly smoothed out by ohmic diffusion at small length scales" (see also Cowling's (1981) discussion of the induction equation for $\underline{B}''$, the part of the fluctuating magnetic field correlated with $\underline{u}'$).
- The role of helicity and turbulent diffusion should be clarified on both large and small scales, leading to more plausible intrinsic determinations of $\alpha$ and $\beta$ in the context of turbulence theory. In particular the present estimates of $\alpha$ seem to be too large. Note, however, that in the case of strong toroidal flux concentrations, a strong $\alpha$ effect should be needed against the no more negligible Lorentz force (Gilman and Miller, 1981).
- The level at which dynamo operates is still matter of discussion: spatial separation of the $\omega$-effect and the $\alpha$-effect is not plausible and would give rise to problems of upward and downward field transport which are not easily overcome even invoking turbulent diffusion. Therefore, since magnetic buoyancy arguments and stability of flux tube configurations (see later) suggest that the $\omega$-effect operates deeply in the c.z. or in the overshooting layer, also the $\alpha$-effect is expected to occur mainly at deep levels. This expectation is supported too by the argument that the $\alpha$-effect on rapidly rising flux tubes should be ineffective (Golub et al 1981).
- The feedback of the magnetic field on the velocity field is not considered in the linear theory, but the Lorentz force is expected to be relevant when strong flux concentrations occur as in the plausible case of toroidal field ropes wound by differential rotation at the bottom of the c.z., or in the observed case of strong filamentary fields at the edges of cellular patterns.
- The magnetic buoyancy force on flux tubes and the problem of their

stability should be studied further in order to get a reasonable estimate of the float-up times, which should be comparable with the amplification time and the diffusion time, both expected of the order of the cycle period.

The two latter points lead to the need of including the Lorentz force, the magnetic buoyancy force and possible stabilizing forces as the hydrodynamical drag and the Coriolis force into the framework describing the interaction of the magnetic and velocity fields. At the present this is done gradually taking into account the different effects separately.

## 2.3 Non-linear dynamo theory and magnetic flux concentrations

A non-linear analysis of dynamo in the Boussinesq approximation has been done by Cuong and Busse (1981), but seems perhaps too idealized to be applicable to the solar case. Another non-linear compressible dynamo model has been worked out by Schüssler (1979) in cartesian geometry. He finds that the growth of the magnetic field is limited by the Lorentz force and the magnetic buoyancy, but not to such an extent to inhibit dynamo action. No considerable differences from the linear case are found for the magnetic field geometry and the period of the $\alpha$-$\omega$ dynamo. Some characteristics of the observed solar cycle (e.g. equatorwards migration and polarity reversals) are reproduced, but the model suffers from idealized geometry and arbitrary spatial distribution of the induction effects.

Dynamo problem as a problem in magnetohydrodynamic turbulence has been studied by Frisch et al (1975), Pouquet, Frisch and Léorat (1976), Meneguzzi, Frisch, Pouquet (1981), and Léorat et al (1982). One of the most interesting results, obtained by numerical simulation, is that magnetic helicity (scalar product of magnetic field and its vector potential) can give rise to a reverse energy cascade, generating magnetic energy on large scales in competition with what the $\alpha$-effect does from kinematic helicity.

Dynamo of small scale fields has recently been investigated by Vainshtein (1980), deriving an equation for the dynamics of the magnetic field in the Lagrangian statistical description of turbulence. It is found that for $\eta \ll \nu$ ($\nu \equiv$ kinematic viscosity) a positive growth rate solution for the magnetic field exists. This may explain the origin of fine structure fields.

Non-linear magnetoconvection studies have been carried out to explain the presence of intense intermittent fields ($\sim$1500 Gauss) at the solar surface through mechanisms for formation of isolated flux tubes in the convection zone. Galloway and Weiss (1981) have recently done a further Boussinesq study of convection in the presence of magnetic fields and found that magnetic flux is rapidly concentrated into sheets at the lateral boundaries of the convective cells, while flux expulsion from the cell interior takes a longer time of the order of a few turnover times. Turbulent convection concentrates magnetic flux until the equipartition value $B_e$ ($\ll B_p$, the pressure equilibrium value, unless at the top of the c.z.) is reached. Rapid evacuation of matter from the flux tubes ('collapse') is then expected, to get the pressure equilibrium field $B_p$. Weaker flux ropes are shredded and dispersed giving rise to smaller size activity features. The observed total flux at the Sun's surface should be compatible, in those authors' opinion, with the toroidal flux contained in a shallow layer located at the bottom of the c.z. This agrees with what is generally speculated on theoretical grounds and with the observational evidence of the coronal hole field corotating nearly uniformly as it were anchored to a deep level in the c.z. (Golub et al 1981).

Non-linear three-dimensional magneto-convection and magnetic field spectrum have been studied by Knobloch (1981 a,b). These works are related to Knobloch and Rosner's (1981) conclusions that the kinematic approach is not sufficient and to Galloway, Proctor and Weiss (1977), who identified different regimes depending on the increasing value of the magnetic field (the kinematic regime, the hydromagnetic regime with fields that can overcome the equipartition value, the overstable regime in which convection is inhibited and no further field concentration occurs).

Knobloch finds that non-linear concentration of magnetic flux by turbulent motions occurs at the cell edges, in agreement with the previous authors (see also Peckover and Weiss 1978), and that different scales of flux tubes arise as a result of different scales of motion. Also a theoretical prediction of size and spatial distribution of the flux tubes is given, as well as the field strength as a function of the tube radius, on the assumption that flux tubes are formed from an initial uniform field in the presence of a given turbulence spectrum. Agreement with the observations is reasonable.

The problem of stability of flux tubes in the c.z. under the action of vigorous turbulent convective motions, magnetic buoyancy, hydrodynamical drag and rotational forces is one of the most debated in the recent years. We refer for a general overview to Parker (1979, chapters 8,10,13) and (Spruit 1981 a,b). This problem is strictly connected with the time scale of the magnetic flux rise to the solar surface to form active regions.

Parker (1975) suggested that toroidal flux generation should occur deeply in the c.z. in order the rise time of the tubes to be comparable with the time scale of the solar cycle. However, some difficulties exist: tubes in thermally equilibrium with the surrounding would float upwards as a result of magnetic buoyancy. But the requirement that they should remain in the c.z. for some years leads to an upper limit of ∿200 Gauss to the field, too small in comparison to the equipartition value. On the other hand, if the tubes were neutrally buoyant, their internal temperature should be lower than the surroundings, and it is not clear what mechanism could maintain the temperature difference. The suggestion of the lower part of the c.z. (or even the transition layers) as a site for magnetic flux storage is also founded on stability requirements. Indeed the strict polarity rules of active regions and, in general, the well definite laws which surface manifestations of activity do obey, seem to imply that the sub-surface magnetic field is highly organized and not subjected to strong deformation by convective motions (Van Ballegooijen 1982b).

Van Ballegooijen (1982a) has studied the stability of adiabatic flux tubes in the layers below the convective zone, where the equipartition magnetic field strength should be $\sim 10^4$ Gauss. In his model buoyancy forces are balanced by hydrodynamical drag and emergence of flux loops from the system, driven by instabilities, should take a few days.

Spruit and Van Ballegooijen (1982) have analyzed the conditions of stability of thin adiabatic flux tubes in the equatorial plane of a convective star, without including rotational effects (for the influence of rotation, see Acheson 1979). They point out that toroidal tubes are subjected not only to buoyant rise, but also to buoyant instabilities such small upward displacement or wave-like disturbances - in which fluid can flow from crests to troughs - if the superadiabaticity of the layer is large enough. In addition tubes are not stable against poleward motion (ribbon slip instability).

They conclude that flux tubes in a stellar convective zone seem to be unstable for all field strengths. Therefore they suggest that toroidal field accumulation occurs in the more stable interface layers between the c.z. and the radiative interior. This idea is also in Van Ballegooijen (1982b), who concludes that an additional force is needed to keep the flux tubes in the overshoot layer against the buoyancy force. He suggests that the Coriolis force, acting on a flow along the toroidal tubes, induced by angular momentum conservation during equatorward meridional motion, would do the job.

A flux tube dynamo model of the solar cycle has been proposed by Schüssler (1980), where attention is devoted to the regeneration process operating on flux tubes in a way similar to Leighton's (1969) model . Further he points out that tubes generated by a weaker poloidal field, therefore shredded by violent convective motions and subjected to smaller magnetic buoyancy, take a longer time to reach the surface (older tubes) where they reveal as Ephemeral Active Regions (EAR) and X-Bright Points (XBP); on the contrary stronger tubes take a shorter time (young tubes) and form large active regions. This way, the antiphase of the cyclic variations of the Wolf number and the EAR (and XPB) number is explained. Schussler's somewhat heuristic model is to be considered a reasonable attempt to incorporate flux tube dynamics into global dynamo theory, but it is clear that more efforts are necessary for a more complete description of the basic interaction between convection, rotation and magnetic fields.

An attempt to do this job is in Gilman and Miller (1981), who consider the fully non-linear hydromagnetic problem of dynamos driven by non-axisymmetric convection in a rotating spherical shell (Gilman 1976, 1977, 1978, 1980, 1982 ; Glatzmaier and Gilman 1982). They point out that previous $\alpha$-$\omega$ models owe part of their success to independently choosing the magnitude and profiles of helicity and differential rotation, and suspect incompatibility with the fluid dynamics laws. Unfortunately, their results are not encouraging with regard to the real Sun. Neither equatorward migration is present, nor polarity reversals, nor preferred symmetry. An excessive $\alpha$-effects is proposed as a cause of the chaotic magnetic field behaviour. The problem of the non-linear global interaction of dynamo magnetic fields and the dynamics of the inducing fluid motions has very recently been reviewed by Gilman (1983), where the dependence of models upon physical parameters as c.z. depth, rotation rate, heating rate, viscous and magnetic diffusivities, compressibility is analyzed, following the results of a large series of numerical experiments . An interesting point is the "regime diagram" which attempts to predict what kind of dynamo action is to be expected, as a function of the electrical conductivity and the influence of rotation on the dynamics of the system. This way, three fundamental regimes are identified: no dynamos, dynamos without cycle, dynamos with cycles, the last corresponding to intermediate influence of rotation. This diagram could be of some relevance in connection with the bimodal dynamo behaviour put in evidence by the Vaughan-Preston gap (see part 3 of the present review ). Another interesting point is in the suggestion, coming out from calculations, that "the seat for cyclic dynamo action is in low latitudes outside the tangent cylinder to the inner boundary of the c.z.". However, due to some difficulties and ambiguities in reproducing the observations, Gilman's models are not fully convincing, even if the position of the problem appears to be correct and fruitful.

## 3 . MODELS OF STELLAR ACTIVITY

It is well known that main sequence stars later than F5 and giant stars later than GO have outer convective envelopes whose extension increases with decreasing effective temperature. The interaction between rotation and convection, leading in the Sun to the observed differential rotation and (dynamo driven) activity cycle, should in principle do the same job in other stars with convective envelopes, even if with different efficiency and mode characteristics, depending on basic parameters as rotation, luminosity class, spectral type and age. In this framework Durney and Latour (1978) stressed the importance of a dynamo generated magnetic field in sustaining angular momentum loss in late main sequence stars, with an efficiency sharply decreasing at F6,where outer convection is practically absent (+). Belvedere et al (1980 c,d; 1981, 1982) made the first attempt to compute differential rotation and magnetic cycle dynamo models for main sequence and giant stars, in analogy to the Sun (Belvedere and Paternò 1977, Belvedere et al 1980a). The results show that differential rotation, magnetic field strength, latitude extension of the activity belt and cycle period length do increase with the advancing spectral type. It has been pointed out in these papers that the ratio of the global convection turnover timescale to the rotational timescale, namely $\omega d/U \approx \omega d^2/\nu$ (where d is the thickness of the c.z. and $\nu$ the kinematic viscosity), which regulates the strength of the interaction of rotation and convection, does increase with the advancing spectral type, leading to increasing differential rotation. This in turn implies a larger $R_\omega$, thus a smaller $R_{\alpha c}$, that means a more favoured dynamo action for later spectral types (see the footnote at p. ). The toroidal magnetic field strength, estimated in the assumption of energy equipartition between the magnetic field and the velocity field (Belvedere et al 1981), is expressed by $B \sim (R^2/\beta)^{\frac{1}{2}}(\omega/R_{\alpha c})$,thus being,for a given $\omega$,of the form $\omega$ x(an increasing function of the spectral type). The cycle period length essentially reproduces the turbulent diffusion timescale which, in the marginal dynamo instability, should be comparable with the period of the dynamo wave. The theoretical predictions of Belvedere et al. have received indirect support from the EINSTEIN X-ray flux observations in main sequence and giant stars (see e.g. Vaiana 1980, 1981; Pallavicini et al 1981) and agree well with the general observational background suggesting that activity increases towards the later spectral types, and the current conviction that angular velocity and depth of the convection zone are the basic ingredients for interpretation of stellar activity (see e.g. Rosner 1980, Durney and Robinson 1982).

The recent observations of chromospheric Ca II emission in stars (Vaughan and Preston 1980), the direct angular velocity measurements derived from its modulation (Vaughan et al 1981), and the comparison

(+) A recent study of dynamo in convective zones of declining thickness and efficiency is in Parker (1981).

of activity cycles in old and young stars (Vaughan 1980) have added to our knowledge of stellar activity, in the same line of Wilson (1978 and references therein).

The following scenario has emerged:

-Ca II emission flux $F$(however, measured against the continuum)increases with the advancing spectral type, while for a fixed spectral type, it increases with $\omega$ and decreases with the age t. The latter result is consistent with Skumanich's (1972) relation $\omega \sim t^{-\frac{1}{2}}$, the former with Skumanich and Eddy's (1981) result that the total magnetic flux erupted increases with the angular velocity, the Ca II emission flux being a magnetic activity indicator.

- There exists a gap in the F-G region of the log $F$ vs.(B-V) diagram, which evidences two well distinct behaviours:

-Stars above the gap (younger and faster rotators , show a larger emission flux, thus stronger activity, but no definite cycles (chaotic behavior);

-Stars under the gap (older and slower rotators) show a smaller flux, thus weaker activity, but well defined cycles (cyclic behavior);

This bimodal behavior has open new problems to dynamo theory, as the discovery of Maunder minima had previously. Also the bimodal behavior can however be accounted for in the strange attractor framework, which, as we have already seen, allows for different trajectories from a bifurcation point, this being a characteristic of non-linear systems. Thus, multimode dynamos are in principle possible, depending on stellar parameters and age.

A different explanation of the Vaughan-Preston gap is given in Durney, Mihalas and Robinson (1981), who derive a relation among the dynamo number, the colour index and the angular velocity in the range F5-MO and make an attempt to reproduce the log $F$ vs.(B-V) diagram. In their opinion a transition from a single-mode dynamo to a multiple-mode dynamo should occur at some critical dynamo number, leading to chaotic field behavior in rapidly rotating young stars, owing to the superposition of several coexcited and interfering modes. This has some analogy with Parker's (1971) result that large dynamo numbers lead to small scale fields varying rapidly and irregularly in time.

Another alternative is that proposed by Knobloch, Rosner and Weiss (1981) who suggest that, as the reciprocal of Rossby number $\sigma \sim \omega\ell/U$ increases over a critical value $\sigma_c$, convection in rolls nearly aligned with the rotational axis is favoured, this decreasing the mean helicity $\underline{u} \cdot \nabla x \underline{u}$. The consequent weakening of the $\alpha$-effect would lead to a more difficult regeneration of the poloidal field, may be no dynamo at all, if the poloidal field does not reverse. Therefore different dynamo mechanisms would operate in the high and the low angular velocity regimes. Another result of these authors is that the mean strength of the magnetic field should be larger for lower mass stars (later spectral types).

A similar result is found by Durney and Robinson (1982) who have estimated the magnetic field strength in the assumption that the rise time of the flux tubes is of the order of the amplification time.

The main result is that, for fixed $\omega$, both the magnetic field strength and its extension over the stellar surface increase with (B-V). This is in the same line as the results of Belvedere et al (1980d), suggesting that the present models, although subjected to different assumptions, do converge in predicting some basic features of stellar activity. This may indicate that the essential points have been captured. However, a difference between Belvedere's et al (1980d) paper and Durney and Robinson's (1982) is in the cycle period length, increasing with (B-V) in the former, decreasing in the latter. These theoretical estimates are indeed sensitive to the characteristic time scales chosen in different models.

Durney and Robinson's (1982) results are essentially confirmed in a more recent work of the same authors (Robinson and Durney 1982), where a relatively simplified local system of dynamo equations, including the magnetic buoyancy term, is solved in the lower part of the convection zone where the magnetic field generation is assumed to occur. Arguments in favour of the latter point are given by Hathaway (1982) in the framework of an analytical model of turbulence in rotating convective zones. A relevant point in this paper is the derivation of the turbulent stress tensor, to which the (pseudo)-tensorial forms of $\alpha$ and $\beta$ are related. The resulting stresses, in the presence of rotation, are expected to be larger at the bottom of the convective zones.

We conclude this review hoping that the comparison of dynamo theory with the new stellar activity data will extend our understanding of the dynamo operation modes, which probably depend on stellar structure, rotation and age.

AKNOWLEDGEMENTS

I am very grateful to the following colleagues who kindly sent me published or preprint material: F.H. Busse, T.G. Cowling, U. Frisch, P. Foukal, P.A. Gilman, D.H. Hathaway, E. Knobloch, H.K. Moffat, E.N. Parker, J.H. Piddington, A. Pouquet, A.A. Ruzmaikin, A. Skumanich, H.C. Spruit, M. Stix, S.I. Vainshtein, A.A. Van Ballegooijen, H. Yoshimura.

I acknowledge also useful discussions with S. Catalano, L. Paternò, M. Rodonò and M. Stix.

REFERENCES

Acheson, D.J. : 1979, Solar Phys. 62, 23.

Belvedere, G., Paternò, L.: 1977, Solar Phys. 54, 289
Belvedere, G., Paternò, L., Stix, M.: 1980a, Geophys. Astrophys. Fluid Dyn.14, 209.
Belvedere, G., Paternò, L., Stix, M.: 1980b, Astron. Astrophys. 86, 40
Belvedere, G., Paternò, L., Stix, M.: 1980c, Astron. Astrophys. 88, 240.
Belvedere, G., Paternò, L., Stix, M.: 1980d, Astron. Astrophys. 91, 328
Belvedere, G., Chiuderi, C., Paternò, L.: 1981, Astron. Astrophys, 96, 369.
Belvedere, G., Chiuderi, C., Paternò, L.: 1982, Astron. Astrophys. 105, 133
Biermann, L.: 1950, Zs. Naturforschung 5a, 65
Busse, F.H., Miin, S.W.: 1979, Geophys. Astrophys. Fluid Dyn. 14, 167
Cowling, T.G.: 1934, Monthly Notices Roy. Astron. Soc. 94, 39
Cowling, T.G.: 1981, Ann.Rev. Astron. Astrophys. 19, 115
Cuong, P.G., Busse, F.H.: 1981, Phys. Earth Planet. Inter. 24, 272
Durney, B.R., Latour, J.: 1978, Geophys. Astrophys. Fluid Dyn. 9, 241
Durney, B.R., Mihalas, D., Robinson, R.D.: 1981, Publ. Astron. Soc. Pacific. 93, 537
Durney, B.R., Robinson, R.D.: 1982, Astrophys. J. 253, 290
Foukal, P.: 1980, Proc. Conf.Ancient Sun, eds. R.O. Pepin, J.A. Eddy, R.B. Merrill, p. 29
Foukal, P.: 1982, "Variations in solar luminosity", preprint
Frisch, U., Pouquet, A., Léorat, J., Mazure, A.: 1975, J. Fluid Mech. 68, 769
Galloway, D.J., Proctor, M.R.E., Weiss , N.O.: 1977, Nature 266, 686
Galloway , D.J., Weiss, N.O.: 1981, Astrophys.J. 243, 945
Gershberg, R.E.: 1978, Astrophysics 13, 310
Gilman, P.A.: 1976, in IAU Symp. 71, Basic Mechanisms of Solar Activity, eds. V.Bumba and J. Kleczeck, Reidel (Dordrecht), p.207
Gilman, P.A.: 1977, Geophys. Astrophys. Fluid Dyn. 8, 93
Gilman, P.A.: 1978, Geophys. Astrophys. Dyn. 11, 157
Gilman, P.A.: 1980, in IAU Coll. 51, Stellar Turbulence, D.F. Gray and J.L. Linsky eds., Reidel (Dordrecht), p. 19.
Gilman, P.A.: 1982, "Convective Dynamos for Rotating Stars" in Proc. 2nd Cambridge Workshop on Cool Stars, Stellar Systems and the Sun, eds. M.S. Giampapa and L. Golub, in press, preprint
Gilman, P.A.: 1983, "Dynamos of the Sun and Stars and Associated Convection Zone Dynamics", IAU Symp. 102, J. Stenflo (ed.), preprint.
Gilman, P.A., Miller, J.: 1981, Astrophys. J. Suppl. 46, 211
Glatzmaier, G.A., Gilman, P.A.: 1982, Astrophys.J., in press
Golub, L., Rosner, R., Vaiana, G.S., Weiss, N.O.: 1981, Astrophys. J. 243, 309
Hall, D.S.: 1981, in Solar Phenomena in Stars and Stellar Systems, eds. R.M. Bonnet and A.K. Dupree, Reidel (Dordrecht), p. 431

Hathaway, D.H.: 1982, "A Convective Model for Turbulent Mixing in Rotating Convection Zones", preprint submitted to Astrophys. J.
Hartmann, L.: 1981, in Solar Instrumentation: what's the next? ed. R.B. Dunn, Sac. Peak Nat. Obs. Sunspot NM, p. 170
Howard, R., La Bonte, B.J.: 1980, Astrophys.J. 239, L33
Knobloch, E.: 1977, J. Fluid Mech. 83, 129
Knobloch, E.: 1981a, Astrophys.J. 247, L93
Knobloch, E.: 1981b, Astrophys. J. 248, 1126
Knobloch, E., Rosner, R.: 1981, Astrophys. J. 247, 300
Knobloch, E., Rosner, R., Weiss, N.O : 1981, Monthly Notices Roy. Astron. Soc. 197, 45p
Köhler, H.: 1973, Astron. Astrophys. 25, 467
Kraichnan, R.H.: 1976, J. Fluid Mech. 75, 657
Krause, F., Rädler, K.H.: 1980, Mean Field Magnetohydrodynamics and Dynamo Theory, Pergamon (Oxford)
La Bonte, B.J., Howard, R.: 1982, Solar Phys. 75, 161
Layzer, D., Rosner, R., Doyle, H.T.: 1979, Astrophys. J. 229, 1126
Leighton, R.B.: 1969, Astrophys. J., 156, 1
Léorat, J., Grappin, R., Pouquet, A., Frisch, U.: 1982, "Turbulence and magnetic fields", to appear in Geophys. Astrophys. Fluid Dyn.
Linsky, J.L.: 1981, in Solar Instrumentation: what's the next?, ed.. R.B. Dunn, Sac. Peak. Nat. Obs., Sunspot NM, p. 180
Meneguzzi, M., Frisch, U., Pouquet, A.: 1981, Phys. Rev. Lett. 47, 1060
Mihalas, D.: 1981, in Solar Instrumentation: what's the next?; ed. R.B. Dunn, Sac.Peak Nat. Obs. , Sunspot NM, p.193
Moffatt, H.K.: 1978, Magnetic Field Generation in Electrically Conducting Fluids, Cambridge Univ. Press
Noyes,R.W.: 1981, in Solar Phenomena in Stars and Stellar Systems, eds. R.M. Bonnet and A.K. Dupree, Reidel (Dordrecht), p. 1
Pallavicini, R., Golub, L., Rosner, R., Vaiana, G.S., Ayres, T., Linsky, J.L.: 1981, Astrophys. J. 248, 279
Parker, E.N.: 1955, Astrophys. J. 122, 293
Parker, E.N.: 1971, Astrophys.J. 165, 139
Parker, E.N.: 1975, Astrophys. J. 198, 205
Parker, E.N.: 1979, Cosmical Magnetic Fields, Clarendon Press (Oxford)
Parker, E.N.: 1981, Geophys. Astrophys. Fluid Dyn. 18, 175
Peckover, R.S., Weiss, N.O.: 1978, Monthly Notices Roy. Astron. Soc. 182, 189
Piddington, J.H.: 1981, Astrophys. J. 247, 293
Piddington, J.H.: 1982, "Dynamo Action in Cosmic Bodies", preprint
Pouquet, A., Frisch, U., Léorat, J.: 1976, J. Fluid Mech. 77, 321
Robinson, R.D.: 1980, Astrophys. J. 239, 961
Robinson, R.D., Durney, B.R.: 1982,"On the generation of magnetic fields in late type stars: a local time-dependent dynamo model", to appear in Astron. Astrophys.

Rodonò, M.:1981, " Progress and Problems in RS CVn Star Research"in Photometric and Spectroscopic Binary Systems, Eds.E.B. Carling and Z. Kopal, Reidel (Dordrecht), p. 285

Rodonò, M.: 1982, "Active Stars and Systems", in XXIV COSPAR Topical Meeting on Stellar Chromospheres and Coronae, Ottawa, eds. J. L. Linsky and G.S. Vaiana, Academic Press (Oxford), in press

Rosner, R.: 1980, in Proc. 1st Cambridge Workshop on Cool Stars, Stellar Systems and the Sun, A.K. Dupree ed., Smithsonian Astrophys. Obs. SP 389, p.79

Ruzmaikin, A.A.: 1981, Comments on Astrophys. 9, 85

Schüssler, M.: 1979, Astron. Astrophys. 72, 348

Schüssler, M.: 1980, Nature 228, 150

Skumanich, A.: 1972, Astrophys.J. 171, 565

Skumanich, A. and Eddy J.A.: 1981, in Solar Phenomena in Stars and Stellar Systems, eds. R.M. Bonnet and A.K. Dupree, Reidel (Dordrecht) p. 349

Spruit, H.C.: 1981a, in Solar Phenomena in Stars and Stellar Systems, ed. R.M. Bonnet and A.K. Dupree, Reidel (Dordrecht), p. 289

Spruit, H.C.: 1981b, in The Sun as a Star, ed. S.D. Jordan, CNRS-NASA NASA-SP-450, p.385

Spruit, H.C., Van Ballegooijen, A.A.: 1982, "Stability of Toroidal Flux Tubes in Stars", preprint

Stix, M.: 1976a, in IAU Symp. 71, Basic Mechanisms of Solar Activity, eds. V. Bumba and J. Kleczek, Reidel (Dordrecht) p. 367

Stix, M.: 1976b, Astron. Astrophys. 47, 243

Stix, M.: 1981, Solar Phys. 74, 79

Vaiana, G.S.: 1980, in Proc. 1st Cambridge Workshop on Cool Stars, Stellar Systems and the Sun, ed. A.K. Dupree, Smithsonian Astrophys. Obs. SP 389, p.195

Vaiana, G.S.: 1981, Space Sci. Rev. 30, 151

Vainshtein, S.I.: 1980, Sov. Phys. JETP 52, 1099

Van Ballegooijen, A.A.: 1982a, Astron. Astrophys. 106, 43

Van Ballegooijen, A.A.: 1982b, "The Overshoot Layer at the Base of the Solar Convective Zone and the Problem of Magnetic Flux Storage", preprint submitted to Astron. Astrophys.

Vaughan, A.H.: 1980, Publ. Astron. Soc. Pacific 92, 392

Vaughan, A.H., Preston, G.W.: 1980, Publ. Astron. Soc. Pacific 92, 385

Vaughan, A.H., Baliunas, S.L., Middelkoop, F., Hartmann, L.W., Mihalas, D., Noyes, R.W., Preston, G.W.: Astrophys. J. 250, 276

Weiss, N.O.: 1981, in Solar Phenomena in Stars and Stellar Systems, eds. R.M. Bonnet and A.K. Dupree, p.499

Wilson, O.C.: 1978, Astrophys.J. 226, 379

Worden, S.: 1981, in Solar Instrumentation: what's the next?, ed. R.D. Dunn, Sac. Peak Nat. Obs., Sunspot NM, p.201

Yoshimura, H.: 1975a, Astrophys.J. Suppl. 29, 467
Yoshimura, H.: 1975b, Astrophys. J. 201, 740
Yoshimura, H.: 1978a, Astrophys. J. 220, 692
Yoshimura, H.: 1978b, Astrophys. J. 226, 706
Yoshimura, H.: 1981, Astrophys. J. 247, 1102

RECENT REVIEWS ON STELLAR DYNAMO (not quoted in the text)

Gilman, P.A.,: 1981, in The Sun as a Star, ed. S.D. Jordan, CNRS-NASA, NASA SP-450, p.231
Gilman, P.A.: "The Solar Dynamo; Observations and Theories of Solar Convection, Global Circulation and Magnetic Fields", Physics of the Sun, Space Science Board, National Acad. of Sciences, USA, in press
Parker, E.N.: 1981, in Plasma Astrophysics, ed. T.D. Guyenne and G.Lévy, ESA SP-161, p.23
Stix, M.: 1982, "The Rotation-Magnetism-Convection Coupling in the Sun", in Proc. 6th European Regional Meeting in Astron., Dubrovnik, in press
Yoshimura, H.: 1981, in Proc. of the Japan-France Seminar on Solar Physics, eds. F. Moriyama and J.C. Hénoux, p.19
Yoshimura, H.: 1982, 'Theory of Solar Cycle", Progress Report for the Section of the IAU Commission 10 for "Reports on Astronomy 1982".

DISCUSSION

Mullan: (most of question lost)....dynamo?

Belvedere: You mean the work by Durney and Robinson. This point is not very clear in this work. I am not in agreement with what they propose. They offer a parameterization of these effects in terms of depth of the layer where the interaction between rotation and convection occurs. They are more in favour of a location at the base of the convection zone.

Rosner: If I may make a comment, I would like to add that they use dynamo numbers which come from classical $\alpha$-$\omega$ dynamo theory which assume that field production occurs throughout the convection zone.

Serio: The evidence presented at this meeting suggests that the period of the magnetic cycle is independent of the rotation period of a star. Would you comment on this?

Belvedere: There are some such observational results but I believe that is not very clear. Dynamo action must depend on the rate of rotation. But there is no connection in theoretical work between cycle period and rotation rate except in the work by Robinson and Durney, who found such a relation. I think we must wait to see what happens in the next few years.

Weiss: I should like to make a quick comment on Peter Gilman's results. At the meeting in Zurich he presented results more recent than those which have been published. He has got cyclic behaviour with his dynamo model giving dynamo waves. Unfortunately they progress from equator to poles. Nevertheless they are far more like stellar dynamos than anything that model has hitherto produced. Another point about his models is that as the dynamo number is increased differential rotation is suppressed to the extent that an entirely different mode of dynamo action occurs. He believes that this is a good explanation for the change in the pattern of magnetic activity in more rapidly rotating stars. If I may add my own opinion to that, I believe that this is the best explanation of that effect which we have at the moment, despite my attachment to work in which I was involved.

Linsky: I would like to express a note of caution concerning comparison between dynamo calculation and observed X-ray emission. That is that there are a great many steps between the generation of magnetic field by the dynamo and X-ray emission. The magnetic field has to make it to the surface, it has to be thermalized, etc. So whether the magnetic field is in open or in closed structures may make a world of difference in terms of the observed X-ray emission.

Belvedere: Yes, you are right. However, we have to proceed step-by-step with both new theory and new observations. As Dr. Weiss has pointed out Gilman's new results are not a matter of observations. Nevetheless new theoretical directions are suggested by new observational data.

Paternò: Perhaps I can comment on the relationship between $\omega$ and rotational period. The linear theory for a marginally critical dynamo provides for having the diffusion time equal to the dynamo wave period. So since the dynamo wave period contains a measure of differential rotation and since the model of differential rotation indicates $\delta\omega$ is proportional to $\omega$, I can suppose that larger rotation rates will produce shorter cycle periods. The dependence should be $\propto \omega^{\frac{1}{2}}$.

Belvedere: This is what I said. The problem is different however. This is an evaluation which results in the particular context of the Parker dynamo wave. We can obtain other relationship also equating amplification times of rise with other things but the length of the cycle as a function of spectral type does depend on the basic assumption you make about the physics of the convection-rotation interaction.

Venugopal: What is your estimate of the thickness of the shell in which the dynamo is working?

Belvedere: This depends on the assumption made. Normally it is a fraction of the scale height of the base of the convection zone. Whereas previously people believed it was of the order of a pressure scale height in this same region.

Rosner: Perhaps I could point out here that the first person to carry out calculations of flux stability at the base of this zone was Acheson.

Belvedere: Yes, I quoted this in the references to my review.

Rosner: It is interesting that the calculations of the Dutch group (van Ballegooijen et al) assume that there are already flux tubes at the base. Acheson, however, assumed the field was initially uniform and posed the question how does one form flux tubes? A number of other people have recently carried out similar calculations viz. Jurgen Schmitt at Harvard and Nigel Weiss and his students.

Belvedere: You mean for the influence of rotation on the stability of flux tube concentration.

Rosner: Yes.

# PHOTOSPHERIC SOURCES OF MAGNETIC FIELD ALIGNED CURRENTS

Ake Nordlund
Copenhagen University Observatory, Denmark

Possible (small-scale) photospheric sources of coronal magnetic field aligned currents are discussed. Such currents are equivalent to local (small-scale) twists of the coronal magnetic field, and may cause field topologies that are (MHD or resistively) unstable, and thus contribute to the (small-scale) coronal activity.

One electro-motive force associated with photospheric magnetic fields is due to the asymmetry between ions and electrons as agents of conductivity and as agents of momentum transfer to the (dominant) neutral gas component. In the solar photosphere, where only some of the easily ionized heavier elements (Mg, Si, Fe, ...) are significantly ionized, the ratio of number density of electrons and ions to neutrals is very small, of the order of the total abundance of these elements, which is of the order $10^{-4}$, by number. Due to the larger electron than ion mobility, electric currents are mainly carried by the electron component of the plasma whereas, because of the ions greater momentum transfer to the neutrals in collisions, the friction between the charged and the neutral components of the plasma is mainly due to the ions. Thus, the Lorenz force $\underline{i} \times \underline{B}$ acts mainly on the electron component of the plasma, whereas the balancing gas pressure gradient and gravity terms act mainly on the neutral component. A consideration of the momentum balance for each component of the plasma shows that the forces acting on the neutral component must be communicated to the electron component through friction to the ions and then, through an electric field to the electron component. Taken per unit volume, the forces on the three components of the plasma have similar magnitudes but, taken per particle, the forces on the charged components must be a factor n(neutral)/n(charged) larger than the forces on the neutrals. This leads to the presence of an (ambi-polar diffusion like) electric potential difference between the inside and the outside of magnetic flux concentrations (inside positive). The potential difference is of the order of the volt-equivalent of the temperature, times the number



*P. B. Byrne and M. Rodonò (eds.), Activity in Red-Dwarf Stars, 601–603.*

ratio of neutrals to ions; i.e., of the order of 10 keV.

Because of the vertical stratification of the photospheric plasma and because the degree of ionization varies strongly with height (temperature), the emf across the interface varies strongly with height. At chromospheric levels and in the subphotosphere, the emf is much reduced due to hydrogen ionization, and a current-free equilibrium is not possible. In the limit where the current system is sufficiently efficient to effectively short out the photospheric emf, the momentum balance between the three plasma components can only be maintained by an additional $\underline{v} \times \underline{B}$ electric field, that is; the plasma has to rotate perpendicular to B and the interface. The significance of this effect depends strongly on the thickness of the boundary layer between field-free plasma and magnetic plasma: a smaller thickness implies a larger electric field and therefore a higher velocity, and a smaller thickness also implies that the plasma has to move a smaller distance for the topology of the magnetic field to become significantly distorted in the boundary layer. The boundary layer thickness is a result of the energy equilibrium in the boundary layer and is mostly controlled by radiative effects. Spruit (1976) has estimated the thickness to be of the order 10 km. This implies a typical plasma velocity of some 10 $ms^{-1}$ for magnetic fields of the order $10^{-1}$ T. A significant distortion of the boundary layer magnetic field topology then takes some $10^3$ seconds.

Another source of field-aligned currents is the small-scale vorticity associated with the horizontal velocity components of the granular velocity field. Numerical 3-D simulations of the interaction of granulation with a magnetic field on a small scale (Nordlund 1982), have shown that the magnetic flux concentrations in the photosphere continually readjust to fill out downdrafts created by the granular convection in the field free areas.

In summary, distortions of the photospheric magnetic field topology in the photosphere cause twists (field aligned currents) to propagate along field lines up into the coronal magnetic field. For small scale magnetic loops, these currents have a duration which is long compared to the propagation time of Alfven waves along the loop. This results in quasi-static twists of the coronal field lines, rather than propagating Alfven waves. The magnetic field aligned currents associated with such twisted fields may cause resistive MHD instabilities similar to Tokamak instabilities (Waddell et al 1979, Carreras et al 1980), and my thus be of importance for the small scale chromospheric and coronal activity.

REFERENCES

Carreras, B., Hicks, H.R., Holmes, J.A. and Waddell, B.V. 1980, Phys. Fluids 23, 1811.

Nordlund, A. 1983, in IAU Symposium No. 102, J. Stenflo (ed.).
Spruit, H. 1976, Solar Phys. 50, 269.
Waddell, B.V., Carreras, B., Hicks, H.R. and Holmes, J.A. 1979, Phys. Fluids 22, 896.

# TURBULENT SOUND GENERATION IN RED DWARF STARS

H.U. Bohn
Inst. f. Astron. u. Astrophys. d. Univ., Am Hubland,
D-87ØØ Wurzburg, FRG

## INTRODUCTION

Since the acoustic heating theory (c.f. Ulmschneider 1979) has been proven successful for the solar chromosphere, it was common practice to extend this concept to other stars. However, as it appeares from observed chromospheric and coronal emissions, the usual theoretical acoustic fluxes for red dwarf star, particularly, are too small to account for the heating of chromospheres and coronae (e.g. Blanco et al 1974; Vaiana et al, 1981). It is therefore the intention of this paper to discuss improvements on the current model calculations for turbulent sound generation from outer convection zones.

## IMPROVED MODEL CALCULATIONS

A grid of convection zone models with effective temperatures between 2500 K and 9500 K and log g=4 and 4.5 representing the physical parameters of red dwarf stars has been constructed using the standard mixing length theory (Cox and Giuli, 1968). The chcmical composition has been chosen as X : Y : Z = 0.68 : 0.30 : 0.02 by mass as for population I stars. The ratio of mixing length to pressure scale height, $\alpha$, was assumed to be 1.0 for all models.

The opacity was interpolated from the third generation Los Alamos tables by Cox and Tabor (1976). For temperatures below 5500 K, where absorption by dust and molecules becomes dominant, the corresponding tables by Alexander (1976) were taken instead. At low temperatures this absorption coefficient is several orders of magnitude larger than the Cox and Tabor (1976) opacity.

The thermodynamical quantities have been calculated by the simultaneous solution of Saha equations for particle densities and an equivalent equation for the $H_2$-dissociation. Due to this molecular dissociation the adiabatic temperature gradient reaches a low value



*P. B. Byrne and M. Rodonò (eds.), Activity in Red-Dwarf Stars, 605–608.*

(< 0.4) at low temperatures and high pressures which facilitates convective instability.

Previous calculations of stellar acoustic fluxes rely on the Lighthill - Proudman theory (Lighthill, 1952; Proudman, 1952). Therefore a homogeneous atmosphere without mechanical boundaries is assumed which implies a pure quadrupole source term in the inhomogeneous wave equation. It has been known for some time, however, that for gravitationally stratified stellar atmospheres dipole and monopole sound generation can also be expected (Unno, 1964).

Stein (1967) generalized Lighthill's theory for density stratifications due to stellar gravity. With his method it is possible to solve the inhomogeneous wave equation numerically after a multipole expansion of the source term. Stein's (1967) code was used to calculate the turbulent sound generation in the convection zones, so monopole- and dipole-terms superpose the usual quadrupole emission. A comparison of the pure quadrupole-, dipole-, and monopole- fluxes revealed an increasing importance of monopole and dipole sound generation for stars of later than solar type. Monopole emission even becomes dominant for dwarfs cooler than 4000 K.

Figure 1 exhibits the total acoustic fluxes from this work (drawn line) in relation to the results of DeLoore (1970; broken line) and the chromospheric CaII fluxes from Blanco et al. (1974). It is evident that the revised acoustic fluxes are comfortably larger than the absolute CaII emissions and the claim that chromospheres are heated by sound waves can be maintained. In the same way, a comparison of the total acoustic flux with observed X-ray emissions of main sequence stars in Figure 2 shows that even acoustic heating of coronae can no more be ruled out by mere insufficiency arguments (thin lines: results of Renzini et al., 1977; fully drawn histogram: observations by Vaiana et al. 1981).

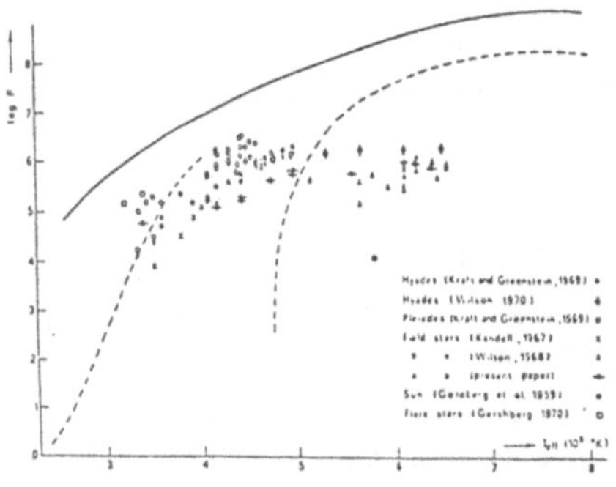

Figure 1

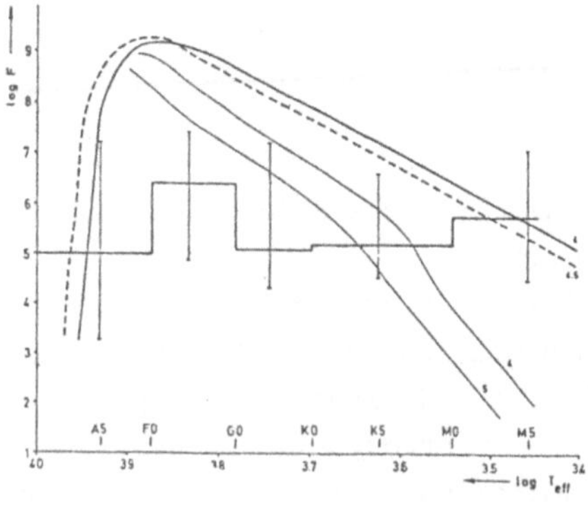


Figure 2

## CONCLUSIONS

The refined calculation of acoustic energy generation from red dwarf stars carried out in this work appears to have removed the insufficiency argument (Linsky, 1980) against the theory of acoustic heating of stellar chromospheres and coronae. This is especially true in view of the conservative assumption which led to these results, as the assumed value of $\alpha = 1.0$ in the mixing length theory is at the lower end of the usual range. It has to be pointed out that the enhancement of acoustic energy generation for the latest dwarfs is a consequence of physical effects which have been neglected in former work and is in no way maximized by adjusting free parameters in the convection theory. Therefore, these calculated fluxes represent lower bounds for possible acoustic emission from stellar convection zones.

## REFERENCES

Alexander, D.R.: 1975, Astrophys. J. Suppl. Ser. 29, 363.
Blanco, C., Catalano, S., Marilli, E., Rodonò, M.: 1974, Astron. Astrophys. 33, 257.
Cox, A.N., Tabor, J.E.: 1976, Astrophys. J. Suppl. Ser. 31, 271.
Cox, J.P., Giuli, R.T.: 1968, "Principles of Stellar Structure", Vols. 1,2 Gordon and Breach.
DeLoore, C.: 1970, Astrophys. Space Sci. 6, 60.
Lighthill, M.J.: 1952, Proc. Roy, Soc. A 211, 564.
Linsky, J.L.: 1980, SAO Spec. Report 389.
Proudman, J.: 1952, Proc. Roy.Soc. A 214, 119.
Renzini, A., Cacciari, C., Ulmschneider, P., Schmitz, F.: 1977, Astron. Astrophys. 61, 39.
Stein, R.F.: 1967, Solar Phys. 2, 385.
Ulmschneider, P.: 1979, Space Sci. Rev. 24, 71.
Unno, W.: 1964, Transactions of the IAU XII, 555, J.C. Pecker, ed.
Vaiana, G.S., Cassinelli, J.P., Fabbiano, G., Giacconi, R., Golub, L., Gorenstein, B., Haisch, B.M., Harnden, F.R. Jr., Johnson, H.M., Linsky, J.L., Maxson, C.W., Mewe, R., Rosner, R., Seward, F., Topka, K., Zwaan, C.: 1981, Astrophys. J. 244, 163.

## DISCUSSION

Vaiana: Did I understand correctly that you said in the case of the Sun the acoustic heating succeeds in explaining the solar atmosphere or did you say that the energy problem does not exist for the Sun?

Bohm: I said that I can balance the solar atmosphere numerically in energy terms with acoustic heating. There are problems however with the mechanism of bringing mechanical energy to the chromosphere and corona.

Vaiana: The OSO IV data would also point in that direction.

# A NEW TRIGGERING MECHANISM FOR RED DWARF FLARES

M.K. Das[1] and J.N. Tandon[2]
[1]Department of Physics, Sri Venkateswara College, Univ. of Delhi, Dhaula Kuan, New Delhi-110 021.
[2]Department of Physics and Astrophysics, Univ. of Delhi, Delhi-110 007

The flare phenomenon associated with dMe stars has received much attention in recent years (Gershberg 1975). Most of the flares have been detected in both optical and radio band (Lovell 1969; Kunkel 1974; Karpen et al, 1977). But as expected (Tandon 1976) only a few display weak soft X-ray emission (Karpen et al, 1977; Haisch and Linsky 1978). Simultaneous X-ray, optical and radio observations of YZ CMi by Karpen et al (1977) shows no X-ray emission above 3σ level accompanying minor flares. Even coincident X-ray coverage with seven radio bursts shows no enhanced X-ray emission. Recently Haisch et al (1981) detected one well resolved X-ray flare on dM5e flare star Proxima Centauri and one coincident optical and radio flare out of five optical and twelve radio flare events. However, the X-ray flare on Proxima Centauri is not accompanied by any ultraviolet, optical or radio emission. Observations on flare stars show that they are more energetic, $10^2$ - $10^3$ times, than the corresponding solar flares. Considering the flare activity in dwarf M-stars to be similar but more energetic to that of a large solar flare, Tandon (1961) proposed red dwarf flares to be the source of low energy galactic cosmic rays. This hypothesis has been reexplored recently by Lovell (1974).

Generally a solar flare type mechanism is envisaged to explain the red dwarf flare phenomenon (Gershberg 1975). It may be noted, however, that during the flare the ratio of energy output in the radio wave band to optical continuum in general is much greater than from the Sun. To explain the enhancement in the total optical magnitude during the flare through a solar type mechanism, Kahn (1974) showed that one requires extremely relativistic electrons ∿300-500 MeV gyrating in a magnetic field ∿ 900 gauss on the star's surface. On similar lines Mullan (1974) linked the stellar flare on YZ CMi with stellar spots carrying magnetic fields of approximately 20 kilogauss: If such a strong magnetic field prevails on the star, one must be able to measure the



*P. B. Byrne and M. Rodonò (eds.), Activity in Red-Dwarf Stars, 609–611.*

optical polarization. However, recent observations on two spotted BY Dra and EV Lac by Petterson and Hsu (1981) and on five UV Ceti and BY Dra by Clayton and Martin (1981) show no linear or circular polarization. It may be remarked that the theories of red dwarf flares based on solar flare mechanism envisages detectable enhanced X-ray emission. Since so far there are no definite evidence of accompanying X-ray flares on these stars, a reconsideration of the problem is called for. In this small note, we therefore, propose an alternative model for the flare activity in dwarf M-stars.

Most of the M dwarf stars are known to be low mass stars. The star must have begun its life from a contracting, gravitating interstellar mass and that the proto-star must have acquired both rotation and magnetic field of the interstellar matter from which it evolved. We assume the magnetic field energy in the interior to be sufficiently large so as to survive the Hayashi turbulence. From the evolutionary track of low mass stars (0.3 - 0.4 $M_o$) (Ezer and Cameron 1967) it can be seen that these stars cease to be wholly convective and a radiative core (approximated as a polytrope of index n=3) forms after about 20-50 million years. With contraction the central temperature rises to about (5-8) x $10^6$K exciting thermonuclear processes inside the newly formed small central convective core. The equilibrium magnetic field inside the core can therefore be approximated as that of a magneto-rotating polytrope of index n=1.5. With subsequent rise in temperature (10-13)x $10^6$K due to enhanced energy production the size of the convective core increases to its maximum and further the interaction of rapid mass-motion or differential rotation with the primeval magnetic field results in the generation of a mainly toroidal component $H_\phi$ of the magnetic field inside the core. Therefore, the ratio of toroidal to poloidal of the magnetic field in the stellar core increases. Due to this and increased core size the magnetic field in the thin radiative region cannot remain in equilibrium with the magnetic field inside the core and subsequently with that existing in the outer convective zone, since the equilibrium magnetic fields for n=1.5 and n=3 are different (Das and Tandon 1976). In order to restore equilibrium, strong pressure waves propagating from the interior along the magnetic field lines will plunge into the outer convective layers of the star and might lead to an outburst. The mass ejecta of the order of $10^{-8}$ to $10^{-6}$ $M_o$ associated with the moving pressure waves with velocities of a few km per sec. will release energy of the order of $10^{35}$ - $10^{37}$ ergs. It may be added that these strong pressure waves derive their energy as a result of an increased rate of release of nuclear energy inside the core. This process is expected to be a piecemeal one with one to several flares at its initial stage. A fraction of the energy released through mass ejecta is capable of enhancing the total luminosity of these red dwarf stars by a few magnitudes and exciting a shock wave in the stellar chromosphere leading to accompanied enhanced radio emission. Since the size of the radio flare will be much larger than that of the optical flare, the ratio of radio to optical output of energy will be much larger. This mechanism does not envisage associated large X-ray emission during the flare and further does not require large surface magnetic fields on these stars.

REFERENCES

Clayton, G.C. and Martin, P.G.: 1981, Astron. J.,86, 1518.
Das, M.K. and Tandon, J.N.: 1976, Astrophys. J.,209, 233.
Ezer, D. and Cameron, A.G.W.: 1967, Can. J. Phys. 45, 3461.
Gershberg, R.E.: 1975, "Variable stars and stellar Evolution", Sherwood and Plaut (eds.), IAU Symp. 67, 47.
Haisch, B.M. and Linsky, J.L.: 1978, Astrophys. J. Letters, 225.
Haisch, B.M. et al.: 1981, Astrophys. J. 245, 1009.
Kahn, F.D.: 1974, Nature, 250, 125.
Karpen, J.T. et al.: 1977, Astrophys. J., 216, 479.
Kunkel, W.E.: 1974, Nature, 248, 571.
Lovell, B.: 1969, Nature, 222, 1126.
Lovell, B.: 1974, Phil. Trans. R. Soc., London, 277, 489.
Mullan, D.J.: 1974, Astrophys. J., 192, 149.
Petterson, B.J. and Hsu, J.C.: Astrophys. J. 247, 1013.
Tandon, J.N.: 1961, Nature, 190, 246.
Tandon, J.N.: 1976, Electricals, Electronics and Telecommunication, March issue, 23.

DISCUSSION

Rosner: I do not understand exactly what you are trying to explain. Are you trying to explain the white-light emission in the flares or what?

Das: We are trying to predict the optical continuum which is being generated. The bulk of the star is involved because pressure waves propagate in all directions in the envolope.

# THEORY AND OBSERVATIONS OF NEGATIVE PREFLARES IN UV CET STARS

V.P. Grinin
Crimean Astrophysical Observatory, USSR

Negative preflares (NPFs) of UV Cet stars, which were first observed by Italian astronomers |1|, are a highly unusual type of preflare activity without direct analogy in solar flares. The rarity of these events in the visual region leads to difficulties in statistical investigations. At present one can consider as reliably established only that the mean NPF-amplitudes increase and the probability of their appearance decreases with a shift toward blue |2|. According to |3| NPFs are observed mainly before flares of smaller amplitudes. Beginning in 1974 at the Crimea and the Astronomical Institute of Tashkent a series of works on the theory and observation of NPFs was carried out. A short review of these is given below.

As shown in |4| the preflare depression of light below the quiescent level may be a consequence of a weak impulsive heating of the stellar atmosphere. This anomalous response is due to the strong temperature dependence of the ionization of metals, the main source of free electrons. As a result even a small heating leads to noticeable increase of the $H^-$ opacity, that in turn leads to the temporary decrease of the flux. The mean duration of NPF is determined by the characteristic time of a temperature relaxation at optical depth $\tau \simeq 1$ and is of the order of a few tens of seconds; its amplitude is maximal in the center of the stellar disc and is negligible at the limb. More detailed calculations |5| accounting also for opacity variations due to molecular lines and bands, showed that this effect takes place in a wavelength region excluding the strong molecular bands of TiO (Fig. 1). Then the amplitude of NPF increases toward long wavelengths, which agrees with the observations. At the same time the observations of such fine effects are especially difficult in the U band due to the veiling of NPFs by the flare itself. In this respect the spectral region $\lambda \sim 1\mu$, free from the strong molecular bands, was proposed in |5| as the most suitable for the monitoring of NPFs.



*P. B. Byrne and M. Rodonò (eds.), Activity in Red-Dwarf Stars, 613–615.*

In 1975 synchronous photoelectric observations of flare stars in the visual and near IR-regions were initiated in the Crimea and then continued in Tashkent. Up to the present more than one hundred flares of UV Cet, EV Lac and AD Leo have been observed |6-8|. These observations fully confirmed the prediction of a sharp increase in the rate of occurrence of NPFs in the near IR-region: on average for two observational seasons, about 40% of flares observed in the i-filter ($\lambda_{eff}$=0.8μ) manifestes NPFs. Most of them are similar to the UV Cet flare 24.8.76 (Fig.2) when a single NPF just before the positive flare was observed. Sometimes the NPF and the following flare are separated by a time interval of up to 1/min. In a few cases more complicated preflare activity was observed. For example in the flare of UV Cet on 23.8.76 three successive NPFs were observed in i: one of them proceded the main flare, two others preceded two earlier bursts in U band |6|.

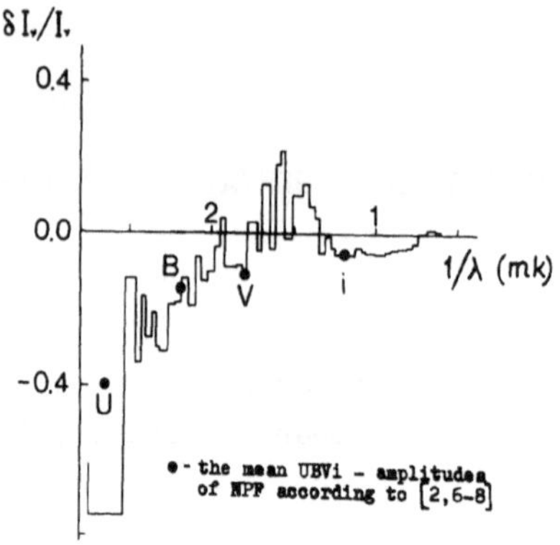


Figure 1. The amplitude of the intensity perturbation $\delta I_\nu/I_\nu$ in an atmosphere with $T_{eff}$=3500K, log g= 5 and $\delta T/T=\exp(-k\tau)$, k=5.

During these observations were obtained two new results related directly to flares themselves. 1) A correlation between the (U-i) colour at light maximum and flare amplitudes was found, which shows that the temperatures of flares increase with increasing amplitude. 2) The systematic reddening of the (U-i) colour was observed along the descending branch of the light curve of the flares |8| due to the relaxation of the heated atmosphere |4|.

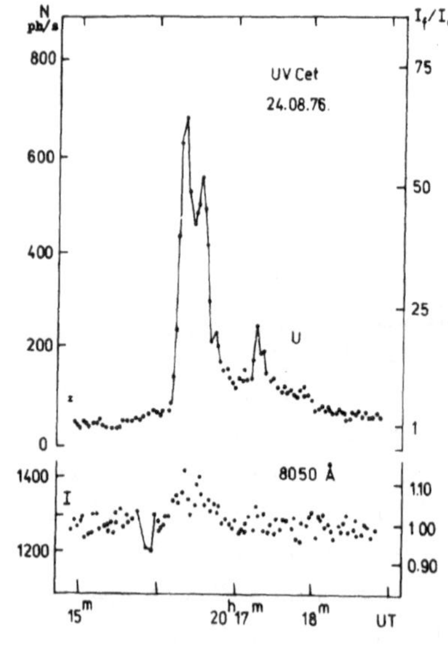


Figure 2. Simultaneous U and i observations of an UV Cet flare.

In total these observations confirm the proposed NPF model. However for the final test of this theory it is necessary to carry out photoelectric observations in the i band and in any narrow bandpass centered on TiO-bands. It follows from Fig.1 that the preflare variations of light in these two regions caused by opacity changes are expected

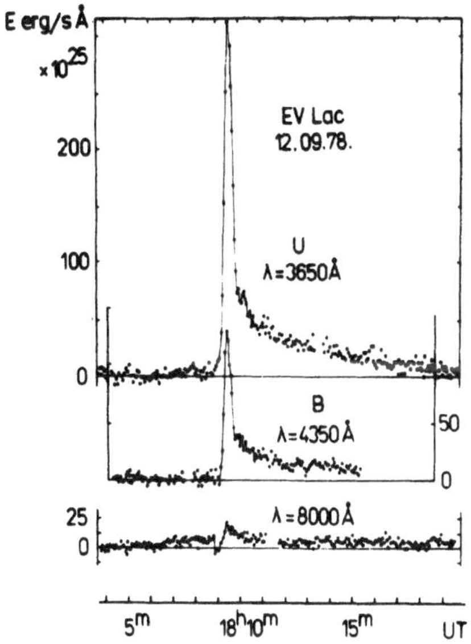


Figure 3. Simultaneous U,B and i observations of an EV Lac flare.

to be in antiphase.

Some other mechanism of NPFs were also briefly discussed in |9, 10|. One of them |9| is based on the assumption that the continuous emission from flare stars consists of two components: photospheric emission and continuous emission formed either in compact active regions or as a result of the superposition of a successive microflares. The decrease of this additional noise-emission related to the preflare variations of the magnetic field might be the reason for the observed preflare depression of light. This hypothesis might be confirmed if the relation between the NPF-amplitudes and the level of additional continuous emission were established. The EV Lac flare of 12.9.78 (Fig.3) might serve as a hint to this relation. Some minutes before the flare the light in i band slowly increased. Then the following NPF reduced the excessive emission returning the light to its initial level.

REFERENCES

1. Cristaldi, S., Rodonò, M.: 1969, in "Non Periodic Phenomena in Variable Stars", IAU Coll. , ed. L. Detre, Acadamic Press, Budapest, p. 149.
2. Cristaldi, S., Gershberg, R., Rodonò, M.: 1980, Astron. Astrophys. 89, 123.
3. Shevchenko, G.G.: 1973, Astron. Circ., No.792, 2.
4. Grinin, V.P.: 1973, Izv. Frym. Astr. Obs. 48, 58.
5. Grinin, V.P.: ibid. 55, 179.
6. Brujevich, V.V. et al.: 1980, ibid. 61, 90.
7. Kiljachkov, N.N., Shevchenko, V.S.: 1976, Soviet. Astron. Letters., 2, 494.
8. Kiljachkov, N.N., Shevchenko, V.S.: 1980, in "Flare Stars, Fuors and Herbig-Haro Objects", ed. L.V. Mirzoyan, Publ. Acad.Sci., Erevan, p. 31.
9. Grinin, V.P.: 1980, ibid., p. 23.
10. Giampapa, M.S., Africano, J.L., Klimke, A., Parks, J., Quiglay, R.J. Robinson, R.D., Worden,S.P.: 1982, Astrophys.J. 252, L39.

# FLARES ON RED DWARF STARS AS A RESULT OF THE DYNAMICAL RESPONSE OF THE CHROMOSPHERE TO THE HEATING

M.M. Katsova
Sternberg Astronomical Institute, Moscow State University, Moscow, USSR
M.A. Livshits
Institute of Terrestrial Magnetism, Ionosphere and Radio Wave Propagation of the Academy of Sciences, Moscow, USSR

The impulsive hard X-ray, EUV and microwave radio bursts always start at or after the onset of a soft X-ray increase - the main phase of a solar flare - before the Hα flare maximum. This hard phase of a solar flare lasts ∿ 100 s and consists of several elementary bursts of 10 s duration. In this time electrons are accelerated up to energy 20-100 keV, then they are injected vertically downwards into denser chromospheric layers. The electron heating as well as the thermal conductivity heating leads to the appearance of the secondary process, in particular, to gasdynamical motions (Kostyuk and Pikel'ner, 1974).

The above mentioned electron event and corresponding secondary process are considered for a red dwarf atmosphere similarly to the solar case. The energy flux of a hard electron beam (> 15 keV) is supposed to be $10^{12}$ erg/cm$^2$s. The numerical solution of the one-dimensional system of gasdynamical equations was given by Katsova et al (1981) and by Livshits et al (1981). Radiative losses balance the energy input at each point taking into account the resonance line opacity. The adopted pressure at the coronae base was P=0.3 dyne/cm$^2$ in the initial model. Due to the initial heating, a region of higher pressure is formed, from which disturbances propagate upwards and downwards - see Figure 1. A shock wave is formed in front of the temperature jump propagating downwards. Between the temperature jump and this shock front a condensation with $\Delta z \approx 1$ km at the beginning and $\approx$ 10 km at the end of this process appears. The optical depth at λ4500 Å for this condensation (with n ∿ $10^{15}$ cm$^{-3}$ and T=8500-9000K) is $\tau \sim 1$. On a star with $T_{eff}$=3250K , R =0.3$R_\odot$ and a flare area S=3x$10^{18}$cm$^2$ the ratio of the flare B-filter-luminosity to the stellar bolometric luminosity is $L_{fl}/L \approx 0.7$; a smaller radius, $\approx 0.1 R_\odot$, leads to $L_{fl}/L \approx 6.3$. Grinin and Sobolev (1977), who were the first to suggest that the



*P. B. Byrne and M. Rodonò (eds.), Activity in Red-Dwarf Stars, 617–619.*

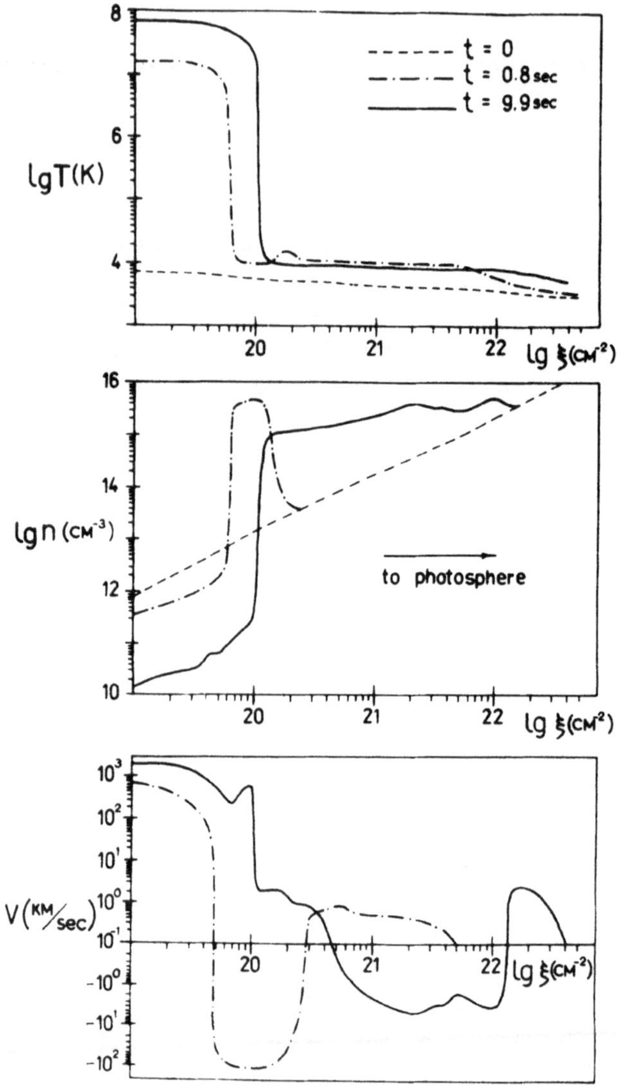


Figure 1. Distributions of the temperature, density and velocity calculated for the response of the red dwarf chromosphere to heating by a non-thermal electron beam with a power-law energy spectrum $vN \sim E^{-3}$. The positive velocities correspond to the gas flowing upwards. The dashed line shows the initial model.

optical flares on UV Cet-stars are located between the photosphere and the chromosphere, give U-B=-1 and B-V=0.5 for the low temperature emission source in the flare. The photometric parameters of this condensation are found to be in a good agreement with observations, in particular, with $T_{fl}$=8500K (Mochnacki and Zirin, 1980).

If the evaporated hot plasma is kept in closed magnetic loops near the stellar surface, the X-ray to optical luminosity ratio $L_X/L_{opt}$ is equal to 0.02 for red dwarf flares. When the hot plasma flows from the star, this ratio will decrease. On the other hand, if the flare optical continuum is absent or is emitted from a small area (as in a solar white flare), $L_X/L_{opt}$ will increase up to the solar value that is equal to several units.

We consider only the electron event without the second main phase of the solar flare, when the solar cosmic rays are sometimes accelerated. Such a second phase may be absent in red dwarf flares, and thus the gamma-radiation and nuclear line fluxes may be less than suggested by data on large solar flares.

REFERENCES

Grinin V.P. and Sobolev V.V.: 1977, Astrofizika, Vol. 13, p. 587.

Katsova M.M., Kosovichev A.G. and Livshits M.A.: 1981, Astrofizika, Vol. 17, p. 285.

Kostyuk N.D. and Pikel'ner S.B.: 1974, Astron. J. USSR, Vol. 51, p.1002.

Livshits M.A., Badalyan O.G., Kosovichev A.G. and Katsova M.M.: 1981, Solar Phys. Vol. 73, p. 269.

Mochnacki S.W. and Zirin H.: 1980, Astrophys. J. Letters, Vol. 239, p. 27.

# MHD THERMAL INSTABILITIES IN COOL INHOMOGENEOUS ATMOSPHERES

G. Bodo[1], A. Ferrari[2,3], S. Massaglia[3] and R. Rosner[4]

1 Osservatorio Astronomico di Torino
2 Istituto di Cosmo-geofisica del C.N.R., Torino
3 Istituto di Fisica Generale, Universita' di Torino
4 Harvard-Smithsonian Center for Astrophysics,Cambridge, MA, USA

## 1. THE MODEL

The Einstein Observatory survey of stellar coronae (Vaiana et al. 1981) and, specifically, the results on cool, low luminosity stars has suggested a correlation between stellar X-ray luminosity and stellar rotational velocity (Pallavicini et al. 1982, Walter 1981, Vaian et al. 1981). In addition the Skylab observations of the solar corona have demonstrated a tight correlation between photospheric surface magnetic structures , which emerge from the interior in the form of "loops" above the photosphere by viz., buoyancy instabilities, (Parker 1979, see also Acheson 1979, Schmitt & Rosner 1982 and references therein), and coronal X-ray emission (Golub et al. 1980). It therefore becomes important to ask how a coronal state (i.e. low density and high temperature plasma) of a stellar atmosphere is formed, presumably from a pure radiative equilibrium configuration.

In principle, a corona may be thought of as a metastable configuration evolved from a cool equilibrium due to some instability in which magnetic fields and currents play a basic role (Ferrari, Rosner & Vaiana 1982). For example these authors have shown that a plane parallel, homogeneous, infinite atmosphere in radiative equilibrium threaded by (re-entmant) magnetic lines, is unstable to MHD thermal instabilities excited by photospheric motions which produce field aligned electric currents. The heating process is related to dissipation of these currents in the presence of filamentation effect; the resulting sharp inhomogeneities lead to quasi-stationary heating processes such as tearing mode reconnection, surface wave dissipation, fast mode damping, which were presumed to ultimately limit the filamentation process by enhancing the current diffusion rate. However, the linear instability analysis in the form presented by Ferrari, Bosner and Vaiana employed a number of simplifying approximations; one of these, namely the spatial uniformity of plasma parameters and concomitant infinite extension of magnetic structures, is relaxed in the present work.

We start from a cool plasma configuration in radiative and hydrostatic equilibrium, and assume that buoyancy instabilities have brought magnetic loops to the solar surface whose spatial scales are large compared to the instability scales considered here, and which remain "anchored" to the photosphere (see Fig. 1). We consider an initially isothermal atmosphere, whose density and pressure have scale heights fixed by the hydrostatic equations.



*P. B. Byrne and M. Rodonò (eds.), Activity in Red-Dwarf Stars, 621–623.*

Linear instability calculations, using the system of equations of motion of Ferrari, Rosner and Vaiana (1982), applied to a non-homogeneous configuration, do not yield simple analytical dispersion relations. Therefore, we have developed a numerical code which allows us to solve for oscillatory perturbations as eigenfunctions of a boundary value problem. With this code we have investigated the effects of atmospheric gradients and finite loop dimension on the scale of unstable perturbations. We want to caution that we have considered gradients for the initial equilibrium atmosphere in the upward direction (i.e. along B) only; no gradients are assumed across magnetic lines, i.e. parallel to the stellar surface. The mathematical problem is hence one-dimensional as far as eigenfunctions are concerned.

## 2. THE RESULTS

The most relevant mode for our problem is the Joule mode (Heyvaerts 1974, Ferrari et al. 1982). This mode is essentially transverse to the magnetic field, and is connected with current filamentation; the complete dispersion relation can be found in Ferrari et al. (1982). In the low frequency limit, such that the plasma can rapidly diffuse across the filaments (so that pressure build-up that could stabilize the mode can be avoided), $\omega \ll c^2 k^2/4\pi\sigma_o$, and

$$\omega = -\, i(\gamma-1)[-(\kappa_{\|} k_{\|}^2 + \kappa_{\perp} k_{\perp}^2) T_o/p_o + (dh/dT)_o\, T_o/p_o + J_o^2\, (d\ln\sigma/d\ln T)_o\, (\sin^2\theta - \cos^2\theta)]/\sigma_o p_o$$

where $\kappa_{\|}$ and $\kappa_{\perp}$ are the thermal parallel and perpendicular conductivities, $h = H - E_R$, $\theta$ is the angle between k and $B_o$ and the subscript "o" denotes parameters evaluated in the equilibrium state. For instability, Im $\omega < 0$, which requires that the last term be positive and exceed the two preceding ones in absolute value.

In Fig. 2, we plot the dispersion relation of Joule modes in the (Im $\omega$, $k_x$) plane. The physical parameters used in this example apply to a "classical" stellar atmosphere, initially without any cromosphere or corona. Joule dissipation is energetically dominant in the optically thin region of the atmosphere, where $\tau \ll 1$ and $\beta \ll 1$.

As Fig. 2 shows, we find that Joule mode instability is an efficient mechanism for current filamentation and subsequent heating in initially cool atmospheres. In particular, this instability is mainly effective at the top of magnetic loops and is not suppressed by thermal conduction.

Typical scales of current filaments (L = 10 ÷ $10^3$ cm for the case of Fig. 2) are too small to be detected by currently available direct observations. However, they are sufficiently small to satisfy the theoretical requirements of steady creation of current inhomogeneities and steep magnetic gradients imposed, but generally left unexplained, in many coronal heating models.

Acknowledgments. This work was supported by NASA under the Solar Terrestrial Theory Program grant NAGW-79 to the Harvard College Observatory (RR) and by the Servizio Attivita' Spaziali del CNR (GB, AF and SM).

Acheson, D.J. 1979 Solar Physics, 62, p.23
Ferrari, A., Rosner, R., and Vaiana, G. S. 1982, Ap. J. in press
Golub, L., et al. 1980, Ap. J., 238 , p. 343
Heyvaerts, J. 1974, Astron. Astrphys. 37, p. 65

Pallavicini, R., et al. 1982, in 2nd Cambridge Cool Stars Workshop, ed. L. Golub and M. Giampapa
Parker, E. N. 1979, Cosmical Magnetic Fields (Oxford, Clarendon Press)
Schmitt, J. and Rosner, R. 1982, Ap. J. in press
Vaiana, G. S. et al. 1981, Ap. J. 244, p. 163
Walter, F. G.. 1981, Ap. J. 245, p. 677

Figure 1 – Model atmosphere:
a) realistic configuration,
b) model configuration.

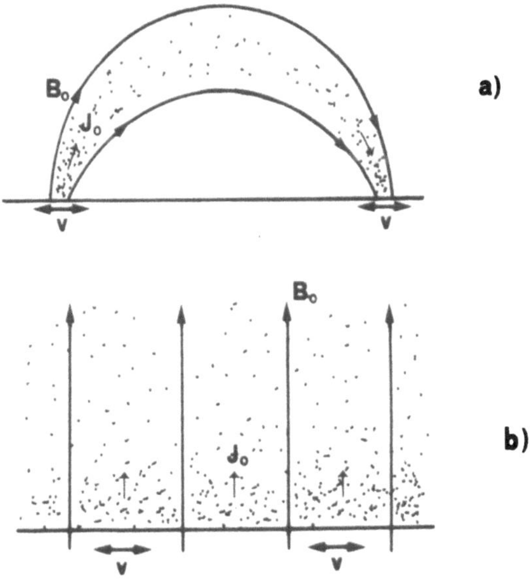


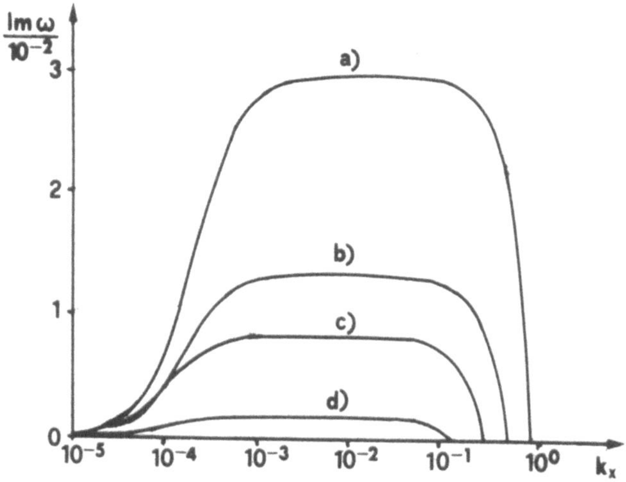


Figure 2 – a) and c): analytic solutions of the homogeneous, infinite atmosphere ($k_{//} \to 0$) dispersion relation; a) refers to plasma conditions at the top of the loop, n=2.9 × $10^6$ $cm^{-3}$, c) to the bottom of the loop, n=$10^7$ $cm^{-3}$.
b): dispersion relation for the case of a model atmosphere with longitudinal gradients.
d): growth rate for a homogeneous atmosphere, in which the loop has finite dimensions, and boundary conditions become important in defining the eigenfunction.

# MAGNETIC ACCRETION MODEL FOR THE ACTIVITIES OF VERY YOUNG STARS

Yutaka UCHIDA
Tokyo Astronomical Observatory, University of Tokyo,
Japan

ABSTRACT : A magnetodynamic model is proposed to interpret the observed inflow-outflow-X-ray emitting region complex in the atmospheres of T Tau stars.

Rapidly varying X-ray emissions ( Feigelson and de Campli 1981 ) as well as the variation of the optical line emissions with widely separated components toward blue and red ( eg., Mundt and Giampapa 1982 ) immediately suggest a highly dynamical character of what is occurring in the atmospheres of T Tau stars. For example, Hα line profile of RW Aur with distinct emission components, one with a blueshift of about 300 km/s and the other with a redshift of about 230 km/s separated by a minimum of near-zero intensity ( Hartmann 1982 ), is difficult to explain, eg., by the rotation effect. Several models with stellar wind or mass accretion have been discussed ( Kuhi 1964, Lynden-Bell and Pringle 1974, Ulrich 1976, Bertout 1979 ), but these models have encountered difficulties in explaining, eg., how in- and outflow can coexist without degenerating into turbulence, or the fact that the outflow is not of the coronal temperature as expected in the case of stellar winds from dwarf stars. Queer enough, further, some observations suggest that the X-ray emitting region is located below Hα emitting region.

We here present a dynamical model to resolve these difficulties. Since the magnetic moment of a new-born star formed from the nebulosity is expected to be parallel to the field of the nebulosity, a magnetically neutral ring is produced in the equatorial plane after the star formation ( Uchida and Low 1981 ). The balance of the Lorentz force exerted by the stellar dipolar field and the gravity acting on the accreted mass gives the distance of the balancing point in the equatorial plane,

$$\frac{r}{r_*} \sim \left( \frac{B_p^2/8\pi\Delta r}{\rho \; GM/r_*^2} \right)^{1/4} \sim 6 \, ,$$

if we take $r_* \sim 7 \times 10^{10}$ cm, $M \sim 2 \times 10^{33}$ g, $B_p \sim 5 \times 10^2$ G, $\Delta r \sim 10^{10}$ cm, and the density of the suspended mass $\rho \sim 10^{-13}$ g/cm$^3$. The magnetic field of the nebulosity condensed to $\rho \sim 10^{-13}$ g/cm$^3$ can be of the same order as the stellar field there, and opposite in sign. The suspension



*P. B. Byrne and M. Rodonò (eds.), Activity in Red-Dwarf Stars, 625–627.*

of mass is equivalent to the storage of energy in the form of potential energy per unit mass, $\varepsilon \sim GM(r_*^{-1} - r^{-1}) \sim 2 \cdot 10^{14}$ erg/g, and the flare energy of these stars may be explained if a release of the mass in a shell of thickness $d \sim 10^9$ cm at the innermost edge of the disk can take place. The release of the mass may be effected through the magnetic reconnection at the neutral ring when the increase of loaded mass distorts the field and compresses the neutral ring region and the critical condition for the magnetic reconnection may be violated ( cf. Hayashi and Sato 1978 ). The reconnection transfers the mass to the stellar field, and the mass can fall almost freely to the stellar surface. The free fall time is $\tau_{ff} \sim (r^3/GM)^{1/4}$ s, and the terminal velocity $v_{ff} \sim (GM/r_*)^{1/2} \sim 3 \times 10^7$ cm/s. The temperature behind the shock produced in the crash amounts to $T_2 \sim (\mu m_H/3k)\, v_{ff}^2 \sim 3 \times 10^6$ K. These explain the infall velocity of the Hα-emitting gas and the production of X-ray emitting region closer to the stellar surface with a suitable time scale for the observed variation.

The problem of the ejection of the cool gas with high velocity requires some more elaboration. By using a tube with varying cross section and effective gravity simulating the critical flux tube in the equilibrium model by Uchida and Low (1981), Shibata and the present author have recently performed a gasdynamic simulation calculation by releasing a lump of mass from the location of the neutral ring. The cool gas falls along the tube with a terminal velocity of the order of 200 km/s, and the temperature of the region just above the stellar surface rises instantaneously to $10^7$ K as the shock hits the surface, and drops to $3 \times 10^6$ K and maintained as the expansion of the heated gas takes place with the shock recoiled at the surface crashing into the tail of the infalling gas. The strength of the shock rapidly increases as it sweeps along the tail of the infalling gas, the gas is driven out with increasing velocity, and the expanded cool part of the driven gas attains a few hundred km/s after the shock has propagated. It is noted that the process of outflow takes place along the reconnected tube while the inflow takes place along the inner stellar flux tube when it comes into contact with the following flux tube of the nebular field. The reconnection separates the in- and outflows and thus both can occur without hindering each other.

## REFERENCES

Bertout, C., 1979, Astr. Astrophys., 80, 138.
Feigelson, E. D., and de Campli, W. M., 1981, Astrophys. J., 243, L89.
Hartmann, L., 1982, CfA preprint No. 1518.
Hayashi, T., and Sato, R., 1978, J. Geophys. Res., 83, 217.
Kuhi, L. V., 1964, Astrophys. J., 140, 1409.
Lynden-Bell, D., and Pringle, J. E., 1974, Mon. Not. R. Astr. Soc., 168, 603.
Mundt, R., and Giampapa, M. S., 1982, CfA preprint No. 1558.
Uchida, Y. and Low, B. V., 1981, Astrophys. Astr., 2, 405.
Ulrich, R. K., 1976, Astrophys. J., 210, 377.

DISCUSSION

Basri: The model seems quite nice for the time when the magnetic field is largely determined by the configuration of the interstellar magnetic field as slipped in by interstellar accretion. But at that stage, if I recall correctly the star would not be visible to us anyway. It would, in fact, look like an infrared cocoon source. At the time when the star is visible and emitting X-rays the star will have started generating magnetic field and associated activity and probably stellar wind as well. Would that not have a large effect on the configuration you have calculated.

Uchida: I have not loocked into this possibility. Anyway in this stage of condensation a large fraction of the nebula's magnetic field escapes, probably at the molecular cloud stage. The magnetic field which I assumed is about 500-1000G. After settling it has a dipolar field and after that the nebular remnant with its magnetic field is accreted to the star causing this particular process. So this process may be applicable only to the very young stars which still 'remember' the nebular field.

# INTERACTING MAGNETOSPHERES IN RS CVn BINARIES - CORONAL HEATING AND FLARES

Y. Uchida[1] and T. Sakurai[2]
1 Tokyo Astronomical Observatory, University of Tokyo, Japan
2 Department of Astronomy, University of Tokyo, Japan

ABSTRACT

Coronae and flares of RS CVn systems are interpreted as due to gradual and sudden releases of magnetic free energy which is built up throught the interaction of magnetic fields of stars in these close binary systems.

## X RAY AND RADIO OBSERVATIONS

Recent observations of RS CVn binaries in soft X-rays (Walter et al 1980) and in radio wavelengths (Gibson et al 1978, Feldman et al 1978) indicate the presence of vigorous coronae and the occurrence of flares. These, combined with the starspot hypothesis based on the "Photometric Wave" (PW) in the light curves (Hall 1972, Eaton and Hall 1979), strongly suggest that these high temperature or high energy phenomena may be due to activated magnetic fields in the outer atmospheres of the component stars, as already found in the case of the Sun (e.g., Sheeley et al 1974, Sakurai and Uchida 1977). In the present paper, we look into this possibility by examining the magnetic field configuration and various distorting effects.

## ACTIVE-LONGITUDE-BELT MODEL

In presenting our argument we first introduce a new aspect into the starspot hypothesis. We propose that the PW may correspond to the stellar analogue, in an extreme form, of an "Active-Longitude-Belt" on the Sun (ALB) in which active spot-groups are seen to emerge, drift across and disappear, rather than being due to a gigantic, long-lived starspot, or aggregate of spots. Introduction of the notion of an ALB relieves us of the somewhat unnatural, though very fascinating, assumption of a gigantic spot or aggregate of spots, staying almost fixed on



*P. B. Byrne and M. Rodonò (eds.), Activity in Red-Dwarf Stars, 629–632.*

the surface of a rapidly rotating star. The 8-10 year recurrence period of the PW means an extremely slow drift or migration of the spot on the surface! Much faster differential rotation may be expected even if the synchronization tends to suppress it. Thus, if we are freed from the restriction that the differential rotation is traced by the PW drift, we are allowed to assume a larger range of differential rotation on K and F stars about the average rotation which is assumed synchronous with the orbital revolution. Polar regions, with latitudes higher than the synchronous latitude, will be rotating retrogressively, while the equatorial regions rotate progressively with respect to orbital motion. The notion of an ALB is also better for interpreting some other observed features of the PW, like the presence of forward or backward drifts and the smooth reversal of the direction of drift in some systems. ALB's on the Sun are known to drift very much more slowly than the individual sunspots and the drift can be in either direction, or even change direction by a mechanism not yet understood (Bumba and Howard 1969, Gaizauskas et al 1982).

It is clear that the notion of an ALB introduces much dynamism into the picture. The size of the spot-pair may now be of the order of the depth of the convection zone rather than of the stellar radius and a much faster differential rotation of spot-pairs is now admissible. The possibility of differential motions of the footpoints of magnetic flux tubes (some of them connecting both stars) introduces a means of energizing the magnetic field, e.g. by induction of field-aligned currents by twisting or the formation of current sheets or in connections of spot-pairs, as well as pole-pole or pole-spot connections extending to the scale of the binary system.

By applying the method of Sakurai and Uchida (1977) to the present situation, we can calculate the magnetic field in the system. Since no direct measurement of magnetic field is yet avilable, we have to rely on an appropriately assumed set of photospheric magnetic field data for both the global and the starspot fields. The example shown in Figure 1 is for RS CVn itself (radii 1.9 $R_\odot$(F4V) and 4.1 $R_\odot$(KOIV), separation 16.8 $R_\odot$). Global magnetic fields are tentatively assumed to be due to dipoles with polar intensitied of 10 G (K star) and −100 G (F star), respectively, with four pairs of spots assumed in the ALB which as a 120° width in longitude on the K star. The spot parameters are: $a_p$= 0.5 $R_\odot$, $B_p$= 3000 G (preceding spots); $a_f$= 0.7 $R_\odot$, $B_f$= −1350 G (following spots), i.e. the area of the following spots are twice that of the preceding, and their fluxes are 10% less than those of the preceding ones as in the solar case.

## HEATING OF THE CORONA

A mechanism for energizing the magnetic field is now introduced in our picture by the notion of ALB's. Winding up of magnetic flux tubes in pole-pole or pole-spot connections, due to differential rotation builds up currents in the corona (Sakurai 1979, 1981), and heating is expected as a result of their dissipation (cf. Kuperus et al 1981). An alternative mechanism is due to the interaction between pole-pole and pole-spot connection which execute a different mode of motion. Flux tubes are strongly disturbed and the setting-up of current sheets at the interface, sudden restoration of the distortion of flux tubes, etc., may take place (Uchida and Sakurai 1977), leading to the subsequent heating of plasmas.

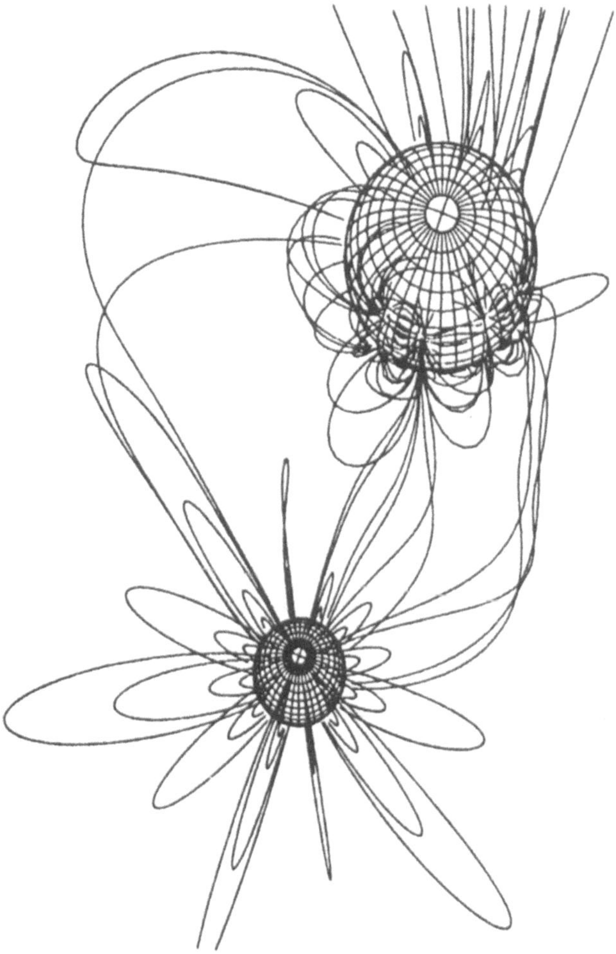

Figure 1. Possible magnetic field configuration in RS CVn system. For details, see the next.

Convective shaking of flux tube footpoints may also result in Alvèn wave heating (e.g. Uchida and Kaburaki 1974) or heating due to current dissipation (Sturrock and Uchida 1981).

## FLARES

We may, as an example, attribute the occurrence of flares to the drastic liberation of magnetically accumulated energy through reconnection (e.g., Priest 1982). As the spots drift in the ALB by differential rotation, the pole-spot connections are streched and interfere with other spot-spot connections. When reconnection takes place, e.g., as old spots disappear near the leading edge and new ones appear near the rearedge of the ALB, the stressed field can relax to a configuration of lower stress and the accumulated energy in the stressed field is liberated into dynamical or thermal modes of energy. In the solar analogy, such phenomena as i) the acceleration of electrons yielding radio and hard X-ray bursts, ii) the heating of magnetic loops to $10^8$ K, iii) the ejection of hot gas into connections created by the reconnection (cf. Simon et al 1980) and so on, may take place vigorously through magnetic energy liberation.

## REFERENCES


Bumba, V., and Howard, R., 1969, Solar Phys., 1, 28.

Eaton, J. A., and Hall, D. S., 1979, Astrophys. J., 227, 907.

Feldman, P. A., Taylor, A. R., Gregory, P. C., Seaquist, F. R., Balonek, T. J., and Cohen, N. L., 1978, Astron. J., 83, 1471.

Gaizauskas, V., Harvey, K. L., Harvey, J. W., and Zwaan, C., 1982, Herzberg Institute Preprint.

Gibson, D. M., Hickes, P. D., and Owen, F. N., 1978, Astron. J., 83, 1945.

Hall, D. S., 1972, Publ. Astron. Soc. Pacific, 84, 323.

Kuperus, M., Ionson, J. A., and Spicer, D. S., 1981, Ann. Rev. Astron. Astrophys., 19, 7.

Priest, E. R., 1983, in this Colloquium.

Sakurai, T., 1979, Publ. Astron. Soc. Japan, 31, 209.

Sakurai, T., 1981, Solar Phys., 69, 343.

Sakurai, T., and Uchida, Y., 1977, Solar Phys., 52, 397.

Sheeley, N. R., Bohlin, J. D., Brueckner, G. E., Purcell, J. D., Scherrer, V., and Tousey, R., 1974, Solar Phys., 40, 103.

Simon, T., Linsky, J. L., and Schiffer, F. H., 1980, Astrophys. J., 239, 911.

Sturrock, P. A., and Uchida, Y., 1981, Astrophys. J., 246, 331.

Uchida, Y., and Kaburaki, O., 1974, Solar Phys., 35, 451.

Uchida, Y., and Sakurai, T., 1977, Solar Phys., 51, 413.

Walter, F., Charles, P., and Bowyer, S., 1980, Proc. Workshop "Cool Stars, Stellar Systems, and the Sun", ed. A. K. Dupree.

# BASIC PARAMETERS DETERMINING X-RAY EMISSION LEVEL IN STARS OF SPECTRAL TYPE LATER THAN F5

Lucio Paternò and Francesca Zuccarello

Osservatorio Astrofisico di Catania, Italy

This is a merely exploratory search for estimating the importance of those stellar parameters which we believe to be relevant to the X-ray emission from stars possessing outer convective envelopes.

Belvedere et al.(1981,1982) have shown that a mechanism which converts magnetic into thermal energy is plausible for explaining the X-ray emission level in late type main sequence and giant stars. One of the main conclusions of their work is that the average surface X-ray flux $F_X$ depends on the square of the average surface magnetic field strength B. The surface magnetic activity, in stars possessing outer convective envelopes, is likely due to the interaction of rotation with convection which produces both differential rotation (Belvedere et al. 1980a) and dynamo action by the α-effect (Belvedere et al. 1980b). Therefore we would expect that the level of magnetic field intensity depends on star's rotation Ω and the depth of the convection zone(c.z.) D, because both parameters are very important in determining the strength of the interaction.

In order to estimate the dependence of B on Ω and D, we assume that the amplitude of B is limited by buoyancy, thus comparing the rise time of a flux tube at the bottom of the c.z. with the amplification time of the dynamo process (Durney and Robinson 1982). This leads to the following expression:

$$B \propto \rho_b^{1/2} \Omega H_b^{5/2} R_b^{-3/2} \qquad (1)$$

where $\rho_b$, $H_b$ and $R_b$ are respectively the density, the pressure scale height and the radial distance from the star's center at the bottom of the c.z. Assuming the c.z. adiabatically stratified, both $\rho_b$ and $H_b$ can be expressed in terms of fractional depth $d=D/R_S$ ($R_S \equiv$ stellar radius) of the c.z. We thus obtain from expression (1), neglecting the very slowly varying factor $(R_b D/R_S)^{1/4}$:

$$B \propto d^3 \Omega \qquad \text{and} \qquad F_X \propto d^6 \Omega^2 \qquad (2)$$

In the following figure we plot the X-ray surface flux $\log F_X$ of 41 stars observed by Pallavicini et al.(1981), Stern et al.(1981) and Walter (1981) vs. their rotational velocities $v_e = \Omega R_S \simeq v \sin i$ (see Uesugi and Fukuda, 1982, for Stern's stars). The straight lines show the behavior of the theoretical relationship (2) calibrated with the Sun, taking the flux



*P. B. Byrne and M. Rodonò (eds.), Activity in Red-Dwarf Stars, 633–635.*

level of $\log F_{XO}=4.7$, for different spectral type and luminosity class stars.

Besides the uncertainties due to v sin i, here assumed as $v_e$, and the observed $F_X$ values, if we take into account that the rotational velocities of K stars are probably overestimated (Catalano, private communication), we see that the positions of most stars in the Figure can be accounted for in the framework of the present discussion.

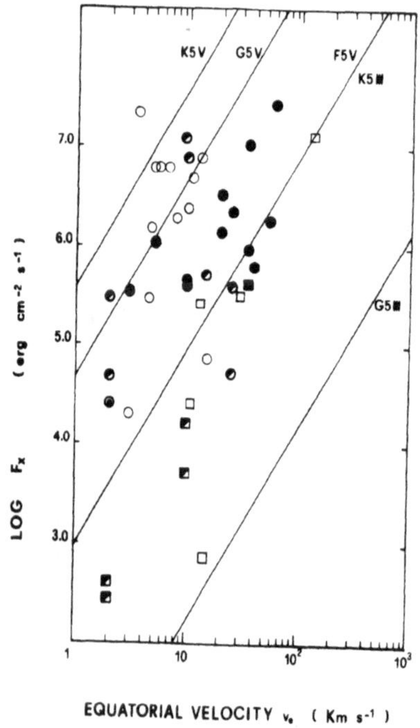


In the figure, circles and squares represent respectively luminosity class V and III stars. Filled, empty and half filled symbols represent respectively F, G and K spectral type stars.

References

Belvedere, G., Paternò, L., Stix, M.: 1980a, Astron. Astrophys. 88, 240
Belvedere, G., Paternò, L., Stix, M.: 1980b, Astron. Astrophys. 91, 328
Belvedere, G., Chiuderi, C., Paternò, L.: 1981, Astron. Astrophys. 96, 369
Belvedere, G., Chiuderi, C., Paternò, L.: 1982, Astron. Astrophys. 105, 133
Durney, B.R., Robinson, R.D.: 1982, Astrophys. J. 253, 290
Pallavicini, R., Golub, L., Rosner, R., Vaiana, G.S., Ayres, T., Linsky, J.L.: 1981, Astrophys. J. 248, 279
Stern, R.A., Zolcinski, M.R., Antiochos, S.K., Underwood, J.H.: 1981, Astrophys. J. 249, 647
Uesugi, A., Fukuda, I.: 1982, Revised Catalogue of Stellar Rotational Velocities, Astron. Dept. Univ. Kyoto, Japan.
Walter, F.: 1981, Astrophys. J. 245, 677

DISCUSSION

Haisch: Comparing time-averaged surface He II fluxes is really the only sensible kind of comparison you can make between theory and observation. The problem is that I do not know what that means physically. We know that the X-ray flux is highly structured and originates from only a small fraction of the stellar surface. So I do not know what it means physically to compare the surface-averaged heating with the average X-ray emission.

Zuccarello: All of these calculations have been made in the framework of dynamo theory. We suppose that the magnetic field intensity depends on the interaction between rotation and convection. So we have used this data to make a comparison between theory and observation. I do not know if it is possible at the moment to make any other such comparison.

Kodaira: I have a comment concerning the small radius involved in the flare. I have shown that this may be up to $10^{19}$ and this is only a few percent of the stellar surface. This is the size of the white-light flare but may be the Hα-emitting region can be smaller than that. The ratio of the size of the white-light kernel to the total volume involved in the flare is of the order 100. If we take that same ratio here a very large volume may be involved, extending to a global instability. So I am not so sure that only a small fraction of the stellar surface may be involved in a stellar flare.

Rosner: I would like to make a comment here. This paper was mostly concerned with quiescent activity and not flaring. So I think that Dr. Haisch's comment was appropriate i.e. that solar X-rays arise in relatively compact active regions.

# SESSION V

## FUTURE RESEARCH DIRECTION AND CONCLUSION

FUTURE RESEARCH DIRECTIONS: THEORETICAL APPROACH AND PERSPECTIVE

N.O. Weiss
Dept. of Applied Mathematics and Theoretical Physics
University of Cambridge
England

Abstract:

It is difficult to provide a comprehensive theoretical explanation for the activity of red dwarf stars. Among particular problems that are ripe for further investigation are: the production of steady, cyclic or irregular patterns of activity by nonlinear dynamo action in stars; the effect of magnetic buoyancy in producing photospheric magnetic fields; the formation of isolated flux tubes and their interaction with convection. These topics are discussed and some future lines of research are suggested.

## 1. THE ROLE OF THEORY IN ASTROPHYSICS

In astronomy, theory is generally led by observations. Even for the sun, where magnetic fields have been observed in great detail over many years, a coherent theoretical description of their structure is only beginning to emerge. The papers presented at this meeting have shown how far theoreticians are from providing an adequate explanation of magnetic activity in red dwarfs. In this review, therefore, I shall attempt to make some general points, and to illustrate them with specific calculations. My choice of topics is somewhat arbitrary, and slanted towards my own interests. I have tried to avoid excessive overlap with other papers in these Proceedings and to focus on the interiors, rather than the atmospheres, of stars. Much of importance is therefore ignored.

It may be helpful to preface these examples with some remarks about the function of theory in astrophysics. I assume that we are dealing with situations where the basic physics is understood (thus we are not, for instance, considering the first $10^{-36}$ seconds after the big bang). The difficulty lies in applying known laws to a particular configuration. To proceed, we have to construct models that are drastically simplified in order to render them mathematically tractable. It is important to distinguish here between two different activities: producing models to



*P. B. Byrne and M. Rodonò (eds.), Activity in Red-Dwarf Stars, 639–650.*

rationalise the observations and using models to provide a basis for detailed theoretical investigations.

"Modelling" frequently implies the construction of the simplest model that fits the observational data. This is a problem in constrained optimization, shown schematically in the following flowchart.

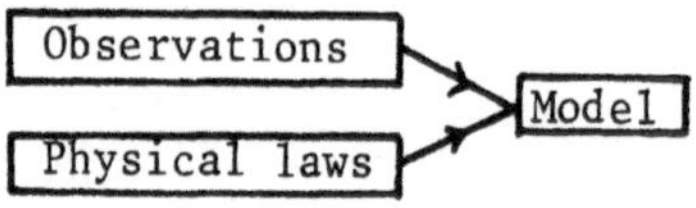


The aim is to find the simplest description that is compatible with the relevant observations, subject to the constraint that some subset of the appropriate laws should be satisfied. For example, one may constuct model atmospheres, models of the structure in magnetic loops or models of the flaring process. Such models are obviously important but I shall not discuss them further here.

A more fundamental approach is to introduce models as an aid to understanding physical processes. This leads to an iterative process, as illustrated below:

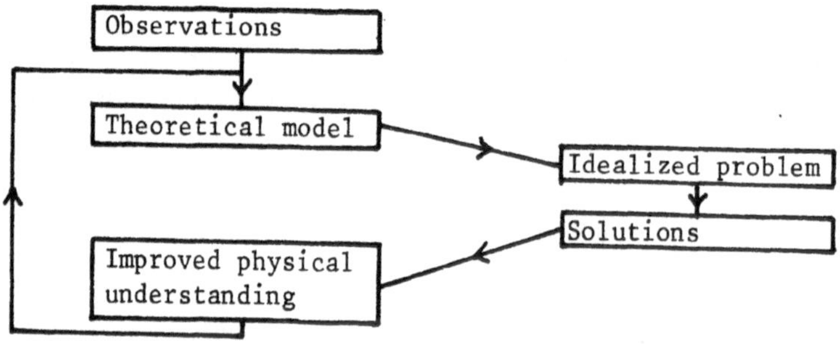


The observations are used to construct a simplified theoretical model, which provides the basis for a properly posed mathematical problem that can be rigorously solved (whether analytically or numerically). The solutions help to improve one's physical understanding and so enable one to improve the theoretical model, thereby generating further (often more complicated) idealized problems and so forth. The success of this procedure depends critically on the interaction between theory and observation. It is vital that theoreticians should listen carefully and sympathetically to observational results but it is also important that observers should appreciate the theoretical significance of their data and organize material so that theoreticians can assimilate it. Above all, effective communication between theoreticians and observers is needed to ensure that new and fruitful observing programs will continue to emerge.

## 2. PHENOMENOLOGICAL DESCRIPTION

The first stage in producing a theoretical model is the development of plausible phenomenological description, which is essentially qualitative. Theoretical approaches to understanding activity in red-dwarf stars have scarcely got beyond this stage. It is not obvious how to extract from the wealth of observations those features that provide the key to understanding the basic mechanisms that are involved. What is clear, however, is the need to combine results obtained from different approaches.

Studies of the sun's magnetic field have, for example, suggested that the solar cycle is driven by a dynamo located at the base of the convective zone. The most likely site seems to be the region of convective overshoot at the interface between the radiative and convective zones (e.g. Spiegel and Weiss 1980; Schüssler 1982), where a magnetic layer may be formed, as sketched in Fig. 1(a). Detailed properties of

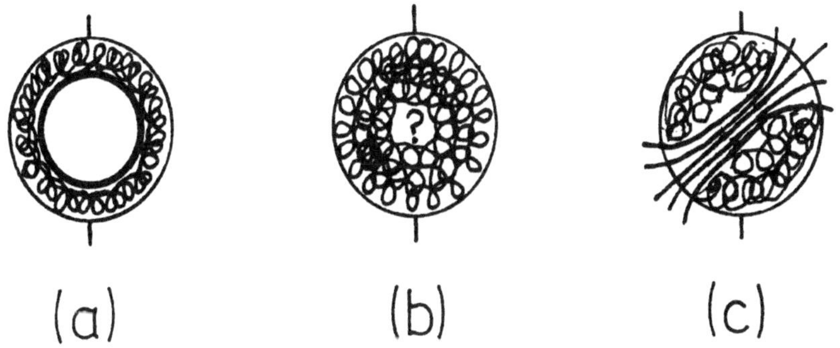


Fig. 1 Field configurations in magnetic stars. (a) Magnetic layer at the base of the convective zone in a G star. (b) Almost fully convective M star. (c) Possible configuration for rapidly rotating RSCVn star.

such a layer have yet to be worked out. Stars of later spectral type than the sun have deeper convective zones and eventually (around M5) become fully convective as in Fig. 1(b). Thus a shell dynamo model is inappropriate for a typical red-dwarf. Is there then a flux-rope dynamo that extends throughout the star, and, if so, where does the transition between the different magnetic configurations take place? Furthermore, solar activity occurs mainly in active regions, which are ephemeral and relatively small; starspots, on the other hand, are long-lived and occupy a greater fraction of the surface. Could the fields in RS CVn variables, for instance, have the structure sketched in Fig.

1(c)? Such basic questions have to be answered before detailed models of magnetic stars can be investigated, and progress here depends on interpreting observational results in the light of whatever theoretical insight is available.

## 3. NUMERICAL SIMULATIONS

Although the full problem of explaining stellar magnetic fields remains intractable, insight can be gained by exploring particular aspects in some detail. Here two approaches are available: one may attempt to simulate some process or one may solve a sequence of simplified model problems. Simulations of nonlinear processes are essentially numerical and, as access to increasingly powerful computers becomes more readily available, increasingly elaborate models can be studied. To illustrate recent progress, I have chosen three examples.

Stellar dynamos. The process by which solar and stellar magnetic fields are maintained is only partially understood, as Belvedere (1982) has explained. Gilman has used the Cray computer at Boulder to study dynamo action in a simplified configuration. He models three-dimensional behaviour in a Boussinesq fluid contained between concentric spheres. The system rotates and is heated from below, so that convective motions are driven in the fluid within the region, as in the outer layers of the sun. The fluid is electrically conducting and small seed magnetic fields can be amplified and maintained (Gilman and Miller 1981). In certain parameter ranges, cyclically varying fields are produced (Gilman 1982), confirming that stellar cycles can be maintained by dynamo action, though dynamo waves progress in a poleward rather than an equatorward direction. Work on a compressible model is already under way.

Turbulent magnetic fields. The laminar diffusivities used in the above calculation have values that can only be related to stellar convective zones if they are interpreted as eddy diffusivities. The behaviour of magnetic fields in a turbulent fluid, when both the ordinary Reynolds number and the magnetic Reynolds number are extremely large, has been simulated by Meneguzzi et al. (1981), using the same Cray computer. They find that the field structure is extremely intermittent: that is to say, magnetic flux is largely confined to isolated regions, where the fields are extremely strong. Intermittent fields are to be expected in stellar convective zones (Galloway and Weiss 1981), though it is difficult to treat them explicitly in any numerical simulation of the dynamo.

Interaction of magnetic fields with granular convection. The most ambitious simulation so far attempted is Nordlund's (1981, 1982) model of intergranular magnetic fields. His code incorporates the full equation of state (including ionization) and the anelastic continuity equation (so that sound waves are filtered out). The energy equation is coupled to radiative transfer in the gas and small scale turbulent dissipation is represented by additional (sub-gridscale) diffusivities. With the aid of a Cray, magnetic fields are again shown to be intermittent, with isolated flux concentrations in the downdrafts between the granules. These calculations point the way towards even more elaborate simulations in the future.

## 4. SIMPLIFIED MODEL PROBLEMS

As an alternative to large scale simulations one may pose a simplified problem that can be properly solved by some combination of analytical or numerical techniques. The main purpose of such a calculation is to investigate systematically the effect of varying certain key parameters: for instance, we would like to know how stellar activity depends on the rate of rotation of the star. Here one must be careful to distinguish between idealized models, which can have both qualitative and quantitative predictive capacity, and parametrized models, which may be illuminating but have no quantitative predictive capacity. My first example is of the latter type.

$\alpha\omega$-dynamo models. As has already been explained in these Proceedings (Belvedere 1982) one can construct an idealized description of a turbulent hydromagnetic dynamo in which the magnetic field has a poloidal component described by the vector potential A, and a toroidal component B. The poloidal field is regenerated by the $\alpha$-effect, depending upon the helicity of the flow (Moffatt 1978, Parker 1979), while the toroidal field is maintained by differential rotation. The growth of the field is governed by the equations

$$\frac{\partial A}{\partial t} = \alpha B + \eta D^2 A, \qquad \frac{\partial B}{\partial t} = r\underset{\sim}{B}.\nabla\Omega + D^2 B \tag{1}$$

and the crucial parameter is the dimensionless dynamo number

$$\mathfrak{D} = \frac{\alpha v' L^3}{\eta^2} . \tag{2}$$

Here $\Omega$ is the angular velocity and v' a typical value of the velocity gradient, $\eta$ is a turbulent diffusivity, L a characteristic length and $D^2$ the Stokes diffusion operator. Linear theory tells us that the field grows exponentially if $\mathfrak{D}$ is sufficiently large. What behaviour can be expected in the nonlinear domain?

The last few years have seen a rapid advance in understanding different patterns of nonlinear behaviour. To explore the behaviour of nonlinear dynamos it is convenient to take a simple parametrized model. Consider one-dimensional dynamo waves, travelling along the x-axis of Cartesian coordinates, with a magnetic field $\underset{\sim}{B} = (0, B, \partial A/\partial x)$ and a velocity $\underset{\sim}{V} = (0, v(z), 0)$ (Parker 1979). Then nonlinear waves can be modelled by a system of three complex ordinary differential equations,

$$\dot{A} = 2\mathfrak{D}B - A , \tag{3}$$

$$\dot{B} = iA - \tfrac{1}{2}iA^*\omega - B ,$$

$$\dot{\omega} = -iAB - \nu\omega \tag{3}$$

(Cattaneo et al. 1982), where $\omega$ represents the nonlinear effect of the Lorentz force on the velocity shear and $\nu$ is a constant. The system (3) has a static solution $A = B = \omega = 0$, which becomes unstable to oscillatory solutions (dynamo waves) when the parameter $\mathfrak{D} = 1$. For $\mathfrak{D} > 1$ there are periodic solutions which are themselves unstable for $\nu < 1$. When $\nu = 0.5$ there are successive bifurcations to doubly and to triply periodic solutions, followed by a period-doubling cascade and chaotic (aperiodic) behaviour. Fig. 2 shows an example of an aperiodic solution, for $\mathfrak{D} = 16$; the recurrent episodes of reduced activity correspond to Maunder minima. Successive bifurcations, leading to chaos, are a feature of many nonlinear systems and similar behaviour must be expected in more elaborate models of stellar dynamos.

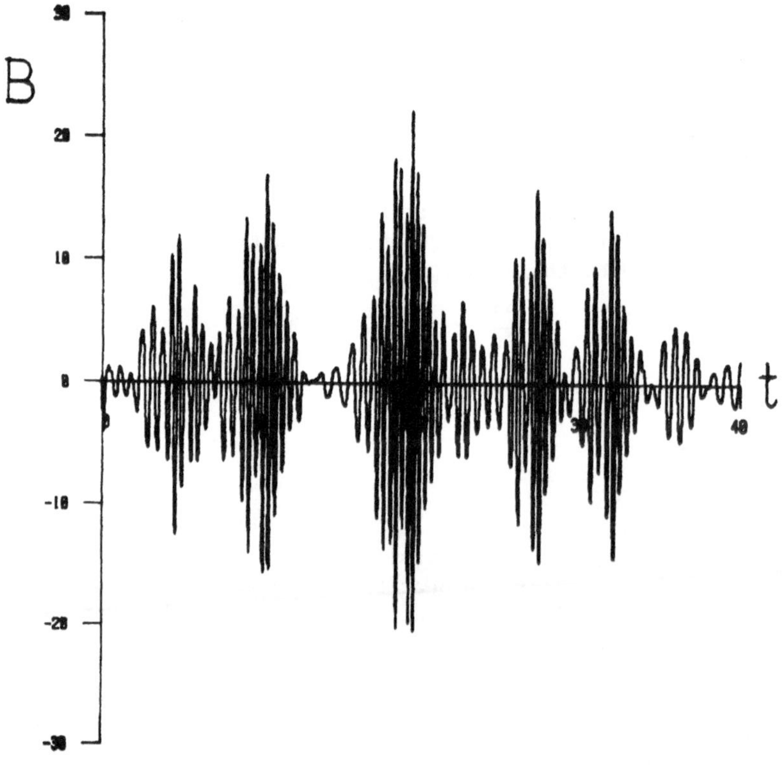


Fig. 2 Chaos in solutions of the model dynamo equations. The toroidal magnetic field B as a function of time for $\nu = 0.5$, $\mathfrak{D} = 16$, showing aperiodic oscillations with episodes of reduced activity (Cattaneo et al. 1982).

Formation of flux ropes by magnetic buoyancy. In the sun, magnetic flux tubes emerge through the photosphere to form active regions. These flux tubes could be formed if the magnetic layer in Fig. 1(a) becomes unstable, as shown schematically in Fig. 3 (Parker 1979). The instabilities can be studied for an idealized model: suppose that there is a

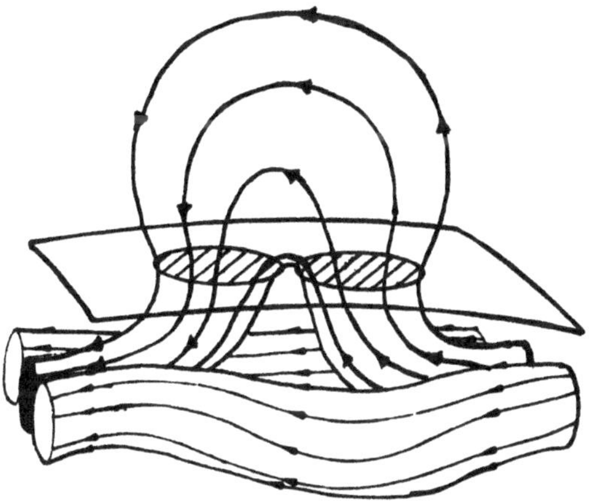

Fig. 3 Sketch showing the formation of a flux rope as a result of instabilities driven by magnetic buoyancy in a magnetic layer (after Parker 1979).

field $\underset{\sim}{B}_o = (0, B_o(z), 0)$ referred to Cartesian coordinates with the z-axis pointing upwards. Such a field is unstable to modes varying as $\exp(i\underset{\sim}{k}.\underset{\sim}{r})$, with $k_x \geq k_z \gg k_y$, if $d(B_o^2)/dz < 0$. More generally, there are complicated doubly-diffusive instabilities, depending on the relative rates at which magnetic fields, heat and momentum diffuse through the medium (Acheson 1978, 1979; Schmitt and Rosner 1982; Hughes 1982), which can be studied within the Boussinesq approximation (Spiegel and Weiss 1982; Hughes 1982). Typically the most unstable modes have a much longer wavelength in the direction parallel to $\underset{\sim}{B}_o$ than in the transverse direction and they might therefore be expected to develop into the configuration sketched in Fig. 3. Thus magnetic buoyancy is an essential ingredient of any theory that relates surface activity to deep-seated dynamos. Work on the nonlinear problem has, however, only just begun (R. Rosner, private communication).

Isolated flux tubes. In the solar photosphere magnetic fields are extremely intermittent, as might be expected from the simulations that were described above. Intense magnetic fields are confined to slender isolated flux tubes, which are partially evacuated so that the magnetic pressure within is contained by the pressure of the external gas (Parker 1979). Thin tube models have been investigated by Spruit (1981), who identified an instability that leads to collapse and partial evacuation;

and the structure of the field within the tubes has been described by Deinzer et al. (1982). These models adopt a parametrized representation of convection in the ambient gas. A proper treatment would combine these effects (important when the magnetic field exerts a pressure comparable to that of the ambient gas) with a better description of convection in the surrounding region, as in Nordlund's simulation.

The kinematic interaction of weak magnetic fields with convection has been investigated in some detail. In two dimensions, or with axial symmetry, flux is expelled from the convection cells and concentrated into isolated sheets or tubes (Galloway and Weiss 1981). In three dimensions the story is, however, much more complicated (Galloway and Proctor 1982; Arter et al. 1982). When there is an imposed horizontal field, Arter (1982) has shown that closed magnetic loops are formed. As a result, the horizontally averaged field is reversed in the upper part of a convecting layer, as shown in Fig. 4. Moreover, flux associated with the reversed field may be greater than that originally imposed. This result upsets many preconceptions about the structure of magnetic fields in stellar convection zones.

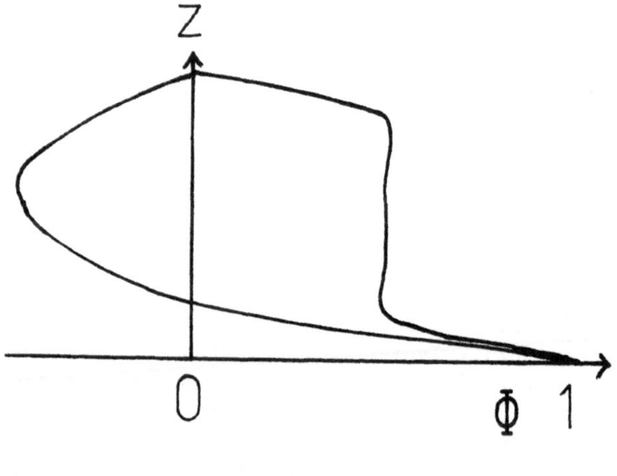


Fig. 4 The horizontally averaged magnetic flux Φ as a function of height, for three-dimensional convection (left-hand curve) and two-dimensional convection (right-hand curve). Note the field free region which is present for the two-dimensional case, and the reversed field in the three-dimensional case (Arter 1982).

Flux concentration produces strong magnetic fields and the Lorentz force becomes important in the equation of motion. Dynamical calculations for convection with two-dimensional or axial symmetry show that isolated tubes and sheets of flux persist (Proctor and Weiss 1982), and some preliminary results are now available for three-dimensional magnetoconvection in a Boussinesq fluid (W. Arter, private communication). Cattaneo (1982) has investigated compressible magnetoconvection and confirmed that flux sheets are partially evacuated in a simple two-dimensional model. A systematic study of three-dimensional compressible magnetoconvection is, however, yet to come. After that, it may be possible to combine treatments of convection in an imposed magnetic field with a description of the dynamo that generates the basic field. At present, owing to the intermittent structure of the field, any such computation seems beyond the bounds of feasibility.

## 5. FUTURE PROSPECTS

The examples that I have described illustrate the different approaches taken by theoreticians, as well as the difficulties that they face. Viewing the problem as a whole, one can also glimpse the way ahead. The one certain feature is that we shall see much more of the same, or similar, investigations. I believe that the emphasis will move from linear to nonlinear behaviour, and so from analysis to computation. Fortunately, advances in bifurcation theory and the theory of dynamical systems are making it easier to interpret the results of such numerical experiments.

As the subject progresses, connections will be established between the different types of theoretical model that I have described. More physics will be fed into the idealized models; for instance, I expect to see fully compressible, three-dimensional computations, utilizing experience gained from plasmas in fusion research. Eventually, the idealized models will merge with simulations, so that we can understand the effect of varying parameters as well as the detailed behaviour of a realistic system. Finally, feedback to phenomenological descriptions may allow us to clarify what happens in a star. Then a coherent dynamo will emerge, incorporating intermittent fields; the role played by rotation will be understood and the origin of the sun's magnetic field, at any rate, will be explained.

Beyond that, we need a global description of the evolution of main sequence magnetic stars. We know that the surface rotation rate is correlated with activity and that angular momentum is lost owing to magnetic braking. The internal rotation rate is much less certain, though non-uniform rotation is to be expected. Indeed, solar oscillations suggest that the core of the sun rotates from 2 to 10 times faster than the surface (Claverie et al. 1981; Gough 1982). So how is the all-important surface rotation rate related to the angular velocity within the star? In the radiative zone, the gradient of $\Omega$ is limited by instabilities driven by differential rotation. The linear theory of these instabilities has been studied in some detail (e.g. Knobloch and Spruit 1982) but their nonlinear development is unclear. Thus we do not know how the angular velocity evolves. Nor do we understand how magnetic behaviour changes with spectral type. What is the effect of deep convective zones on the distribution of angular momentum? Are M dwarfs more or less uniformly rotating? These questions have to be answered in the future.

REFERENCES

Acheson, D.J. : 1978, Phil. Trans. Roy. Soc. A 289, pp.459-500.
Acheson, D.J. : 1979, Solar Phys. 62, pp.23-50.
Arter, W. : 1983, in "Solar and Stellar Magnetic Fields" (IAU Symp. No.102), ed. J.O. Stenflo, Reidel, Dordrecht.
Arter, W., Galloway, D.J. and Proctor, M.R.E. : 1982, Mon. Not. Roy. Astr. Soc., in press.
Belvedere, G. : 1983, in these Proceedings.
Cattaneo, F. : 1982, in preparation.
Cattaneo, F., Jones, C.A. and Weiss, N.O. : 1983, in "Solar and Stellar Magnetic Fields" (IAU Symp. No.102), ed. J.O. Stenflo, Reidel, Dordrecht.
Claverie, A., Isaak, G.R., McLeod, C.P. and van der Raay, H.B. : 1981, Nature 293, pp.443-445.
Deinzer, W., Hensler, G., Schmitt, D., Schüssler, M. and Weisshaar, E. : 1983, in "Solar and Stellar Magnetic Fields" (IAU Symp. No. 102), ed. J.O. Stenflo, Reidel, Dordrecht.
Galloway, D.J. and Proctor, M.R.E. : 1982, in "Planetary and Stellar Magnetism", ed. A.M. Soward, Gordon and Breach, London.
Galloway, D.J. and Weiss, N.O. : 1981, Astrophys. J. 243, pp.945-953.
Gilman, P.A. : 1983, in "Solar and Stellar Magnetic Fields" (IAU Symp. No. 102), ed. J.O. Stenflo, Reidel, Dordrecht.
Gilman, P.A. and Miller, J. : 1981, Astrophys. J. Supp. 46, pp.211-238.
Gough, D.O. : 1982, Nature 298, pp.334-339.
Hughes, D.W. : 1982, in preparation.
Knobloch, E. and Spruit, H.C. : 1982, Astron. Astrophys., in press.
Meneguzzi, M., Frisch, U. and Pouquet, A. : 1981, Phys. Rev. Lett. 47, pp.1060-1064.
Moffatt, H.K. : 1978, "Magnetic Field Generation in Electrically Conducting Fluids", Cambridge University Press.
Nordlund, A. : 1981, Astron. Astrophys. 107, pp.1-10.
Nordlund, A. : 1983, in "Solar and Stellar Magnetic Fields" (IAU Symp. No. 102), ed. J.O. Stenflo, Reidel, Dordrecht.
Parker, E.N. : 1979, "Cosmical Magnetic Fields", Clarendon Press, Oxford.
Proctor, M.R.E. and Weiss, N.O. : 1982, Rep. Prog. Phys., in press.
Schmitt, J. and Rosner, R. : 1982, Astrophys. J., in press.
Schüssler, M. : 1983, in "Solar and Stellar Magnetic Fields" (IAU Symp. No. 102), ed. J.O. Stenflo, Reidel, Dordrecht.
Spiegel, E.A. and Weiss, N.O. : 1980, Nature 287, pp.616-617.
Spiegel, E.A. and Weiss, N.O. : 1982, Geophys. Astrophys. Fluid Dyn., in press.

Spruit, H.C. : 1981, in "The Sun as a Star", ed. S.D. Jordan, pp.385-412, NASA SP-450, Washington.

DISCUSSION

van Leeuwen: Do you consider it possible that by means of magnetic braking a K star could lose 99% of its angular momentum in 10-20 million years as we observe for the Pleiades?

Weiss: I do not know of anyone having and adequate model of magnetic braking or understanding how it could be as efficient as that. Clearly, however, the picture that we have implies a rather gradual loss of angular momentum through magnetic braking. The timescale you mention is not inconceivable but it would require a very vigorous form of magnetic activity so that the lever arm for magnetic braking would be much greater.

Rosner: I would like to comment that it is probably difficult to know something about the angular momentum of a star. All you have is a measure of the spindown of the surface. So one does not know anything about the angular momentum of the whole star unless one assumes something about differential rotation as a function of depth. So it is very model independent.

Weiss: It is possible that the magnetic field would be confined to a skin near the surface which is, as a result, rotating very slowly, while the core is whizzing round. However one is reluctant to assume this unless it is forced on one by observations.

Lang: One of your first diagrams showed observations leading to models and so on. I wonder if you should not complete that feedback loop. For instance, does the production of flux ropes by magnetic buoyancy reproduce any observable properties of sunspots? For instance, is an isolated flux tube observable? How thin would it be? What is its magnetic field strength?

Weiss: I think that at present the argument goes only the one way. We have gone from the observations to a phenomenological description. We observe flux tubes emerging as active regions at the surface of the Sun which suggests that some process of this type is occurring. For a variety of reasons some of us think that they originate very down in the convective zone. The only mechanism that one can imagine for driving this process is an instability driven by magnetic buoyancy. Although it was understood that convection could produce locally magnetic fields through the flux expulsion mechanism, nobody has supposed that intense magnetic fields, virtually completely evacuated, could exist at the surface of the Sun until such fields were discovered. Intense magnetic fields were discovered 10 years ago and they have been intensively investigated. We know there are isolated flux tubes with fields of 1500G or so and magnetic pressure is comparable with external gas pressure at the Sun's surface. This has led Spruit and others to seek mechanism for an instability... (part of recording lost)... But I think it is true to say that in these cases observations led theory.

Catalano: You posed the question how does an M star rotate? This arises because they may be fully convective and so on. We have evidence that the rotation period of a main sequence star changes with some power of the mass and indirect evidence, from Ca K-line behaviour (Catalano and Marilli, this volume) in the Hyades and Pleiades on the decay of rotation with time, that the rotation speed changes exponentially with time and with a powerlaw dependence on mass. So perhaps this could be an answer to the question you pose?

Weiss: This is very helpful. What one would expect from simple considerations is that magnetic braking might be less effective for stars with deeper convective zones. Since one could not decelerate an outer shell. Rather one would have to decelerate to whole star.

Gibson: Would not tidal forces in the RS CVn stars have eliminated differential rotation, at least while they were on the main sequence?

Weiss: Indeed yes. I would have thought that in close binaries one would have thought that tidal forces would have forced at least rough co-rotation, although there might be a modest degree of differential rotation within the convective zone.

Gibson: So as much an object evolves off the main sequence it is possible that the inner parts would rotate more rapidly again than the surface. Would this rapidly rotating region incorporate some of the convective zone or would it be in the radiative zone?

Weiss: That is a question which is beyond my capacity to answer. One should need to know more about stellar structure than I do.

Rodonò: I would say that certainly there would be competition between differential rotation and gravitational effects in close binaries. There are two papers by Scharlemann in Astrophysical Journal (1981 and 1982) which treat the effect of tidal coupling on the differential rotation of RS CVn stars.

# OBSERVATIONAL APPROACH AND PERSPECTIVE

G. S. Vaiana
Astronomical Observatory, University of Palermo, Italy
and Center for Astrophysics, Cambridge, USA

Paper not received.

## DISCUSSION

Goldberg: Well you did not cover more than half of my planned talk! (laughter). Let me comment on interferometric techniques, in particular speckle imaging which you mentioned. Doing speckle imaging with the largest telescopes now available will not give you better than the theoretical resolving power of the telescope. With a 4m telescope that is about 30 marc sec in the visible. That happens to be the radius of the supergiant Betelguese. So you are not going to achieve much with speckle imaging on these stars. One technique which has not been adequately exploited is that of lunar occultation which can give much better angular resolution than speckle, of the order of 2-3 marc sec. By using suitably chosen filters it may be possible to see structure on the disks of stars.

Vaiana: I was aware of this limitation but I was speculating on the very distant future with the availability of very large telescopes. With regard to the second possibility, it might be very difficult to do speckle of accretion disks in binaries because of the high luminosity of one of the components. In the case of the M stars the situation might be more favourable.

Worden: I do not wish to disagree with Dr. Goldberg too much but it will be possible to use speckle interferometry in the next few years. It has been possible both in the work of Labeyrie and in Arizona to phase two spatially separated mirrors. It is possible to increase resolution in this way. An even more interesting possibility is the suggestion of Jacques Becker (?) which is a little related to Vogt and Penrod's work (this volume), i.e. speckle imaging in absorption lines. As you pass through the



*P. B. Byrne and M. Rodonò (eds.), Activity in Red-Dwarf Stars, 651–652.*

absorption line with rotation you can get a one-dimensional image of the star. He believe he can get very interesting information, even on some of the dwarf stars. There is also a proposal from Harvard to build a space interferometer which would give direct images. In practise these ideas may run into difficulty but theoretically they should work.

de la Reza: (part of question lost) ... they have low lives, low mass rates and no specially peculiar abundance. Perhaps the abundance may be important in order to have a measure of the convection.

Vaiana: Yes, this need to be pursued with great care. I do not think we have measured mass loss from those systems and so it does not appear that they have very large mass loss. Given that they are so active both in the "quiescent" and flaring states I think it is worth re-examining these objects in the light of what we are learning in the UV and X-ray ranges.

Walter: I would like to put in a plug for the eclipsing binaries because they give you an opportunity to see surface structure as one star passes in front of the other. All you need is adequate time resolution. As we showed with Einstein with 500 sec resolution sizes of order $10^9$cm in stellar coronae have been resolved. I believe it is possible to do 1 or 2 orders of magnitude better. This is very promising for the very near future in studying stellar chromospheres and coronae.

Vaiana: You need high photon counts for this so it can only be done for bright sources or with large aperture detectors.

Walter: Well although Einstein is a small aperture instrument the Space Telescope will be able to do much better.

# SUMMARY OF THE COLLOQUIUM

L. Goldberg
Kitt Peak National Observatory*

I have rarely had the pleasure of attending a conference in which the sheer volume of information has been so impressive and the contributions have been so diversified. Ordinarily, an I.A.U. colloquium is quite specialized, but this one might easily have been sponsored by half a dozen commissions and several international scientific unions. Astronomers from many sub-disciplines, atomic physicists, plasma physicists, aerodynamicists and applied mathematicians have come together to unite solar and stellar physics and to learn one another's scientific language. In principle, my task is to summarize more than 100 papers, which is clearly impossible even if I had had two evenings to prepare instead of one. The best I can do is to mention a few of the papers which seem to me to have conveyed the flavor of the conference and to ask the forgiveness of the remaining authors.

The idea of using solar guidelines to investigate other stars is an ancient one. In 1905, G. E. Hale gave as one reason for founding the Mt. Wilson Solar Observatory "the investigation of the Sun as a typical star in connection with the study of stellar evolution," and in the year I was born, 1913, Eberhard and Schwarzschild pointed out that the emission reversals of the H and K lines in stellar spectra implied "the same kind of eruptive activity that appears in sunspot, flocculi and prominences." They suggested that the intensities of the emission lines might vary analogously to the sunspot period. These early suggestions could not be implemented immediately because the observational capabilities were lacking. Photographic techniques were too crude, detectors too inefficient, gratings too imperfect and most important, we did not have access to the radio, ultraviolet and X-ray regions of the spectrum.

This conference has focussed on the red dwarf stars, objects at the lower end of the main sequence, which are much harder to observe than their more flamboyant fellows in the upper half of the H-R diagram. They are a unique class of objects because their ages are spread over a

*Kitt Peak is operated by AURA, Inc. under contract with the National Science Foundation.



*P. B. Byrne and M. Rodonò (eds.), Activity in Red-Dwarf Stars, 653–661.*

factor of 100 or more, from those approaching the age of the galaxy to others as young as a few tens of millions of years. This diversity enables us to examine how various important parameters vary with age-rotation, magnetic fields, chromospheric and coronal activity, chemical composition, etc.

A widely used conceptual model for red dwarf stars is closely linked to current ideas on how activity is generated in the Sun. These ideas have crystallized in the last decade, although their genesis occurred much earlier. In giving my impressions of this colloquium, I should like to refer briefly to this model and its solar inspiration.

It is generally accepted that rotation and convection are the principal ingredients in the recipe for magnetic activity in the Sun. First, as Belvedere has described in his thorough and lucid review, differential rotation in the convection zone amplifies a weak, poloidal field by dynamo action. Convection then carries the field to the surface, where it appears in the form of magnetic flux tubes, some of which cluster tightly together to form plages and sunspots. Turbulent, convective motions store some of their energy in the magnetic fields by bending and twisting lines of force, after which a number of possible mechanisms may dissipate the stored non-thermal energy and thereby heat the chromosphere and corona, throw off prominences, and generate solar flares and the solar wind. Energy is lost from closed magnetic regions by radiation and from open regions or coronal holes by mass transport.

The same scenario is extended to other late-type stars on the main sequence by postulating that young stars are rotating rapidly and therefore that magnetic activity with all its consequences is extremely intense. Objects like T Tau and BY Dra may be extreme examples. In some fashion, the magnetic activity generates stellar winds, which carry away angular momentum and put a brake on stellar rotation, resulting in a decrease of activity with stellar age. Close binaries are of special interest because synchronism of orbital and rotational periods causes rapid rotation and excessive magnetic activity on relatively old stars. The RS CVn and W UMa stars are thought to be such objects. We have heard much in this meeting about the behavior of stars in close binaries from Bopp, Catalano, Charles, Dupree and many others.

Clues to the understanding of red dwarfs can also come from the study of red giants and supergiants, in which convection occurs on enormously greater scales and is therefore more easily observed. For example, granules should have typical diameters of a sizable fraction of a stellar radius. Perhaps a dozen convection elements cover the entire surface; temperature fluctuations may be on the order of 1000 K and convective time scales may be 200 days or longer.

Turning now to the subject matter of the colloquium, we were introduced to the red dwarf stars by several comprehensive reviews covering their global and photospheric physical parameters, the properties of their quiescent chromospheres, transition regions and coronae and the methods used to derive them. Petterson showed how the fundamental parameters, L, and $T_{eff}$ can be determined empirically and used in conjunction with calculated evolutionary tracks to infer the evolutionary states of individual red dwarf stars. Masses as low as 0.02 $M_\odot$ have been deduced in this way. The chemical abundances in flare stars appear to be the same as those of other dwarfs and the Sun. Surprisingly, Li has been seen in only one flare star. As in the sun, the detection of oscillations in red dwarfs would give information on the structure of the interior. Periods of 14-60 minutes have been predicted but there has been no positive detection as yet. The detection of oscillations should have high priority. Long periods of observation will be required either from space platforms or perhaps at the South Pole, if the hazards of the Antarctic winter can be dealt with. Paterno and Zuccarello suggested that the depth of the convection zone may be an important parameter in the determination of X-ray fluxes from red dwarfs. It is not yet clear whether or not oscillations with high $\ell$-values, which would give information on the depths of convection zones, can be detected in integrated starlight.

It is now well established that stellar rotation is a fundamental parameter underlying red dwarf activity. Petterson described several methods for inferring rotational velocities, including rotational broadening of absorption lines through the use of cross-correlation techniques, which gives V sin i, and rotational flux modulation by spots (photometric light variations) and plages (Ca + emission intensity variations), which yields equatorial velocities. Binaries tend to rotate faster than single stars and spotted stars much faster than plage stars. A remarkable new technique, called Doppler imaging of starspots, has been invented by Vogt and Penrod. The method gives not only rotational periods, but the sizes of starspots, information on their migration across the star, and, when supplemented by V-R photometry, their temperatures. The method can only be applied to the very fast rotating stars like BY Dra and RS CVn. An important conclusion from the work of Vogt and Penrod is that starspots are not morphologically similar to sunspots - they are quite large, covering up to 20% of the stellar surface and their temperatures indicate they resemble giant umbrae, with little or no penumbrae. In the BY Dra stars, the spots have been found to migrate to the poles where they dissolve into remnants, much like high latitude solar coronal holes, and to exhibit period variations implying differential rotation in the same sense as in the sun. During the past 8 years, Rodono and associates at the Catania Observatory have been studying the light variations of BY Dra and II Peg, from which they have inferred cyclic migration of spots in latitude and lower limits to rates of differential rotation. In the RS CVn stars, the spots appear to migrate in all directions, N or S and E or W. Estimates of the sizes, temperatures and migration rates of spots may also be inferred from photometric measurements. Vogt and Penrod have applied the Doppler-imaging technique in some detail to the RS CVn star HR 1099. The behavior of a spot group on this star was more like that of a solar complex of activity than a sunspot group. Blanco and his collaborators reported the results of twenty years of dedicated photoelectric photometry of RS CVn at Catania Observatory. Analysis of the light curves suggests a qualitative model in which huge spotted regions migrate in both latitude and longitude in a differentially rotating star.

Long-period cycles similar to the 11-year sunspot cycle have been found from Ca II emission and from photometric waves. Vogt's compilation of these cycle periods for 11 active, single dwarfs reveals that for $P_{rot} > 7$ days, $P_{cycle}$ is approximately constant at 9 ± 2 years. Stars in this category do not appear to show large spots. However, from examination of 6 binaries with $P_{rot} < 7$ days, $P_{cycle}$ was found to be proportional to $1/P_{rot}$ and, moreover, the stars in this group display large, dark spots. This remarkable, apparent discontinuity in the cycle period at $P_{rot} = 7$ days needs to be investigated for a larger sample of stars and would obviously be of great significance if confirmed.

Other than the evidence from the poleward migration of spots in BY Dra stars, to which I have already alluded, no sure evidence of differential rotation in solar-type stars was reported, although Baliunas offered the possibility that the rotation periods derived from Ca II emission intensities might vary with cycle phase. On the subject of Ca II emission intensities, Soderblom cast doubt on the evolutionary significance of the Vaughan-Preston gap in the relation between Ca II intensities and B-V, by pointing out that the intensity is not a linear function of age, and that there is very little difference in the H-K intensities between stars in the Hyades and those in the Pleiades. Catalano and Marilli showed that the K-line emission decays exponentially with the square root of the age.

The key to interpreting stellar activity is the direct measurement of magnetic fields. The Robinson method of deconvolving a Zeeman sensitive line by use of an insensitive line, which yields both the field strength and the area covered by the field, is beginning to be widely used. Marcy, at Lick Observatory, has observed that 10 of 29 G0V - K5V stars have obvious fields in the range 800-1500 gauss. Area filling factors of 60-80% are common. In principle, the Robinson method ought to be applied at the longest possible wavelengths, since the ratio of Zeeman splitting to the Doppler width is proportional to wavelength. For example, the 12$\mu$ emission lines observed in the Sun, by Brault and Noyes, which I have discussed in a contributed paper, would be ideal, but their detectability in other stars remains to be demonstrated.

The enormous areas covered by starspots will not be easy to explain. Where does the energy go? Gershberg told us that the energy of X-ray radiation detected from the quiet coronae of flare stars is comparable with the radiation missing in dark spots. He made the novel suggestion that the energy is converted into hydromagnetic waves which heat the coronae. Mullan suggested that the missing energy is stored in the form of trapped Alfvén waves and that the escape of as little as 1% of the stored energy would provide enough energy to power large solar proton flares.

Quiescent coronae and chromospheres of red dwarf stars have been studied principally with the Einstein and IUE satellites, and the results were summarized here by Golub, Linsky and Johnson. Beginning at late F spectral classes, the X-ray flux increases with increasing rotation rate, as reported by both Golub and Johnson. The M-dwarf coronae are generally much more active than the solar corona, exhibiting greater variability and temperatures up to $10^7$ deg. K. Magnetic fields inferred from X-ray fluxes are also much stronger than in the sun, by one or two orders of magnitude. The fact that the X-ray flux decreases strongly with B-V may imply, as Giampapa has tentatively suggested, that the entire star is becoming convective and that the stellar dynamo has no room in which to operate. New computations on the generation of acoustic energy from convection zones of late-type stars were reported by Bohn, who finds that the flux of acoustic energy from the very cool stars is nearly five orders of magnitude greater than that given by previous calculations. It now seems that more than enough acoustic energy is being generated to account for the X-ray emission of late-type stars but this does not alter the prevailing view that the energy will be dissipated in the chromospheres before reaching the coronas of M stars.

Evidence for red dwarf activity in chromospheres and transition regions, based on IUE data, is beginning to pile up. Linsky emphasized the all-pervasive influence of magnetic fields in his summary, while pointing out that rotational modulation by plages can be seen in transition-region lines. In II Peg, the fluxes in these lines are anti-correlated with the photometric variations that indicate the transit of dark starspots. Stellar plages may be much brighter than in the sun, up to a factor of 50 in the CIV lines. The CIV flux shows variations of up to 70%.

In no aspect of red dwarf research has progress in the past 2-3 years been so dramatic as in the study of flares. Where previously there was a scattering of visible light curves and radio detections, only occasionally simultaneously, there is now a well-coordinated program of observations in visible, UV, X-ray and radio wavelengths and a sophisticated approach to the analysis and interpretation of data based on solar experience. Optical photometry, optical and UV spectroscopy and X-ray and radio data are combining to give a clear picture of the similarities and differences between solar and stellar flares.

Byrne reviewed the morphology of flare light curves and the statistics of times of occurrence of optical flares. Among the interesting questions he discussed were the extent to which the parameters of the light curves can be systematized, whether there is evidence for precursor events and whether there may be periodicities in the occurrence of flares, related perhaps to the rotation period of the underlying star.

The energy released in dMe flares sometimes exceeds that from solar flares by $10^3$. Yet, Worden showed that optical and UV stellar flare spectra are surprisingly like their solar counterparts. A major handicap at present is the cutoff of IUE spectra at 1100 Å. Plans are being drawn to fill the gap between 1100 Å and X-rays. I was pleasantly surprised to learn that EUV radiation, at least below 500 Å, can penetrate the interstellar medium to the sun from most known flare stars.

Gibson and Lang gave impressive demonstrations of the power of the VLA in flare studies. Such parameters as temperature, electron density, magnetic field and emission measure are readily derived and polarization measurements give unique information on non-thermal processes and mechanisms.

Remarkable progress has been made in the theoretical modeling of solar flares, guided by laboratory experiments with such devices as Tokamak, as we heard from Priest and from Spicer. Very impressive simulations of solar flare loop models were carried out by Pallavicini and found to be in good agreement with observations from SMM. Similar simulations may now be used to interpret stellar flare observations from the Einstein satellite.

It is now possible to analyze optical and X-ray data on stellar flares with the help of models developed for the sun, as Giampapa and Haisch have shown. Haisch's detailed examination of the X-ray observations in the framework of the solar loop model was most impressive. It is remarkable how one can derive estimates of such parameters as temperature, density, pressure, loop lengths, magnetic field strengths, etc. The first results show that the loop model is applicable to some but not all types of flares or stars. Kodaira has constructed empirical models of flares which will help to guide further theoretical developments.

Finally, I should like to give my view of what seem to be the major conclusions of the conference:

1) The connection between rotation and magnetic activity seems well established. For the M dwarfs, the depth of the convection zone may be a contributing factor to the magnitude of the X-ray flux.

2) Acoustic waves provide more than enough energy to heat the coronas of M stars, but their heating effect, if any, is probably limited to the chromospheres.

3) Activity on red dwarf stars is remarkably similar to that on the sun, but there are important differences, e.g., stars with $P_{rot} < 7$ days seem to fall in a separate class: they have very large, dark spots and the periods of their cycles are inversely proportional to $P_{rot}$.

4) Starspots last much longer than sunspots and their average fields are much stronger, perhaps by factors of 20-30.

5) Flares on dMe stars are similar on the whole to those on the sun, but the energetics are different. The strongest stellar flares emit perhaps $10^3$ times as much energy as the most energetic solar flares.

Before closing, may I add some comments on the future of research in stellar physics, as a supplement to the excellent presentations by Vaiana and Weiss. We are now clearly on the linear branch of the curve of growth so far as knowledge of red dwarf stars is concerned and it is at this point that we ought to look ahead to the developments that are needed to sustain and accelerate the growth. It seems to me that the key to the future of stellar physics research is high angular resolution, but most of the emphasis in long-range planning, by optical astronomers at least, is on modest increases in the diameters of filled apertures rather than on systems that yield quantum jumps in spatial resolution. While the radio astronomers are planning a very long baseline array with resolution of a few tenths of a millisecond, optical astronomers are looking forward to a resolving power of 0.1 arc sec with the ST and even the proposed New Technology Telescopes can increase the available resolution by factors of 2-3 at most. Such telescopes will be capable of resolving the disks of a few of the brighter red giants, e.g., by speckle interferometry, but only just barely, and the red dwarfs will remain far beyond resolution.

What is needed is a much greater concentration of effort on long base-line optical interferometers, which are now being pioneered for use on the ground by Labeyrie and others, but which must be put in earth orbit for really effective performance. Interferometers with a base-line of 10-15 meters could probably be launched by the space shuttle within a few years, and a base-line of 200 meters, yielding an angular resolution of 0".001 in the visible, should be in reach by the end of the century. That is only 18 years off and I remind you that at least that much time will have elapsed between the start of serious planning for the ST and its actual launch. A gain in optical resolving power by a factor of 100 or more is bound to have an impact on astronomy comparable to that of the large reflecting telescopes that Hale pioneered at the beginning of this century.

DISCUSSION

Linsky: I would like to make a modest comment and suggestion.

Anon: Impossible! (Lots of laughter).

Linsky: Now that I have everybody's attention I will say that a great deal of what has been said at this meeting tells us that we need to monitor objects for long periods of time. The allocation of telescope time, both in space and on the ground, is not usually made by people who recognize that elementary fact. I would like to give a particular example of that. Shortly after Olin Wilson first began his study which led to the discovery of activity cycles he applied to the U.S.National Science Foundation for a grant of a small amount of money to continue his work. Even though he was a well-known astronomer and wrote a reasonable proposal the referees said in effect that it was not very exciting science. So it should not be founded. It was only due to the persistence of Wilson himself that he managed to continue and we all know the result. So the moral is that there is a great virtue in long-term monitoring and this should be recognized.

Golub: There is now the possibility of making normal incidence optics in the X-ray region. This raises the possibility, working at 50 Å for instance, of getting down to the diffraction limit, which would give rise to resolution competitive with present radio resolution.

Goldberg: This would be great but it does not remove the need for optical telescopes of the same resolution, of course.

Charles: I would like to refer back to Prof. Vaiana's talk and remind people of an instrument which we hope will become available very soon i.e. EXOSAT. The gratings on board will be very much more sensitive than Einstein's and should extend out to 300 Å. After six months of operation all of the time will be available for guest observers.

Paternò: I would like to make a much more modest comment than Jeff (Linsky) (laughter). It concerns oscillations of these stars. Oscillations are a beatiful tool for determination of the depth of the convective zone, which is in turn ultimately important for the generation of X-rays. For this purpose it is important to detect high-mode p-type oscillations, since it is these which are trapped in the convective zone. It is they which determine the depth of the convective zone since the latter is one of the model parameters which must be fitted to get the best model fit for any observed spectrum. Observationally this is a very difficult task. Incidentally this is the topic of the next meeting in Catania, next June viz. the study of oscillations in order to probe the interior of the stars.

Mullan: I was very glad to hear mention of the South Pole as a site for prolonged observations. At the present time the Bartol Foundation has an observatory there equipped with a 3-inch telescope. You may laugh and ask what astronomy is it possible to do with a 3-inch telescope? Well, solar seismology was revolutionized by that telescope when it was used by Hussak(?), Gregg and Pomeranz. They probed the 5-mins oscillations with a resolution of 1 Hz. The achieved 120 hours of continuous observation of the Sun. We now have an 8-inch telescope giving a 20 cm solar image. This however is just the beginning.

van Leeuwen: (part of question lost) ... you all to observe these stars in open clusters as much as possible in order to constrain the parameters of age and mass. That way I think you can see much more detail in the parameters, in the physics of the stars.

Popper: I have been listening, throughout today's presentations of theoretical interpretations of stellar activity, for some discussion of what appears to me to be one of the primary observational facts calling for theoretical understanding in terms of the spottedness hypothesis. That is the very great amplitudes of phase-related light variations in some active stars. In some cases, modulation over one orbital period of the visual light of the larger, cooler component in a binary may be as great as 30% (e.g. RS CVn, RW UMa). It would seem that a satisfactory theory of stellar activity must show that the amount and distribution of spottedness over the surface, required to account for such large light variation, follows naturally from the theory. I failed to hear any serious reference to this outstanding problem. Four possible explanations occur to me. 1) I didn't listen attentatively. 2) The explanation is so simple and obvious that no comment is needed. 3) The speakers are unaware of the phenomenon. 4) The speakers are aware of it, but have no explanation.

SUBJECT INDEX


Abundances 22,23,24,77,420
Acoustic Flux
vs Spectral type 606
Activity Cycles 146,147,148,166,169,350,351,352,399
Model of 591,642
vs (B-V) 593
vs Rotation period 148,196,599
Activity Parameters (cf specific parameters)
Angular Momentum (Pleiades K stars)
vs Mass 191
Atmospheric Structure
RS CVn stars, in 443
Chromospheric emission line variability, from 41,61,443
Photometric variability, from 41
W UMa stars, in 455
Balmer Spectrum
Decrement in flares 214,227,229
Diagnostic use in flares 227,377
Flare spectrum 203,214,227,239,368,562
Formation of 177
$H_{\alpha}$
RS CVn stars, in 368
Flare-like variability 369
vs Photometric wave 369
FK Com stars, in 374
Flare, in 369
Reversal in flares 229
T Tauri stars, in 229
Model of profile 625
Variability 501
vs Polarization 515
$H_{\beta}$
Flare, in 237
Preflare dips, in 237
$L_X$, derived from 228

Profiles in flares 216,229,238,369,513
Search for, in emission, in low-mass dwarfs 92
Bolometric Luminosity ($L_{bol}$) 19
Bolometric correction (BC) 19
vs Colour 21
vs Radius 21
CA II H & K 71
RS CVn stars, in 363
FK Com stars, in 372
Diagnostic use 377
Flarelike variations 197,367
Solar flares, in 210
Stellar flares, in 216
Surface flux F(Ca II) in RS CVn's 75,364,366
vs $F_X$ 419
Variation, cyclic 146,195
vs Age 71
vs Acoustic flux 606
vs (B-V) (=Vaughan-Preston diagram) 67
vs Binary phase 367
vs He I $\lambda$10830 372
vs Mass 72
vs (R-I) 67
vs Rotation 71,196,367
Wilson intensities, calibration 365
Chromospheres
Emission lines vs spots 145
UV line fluxes in Hyades stars 131
Close Binaries-Period Distribution 391
Coronae
Late-type stars 86
$L_X$ vs L(NV) 456
Models 423
Physical parameters 61,97,127,261,282,332,422,443
Solar 84
Structure of 443
Temperature vs $L_X$ 98
Transients 301
Diagnostic Spectral Lines 62,64,234,372,377,439
Differential Rotation 168,351,353,387,650
Dynamo 531,579,642
Bimodal (Vaughan-Preston gap) 592
"Maunder" minima 585
Saturation 458
Timescales 583
Turn-of in low-mass stars 97
Waves 598,642
Emission Measure
Distributions 62,509
Flares, in 228,230,271,296,316
X-ray EM in RS CVn's 422

EUV Emission
Solar flares, in 211,565
Stellar flares, in 217
Sunspots, in 325
Filling Factors
Magnetic fields 103,150
Starspots 421
Flares, Solar
Coronal heating, cf 560
Coronal transients 301
Electron beams 298,318,321
Flash-phase 208,291,293,309,321,340
Gamma-ray spectra 295,318
General description 265,289,546,561
Homologous flares 301,560
Light curves 562
Mass motions in 297,299,313,322
Microwave emission 297
Optical flare spectra 209
Proton flares 528
Radio interferometry of flaring regions 331,335,343
Stellar flares, cf 247,264,561
Thermal phase 291,321
Two-ribbon flares 309
X-ray data 289,307
White light flares 562,565
Flares, Stellar
Colours of 159,239,563
Continuous heating in 263,269
RS CVn stars, in 349,368,411,420,424,430,443
Energy in optical 159,163,165,224,412
General description 157,265,561,573
Light curves 157,163,166,203,218,239,562
Mass motion 218,238,617
Microflaring 162,198
Models 527,554,561,567,609,617,625,628
Optical photometry 157,203,239,246,412
Optical spectroscopy 197,203,211,214,216,224,229,237,239,246
Preflare dips/rises 168,204,613
Radio data 273,430
Solar flares, cf 247,264
Time distribution 161-167, 432
T Tau stars, in 503
W UMa stars, in 481
UV observations 246,249,443
vs Spots 145,172,529
X-rays, in 43,93,118,127,131,255,420,424
Flarelike Activity
CA II 197,367
RS CVn stars, in 367,411
$H_\alpha$ 369

Red giants, in 251,254
T Tau stars, in 501
X-rays, in 97
He λ1640 Line 59,64,440
He I λ10830 Line 371,377
Heating of Chromospheres/Coronae 10,44,49,101,538,605,618,625
Hydromagnetic Waves 487,537
IR Unidentified 12μ Emission Lines 327
Light Curves 249,388,393,395,399,401,403,408,409,443,481,486,503
Amplitude vs period 189,346
Variability, of 176,179,189,346,348
vs UV line emission 444
Loop
Coronal heating, by 101
Model 100,260,266,423,475
Parameters 98,102,261
Solar/stellar flares 268,316,551
$L_{opt}/L_X$
Flares, in 228,264,291,565,619
$L_X/L_{bol}$ 126,453
Magnetic Cycles 42
Model for 579
Magnetic Fields, Stellar 102,150,151,261,282,384,529,584,601
Braking 649
Buoyancy 645
Concentration of 530
Configurations 629,641
Filling factor 103,150
Flux vs equatorial velocity 152,633
Flux tubes 10,528,531,532,533,546,552,554,588,617,645
Instability 533,545,625
Variability 152,485,582
vs Convection 633,642,646
Mass 22
Mass-luminosity relation 22
vs Angular momentum 191
vs Ca II emission 72
Oscillations 660
Magnetic flux, of 485
Radio flares, in 280
Red-dwarfs, in 28
TR (over sunspots), in 301
Physical Parameters of Dwarf-Stars 17,19,48,49,112
Plages 43,249,443,529,535
Polarization
Optical, in flare stars 169
Radio flares, in 277,279,331,335,573
T Tau stars, in 515
Preflare Dips/Increases 168,169,204,237,613
Radio
Emission from

RS CVn stars 429,444
BY Dra stars 429
Flare stars 273,282,562
T Tauri stars 505
W UMa stars 429
Emission
Parameters, definition 274
Interferometry of solar AR 331,335
Mechanisms 279,430
Noise storms 300
Oscillations 280
Polarization 277,279,331,335,515,573
Solar flares 296,297
Stellar flares 276,278,368,430,431,565
Cumulative frequency diagrams 432,433
Rotation (see also Differential Rotation) 27,28,29,148,176
Equatorial velocity vs magnetic flux 19,152,633
Periods vs activity cycle 148,196
Periods of RS CVn's 417
Periods vs equatorial velocity 113
Spindown 9
Velocity of T Tauri's 504
vs Ca II emission 71,196,366
vs Light curve amplitude 190
vs X-ray luminosity 416,633
Scaling Laws 62,84,101,107,271
Speckle Imaging 505,651
Spectral Types
TiO band photometry, from 35
Spots, of 141
vs Colour 35
Spot
Cooling 536,538
Cycles 146,169,350
Energy storage 536
EUV spectra in sunspots 325
Filling factor 423
Magnetic flux, in 382
Migration 144,168,185,350
Missing flux vs coronal heating 528,536,538
Models 139,153,185,326,379,382,403,423,483
Photometric and spectroscopic features 141,153,175,344,379
Physical parameters 139,141,143,144,325,379,384,527,532,539
Pleiades stars, in 189
Stability 532,534
Stars vs plage stars 535
T Tauri stars, in 503
W UMa stars, in 471,481
vs Chromospheric emission 145
vs Flaring 145,174,539
Stars and Stellar Systems

RW Aur 513
44 Boo 451
WY Cnc 399
YZ CMi, flare of 25.10.1979 561
RS CVn and RS CVn-stars 343,363,387,415,650
    Balmer emission in 368
    Ca II H & K emission in 363
        vs Binary phase 367
        vs Rotation rate 367
    Cycles 352,400
    Evolutionary status 463
    Evolution of light curve 345,348,351,387,395,399,403,409
    Flares on 349,411,420,429,445
    Mass loss from 354
    Migration of photometric wave 350,387,395,409
    Orbital period variations 353,388
    Photometric variability 75,344,387,393,395,399
    Rotation periods 399,403,417
    X-ray observations 417,454
DI Cep 497,515
VW Cep 451,454,479,481,485
FK Com stars 372
ε CrA 451
BY Dra and BY Dra-stars
    Cycles 168,352
    Flares on 249
    Radio emission 429
Gliese 182 77
HR 1099 (see V711 Tau)
AR Lac 401,443,445
HK Lac 403
EV Lac
    Flare/preflare dip 203
    Unseen companion 201
AU Mic 249
II Peg 168,185,443
AI Phe 407
T Tau and T Tau-stars 505,515
    Accretion model 625
    Activity 487,501,503
    Colour vs magnitude 501
    Evolutionary status 488
    Flares, FU Ori 490
    Nebulae, circumstellar extinction 498
        Formation of molecules 512,523
        Physical processes, in 519
    Polarization model 515
    Radio observations 505
    Stellar wind 507,509
    UV Spectra 508
V711 Tau = HR 1099 383,409,498

XI UMa 411
W UMa stars 447
Atmospheric structure 415
Activity indicators 469
Evolution of 447,463,465
Flare on 481
Orbital period variations 465,485
Orbital period, correlations with 454,457,475
Radio emission 429
Spectral type, correlations with 470,475
Spots on 471,482
UV spectra 448
X-ray data 452
Stellar Structure and Evolution 25
Transition Region
Flux vs Orbital period 457,470,474
vs Spectral type 470,474
vs X-ray 131,266,456,457,474
Flux ratios of UV line emissions 79
Hyades stars, in 131
Model 80
UV Emission
RS CVn stars, in 443
Fluxes at Earth 115,131,443
Flux ratios 79
Line variability 60,249
vs photometric wave 444,451
Stellar flares, in 207,230,245,249,562
Surface fluxes 51,181,450
T Tau stars, in 509
W UMa stars, in 448
"Vaughan-Preston" gap 8,67,70,592
Model for 592
vs X-rays 266
X-Rays
Abundances from 420
Coronal structure/parameters from 5,445
Flares, in 93,118,127,255,269,289,420,424,562,565
Fluxes at Earth 113,446
Hyades stars 131
$L_X$ in flare stars 89,113,125,261
$L_X$, predicted from Balmer EM 228
$L_X$ vs stellar parameters 87,92,98,115,116,453,457,470
Solar flares, in 289,307
Surface fluxes vs stellar parameters 51,87,117,416,419,604,633
Variability, short-term 92
vs He I 10830 372
vs UV line fluxes 266
W UMa stars, in 452